Studies in Surface Science and Catalysis 27

CATALYTIC HYDROGENATION

Studies in Surface Science and Catalysis
Advisory Editors: B. Delmon and J.T. Yates

Vol. 27

CATALYTIC HYDROGENATION

Editor

L. Červený

Department of Organic Technology, Prague Institute of Chemical Technology, Suchbátarova 5, 166 28 Prague 6, Czechoslovakia

ELSEVIER

Amsterdam — Oxford — New York — Tokyo 1986

ELSEVIER SCIENCE PUBLISHERS B.V.
Sara Burgerhartstraat 25
P.O. Box 211, 1000 AE Amsterdam, The Netherlands

Distributors for the United States and Canada:

ELSEVIER SCIENCE PUBLISHING COMPANY INC.
52, Vanderbilt Avenue
New York, NY 10017, U.S.A.

ISBN 0-444-42682-5 (Vol. 27)
ISBN 0-444-41801-6 (Series)

© Elsevier Science Publishers B.V., 1986

All rights reserved. No part of this publication may be reproduced, stored in a retrieval system or transmitted in any form or by any means, electronic, mechanical, photocopying, recording or otherwise without the prior written permission of the publisher, Elsevier Science Publishers B.V./Science & Technology Division, P.O. Box 330, 1000 AH Amsterdam, The Netherlands.

Special regulations for readers in the USA — This publication has been registered with the Copyright Clearance Center Inc. (CCC), Salem, Massachusetts. Information can be obtained from the CCC about conditions under which photocopies of parts of this publication may be made in the USA. All other copyright questions, including photocopying outside of the USA, should be referred to the publisher.

Printed in The Netherlands

CONTENTS

Studies in Surface Science and Catalysis (other volumes in the series) . . XVII
Preface XIX
List of Contributors XXIII

PART I: KINETICS AND MECHANISM OF HYDROGENATION AND HYDROGENOLYTIC REACTIONS

Chapter 1: Some problems of chemical kinetics in heterogeneous hydrogenation
    catalysis (S.L. Kiperman) 1
    1.1 Introduction 1
    1.2 Specific kinetic aspects of catalyzed hydrogenation reactions . . . 1
    1.3 Mechanism and kinetics of hydrogenation. General concepts 3
    1.4 Kinetic models of hydrogenation in the gas phase 10
        1.4.1 Hydrogenation of olefins 10
        1.4.2 Hydrogenation of diene hydrocarbons 14
        1.4.3 Hydrogenation of acetylene compounds 15
        1.4.4 Hydrogenolysis of aliphatic and alicyclic hydrocarbons . . . 16
        1.4.5 Hydrogenation of aromatic compounds 18
        1.4.6 Hydrogenation of oxygen- and nitrogen-containing compounds . 23
        1.4.7 Interaction of heterocyclic compounds with hydrogen 28
    1.5 Kinetic behaviour of liquid-phase hydrogenation 31
    1.6 Hydrogenation in systems containing several unsaturated compounds . 35
    1.7 Selectivity of hydrogenation 38
    1.8 Conclusion 43
    References 45

Chapter 2: Synergy in catalytic reactions involving hydrogen: possible role
    of surface-mobile species (B.K. Hodnett and B. Delmon) 53
    2.1 Introduction 53
    2.2 Literature survey of synergistic effects in catalytic reactions
        involving hydrogen 54
        2.2.1 Hydrogenation on catalysts composed of noble metals (Pt and
            Pd) and inorganic oxides (SiO_2 and Al_2O_3) 56
        2.2.2 Hydrogenation and hydrodesulphuration over sulphide catalysts 58
        2.2.3 Coke formation on reforming catalysts 59
    2.3 Possible origins of synergy in reactions involving hydrogen 59
        2.3.1 Classical bifunctional catalysis 59
        2.3.2 Formation of compounds between two phases 60

    2.3.3 Contamination of the surface of one phase by elements from the other phase 61
    2.3.4 Reaction of spill-over hydrogen produced on one phase with a reactant adsorbed on another phase 62
    2.3.5 Creation or regeneration of catalytic centres on one phase by spill-over hydrogen emitted by the other phase: the "remote control" concept 65
2.4 Special discussion of mechanisms involving surface-mobile species: possible outlook; control of selectivity 67
    2.4.1 Extent of formation of spill-over hydrogen on SiO_2 and Al_2O_3 .. 68
    2.4.2 Factors influencing spill-over: possible mechanisms 69
    2.4.3 Creation and regeneration of surface sites by spill-over hydrogen 70
    2.4.4 The role of surface-mobile species in determining selectivity ... 74
2.5 Conclusions 76
References 76

Chapter 3: Adsorption and hydrogenation of carbonyl and related compounds on transition metal catalysts (K. Tanaka) 79

3.1 Introduction 79
3.2 Comparison with olefin hydrogenation 80
    3.2.1 Complexity in carbonyl compound hydrogenation 80
    3.2.2 Equilibrium positions 80
    3.2.3 Reaction pathways 81
3.3 Deuterium tracer studies 83
    3.3.1 Aliphatic ketones and aldehydes 83
    3.3.2 Alicyclic ketones 85
    3.3.3 Aromatic ketones 89
3.4 Stereochemistry of alicyclic ketone hydrogenation 90
3.5 Kinetics, substituent effects and related subjects 91
    3.5.1 Kinetic studies 91
    3.5.2 Substituent effects, competitive reactions 92
3.6 Characterization of adsorbed species 94
    3.6.1 IR spectroscopy 94
    3.6.2 Thermal desorption 98
    3.6.3 Ultraviolet photoelectron spectroscopy 99
    3.6.4 Extrapolation from UHV-low temperature conditions 99
3.7 Conclusions 100
References 101

Chapter 4: Hydrogenation of nitriles (J. Volf and J. Pašek) 105
   4.1 Introduction 105
   4.2 Reaction scheme in the formation of primary, secondary and tertiary amines 106
   4.3 Catalysts for hydrogenation of nitriles 111
   4.4 Properties of the main types of catalysts for hydrogenation of nitriles 114
      4.4.1 Nickel and cobalt catalysts 114
      4.4.2 Copper catalysts 119
      4.4.3 Catalysts from the platinum group metals 121
   4.5 Effect of nitrile structure on hydrogenation 125
   4.6 Effect of reaction conditions on hydrogenation of nitriles . . . 128
      4.6.1 Temperature 128
      4.6.2 Hydrogen pressure 130
      4.6.3 Ammonia 132
      4.6.4 Water 135
   4.7 Conclusion 140
   References 141

Chapter 5: Hydrogenolysis of C-C bonds on platinum-based bimetallic catalysts (F. Garin, L. Hilaire and G. Maire) 145
   5.1 Introduction 145
   5.2 General properties of bimetallic catalysts 146
      5.2.1 Geometric and electronic factors 146
      5.2.2 The rigid band model 146
      5.2.3 The geometric effect 147
      5.2.4 The electronic factor (ligand effect) 148
      5.2.5 Synergistic effects 148
      5.2.6 Surface segregation 149
      5.2.7 Small particles 149
   5.3 Hydrogenolysis reactions 150
      5.3.1 Thermodynamic and kinetic data 150
      5.3.2 Mechanisms of hydrogenolysis reactions on platinum catalysts 154
      5.3.3 Hydrogenolysis of C-C bonds on alloys and bimetallic systems 157
   Concluding remarks 194
   References 195

Chapter 6: Hydrogenative denitrogenation of model compounds as related to
the refining of liquid fuels (H. Schulz, M. Schon and N.M. Rahman) . 201
  6.1 Organic nitrogen compounds in tars, oils from coal, shale and
    petroleum 201
  6.2 Evaluation of catalyst selectivity for HDS, HDO and HDN reactions
    and hydrogenation of multiple bonds 204
    6.2.1 Conversion of a four compound model mixture 205
    6.2.2 Evaluation of specific HDS activity 207
    6.2.3 Evaluation of specific HDO activity 209
    6.2.4 Evaluation of specific HDN activity 210
    6.2.5 Concluding evaluation of specific catalyst activity 212
    6.2.6 Variation of the compounds of the mixture 213
    6.2.7 Variation of the partial pressure of one of the compounds
       of the model mixture 213
  6.3 Simultaneously proceeding hydrogenation reactions during refining
    of oils from "Sumpfphase" coal hydrogenation 216
  6.4 Reaction steps in denitrogenation 218
    6.4.1 Nitrogen in aliphatic amines 218
    6.4.2 Nitrogen in amino groups attached to aromatic rings: aniline
       and homologues 222
    6.4.3 Nitrogen in aromatic six-membered heterocycles: pyridine,
       quinoline, isoquinoline, acridine 225
    6.4.4 Nitrogen in aromatic five-membered heterocycles: pyrrole,
       indole, carbazole 239
    6.4.5 Nitrogen in saturated monocyclic five- and six-membered
       rings: pyrrolidine and piperidine 244
  6.5 Generalization for fast and slow steps in HDN reaction networks . 246
  Appendix 252
  References 253

Chapter 7: Effect of catalyst composition on reaction networks in
hydrodesulphurization (M. Zdražil and M. Kraus) 257
  7.1 Introduction 257
  7.2 The chemistry of hydrodesulphurization 257
  7.3 Kinetic consequences 259
  7.4 Effects of structure, temperature and hydrogen pressure 262
  7.5 Synergic effects on the distribution of intermediates 262
    7.5.1 Non-aromatic cyclic sulphur compounds 262
    7.5.2 Unsaturated hydrocarbons 269
  7.6 Synergic effects on individual reactions 272
  7.7 Conclusions 273
  References 275

PART II: HETEROGENEOUS HYDROGENATION CATALYSTS. NEW ASPECTS

Chapter 8: Carrier effect on hydrogenation properties of metals (G.M. Pajonk and S.J. Teichner) 277
    8.1 Introduction 277
    8.2 Some previous data 278
    8.3 Recent views on the metal-support effects 283
    8.4 Hydrogenation of carbon oxides 284
        8.4.1 Nickel-based catalysts 284
        8.4.2 Platinum-based catalysts 289
        8.4.3 Palladium-based catalysts 290
        8.4.4 Ruthenium-based catalysts 291
        8.4.5 Rhodium-based catalysts 292
        8.4.6 Cobalt-based catalysts 293
        8.4.7 Iron-based catalysts 293
        8.4.8 Molybdenum-based catalysts 295
    8.5 Hydrogenation of benzene and other aromatics 295
        8.5.1 Metals supported on TiO_2 catalysts 296
        8.5.2 Metals supported on other inorganic supports 297
        8.5.3 Metals supported on organic carriers 299
    8.6 Hydrogenation of mono- and diolefins 299
    8.7 Miscellaneous hydrogenations 301
        8.7.1 Organic reactants 301
        8.7.2 Inorganic reactants 304
    8.8 Influence of the metal-support interaction on the deactivation of supported catalysts 305
    8.9 Conclusions 307
    References 307

Chapter 9: Role of bimetallic catalysts in catalytic hydrogenation and hydrogenolysis (L. Guczi and Z. Schay) 313
    9.1 Introduction 313
    9.2 Ensemble size effect and surface segregation 315
        9.2.1 Ni-Cu alloys 315
        9.2.2 Pt-Au alloys 318
        9.2.3 Ru-Cu alloys 319
        9.2.4 Pt-Pd alloys 320
    9.3 Ensemble size effect accompanied by secondary effects 322
        9.3.1 Matrix effect 322
        9.3.2 Hydrogen effect 325
        9.3.3 Metal-support interaction and segregation 330

9.4 Conclusions 333
References 333

Chapter 10: Supported mono- and bimetallic catalysts in hydrocarbon
conversions (J. Völter) 337
   10.1 Introduction 337
   10.2 Preparation and structure 338
       10.2.1 Monometallic systems (Pt/Al_2O_3) 338
       10.2.2 Bimetallic systems 343
       10.2.3 Further bimetallic systems of Pt, Pd, Rh and Ni 348
   10.3 Catalysis with bimetallic systems 348
       10.3.1 Hydrogenation 348
       10.3.2 Dehydrogenation 349
       10.3.3 Dehydrocyclization (reforming) 351
       10.3.4 Dehydroisomerization 352
       10.3.5 Hydrogenolysis 353
   10.4 Coking 353
       10.4.1 Initial coking 353
       10.4.2 Long-term coking 355
   10.5 General discussion 357
       10.5.1 Preparation and structure 357
       10.5.2 Classification of reactions 358
       10.5.3 Bimetal effect in selective conversions 360
       10.5.4 Bimetal effect in non-selective conversions 362
   10.6 Conclusions 367
       10.6.1 Preparation and structure 367
       10.6.2 Coking 367
       10.6.3 Catalysis 368
   References 369

Chapter 11: Supported bimetallic catalysts prepared by controlled surface
reactions (J. Margitfalvi, S. Szabó and F. Nagy) 373
   11.1 Introduction 373
   11.2 The mode of control of surface reactions involved in the preparation of supported bimetallic catalysts 374
       11.2.1 General aspects 374
       11.2.2 Preparation via anchored ionic surface complexes 375
       11.2.3 The use of bimetallic complexes or clusters 376
   11.3 New approaches 376
       11.3.1 Electrochemical approach 376
       11.3.2 Organometallic approach 379

11.4 Preparation of supported bimetallic catalysts via metal adsorption 380
    11.4.1 Technical aspects of the preparation 380
    11.4.2 Preparation of bimetallic Pd-Pt/Al_2O_3 catalysts 381
    11.4.3 Preparation of Pt/Al_2O_3 catalysts modified by adsorbed Re and Sn 385
11.5 Preparation of supported bimetallic catalysts by using organometallic compounds 385
    11.5.1 The experimental technique and procedures used 385
    11.5.2 Preparation of supported Sn-Pt catalysts 386
    11.5.3 Preparation of lead-modified alumina-supported nickel catalysts 390
11.6 Catalytic properties of supported bimetallic catalysts prepared via controlled surface reactions 391
    11.6.1 General aspects 391
    11.6.2 Benzene hydrogenation 392
    11.6.3 n-Hexane conversion 393
    11.6.4 Hydrogenation of acrylonitrile 405
11.7 Conclusions 407
References 407

Chapter 12: New supported metallic nickel systems (J.M. Marinas, J.M. Campelo and D. Luna) 411
12.1 Introduction 411
12.2 Preparation and characterization of nickel catalysts 412
    12.2.1 Introduction 412
    12.2.2 New preparation methods 412
    12.2.3 New materials 414
12.3 Metal-support interaction 434
12.4 Catalyzed reactions 440
12.5 Poisoning and deactivation 442
12.6 Promoting effects 447
12.7 Conclusions 450
References 450

PART III: ADVANCES IN HOMOGENEOUS HYDROGENATION

Chapter 13: Supported metal complexes as hydrogenation catalysts (Yu.I. Yermakov and L.N. Arzamaskova) 459
   13.1 Introduction 459
      13.1.1 General information on catalysts prepared by anchoring of metal complexes 459
      13.1.2 The stability of anchored metal complexes 465
   13.2 Hydrogenation of unsaturated hydrocarbons in the presence of anchored metal complexes 466
      13.2.1 The problem 466
      13.2.2 Parameters influencing the accessibility of active centers 467
      13.2.3 Substrate activation in the coordination sphere of anchored palladium complexes 472
      13.2.4 Some conclusions 478
   13.3 Hydrogenation of unsaturated hydrocarbons in the presence of transition metal ions on non-functionalized oxides 479
      13.3.1 Catalysts prepared from anchored π-allyl complexes of Group 6 metals 479
      13.3.2 Catalysts prepared from anchored complexes of Group 4 metals 480
      13.3.3 Catalysts prepared from anchored organometallic complexes of actinides 481
   13.4 Anchored organometallic complexes as precursors for the preparation of highly dispersed metallic particles 481
      13.4.1 Preparation of metallic catalysts by decomposition of surface organometallic complexes 482
      13.4.2 Low-valent surface ions as anchoring sites for stabilization of dispersed metals and the problem of the strong metal-support interaction 483
      13.4.3 The preparation of metallic catalysts by decomposition of anchored metal carbonyls 484
      13.4.4 Some conclusions 486
   13.5 Conclusion 487
   References 487

Chapter 14: Supported asymmetric hydrogenation catalysts (J. Hetflejš) . 497
   14.1 Introduction 497
   14.2 Polymer-supported asymmetric hydrogenation catalysts 498
   14.3 Asymmetric hydrogenation catalysts anchored to inorganic supports 509
   14.4 Concluding remarks 512
   References 512

PART IV: CATALYTIC HYDROGENATION REACTORS AND TECHNOLOGIES

Chapter 15: Liquid-phase hydrogenation: the role of mass and heat transfer in slurry reactors (G. Gut, O.M. Kut, F. Yuecelen and D. Wagner) . . . 517
    15.1 Introduction 517
    15.2 Overall model 517
        15.2.1 Surface reaction 518
        15.2.2 Poisoning and inhibition 521
    15.3 Mass transfer and kinetics 523
        15.3.1 Transport of hydrogen 523
        15.3.2 Transport of substrate 524
        15.3.3 Estimation of the absorption coefficient of hydrogen . . . 525
        15.3.4 Estimation of the transfer coefficient to the catalyst . . 527
        15.3.5 Checking for effects of mass transfer on the kinetics . . 530
        15.3.6 Pore diffusion 530
    15.4 Examples of application and simulation 531
        15.4.1 Batch reactor: effects of hydrogen and substrate transfer 531
        15.4.2 Continuous stirred tank slurry reactor: basic concept . . 533
        15.4.3 CSTR: combination of mass transfer with chemical reaction 534
        15.4.4 CSTR: simulation of conversion profiles 535
        15.4.5 Batch reactor: effects of the rates of heat production and heat transfer 537
        15.4.6 Batch reactor: effects of film and pore diffusion 538
    15.5 Notation 542
    References 544

Chapter 16: Application of fixed-bed reactors to liquid-phase hydrogenation (J. Hanika and V. Staněk) 547
    16.1 Introduction 547
    16.2 Literature survey 548
    16.3 Parameters affecting the productivity of multi-phase fixed-bed reactors 548
    16.4 Mass and heat transfer in three-phase fixed-bed reactors . . . 549
    16.5 Structural properties of fixed catalytic beds 550
    16.6 Mathematical modelling in the design and operation of trickle-bed reactors 552
    16.7 Hydrogenation of cyclohexene in a trickle-bed reactor 557
        16.7.1 Experimental set-ups and reaction conditions 558
        16.7.2 Steady-state operation of the trickle-bed reactor 559
        16.7.3 Dynamic behaviour of the trickle-bed reactor 566
        16.7.4 Hysteresis properties of trickle-bed reactors 569

16.8 Conclusion 574
Symbols 575
References 576

Chapter 17: Control of hydrogenation autoclaves (J. Horák) 579
  17.1 Introduction 579
  17.2 Formulation of the control task 581
    17.2.1 Control objectives 581
    17.2.2 Characteristics of the reaction regime 581
    17.2.3 Optimization of the reaction temperature 583
    17.2.4 Control constraints 584
  17.3 Control of small laboratory autoclaves 585
    17.3.1 Temperature control 586
    17.3.2 Pressure control 586
  17.4 Industrial and pilot-plant batch autoclaves 588
    17.4.1 Process safety 588
    17.4.2 Temperature control 591
    17.4.3 Dynamic properties of autoclaves 597
    17.4.4 Conversion control 602
  17.5 Industrial and pilot-plant continuous autoclaves 602
    17.5.1 Temperature control 603
  17.6 Early recognition of hazardous states 605
  List of symbols 606
  References 608

Chapter 18: Selective hydrogenation applied to the refining of petrochemical
raw materials produced by steam cracking (M.L. Derrien) 613
  18.1 Introduction 613
  18.2 C_2 cut hydrorefining 616
    18.2.1 The different processes for the hydrorefining of the
           ethylene-rich cut (C_2 cut) 616
    18.2.2 Industrial catalysts 618
    18.2.3 Selective hydrogenation of acetylene in the C_2 cut, detailed
           study 618
  18.3 C_3 cut hydrorefining 625
    18.3.1 Purpose 625
    18.3.2 Thermodynamic and kinetic aspects of hydrogenation . . . 626
    18.3.3 Technological aspects of C_3 hydrorefining; liquid-phase or
           gas-phase operation 629
    18.3.4 Industrial catalysts 630
    18.3.5 Industrial liquid-phase processes 631
    18.3.6 Conclusions on C_3 hydrogenation 634

18.4 C_4 cut hydrorefining	635
18.4.1 Hydrorefining of the C_4 butadiene-rich cut (selective hydrogenation of vinylacetylene)	636
18.4.2 Hydrorefining of 1-butene-rich cuts (selective hydrogenation of butadiene)	643
18.4.3 Hydroisomerization of 1-butene cuts	648
18.5 Gasoline cut hydrorefining	652
18.5.1 Characteristics of steam-cracked gasolines and specifications of hydrogenated products	652
18.5.2 Refining by selective hydrogenation	653
18.5.3 Processes	656
18.5.4 Industrial performance	659
18.6 Conclusion	665
References	665
Subject Index	667

STUDIES IN SURFACE SCIENCE AND CATALYSIS

Advisory Editors: B. Delmon, Université Catholique de Louvain, Louvain-la-Neuve, Belgium
J.T. Yates, University of Pittsburgh, Pittsburgh, PA, U.S.A.

Volume 1	**Preparation of Catalysts I.** Scientific Bases for the Preparation of Heterogeneous Catalysts. Proceedings of the First International Symposium held at the Solvay Research Centre, Brussels, October 14–17, 1975 edited by **B. Delmon, P.A. Jacobs and G. Poncelet**
Volume 2	**The Control of the Reactivity of Solids.** A Critical Survey of the Factors that Influence the Reactivity of Solids, with Special Emphasis on the Control of the Chemical Processes in Relation to Practical Applications by **V.V. Boldyrev, M. Bulens and B. Delmon**
Volume 3	**Preparation of Catalysts II.** Scientific Bases for the Preparation of Heterogeneous Catalysts. Proceedings of the Second International Symposium, Louvain-la-Neuve, September 4–7, 1978 edited by **B. Delmon, P. Grange, P. Jacobs and G. Poncelet**
Volume 4	**Growth and Properties of Metal Clusters.** Applications to Catalysis and the Photographic Process. Proceedings of the 32nd International Meeting of the Société de Chimie Physique, Villeurbanne, September 24–28, 1979 edited by **J. Bourdon**
Volume 5	**Catalysis by Zeolites.** Proceedings of an International Symposium organized by the Institut de Recherches sur la Catalyse – CNRS – Villeurbanne and sponsored by the Centre National de la Recherche Scientifique, Ecully (Lyon), September 9–11, 1980 edited by **B. Imelik, C. Naccache, Y. Ben Taarit, J.C. Vedrine, G. Coudurier and H. Praliaud**
Volume 6	**Catalyst Deactivation.** Proceedings of the International Symposium, Antwerp, October 13–15, 1980 edited by **B. Delmon and G.F. Froment**
Volume 7	**New Horizons in Catalysis.** Proceedings of the 7th International Congress on Catalysis, Tokyo, June 30–July 4, 1980. Parts A and B edited by **T. Seiyama and K. Tanabe**
Volume 8	**Catalysis by Supported Complexes** by **Yu.I. Yermakov, B.N. Kuznetsov and V.A. Zakharov**
Volume 9	**Physics of Solid Surfaces.** Proceedings of the Symposium held in Bechyňe, September 29–October 3, 1980 edited by **M. Láznička**
Volume 10	**Adsorption at the Gas–Solid and Liquid–Solid Interface.** Proceedings of an International Symposium held in Aix-en-Provence, September 21–23, 1981 edited by **J. Rouquerol and K.S.W. Sing**
Volume 11	**Metal-Support and Metal-Additive Effects in Catalysis.** Proceedings of an International Symposium organized by the Institut de Recherches sur la Catalyse – CNRS – Villeurbanne and sponsored by the Centre National de la Recherche Scientifique, Ecully (Lyon), September 14–16, 1982 edited by **B. Imelik, C. Naccache, G. Coudurier, H. Praliaud, P. Meriaudeau, P. Gallezot, G.A. Martin and J.C. Vedrine**
Volume 12	**Metal Microstructures in Zeolites.** Preparation – Properties – Applications. Proceedings of a Workshop, Bremen, September 22–24, 1982 edited by **P.A. Jacobs, N.I. Jaeger, P. Jirů and G. Schulz-Ekloff**
Volume 13	**Adsorption on Metal Surfaces.** An Integrated Approach edited by **J. Bénard**
Volume 14	**Vibrations at Surfaces.** Proceedings of the Third International Conference, Asilomar, CA, September 1–4, 1982 edited by **C.R. Brundle and H. Morawitz**

Volume 15 **Heterogeneous Catalytic Reactions Involving Molecular Oxygen**
by G.I. Golodets

Volume 16 **Preparation of Catalysts III.** Scientific Bases for the Preparation of Heterogeneous Catalysts. Proceedings of the Third International Symposium, Louvain-la-Neuve, September 6–9, 1982
edited by G. Poncelet, P. Grange and P.A. Jacobs

Volume 17 **Spillover of Adsorbed Species.** Proceedings of the International Symposium, Lyon-Villeurbanne, September 12–16, 1983
edited by G.M. Pajonk, S.J. Teichner and J.E. Germain

Volume 18 **Structure and Reactivity of Modified Zeolites.** Proceedings of an International Conference, Prague, July 9–13, 1984
edited by P.A. Jacobs, N.I. Jaeger, P. Jirů, V.B. Kazansky and G. Schulz-Ekloff

Volume 19 **Catalysis on the Energy Scene.** Proceedings of the 9th Canadian Symposium on Catalysis, Quebec, P.Q., September 30–October 3, 1984
edited by S. Kaliaguine and A. Mahay

Volume 20 **Catalysis by Acids and Bases.** Proceedings of an International Symposium organized by the Institut de Recherches sur la Catalyse–CNRS–Villeurbanne and sponsored by the Centre National de la Recherche Scientifique, Villeurbanne (Lyon), September 25–27, 1984
edited by B. Imelik, C. Naccache, G. Coudurier, Y. Ben Taarit and J.C. Vedrine

Volume 21 **Adsorption and Catalysis on Oxide Surfaces.** Proceedings of a Symposium, Brunel University, Uxbridge, June 28–29, 1984
edited by M. Che and G.C. Bond

Volume 22 **Unsteady Processes in Catalytic Reactors**
by Yu.Sh. Matros

Volume 23 **Physics of Solid Surfaces 1984**
edited by J. Koukal

Volume 24 **Zeolites: Synthesis, Structure, Technology and Application.** Proceedings of the International Symposium, Portorož-Portorose, September 3–8, 1984
edited by B. Držaj, S. Hočevar and S. Pejovnik

Volume 25 **Catalytic Polymerization of Olefins.** Proceedings of the International Symposium on Future Aspects of Olefin Polymerization, Tokyo, July 4–6, 1985
edited by T. Keii and K. Soga

Volume 26 **Vibrations at Surfaces 1985.** Proceedings of the Fourth International Conference, Bowness-on-Windermere, September 15–19, 1985
edited by D.A. King, N.V. Richardson and S. Holloway

Volume 27 **Catalytic Hydrogenation**
edited by L. Červený

PREFACE

Catalytic hydrogenation was discovered at the end of the last century, and for several decades has aroused the interest of a vast number of academic and industrial research centres. It is used as a means of obtaining a great variety of products and semiproducts whose synthesis would otherwise be difficult. The number of original papers and books devoted to catalytic hydrogenation reflects its importance. Some examples are: "Catalytic Hydrogenation" by Augustine (Marcel Dekker, 1965), "Metally-Katalizatory Gidrogenizatsii" by Sokolskii and Sokolskaya (Nauka, 1970), "Homogeneous Hydrogenation" by James (Wiley, 1973), "Catalytic Hydrogenation in Organic Synthesis Procedures and Commentary" by Freifelder (Wiley, 1978), "Catalytic Hydrogenation in Organic Syntheses" by Rylander (Academic Press, 1979).

Why, then, yet another book entitled "Catalytic Hydrogenation"? The publication explosion of the last fifteen years has considerably impeded the efforts of potential authors to produce up to date articles on broad areas of scientific knowledge. On the other hand, specific areas can be dealt with much more quickly, easily and efficiently. I therefore requested some renowned specialists to contribute articles in their respective fields of interest. In this way, a collection of eighteen chapters was obtained which represent the most recent state of the art, summaries of published results being supplemented by results, mostly unpublished, obtained by the authors themselves. The book is divided into four parts:

Part I. Kinetics and mechanism of hydrogenation and hydrogenolytic reactions.
Part II. Heterogeneous hydrogenation catalysts. New aspects.
Part III. Advances in homogeneous hydrogenation.
Part IV. Catalytic hydrogenation reactors and technologies.

For an indication of the topics covered in the individual chapters I use the words of the authors themselves:

In Chapter 1 an attempt is made to give a concise account of the available data on the kinetics of hydrogenation of organic compounds and related kinetic problems of hydrogenation catalysis (Kiperman).

Chapter 2 considers the evidence for the role played by surface-mobile hydrogen in various heterogeneously catalysed reactions where synergy occurs. The reaction of spill-over hydrogen with organic molecules or inorganic solids is also discussed (Hodnett and Delmon).

Chapter 3 is confined predominantly to the transition-metal catalysed hydrogenation of isolated carbonyl functions (Tanaka).

Chapter 4 summarizes existing knowledge on the hydrogenation of nitriles in the liquid phase, and supplements it by some, mostly unpublished results from the authors' own research (Volf and Pašek).

Chapter 5 deals with the hydrogenolysis reactions of saturated hydrocarbons performed on platinum-based bimetallic catalysts (Garin, Hilaire, Maire).

Chapter 6 discusses hydrogenative denitrogenation of model compounds as related to the refining of liquid fuels (Schulz, Schon and Rahman).

Chapter 7 summarizes the literature data on the influence of the catalyst composition on the product distribution in the hydrodesulphuration of model sulphur compounds (Zdražil and Kraus).

In Chapter 8 some ancient data on the hydrogenation of ethylenic and acetylenic bonds in linear hydrocarbons, of aromatics or cycloolefins, and of other functions are first considered. It is then shown that, when account is taken of the metal-support interaction, the interpretation of the role of the support in the catalytic activity becomes easier (Pajonk and Teichner).

Chapter 9 demonstrates how the ensemble size, dispersion, nature of adsorbed hydrogen, metal-support interaction and other factors control the behaviour of bimetallic catalysts in some hydrocarbon transformations (Guczi and Schay).

In Chapter 10 a new attempt is made to elucidate the relationships between the structure of bimetallic catalysts and their specific catalytic properties, as evidenced by the conversion of hydrocarbons with hydrogen (Völter).

The aim of Chapter 11 is to demonstrate that the principle of Controlled Surface Reactions is a very powerful approach, and that supported bimetallic catalysts prepared by these methods possess unique catalytic properties, in hydrocarbon conversion and in the hydrogenation of organic compounds (Margitfalvi, Szabó, Nagy).

Chapter 12 reviews new supported nickel catalysts, revealing the significant progress in the characterization of such catalysts and their use in catalytic processes (Marinas, Campelo, Luna).

Chapter 13 analyzes the results of the anchoring of metal complexes in the preparation of hydrogenation catalysts (Yermakov and Arzamaskova).

In Chapter 14 studies on asymmetric hydrogenation, homogeneous catalysts attached to organic polymers and inorganic supports are surveyed (Hetflejš).

In Chapter 15, hydrogenation is modelled in terms of the Langmuir-Hinshelwood type kinetic behaviour, emphasis being placed on the effects of heat and mass transfer on kinetics (Gut, Kut, Yücelen and Wagner).

Chapter 16 concerns the application of fixed-bed reactors to liquid-phase hydrogenation. The simple pseudohomogeneous reactor model is recommended for the description of the majority of hydrogenations in organic technology (Hanika and Staněk).

No monographs or specialized chapters in textbooks have been published on hydrogenation autoclave control. In Chapter 17 guidance is provided for the design of control systems, limited to smaller plants with changing product spectra (Horák).

The final Chapter (Derrien) is beyond the scope of this book. It deals mostly with technological aspects (of hydrorefining of C_2, C_3, C_4 and gasoline cuts). However, because of the immense importance of such matters in the petrochemical industry, I believe that the reader will find this chapter useful.

Unfortunately, no space was left for chapters devoted to the effect of reactant structure and solvent properties on the kinetics of the liquid phase hydrogenation. To those interested in these problems, our papers in the Advances in Catalysis (1981) or in the Catalysis Reviews (1982) can be recommended.

It is my pleasant duty to thank all those who have found time to contribute to this book.

Libor Červený

LIST OF CONTRIBUTORS

L.N. Arzamaskova, Institute of Catalysis, Siberian Branch of the Academy of
    Sciences of the U.S.S.R., Novosibirsk 630090, U.S.S.R.

J.M. Campelo, Department of Organic Chemistry, Faculty of Sciences, Córdoba
    University, E-14004 Córdoba, Spain

B. Delmon, Groupe de Physico-Chimie Minérale et de Catalyse, Université
    Catholique de Louvain, Place Croix du Sud 1, 1348 Louvain-la-Neuve, Belgium

M.L. Derrien, Institut Français du Pétrole, B.P. 311, 92506 Rueil Malmaison
    Cedex, France

F. Garin, Laboratoire de Catalyse et Chimie des Surfaces, U.A. 423 du CNRS,
    Université Louis Pasteur - Institut Le Bel, 4 Rue Blaise Pascal, 67070
    Strasbourg Cedex, France

L. Guczi, Institute of Isotopes, Hungarian Academy of Sciences, P.O. Box 77,
    H-1525 Budapest, Hungary

G. Gut, Swiss Federal Institute of Technology (ETH), Universitätstrasse 6,
    CH-8092 Zurich, Switzerland

J. Hanika, Department of Organic Technology, Prague Institute of Chemical
    Technology, Suchbátarova 5, 166 28 Prague 6, Czechoslovakia

J. Hetflejš, Institute of Chemical Process Fundamentals, Czechoslovak Academy
    of Sciences, Rozvojová 135, 165 02 Prague 6 - Suchdol, Czechoslovakia

L. Hilaire, Laboratoire de Catalyse et Chimie des Surfaces, U.A. 423 du CNRS,
    Université Louis Pasteur - Institut Le Bel, 4 Rue Blaise Pascal, 67070
    Strasbourg Cedex, France

B.K. Hodnett, Department of Materials Engineering and Industrial Chemistry,
    NIHE, Limerick, Ireland

J. Horák, Department of Organic Technology, Prague Institute of Chemical
    Technology, Suchbátarova 5, 166 28 Prague, Czechoslovakia

S.L. Kiperman, Institute of Organic Chemistry, Academy of Sciences of the U.S.S.R., Moscow, U.S.S.R.

M. Kraus, Institute of Chemical Process Fundamentals, Czechoslovak Academy of Sciences, Rozvojová 135, 165 02 Prague 6 - Suchdol, Czechoslovakia

O.M. Kut, Swiss Federal Institute of Technology (ETH), Universitätstrasse 6, CH-8092 Zurich, Switzerland

D. Luna, Department of Organic Chemistry, Faculty of Sciences, Córdoba University, E-14004 Córdoba, Spain

G. Maire, Laboratoire de Catalyse et Chimie des Surfaces, U.A. 423 du CNRS, Université Louis Pasteur - Institut Le Bel, 4 Rue Blaise Pascal, 67070 Strasbourg Cedex, France

J. Margitfalvi, Central Research Institute for Chemistry, Hungarian Academy of Sciences, P.O. Box 17, H-1525 Budapest, Hungary

J.M. Marinas, Department of Organic Chemistry, Faculty of Sciences, Córdoba University, E-14004 Córdoba, Spain

F. Nagy, Central Research Institute for Chemistry, Hungarian Academy of Sciences, P.O. Box 17, H-1525 Budapest, Hungary

G.M. Pajonk, Laboratoire de Thermodynamique et Cinétique Chimiques, L.A. 231 du CNRS, Université Claude Bernard LYON I, 43 Boulevard du 11 novembre 1918, 69622 Villeurbanne Cedex, France

J. Pašek, Department of Organic Technology, Prague Institute of Chemical Technology, Suchbátarova 5, 166 28 Prague 6, Czechoslovakia

N.M. Rahman, Engler-Bunte-Institut, Universität Karlsruhe, 7500 Karlsruhe, F.R.G.

Z. Schay, Institute of Isotopes, Hungarian Academy of Sciences, P.O. Box 77, H-1525 Budapest, Hungary

M. Schon, Engler-Bunte-Institut, Universität Karlsruhe, 7500 Karlsruhe, F.R.G.

H. Schulz, Engler-Bunte-Institut, Universität Karlsruhe, 7500 Karlsruhe, F.R.G.

V. Staněk, Institute of Chemical Process Fundamentals, Czechoslovak Academy of Sciences, 165 02 Prague 6, Czechoslovakia

S. Szabó, Central Research Institute for Chemistry, Hungarian Academy of
    Sciences, P.O. Box 17, H-1525 Budapest, Hungary

K. Tanaka, Institute of Physical and Chemical Research (Riken), Wako, Saitama
    351-01, Japan

S.J. Teichner, Laboratoire de Thermodynamique et Cinétique Chimiques, L.A. 231
    du CNRS, Université Claude Bernard LYON I, 43 Boulevard du 11 novembre
    1918, 69622 Villeurbanne Cedex, France

J. Völter, Central Institute of Physical Chemistry, Academy of Sciences of the
    G.D.R., Rudower Chaussee 5, DDR-1199 Berlin, G.D.R.

J. Volf, Department of Organic Technology, Prague Institute of Chemical
    Technology, Suchbátarova 5, 166 28 Prague 6, Czechoslovakia

D. Wagner, Swiss Federal Institute of Technology (ETH), Universitätstrasse 6,
    CH-8092 Zurich, Switzerland

Yu.I. Yermakov, Institute of Catalysis, Siberian Branch of the Academy of
    Sciences of the U.S.S.R., Novosibirsk 630090, U.S.S.R.

F. Yuecelen, Swiss Federal Institute of Technology (ETH), Universitätstrasse 6,
    CH-8092 Zurich, Switzerland

M. Zdražil, Institute of Chemical Process Fundamentals, Czechoslovak Academy
    of Sciences, Rozvojová 135, 165 02 Prague 6 - Suchdol, Czechoslovakia

Chapter 1

SOME PROBLEMS OF CHEMICAL KINETICS IN HETEROGENEOUS HYDROGENATION CATALYSIS

S.L. KIPERMAN
N.D. Zelinsky Institute of Organic Chemistry, Academy of Sciences of the USSR, Moscow (USSR)

1.1. INTRODUCTION

Chemical kinetics is of increasing significance in heterogeneous catalysis, acting as a bridge between the theory and practice of catalytic processes. Kinetic investigations have proven to be essential for studying both the generalities and the particularities of catalysis.

The heyday of universal theories, which attempted to explain the complex catalytic phenomena from a general standpoint, is now over. It has become clear that any further generalisations will require a gradual accumulation of new data. Today, chemical kinetics provides a theoretical basis, experimental techniques and calculative methods, which allow the construction of accurate and reliable models for qualitative and quantitative description of catalytic processes (refs. 1-9).

When using the results of kinetic investigations one must make sure there are no distortions due to the retarding effects of mass and heat transfer; i.e., that the reaction proceeds in the kinetic region. For this there is a number of experimental and theoretical check procedures (refs. 1, 10, 11). The calculations of kinetic data see *e.g.* refs. 12-16.

An attempt will be made in this chapter to give a concise account of the available data on the kinetics of hydrogenation of organic compounds and related problems in the kinetics of hydrogenation catalysis. Whenever expedient, other interactions involving hydrogen and reactions with non-stoichiometric involvement of hydrogen will also be discussed.

1.2 SPECIFIC KINETIC ASPECTS OF CATALYZED HYDROGENATION REACTIONS

Certain specific features of hydrogenation reactions are very important in kinetic studies:

(1) The diversity of catalytic systems, and hence the possible diversity of intermediates and mechanisms, giving rise to different kinetic behaviours. This is the reason why any attempt to give a unified kinetic description of hydroge-

nation and to associate it with a single kinetic equation and mechanism is futile from the outset.

(2) The complex nature of these reactions, giving rise to multi-step mechanisms.

(3) The complexity of reactions, leading in various directions. Under stationary conditions this results in different independent reaction routes, including stoichiometric routes leading to different products.

(4) The variety of adsorbed forms of hydrogen under different conditions, with possible penetration into the subsurface layer of the catalyst, complicating the kinetics.

(5) The high reaction rates already at rather low temperatures leading to almost complete conversion.

(6) The sensitivity of kinetics towards relatively small changes in the conditions, leading to new routes and new mechanisms and the replacement of one equation by another. Hence the necessity of thoroughly exploring the borderline kinetic regions.

(7) The existence of temperature maximum of hydrogenation rate (near 200°C at atmospheric pressure (see $e.g.$ ref. 17).

The kinetic data and their interpretation in the case of liquid phase hydrogenation may be complicated by the following effects:

(1) The three-phase nature of the system means that hydrodynamic factors and the transport to the boundary must be taken into account (refs. 18-20).

(2) Numerous influences of the liquid reactants and solvent, due to solvation of the reactive system, catalyst and active complexes, to ionic and dipole interactions, to the change in reactivity in concentrated solutions, to cell effects and the combination of all these (see, $e.g.$ refs. 21-23). All these effects and their mechanisms have not yet been quantitatively included into kinetic models, although such a possibility is being discussed (see, $e.g.$ refs. 24-26).

(3) Increased likelihood of inhibition by mass and heat transfer and by dissolution of the components.

(4) The possibility of transition of the reaction into the volume (similar to liquid-phase oxidation, see refs. 27-32). But these effects in hydrogenation reactions have not been found yet.

(5) The possibility of transition from one stationary or quasi-stationary regime to another - in particular, because of abrupt changes in the concentration ratios of components in closed systems, which may give rise to drastic changes in kinetic relationships (refs. 33, 34).

Thus the researcher must be prepared for possible surprises in the interpretation of kinetic experiments and for increased complexity of the kinetic models of hydrogenation.

1.3 MECHANISM AND KINETICS OF HYDROGENATION. GENERAL CONCEPTS

The practice of catalyzed hydrogenation involves various concepts about its nature, mechanism and sequence of steps, as well as of the relevant kinetic relationships.

For instance, the Langmuir-Hinshelwood mechanism (ref. 35) assumes that the species to be hydrodenated and hydrogen react in the adsorbed state in the ideal layer (refs. 1, 6). Given that the slow step is the surface interaction, the general kinetic equation follows

$$r = k \frac{P_1^{n_1} P_2^{n_2}}{(L + k_1 P_1 + k_2 P_{H_2}^m + k_3 P_2^n)^l} \quad (1.1)$$

where P_1, P_{H_2} and P_2 are the partial pressures of the original substance, hydrogen and the product respectively; r is the reaction rate, the remaining symbols denote constants. For this mechanism it is usual that $n_1 = 1$, and $L = 0$ or 1. For low coverages $l=0$, and eqn. 1.1 becomes

$$r = k \, P_1^{n_1} P_{H_2}^{n_2} \quad (1.2)$$

For atomic and molecular adsorption of hydrogen, the index m has values of 0.5 and 1 respectively, and l=2 corresponds to the interaction of two particles in the surface layer.

In a real surface layer (refs. 1, 6, 36), the reaction rate may be given by an equation similar to 1.1, but with different meanings of the constants. Eqn.1.1 may also be valid for other slow steps, perhaps with minor amendments (ref. 6).

The Eley-Rideal mechanism (ref. 37) assumes that one of the components enters the reaction from a chemisorbed state, and the other comes from a state of physical adsorption or directly from the gas phase (collision mechanism). This mechanism can also be represented by eqn. 1.1, although some constants in the denominator will vanish. Such a mechanism has recently been criticised (ref. 38).

Clearly, these notions do not consider the details of the nature and the sequence of the steps of hydrogenation, but treat the reaction in a rather general way. A more detailed picture is given by other theories, dealing directly with such reactions or with related processes involving hydrogen. Let us first focus on the nature of adsorption of reactants. The relevant data can be found in a number of monographs and reviews (see *e.g.* refs. 39-48).

In most cases here we shall restrict ourselves to metallic catalysts, not excluding others, of course, as long as they are backed with reliable kinetic data (which is rare). Some forms of hydrogen adsorption have been observed on hydrogenation catalysts. At not too low temperatures (above room temperature) equilibrium may be established between different forms of adsorption. Then for the kinetics of reaction it is not too important which of these forms exhibits the highest reactivity.

Unsaturated initial substances and products may be attached to the catalyst surface by different kinds of bonds: δ- and π-bonds, donor-acceptor bonds or (at not-too-high temperatures) as physically adsorbed forms. Dissociative adsorption with loss of hydrogen is possible, also with the formation of strong δ-bonds. Reactive π-surface complexes as well as π-allyl ones may arise (refs. 40, 49, 50). In some cases the existence of such surface structures has been demonstrated *e.g.* for benzene on platinum (ref. 51).

The saturated products of hydrogenation may be held on the catalyst surface by donor-acceptor bonds, by physical or dissociative adsorption. Accordingly, one cannot directly exclude the possibility of adsorption and desorption being the slow steps in the overall process of hydrogenation.

Balandin's multiplet theory (ref. 52) assumes that hydrogenation reactions involve the formation and decomposition of a so-called multiplet complex. In the simple cases of hydrogenation of multiple bonds the doublet mechanism is assumed, which involves four catalyst surface sites; in the more complicated cases of hydrogenation of aromatic bonds, in accordance with the principle of microscopic reversibility, twelve active sites are required (refs. 1, 6). Here the theory assumes a single two-step mechanism, which is not in accord with the variety of kinetic data available on hydrogenation.

Horiuti and Polanyi (ref. 53) proposed an associative mechanism of hydrogenation, based on consecutive interaction between the adsorbed unsaturated species and hydrogen atoms with the formation of intermediate half-hydrogenated forms. Hence different kinetic relationships follow, depending on the adsorptivity of the reactants, the surface coverage, the ratio of rates of the individual steps and the nature of the adsorbed layer. For instance, if the hydrogen adsorption is predominant and slow (which, however, is not very realistic), or if the half-hydrogenated form interacts slowly with hydrogen when the surface is predominantly occupied by the initial species, we get an equation

$$r = k\, P_{H_2} \tag{1.3}$$

which often fits the experimental results, but has little to say about the actual mechanism of the process.

Farkas and Farkas (ref. 54) considered the dissociative mechanism, which assumes adsorption of the original substance with rupture of a multiple bond, dissociative adsorption of hydrogen and interaction without the formation of a half-hydrogenated form.

The newly developed concepts of adsorption of organic compounds by metals in the form of surface π-complexes (ref. 50), prompted modifications of the mechanisms in refs. 53 and 54 in the form of associative and dissociative π-complex substitution (ref. 49) for isotope exchange in aromatic compounds. The authors believe that this mechanism includes hydrogenation as well. The mechanism in ref. 54 also leads to a kinetic equation similar to eqn. 1.1.

Twigg and Rideal (ref. 55) discussed the mechanism of hydrogenation of hydrocarbons using the interaction of ethylene with hydrogen as an example. It was assumed to lead partly to a half-hydrogenated form, and partly to a slow direct formation of saturated hydrocarbon. This mechanism was further modified by Twigg (ref. 56), who took for the slow step the interaction between the half-hydrogenated form and adsorbed atomic hydrogen. Hence, assuming that the main portion of the catalyst surface is covered by adsorbed hydrocarbon, eqn. 1.3 follows. This mechanism was confirmed (ref. 57), using as the catalyst a palladium membrane admitting hydrogen into the reaction zone.

The theory elaborated by Balandin (ref. 58) provides a comprehensive generalisation of various types of hydrogenation processes, including their kinetic aspects. All sites of the catalyst surface are assumed to be capable of adsorbing the components, although some of them activate only the initial species, while others activate hydrogen (assuming homogeneity or quasi-homogeneity of the surface, refs. 6, 59). This concept is not in accord with the earlier notion of different kinds of active sites, each of which selectively adsorbs a particular substance. Assuming adsorption and desorption to be in equilibrium, and assuming the slow interaction of the half-hydrogenated form with hydrogen, Balandin obtained a general kinetic equation

$$r = k \frac{P_1 P_{H_2}^n}{(1+k_1 P_1 + k_2 P_2 + k_3 P_{H_2}^m)(1+k_1' P_1 + k_2' P_2 + k_3' P_{H_2}^{m'})} \qquad (1.4)$$

where the constants in the denominator characterise the coefficients of adsorption of the reactants on each type of surface, and the values of the indices m, m' and n, depending on the form of adsorption of hydrogen (either atomic or molecular) may be 0, 5 or 1. Various kinetic relationships may be obtained depending on the relative values of the partial pressures and constants in the equation, fitting the vast experimental data quoted by Balandin.

It is clear that this theory, despite its alleged generality, concerns only one particular mechanism of reaction and the corresponding kinetic equation. Its universality is contradicted by experimental evidence, and the observed agreement between the experimental results and eqn. 1.4 with appropriate choice of constants does not prove the validity of underlying assumptions, since the same relationships may be derived from different assumptions and by different reasoning.

Jenkins and Rideal (ref. 60) considered the relative rates of dissociative adsorption of hydrogen and its interaction with the initial substance for the case of hydrogenation of ethylene. The process is presumed to go via the interaction of organic species with two adsorbed hydrogen atoms. This step is assumed to be the slow one at temperatures above that corresponding to the maximum reaction rate, while below this temperature the slow step is represented by the adsorption of hydrogen. These assumptions also comply with eqns. 1.3 and 1.1 with appropriate simplifications.

A number of works have dealt with the surface chain hydrogenation mechanisms, mainly in the hydrogenation of ethylene. For instance, Wagner and Hauffe (ref. 61) suggested that the reactions may proceed via the interaction of adsorbed atomic hydrogen with the initial species leading to a half-hydrogenated form, which in its turn reacts with molecular hydrogen, yielding the final product and an adsorbed hydrogen atom which serves as the chain carrier.

Similarly, Taylor and Thon (ref. 62) considered the adsorbed atoms or radicals to be the "active sites" of the reactions, and also assumed these reactions to be of the chain type. For the hydrogenation reactions it is assumed that the "active sites" are the surface hydrogen atoms, whereas the other surface compounds compete with the latter. The constant density of such sites is believed by the authors to ensure that the reaction follows a zero-order equation. Variations in the order of reaction and other complications in the kinetic behaviour are explained by a change in the density of active sites. This hypothesis, however, does not seem to have a sound physical foundation and does not pay due attention to the catalyst nature.

Similar concepts have been developed by Voevodskyi, Volkenstein and Semenov (ref. 63, 64) for hydrogenation on oxide and metallic catalysts. They considered the chain carriers to be the surface compounds resulting from the interaction of dissociatively adsorbed hydrogen and organic radicals with free surface valencies.

The feasibility of a chain mechanism in the hydrogenation and hydrogen-deuterium exchange of ethylene was tested in experiments with palladium membranes (refs. 65-67). The tests revealed that, although such a mechanism is not precluded, the chain length does not exceed 0.5, and hence the contribution from the chain mechanism is negligible.

Horiuti and Miyahara (refs. 68, 69) carried out a consistent treatment of hydrogenation rates using the method of quasi-stationary concentrations and the Horiuti-Polanyi mechanism. They claim that the kinetic parameters derived confirm their basic assumptions.

Some theories deal specifically with the hydrogenation of aromatic compounds. The mechanism by Kagan, described in ref. 70 assumes the reaction to proceed via slow attachment of hydrogen molecules and subsequent fast disproportionation of the resulting cyclic diene. This, however, does not satisfy the principle of microscopic reversibility (see ref. 6).

The mechanism of Parravanno (ref. 71) consists in simultaneous joining of initial species from the gas phase to the surface hydrogen atoms, the adsorption of which is assumed to be slow. The latter assumption is not in accord with experiment, and the author gives no kinetic relationships.

Sokolskyi (refs. 72, 73) considers four possible mechanisms of liquid-phase hydrogenation: (1) activation of organic species on the surface covered by hydrogen, i.e., slow interaction between hydrogen and initial species or displacement of hydrogen by the latter; (2) activation of both components, i.e., slow surface interaction; (3) activation of hydrogen on the surface covered by initial species, i.e. slow interaction in the surface layer with hydrogen from the solution or interaction of hydrogen with the surface; (4) slow transfer of an electron from the catalyst to the initial species (quinone, nitrobenzene, etc.). The author does not include any accurate kinetic data but advocates studying the kinetics with simultaneous monitoring of the catalyst potential in the course of the process.

Various mechanisms of hydrogenation in the liquid phase, taking due account of the processes occurring in real surface layers, have been considered in refs. 33, 34, 74. The author proposed a unified scheme (see Table 1.1)

TABLE 1.1.
Various possible routes of the liquid phase hydrogenation.

Steps: / Routes:	Stoichiometric numbers:				
	I	II	III	IV	V
1. $A_{(s)} = A_{(a)}$	1	0	0	1	1
2. $H_{2(s)} = 2H_{(a)}$	0	1	0	0	1
3. $A_{(a)} + H_{2(s)} = AH_{2(s)}$	1	0	0	0	0
4. $A_{(s)} + 2H_{(a)} = AH_{2(s)}$	0	1	0	0	0
5. $A_{(s)} + H_{2(s)} + \text{catalyst} = A(cat) \cdot H_2$	0	0	1	0	0
6. $A(cat) \cdot H_2 = AH_{2(s)} + \text{catalyst}$	0	0	1	0	0
7. $A_{(a)} + H_{2(s)} = AH_{(a)} + H_{(a)}$	0	0	0	1	0
8. $A_{(a)} + H_{(a)} = AH_{(a)}$	0	0	0	0	1
9. $AH_{(a)} + H_{(a)} = AH_{2(s)}$	0	0	0	1	1

embracing five routes for the reaction $A_{(s)}+H_{2(s)}=AH_{2(s)}$, where the subscript (s) denotes solution. These routes actually characterise different possible mechanisms of the reaction. They include interaction of the adsorbed organic component with adsorbed atomic hydrogen or with molecular hydrogen from the solution, with or without the formation of the intermediate half-hydrogenated species and also simultaneous interaction of both components from the solution with the catalyst. The latter mechanism, although it requires a triple collision, is quite plausible in liquid-phase reactions (ref. 75).

In Table 1.1 the subscript (a) denotes adsorbed particles. If the first two steps are equilibrium ones (which is very likely) and the reaction generally proceeds via all these routes simultaneously (i.e. various mechanisms are realised at the same time), then the following kinetic equations will correspond, respectively, to low, medium and high coverages of the catalyst surface (ref.74):

$$r = k\, c_1 c_{H_2} (1 + k' c_{H_2}^{-0.5}) \tag{1.5}$$

$$r = k\, \frac{c_1 c_{H_2}}{M^{\alpha'}} (1 + \frac{1 + k' c_{H_2}^{-0.5}}{k\, M^{\alpha'}}) \tag{1.6}$$

$$r = k\, \frac{c_1 c_{H_2}}{M} (1 + \frac{1 + k' c_{H_2}^{-0.5}}{k'' M}) \tag{1.7}$$

$$M = a_1 c_1 + a_H c_H + a_{A(cat)H_2} c_{A(cat)H_2} + a_{AH} c_{AH} \tag{1.8}$$

where symbol C denotes reactant concentration in solution and a is a constant.

The symbols c_1, c_H, $c_{A(cat)H_2}$ and c_{AH} in eqn. 1.8 denote the surface concentrations of reactants and intermediates, which can be expressed by the reactant concentrations in solution. In these equations, α' characterises the relationship between the activation energy and the heat of formation of surface compounds at different sites of the real adsorbed layer ($0 < \alpha' < 1$). The values $\alpha'=0$ and $\alpha'=1$ formally correspond to low and high coverages of the surface. For low coverages

$$k = k_3 + k_4 + k_6 + k_7$$
$$k' = k_8 \tag{1.9}$$

and otherwise

$$k = k_3 + k_6$$

$$k' = \frac{k_8}{k_4 + k_7} \qquad (1.10)$$

$$k'' = \frac{k_3 + k_6}{k_4 + k_7}$$

where the subscripts correspond to the steps in the Table 1.1.

For medium and high coverages the form of the equation depends essentially upon whether a single adsorbed form is predominant; for low coverages there is no such dependence.

With the realisation of the first three routes I-III and medium coverages

$$r = kC_1^{1-\alpha'} C_{H_2} \qquad (1.11)$$

and at saturation

$$r = kC_{H_2} \qquad (1.12)$$

The last equation is equivalent to eqn. 1.3 owing to Henry's law.

If the catalyst surface predominantly adsorbs hydrogen, then for medium and high coverages, respectively,

$$r = kC_1 C_{H_2}^{1-\alpha'} \qquad (1.13)$$

$$r = kC_1 C_{H_2}^{0.5} \qquad (1.14)$$

and for a sufficiently high concentration of hydrogen, when the surface is covered predominantly by the complex $A \cdot (cat) \cdot H_2$, the reaction at saturation level will exclusively take route III. The rate of the process will then be given by

$$r = k^* \qquad (1.15)$$

i.e., zero-order reaction. This general scheme is obviously rather simplified; it does not account for other possible mechanisms of reactions. Nevertheless, it covers the most common cases, although requires verification in each particular case.

This was a brief account of the general views on the mechanisms and kinetics of hydrogenation without reference to particular processes. Unfortunately, a considerable number of investigations, especially those dealing with processes in the liquid phase, failed to present careful experimental tests of the kinetic models. The available data on the kinetics of particular reactions of catalytic hydrogenation will be discussed later.

1.4 KINETIC MODELS OF HYDROGENATION IN THE GAS PHASE

Here and elsewhere the most characteristic and complete data on the kinetics of hydrogenation by various catalysts will be quoted. For more detailed information the reader will be referred to the summaries included in reviews, *e.g.* refs. 39, 40, 68, 74.

1.4.1 Hydrogenation of olefins

The kinetics of these reactions at not-too-high temperatures (usually below 150-170°C) is described by simple equations (either 1.2 or 1.3), although sometimes, as we shall see later, more complicated relationships may be encountered. The activation energy E is usually 40-44 kJ/mol or somewhat lower. Some authors obtained negative values for n_2, which demonstrate the approximative nature of kinetic equations of this kind.

Eqn. 1.3 was obtained in ref. 57 by the gradientless technique (refs. 3, 6-9, 76) for the hydrogenation of ethylene on palladium membranes. In the temperature range from 0 to 42°C the activation energy is 42,6 kJ/mol, in accordance with Beeck's data (ref. 77) for various metals. At higher temperatures (102-209°C), $E \approx 5$ kJ/mol (near the temperature at which the reaction rate is maximum), while at 176°C the kinetics complies with the equation (ref. 57)

$$r = k\, P_1^{0.5}\, P_{H_2} \tag{1.16}$$

Within the framework of Twigg's mechanism, the fulfilment of this equation indicates the transition from a large to medium coverage of the catalyst surface by ethylene. Eqn. 1.3 was also obtained in ref. 78 for hydrogenation of ethylene in the gradientless system, on a nickel catalyst suspended in an inert liquid (n-octane) through which ethylene passed.

The kinetics of ethylene hydrogenation, also in the gradientless system (ref. 79), on a rhenium catalyst, is different and follows the equation:

$$r = kP_1 \tag{1.17}$$

The authors considered this to indicate that most of the catalyst surface was covered by hydrogen.

The kinetics of hydrogenation of ethylene on zinc oxide in a closed system was claimed (ref. 80) to be described by eqn. 1.2, the values of n_1 and n_2 varying with the composition of the mixture and even with the time, i.e., the kinetic dependence was not actually established. The authors reported irreversible poisoning of the surface by ethylene.

The available data on the kinetics of hydrogenation of C_3-C_5 olefins have been reviewed (refs. 39, 40); the kinetic behaviour is described by eqn. 1.2.

A kinetic study of the hydrogenation of propylene was carried out (ref. 81) at about $100^\circ C$ on various Group 8 metals. The kinetics is characterised by eqn. 1.3; the catalytic activity of the metals was compared with that for the hydrogenation of ethylene. The most active catalyst was found to be rhodium, as in the hydrogenation of ethylene. In contrast, iridium exhibited a high catalytic activity in the hydrogenation of propylene but not in that of ethylene. This has been explained (ref. 82) in terms of the hypothesis of vibration resonance in catalysed reactions, i.e., the matching of the frequencies of vibration of the metal-hydrogen bonds in the surface layer and vibrations in the species to be hydrogenized. The activation energies were in the interval 26-64 kJ/mol.

Fott and Schneider (ref. 83) compared the kinetics of hydrogenation of ethylene and butylene on a cobalt-molybdenum catalyst at $360^\circ C$ in a circulating flow gradientless reactor. The rates of both reactions were described by the equation

$$r = k \frac{P_1 P_{H_2}}{1+k'P_1} \tag{1.18}$$

Kraus and co-workers (ref. 84) obtained kinetic data on the hydrogenation of C_2-C_{10} olefins on a similar cobalt-molybdenum catalyst in a flow reactor at $300^\circ C$. The initial reaction rates can be expressed by

$$r^o = k \frac{P_1 P_{H_2}}{(1 + k_1 P_1 P_{H_2}^{0.5} + k_2 P_{H_2}^{0.5})^2} \tag{1.19}$$

corresponding to the slow interaction between hydrogen and the half-hydrogenated form. The back reaction is neglected, since the initial rates are being considered. The authors point attempted to describe the effects of olefin structure on the rate of hydrogenation by use of Taft's equation.

A subsequent work (ref. 85) gave a kinetic description of the hydrogenation of C_2-C_8 olefins on a sulphurated Al-Ni-W catalyst at $300^\circ C$ in a flow reactor. The initial reaction rates can be expressed by

$$r^o = k \frac{P_1 P_{H_2}}{(1 + k_1 P_1 + k_2 P_{H_2})^2} \tag{1.20}$$

equally well with $l=2$ and $l=3$, and by:

$$r^o = k \frac{P_1 P_{H_2}}{(1 + k_1 P_1 + k_2 P_{H_2}^{0.5})^3} \tag{1.21}$$

The constants characterising the adsorption of olefins increase with increasing molecular weight. The rate constants correlated with the steric hindrances in the respective half-hydrogenated forms.

The kinetics of hydrogenation of C_6-C_9 olefins on a palladium sulphide catalyst in a gradientless system was studied in the gas phase at 40-80°C (ref. 86), with and without an excess of aromatic hydrocarbons. The isotopic and IR spectroscopic investigations were simultaneously carried out. Under these conditions the aromatic compounds were not hydrogenated, and the process was selective. The rate of hydrogenation of each olefin was given by

$$r = k \frac{P_1 P_{H_2}}{M} \tag{1.22}$$

$$M = P_{H_2}^{0.5} + k_1 P_1 + k_1' P_1 P_{H_2}^{0.5} + \sum_j k_j P_j \tag{1.23}$$

Here k, k_1, k_1', k_j are constants; subscript j refers to the aromatic hydrocarbons. At the same time, isomerization with transposition of double bounds takes place according to the equation

$$r_i = k_i \frac{P_1 P_{H_2}^{0.5}}{M} \gamma \tag{1.24}$$

where γ is a correction for the back reaction. When the pressure was raised to 0,5 MPa, an additional term, $(1 + k' P_{H_2}^{0.5})$ in the denominator had to be employed to account for the increase in the concentration of the half-hydrogenated form.

The kinetic isotope effect, when hydrogen is replaced by deuterium, was found to be equal to 1.5-1.6 both for hydrogenation and isomerization. IR spectroscopy indicated that the role of the aromatic hydrocarbons consists in adsorptive displacement of the olefins to the sites of weaker adsorption; the sulphuration of the catalyst tends to weaken the bonding between its surface and the olefins. Based on the data accumulated the scheme of the process was supposed. This scheme includes two slow steps: the formation of a half-hydrogenated form and its conversion either into a paraffin upon addition of hydrogen, or into an olefin isomer after the loss of another hydrogen atom.

Studies of the kinetics of dehydrogenation of higher $C_{10}-C_{12}$ linear paraffins into olefins, diolefins and aromatic hydrocarbons on platinum catalysts in a gradientless systems provided information regarding also the kinetics of the back reaction, i.e. the hydrogenation of higher olefins at 420-470°C (refs. 87-89). As indicated by isotope investigations and data on changes in selectivity, the dehydrogenation of paraffins proceeds consecutively, while the hydrogenation and dehydrogenation of olefins occur in parallel. The rate of hydrogenation is given by

$$r = k \frac{P_1 P_{H_2}^\gamma}{P_{H_2}^{1.5} + k_1 P_1 + k_2 P_2'} \qquad (1.25)$$

where P_2' is the partial pressure of simultaneously formed dienes. The kinetic isotope effect is absent, and the kinetic model corresponds to a scheme in which the slow step is the adsorption of the olefins. The reaction also yields coke, and when its concentration on the catalyst surface reaches a certain level C, the process becomes non-stationary; its rate is given (ref. 90) by

$$r^* = k^* \frac{P_1 P_{H_2}^\gamma}{C^{2/3} + k' P_1 + k_2' P_2'} \qquad (1.25a)$$

The hydrogenation of cyclic olefins on six samples of a sulphurated cobalt-molybdenum catalyst prepared by different methods was studied (ref. 91) in connection with the hydrodesulphurization of thiophene in a flow reactor at 350°C. The initial rate can be expressed by

$$r^\circ = k \frac{P_1 P_{H_2}^2}{(1 + k_1 P_1)^2} \qquad (1.26)$$

while on the samples with the highest cobalt content (0.5 and 1%) another equation was also valid

$$r^\circ = k \frac{P_1 P_{H_2}}{(1 + k_1 P_1^{0.5})^2} \qquad (1.27)$$

The authors associate these equations with the Langmuir-Hinshelwood mechanism; the constants accounting for the adsorption of hydrocarbons were the same for different samples of catalyst, while the reaction rate constants were correlated with the amount of hydrogen adsorbed.

1.4.2 Hydrogenation of diene hydrocarbons

There have been few investigations in this field; some kinetic data have been reviewed (refs. 39, 40).

The kinetics of hydrogenation of allene was studied in a closed system on supported nickel, platinum and palladium in the range 0-200°C (ref. 92). The reaction rate can be described by eqn. 1.3 which, in the authors' view, corresponds to the strong adsorption of allene. The selectivity is close to unity on Pd and equals 0.8 on Pt.

The kinetics of this reaction was studied (ref. 93) also on various Group 8 metals supported on silica gel, with monitoring the surface compounds by IR spectroscopy. The reaction rate is also given by eqn.1.3 and partly by eqn.1.16. The activity of the catalysts decreases in the order Pd>Rh>Pt>Ni>Co>Ru>Ir. The reaction proceeds via the formation of a double-bonded compound at the surface, which reacts with hydrogen and passes into the gas phase.

The results of kinetic studies of the hydrogenation of butadiene in a closed system on all Group 8 metals have been reviewed in ref. 40 (also including data for copper). The reaction kinetics follows either eqn. 1.3 or eqn. 1.2: for iridium $n_1=0$, $n_2=0.8$; for nickel $n_1=-0.3$, $n_2=1$. The selectivity varies from 0.35 on Ir to unity on Pd, Ni, Co, Fe and Cu.

The kinetics of hydrogenation of isoprene on Pd-Pb catalyst was studied in a gradientless system at 57-118°C together with the hydrogen-deuterium exchange in the course of the reaction, the nature and transformations of surface compounds by IR spectroscopy and thermal desorption of hydrogen (refs. 94, 95). The rate of formation of each of the isoamylene isomers complied with eqn. 1.16. The slow steps are the formation of the half-hydrogenated form and its conversion. The high selectivity of this catalyst is apparently due to Pb which blocks the surface sites having the strongest bonds with carbon (as confirmed by IR spectroscopy).

Selective hydrogenation of cyclopentadiene on palladium and platinum catalysts as compared with the hydrogenation of alkadienes (1,3-butadiene and isoprene) was studied by Kripylo and co-workers (refs. 96-101), using a gradientless reactor (ref. 101) or circulation system (ref. 102). In the range 80-140°C the results could be described by the equation

$$r = k \frac{P_1 P_{H_2}}{(1+k_1 P_1 + k_2 P_2)(1+k' P_{H_2})} \tag{1.28}$$

Up to certain concentrations, hydrogen did not influence the process rate, then the kinetic equation becames more simple. Consecutive hydrogenation of the forming alkenes and cycloalkenes was characterised by a similar kinetic equation.

On the basis of their data, including the results obtained with deuterium, the authors concluded that the reaction proceeds in two steps, with the formation of an intermediate π-allyl compound, whose stability determines the selectivity of the process. This compound is unstable on platinum, which accounts for the poor selectivity of the latter.

Grjaznov and co-workers (refs. 103-105) studied the hydrogenation of cyclic polyenes on hydrogen-penetrable membranes of palladium-ruthenium alloys in gradientless reactors at temperatures up to $220°C$. The starting compounds were cyclopentadiene, cyclooctadiene, cyclooctatetraene and cyclodecatriene. The rate is given by the general eqn. 1.2 with n_1 and n_2 varying from 0 to 1 depending on the initial proportions of the mixture. This indicates that eqn. 1.2 has only approximate validity in this case. For the hydrogenation of cyclopentadiene, $n_1=1$ and $n_2=0.5$; the same values are used to describe the hydrogenation of the forming cyclopentene. The authors assumed slow interaction with hydrogen atoms passing through the membrane, although simultaneous interaction with molecular hydrogen cannot be excluded.

For the hydrogenation of the C_7-C_{12} cyclic hydrocarbons the authors gave the following general kinetic equation

$$r = k \frac{P_1 P_{H_2}}{(1+k_1 P_1)(1+k_2 P_{H_2}^{0.5})^2} \tag{1.29}$$

derived under the assumption of the slow joining of the adsorbed hydrogen atom to the hydrocarbon in the surface layer. The hydrogenation rate was found to decrease with increasing number of carbon atoms in the original hydrocarbon, providing the number of double bonds was the same, and to pass through a maximum when the number of double bonds was varied.

1.4.3 Hydrogenation of acetylene compounds

The kinetics of hydrogenation of acetylene hydrocarbons has been studied by many researchers; however, accurate kinetic models are few. Early kinetic data have been reviewed (refs. 39, 40). On most Group 8 metals the reaction rate is given either by eqn. 1.3 or by eqn. 1.2, with a zero or formally negative order in the hydrocarbon pressure.

Takeuchi and Miyahara (ref. 106) studied the hydrogenation of acetylene on a nickel filament in a closed system. The process simultaneously led to the formation of ethane, propylene and C_4 hydrocarbons. The initial reaction rate at $120°C$ in all these directions was given by eqn. 1.2 with n_1 in the interval 1,2÷1,5 and formally negative values of n_2 from -0.02 to -0.3; the products were found to have practically no effect on the kinetics. Up to a pressure of $75 \cdot 10^4$ Pa the slow step is presumably the adsorption of hydrogen, while at higher partial pressures it is represented by the formation of intermediate surface compounds.

Data on the kinetics of hydrogenation of methylacetylene in a closed system on supported Group 8 metals have been reported (ref. 107). The results can be described by eqn. 1.2; the reaction orders depended upon the experimental conditions.

The hydrogenation of heptyne-1 on a pumice-supported platinum catalyst was studied in a gradientless system at 67°C (ref. 108). The rates of build-up of heptene and heptane, following a consecutive-parallel scheme, are expressed, respectively, by

$$r_1 = \frac{k_1 P_1 P_{H_2} - k_{III} P_2 P_{H_2}}{P_{H_2}^{0.5} + k_1 P_1 P_{H_2}^{0.5} + k_2 P_1 + k_3 P_2 P_{H_2}^{0.5}} \tag{1.30}$$

$$r_2 = \frac{k_{II} P_1 P_{H_2} + k_{III} P_2 P_{H_2}}{P_{H_2}^{0.5} + k_1 P_1 P_{H_2}^{0.5} + k_2 P_1 + k_3 P_2 P_{H_2}^{0.5}} \tag{1.31}$$

The slow step is presumed to be the attachment of hydrogen to surface intermediates.

The hydrogenation of 2-methyl-3-butyn-2-ol on a palladium-lead catalyst proceeds selectively, without involving the hydroxyl group and the emerging double bond. A study of the kinetics in a gradientless system (ref. 109) and the use of other physicochemical techniques (isotopic exchange, isotopic effects, IR and NMR spectroscopy ref. 110), led to the equation

$$r = k \frac{P_1 P_{H_2}}{(P_1 + k P_{H_2}^{0.5})^{2.1}} \tag{1.32}$$

and to a scheme involving slow attachment of two hydrogen atoms or an hydrogen molecule to the adsorbed alcohol and direct release of 2-methyl-3-butene-2-ol into the gas phase.

1.4.4 Hydrogenolysis of aliphatic and alicyclic hydrocarbons

The hydrogenolysis of hydrocarbone catalysed by Group 8 metals has been studied in detail, and the kinetic data was reviewed by Sinfelt (ref. 111). The rates of these reactions can usually be expressed by eqn. 1.2 with a negative order with respect to hydrogen pressure, or by eqn. 1.3. This is believed to represent an approximation of non-power equations, which follow from the assumption that the dissociative adsorption of the hydrocarbon is an equilibrum step but the interaction of the hydrocarbon fragments with hydrogen proceeds slowly. For instance, for hydrogenolysis of ethane on nickel (refs. 112-116)

$$r = k \frac{P_{H_2}^{(1-x/2)} P_{C_2H_6-y}}{1 + k_1 P_{C_2H_6-y}} \tag{1.33}$$

$$P_{C_2H_6-y} = K' \frac{P_{C_2H_6}}{P_{H_2}^{y/2}} \tag{1.34}$$

where y is the number of hydrogen atoms lost in the course of dissociative adsorption of the hydrocarbon, $P_{C_2H_6-y}$ is the fugacity of the surface layer of dissociated fragments and K' is the equilibrium constant for dissociative adsorption of the hydrocarbon. These results were recently confirmed (ref. 116).

The kinetics and mechanism of hydrogenolysis of n-pentane on nickel and platinum in a gradientless system were studied (refs. 117-119) with simultaneous use of isotope techniques. The results once again complied with eqn. 1.2 where $n_1=1$ and $n_2= -0.5$, although they could be expressed with higher accuracy by

$$r = k \frac{P_1}{P_1 + k'P_{H_2}} \tag{1.35}$$

which implies equilibrium dissociative adsorption of pentane and slow splitting of fragments $C_5H_{11(ads)}$. At the same time, isomerization of n-pentane into isopentane takes place on platinum in the presence of hydrogen. With a Pt-Pb-alumina catalyst its rate is given by (ref. 120).

$$r = k \frac{P_1 P_{H_2}^{0.25} \gamma}{(P_{H_2} + k_1(P_{n-C_5} + P_{iso-C_5}))^{0.5}} \tag{1.36}$$

which follows from a similar assumption of equilibrium adsorption of pentane with subsequent fast migration to the catalyst surface sites adjoining the carrier, and slow interaction with hydrogen. The measurements of isotope effects confirm the participation of hydrogen in the slow steps of isomerization, contrary to hydrogenolysis.

The kinetics of hydrogenolysis of cyclopropane on nickel is described by (ref. 121)

$$r = k \frac{P_1}{1 + k'P_1} \tag{1.37}$$

derived under the assumption of equilibrium adsorption of cyclopropane and the slow rupture of a C-C- bond after adsorption. This assumption is based on adsorption and isotopic data.

Beránek and co-workers (ref. 122) studied the hydrogenolysis of methylcyclopentane over platinum-alumina and obtained the following expression for the rates of formation of 2- and 3-methylpentane in a flow reactor at 230°C:

$$r = k \frac{P_1 P_{H_2}}{(1 + k_1 P_1^{0.5} + k_2 P_{H_2})^3} \tag{1.38}$$

The constants k_2 for both directions are similar, while the values of k_1 differ by a factor of two. This was taken to indicate that the conversion in each direction takes place on different surface sites, the slow steps being the same.

An investigation of the kinetics of dehydrocyclization of isooctane on platinized carbon (ref. 123) led to the following equation for the rate of hydrogenolysis of 1,1,3-trimethylcyclopentane into isooctane

$$r = k \frac{P_1 P_{H_2}^{1.5}}{P_1 + k_1 P_{H_2} + k_2 P_1 P_{H_2} + k_3 P_2 P_{H_2}^{-0.5n}} \tag{1.39}$$

where $n \geq 1$. The slow step is the dissociation of the intermediate cyclic compound on the surface upon interaction with hydrogen (as shown by isotopic data).

1.4.5 Hydrogenation of aromatic compounds

The kinetics of hydrogenation of benzene has been studied extensively, although reliable results are scarce. Some studies are quoted in monograph (ref. 39). The majority of investigations were carried out in flow or static systems. Some results can be described by eqn. 1.2, *e.g.* with $n_1=0$ and $n_2=(0.4:0.5)$ for the reaction on nickel films at 20-60°C in a static system (ref. 124), and with $n_1=0$ and $n_2=2$ over nickel in a flow system at 40-100°C (ref. 125). In ref. 126 a value of $n_2=3$ is quoted for a platinum catalyst at high pressures. Japanese researchers (ref. 127) found $n_1=1$ and $n_2=3$ for low conversions on a thin nickel film (a differential reactor) at 140-200°C, and other work (ref. 128) quoted $n_1=1$ and $n_2=2$, also with a nickel catalyst in a differential reactor. On nickel catalysts at room temperature (ref. 129) the following values were found $n_1=(0.06:0.14)$ and $n_2=0.5$. As the temperature increases, the process of hydrogenolysis

$$C_6H_6 + 9H_2 = 6CH_4 \tag{1.40}$$

becomes considerable; at 247°C the results also comply with eqn. 1.2 when $n_1= -0.4$ and $n_2= 0.3$.

Snagovskyi and co-workers (ref. 130) thoroughly investigated the kinetics of hydrogenation of benzene on a nickel catalyst in a gradientless system at normal and high pressures. They proposed a scheme involving six slow steps in which hydrogen is attached to benzene in the surface layer. This is described by a complicated kinetic equation, which after certain simplifications, becomes eqn. 1.2 with $n_1=(0:0.5)$ and $n_2=(0.5:1.2)$, both n_1 and n_2 increase with increasing temperature.

The rate of interaction of benzene with deuterium catalyzed by platinum, palladium and ruthenium is also given by eqn. 1.2 with $n_1=(0.1:0.3)$ and $n_2=(0.6:0.9)$ (ref. 131). The slow step is presumably the attachment of adsorbed hydrogen to benzene in the surface layer.

Other studies reported the validity of eqn. 1.3 for the reaction over a nickel catalyst at up to 50°C (ref. 132) and 175°C (ref. 133) and over ruthenium at 40°C (ref. 134). According to ref. 135, eqn. 1.3 is valid for nickel at up to 180°C, whereas at higher temperatures, eqn. 1.17 is appropriate.

Many authors realise that power equations of this kind may be only approximations of more complicated relationships, *e.g.* those in refs. 127, 128, 130, 134-136. At the same time, in certain cases (for particular reaction mechanisms and particular relative adsorptivities of reactants), the power equations may give a true picture of the reaction kinetics. It is also possible that the obtained non-power relationships can actually be expressed by simpler equations, when account is taken of the reaction mechanism. This can be done, however, only on the basis of detailed investigations.

For example, for the hydrogenation of benzene on supported platinum, palladium and nickel in a flow system the following equation was proposed (ref. 137)

$$r = k \frac{P_{H_2}}{1 + k'P_{H_2}} \tag{1.41}$$

The authors, in accordance with experiment, considered the independence of r from P_1 and the Langmuir isotherm of hydrogen adsorption, without any speculations regarding the reaction mechanism.

The temperature dependence of the rate of benzene hydrogenation varies with the reaction conditions. Some researchers report a maximum rate at 170-190°C (see *e.g.* refs. 17, 127, 128, 138-140) or at 150°C (ref. 129). Such a maximum is apparently not associated with the approach to the equilibrium, since the rate of the back reaction becomes considerable only at higher temperatures. This is obviously caused by a change either in the reaction mechanism or in the adsorption of the reactants, which ought to be taken into account in the analysis of the kinetic data and in planning the experiments.

A series of works, summarised in refs. 17, 140, dealt with a large-scale investigation of the kinetics and mechanism of benzene hydrogenation on a Ni·ZnO catalyst over a wide temperature range, using isotopic and adsorption techniques. The kinetics of benzene hydrogenation, as well as the kinetics of the back reaction, were thoroughly studied in a flow-circulating system. Also studied were the possible intermediate reactions (cyclohexene and cyclohexadiene hydrogenation), the adsorption of the reactants, the relative rates of the individual steps, the isotopic exchange effects in the course of the reaction and the kinetic isotopic effects. In the reversibility region (above the maximum rate, at 210-270°C) the rate of benzene hydrogenation is given by

$$r = k_1 P_1^{0.5} P_{H_2}^3 - k_2 P_2 P_1^{-0.5} \qquad (1.42)$$

where P_2 is the partial pressure of cyclohexane. In this region the reaction rate decreases with increasing temperature and the energy of activation is formally negative, $E = -88$ kJ/mol.

In the regions of practical irreversibility, at 90-135°C and 150-185°C (below the maximum rate), the reaction rate is described respectively by

$$r = k P_{H_2}^{0.5} \qquad (1.43)$$

and by

$$r = k P_1 P_{H_2} \qquad (1.44)$$

The comparison with the hydrogenation of cyclohexene

$$C_6H_{10} + H_2 = C_6H_{12} \qquad (1.45)$$

as well as with the hydrogenation and dehydrogenation of cyclohexadiene showed the rates of these reactions to be much higher than the rate of hydrogenation of benzene. Reaction 1.45 in the presence of an excess of hydrogen can be described by the equation

$$r = k P_{C_6H_{10}} P_{C_6H_{12}}^{-0.5} \qquad (1.46)$$

The rate of hydrogen-deuterium exchange in benzene and cyclohexane in the course of the reaction exceeds the respective rates of hydrogenation and dehydrogenation, but the isotopic exchange in cyclohexene was much slower than reaction 1.45. The comparison with para-ortho hydrogen conversion in the course of the process indicated that the rate of this reaction is much higher than the rates of hydrogenation and dehydrogenation. All this demonstrates that the rates of adsorption of the initial reactants and of desorption of the final products of benzene hydrogenation and cyclohexane dehydrogenation cannot be slow under these conditions.

Furthermore, labelling with ^{14}C indicated that the label is not incorporated into cyclohexene or cyclohexadiene. This confirms the truly intermediate nature of these compounds, which are not desorbed in the course of the reaction. The kinetic isotope effects resulting from the replacement of protium by deuterium in hydrogen, initial and final hydrocarbons in the reactions of benzene hydrogenation and cyclohexane dehydrogenation are noticeable, but not in cyclohexene hydrogenation.

The measured kinetic isotopic effects for the direct and back reactions allowed calculation of the stoichiometric number of the limiting step (ref. 141), which was equal to unity (ref. 140). This result also precludes the possibility of the adsorption of hydrogen or the simultaneous reaction with three molecules of hydrogen being the slow steps.

The study of the adsorption of benzene in the course of the reaction revealed a considerable chemisorption of benzene on the nickel phase and its parallel orientation to the surface. The ortho-para conversion of hydrogen was noticeably accelerated in the presence of cyclohexene and cyclohexane. It indicates the dissociative nature of their adsorption. This was verified by IR spectroscopy of surface compounds and by ferromagnetic resonance studies.

Quantum chemical calculations indicate the feasibility of formation of surface π-complexes of benzene in the course of the reaction; their stability depends on the coverage of the catalyst surface (ref. 142). Other experiments imply that the isotopic exchange in benzene and cyclohexane on nickel requires chemical adsorption of hydrogen, whereas even at low temperatures the hydrogenation of benzene proceeds well with hydrogen from the gas phase or from the physically adsorbed state (ref.143).It was recently confirmed by the existence of a temperature threshold in the hydrogen-deuterium exchange of benzene on nickel-molybdenum catalyst in contrast with the benzene hydrogenation (ref. 144).

On the whole, the results indicate that the reversible hydrogenation of benzene can include neither the consecutive nor the simultaneous slow attachment of hydrogen with direct production of cyclohexane. A scheme that fits well the bulk of experimental data presumes fast adsorption of benzene with the formation of a surface π-complex, followed by a slow formation of a kind of reactive complex involving three hydrogen molecules, physically adsorbed nearby. Although the hydrogen is not bound chemically, such an unstable complex can be treated by analogy with known complexes such as arene complexes with weakly bound hydrogen, or complexes encountered in homogeneous catalysis by Group 8 metals. The next step is fast consecutive isomerization of the reactive complex, with hydrogen gradually passing from the outer into the inner sphere, which ultimately results in the formation of cyclohexane.

The investigation also revealed that in reaction 1.45, in contrast to hydrogenation of benzene, the slow step is the adsorption of cyclohexene.

At low temperatures ($85^{\circ}C$), isotopic effects are absent in the hydrogenation of benzene over platinum catalysts (ref. 145) and are negligible over nickel on silica gel, but at the same time are prominent in the isotope exchange. These results were related to the dissociative mechanism of π-complex substitution. It is clear, however, that at such low temperatures the mechanism of reaction must be different, although the reason why hydrogen takes no part in the slow steps is hard to explain.

The hydrogenation of benzene on nickel-alumina catalyst in a gradientless reactor is described in ref. 146 by

$$r = k \frac{P_1 P_{H_2}^{0.5}}{(1 + k'P_1)(1 + k''P_{H_2}^{0.5})} \quad (1.46a)$$

where k and k'' are constants, k' is a complicated function of P_{H_2}. The authors supposed the reaction to proceed via π- and δ-bonded benzene fragments on nickel.

A detailed investigation of the hydrogenation of toluene over a nickel-zinc oxide catalyst (for a summary of the results see ref. 147) showed its complete analogy with the hydrogenation of benzene, which allowed the proposed mechanism to be extended to this reaction as well.

A comparison of the rates of hydrogenation of different alkyl derivatives of benzene on nickel, cobalt and rhodium catalysts was undertaken by Völter and co-workers (refs. 148, 149). They used eqn. 1.17 and compared the values of its constants in juxtaposition with other characteristics. The relative hydrogenation rates on nickel were as follows: benzene:toluene:ethylbenzene:p-xylene:mesitylene = 1:0.87:0.54:0.54:0.29, consistent with the variations in the ionisation potentials of the hydrocarbons and with the increasing stabilities of the π-complexes. This led to the conclusion that these reactions involve the formation of surface π-complexes, whose stabilities increase with increasing number of substituents.

Accurate values of the rates of hydrogenation and isotopic exchange for a series of alkyl-substituted aromatic hydrocarbons over a nickel-zinc oxide catalyst at, respectively, 225 and 30°C have been compared (refs. 147, 150). The rates of hydrogenation were: benzene:ethylbenzene:toluene:m-xylene:p-xylene:o-xylene:mesitylene = 1:0.59:0.33:0.20:0.18:0.08. No consistent variations, however, could be detected in the rates of hydrogen-deuterium exchange in these hydrocarbons, whether in the aromatic ring or in the substituents. Thus it was concluded that different intermediates are responsible for the hydrogenation and isotope exchange in these systems (taking into account the difference in the temperatures of the reactions).

This conclusion was later independently verified by Gudkov (ref. 151) who in the same catalytic system studied the simultaneous reactions of hydrogenation of benzene and isotope exchange at 30°C. The combination of the two reactions opened up new opportunities for developing and interpreting kinetic models. The rates of isotope exchange and of hydrogenation (deuteration) of benzene were expressed, respectively, by

$$r_I = k_1 \frac{P_1 P_{D_2}^{0.5}}{M} \qquad (1.47)$$

$$r_{II} = k_{II} \frac{P_1' P_{D_2}}{M^2} \qquad (1.48)$$

$$M = P_{D_2}^{0.5} + k'P_1' \qquad (1.49)$$

where P_1' is the partial pressure of deuterium-substituted benzene. These equations correspond to a scheme, in which the exchange must proceed through the dissociation of benzene on the catalyst surface upon interacting with deuterium and loss of an HD molecule, and the hydrogenation is associative, through consecutive attachment of deuterium in the surface layer.

1.4.6 Hydrogenation of oxygen- and nitrogen-containing compounds

A detailed investigation of the dehydrogenation of isopropanol over a nickel catalyst (refs. 152, 153), which included a study of the kinetics of the direct and back reactions, isotope exchange in the components, IR and NMR spectroscopy and kinetic isotope effects, led to the following kinetic equation for acetone hydrogenation in the range 130-200°C (ref. 154)

$$r = k \frac{P_1 P_{H_2}(1 + k P_{H_2}^{0.5})\gamma}{1 + k_1 P_1 + k_2 P_{H_2}^{0.5} + k_3 P_1 P_{H_2}^{0.5}} \tag{1.50}$$

where γ is the correction for the back reaction. This equation corresponds to a two-route scheme, based on plenty of data, which presumes that the slow stages of formation of the hydroxyl-containing surface compounds are due to the interaction of adsorbed acetone with both adsorbed and molecular hydrogen. The scheme relies on the unusual direct participation of hydrogen in the back dehydrogenation reaction, the accelerating it, as well as the participation of the hydroxyl hydrogen in the slow steps.

In the region of practical irreversibility (50-90°C), the rate of acetone hydrogenation over a nickel catalyst in a gradientless system is given by (ref. 155).

$$r = k \frac{P_1 P_{H_2}^{0.5}}{(1 + k_1 P_1 + k_2 P_2)(1 + k' P_{H_2}^{0.5})} \tag{1.51}$$

corresponding to the adsorption of the organic compound and hydrogen on different kinds of surface sites and their subsequent slow interaction to produce the half-hydrogenated form. The authors do not preclude the alternative mechanism of slow interaction between adsorbed hydrogen and the half-hydrogenated form, which is described by a more complicated equation (ref. 156):

$$r = k \frac{P_1 P_{H_2}}{(1 + k_1 P_1 + k_2 P_2 + k_3 P_1 P_{H_2}^{0.5})(1 + k' P_{H_2}^{0.5})} \tag{1.52}$$

The mechanism proposed for the copper-catalysed reaction in the same region presumes slow interaction with molecular hydrogen, as well as the adsorption of atomic hydrogen by other sites, resulting in a half-hydrogenated form which subsequently interacts slowly with hydrogen. This is described by the kinetic equation (ref. 156)

$$r = k \frac{P_1^2 P_{H_2}}{(1 + k_1 P_1 + k_2 P_2 + k_3 P_1 P_{H_2}^{0.5})^2} \tag{1.53}$$

The kinetics on this reaction on other copper catalysts was recently studied by

Golodets and co-workers (ref. 157) and described by

$$r = k \frac{P_1 P_{H_2}}{(1 + k_1 P_1 + k_2 P_2)(1 + k' P_{H_2})} \quad (1.53a)$$

On nickel-alumina and cobalt-alumina these authors earlier found the equation 1.51 with $k_2=0$.

Šimoník and Beránek (refs. 158, 159) studied the kinetics of transformations in a set of simultaneous and consecutive reactions in the course of the interaction of crotonaldehyde with hydrogen on a platinum-iron catalyst at 160°C in a flow reactor. This process results in butyraldehyde (reaction I) and crotyl alcohol (reaction II) which isomerizes to aldehyde (reaction III), whereupon both products are hydrogenated to butanol (reactions IV, V). The initial rates are described by

$$r = k \frac{P_1 P_{H_2}}{(1+k_1 P_1^{0.5}+k_2 P_2^{0.5}+k_3 P_3+k_4 P_4+k_{H_2} P_{H_2}^{0.5})^l} \quad (1.54)$$

where $l=4$ for reactions I, II and IV, and $l=2$ for reaction V. These equations are the best fits under the assumption of surface interactions between the components.

The kinetics of hydrogenation of phenol on nickel and palladium at 110-130°C was studied (ref. 160) using the gradientless technique. Kinetic relationships were derived on the basis of a stepwise scheme, which assumes the attachment of adsorbed hydrogen to phenol in four slow steps, with the formation of cyclohexanone being followed by simultaneous attachment of two adsorbed hydrogen atoms to form cyclohexanol, the rates being denoted, respectively, by r_I and r_{II}

$$r_I = k_I \frac{P_1 P_{H_2}^2 \gamma}{(1 + k^* P_{H_2}^{0.5}) Y} \quad (1.55)$$

$$r_{II} = k_{II} \frac{P_2 P_{H_2} (1+k' P_{H_2}^{0.5}+k'' P_{H_2}^{1.5}+k''' P_{H_2}) \gamma'}{(1 + k^* P_{H_2}^{0.5}) Y} \quad (1.56)$$

where γ, γ' are corrections for the back reactions and Y is a complicated function of the partial pressures of all the components; $r_I \ll r_{II}$ on nickel, and $r_I \gg r_{II}$ on palladium.

The selective hydrogenation of phenol to cyclohexanone on a palladium catalyst in a flow system has been studied (ref. 161). The results at 145-175°C are best described by the equation

$$r = k \frac{P_1 P_{H_2}}{(1 + k' P_{H_2}^{0.5})^2} \qquad (1.57)$$

which corresponds to the assumption of a fast interaction of phenol from the gas phase with an adsorbed hydrogen atom, slow attachment of another surface hydrogen atom and, in further fast steps, attachment of extra surface hydrogen and rearrangement of cyclohexenol into cyclohexanone. The reaction is assumed to proceed in the ideal surface layer, as in the previously cited studies on the hydrogenation of phenol.

The kinetics of hydrogenation of phenol on a platinum catalyst (refs. 162, 163), studied in a flow reactor at 150°C, is more or less equally well described by three equations, of which the authors chose the one which corresponds to the assumption of a slow interaction between the initial species and adsorbed hydrogen

$$r_I = k_I \frac{P_1 P_{H_2}^2}{(1 + k_1 P_1 + k_2 P_2 + k_3 P_{H_2})^3} \qquad (1.58)$$

The hydrogenation of cyclohexanone formed to yield cyclohexanol is described by

$$r_{II} = k_{II} \frac{P_2 P_{H_2}}{(1 + k_1 P_1 + k_2 P_2 + k_3 P_{H_2})^2} \qquad (1.59)$$

which presumes a slow surface reaction with molecular hydrogen.

The initial rates of hydrogenation of various alkylphenols on nickel in a flow system at 160°C have been compared (ref. 164). The results are formally described by

$$r^o = k \frac{P_1 P_{H_2}^2}{(1 + k_1 P_1 + k_2 P_{H_2}^{0.5})^5} \qquad (1.60)$$

assuming slow interactions between the components in the surface layer. The relative initial rates of hydrogenation of different phenols were:phenol:o-cresol:

ethylphenol:*o*-isopropylphenol:*o*-propylphenol:*m-sec*-butylphenol:*o-tert*-butylphenol = 1:0.47:0.67:0.32:0.58:0.28:0.98.

A thorough investigation of the dehydrogenation of cyclohexanol by copper catalysts in a gradientless system, combined with a study of isotopic exchange in the reactants and measurement of kinetic isotopic effects in the direct and back reactions (ref. 165-167), led to a kinetic equation for the hydrogenation of cyclohexanone to cyclohexanol

$$r = k \frac{P_1 P_{H_2}}{(P_1 + k'P_2)^l} \qquad (1.61)$$

where $l = (0.5 \div 0.6)$. It describes the fast rearrangement of the surface compound of cyclohexanone into the enol form, with fast attachment of adsorbed H at the α-position of the ring, followed by a slow attachment of $H_{(ads)}$ at the β-position of the ring on the inhomogeneous catalyst surface.

From data on the dehydrogenation of cyclohexanol on alloyed copper catalysts in a flow reactor at 190-310°C (ref. 168), a kinetic equation for the hydrogenation of cyclohexanone to cyclohexanol was obtained

$$r = k \frac{P_1 P_{H_2}}{1 + k_1 P_1 + k_2 P_2} \qquad (1.62)$$

assuming the interaction of adsorbed ketone with molecular hydrogen on the homogeneous catalyst surface, and weak adsorption of hydrogen at the steady state.

Reduction of nitrobenzene to aniline is an important industrial process; reliable kinetic data, however, are quite scarce. The kinetics of this reaction have been studied in a flow-circulating reactor at 120-160°C over copper and nickel catalysts (supported, in the form of foil, as well as alloyed, ref. 169). The rate of reaction catalysed by copper is given by

$$r = k \frac{P_1 P_{H_2}}{1 + k^* P_1} \qquad (1.63)$$

which, as suggested by the authors, follows from a scheme presuming slow interaction of the adsorbed nitrobenzene with molecular hydrogen on a homogeneous surface. With nickel catalysts, the results can be described by eqn. 1.2 where $n_1 = 0.3$ and $n_2 = 0.7$, which is attributed to the effects of the inhomogeneity of the catalyst surface, both the reactants being adsorbed at different sites.

An investigation of the kinetics of reduction of nitrobenzene on modified nickel catalysts was carried out (ref. 170) in a flow-circulating reactor at

130-370°C. The process is accompanied by catalyst deactivation. The results were described by eqn. 1.2, treated in the same way as in ref. 169. The authors also described the change in the activity of the catalyst, taking into account the inhomogeneity of its surface (see also ref. 171).

1.4.7 Interaction of heterocyclic compounds with hydrogen

The results of a study of the kinetics of the desulphurization of thiophene and its derivatives have been reviewed (ref. 172); here we shall deal only with the studies not included in this review.

Data on the kinetics of hydrogenolysis of thiophene on cobalt-molybdenum catalysts, with simultaneous monitoring of the changes in the catalyst by XPS, have been obtained (ref. 173). The rate of the process in the presence of an excess of hydrogen can be expressed by:

$$r = k \frac{P_1}{1 + k_1 P_1 + k_2 P_{H_2S}} \qquad (1.64)$$

Fott and Schneider (ref. 174) studied the kinetics of this reaction in a gradientless system at 360°C and found the results to be described, to about the same accuracy either by equations characterising the production of hydrogen sulphide and butylene (r_I) and butane (r_{II})

$$r_I = k_I \frac{P_1 P_{H_2}}{1 + k_1 P_1 + k_2 P_2 + k_3 P_{H_2S}} \qquad (1.65)$$

$$r_{II} = \frac{k' r_I P_{H_2} + k'' P_2 P_{H_2}}{1 + k_I' P_2} \qquad (1.66)$$

which correspond to the consecutive hydrogenation of butylene to butane, or by a combination of eqn. 1.66 and

$$r_{II} = k^° r_I + \frac{k'' P_2 P_{H_2}}{1 + k_1 P_1 + k_2 P_2 + k_3 P_{H_2S}} \qquad (1.67)$$

corresponding to a parallel-consecutive scheme. Both versions presume the conversion of thiophene and butylene to take place on different sites of the catalyst surface, with slow interaction according to the Langmuir-Hinshelwood mechanism.

The following equations have been proposed (refs. 175, 176) for the kinetics of thiophene hydrogenolysis according to the Langmuir-Hinshelwood mechanism

$$r = k \frac{P_1 P_{H_2}}{1 + k' P_{H_2}^{0.5}} \qquad (1.68)$$

and

$$r = k \frac{P_1 P_{H_2}}{1 + k' P_{H_2}} \qquad (1.69)$$

Little is known about the kinetics of interaction of other heterocyclic compounds with hydrogen. The hydrogenation of furfural in a flow system on a copper chromite catalyst has been studied (ref. 177). The rates of formation of 2-methylfuran (sylvan), tetrahydrosylvan and side products in the presence of an excess of hydrogen can be described by first-order equations under the assumption of weak adsorption of all the components.

A detailed investigation of the transformations of 5-methylfurfural in the presence of hydrogen on a palladium catalyst in a gradientless system has been carried out (refs. 178-180). The transformations were studied in the directions of hydrogenation to dimethylfuran (reaction I), decarbonylation to sylvan (reaction II), as well as further conversions of sylvan in the absence of the original heterocyclic compound: hydrogenation to tetrahydrosylvan (reaction III), hydrogenolysis of sylvan (reaction IV) and isomerization of tetrahydrosylvan into methyl propyl ketone (reaction V). This was combined with a study of the hydrogen-deuterium exchange, measurement of kinetic isotope effects and IR spectroscopy of some of the surface compounds formed in the process.

In the course of the conversion of methylfurfural the catalyst steadily loses activity due to the blocking of its surface by the initial heterocyclic compound, which partly forms polymer-like substances and forces out other components formed in the process. Reaction I proceeds through slow consecutive reduction of the aldehyde group, bound directly to the catalyst, while reaction II goes through the slow split-off of this group in a flat-faced orientation, followed by fast introduction of externally adsorbed hydrogen into the furan ring. The rates of these reactions are expressed, respectively, by

$$r_I = k_I \frac{P_1 P_{H_2}}{P_{H_2}^{0.5} + k' P_1} \sigma \qquad (1.70)$$

$$r_{II} = k_{II} \frac{P_1}{P_{H_2}^{0.5}+K'P_1} \sigma + k_{II} \frac{P_1}{1 + k''P_1} (1 - \sigma) \qquad (1.71)$$

$$\frac{d\sigma}{dt} = k* \frac{P_1}{P_{H_2}^{0.5} + k'P_1} \sigma \qquad (1.72)$$

where σ is the proportion of unblocked catalyst surface. Further transformations of sylvan proceed via slow attachment of hydrogen to the dihydrosylvan formed, and hydrogenolysis and isomerization through slow splitting of the ring, while tetrahydrosylvan is retained by the surface by virtue of the unshared pair of electrons on the oxygen atom.

The rates of reactions III-V are described by

$$r_{III} = k_{III} \frac{P_1' P_{H_2}^2}{P_{H_2}^{0.5} + k_1'P_1' + k_2'P_2'} \qquad (1.73)$$

$$r_{IV} = k_{IV} \frac{P_1' P_{H_2}}{P_{H_2}^{0.5} + k_1'P_1' + k_2'P_2'} \qquad (1.74)$$

$$r_V = k_V \frac{P_2'}{P_{H_2}^{0.5} + k_1'P_1' + k_2'P_2'} \qquad (1.75)$$

where P_1' and P_2' are the partial pressures of sylvan and tetrahydrosylvan respectively. We see that the rate of isomerization depends not upon hydrogen(in the numerator of eqn.1.75),which comes to play only in quick steps of the reaction. Different forms of the non-stoichiometric hydrogen in various processes see refs. 181, 188.

* * *

The results presented in this section indicate that both the kinetics and the mechanisms of interaction of hydrogen with various compounds may be quite different. In most cases, accurate kinetic investigations lead to kinetic models of the non-power fractional form.

1.5 KINETIC BEHAVIOUR OF LIQUID-PHASE HYDROGENATION

From the arguments developed above it follows that in the study of the kinetics of catalysed hydrogenation the transition from gaseous to liquid phase cannot be done in a formal way, but detailed and comprehensive analysis is necessary. This gives rise to two major problems. The first lies in the design of kinetic experiments, suitable for acquiring reliable information regarding the liquid-phase processes. Secondly, there is the actual performance of the experiment and the analysis of the results obtained, taking due account of all liquid-phase effects, which might well complicate the kinetics. To these belong interactions in solutions, solvation of reactants and activated complexes, changes in the rates and even the natures of the individual steps, as well as other factors whose significance may vary with the circumstances.

The solution of these problems presents the researcher with a formidable task. The kinetics of liquid-phase reactions is studied almost exclusively in closed systems, which possess all the drawbacks of integral reactors (refs. 6, 7). Liquid-phase modifications of gradientless reactors have been proposed (refs. 7, 30, 183-186); but are of limited use because of the complications which arise, in particular, due to blowing out of finely divided catalyst. In addition, it is very hard to embrace all the factors which influence the kinetics in a three-phase system, including the solution, the solid-state catalyst and gaseous hydrogen. To make things easier it is either assumed that the common liquid-phase effects are of secondary importance compared with the effects arising from the presence of boundary surfaces (which is not always self-evident), or they are assumed to be accounted for by constants, which implies their independence of the actual experimental conditions.

The effects of the solvent in hydrogenation reactions have been studied in detail by Červený and Růžička (see, e.g., refs. 187-194). These authors demonstrated the feasibility of quantitative assessment of the effects of the reactant structure and the nature of the solvent on the kinetic behaviour, using linear free-energy relationships and correlation formulae. However, most published investigations dealing with various liquid-phase effects have not been pursuid to the extent of the construction of comprehensive and soundly based kinetic models for various processes of hydrogenation. Such models are quite rare indeed, probably for the reasons already indicated above.

As regards reactions in open systems, one also cannot exclude the possibility of direct interaction with the hydrogen flow passing through the catalyst, so that the hydrogen concentration exceeds that of its solubility. In this case, however, tiny bubbles of hydrogen must be able to penetrate into the thin boundary layer of liquid surrounding each particle of catalyst. This is less likely than dissolution. If the rates of dissolution of gases are high enough, the concentration of hydrogen in the solution in the course of reaction depends only on its solubility under the given conditions, providing the rate of reaction does not exceed the rate of dissolution (for instance, when the amounts of catalyst are very large). For this reason the concentration of hydrogen is usually low, but in the presence of an excess of solvent and at constant temperature and pressure it should remain unchanged throughout the process. In the absence of solvent, the concentration of hydrogen must depend on its relative solubility in the reactants.

From eqns. 1.5-1.15 it is evident that the overall rates of reactions taking place in the liquid phase at various step velocity ratios and different surface coverages should be described by different kinetic relationships; in particular, by equations of first or fractional order, with product inhibition and also perhaps by zero-order equations. Such relationships have actually been observed and described, *e.g.* refs. 34, 195, 196. It is important that the same equations could correspond to different mechanisms and step rate relations.

Bearing in mind all these arguments, we shall now discuss some specific kinetic models of liquid-phase hydrogenation with reference to specialised summaries and reviews. For economy of space, no difference between the symbols for partial pressure and for concentration will be made when referring to the earlier equations.

Hartog et al. (ref. 197) have described the kinetics of hydrogenation of benzene in a liquid-phase closed system over Raney nickel at atmospheric pressure and room temperature by use of eqn. 1.2 with $n_1=0.3$ and $n_2=0.6$. In a later work (ref. 131), concerning both liquid and gas phases on platinum, palladium and ruthenium catalysts and involving deuterium, the same equation was used with $n_1=0.1$ and $n_2=(0.7:0.9)$. It was assumed that the slow step was the attachment of adsorbed atomic hydrogen to benzene in the surface layer.

The kinetics of this reaction on supported nickel has been studied at high pressures (up to 5 MPa) in the temperature range of 175-275°C (ref. 198). The reaction rate is described by eqn. 1.3; at the same time the reaction was found to be slightly inhibited by the products, which suggests a more complicated model. According to ref. 199, the rate of this reaction at 0.9-14 MPa and 75-230°C is also described by eqn. 1.2 with $n_1=0$ and n_2 increasing from 0 to 1 with increasing temperature.

A similar change in the reaction order with temperature was observed by other researchers (see, *e.g.* ref. 200). This must be seen as an indication that the mechanism proposed is not fully adequate. On nickel and platinum as catalysts, the reaction orders were also found to vary with the temperature, n_1 ranging from 0.3 to 1, and n_2 as widely as from 0.6 to 2.5 in the temperature range 100-190°C (refs. 201, 202).

The kinetics of hydrogenation of acetone, benzene, its homologues and derivatives, as well as of multinuclear compounds has been studied (ref. 203). The reaction rates are also in accord with eqn. 1.2 with, in most cases, powers close to zero, although they vary with the experimental conditions.

Kaplan and Kiperman (ref. 78) studied the kinetics of hydrogenation of acetone on nickel in a liquid-phase gradientless system in the range 50-95°C. The reaction rate may be expressed by the equation

$$r = k \frac{c_1 c_{H_2}^{0.5}}{c_2^{0.5}} \tag{1.76}$$

which is interpreted in accordance with a two-step scheme taking place on the inhomogeneous catalyst surface, with slow adsorption of acetone and assuming equilibrium between the acetone surface compound, hydrogen and isopropanol in solution.

Temkin and co-workers (refs. 204, 205) studied the kinetics of hydrogenation of benzoic acid on palladated carbon at 130°C and pressures between 0,1 and 6 MPa. The experimental data could be fitted with the equation

$$r = k \frac{c_1 P_{H_2}}{1 + k' c_1 P_{H_2}} \tag{1.77}$$

which is viewed as deriving from the reaction on the homogeneous catalyst surface, via the formation of a surface complex between the reactants which undergoes further isomerization and fast hydrogenation to the final product. This scheme is in fact equivalent to that proposed earlier (refs. 34,74) to explain the transition to zero reaction orders (see below).

A similar equation was proposed (refs. 206, 207) for the hydrogenation of molten benzoic acid and phenol on nickel and palladium.

Konyuchov and Zyskin (ref. 208) drew recently attention to the possibility to describe the benzene hydrogenation data (ref. 146) also by eqn. 1.77 as well as in ref. 206 for this reaction on nickel in gas phase. But as pointed out Prasad (ref. 209) eqn. 1.46a reflected the mechanistic features of the reaction, therefore it was more preferable.

The rate of the sulpholene and methylsulpholene hydrogenation on nickel at 20°C and 0.7-20 MPa is described by eqn. 1.17 (ref. 210).

Gostikin (ref.211) considered the kinetics of various reactions by:

$$r = k \frac{c_1}{1 + k_1 c_1} \frac{c_{H_2}}{1 + k_2 c_{H_2}} \tag{1.78}$$

which assumes slow interaction between the two components occupying different sites of the homogeneous surface of the catalyst. The surface is supposed to contain two types of sites, which selective adsorb the components (similar to ref. 155). In particular, this kinetic treatment has been applied to the hydrogenation of nitrogen-containing compounds. As indicated by the author, it is necessary to take into account the possibility of non-stationary phenomena and diffusive hindrances.

A monograph (ref. 212) gives a comprehensive treatment of various theoretical models of the kinetics of hydrogenation in the ideal surface layer under different assumptions regarding the mechanism and relationships between the individual step rates. The authors associate the kinetic changes also with the variations in the catalyst potential, although they do not go so far as to give specific applications and accurate experiment-based models.

The literature occasionally contains kinetic data on enantioselective hydrogenation catalysed by metals, i.e., reactions leading to optical isomers. These studies are quoted in the monograph (ref. 213) and will be mentioned only briefly. The hydrogenation of ethyl acetoacetate over copper-ruthenium catalysts at 70-130°C and 0,5-2 MPa may be described as a function of P_{H_2} by equation 1.3. The reaction is assumed to take either of two independent routes: via interaction of the adsorbed ketoester with H_{ads}, or via the interaction of the half-hydrogenated form with hydrogen from solution, the initial reactants being adsorbed by different kinds of sites. Enantioselective hydrogenation of ethyl acetoacetate over modified ruthenium catalysts in the range 1-7 MPa occurs with various orders with respect to P_{H_2}. This is explained by the slow adsorption of hydrogen and subsequent slow interaction of the components in the surface layer, the kinetic equation having the form:

$$r = k \frac{P_{H_2}}{(1 + k'P_{H_2}^{0.5})^2} \qquad (1.79)$$

The authors believe that the variable reaction orders arise from the incorrect approximation of this equation by a power-type relationship. On the basis of the little data available, the transition from ordinary to enantioselective hydrogenation does not introduce any essential changes into the general kinetics of the reaction.

As noted above, some researchers observed changes in the reaction orders of liquid-phase hydrogenation upon reaching certain hydrogen pressures. These changes consist in a sharp transition to zero reaction order (usually at 6 to 8 MPa although sometimes the threshold pressure may be different), while below the threshold pressure the reaction is of first order with respect to P_{H_2} and of zero, first or fractional order with respect to other reactants (refs. 74, 195, 214). Various explanations of this effect have been given which treated the observed relationships in a not very rigorous fashion (*e.g.* in ref. 196), from the standpoint of Balandin's theory of hydrogenation, taking into account only the surface concentrations of the reactants (the discussion see in ref. 74).

It was shown (ref. 74) that, in general, the abrupt transitions from one kinetic pattern to another may be due to a change in the mechanism of the process when the partial pressure of hydrogen reaches a certain limit, whereupon hydrogen displaces all the other components from the surface. Then the reaction proceeds predominantly via the formation of a surface complex with the two components coming from the solution (see Table 1.1 and eqn. 1.15). A similar treatment of a special case (the transition from first to zero order with respect to both components) is based on eqn. 1.77 (see refs. 204-206). In this case, however, the transition from one equation to the other should be gradual rather than abrupt.

This discussion of the kinetics of liquid-phase hydrogenation leaves us with the impression that, despite the abundance of investigations, only few of the processes have been describe by accurate and reliable kinetic models, although general kinetic models and their implications have been discussed at length.

1.6 HYDROGENATION IN SYSTEMS CONTAINING SEVERAL UNSATURATED COMPOUNDS

Among the main problems of the kinetics of catalyzed hydrogenation are those connected with the hydrogenation of mixtures of organic compounds. The main question is, how to derive from the known kinetics of transformations of individual compounds the unknown kinetics of reactions of their mixtures under the same conditions. This problem has been considered (refs. 215, 216) with due regard to the possible differences both in the kinetics and in the mechanisms of individual reactions.

Consider the reactions

$$a_i A_i + b_i B_i = c_i C_i + d_i D_i \tag{1.80}$$

where a_i, b_i, c_i, d_i are stoichiometric coefficients, some of which may equal zero. Regardless of their mechanism, these reactions can be described by kinetic equations of the general form:

$$r_i = k_i \frac{P_{A_i}^{m_i} P_{B_i}^{n_i} \gamma_i}{M_i^{l_i}} \tag{1.81}$$

$$M_i = L_i + k_{A_i} P_{A_i} + k_{B_i} P_{B_i} + k_{C_i} P_{C_i} + k_{D_i} P_{D_i} + \sum_{jj} \alpha_{jj} P_j^{\beta_j} P_{j'}^{\beta_{j'}} \qquad (1.82)$$

In these equations γ_i is a correction for the back reaction, and the meaning of the constants m_i, n_i, l_i, k_{A_i}, k_{B_i}, k_{C_i}, k_{D_i}, $\alpha_{jj'}$, β_j, $\beta_{j'}$ depends on the mechanism of the process (some of them may be zero); the constants in the denominator do not necessarily account for the adsorption coefficients (ref. 217). The majority of the kinetic equations proposed for systems of this kind may be viewed as particular cases of eqn. 1.81.

When two or more reactions are combined, the change in the kinetics will be due to the mutual influence of the components upon the catalyst surface. This influence may be totally absent, or be simple, comprising only adsorptive competition, or it may be complicated, altering the nature of the intermediates and even changing the direction of the reaction. These cases may be classified as follows:

(1) Before and after superposition, the reactions take place in the region of low coverage of the catalyst surface. Then there will be no mutual influence and the kinetics of the reactions will remain the same, as will their velocities ($r_i' = r_i$); slight changes only may result due to variations in the values of γ_i (corrections for back reactions).

(2) The combined reactions take place on different sites of the catalyst surface. The outcome is the same as in (1), provided that there is no interaction between the surface compounds on the borders of these areas.

(3) The species A_1, engaged in one of the reactions, almost entirely displaces the initial substances from other reactions (A_2, ..., A_i), as well as the components B_2, ..., B_i; or, alternatively, the components B_2, ... B_i may be common to the combined reactions (in hydrogenation, for instance, $B_1 = B_2 = ... = B_i \equiv H_2$). Then $r_1' = r_1$; $r_2' = ... = r_i' = 0$.

(4) Combination leads to adsorptive competition. Then, in general, for each of the individual reactions $r_i' < r_i$, and

$$r_i' = k_i \frac{P_{A_i}^{m_i} P_{B_i}^{n_i}}{(M_i + \Delta M_i)^{l_i}} \qquad (1.83)$$

where ΔM_i denotes additional terms in eqn. 1.82 which depend on the mechanisms of the reactions combined with the given one and on adsorptive relationships.

(5) As in (4), but by virtue of superior adsorptivity of the components of one of the reactions we obtain $r_1' = r_1$; $r_2' < r_2$; $r_i' < r_i$. Then $M_1 + \Delta M_1 \approx M_1 = M_2 + \Delta M_2 = ... = M_i + \Delta M_i$.

(6) Combination leads to the appearance of new intermediates, new directions and pathways. This opens the way for arbitrary changes in the reaction rates, and in the form of the kinetic equations and parameters. This is more likely when the initial compounds are of starkly different natures.

(7) Combination leads to mutual interaction of the components on the catalyst, resulting in the creation of new stationary states with any arbitrary changes in the reaction rates.

Some studies of the processes of hydrogenation have actually reported cases of mutual influence, where knowledge of the kinetics of individual reactions allowed the prediction of the kinetics of transformations of the mixtures. Kinetic equations which describe this simple interference and are verified experimentally have been given, for example, by Beránek and co-workers (ref. 218) for the hydrogenation of mixtures of phenol and cyclohexanone (refs. 162, 163), and for the parallel-consecutive hydrogenation of crotonaldehyde (ref. 158). Simple mutual influence is also observed in the hydrogenation of heptyne and heptene (ref. 108), as well as mixtures of C_6-C_9 olefins (ref. 86), and the predictions are in accord with the actual kinetics observed.

The essential point here is that the description of the kinetics of hydrogenation of mixtures does not require any new constants apart from those already known from the kinetic equations for hydrogenation of the individual substances. This enables the simple "design" of the required kinetic models.

Similar opportunities arise also for the reactions of liquid-phase hydrogenation, insofar as there exists ample experimental evidence of such simple mutual effects (see the monographs in refs. 58, 195, 219). One such case is the total displacement of one species by another from the surface, as in the consecutive selectivity scheme. Selectivity of this kind is observed, as indicated above, in the hydrogenation of triple-bond compounds and dienes, in the course of which the olefine are completely expelled from the surface of selective metal catalysts (or remain on non-reactive sites), and undergo hydrogenation only after the complete conversion of the initial substances (case 3 above). Extensive experimental results on liquid-phase hydrogenation reactions, combined and competing, have been reviewed by Červený and Růžička (ref. 220).

The available data on combined transformations indicate that in the hydrogenation of mixtures in the case of simple mutual influence the decisive factor is not the initial rates of the individual reactions, but rather the relative adsorptivity of the components, which ultimately determines the degree and nature of these effects. The form 1.83 of the kinetic equations for combined reactions is, obviously, not the only possible one. For example, the hydrogenation of a mixture of hexene and benzene has been found described by ref. 221 to be

$$r = k \frac{P_1 P_{H_2}^2}{(P_1' P_2')^{0.5}} \tag{1.84}$$

where P_1' and P_2' correspond to the products of hydrogenation of each species. This equation implies the displacement of benzene by hexene from the inhomogeneous catalyst surface, similarly to the influence of hydrogen on the hydrogenation of nitrogen (ref. 222). By contrast, in the hydrogenation of olefins on a palladium sulphide catalyst (ref. 86), it is the aromatic hydrocarbons that displace the olefins, including hexene.

When the natures of the hydrogenated species are starkly different, a simple mutual influence may be superseded by complex effects. Then new intermediates and new routes are quite likely to arise, and the form of the kinetic relationships may be drastically changed (refs. 215, 216). In multicomponent mixtures both types of mutual influence may occur simultaneously; then the changes in the rates of the combined processes may follow a predictable pattern in the region where they are governed by the simple interference. For example, it was demonstrated (ref. 83) that although the hydrogenation of olefins and thiophene on a cobalt-molybdenum catalyst takes place at different sites of the surface, the combination of these reactions results in the attack by sulphurous compounds on the surface sites responsible for catalysing the hydrogenation of olefins. This is exactly what is meant by a change in the state of the catalyst as a result of complex influences.

Hopefully, the study of the kinetics of hydrogenation of various mixtures of organic compounds, and the construction of accurate kinetic models, will throw light on the details of the mechanism of hydrogenation catalysis.

1.7 SELECTIVITY OF HYDROGENATION

The selectivity of a process is an important kinetic function, whose magnitude and variation characterise the reaction under consideration. The selectivity aspects of hydrogenation have been discussed in a number of monographs and reviews (see, e.g. refs. 58, 216, 220, 223-225). A kinetic approach to selectivity in the light of the research done in refs. 1, 13, 74, 223 will be discussed below.

Selectivity is defined as the ratio of the rate of accumulation of the desired product to the overall rate of transformation of the initial reaction mixture in all possible directions, all rates being related to the same conditions and conversions. For the sake of simplicity, we consider two reactions of selective and exhaustive hydrogenation according to the parallel-consecutive scheme

$$A \xrightarrow[I]{H_2} AH_2 \xrightarrow[III]{H_2} AH_4$$
$$A \xrightarrow[II]{2H_2} AH_4$$

where AH_2 is assumed to be the desired product.

This scheme envisages the formation of the desired product AH_2 by reaction I, and the waste product AH_4 by reactions II and III. Then the selectivity of the process can be expressed as

$$S_{AH_2} = \frac{r_I - r_{III}}{r_I + r_{II}} \qquad (1.85)$$

If $r_{II} = 0$, the scheme becomes consecutive, and parallel if $r_{III} = 0$.

The reaction rate can always be presented as a product of the initial rate r^o and a function $\phi(x)$, which defines the dependence of the rate upon the conversion x and characterises the properties of the kinetic model (ref. 226)

$$r = r^o \phi(x) \qquad (1.86)$$

Hence

$$S = \frac{1 - \beta \dfrac{\phi_{III}(x)}{\phi_I(x)}}{1 + \alpha \dfrac{\phi_{II}(x)}{\phi_I(x)}} \qquad (1.87)$$

where

$$\alpha = r_{II}^o / r_I^o \qquad (1.88)$$

$$\beta = r_{III}^o / r_I^o \qquad (1.89)$$

The cases $\alpha=0$ and $\beta=0$ correspond, respectively, to the consecutive and the parallel schemes. Functions ϕ are normalised to unity; for x=0 and x=1 they equal 1 and 0 respectively.

Consistent with the earlier discussion, the reaction rates can be expressed by

$$r_1 = k_1 \frac{P_1^{m_1} P_{H_2}^{n_1}}{M_1^{l_1}} \;;\quad r_{11} = k_{11} \frac{P_1^{m_{11}} P_{H_2}^{n_{11}}}{M_{11}^{l_{11}}} \;;\quad r_{111} = k_{111} \frac{P_2^{m_{111}} P_{H_2}^{n_{111}}}{M_{111}^{l_{111}}} \qquad (1.90)$$

Coincidence of the form of these equations is, obviously, possible only for the first two reactions when $m_I = m_{II}$, $n_I = n_{II}$, $l_I = l_{II}$ and $M_I = M_{II}$, or for all three reactions when, in addition, $m_I = m_{II} = m_{III} = 0$ (zero order with respect to the initial species). Casting aside the latter case as being rather unlikely, we see that similar kinetic patterns can be realised only in the parallel scheme. Those reactions which exhibit similar kinetic patterns, differing only in the values of the constants, are called isokinetic reactions (refs. 1, 13, 223).

Thus, if the reaction order with respect to the initial species is non-zero, and the reactions are isokinetic, it means that they should follow the parallel scheme. Expression 1.87 shows the influence of the various parameters and their changes on the selectivity of the process. In general, the selectivity will strongly depend on the reaction conditions and will vary in the course of reaction. For this reason it cannot be correctly represented as a ratio of reaction rate constants or, for that matter, of any other constant factors. It follows from eqn. 1.87 that the selectivity will remain constant under varying conditions if the values of α and β remain unchanged. Also, constancy of selectivity is possible when the ratios of the functions ϕ don't change throughout the course of the reaction. Such conditions can be satisfied by isokinetic reactions, and, as follows from the above arguments, by the parallel scheme (with a minor exception). Independence of β of variations in the initial conditions is not likely (besides the case of zero order reactions), since this quantity includes the initial concentrations of the primary substances, which change continuously in the consecutive scheme and in different manners under different conditions.

Consequently, if the selectivity remains constant in the course of the process or when the initial conditions are altered, the reaction probably follows the parallel scheme. On the other hand, extrapolation of the selectivity towards its initial value, S_o, at x=0 may serve as an indication of the consecutive scheme when $S_o = 1$ or the parallel-consecutive (or parallel) scheme when $S_o < 1$. In this way, analysis of the selectivity pattern allows one to deduce the properties of kinetic models. This is reflected in the kinetic behaviour of catalyzed hydrogenation.

In the paper (ref. 223) some results are quoted which actually indicate that for isokinetic reactions the selectivity remains constant when the partial pressures of hydrogen or of the initial substance are varied, as well as in the course of the process (see also ref. 13). This specific features provides a reliable means for assessing the character of kinetic models. For instance, for the hydrogenation of higher olefins (refs. 87-89) the selectivity of formation of paraffins and of dehydrogenation to dienes does not change in the course of the process. This points to the isokinetic nature of the reactions and to the parallel selectivity scheme of transformations of olefins. It also follows that the inverse process of dehydrogenation of paraffins follows a consecutive scheme, evidence in support of which is provided by isotopic data and by the fact that the selectivity of production of olefins decreases with increasing conversion (ref. 13).

For the hydrogenation of heptyne on platinum (ref. 108) the initial selectivity, extrapolated towards $x=0$, was found to be less than unity ($S_o \approx 0.8$). For small values of x the selectivity is practically independent of x (low P_{H_2}), although it decreases with increasing x for large excesses of hydrogen. This is characteristic of the parallel-consecutive selectivity scheme, which was proposed in the kinetic model (see eqns. 1.30, 1.31). A similar pattern was observed for the hydrogenation of vegetable oil (ref. 196).

On the contrary, for the hydrogenation of cyclopentadiene on palladium the extrapolation of selectivity yields $S_0=1$, therefore the consecutive scheme is realised (refs.13,97,101). In the hydrogenation of isoprene (ref.94,95) the selectivity remains high ($S>0.92$) throughout the process, which points to the displacement of isoamylenes from the surface, as confirmed by the constant proportion of their isomers and by the absence of isotope exchange in the isomers.

The hydrogenolysis of n-pentane on a palladium catalyst (refs. 117-119) also gives rise to isopentane, which simultaneously undergoes hydrogenolysis, as described by a similar kinetic model. The rate of isomerization complies with a different model. Accordingly, in this case we are dealing with the constant selectivity of hydrogenolysis of isopentane with respect to the hydrogenolysis of n-pentane with increasing conversion, and with decreasing selectivity of isomerization with respect to hydrogenolysis (ref. 13). Similar patterns are also observed in other cases (ref. 224). Thus we see that the nature of selectivity variations in reactions involving hydrogen correlates with the properties of kinetic models and can be useful for establishing and verifying the latter.

Spectroscopic studies of surface compounds, in combination with kinetic data, allowed the selectivity series for various metallic catalysts in reactions involving hydrogen to be ascertained (ref. 227)- for hydrogenolysis:

$$Ni, Co, Fe, Os > Rh, Ir \gg Pt, Pd$$

and for selective hydrogenation of dienes and alkynes:

$$Ru > Pd > Rh > Pt.$$

The rates of hydrogenation of various C_5 hydrocarbons at $86^{\circ}C$ and $x=0.2$ in the vapour phase on a palladium-lead catalyst, which characterise the selectivity of the conversions, run in the following order (ref. 228):

$$\text{isoprene : piperylene : pentyne-1 : pentene-1 : isoamylene =}$$
$$= 1 : 1.98 : 1.05 : 0.04 : 0.02$$

under conditions of incomplete removal of isoprene from the catalyst surface after hydrogenation. For the hydrogenation of C_5 hydrocarbons (alkynes) in the liquid phase, Bond and Rank (ref. 229) reported the following selectivity series for Group 8 metals

$$Pd > Pt > Ru > Rh > Ir$$

and for stereoselectivity:

$$Pd > Rh > Pt > Ru > Ir$$

Kripylo and co-workers (refs. 96-101) examined selectivity in terms of the variations in the activation entropy with different metal catalysts, which allegedly account for the observed differences. They assumed, that the kinetic models in all these cases are similar, for which there actually are no grounds. A similar approach to the selectivity of hydrogenation, based on a single kinetic equation and single mechanisms, is typical of Balandin's hydrogenation theory (ref. 58). The selectivity here is characterised by the "selectivity index", which is a ratio of the product of the reaction rate constants to the product of the coefficients of adsorption of the initial species. A similar characteristic was introduced by Wauquier and Jungers (ref. 230). The universality of this approach, however, is not in accord with the diversity of kinetic models and mechanisms for the processes of hydrogenation.

Extensive results on liquid-phase hydrogenation, collated by Červený and Růžička and reviewed (ref. 221), allowed the authors to relate selectivity to other characteristics, once again on the basis of a single kinetic model. As observed by the authors, the extreme cases of zero and first order (possibly arising from different kinetic models), often encountered in liquid-phase reactions, complicate the use of relationships of this kind.

Ipatiev Jr. and co-workers (ref. 231) adopted purely thermodynamic approach to selectivity, and derived the latter from the ratio of the equilibrium constants for the reactions of hydrogenation after equating their rates. This, however, is also only a particular case of reactions advancing at the same rate without violating the required parity between the changes in the free energies. This approach was discussed in ref. 223, see also ref. 195. Bond (ref. 39) defined the thermodynamic selectivity factor in the form of relationships between the kinetic and adsorptive characteristics, together with the mechanistic factor, depending on the particular properties of the catalyst which determine the mechanism of the process.

Clearly, selectivity is such an important characteristic of the kinetic pattern of reactions that its assessment in catalysed hydrogenation is no less essential than detailed studies of the kinetics and mechanisms of these processes.

1.8 CONCLUSION

The kinetic behaviour of catalysed hydrogenation has formed the subject of many detailed investigations, some of which are mentioned in this chapter. Such an abundance of works reflects the increasing interest in the kinetics of catalyzed reactions and the growing recognition of the fact that the mechanism of hydrogenation cannot be understood without a proper kinetic basis.

At the same time it must be admitted that only a minor proportion of the kinetic data available can be used for constructing accurate, complete and reliable kinetic models. This is why the construction of such models for both theoretical and practical purposes remains the central problem of chemical kinetics in hydrogenation catalysis. These models are required to stand up to quite high standards as regards their accuracy, sensitivity and comprehensiveness, since they serve as a basis for the design and optimisation of industrial catalysed processes in order to achieve the highest possible efficiency.

The problem of determining the optimum parameters and their relation to the observed phenomena is of great importance for the kinetics of catalysed processes. The optimum characteristics of liquid-phase hydrogenation have formed the subject of special research (ref. 34), which took account of the inhomogeneity of the catalyst surface, the changes in surface coverages and in the reaction conditions. It was shown that the optimal formation heats of intermediates may essentially change in the course of reactions.

Variations in the temperature and in the reactant concentrations may cause a transition to another region of the catalyst surface coverages, resulting in changes in relationships between the individual step rates, which will strongly affect the kinetics. The influence of these factors is far from being clear, and the investigation of these effects and their relationships to the properties of kinetic models still remains an important problem.

The kinetic data on hydrogenation correspond mostly to steady-state conditions. Few investigations have been concerned with possible non-stationary regimes (see, *e.g.* Gostikin and co-workers, refs. 212, 232-238). They emphasize the need to pay great attention to the non-stationary effects likely to arise in liquid-phase reactions dominated by the processes of dissolution and diffusion. On the other hand, the use of non-stationary techniques for studying the transition processes, critical phenomena and other effects will throw light on the mechanism and kinetics in steady-state regimes. Non-stationary regimes may prove to be very advantageous and informative. In particular, their selectivity may be higher compared to that under steady-state conditions. An important direction of future kinetic investigations consists, therefore, in the use of non-stationary methods and techniques for studying the fine details of the reaction mechanisms.

One cannot avoid the impression that the kinetics of liquid-phase hydrogenation is still to a large extent unexplored. This situation is due both to the lack of reliable liquid-phase kinetic techniques, and to insufficient knowledge of the detailed effects of the solvents, the solvation of reactants, active complexes and catalysts, a lack of reliable correlation between precise physicochemical measurements and substantiated kinetic descriptions. One of the major problems consists, therefore, in further development of accurate kinetic techniques for liquid-phase reactions and in the realisation of thorough complex investigations of systems of this kind.

Despite much attention paid to the selectivity aspects of catalytic reactions, the changes in selectivity have not been reliably correlated to the properties of the kinetic models of hydrogenation, and there still are many gaps in our knowledge of the hydrogenation selectivity and its dependence on various factors. Prime importance is, therefore, attached to investigations of all aspects of the selectivity of hydrogenation, and, in particular, of the prerequisites and conditions for selective hydrogenation of triple and conjugated bonds.

Finally, a no less urgent problem is the study of the macrokinetics of hydrogenation processes - the more so, since under practical conditions the diffusion resistance is more of a nuisance than one would like.

As pointed out by Gostikin (ref. 211), most kinetic data on liquid-phase hydrogenation are liable to be distorted by the transfer effects including the dissolution of hydrogen.

Hydrodynamic effects in the kinetic studies of liquid-phase hydrogenation were examined by Sokolskyi and co-workers (ref. 239). Experimental technique to monitor the hydrogen solubility has been proposed in ref. 240. The problem of eliminating the external diffusion resistance in liquid-phase reactors is especially urgent since the diffusion coefficients in liquids are small. The intensity of stirring required to eliminate the external diffusion effects is estimated by Temkin and co-workers (refs. 10, 241-243).

Of no less importance is the elimination of internal diffusion resistance in liquid-phase reactions. It was found for some reactions that the kinetic region is achieved only on very fine-dispersion catalyst samples(refs.244-247),see also refs. 18, 211. Diffusion effects in some hydrogenation reactions have been considered *e.g*. in refs. 170, 248, 249. The appropriate techniques for calculating the kinetic parameters taking into account the inner diffusion effects have been proposed in refs. 232-238, 250-254.

Kinetic investigations obviously represent just one branch of the comprehensive study of catalytic processes, aimed at discovering the nature of catalysis. Nevertheless, they play an important and distinguished role and their usefulness will be much greater if they are combined with the use of other techniques.

REFERENCES

1. S.L. Kiperman, Foundations of Chemical Kinetics in Heterogeneous Catalysis, Chimiya, Moscow, 1979.
2. S.L. Kiperman, Usp. Khimii, 47 (1978) 3-38.
3. S.L. Kiperman, in S.L. Kiperman (Editor), Kinetics of Heterogeneous Catalytic Reactions. Methodics Problems in Kinetica, Chernogolovka, 1983, p. 5.
4. S.L. Kiperman, in G.K. Boreskov and T.V. Andrushkevich (Eds.), Mechanism of Catalysis. Part I, Nature of Catalytic Action, Nauka, Novosibirsk, 1984, p. 159.
5. G.M. Schwab, Disc. Faraday Soc., 41 (1966) 415-417.
6. S.L. Kiperman, Introduction to Kinetics of Heterogeneous Catalytic Reactions, Nauka, Moscow, 1964.
7. S.L. Kiperman, Int. Chem. Eng., 11 (1971) 513-521.
8. S.L. Kiperman, Kinet. Katal., 13 (1972) 625-639.
9. S.L. Kiperman, Chem. Technik. 22 (1970) 327-334.
10. M.I. Temkin, N.V. Kulkova, V.Yu. Konyuchov and V.L. Lopatin, in S.L. Kiperman (Editor), Kinetics of Heterogeneous Catalytic Reactions. Methodics Problems in Kinetics, Chernogolovka, 1983, p. 27.
11. R.P. Chambers and M. Boudart, J. Catalysis, 6 (1966) 141-145.
12. N.I. Koltsov and S.L. Kiperman, J. Res.Inst. Catalysis Hokk. Univ., 26(1978) 85-99.
13. S.L. Kiperman, Kinet. Katal., 22 (1981) 30-44.
14. M. Boudart, Chem. Eng. Sci., 22 (1967) 1397-1399.
15. G.F. Froment and K.B. Bischoff, Chemical Reactor Analysis and Design, Wiley, N.Y. 1979.
16. G.F. Froment, Amer. Inst. Chem. Eng. Journ., 21 (1975) 1041-1062.
17. S. Kiperman, D. Shopov, A. Andreev, N. Zlotina and B. Gudkov. Comm. Dptm. Chem. Bulg. Acad. Sci., 4 (1971) 237-266.
18. G.C. Bond, in D.V. Sokolskyi (Editor), Catalytic Reactions in Liquid Phase, Nauka, Alma-Ata, 1972, p. 3.
19. I.I. Ioffe, in D.V. Sokolskyi (Editor), Catalytic Reactions in Liquid Phase, Nauka, Alma-Ata, 1976, p. 76.
20. M.G. Slin'ko, A. Ermakova, G.K. Siganshin, E.F. Stefoglo and A.N. Garzman, in Simulation of Chemical Processes and Reactors, Vol. 2, Novosibirsk, 1971, p. 1095.
21. J.E. Leffler and E. Grunwald, Rates and Equilibria of Organic Reactions, Wiley, New York, London, 1963.
22. E.S. Amis, Solvent Effects on Reaction Rates and Mechanisms, Academic Press, New York, London, 1966.
23. S.G. Entelis and R.P. Tiger, Kinetics of Liquid Phase Reactions. Quantitative Consideration of Media Effects, Chimiya, Moscow, 1973.
24. D.V. Sokolskyi, Ya.A. Dorfman, M.N. Anchevskaya and T.L. Rakitskaya, in Ya.A. Dorfman (Editor), Homogeneous Catalysis. Proc. of Institute of Org. Catalysis and Electrochemistry, Vol. 8, Nauka, Alma-Ata, 1974, pp. 3-35.
25. D.V. Sokolskyi and Ya.A. Dorfman, in N.P. Maslov (Editor),"Kinetics-2", Proc. 2nd Conference of Kinetics of Catalytic Reactions, Vol. 2, Institute of Catalysis, Acad. Sci. USSR, Novosibirsk, 1975, pp. 144-153.
26. D.V. Sokolskyi, Ya.A. Dorfman and M.N. Anchevskaya, Doklady Akad. Nauk SSR, 217 (1974) 1366-1369.
27. I.I. Ioffe, in D.V. Sokolskyi (Editor),Catalytic Reactions in Liquid Phase, Nauka, Alka-Ata, 1967, p. 36.
28. N.M. Emanuel, G.E. Saikov and S.K. Maisus, Role of Media in Radical-Chain Oxidation Reactions, Nauka, Moscow, 1973.
29. Ya.B. Gorochovatskyi, in D.V. Sokolskyi (Editor),Catalytic Reactions in Liquid Phase, Nauka, Alma-Ata, 1962, p. 85.
30. S.Z. Roginskyi, in V.N. Kondratiev (Editor), Chemical Kinetics and Chain Reactions, Nauka, Moscow, 1966, p. 483.
31. I.I. Ioffe and L.M. Pismen, Chemical Engineering in Heterogeneous Catalysis, Chimiya, Leningrad, 2nd ed., 1972.

32 Ya.B. Gorochovatskyi, T.M. Kornienko and V.V. Shalja, Heterogeneous-Homogeneous Reactions, Technika, Kiev, 1972.
33 S.L. Kiperman, in D.V. Sokolskyi and L.I. Zueva (Editors), Hydrogenation of Fats, Sugars and Furfural, Kazachstan, Alma-Ata, 1967, p. 14.
34 S.L. Kiperman, Comm. Dptm. Chem. Bulg. Acad. Sci., 1 (1968) 73-91.
35 C.N. Hinshelwood, Kinetics of Chemical Change, Clarendon Press, Oxford, 1942.
36 M.I. Temkin, Adv. Catal., 28 (1979) 173.
37 E.K. Rideal, Concepts in Catalysis, Academic Press, London, New York, 1968.
38 G. Ertl, Plenary Lecture on 7th International Congress on Catalysis, Tokyo, 1980.
39 G.C. Bond, Catalysis by Metals, Academic Press, London, New York, 1962.
40 G.C. Bond and P.B. Wells, Adv. Catal. 15 (1964) 91.
41 J. Horiuti, and T. Toya, in M. Green (Editor), Solid State Surface Science, Vol. 1, Marcel Dekker, New York, 1969, p. 5.
42 D.V. Sokolskyi and A.M. Sokolskaya, Metals - Hydrogenation Catalysts, Nauka, Alma-Ata, 1970.
43 G.D. Zakumbaeva, N.A. Zakarina, L.A. Beketaev and V.A. Naidin, Metal Catalysts, Nauka, Alma-Ata, 1982.
44 G.D. Zakumbaeva. Interaction of Organic Compounds with the surface of the Group 8 Metals, Nauka, Alma-Ata, 1978.
45 D.V. Sokolskyi and G.D. Zakumbaeva, Adsorption and Catalysis on the Group 8 Metals in Solutions, Nauka, Alma-Ata, 1973.
46 G.A. Somorjai, Chemistry in Two Dimensions Surfaces, Cornell University Press, Itaka, London, 1981.
47 N.M. Popova, L.V. Babenkova and G.A. Saveljeva, Adsorption and Interaction of the Simplest Gases with the Group 8 Metals, Nauka, Alma-Ata, 1979.
48 P.B. Moyes and P.B. Wells, Adv. Catal., 23 (1973) 121.
49 J.L. Garnett and W.A. Sollich-Baumgartner, Adv. Catal. 16 (1966) 95.
50 J.J. Rooney, J. Catal. 1 (1962) 255-274.
51 D. Shopov and A. Palazov, Compt. Rend. Acad. Sci. Bulg., 22 (1969) 181-184.
52 A.A. Balandin, Modern State of Multiplet Theory in Heterogeneous Catalysis, Nauka, Moscow, 1968.
53 J. Horiuti and M. Polanyi, Trans. Faraday Soc., 30 (1934) 1164-1172.
54 A. Farkas and L. Farkas, J. Am. Chem. Soc., 60 (1938) 22-32.
55 G.H. Twigg and E.K. Rideal, Proc. Roy. Soc., A171 (1939) 55-69.
56 G.H. Twigg, Disc. Faraday Soc., 8 (1950) 152-159.
57 L.O. Apel'baum and M.I. Temkin, Zh. Phys. Khim. 33 (1961) 2060-2070.
58 A.A. Balandin, Multiplet Theory of Catalysis, Part 3, Moscow Univ. Publ., Moscow, 1970.
59 A.A. Balandin, Doklady Akad. Nauk SSSR, 93(1953) 55-58.
60 G.I. Jenkins and E.K. Rideal, J. Chem. Soc. (1955) 2490-2500.
61 C. Wagner and K. Hauffe, Z. Elektrochem. 45 (1939) 409-425.
62 H.A. Taylor and N. Thon, J. Am. Chem. Soc., 75 (1953) 2747-2750.
63 V.V. Voevodskyi, F.F. Vol'kenstein and N.N. Semenov. In V.N. Kondratiev and N.M. Emanuel (Editors), Problems of Chemical Kinetics, Catalysis and Reactivity, Acad. Sci. USSR Publ., Moscow, 1955, p. 423.
64 N.N. Semenov and V.V. Voevodsky, in G.K. Boreskov et al. (Editors), Heterogeneous Catalysis in Chemical Industry, Goskhimizdat, Moscow, 1955, p. 233.
65 M.I. Temkin, Comm. Dptm. Chem. Bulg. Acad. Sci., 1 (1968) 65-72.
66 M.I. Temkin and L.O. Apel'baum, in Ya.M. Varshavskyi (Editor), Problems of Physical Chemistry, No. I, Goskhimizdat, Moscow, 1958, p. 94.
67 V.B. Kazanskyi and V.V. Voevodskyi, in S.Z. Roginskyi and O.V. Krylov (Editors), Problems of Kinetics and Catalysis, Vol. 10, Acad. Sci. USSR Publ., Moscow, 1960, p. 398.
68 J. Horiuti and K. Miyahara, Hydrogenation of Ethylene on Metallic Catalysts, U.S. Nat. Bur. Stand., 13 (1971).
69 K. Miyahara, in Theoretical Problems of Catalysis, Proc. of 1st Sovjet-Japan Seminar on Catalysis, Novosibirsk, July 4-7, 1971, Preprint I, pp. 1-22.
70 M.Ya. Kagan, in S. Z. Roginskyi et al. (Editors), Problems of Kinetics and Catalysis, Vol. 6, Acad. Sci. USSR Publ., Moscow, Leningrad, 1949, p. 232.

71 G. Parravanno, Chim. e l'Industria 51 (1969) 937-943.
72 D.V. Sokolskyi, in D.V. Sokolskyi (Editor), Proc. of Inst. of Catalysis and Electrochemistry, Vol. 9, Nauka, Alma-Ata, 1974, p. 3.
73 D.V. Sokolskyi, in D.V. Sokolskyi (Editor), Catalytic Reactions in Liquid Phase, Part I, Acad. Sci. Kaz. SSR Publ., Alma-Ata, 1974, p. 4.
74 S.L. Kiperman, Int. Chem. Eng. 18 (1978) 59-74.
75 N.M. Emanuel, Izv. Akad. Nauk SSSR, Ser. Khim., (1974) 1056-1098.
76 I.M. Berty, Catal. Rev.-Sci. Eng., 20 (1979) 75-95.
77 O. Beeck, Disc. Faraday Soc., 8 (1950) 118-126.
78 S.L. Kiperman and G.I. Kaplan, Kinet. Katal., 5 (1964) 888-892.
79 D.A. Kondratiev, T.N. Bondarenko, A.A. Dergachev and Ch.M. Minachev, Kinet. Katal., 18 (1977) 1591-1595.
80 F. Bozon-Verdura and S.J. Teichner, in Proceedings of the 4th Intern.Congress on Catalysis, Moscow, 23-29 June 1968, Akademiai Kiado, Vol. I, Budapest, 1971, p. 91.
81 R.S. Mann and T.P. Lien, J. Catal., 15 (1969) 1-9.
82 R.S. Mann and T.P. Lein, J. Catal., 88 (1984) 509-512.
83 P. Fott and P. Schneider, Collect. Czech. Chem. Commun., 45 (1980) 2742-2750.
84 J. Uchytil, E. Kočová and M. Kraus, Collect. Czech. Chem. Commun., 46 (1981) 2076-2082.
85 J. Uchytil, E. Jakubičkova and M. Kraus, J. Catal., 64 (1980) 143-149.
86 T.M. Matveeva, N.V. Nekrasov and S.L. Kiperman, in D. Shopov et al (Editors), Heterogeneous Catalysis , Proc. 5th Intern. Symposium on Heterog. Catalysis, Varna, October 3-6, 1983, Bulg. Acad. Sci. Publ., Varna, 1983, Part I, p.207.
87 T.L. Krylova, S.A. Sadychova, N.V. Nekrasov, B.S. Gudkov, S.K. Martjanova, N.A. Gaidaj, V.R. Gurevich and S.L. Kiperman, in M.G. Slin'ko (Editor) "Kinetics-3", Proc. 3rd All Union Conference, Kalinin, March 16-19, 1980, Vol. I, Kalinin, 1980, p. 251.
88 S.A. Sadychova, N.V. Nekrasov, V.R. Gurevich and S.L. Kiperman, Kinet.Katal., 22 (1981) 396-402.
89 S.A. Sadychova, N.V. Nekrasov, V.R. Gurevich and S.L. Kiperman. Kinet.Katal., 25 (1984) 593-597.
90 S.L. Kiperman, N.A. Gaidaj, N.V. Nekrasov and M.M. Kostyukovsky, Chem. Eng. Comm., 18 (1982) 39-50.
91 V. Vyskočil and M. Kraus, Collect. Czech. Chem. Commun., 44 (1979) 3676-3678.
92 G.C. Bond and R. Sheridan, Trans. Faraday Soc., 48 (1952) 658-663.
93 C.P. Khulbe and R. Mann, in G.C. Bond, P.B. Wells and E.C. Tompkins (Editors), Proceedings of the 6th International Congress on Catalysis, July 1976, Vol. I, London Chem. Soc. Publ., London, p. 447.
94 S.T. Beisembaeva, N.I. Popov and S.L. Kiperman, Izv. Akad. Nauk SSSR, Ser. Khim., (1976) 37-41.
95 S.T. Beisembaeva, B.S. Gudkov and S.L. Kiperman, Izv. Akad. Nauk SSSR, Ser. Khim., (1984) 525-530.
96 P. Kripylo, F. Turek and R. Oertel, Wiss. Zeitschr. Techn. Hochsch., Carl Schorlemmer, Leuna-Merseburg, 21 (1979) 88-96.
97 P. Kripylo, F. Turek, H.D. Hempe and H. Kirmse, Chem. Techn., 27 (1975) 675-682.
98 P. Kripylo and D. Klose, Chem. Techn., 28 (1976) 22-25.
99 P. Kripylo, P, Münch, T. Borchert, D. Klose, Chem. Techn., 29 (1977) 322-324.
100 P. Kripylo, P. Münch, T. Borchert, D. Klose., Chem. Techn., 30 (1978) 29-32.
101 P. Kripylo, H. Strecha and G. Müller, Chem. Techn., 32 (1980) 531-532.
102 K. Ramhold and R. Zimmerman, J. Prakt. Chem., 316 (1974) 839-850.
103 V.M. Gryasnov, M.M. Ermilova, N.V. Orechova and L.S. Morozova, in G.K. Boreskov (Editor), Mechanism of Catalytic Reactions, Proc. 3rd All Union Conference, Novosibirsk, 20-23 April, 1982, Part 1, pp. 179-182.
104 M.M. Ermilova, N.V. Orechova, L.D. Gogua and L.S. Morozova, in V.M. Gryasnov (Editor), Metals and Alloys as Membrane Catalysts, Nauka, Moscow, 1981, p.82.
105 V.M. Gryasnov, V.S. Smirnov, M.M. Ermilova, V.M. Zhernosek, N.V. Orechova, E.V. Chrapova, L.D. Gogua and N.N. Michaylenko, in O.V. Bragin (Editor), All Union Conference of Mechanism of Catalytic Reactions, Moscow, 1978, Vol. I, Nauka, Moscow, 1978, p. 202.

106 A. Takeuchi and K. Miyahara, J. Res. Inst. Catal. Hokk. Univ., 21 (1974) 132-149.
107 R.S. Mann and K.C. Khulbe, in Proceedings of the 4th International Congress on Catalysis, Moscow, June 23-29, 1968, Vol. 1, Akademiai Kiado, Budapest, 1971, p. 373.
108 I.M. Fialkova, G.N. Shvedova, N.V. Nekrasov, M.V. Hofman and S.L. Kiperman, in M.G. Slinko (Editor), Kinetics-3, Proc. 3rd All Union Conference, Kalinin, 1980, Vol. 2, Kalinin, 1980, p. 478.
109 D.Z. Levin, F.A. Melamed, M.A. Besprozvannyi and S.L. Kiperman, Kinet. Katal. 12 (1971) 1455-1463
110 D.Z. Levin, E.P. Prokophiev, B.S. Gudkov, M.A. Besprozvannyi and S.L. Kiperman, Kinet. Katal., 15 (1974) 98-103.
111 J.H. Sinfelt, Adv. Catal., 23 (1973) 1.
112 J.C. Kempeling and R.B. Anderson, Ind. Eng. Chem., Proc. Des. Dev., 11 (1972) 146-150.
113 P.K. Tsieng, R.B. Anderson, Can. Chem. Eng., 54 (1976) 101-106.
114 C. Machiels and R.B. Anderson, J. Catal., 58 (1979) 260-275.
115 A. Cimino, M. Boudart and H.S. Taylor, J. Phys. Chem., 58 (1954) 796-805.
116 G. Leclercq, L. Leclercq, L.M. Bouleaun, S. Pietrzyk and R. Maurel, J. Catal. 88 (1984) 8-17.
117 E.M. Davydov, M.S. Kharson and S.L. Kiperman, Izv. Akad. Nauk SSSR, Ser. Khim., (1977) 2687-2694.
118 E.M. Davydov, M.S. Kharson, N.I. Koltsov and S.L. Kiperman, Kinet. Katal., 19 (1978) 955-959.
119 E.M. Davydov, B.S. Gudkov, M.S. Kharson and S.L. Kiperman, Kinet. Katal., 19 (1978) 650-653.
120 I.A. Vartanov, M.S. Kharson, M.M. Kostyukovsky, V.G. Lipovich and S.L. Kiperman, Kinet. Katal., 25 (1984) 142-146.
121 T.R. Sridhar and D.M. Ruthven, J. Catal., 24 (1972) 153-160.
122 C. Corolleur, F.G. Gault, L. Beránek, React. Kinet. Catal. Lett., 5 (1976) 459-463.
123 S.A. Krasavin, M.S. Kharson, M.M. Kostyukovsky, O.V. Bragin and S.L. Kiperman, Izv. Akad. Nauk SSSR, Ser. Khim., (1982) 1231-1237.
124 O. Beeck and A.W. Ritchie, Disc. Faraday Soc., 8 (1950) 159-166.
125 R.A. Ross and B.G. Walsh, J. Appl. Chem., 11 (1961) 469-473.
126 P.F. Korbach and W.E. Stewart, Ind. Eng. Chem., Fundam., 6 (1964) 24-27.
127 S. Nagata, K. Hashimato, J. Tanijama, H. Nishida and S. Iwane, Chem. Eng. Japan, 27 (1963) 558-561.
128 J.E. Germain, R. Maurel, J. Burgeois and R. Sinn, J. Chim. Phys., 60 (1963) 1219-1222.
129 G.A. Martin and J.A. Dalmon, J. Catal., 75 (1982) 233-242.
130 Yu.S. Snagovskyi, G.D. Lubarskyi and G.M. Ostrovskyi, Kinet. Katal., 7(1966) 258-265.
131 F. Hartog, J.H. Tebben and C.A.M. Weterings, in W.M.H. Sachtler, G.C.A. Schuit and P. Zwietering (Editors), Proceedings of the 3rd International Congress on Catalysis, Amsterdam, July 20-25th, 1964, Vol. 2, North-Holland, Amsterdam, 1965, p. 1210.
132 W.F. Madden and C. Kemball, J. Chem. Soc., (1961) 302-310.
133 R.M. Flid, in Proceedings Inst. Fine Chem. Technol., No 6, Moscow, 1956, p. 51.
134 A. Amano and G. Parravano, Adv. Catal., 9 (1957) 716.
135 Cl. Herbo, Bull. Soc. Chim. Belg., 51 (1942) 44-49.
136 R.L. Motard, R.F. Burke, L.N. Kanjar and R.B. Beekmann, J. Appl. Chem., 7 (1957) 1-14.
137 J.C. Aben, J.C. Platteeuw and B. Stouthamer, in Proceedings of the 4th International Congress on Catalysis, Moscow, 23-29 June, 1968, Vol. I, Akademiai Kiado, Budapest, 1971, p. 345.
138 P. Sabatier and J. Senderens, Ann. Chim., 4 (1905) 319-488.
139 Cl. Herbo and V. Hauchard, Bull. Soc. Chim. Belg., 54 (1945) 203-235.
140 S.L. Kiperman, B.S. Gudkov and N.E. Zlotina, in O.V. Krylov and V.A. Selesnev (Editors), Problems in Kinetics and Catalysis, Vol. 15, Nauka, Moscow,

1973, p. 110.
141 J. Horiuti, Ann. N.Y., Acad. Sci., 213 (1973) 5-30.
142 S.L. Kiperman, in Proceedings of the 4th International Congress on Catalysis, Moscow, 23-29 June, 1968, Vol. I, Akademiai Kiado, Budapest, 1971, p. 593.
143 N.V. Nekrasov, B.S. Gudkov and S.L. Kiperman, Dokl. Akad. Nauk SSSR, 203 (1973) 1150-1153.
144 A.A. Slinkin, M.V. Chotuleva, B.S. Gudkov, T.N. Kucherova and S.L. Kiperman, Kinet. Katal., 25 (1984) 1014-1015.
145 R.Z.C. van Meerten, A. Morales, J. Barbier and R. Maurel, J. Catal., 58 (1979) 43-51.
146 K.H.V. Prasad, K.B.S. Prasad, M.M.Malekarjunan and R. Vaideswaran, J. Catal., 84 (1983) 65-73.
147 L.M. Mamaladze, N.V. Nekrasov, B.S. Gudkov and S.L. Kiperman, Acta Chim. Acad. Sci. Hung., 92 (1977) 73-83.
148 J. Völter, B. Lang and W. Kuhn, Z. Anorg. Algem. Chem., 340 (1965) 253-260.
149 J. Völter, M. Hermann and K. Heise, J. Catal., 12 (1968) 307-313.
150 B.S. Gudkov, L.M. Mamaladze and S.L. Kiperman, Izv. Akad. Nauk SSSR, Ser. Khim., (1975) 757-764.
151 B.S. Gudkov, Kinet. Katal., 20 (1979) 668-673.
152 T.P. Gegenava and S.L. Kiperman, in All Union Conference on Mechanism of Heterogeneous Catalytic Reactions, Moscow, 1974, Preprint 55, pp. 1-10.
153 T.P. Gegenava and S.L. Kiperman, Bull. Acad. Sci. Georgian SSR, 74 (1974) 613-616.
154 T.P. Gegenava, Proc. Acad. Sci. Georgian SSR, 2 (1976) 259-262.
155 P.B. Babkova, A.K. Avetisov, G.D. Lubarskyi and A.I. Gelbstein, Kinet. Katal., 10 (1969) 1086-1089.
156 P.B. Babkova, A.K. Avetisov, G.D. Lubarskyi and A.I. Gelbstein, Kinet.Katal., 11 (1970) 1451-1456.
157 A.I. Tripolskyi, N.V. Pavlenko and G.I. Golodets, Kinet. Katal., 26 (1985) 1131-1135.
158 J. Šimoník and L. Beránek, Collect. Czech. Chem. Commun., 37 (1972) 353-368.
159 J. Šimoník and L. Beránek, J. Catal., 24 (1972) 348-356.
160 Ju.S. Snagovskyi, M.M. Strelets, V.M. Borisov and G.D. Lubarskyi, in S.L. Kiperman (Editor), Reactions with Participation of Molecular Hydrogen, Novosibirsk, 1973, p. 73.
161 T. Mathe, J. Petro, A. Tungler, Z. Csüros and L. Lugosi, Acta Chim. Acad. Sci. Hung., 103 (1980) 241-248.
162 L. Beránek and V. Hančil, in 4th Intern. Congr. Catalysis, Moscow, 23-29 July, 1968, Symposium Mechanism and Kinetics of Complex Reactions, Preprint 10.
163 V. Hančil, P. Mitschka and L. Beránek, J. Catal., 13 (1969) 435-444.
164 J. Verjosta, V. Klenha and L. Beránek, Collect. Czech. Chem. Commun., 37 (1972) 1097-1105.
165 O.N. Medvedeva, A.S. Badrian and S.L. Kiperman, Kinet. Katal., 17 (1976) 1530-1536.
166 O.N. Medvedeva, B.S. Gudkov, A.S. Badrian and S.L. Kiperman, Izv. Akad. Nauk SSSR, Ser. Khim., (1977) 19-23.
167 V.A. Rovskyi, O.N. Medvedeva, R.I. Bel'skaya, N.I. Koltsov and S.L. Kiperman, Izv. Akad. Nauk SSSR, Ser. Khim., (1981) 519-521.
168 I.V. Orizarskyi, L.A. Petrov and V.M. Petrova, React. Kinet. Catal. Lett., 17 (1981) 427-431.
169 V.V. Pogorelov, F.L. Vigdorovich and A.I. Gelbstein, in N.P. Maslov (Editor), "Kinetics-2", Proc. 2nd Conference on Kinetics of Catalytic Reactions, Vol. 2, Inst. of Catalysis Acad. Sci. Publ., Novosibirsk, 1975, p. 88.
170 F.L. Vigdorovich, V.V. Pogorelov, Ju.K. Kapkov, A.G. Gorelik and A.I. Gelbstein, Kinet. Katal., 21 (1980) 975-982.
171 A.K. Avetisov and F.L. Vigdorovich, in S.L. Kiperman (Editor) Theoretical Problems of Kinetics of Catalytic Reactions, Chernogolovka, 1984, p. 34.
172 M.L. Vrinat, Appl. Catal., 6(1983) 137-158.
173 Y. Okamoto, H. Tomione, I. Imanaki and S. Teranishi, Proceedings of the 7th Intern. Congr. on Catalysis, Tokyo, 1980, Preprint A43.

174 P. Fott and P. Schneider, Collect. Czechosl. Chem. Commun., 45 (1980) 2728-2841.
175 I.M. Kolesnikov, V.A. Poteryachin, V.F. Sabitova, A.G. Rakhimkulov, V.N. Pavlichev and S.A. Akhmetov, Zhurn. Fiz. Khim., 52 (1978) 941-944.
176 J. Devanneaux and J. Mauren, J. Catal., 80 (1983) 491.
177 M. Vrinat and J. de Mourges, Compt. Rend.(II), 292 (1981) 584-591.
178 J.G. Yuskovets, N.V. Nekrasov, M.V. Shimanskaya and S.L. Kiperman, Kinet. Katal., 22 (1981) 1214-1217.
179 J.G. Yuskovets, N.V. Nekrasov, M.S. Kharson, M.M. Kostyukovsky, M.V. Shimanskaya and S.L. Kiperman, Kinet. Katal., 24 (1983) 1134-1139.
180 J.G. Yuskovets, N.V. Nekrasov, M.M. Kostyukovskyi, M.V. Shimanskaya and S.L. Kiperman, Kinet. Katal., 25 (1984) 1361-1364.
181 S.L. Kiperman, Comm. Dptm. Chem. Bulg. Acad. Sci., 16 (1983) 22-31.
182 Z. Paal and P.G. Menon, Catal. Rev.-Sci. Eng., 25 (1983) 229-334.
183 S.L. Kiperman, Kinet. Katal., 3 (1962) 520-522.
184 S.L. Kiperman and N.V. Nikolaeva, Kinet. Katal., 2 (1961) 936-939.
185 G.F. Tichonov, R.M. Flid, G.K. Shestakov and O.N. Temkin, Kinet. Katal., 7 (1966) 1914-1916.
186 F.B. Bizhanov D.V. Sokolskyi and U.I. Yunusov, Hydrolysnaja i Lesokhim. Promyshl., No. 1 (1973) 9-10.
187 L. Červený and V. Růžička, Adv. Catal., 30(1981) 335.
188 L. Červený and V. Růžička, Collect. Czech. Chem. Commun., 34 (1969) 1570-1579.
189 L. Červený and V. Růžička, Kinet. Katal., 24 (1983) 90-93.
190 V. Růžička, L. Červený and J. Pachta, Collect. Czech. Chem. Commun., 34(1969) (1969) 2074-2082.
191 L. Červený, A. Procházka and V. Růžička, Collect. Czech. Chem. Commun., 39 (1974) 2463-2469.
192 L. Červený, J. Bartoň and V. Růžička, Collect. Czech. Chem. Commun., 42 (1977) 3402.
193 L. Červený, J. Bartoň, V. Nevrkla and V. Růžička, Collect. Czech. Chem. Commun., 42 (1977) 3325-3332.
194 L. Červený, R. Junová and V. Růžička, Collect. Czech. Chem. Commun., 44 (1979) 2378-2383.
195 D.V. Sokolskyi, Hydrogenation in Solutions, Nauka, Alma-Ata, 1979.
196 D.V. Sokolskyi and K.A. Zhubanov, Hydrogenation of Vegetable Fats, Nauka, Alma-Ata, 1972.
197 F. Hartog, J.H. Tebben and P. Zwietering, in Actes 2me Congres Intern. Katalyse, Vol. I, Technip, Paris, 1961, p. 1229.
198 M. Magyar and L. Toll, Magyar Kem. Foliorat, 66 (1960) 297-302.
199 J.E. Germain, J. Borgeois, Bull. Soc. Chim. France, (1960) 2127-2130.
200 J. Nicolai, R. Martin and J.C. Jungers, Bull. Soc. Chim. Belg., 55 (1948) 555-568.
201 M.K. Dyakova and A.V. Lozovoyi, Zh. Obshch. Chim., 7 (1937) 2964-2977.
202 R.K. Greenhalgh and M. Polanyi, Trans. Faraday Soc., 35 (1939) 520-542.
203 E. de Ruiter and J.C. Jungers, Bull. Soc. Chim. Belg., 58 (1949) 210-246.
204 M.I. Temkin, V.Yu. Konyukhov and N.V. Kulkova, J. Res.Inst. Catalysis Hokk. Univ., 28 (1980) 363-370.
205 N.V. Kulkova, V.Yu. Konyukhov, I.M. Genkina, D.I. Parazich and M.I. Temkin, Kinet. Katal., 25 (1984) 678-681.
206 P.N. Palkin, V.Yu. Konyukhov, V.P. Gostikin and M.I. Temkin, Kinet.Katal., 24 (1983) 1396-1399.
207 M.I. Temkin, V.Yu. Konyukhov and N.V. Kulkova, Kinet. Katal., 25 (1984) 1257-1259.
208 V.Yu. Konyuchov and A.G. Zyskin, J. Catal., 94 (1985) 319.
209 K.B.S. Prasad, J. Catal., 94 (1985) 322.
210 A.V. Mashkina, Heterogeneous Catalysis in Chemistry of Sulphur Compounds, Nauka, Novosibirsk, 1977.
211 V.P. Gostikin, in M.G. Slinko (Editor), Kinetics-3, Proceedings of the 3rd Conference on Kinetics and Catalytic Reactions, Kalinin, March 1980, Vol. 1, Kalinin, 1980, p. 107.

212 D.V. Sokolskyi, G.D. Zakumbaeva, N.M. Popova, A.M. Sokolskaya and K.A. Zhubanov, Hydrogenation Catalysts, Nauka, Alma-Ata, 1975.
213 E.I. Klabunovskyi and A.A. Vedenyapin, Asymmetric Catalysis. Hydrogenation on Metals, Nauka, Moscow, 1980.
214 F.B. Bizhanov, in Hydrogenation and Oxidation on Heterogeneous Catalysts. Proc. Inst. Org. Catalysis and Electrochem. of Acad. Sci. Kazach. SSR, Vol. 19, Nauka, Alma-Ata, 1978, p. 115.
215 S.L. Kiperman, Izv. Akad. Nauk SSSR, Ser. Khim., (1984) 51-60.
216 S.L. Kiperman, in S.L. Kiperman (Editor), Theoretical Problems in Kinetics, Chernogolovka, 1984, p. 12.
217 A.A. Balandin and S.L. Kiperman, J. Chim. Phys., (1958) 363-369.
218 L. Beránek, Adv. Catal., 24 (1975) 1.
219 D.V. Sokolskyi, Hydrogenation in Solutions, Nauka, Alma-Ata, 1962.
220 L. Červený and V. Růžička, Catal. Rev.-Sci. Eng., 24 (1982) 503-566.
221 A. Palazov, A. Andreev, D. Shopov, Kinet. Katal., 12 (1971) 969-973.
222 S.L. Kiperman, Zhurn. Phys. Khim., 21 (1947) 1435-1448.
223 S.L. Kiperman, in D.V. Sokolskyi (Editor), Catalytic Reactions in Liquid Phase. Proc. 5th All Union Conference Nauka, Alma-Ata, 1980, p. 171.
224 S.L. Kiperman, in T.P. Chochlova (Editor), Methods of Catalytic Activities Testing, Proc. of Coordination Center, No 4, Novosibirsk, 1975, p. 7.
225 L. Beránek, in T.P. Chochlova (Editor), Methods of Catalytic Activities Testing, Proc. of Coordination Center, No. 4, Novosibirsk, 1975, p. 37.
226 S.G. Bashkirova and S.L. Kiperman, Kinet. Katal., 11 (1970) 631-637.
227 M.S. Kharson, E.M. Davydov and S.L. Kiperman, in Travaux du 4me Seminaire Sovietique-Franç. sur la Katalyse, Tbilissi, août 28 - septempre 2, 1978, Tbilissi, 1978, p. 170.
228 S.T. Beisembaeva, B.S. Gudkov, M.S. Kharson, N.I. Popov and S.L. Kiperman, in N.P. Maslov (Editor), Kinetics-2, Proceedings of the 2nd All Union Conference, Novosibirsk, 1975, Vol. 2, Inst. of Catalysis Acad. Sci. Publ., Novosib., 1975, p.104.
229 G.C. Bond and J.S. Rank, in W.M.H. Sachtler, G.C.A. Schuit and P. Zwietering (Editors), Proceedings of the 3rd International Congress on Catalysis, Amsterdam, July 20-25, 1964, Vol. 2, North-Holland, Amsterdam, 1965, p. 1225.
230 J.P. Wauquier and J.C. Jungers, Bull. Soc. Chim. France, (1957) 1280-1288.
231 V.V. Ipatiev, M.I. Levina and A.I. Karlbom, Usp. Khim., 8 (1939) 481-536.
232 L.G. Nishchenkova, V.P. Gostikin and K.N. Belonogov, Izv. Vyssh. Uchebn. Zaved., Khim. i Khim. Technol., 21 (1978) 1310-1313.
233 K.N. Belonogov, V.P. Gostikin, S.V. Klevtsov and V.M. Filatov, Kinet. Katal., 19 (1978) 468-473.
234 V.P. Gostikin, K.N. Belonogov, S.A. Komarov and S.V. Klevtsov, Kinet. Katal., 19 (1978) 474-479.
235 L.G. Nishchenkova, I.N. Leibovich, V.P. Gostikin and K.N. Belonogov, in K.N. Belonogov (Editor), Problems of Kinetics and Catalysis, No 1, Ivanovo, 1973, p. 55.
236 L.G. Nishchenkova, K.N. Belonogov, V.P. Gostikin and I.V. Savinov, in K.N. Belonogov (Editor), Problems of Kinetics and Catalysis, No 1. Ivanovo, 1973, p. 63.
237 A.D. Babneev, V.P. Gostikin, S.N. Dolgov, L.G. Nishchenkova and L.G. Kuvykina, in K.N. Belonogov (Editor), Problems of Kinetics and Catlysis, No 3, Ivanovo, 1976, p. 38.
238 V.P. Gostikin, L.G. Nishchenkova, T.I. Sovina and K.N. Belonogov, in K.N. Belonogov (Editor), Problems of Kinetics and Catalysis, No 3, Ivanovo, 1976, p. 33.
239 F.B. Bizhanov, D.V. Sokolskyi and U.A. Sadykov, in D.V. Sokolskyi (Editor), Proceedings Instit. Catalysis and Electrochem., Vol. 19, Nauka, 1978, Alma-Ata, p. 151.
240 I.I. Ioffe and E.I. Kozlov, in M.G. Slinko (Editor), Kinetics-3, Proceedings of the 3rd All Union Conference, Kalinin, 1980, Vol. 1, Kalinin, 1980, p.33.
241 M.I. Temkin, Kinet. Katal., 18 (1977) 493-496.
242 V.Yu. Konyukhov, N.V. Kulkova and M.I. Temkin, in M.G. Slinko (Editor), Kinatics-3, Proceedings of the 3rd All Union Conference, Kalinin, 1980, Vol.1, Kalinin, 1980, p. 266.

243 V.Yu. Konyukhov, N.V. Kulkova and M.I. Temkin, Kinet. Katal., 23 (1982) 507-510.
244 S.L. Kiperman, A.A. Balandin and I.R. Davydova, Izv. Akad. Nauk SSSR, Otdel. Khim. Nauk, (1957) 1457-1459.
245 T. Freund and H.M. Hulburt, J. Phys. Chem., 61 (1957) 909-912.
246 J. Hoogschagen, Ind. Eng. Chem., 47 (1955) 906-913.
247 L.K. Philipenko, K.N. Belonogov and V.P. Gostikin, Izv. Vyssh. Uchebn. Zaved. Khim. i Khim. Technol., 13 (1970) 553-555.
248 V.M. Safronov, V.I. Vorobyeva and A.B. Fasman, Vestn. Akad. Nauk Kazach. SSR No 10 (1982) 55-60.
249 F. Turek and R. Geike, Chem. Techn., 33 (1981) 24-28.
250 M. Berowian, J. Perak and R. Potocky, Przem. Chem., 51 (1972) 530-532.
251 G.K. Ziganshin, E.F. Stephoglo and A. Ermakova, Kinet. Katal., 14 (1973) 530-532.
252 K. Sporka, J. Hanika, V. Růžička and J. Pouček, Collect. Czech. Chem. Commun., 38 (1973) 166-170.
253 P.K. Anikeev, N.Ch. Valitov and G.M. Panchenkov, Kinet. Katal., 16 (1975) 1310-1313.
254 A. Ermakova, A.S. Umbetov and P. Valko, Hung. J. Ind. Chem., 8 (1980) 77-86.

Chapter 2

SYNERGY IN CATALYTIC REACTIONS INVOLVING HYDROGEN : POSSIBLE ROLE OF SURFACE-MOBILE SPECIES

B.K. HODNETT[1] and B. DELMON[2]

[1] Department of Materials Engineering and Industrial Chemistry, NIHE, Limerick (Ireland)
[2] Groupe de Physico-Chimie Minérale et de Catalyse, Université Catholique de Louvain, Place Croix du Sud 1, 1348 Louvain-la-Neuve (Belgium)

2.1 INTRODUCTION

A great deal of evidence has been accumulated over the past ten years which suggests that surface-mobile species, such as the so-called "spill-over" hydrogen, can play a role in catalytic reactions involving hydrogen (ref. 1). However, spill-over hydrogen is elusive. It has never been detected by physico-chemical means under conditions similar to those prevailing during catalysis. It is therefore difficult to determine the real role of this species in catalytic hydrogenation, hydrogenolysis and other hydrotreating reactions. More precisely, the question is whether the surface mobility of hydrogen is responsible for more than minor phenomena which only occur under conditions far from those under which catalysed reactions occur.

The vast majority of experiments whose rationalization involves surface-mobile hydrogen involve two phases (refs 2-4). One phase can produce the mobile species from molecular hydrogen; it is called the DONOR (D). The second phase is the ACCEPTOR (A). The effects attributed to spill-over are usually chemical, e.g., hydrogenation by spill-over hydrogen of species adsorbed on a support, removal of carbonaceous deposits, occurrence or enhancement of a given reaction and, more generally, change in catalytic activity. These phenomena are frequently anomalous and can often be explained only by invoking surface mobility and spill-over from one phase to another. The most undisputable evidence of mobile or spill-over species meets the following requirements :

(1) The studied system comprises two separate phases, D and A.
(2) The presumed mobile species flows from D to A.
(3) The observed effects result from the reaction of the mobile species on, or with, A.

A conspicuous feature of these systems is that synergy has often been observed. Synergy exists if the observed rate of a catalysed reaction is greater in the presence of a two-phase catalyst than the sum of the individual rates which would be observed with each of the isolated phases present in the same quantities as in the composite catalyst.

This chapter will consider evidence for the role which surface-mobile hydrogen plays in various heterogeneously catalysed reactions where synergy occurs. The reactions of spill-over hydrogen with organic molecules or inorganic solids will be considered. We shall attempt to show that spill-over hydrogen can have important effects on activity, selectivity and deactivation of heterogeneous catalysts. The first part of this chapter will be a short literature survey of synergistic effects in catalysed reactions involving hydrogen; a discussion of the possible origins of synergy will follow. We will conclude with a special discussion of possible mechanisms involving surface-mobile species and how these species might change the activity and selectivity of catalysts.

2.2 LITERATURE SURVEY OF SYNERGISTIC EFFECTS IN CATALYTIC REACTIONS INVOLVING HYDROGEN

An extensive literature now exists on spill-over phenomena (refs 1, 4-6). A recent symposium (ref. 1) was devoted to this subject and for a general outline of the phenomenon the reader is referred to a very comprehensive review presented there (ref. 7).

Here we shall direct our attention mainly to reports dealing with catalysed reactions in which synergy occurs between phases prepared separately and mixed after their preparation. Synergy in hydrogenation and in other reactions involving hydrogen will be considered. Our selection will be made on the basis that the authors invoked the intervention of mobile species in explaining their results. These reports are collected in Table 2.1. To complement these data, a description of phenomena observed with supported catalysts is presented in Table 2.2. Any synergistic effects are less conspicuous with these catalysts than with mixtures of separate phases, but the studies presented in this table strongly suggest a role for surface mobility.

TABLE 2.1
Synergy between separate phases in reactions involving hydrogen

Phases	Reaction where synergy was observed	Activation conditions	Temperature	Conditions	References
Pt/SiO$_2$ + γ-Al$_2$O$_3$	Ethylene hydrogenation	500°C/H$_2$	100 and 153°C	continuous flow measured after 3 min on stream	8,10
Pd/Al$_2$O$_3$ + Al$_2$O$_3$	benzene hydrogenation	200°C/H$_2$	150°C	Slug reactor	11,12
Pt/γ-Al$_2$O$_3$ + γ-Al$_2$O$_3$	benzene hydrogenation	500°C/H$_2$	50-250°C	differential reactor	13
MoS$_2$ + Co/Al$_2$O$_3$ or MoS$_2$ + Co/SiO$_2$	thiophene hydrogenolysis	400°C/H$_2$	400°C	differential reactor	14
MoO$_3$ + Co$_3$O$_4$	cyclohexene hydrogenation and thiophene hydrogenolysis	-	305°C	flow reactor	15
MoS$_2$ + Co$_9$S$_8$	cyclohexene hydrogenation and thiophene hydrogenolysis	-	305°C	flow reactor	16-18
MoS$_2$ + Ni$_2$ or Co on Y-zeolite	cyclohexene hydrogenation and thiophene hydrogenolysis	-	305°C	flow reactor	19
Pt/Al$_2$O$_3$ + Al$_2$O$_3$	removal of coke	-	temp. prog. 25-600°C	flow reactor	20
Ni/SiO$_2$ + H-mordenite	removal of coke	-			21

TABLE 2.2

Synergy probably due to the surface mobility in reactions involving hydrogen (supported catalysts)

Catalyst	Reaction where synergy was observed	Reference
Rh	benzene hydrogenation	22
Ru/Ni-La$_2$O$_3$	oxidation of hydrogen	23
MoO$_3$/Co exchanged Y zeolite	thiophene hydrogenolysis	18
Pt/γAl$_2$O$_3$ (Cl)	removal of coke	25
Pt/TiO$_2$	photocatalytic dehydrogenation of propan-2-ol	26,27

2.2.1 <u>Hydrogenation on catalysts composed of noble metals (Pt and Pd) and inorganic oxides (SiO$_2$ and Al$_2$O$_3$)</u>

The activity of catalysts comprising platinum or palladium supported on SiO$_2$ or Al$_2$O$_3$ and mechanically mixed with further quantities of SiO$_2$ or Al$_2$O$_3$ has been described (refs 8,11). The ratios of Pt supported on SiO$_2$ or Al$_2$O$_3$ to diluent SiO$_2$ or Al$_2$O$_3$ used were in the range 1:1 -100. Neither pure SiO$_2$ nor Al$_2$O$_3$ alone is able to catalyse hydrogenation reactions, but Sinfelt and coworkers (refs 8-10) reported that a simple mixture of 0.05% Pt/SiO$_2$ and γ-Al$_2$O$_3$ was 5-7 times more active in ethylene hydrogenation than when the same Pt/SiO$_2$ catalyst was mixed with SiO$_2$. The implication of these results was that γ-Al$_2$O$_3$, but not SiO$_2$, could be activated as a hydrogenation catalyst. It was postulated that spill-over hydrogen was responsible for this effect. Further details of this work are listed in Table 2.1.

Several years later Sancier (refs 11,12) reported similar results for the hydrogenation of benzene on Pd/Al$_2$O$_3$ mixed with pure Al$_2$O$_3$. He decreased the amount of Pt/Al$_2$O$_3$ in his reactor as the amount of diluent Al$_2$O$_3$ was raised. The overall conversion rate decreased but the metal specific activity increased as a function of the amount of diluent Al$_2$O$_3$ added (Table 2.1). Here also it was postulated that the Al$_2$O$_3$ had been activated by spill-over hydrogen.

The work of Sinfelt and coworkers and that of Sancier was later criticised. Schlatter and Boudart (ref. 28) concluded that the apparent synergistic effects observed previously were due to ready release of contaminants from the SiO_2 diluent to the Pt/SiO_2 catalyst, with subsequent loss of activity. They hypothetised that when Al_2O_3, which held contaminants more strongly, was used as diluent the true activity of the metal surface was measured. Vannice and Neikam (ref. 29) postulated that the effects observed by Sancier were due to inhibition of hydrogenation by the product, cyclohexane, for the high benzene conversions which were reached for low dilutions with Al_2O_3. The Pt/Al_2O_3 system was examined most recently by Antonucci and coworkers (ref. 13) for benzene hydrogenation. This study seems to confirm the previously reported synergistic effects when Al_2O_3 was used as diluent. The magnitude of the synergy is striking (fig. 2.1).

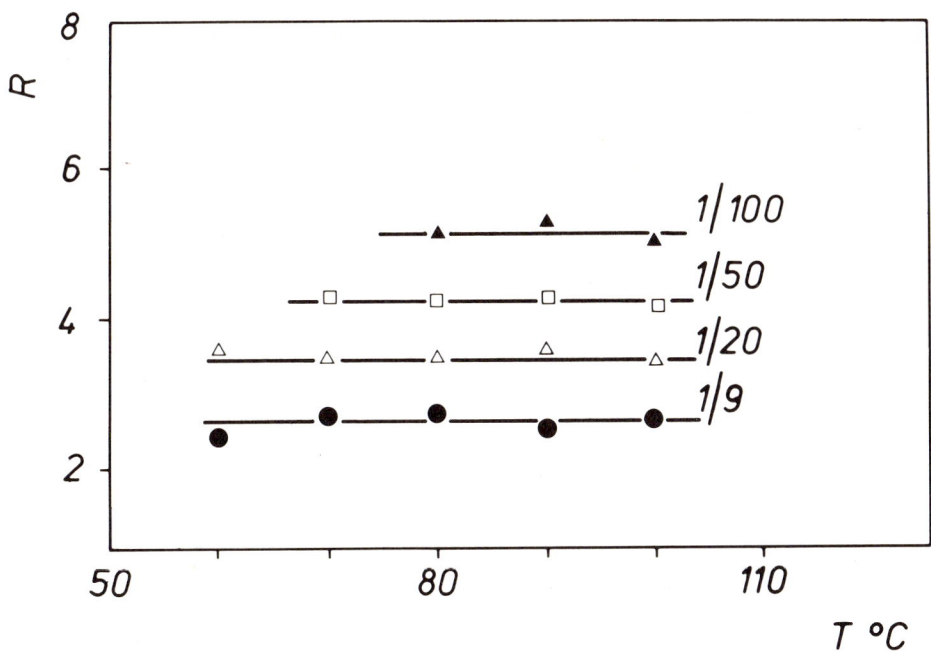

Fig. 2.1. Activity, in benzene hydrogenation, of 1.35 % $Pt/\gamma-Al_2O_3$ catalyst diluted with $\gamma-Al_2O_3$ in proportions $Pt/\gamma-Al_2O_3: \gamma-Al_2O_3$ of 1:9, 1:20, 1:50 and 1:100, compared to the activity of pure $Pt/\gamma-Al_2O_3$. R is the ratio of benzene conversion per mg of Pt on the diluted catalyst divided by the value on the pure catalyst. The contact times with the beds of diluted and undiluted catalyst were identical (W/F = 0.73×10^6 g sec mol^{-1}).

2.2.2 Hydrogenation and hydrodesulphuration over sulphide catalysts

This is another class of catalytic reaction where spill-over hydrogen has been invoked to explain the synergy that occurs when a mixture of a Group 8 (Co or Ni) and a Group 6B (Mo or W) metal is used as catalyst, with each metal in its sulphided state (ref. 30). The investigated reactions are typically hydrogenation of cyclohexene and hydrogenolysis of the C-S bonds in thiophene to produce butene and butane, but many variants have been studied. Synergy is usually observed for different Group 8:Group 6B ratios for hydrogenation and hydrodesulphuration reactions (Fig. 2.2). Although the structure of commercial hydrotreating catalysts has not yet been fully elucidated (refs. 31, 32), synergistic effects similar in all important characteristics are observed when mechanical mixtures of Co_9S_8 and MoS_2 are used.

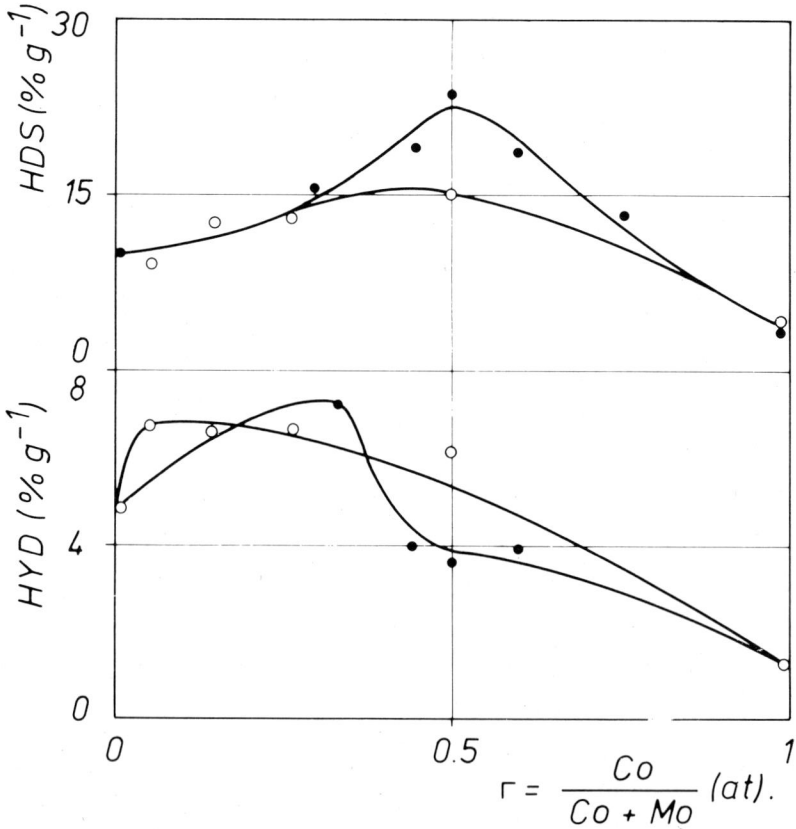

Fig. 2.2. Catalytic activities in hydrodesulphuration (HDS) and hydrogenation (HYD) of catalysts prepared by mechanically mixing MoS_2 with one of two samples of Co_9S_8 which had different specific surface areas (9 m^2 g^{-1}; 27 m^2 g^{-1}) (ref. 18).

2.2.3 Coke formation on reforming catalysts

By direct, very elegant experiments using Pt/Al_2O_3 mixed with diluent Al_2O_3, Parera and coworkers (ref. 20) showed that spill-over hydrogen (as well as spill-over oxygen) could remove coke or coke precursors formed on catalysts under conditions similar to those prevailing during naphtha reforming (or the regeneration of these catalysts). Another set of experiments by Gnep and coworkers (ref. 21) showed that coke formation on H-mordenite during o-xylene isomerization and disproportionation was inhibited by the presence of Ni/SiO_2. They concluded that spill-over hydrogen from Ni/SiO_2 inhibited disproportionation and coke formation by reacting with carbocations. The observed synergy corresponded to an increased resistance of the isomerization reaction to deactivation.

2.3 POSSIBLE ORIGINS OF SYNERGY IN REACTIONS INVOLVING HYDROGEN

In this section we consider how synergy may develop in various catalytic systems including : (a) classical bifunctional catalysis; (b) synergy through the formation of compounds between two phases acting synergistically; (c) through contamination of the surface of one phase by elements from another; (d) reaction of spill-over hydrogen, produced on one phase, with a reactant adsorbed on another phase; (e) the "remote control" concept, i.e.; creation or regeneration of active centres on one phase by spill-over hydrogen emitted by the other phase.

2.3.1 Classical bifunctional catalysis

Catalytic reforming involves the restructuring of hydrocarbon molecules without changing their carbon numbers, with the aim of increasing the octane number of the treated mixtures (ref. 33). Typical catalysts for the process are Pt/Al_2O_3 and Pt-Re/Al_2O_3 and are composed of at least two phases, i.e., supported metals and an acidic support. The metal loading rarely exceeds 0.6 wt % (ref. 34). A complex series of reactions occur during reforming, including isomerization, hydrogenation, dehydrogenation and dehydrocyclization. The last three reactions are catalysed on the metal, whereas isomerization as well as some hydrocracking occurs on the support. Each of the components of these catalysts, metal and acidic support, are necessary in producing a high octane number fuel (ref. 33). Synergy is normally observed if the activities and selectivities of the two-phase catalyst are compared to those of the separate metal and support.

Synergy in this system can readily be explained in terms of Scheme 2.1 which envisages partial transformation of a reactant on the metal followed by further reaction on the support. An example is the hydrogenation of 1-heptene followed by its isomerization to form 2-methylhexene. In appropriate cases the reverse sequence may be envisaged, i.e., the partial reaction occuring on the support, followed by transfer of the primary reaction product to the metal for further reaction.

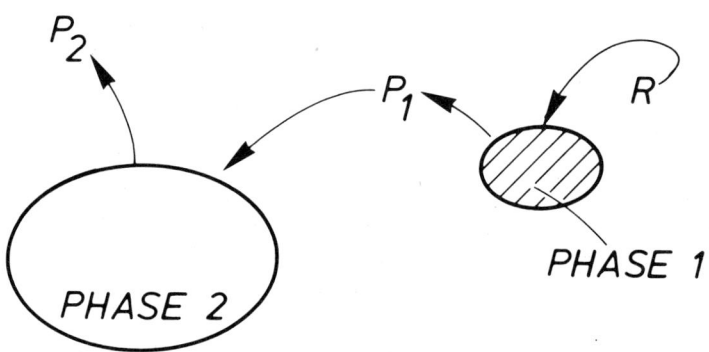

Scheme 2.1. Simple representation of bifunctional catalysis. The reactant (R) is transformed into the intermediate product (P_1) on phase 1; P_1 readsorbs on phase 2 to be transformed into the final product (P_2).

2.3.2 Formation of compounds between two phases

Most promoted and alloy catalysts may be included in this category. These are catalysts in which the second component changes the surface geometry or the electronic environment and many examples have been reported where a distinct new phase forms.

Addition of iridium to a Ni/SiO_2 catalyst increases its resistance to sulphur poisoning and its methanation activity at 282°C. Wentrcek and coworkers (ref. 35) postulated that the added iridium changed the surface structure of the nickel particles. A synergistic effect was observed when an optimum surface concentration of iridium reduced the density of four- and three-fold coordinate sites, preferred for sulphur adsorption, and increased the number of sites (two-fold) available for CO adsorption.

The presence of some potassium metal in iron catalysts used for methanation and Fischer-Tropsch reactions increases the rate of the former and shifts the selectivity of the latter in favour of longer chain hydrocarbons (ref. 36). The effect appears to be electronic, with the alkali metal donating electron density to the iron. When CO adsorbs on this structure, the Fe-C bond is stronger and the C-O bond weaker than on the unpromoted catalyst. It has been argued that carbon, activated in this way, is more prone to attack by hydrogen (ref. 37). If the K:Fe ratio exceeds a critical value these effects are lost, presumably due to the presence of excessive amounts of inactive potassium on the iron surface.

Poels and coworkers (ref. 38) studied $Pd-Mg/SiO_2$ for hydrogenation of CO to methanol. Optimum activity was observed for Pd:Mg = 1:1. Scanning electron microscopy indicated that Pd, Mg and Cl occured simultaneously in many locations in the catalysts. The synergistic behaviour followed exactly the trend in the concentration of unreduced Pd present during catalysis. It was postulated that Mg^{2+} stabilized Pd^{n+}, the active centres for this reactions.

Daage and Bonnelle (ref. 39) studied the hydrogenation of isoprene and 1,3 pentadiene over a series of Cu-Cr-O spinel bronzes which also contained small amounts of supported metallic copper. The Cu:Cr ratios were in the range 0.5-1.4:1, but optimum activity was observed with the 1:1 composition. NMR studies pointed to the presence of a special hydrogen species in the bulk of the spinel phase after treatment in hydrogen at 150°C. This hydrogen diffused from the bulk and participated in hydrogenation, even when gaseous hydrogen was present. Some unusual selectivities were observed : the less subsituted double bond in isoprene was hydrogenated preferentially, but in 1,3-pentadiene the more substituted double bond was the most reactive. These effects were attributed to copper ions in octahedral environments which acted as active sites when combined with hydride ions.

2.3.3 Contamination of the surface of one phase by elements from the other phase

It is often difficult to prove whether a synergistic effect arises because of the interaction of two clearly defined phases or whether small amounts of one phase become detached and contaminate the second phase, hence acting like a classical promoter. For Cu-Zn-O catalysts, often used for the conversion of $CO+H_2$ or CO_2+H_2 into methanol, the weight of evidence to date would favour a model in which small amounts of Cu^+, mixed with ZnO, act as a promoter. The optimum Cu:Zn ratio is usually 1:1. However, a doubt remains because mechanical mixtures of CuO and ZnO can catalyse these reactions nearly as efficiently as coprecipitated catalysts (ref. 40).

Other systems of interest in this category are those which exhibit the so-called "strong metal-support interaction". This term is applied to many metals supported on reducible oxides, such as TiO_2, which, when treated in hydrogen at 500°C, lose the capacity to adsorb hydrogen and catalyse hydrogenation and hydrogenolysis reactions. Originally this effect was thought to be due to electron transfer from the reduced oxide to the metal (ref. 41), but evidence has been accumulated which suggests that the metal particles become contaminated by the support (ref. 42).

2.3.4 Reaction of spill-over hydrogen produced on one phase with a reactant adsorbed on another phase

The phenomenon of interest here is described in Scheme 2.2. Essentially a hydrogen species is formed on one phase (usually a metal) and spills over to react on the other phase. From the many examples of this type we will consider the following in detail : (i) the reactions of ethylene and acetylene over Pd/Al_2O_3; (ii) the role of spill-over hydrogen in the prevention of coke formation; (iii) the influence of spill-over hydrogen in controlling the selectivity of catalysed reaction.

Sarkany and coworkers (ref. 43) studied the hydrogenation of a mixture of 0.29 mole % C_2H_2, 0.44 mole % H_2 and C_2H_4 up to 100 %, a so-called tail-end mixture, on palladium black and several Pd/Al_2O_3 catalysts. Hydrogenation of C_2H_4 increased with time on stream for all the Al_2O_3-supported catalysts; the opposite behaviour was noted with palladium black. Polymer formation was noted for all catalysts studied and also increased with time. It was recognized that a small number of C_2H_4 hydrogenation sites were located on the metal but the majority were on the polymer-covered support. The authors proposed that C_2H_4 adsorbed on the support and was hydrogenated there. Spill-over hydrogen was tentatively identified as the source of hydrogen. Because of the parallelism between polymer formation and ethylene hydrogenation, it was proposed that the surface polymer served as a hydrogen pool or facilitated diffusion of hydrogen from Pd to the support.

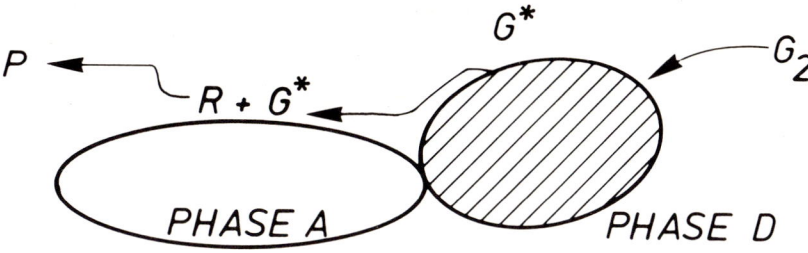

Scheme 2.2. Reaction between a species (G*), formed on phase D, then transferred to phase A by spill-over and surface diffusion, where it reacts with an adsorbed molecule R.

Most reforming catalysts contain small amounts of chlorine located on the support. This influences the activity and selectivity : if too little Cl is present activity is lost; if too much is present selectivity, (i.e., high octane number), is lost. The optimum is usually ca. 0.9 wt % Cl (ref. 34) (Fig 2.3). Parera and coworkers (ref. 25) observed a parallelism between chloride content and the build-up of carbonaceous deposits on these catalysts. Minimum coke build-up occurred when the chlorine concentration was such that half of the surface hydroxyl groups were replaced by chlorine. Maximum hydrogen spill-over was also observed at this concentration. A further publication (ref. 20) by the same group indicated that spill-over hydrogen from Pt could completely eliminate slightly polymerized coke on Al_2O_3, by hydrogenation. In the absence of Pt, decontamination of Al_2O_3 by hydrogen was incomplete. With more highly polymerized coke, hydrogenation with and without Pt succeeded in removing only a small fraction of the coke. However, this fraction was always higher with Pt than without. The proposed mechanism is outlined in Scheme 2.3. The implication of this work is that with an optimum chlorine concentration the rate of coke formation is lower because spill-over hydrogen keeps the support clean by removing coke precursors.

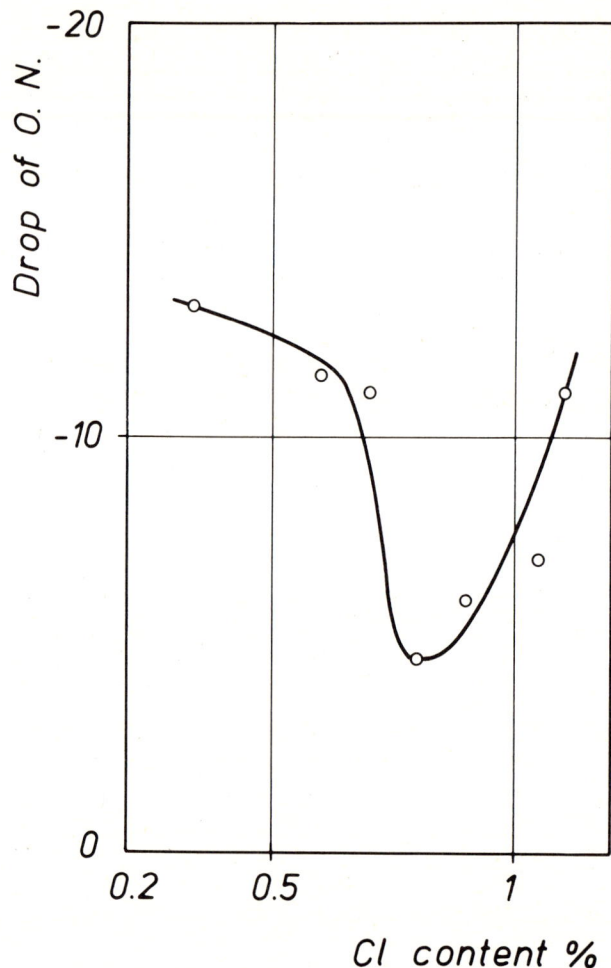

Fig. 2.3. Influence of chlorine content on the deactivation of Pt/Al$_2$O$_3$ reforming catalysts (Ref. 25), measured by the decrease in octane number (O.N.) of the product.

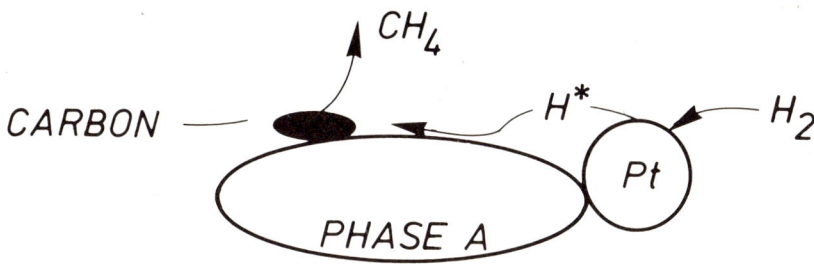

Scheme 2.3. Removal of carbonaceous deposits, more generally, of fouling materials or poisons, by reaction of spill-over hydrogen, leading to restoration of the catalytic site.

Gnep and coworkers (ref. 21) studied the reaction of o-xylene over a series of mordenite catalysts. Isomerization, disproportionation and coke formation occured. These catalysts were more resistant to deactivation in hydrogen when Ni/SiO$_2$ was added. The greatest resistance was observed for mordenite which was treated in spill-over hydrogen which originated from Ni/SiO$_2$ (the latter was removed before exposure of the mordenite to o-xylene). Carbon deposition and disproportionation were both inhibited under these conditions by comparison with untreated mordenite. These workers postulated that spill-over hydrogen was capable of generating new catalytic sites on mordenite; these catalytic sites were believed to be capable of activating hydrogen for reaction with carbocations, which were intermediates in the disproportionation and coke formation reactions. The isomerization of o-xylene did not involve carbocations, but it benefited by not being inhibited by coke deposition.

2.3.5 Creation or regeneration of catalytic centres on one phase by spill-over hydrogen emitted by the other phase : the "remote control" concept

The creation of catalytic centres on phase A by reaction of spill-over hydrogen emitted by phase D is depicted in Scheme 2.4.

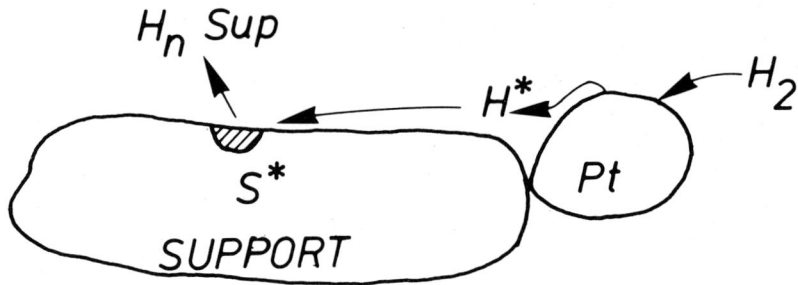

Scheme 2.4. Creation of an active site on phase A by reaction of its surface with spill-over hydrogen (with possible formation of a product H_n Sup.).

Some of the most unambiguous evidence for the creation of catalytic centres by spill-over hydrogen comes from the work of Teichner and coworkers (ref. 44-54). In the early 1970s these workers developed an experimental device whereby an oxide, usually Al_2O_3, could be activated by spill-over hydrogen produced on Pt/Al_2O_3; immediately after activation the donor, i.e., Pt/Al_2O_3, could be removed. This system had the advantage that the oxide alone could be tested for catalytic activity in a number of hydrogenation and isomerization reactions. For our purposes it suffices to recall the principal conclusions :

(1) Al_2O_3 could be activated for hydrogenation by heating in spill-over hydrogen at 430°C;
(2) The sites which formed were catalytic, i.e., they were used more than once in hydrogenating feed molecules;
(3) An induction period was often observed, which was ascribed to the need to desorb spill-over hydrogen before the sites were free for catalytic reactions.

In addition, for the temperature range within which synergetic effects of interest in catalysis occur, transport of the metal has been excluded (ref. 54).

Synergy in catalysis by sulphides has already been alluded to in Section 2.2. For catalysts composed of two clearly defined phases, such as Co_9S_8 and MoS_2, the "remote control" model has been develop by Delmon and coworkers (refs. 18, 55-58). This model postulates that hydrogen is activated on the Co_9S_8 phase and spills over to create or regenerate two types of site on the MoS_2 phase, i.e., a slightly reduced region responsible for hydrogenation, and a more deeply reduced region responsible for hydrodesulphuration. The concept is shown in Scheme 2.5 (ref. 59), which is a detailed representation for hydrodesulphuration catalysts of the process suggested in Scheme 2.4.

Two synergistic effects are observed with these catalysts (see Fig. 2.2). The first is for hydrogenation : its maximum usually occurs at fairly low Co:Mo ratios. Under these conditions little spill-over hydrogen becomes available because little of phase D is present and the MoS_2 only becomes mildly reduced. The second synergistic effect is observed for hydrodesulphuration, and its maximum occurs at higher Co:Mo rations. Under these conditions more spill-over hydrogen should be available from the higher Co_9S_8 content, hence the MoS_2 phase can become more deeply reduced.

The "remote control" hypothesis implies that sites responsible for hydrogenation may be converted into hydrodesulphuration sites by appropriate manipulation of the experimental conditions.

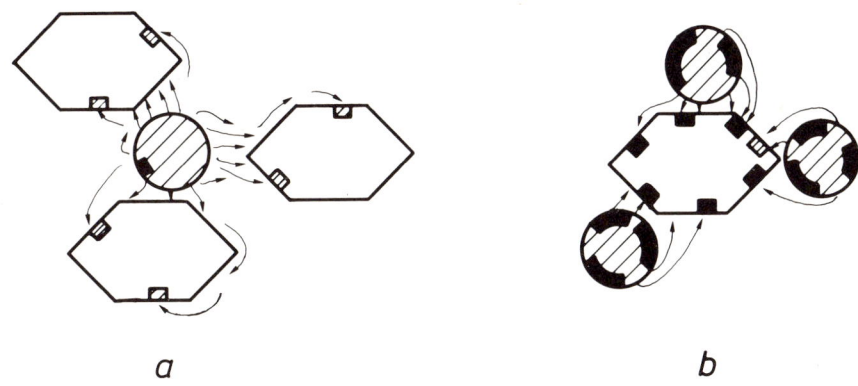

Scheme 2.5. Formation of active sites for hydrogenation and hydrodesulphuration by hydrogen spill-over : (a) low proportion of Co_9S_8, hence small amounts of spill-over hydrogen with limited reduction of the MoS_2 phase; (b) high proportion of Co_9S_8 giving deeper reduction of MoS_2.

2.4 SPECIAL DISCUSSION OF MECHANISMS INVOLVING SURFACE-MOBILE SPECIES : POSSIBLE OUTLOOK; CONTROL OF SELECTIVITY

In view of the lack of physico-chemical techniques capable of detecting spill-over hydrogen and the indirect nature of the evidence for its occurrence, it is not surprising that a consensus has not been reached about the mechanism of its formation and action. Although some spill-over processes occur by transport of the active species through the gas phase (ref. 60), it is now generally believed that surface diffusion is involved in most of the cases of interest in catalysis. There is also a great deal of evidence that spill-over hydrogen is a form of atomic hydrogen, i.e., H^+, H^0 or H^-, and acts as a powerful reducing agent on metal oxides where an oxidation state is available one below that of the acceptor oxide (ref. 4).

2.4.1 Extent of formation of spill-over hydrogen on SiO_2 and Al_2O_3

Some information is available about the extent to which spill-over hydrogen is formed. Kramer and Andre (ref. 61) estimated the concentration of hydrogen spill-over from Pt to Al_2O_3 to be equivalent to 2×10^{12} sites cm^{-2} of Al_2O_3 at 400°C. Thus for a catalyst with 1% Pt the number of metal sites per gram of catalyst was 1.2×10^{19}, compared to $(1.5 - 2) \times 10^{18}$ alumina sites per gram of catalyst activated by spill-over hydrogen. These values are consistent with the many values reviewed by Sermon and Bond (4) who found that, for supported metal catalysts where spill-over has been invoked, the ratio of the number of hydrogen atoms adsorbed to the total number of metal atoms present on the support rarely exceeds 2:1 for non-reducible supports, such as SiO_2 and Al_2O_3.

Experimental data published by the group of Teichner and coworkers (refs. 47,48) enables us to calculate the approximate turnover number for sites activated on Al_2O_3 in the manner outlined in Section 2.3.5. The reaction of interest is the hydrogenation of ethylene at 25, 110 and 300°C. The amounts of spill-over hydrogen produced at these temperatures corresponded to the activation of 5.7×10^{12}, 1.5×10^{13} and 2.9×10^{13} sites cm^{-2}. The turnover numbers (N= number of molecules of C_2H_4 hydrogenated sec^{-1}/number of sites) were close to : 6×10^{-7} sec^{-1} at 25°C, 1.2×10^{-4} sec^{-1} at 110°C and 5.6×10^{-4} sec^{-1} at 300°C. These values compare with turnover numbers of 1-5 sec^{-1} reported by Schlatter and Boudart (ref. 28) for the same reaction catalysed on Pt at 25°C.

Evidence from other sources corroborates these findings. Kramer and André (ref. 61) measured a diffusion coefficient close to 1×10^{-15} $cm^2 sec^{-1}$ for spill-over hydrogen on Al_2O_3 at 400°C. They postulated that surface diffusion was the slow step. Sermon and Bond (ref. 62) reported data for the reaction of SiO_2-held hydrogen, i.e., spill-over hydrogen, with pentene at 100 and 200°C. These workers postulated that reverse spill-over from the support to the metal was the slow step, but the reaction was again very slow by comparison with the same reaction catalysed on Pt. We may therefore conclude that for systems which involve <u>clean</u> SiO_2 or Al_2O_3 as supports the extent of spill-over is small and the activity of the sites for hydrogenation is very low. In order to explain the remarkable synergistic effects in reactions involving hydrogen which have often been attributed to spill-over hydrogen (refs. 8-13), we must therefore consider the phenomenon from another point of view, and examine the nature of the surface sites produced.

2.4.2 Factors influencing spill-over : possible mechanisms

The small amounts of spill-over on SiO_2 and Al_2O_3 cannot be attributed to the metal's inability to produce any further quantities. The formation of hydrogen bronzes (ref. 4), with formulae such as $H_{1.6}MoO_3$ and $H_{0.3}WO_3$, by spill-over of hydrogen from Pt onto and into MoO_3 or WO_3 suggests that the metal alone is capable of producing larger amounts of spill-over hydrogen provided a suitable acceptor phase is available. Facile reduction of some oxides by spill-over hydrogen also implies that large amounts may be produced (ref. 4). This point may be further illustrated by the very large amounts of spill-over hydrogen produced on the Pt/C system (ref. 4).

Of particular interest to this review are studies of the influence of adsorbed molecules on the rate at which spill-over occurs. Neikam and Vannice (refs. 63,64) found enhancement of the rate of spill-over on Y-zeolites of Na, La and Ce when large aromatic or aliphatic molecules were adsorbed at room temperature. Levy and Boudart (ref. 65) followed the rate of reduction of WO_3 as a means of determining the rate of production of spill-over hydrogen in the presence of adsorbed water, aliphatic alcohols or acetic acid. The reduction rate, hence the rate of transfer of spill-over hydrogen, increased as the proton affinity of the adsorbed layer decreased. A mechanism was proposed which involved movement of a solvated proton through the adsorbed layer; the slow step was release of the proton at the reduction site. As the coverage by the adsorbate decreased the reduction rate also decreased. In these circumstances the slow step was thought to be diffusion of spilt-over hydrogen from Pt to WO_3.

A special form of mobile proton-electron pair has been postulated as the spill-over species by Keren and Soffer (ref. 66) on the basis of an electrochemical investigation of the Pd/C system. These workers envisaged : (1) dissociative adsorption of the hydrogen on Pd as (H^+ + e); (2) transfer of the electron to the carbon support; (3) electronic conduction in the support; (4) ionic conduction of the H^+ in the adsorbed layer, such as conductivity through the surface hydroxyl network.

Mechanisms for spill-over which take account of the role of adsorbed layers in the surface diffusion step readily explain the large amounts of spill-over observed with carbon supports (ref. 4), i.e., high electronic conductivity, and lead to the idea that carbonaceous deposits formed in reactions involving hydrocarbons could play a similar role. The hypothesis is outlined in Scheme 2.6 and may have application to the work of Sarkany and coworkers (43).

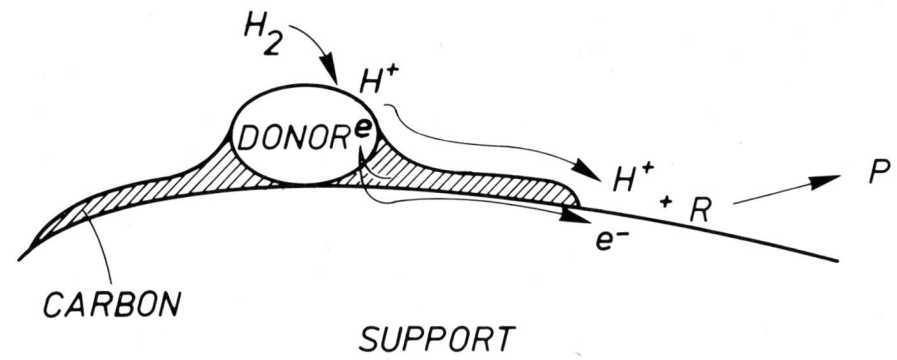

Scheme 2.6. Surface diffusion of a (proton + electron) pair aided by the presence of a partial layer of conducting carbonaceous deposit.

2.4.3 Creation and regeneration of surface sites by spill-over hydrogen.

The creation of catalytic centres with unusual properties (e.g., sites on SiO_2 capable of cracking benzene to acetylene (ref. 54)), observed by Teichner and co-workers suggests an effect on catalyst selectivity. In addition the "remote control" concept of Delmon suggests that spill-over hydrogen, acting through the creation or regeneration of active centres capable of transforming many reactant molecules, would have effects not commensurate with its scarcity, namely that the creation of catalytic centres of high activity would reveal on a larger scale the formation of spill-over hydrogen.

In this section it is convenient to discuss separately the two possibilities by which the number of catalytic centres might be modified. These are :

1) New surface sites created by the action of spill-over hydrogen; these sites are capable of acting catalytically, i.e. capable of aiding the transformation of more than one reactant molecule before being destroyed; they may subsequently be regenerated by spill-over hydrogen. This type of behaviour has been reported by Teichner and co-workers (refs 44-54) for Al_2O_3 and by Delmon (refs. 55-59) for hydrodesulfuration catalysts (Schemes 2.4 and 2.5);

2) Spill-over hydrogen reacts directly with a fouling or contaminating species deactivating an existing surface site thereby removing that contamination. In this category of effects we can include reactions of spill-over hydrogen with coke deposited on Al_2O_3 (Scheme 2.3).

Here we will try to concentrate on those factors which could cause the dramatic magnifying effects which are necessary so that small amounts of spill-over hydrogen can give rise to conspicuous effects in terms of synergy and selectivity.

(a) New surface sites. The spill-over phenomena observed by Teichner and co-workers (refs. 44-54) were revealed thanks to an experimental arrangement which allowed the activity of the support modified by the action of the spill-over to be measured separately. However, the results are not conclusive with respect to the possibility of bringing about large effects due to spill-over species. This is a consequence of the low turnover (see above). These experiments have the exceptional merit of giving direct proof of the creation of active sites. It is understandable that the special conditions under which they had to be conducted did not allow all the facets of the problem to be examined; some of the sites might disappear during the treatments carried out between flooding with spill-over hydrogen and the catalytic experiments. The experimental set-up, which had to allow withdrawal of the donor (Pt/Al_2O_3), could not provide an excellent and intimate contact between the donor and acceptor (SiO_2 or Al_2O_3). However, a point of special interest for this chapter is the fact that some sites can be activated by spill-over to catalyse reactions not normally catalysed by either the metal or the support even under much more severe reaction conditions, i.e. the creation of sites on SiO_2 capable of cracking benzene to acetylene (ref. 54).

The remote control mechanism presented by Delmon (refs. 55-59) to explain the synergy between cobalt and molybdenum in hydrodesulphuration catalysts suggests how a small amount of spill-over hydrogen can control the surface state of a catalyst and hence have a large, even predominant influence in determining its catalytic activity. Scheme 2.7 summarizes this concept. The mechanism of modification of activity and, if applicable, selectivity is best represented by the juxtaposition of two cycles : one (right-hand side) is a classical catalytic cycle, operating as long as the active centre exists, (e.g. operating 10^4 times before destruction or deactivation); the mechanism represented in the second cycle (left-hand side) only operates on non-active or deactivated parts of the surface of the potentially active phase (MoS_2 or Co edge-decorated MoS_2 in hydrodesulphuration catalysts). The most important

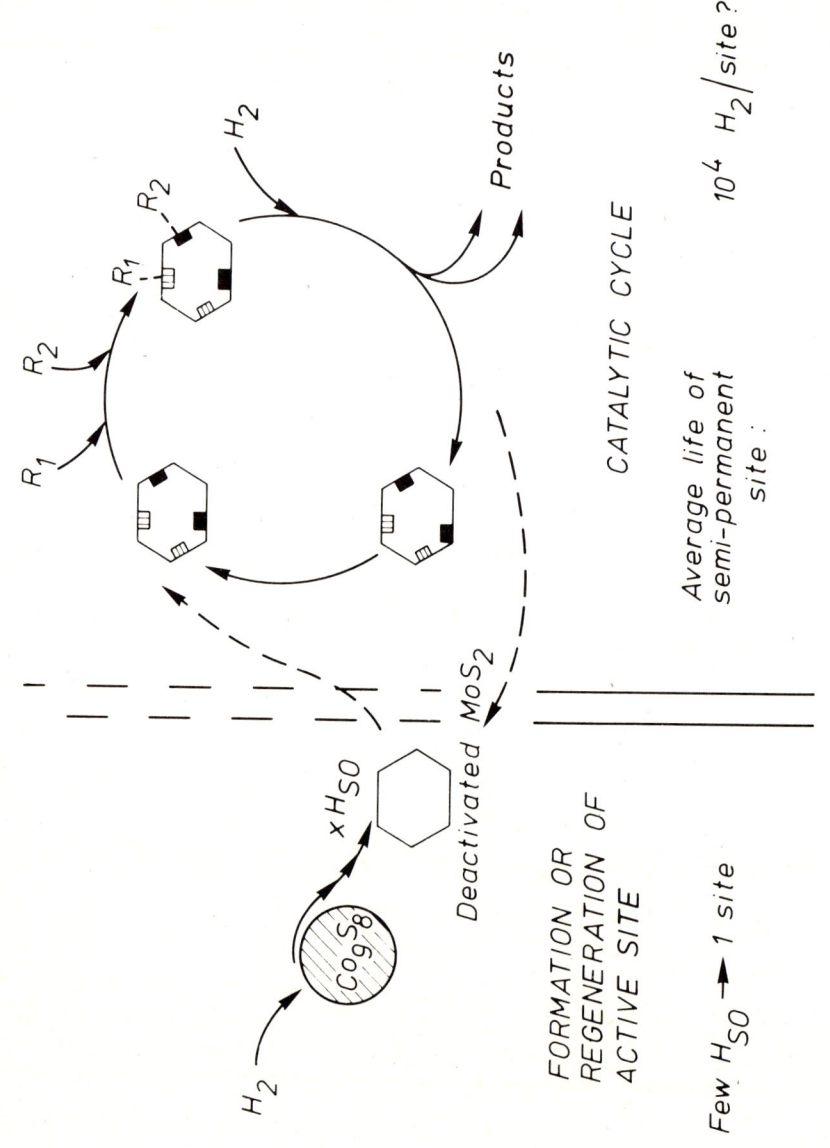

Scheme 2.7. Formation of active sites on MoS_2 and catalytic cycle involved in hydrodesulphuration.

feature is that one or two spill-over atoms create a centre which will transform a very large number of molecules. Although very scarce, the spill-over species can bring about important consequences. If, in addition, the spill-over species can create two different sites with different activities, (e.g., hydrogenation of unsaturated C-C bonds and hydrogenolysis of C-S bonds in hydrodesulphuration), it modulates the selectivity as a function of its abundance. The donor phase, here Co_9S_8, emitting the spill-over species, remotely controls the activity and selectivity of the acceptor phase. Synergy results from the varying numbers of surface sites created on MoS_2.

We have no experimental data for the number of catalytic sites created by spill-over in hydrodesulphuration catalysts because of the high pressures (3MPa) at which the reactions are normally carried out. However, compared to the systems studied by Teichner and co-workers (refs. 44-54), the number of active sites can reasonably be expected to be much greater : MoS_2 is more reactive towards hydrogen than Al_2O_3 or SiO_2 and there is a great deal of carbonaceous material deposited during hydrodesulphuration. Provided this does not become excessive it may aid the transfer of spill-over hydrogen, as suggested by the experiments of Keren and Soffer (ref. 66) and of Neikam and Vannice (refs. 63,64).

With sulphided Co-Mo catalysts (or similar catalysts where Co is replaced by Ni or Fe, and Mo by W or U), it is easy to propose a picture of the active centres on the Group 6B metal sulphide. It would consist of Group 6B ions (or Group 8 ions in the representation of Topsøe and coworkers (ref. 67) whereby the Group 8 ions are associated with the edges of the Group 6B metal sulphide), the coordination of which is reduced, compared to that of normal surfaces. Specifically, spill-over hydrogen would react with surface sulphur ions and remove them, leaving unsaturated coordination sites. It is easy to imagine that the degree of unsaturation achieved depends on the quantity of spill-over hydrogen available. Removal of a moderate number of sulphur ions would result in the formation of hydrogenating centres; a more extensive removal would bring about the formation of hydrodesulphurating centres.

(b) <u>Direct reaction of spill-over hydrogen with contaminating species on existing sites.</u> Many examples appear in the literature where this type of mechanism has been invoked, and some of these are described in Section 2.2 and 2.3 above. A feature of these studies is that the metal was present during exposure of the catalyst to hydrocarbon-hydrogen mixtures. Consequently, spill-over hydrogen could be continuously supplied under the reaction conditions. The presence of metal and support during exposure to hydrocarbon-hydrogen mixtures favours the formation of some carbonaceous deposits on the catalyst. Provided that these are not excessive, i.e., do not completely envelope the metal or the support, spill-over could be enhanced in a mechanism such as that shown in Scheme 2.6. This is the type of behaviour observed by Sarkany and coworkers (ref. 43); ethylene hydrogenation was observed on their support only after a certain amount of carbonaceous residue had built up.

Little evidence is available concerning the optimum amount of carbonaceous deposits which favour spill-over without blocking all the surface sites which produce this species. The possibility that carbonaceous residues themselves can act as catalytic sites is being actively examined at the present time. However, the reports of Parera and coworkers (refs. 20,25) would suggest that when these formations become too graphite-like, they become extremely stable and are unlikely to serve as catalytic centres.

2.4.4 The role of surface-mobile species in determining selectivity

Most of the data available to date concerning the reactivity of spill-over hydrogen indicate that it is a highly reactive species : it readily reduces a large number of oxides and sulphides; it can hydrogenate surface carbon; it appears readily to hydrogenate benzene and many other unsaturated molecules. Consequently, with conventional catalysts such as Pt/Al_2O_3 we cannot be hopeful that it will behave towards molecules with more than one unsaturated centre, such as isoprene, in a selective manner.

These remarks have two series of consequences, both related to the reactivity of spill-over hydrogen. One series concern the reactivity towards organic molecules, the other one concerns reactions with the surface of the catalyst. Very little work related directly to the selectivity of reactions of spill-over hydrogen has been reported. However, Daage and Bonnelle (ref. 39) did observe some curious selectivity effects during hydrogenation of isoprene and 1,3-pentadiene over Cu-Cr-O spinels, which can in certain aspects be considered as bronzes. These effects may have been due more to steric effects than to selectivity associated with the hydrogen species, but were directly commensurable with the amount of spill-over hydrogen produced.

However, spill-over hydrogen can affect selectivity in other more powerful ways. Hydrodesulphuration has already been mentioned (refs. 54-58). A second mechanism concerns reactions with organic species. It involves the removal, by very effective reactions with spill-over hydrogen, of coke precursors or of other species (like carbonaceous oligomers) which poison useful catalytic sites.

Contamination by carbonaceous deposits on useful catalytic sites (S_u) would effect selectivity. By maximizing the number of S_u catalysing the desired reaction, the effects of other sites (S_n) which might catalyse unwanted (noxious) reactions would be diminished, on a relative basis. If S_n were immune to fouling or poisoning by contaminating species then

$$S_u \begin{cases} \rightarrow \text{free } S_u* \\ \rightarrow \text{contaminated } (S_u - S_u*) \end{cases}$$

$$S_n \longrightarrow \text{always free}$$

$$\text{Selectivity} = \frac{S_u*}{S_u* + S_n}$$

i.e., the selectivity would be proportional to the number of decontaminated sites, i.e. S_u*.

The second subcase is that in which elimination of a coke precursor A* is involved (Scheme 2.8). Spill-over hydrogen then inhibits the formation of coke. This scheme encompasses the work of Gnep and coworkers (ref. 21) (section 2.3.4). There A* is simply a carbocation whose elimination by spill-over hydrogen, or hydrogen from a surface site produced by spill-over, inhibits coke formation and disproportionation of o-xylene and increases selectivity for isomerization (ref. 21).

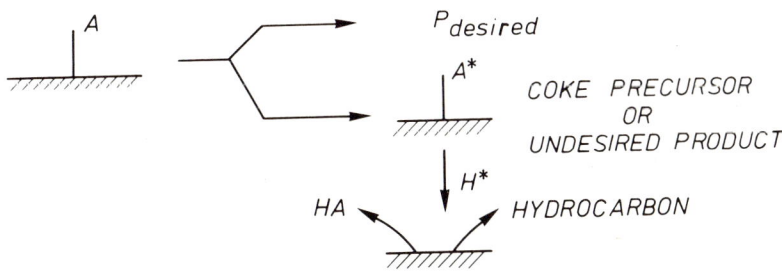

Scheme 2.8. The action of spill-over hydrogen in removing coke precursors or in eliminating an undesired product.

2.5 CONCLUSIONS

Spill-over hydrogen is present in very small amounts and its diffusion is slow on conventional supports, such as SiO_2 or Al_2O_3. Therefore, it can have only minor direct effects in forming products. However, it is extremely reactive; it creates active centres and changes the proportion of active centres with different activities and selectivities. This creation could correspond to the removal of part of the co-ordination shell of active surface elements, leaving a species with more co-ordinative unsaturation, which thus becomes potentially catalytically active. Spill-over hydrogen can also react with contaminating species, thereby regenerating deactivated centres and increasing activity. It can change selectivity by reaction (i) with the surface to increase the proportion of desired to undesired sites, (ii) with organic molecules contaminating desired sites, or (iii) with coke precursors.

These mechanisms would bring about powerful magnifying effects, namely the reaction of one or two spill-over hydrogens would create or regenerate a centre which could, afterwards, transform many molecules catalytically.

REFERENCES

1. G.M. Pajonk, S.J. Teichner and J.E. Germain (Editors) Spillover of Adsorbed Species, Elsevier, Amsterdam, 1983.
2. S. Khoobiar, J. Phys. Chem., 68 (1964) 411.
3. P.A. Sermon and G.C. Bond, J. Chem. Soc. Faraday Trans 1, 72 (1976) 730.
4. P.A. Sermon and G.C. Bond, Catal. Rev. Sci. Eng., 8 (1973) 211.
5. D.A. Dowden in "Catalysis" A specialist periodical report, the Chemical Society, London, 3 (1980) 136.
6. H. Charcosset and B. Delmon, Ind. Chem. Belg., 38 (1973) 481.
7. G.C. Bond, in G.M. Pajonk, S.J. Teichner and J.E. Germain (Editors), Spillover of Adsorbed Species, Elsevier, Amsterdam, 1983, p.1.
8. J.H. Sinfelt and P.J. Lucchesi, J. Amer. Chem. Soc., 85 (1963) 3365.
9. J.H. Sinfelt, J. Phys. Chem., 68 (1964) 856.
10. J.L. Carter, P.J. Lucchesi, J.H. Sinfelt and D.J.C. Yates, in W.M.H. Sachtler, G.C.A. Schuit and P. Zwietering (Editors), Proc. 3rd Int. Congr. Catalysis, Amsterdam 1964, North-Holland, Amsterdam, part 2, 1965, p. 644.
11. K.M. Sancier, J. Catal., 20 (1971) 106.
12. K.M. Sancier, J. Catal., 20 (1971) 404.
13. P. Antonucci, N. van Troung, N. Giordano and R. Maggiore, J. Catal., 75 (1982) 140.
14. V.H.J. de Beer, T.H.M. van Sinfelt, G.H.A.M. van der Steen, A.C. Zwaga and G.C.A. Schuit, J. Catal., 35 (1974) 297.
15. J.M. Zabala, M. Mainil, P. Grange and B. Delmon, C.R. Hebd. Seances Acad. Sci. Ser. C, 280 (1975) 1129.
16. J.M. Zabala, M. Mainil, P. Grange and B. Delmon, React. Kinet. Catal. Lett., 3 (1975) 285.
17. D. Pirotte, P. Grange and B. Delmon, in T. Seiyama and T. Tanabe (Editors), New Horizons in Catalysis, Part B, Proc. 7th Int. Congr. Catalysis, Tokyo, 1980, Kondansha, Tokyo and Elsevier, Amsterdam, 1981, p. 1422.
18. D. Pirotte, J.M. Zabala, P. Grange and B. Delmon, Bull. Soc. Chem. Belg., 90 (1981) 1239.
19. D. Pirotte, Thesis, Université Catholique de Louvain, Louvain-la-Neuve, Belgium, 1983.

20 J.M. Parera, E.M. Traffano, J.C. Musso and C.C. Pieck, in G.M. Pajonk, S.J. Teichner and J.E. Germain (Editors), Spillover of Adsorbed Species, Elsevier, Amsterdam, 1983, p. 101
21 N.S. Gnep, M.L. Martin de Armando and M. Guisnet, in G.M. Pajonk, S.J. Teichner and J.E. Germain (Editors), Spillover of Adsorbed Species, Elsevier, Amsterdam, 1983, p. 309.
22 G. del Anger, B. Log, R. Dutartre, F. Fajula, F. Figueras and C. Leclercq, in G.M. Pajonk, S.J. Teichner and J.E. Germain (Editors), Spillover of Adsorbed Species, Elsevier, Amsterdam, 1983, p. 301.
23 T. Inui, Y. Miyamoto and Y. Takegami, in G.M. Pajonk, S.J. Teichner and J.E. Germain (Editors), Spillover of Adsorbed Species, Elsevier, Amsterdam, 1983, p.181.
24 E. Furimsky and C.H. Amberg, Can. J. Chem., 53 (1975) 2542.
25 J.M. Parera, N.S. Figoli, E.L. Jablonski, M.R. Sad and J.N. Beltramini, in B. Delmon and G.F. Froment (Editors), Catalyst Deactivation, Elsevier, Amsterdam, 1981, p.571.
26 I. Ait Ichou, M. Fromenti and S.J. Teichner, in G.M. Pajonk, S.J. Teichner and J.E. Germain (Editors), Spillover of Adsorbed Species, Elsevier, Amsterdam, 1983, p. 63.
27 H. Courbon, J-M. Hermann and P. Pichat, J. Catal., 72 (1981) 129.
28 J.C. Schlatter and M. Boudart, J. Catal. 24 (1972) 482.
29 M.A. Vannice and W.C. Neikam, J. Catal. 23 (1971) 401.
30 P. Grange, Catal. Rev. Sci. Eng., 21 (1980) 135.
31 B. Delmon, in H.F. Barry and P.C.H. Mitchell (Editors), Proc. Climax 3rd Intern. Congr. Chemistry and Uses of Molybdenum, Climax Molybdenum, Ann Arbor, 1979, p.73.
32 H. Topsøe, in J.P. Bonnelle, B. Delmon and E. Derouane, Surface Properties and Catalysis by Non-Metals, N.A.T.O. A.S.I. Ser. C, D. Reidel, Dordrecht, 1982, p. 329.
33 B.C. Gates, J.R. Katzer and G.C.A. Schuit, Chemistry of Catalytic Processes, McGraw-Hill, New-York, 1979, p. 184.
34 Z. Paal, P.G. Menon, Catal. Rev. Sci. Eng., 25 (1983) 229.
35 P.W. Wentrcek, J.G. McCarty, C.M. Ablow and H. Wise, J. Catal., 61 (1980), 232.
36 G.A. Martin, in B. Imelik, C. Naccache, G. Coudurier, H. Praliaud, P. Mériaudeau, P. Gallezot, G.A. Martin and J.C. Vedrine (Editors), Metal-Support and Metal-Additive Effects in Catalysis, Elsevier, Amsterdam, 1982, p. 315.
37 M.E. Day, T. Shingles, L.J. Boshoff and G.P. Oosthuizen, J. Catal., 15 (1969) 190.
38 E.K. Poels, R. Koolstra, J.W. Geus and V. Ponec, in B. Imelik, C. Naccache, G. Coudurier, H. Praliaud, P. Mériaudeau, P. Gallezot, G.A. Martin and J.C. Védrine (Editors), Metal-Support and Metal-Additive Effects in Catalysis, Elsevier, Amsterdam, 1982, p. 233.
39 M. Daage and J.P. Bonnelle, in G.M. Pajonk, S.J. Teichner and J.E. Germain (Editors), Spillover of Adsorbed Species, Elsevier, Amsterdam, 1983, p. 261.
40 D. Duprez, J. Barbier, Z. Ferhat.Hamida and M. Bettahar, Appl. Catal., 12 (1984) 219.
41 S.J. Tanser, S.C. Fung, R.T.K. Baker, and J.A. Horseley, Science, 211 (1981) 1121.
42 G.L. Haller, V.E. Henrich, M. McMillan, D.E. Resasco, H.R. Sadeghi and S. Sakellson. Proc. 8th Int. Congr. Catalysis, Berlin, 1984, Verlag Chemie, Weinheim, Vol. V, 1984, p. 135.
43 A. Sarkany, L. Guczi and A. Weiss, Appl. Catal., 10 (1984) 369.
44 G.E.E. Gardes, G.M. Pajonk and S.J. Teichner, C.R. Hebd. Seances Acad. Sci., Ser. C, 277 (1973) 191.
45 G.E.E. Gardes, G.M. Pajonk and S.J. Teichner, J. Catal., 33 (1974) 145.

46 G. Hoang-Van, A.-R. Mazabrard, C. Michel, G.M. Pajonk and S.J. Teichner, C.R. Hebd. Séances Acad. Sci. Ser. C, 281 (1985) 211.
47 D. Bianchi, G.E.E. Gardes, G.M. Pajonk and S.J. Teichner, J. Catal., 38 (1975) 135; G.E.E. Gardes, G.M. Pajonk and S.J. Teichner, J. Catal., 33 (1974) 145.
48 S.J. Teichner, A.-R. Mazabrard, G.M. Pajonk, G.E.E. Gardes and C. Hoang-Van, J. Colloid Interface Sci., 58 (1977) 88.
49 M. Lacroix, G.M. Pajonk and S.J. Teichner, Bull. Soc. Chim. Fr., (1981) 87.
50 M. Lacroix, G.M. Pajonk and S.J. Teichner, Bull. Soc. Chim. Fr., (1981) 94.
51 M. Lacroix, G.M. Pajonk and S.J. Teichner, Bull. Soc. Chim. Fr., (1981) 101.
52 M. Lacroix, G.M. Pajonk and S.J. Teichner, Bull. Soc. Chim. Fr., (1981) 258.
53 M. Lacroix, G.M. Pajonk and S.J. Teichner, Bull. Soc. Chim. Fr., (1981) 265.
54 S.J. Teichner, G.M. Pajonk and M. Lacroix, in J.P. Bondelle, B. Delmon and E. Derouane (Editors), Surface Properties and Catalysis by non-metals, N.A.T.O A.S.I. Series C,, D. Reidel, Dordrecht, 1983, p. 457.
55 B. Delmon, Bull. Soc. Chim. Belg., 88 (1979) 979.
56 B. Delmon, React. Kinet. Catal. Lett., 13 (1980) 203.
57 B. Delmon, Ind. Chem. Eng., 20 (1980) 639.
58 B. Delmon, C.R. Hebd. Seances Acad. Sc., Paris, Ser. C, 289 (1979) 173.
59 B. Delmon, Bulg. Acad. Sci., Comm. Dept. Chem., 17 (1984) 107.
60 G.E. Batley, A. Ekstrom and P.A. Johnson, J. Catal., 34 (1974) 368.
61 R. Kramer and M. Andre, J. Catal., 58 (1979) 287.
62 P.A. Sermon and G.C. Bond, J. Chem. Soc. Faraday Trans 1, 72 (1976) 745.
63 N.C. Neikam and M.A. Vannice, J. Catal., 20 (1971) 260.
64 W.C. Neikam and M.A. Vannice, J. Catal., 27 (1972) 207.
65 R.B. Levy and M. Boudart, J. Catal., 32 (1974) 304.
66 E. Keren and S. Soffer, J. Catal., 50 (1977) 43.
67 H. Topsøe, R. Candia, N.-Y. Topsøe and B.S. Clausen, Bull. Soc. Chim. Belg. 93 (1984) 783.

Chapter 3

ADSORPTION AND HYDROGENATION OF CARBONYL AND RELATED COMPOUNDS ON TRANSITION METAL CATALYSTS

KAZUNORI TANAKA

The Institute of Physical and Chemical Research (Riken), Wako, Saitama 351-01 (Japan)

3.1 INTRODUCTION

The hydrogenation of aldehydes and ketones to the corresponding alcohols resembles the hydrogenation of olefins to alkanes in that two hydrogen atoms are added across a C=O or C=C double bond. Olefin hydrogenations catalyzed by transition metals are among the most extensively studied, and most of the data obtained have been explained according to the Horiuti-Polanyi mechanism, a stepwise addition of two adsorbed hydrogen atoms to the adsorbed olefin. One might ask whether the Horiuti-Polanyi mechanism can be extended to include the hydrogenation of C=O double bonds. This question will be examined in this chapter by means of a literature survey.

Several reviews have already been published. Five books (refs. 1-5) and one review (ref. 6) on catalytic hydrogenation include chapters on carbonyl functions. These are especially useful for synthetic purposes because they focus upon the product selectivity for a variety of catalysts and reaction substrates. Reviews based on kinetics and mechanisms were presented by Taylor (ref. 7), Bond (ref. 8), and Kieboom and Rantwijk (ref. 9), and some based on substituent effects by Kraus (refs. 10, 11). The literature up to the 1960s has been reviewed comprehensively from various angles by Mitsui (ref. 12). None of these reviews, however, refers to modern spectral studies on the adsorption of carbonyl compounds. A review including this subject was written in Japanese by Tanaka (ref. 13). This chapter incorporates most of that review, supplemented by publications which appeared afterwards.

This chapter will be confined largely to the transition-metal catalyzed hydrogenation of isolated carbonyl functions. Hydrogenations by other means such as homogeneous catalysis, electrochemical reduction and the use of reducing agents are not included. The asymmetrical hydrogenation of prochiral ketones and the reduction of conjugated carbonyl functions, though very intriguing, are also outside of the scope of this chapter, simply because of space limitations.

3.2 COMPARISON WITH OLEFIN HYDROGENATION

3.2.1 Complexity in carbonyl compound hydrogenation

The hydrogenation of carbonyl compounds, though resembling olefin hydrogenation, may involve further complications owing to (i) the possible participation of the enol form, (ii) the significant reverse reaction, (iii) the existence of two different "half-hydrogenated states", i.e., the C-metal- and O-metal-bonded species, (iv) the polar nature of the carbonyl double bond and (v) the existence of oxygen lone-pair electrons.

3.2.2 Equilibrium positions

In connection with item (ii) let us compare the equilibrium constant K_p for acetone hydrogenation

$$\text{Acetone} + H_2 \rightleftharpoons \text{2-Propanol} \tag{3.1}$$

with that for the analogous olefin hydrogenation

$$\text{Isobutene} + H_2 \rightleftharpoons \text{Isobutane} \tag{3.2}$$

The literature values of K_p for both reactions (refs. 14, 15) are shown in Fig. 3.1A. At room temperature, isobutene hydrogenation (eqn. 3.2) is practically irreversible, but acetone hydrogenation (eqn. 3.1) is rather reversible. This is also seen in Fig. 3.1B, whose ordinate, α, is the fractional equilibrium conversion of 2-propanol or isobutane in the backward reaction of eqns. 3.1 and 3.2. The values for α were calculated from

$$K_p P_t = (1-\alpha^2)/\alpha^2 \tag{3.3}$$

by setting the total pressure, P_t, equal to 1 atm, around which most laboratory experiments are conducted (ref. 16). What is striking is that even at room temperature a manometer-

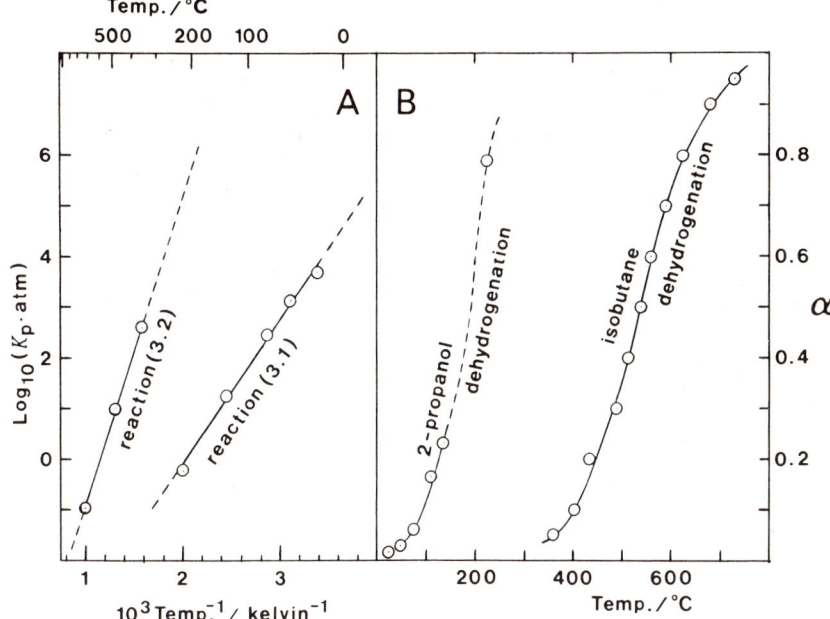

Fig. 3.1. Comparison of equilibrium parameters for reactions (3.1) and (3.2). (A) equilibrium constant, (B) the fractional equilibrium conversion, α, of 2-propanol or isobutane in the reverse reactions of (3.1) and (3.2).

detectable pressure of acetone or hydrogen exists at the equilibrium of reaction 3.1. Bond additivity or group additivity rules for thermodynamic quantities (refs. 17, 18) suggest that the above stark contrast between acetone hydrogenation and isobutene hydrogenation is also true for all ketone-olefin or aldehyde-olefin analogue pairs.

It may be noted that in the above discussion we have been dealing with gas-phase hydrogenation. In liquid-phase hydrogenation, the equilibrium is generally anticipated to shift much toward the alkane or alcohol (ref. 19), due to the severe restrictions on the translational and rotational motions of liquid-phase species. In any reaction of the type A+B ⇌ AB, such motional limitation serves to push the equilibrium toward AB by reducing the entropic gain of A+B relative to AB.

3.2.3   Reaction pathways

The Horiuti-Polanyi mechanism for ethylene hydrogenation is shown in Fig. 3.2, where (a) denotes the adsorbed state. This

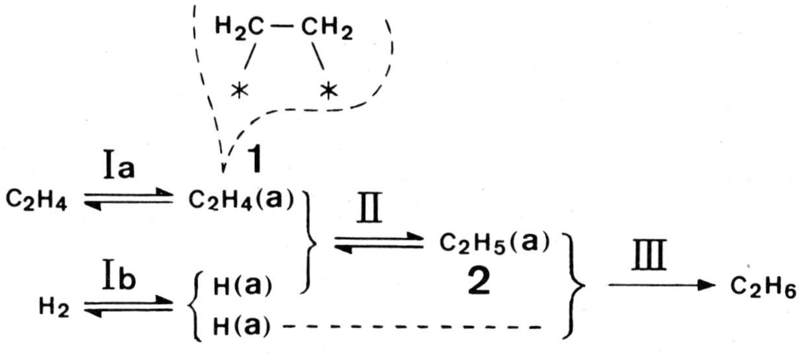

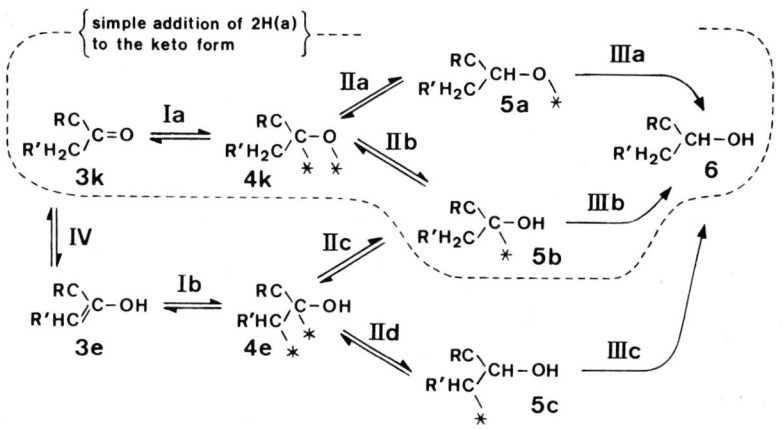

Fig. 3.2. The Horiuti-Polanyi mechanism for ethylene hydrogenation. (a) denotes the adsorbed state, ∗ signifies the adsorption site.

Fig. 3.3. Horiuti-Polanyi mechanism applied to ketone hydrogenation. ∗ signifies the adsorption site (ref. 13).

reaction scheme is characterized by the stepwise addition of H(a) to the associatively adsorbed ethylene **1**. The analogous reaction scheme for ketone or aldehyde hydrogenation is given in Fig. 3.3, where H_2 and H(a) are omitted for the sake of simplicity. The rate-determining step, though tentatively assigned to step III, may be different depending upon the catalyst and the reaction conditions. Items (i) and (iii) in Section 3.2.1 are explicitly taken into account in this reaction scheme. Item (iv) may call for proton, hydride ion or polarized hydrogen molecule (H^+-H^-) as the hydrogen source for hydrogenation. Species **4k**, corresponding to the associative adsorption **1** in the olefin system, is the sole adsorption form for keto-type ketones within the framework of the Horiuti-Polanyi mechanism. Item (v), however, suggests that carbonyl compound hydrogenation may also proceed via other adsorption species such as the π-bonded one (7), the one coordinated through the oxygen lone pairs (8) and the π-oxaallylic intermediate (9). All of these are shown in Fig. 3.4.

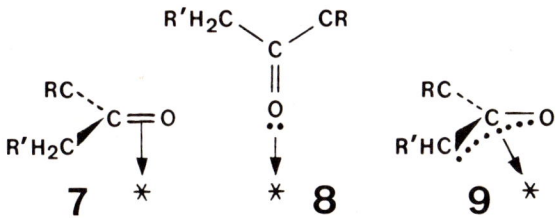

Fig. 3.4. Possible reaction intermediates not involved in the Horiuti-Polanyi mechanism.

Under certain conditions, hydrogenation is accompanied by hydrogenolysis to form hydrocarbons. This side reaction is usually promoted by acids and inhibited by bases (ref. 5), and is often notable on Pt (ref. 12).

3.3 DEUTERIUM TRACER STUDIES

Since the tracer studies up to 1952 have been reviewed in ref. 7 and those up to 1958 in ref. 8, only representative later studies are reviewed here, except for a few earlier ones which deserve particular attention.

3.3.1 Aliphatic ketones and aldehydes

(i) <u>Selected early work up to the 1950s</u>   Simple deuterium addition to the keto form **3k** and to the enol form **3e** leads to the isotopomer alcohol containing the C_αD-OD moiety and to that containing the C_βD-C_αD moiety, respectively. By quantifying the C_βD bond in the product alcohol, therefore, Anderson and MacNaughton (ref. 20) attempted to estimate the contribution of the enol form **3e** in the metal-catalyzed hydrogenation of several carbonyl compounds including acetone. The conclusion reached was that ketonic addition is predominant at low temperatures, and more enolic addition takes place at higher temperatures. The enolic mechanism at higher temperatures, however, is inconclusive because the C_βD bond could also be formed from the keto form **3k** through the migration of the adsorption bond (steps IIb plus IIc of Fig. 3.3) or the formation of adsorbed π-oxaallylic species (**9**, Fig. 3.4).

Friedman and Turkevich (ref. 21) demonstrated a contrast between the gas-phase and liquid-phase reductions of acetone with deuterium. Under otherwise similar conditions, the main reaction product in the gas phase was propane with extensive deuterium exchange, but that in the liquid phase was 2-propanol with only a small amount of exchange. This result seems relevant to the equilibrium shift in the liquid phase, as mentioned in Section 3.2.2.

Catalytic reactions between acetone vapour and deuterium were also examined by Kemball and Stoddart (ref. 22) on a series of evaporated metal films (Rh, Pd, Pt, Ni, Fe, W, Au and Ag). Although the partial pressure of acetone used was as low as 0.2 kPa, the main reduction product was 2-propanol rather than propane. A simple addition mechanism (a series of steps Ia, IIb and IIIb in Fig. 3.3) was suggested from the observed time variation of the deuterium contents of acetone and 2-propanol. The rate-determining step was assigned to step IIIb because the rate of acetone deuteration was comparable with the rate of secondary-hydrogen exchange in 2-propanol, and both processes had roughly the same activation energy. The relative rates of acetone hydrogenation and deuteration, k_H/k_D, and the difference in activation energy between them can also be accounted for by this reaction model. In connection with these studies, Anderson and Kemball (ref. 23) examined the catalytic reaction between aliphatic alcohols and deuterium. The hydroxyl hydrogen exchanged much more rapidly and with a lower activation energy than did the aliphatic hydrogens. Extensive multiple exchange was observed only on Rh and Fe.

(ii) <u>Anomaly on Cu. Intermolecular hydrogen transfer.</u> A series of studies were undertaken by Burwell and co-workers (refs. 24-26) on isotopic exchange reactions involving alcohols, ketones and deuterium on Cu, Ni and Pd. The observed deuterium distributions of the product butanol in butanone deuteration over Cu are shown in Fig. 3.5 (ref. 26) : rather surprisingly, the versatile OH group was only 20% D-labelled while at the α-position the hydrogen was almost completely replaced by deuterium. A similar characteristic distribution pattern was observed in the

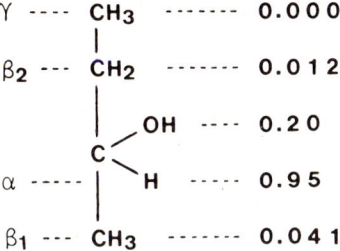

Fig. 3.5. Results of butanone deuteration catalyzed by Cu. The figures indicate the average number of deuterium atoms per molecule of the product 2-butanol (ref. 26).

exchange of 2-butanol with deuterium (ref. 26). These results, in conjunction with other experimental data, were accounted for by considering the nature of the hydrogen atoms in the butanone and 2-butanol molecules. The surface hydrogen on Cu and the α-hydrogen on butanol were regareded as hydride-like, and the β-hydrogen atoms of butanone and butanol as proton-like. If preferential exchange occurs only between like hydrogens and is accompanied by intermolecular proton transfer from alcohol hydroxyl to ketone carbonyl, then the isotopic distribution patterns would be as observed. A mechanism for the deuteration and exchange on Cu was proposed which involves the alkoxy (5a) and the π-oxaallylic (9) adsorbed species. Species 9 was also assumed on Pd (ref. 25).

3.3.2 Alicyclic ketones

(i) <u>Lability of the α-hydrogen of ketones.</u> Catalytic reactions of cyclopentanone vapour with hydrogen and deuterium were investigated by Kemball and Stoddart (ref. 27) using a series of evaporated metal films (Rh, Pd, Pt, W, Ni). Compared with their previous studies on acetone reactions (ref. 22), the activation energies and frequency factors for hydrogenation were broadly similar on each metal. This similarity suggests the same rate-determining step, i.e., step IIIb in Fig. 3.3, for the hydrogenation of both ketones. Exchange data were not obtained over Rh and Pt because the reduction to cyclopentanol and cyclopentane was too rapid. The kinetics of cyclopentanone-deuterium exchange observed on W and Ni was similar to that previously found for acetone-deuterium exchange (ref. 22), but on Pd four of the eight hydrogen atoms in a cyclopentanone molecule, presumably those of the α-methylene groups, were much more readily exchanged. Cornet and Gault (ref. 28) later used as exchange substrates four methyl-substituted cyclopentanones differing in the number of α-hydrogens, and gave further evidence for the lability of α-hydrogen on Pd and also on Ni. α-Hydrogen lability was explained by assuming adsorbed π-oxaallylic species.

  (ii) <u>Stereochemical mechanism.</u> The stereochemistry for the formation of the isomeric alcohol in the hydrogenation of substituted alicyclic ketones is of continuing interest. Two distinct orientations are conceivable for the adsorption of alicyclic ketones, depending upon which side of the molecular ring is directed toward the catalyst surface. Figure 3.6 depicts this

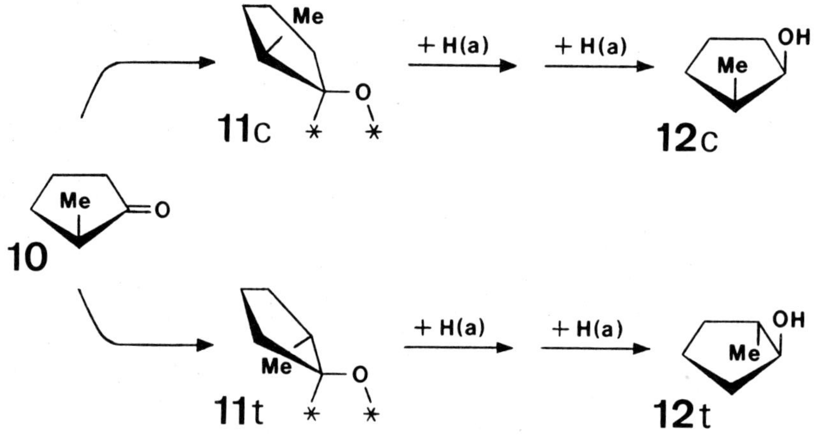

Fig. 3.6. A reaction mechanism for the formation of stereo-isomeric alcohols 12c and 12t by the hydrogenation of compound 10.

situation, taking 2-methylcyclopentanone (10) as an example. The cis-addition of two hydrogen atoms to the adsorbed carbonyl from the catalyst side leads one orientation (11c) to the cis-alcohol (12c) and the other (11t) to the trans-isomer (12t).

  According to this conventional reaction model, isomers 12c and 12t are produced via different reaction pathways. Upon deuteration of compound 10, therefore, one might expect entirely different deuterium distribution patterns for 12c and for 12t. However, the experiments of Cornet and Gault (ref. 29) on Ni, Rh and Pt showed (a) no significant difference in the deuterium distribution between **12c** and **12t**; (b) the existence of fast cis-trans isomerization of **12** not via ketone **10**. Result (b) seems difficult to interpret, but can be accounted for by assuming an **8**-like intermediate through which the rapid cis-trans isomerization occurs:

where "λ-adsorbed" represents coordination to the surface through an oxygen lone pair. This rapid isomerization path could also be used to interpret result (a) basically within the framework of the conventional reaction scheme. The mechanism proposed by Cornet and Gault (ref. 29), however, was quite different, and is characterized by the formation of a π-oxaalyllic species **9** followed by hydrogen addition to the allylic carbons from either side of the adsorbed molecules: the addition of H(a) from the bottom and of an H_2(a) hydrogen atom from the top.

(iii) <u>Direct location of incorporated D atoms.</u> In the deuterium tracer studies described in the preceding sections the carbon position of deuterium incorporation was not directly determined, but inferred from mass spectral data. There was no information at all on the axial and equatorial deuterium contents at each carbon position. Takagi et al. (refs. 30, 31) succeeded in obtaining such more direct and more detailed data using NMR spectroscopy with the aid of paramagnetic shift reagents for the deuteration of 4-tert.-butylcyclohexanone (**13**) catalyzed by

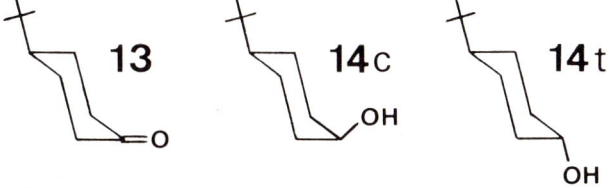

platinum metals. Table 3.1 lists the observed axial and equatorial D contents at each carbon position of the product alcohol, i.e., cis- and trans-4-tert.-butylcyclohexanols (**14c**, **14t**). Apparently, over Ru, Os, Ir and Pt, deuterium incorporation occurs only at the C-1 position (besides that at the oxygen position). This result suggests the simple addition of two

TABLE 3.1

Stereochemical deuterium distributions in species **14c** and **14t** produced by the deuteration of **13**.

Catalyst	Sample	Average number of axial and equatorial D atoms at each carbon position[*]				
		D_1	$D_{2,6a}$	$D_{2,6e}$	D_{other}	D_m
Rh	14c	0.57	0.44	0.44	0.0	1.45
	14t	0.59	0.50	0.50	0.0	1.59
Pd	14c	0.40	0.64	0.64	0.0	1.68
	14t	0.34	0.68	0.68	0.0	1.70
Ru,Os,Ir,Pt	14c	1.0	0.0	0.0	0.0	1.0
	14t	1.0	0.0	0.0	0.0	1.0

[*] D represents the number of D atoms at the carbon position specified. a = axial, e = equatorial, other = any carbon positions other than C(1) + C(2) + C(6), m = mean value for the whole molecule.
Note : $D_{2,6a} = D_{2a} + D_{6a}$, $D_m = D_1 + D_{2,6a} + D_{2,6e} + D_{other}$

D(a) atoms to the adsorbed carbonyl bond. In contrast, over Rh and Pd the C-2 and C-6 positions (mutually identical) are also deuterated. This suggests that these carbon positions as well as the carbonyl group participate in adsorption. Nearly the same deuterium contents for **14c** and **14t** are consistent with result (a) in Section 3.3.2, point ii. Another equality in deuterium content is seen between the axial and equatorial C-2,6 positions. This double equality can be explained by assuming a quasi-equilibrium between the cis-type and trans-type intermediates **5b** (Fig. 3.3) through free ketone **3k** or the coordinatively adsorbed one **8**.

Significant deuterium incorporation at C-2 and C-6 over Rh and Pd was also observed in the deuteration of cyclohexanone (ref. 32) and its 2-methyl (refs. 33, 34) and 2-ethyl derivatives (**15**) (ref. 34). Table 3.2 shows the observed deuterium distribution for ketone **15** and its corresponding alcohols (**16c**, **16t**). A closer

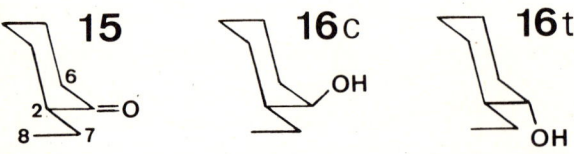

TABLE 3.2

Stereochemical deuterium distributions in species **15**, **16c**, and **16t** in the deuteration of **15**.

Catalyst	Sample	Average number of axial and equatorial D atoms at each carbon position*							
		D_1	D_{2a}	D_{6a}	D_{6e}	D_{3e}	D_{5e}	D_7	D_m
Rh	15		0.26	0.38	0.17	0.0	0.17	0.14	1.12
	16c	0.73	0.14	0.26	0.14	0.0	0.07	0.14	1.48
	16t	0.74	0.16	0.32	0.16	0.0	0.08	0.0	1.46
Pd	15		0.83	0.71	0.74	0.0	0.0	0.0	2.28
	16c	0.60	0.64	0.47	0.44	0.0	0.0	0.0	2.15
	16t	0.48	0.71	0.41	0.38	0.0	0.0	0.0	1.98

* See Table 3.1 for details.

look at the data of Table 3.2 reveals a small but definite difference between Rh and Pd: distribution of deuterium beyond the C-2 and C-6 positions occurs on Rh, but not on Pd. This distinction was also confirmed indirectly by mass spectral analysis.

A possible explanation for this result is that the main adsorbed species responsible for deuterium exchange is π-allylic on Pd but α,β-diadsorbed on Rh. This could be tested by carrying out deuteration of adamantanone (**17**) over these metals and by

examining the deuterium distribution in the product 2-adamantanol. Owing to the rigidity of its molecular framework, compound **17** is unable to undergo tautomerization to the enol form and cannot yield the π-oxaallylic species upon adsorption. Thus it is expected that deuterium incorporation on Pd is limited only to deuterium addition across the carbonyl double bond, whereas on Rh the deuterium is distributed beyond the C-1 and C-3 positions by propagations of the so-called α,β-process (ref. 35). This expectation was confirmed by experiment (ref. 36).

3.3.3 Aromatic ketones

Bonnet et al. (ref. 37) studied the hydrogenation of acetophenone-d₃ (**18**) specifically labelled at the methyl position.

Ph–C(=O)–CD₃ **18**

Their main concern was in establishing whether the enol form plays a role in the metal-catalyzed hydrogenation of acetophenone. Only the d_3-alcohol must be formed if there is no participation of the enol form, whereas if it participates the depleted species d_0-d_2 are also expected, on account of the D transfer from the CD_3 to the carbonyl to form a OD group followed by isotopic exchange between the OD group and the pool of surface hydrogen on the catalyst. By conducting the hydrogenation in cyclohexane using Pt/SiO_2 as catalyst they obtained essentially the d_3-alcohol alone containing only 8% d_2, thus ruling out the keto-enol tautomerization both in the bulk liquid phase and on the catalyst surface.

3.4 STEREOCHEMISTRY OF ALICYCLIC KETONE HYDROGENATION

Two stereoisomers are usually formed in the hydrogenation of substituted cycloalkanones to the corresponding alcohols — the axial and the equatorial alcohols. In the case of alkyl-substituted cyclohexanones, the equatorial alcohol is more stable and hence more abundant at isomeric equilibrium (refs. 38-40). However, metal-catalyzed hydrogenation usually gives more of the axial alcohols (refs. 38, 40, 41-43). This axial selectivity is generally accounted for by making the following assumptions (refs. 38, 44, 45):

  a) Adsorbed cyclohexanones are in the chair form.
  b) The adsorption occurs in the least hindered state, i.e., in such a manner that the carbonyl bond axis becomes parallel to the catalyst surface, with the substituent R inclined away from the catalyst.
  c) Hydrogen atoms add cis to the carbonyl linkage from the catalyst side.

The axial alcohol alone is expected from this mechanism, but in fact the axial selectivity is not complete: the equatorial alcohol is also formed, especially in neutral and basic media. A variety of explanations have been advanced in order to explain the simultaneous formation of the equatorial isomer. Among them are the adsorption of substituted cyclohexanones from the more hindered side (refs. 31, 34, 42), the isomerization of the produced axial alcohol to the equatorial isomer (refs. 28, 38, 44) and the trans addition of hydrogen atoms to the adsorbed π-oxaallylic intermediate (refs. 28, 29). The likelihood of the equatorial alcohol being formed especially in neutral or basic media was explained by assuming the adsorption of the ketone (ref. 46) or the enolate (ref. 47) from the least hindered side, followed by hydride ion transfer from the catalyst and protonation from the solvent. It was reported, however, that the addition of NaOH anomalously enhanced the axial alcohol production in the hydrogenation of substituted cyclohexanones over Raney-Ni (ref. 42) and Pt (ref. 48).

In the reduction of 2-methylcyclohexanone by sodium borohydride, Wigfield et al. (ref. 49) reported evidence of significant participation of the less stable chair conformation with the alkyl group axial. This may also be true for metal-catalyzed hydrogenation. Notable in this connection is the fact that the percentage of the less stable axial-alkyl conformer in the chair-chair equilibrium of alkylcyclohexanones is quite significant, though small; for the 2-alkyl derivatives the reported values (ref. 50) are 10.3% for methyl, 14.5% for ethyl, 23.4% for isopropyl and 7.0% for tert.-butyl.

3.5 KINETICS, SUBSTITUENT EFFECTS AND RELATED SUBJECTS

3.5.1 Kinetic studies

The main papers up to 1959 have been reviewed (ref. 8). Acetone hydrogenation was subsequently intensively studied by several more groups, i.e., by Simonikova et al. (ref. 51), Kishida et al. (refs. 52, 53), Iwamoto et al. (refs. 54, 55, 56), Kiperman and Kaplan (ref. 57), and Babkova et al. (ref. 58). All of these investigations concerned a comparison of the rate expressions determined by experiment with those derived from reaction models (ref. 59). According to Simonikova et al. (ref. 51), on Pt, Pd and Rh, acetone hydrogenation is controlled by the reaction of the adsorbed hydrogen with acetone adsorbed molecularly on a single metal site, whereas on Cu it occurs by the reaction of dissociatively adsorbed acetone with adsorbed molecular hydrogen. Iwamoto et al. (ref. 54) obtained two different rate expressions over Raney-Ni, depending upon the solvent used. They concluded that in methanol or hexane the rate is controlled by addition of H(a) to the half-hydrogenated acetone, whereas in C_2-C_4 alcohols the conversion of the acetone-H_2 complex into 2-propanol is rate determining. Similar results were obtained for hydrogenation of other alkyl methyl ketones (ref. 55).

The hydrogenation of cyclohexanone were similarly studied over copper chromite (ref. 60).

3.5.2 Substituent effects, competitive reactions

For studies up to the 1970s the reader should consult ref. 11. In this reference the hydrogenation of $RCOCH_3$ ketones is correlated with the Taft σ^* constant, and that of $R \cdot C_6H_4 \cdot COCH_3$ ketones with the Hammett-type equations. This indicates the existence of polar effects in metal-catalyzed hydrogenation of ketones. The involvement of steric effects is also discussed.

Though not well known, there should in principle be one more factor in substituent effects: the non-polar, non-steric entropy effects which depend on the mass, m, and the moment of inertia, I, of the reactant molecule. To describe this factor, the term "ponderal" effect was coined in 1955 (ref. 61) although there has been a recent claim that the term "non-potential energy" effect is preferable (ref. 62). The ponderal factor in substituent effects is considered to be especially pronounced in heterogeneous catalysis, owing to the loss of translational and rotational entropies due to the adsorption of the reactant molecule. The larger the substituent, the greater is the entropy loss, thus giving rise to a certain magnitude of ponderal substituent effects. Indeed, relative rate data for the metal-catalyzed competitive hydrogenation of cyclohexanone and its 2-alkyl derivatives (Me, Et, Pr) were satisfactorily accounted for on the basis of ponderal effects (refs. 63, 64), although this term was not explicitly used. The relative hydrogenation rate of 4-isopropyl- to 4-tert.-butylcyclohexanone is also explicable by this concept (ref. 65).

Nishimura et al. (ref. 66) estimated the relative adsorption coefficient, K_B/K_A, of 5α- and 5β-cholestan-3-one (B) to 4-tert.-butylcyclohexanone (A) by the analysis of relative rate data for individual and competitive hydrogenation in tert.-butanol. The K_B/K_A values obtained were 81 (5α) and 63 (5β) for Pd and 1.4 (5α) for Ir. This suggests that the steroid α-face, i.e., the large hydrocarbon moiety in ketone B, interacts very attractively with a palladium surface. The attractions of substituent hydrocarbon moieties toward such a surface were also observed for 3- and 4-

alkyl derivatives of cyclohexanone (refs. 65, 67). Palladium is conspicuous among other metals for this attraction, presumably because of its unusual weakness for the adsorption of carbonyl groups (refs. 66, 68).

Jenck and Germain (ref. 69) measured the relative reactivities of 19 saturated aliphatic and alicyclic ketones in the copper chromite-catalyzed vapour-phase hydrogenation of binary or ternary mixtures at 185-240°C under $2 \times 10^3 - 10 \times 10^3$ kPa total pressure, with an excess of hydrogen. The relative reactivities varied with pressure and temperature, and when plotted against temperature exhibited a sharp break at the point where the liquid phase forms or disappears in the reactor. A linear correlation between the logarithm of the reactivities and Taft's E_s constants indicates the predominance of steric effects, in contrast with the results of Simonikova et al. (ref. 70) for kieselguhr-supported Cu, Rh and Pt, and of Iwamoto et al. (ref. 55) for Raney-Ni. These two groups obtained the best correlation with the Taft σ^* constant, indicative of the inductive effect. The work of Jenck and Germain (ref. 69) was further extended to include aldehydes

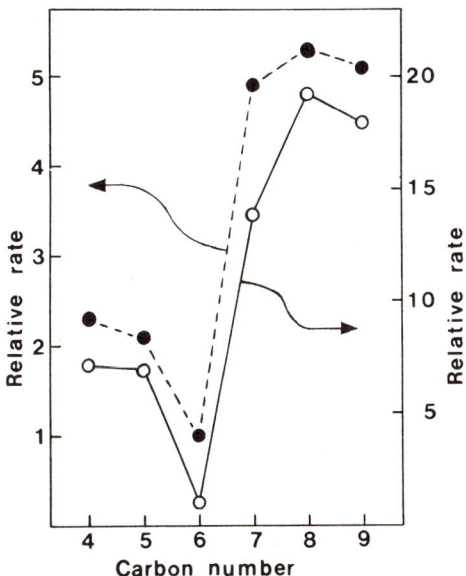

Fig. 3.7. The effects of ring size on reaction rate in the catalytic hydrogenation of cycloalkanones over Pt/SiO$_2$ (Solid line) and also in the chromic acid oxidation of cycloalkanols (dashed line) (ref. 72).

and olefins (ref. 71).

Geneste et al. (ref. 72) measured the relative rates for the hydrogenation of various cycloalkanones on Pt/SiO_2 using cyclohexane as solvent according to the competitive kinetic method. These rates are plotted against carbon number in Fig. 3.7 (solid line). This profile closely parallels the corresponding curve for the reverse reaction, i.e., the chromic acid oxidation of the cycloalkanols (broken line). Since the slow step in the latter reaction is the formation of the ketone (sp^2) from the chromic ester intermediate (sp^3) having an O-Cr bond, it is very likely that the hydrogenation is rate-controlled by the conversion of the π-adsorbed (sp^2) into the half-hydrogenated intermediate (sp^3) having an O-Pt bond:

$$\underset{Pt}{\overset{}{\underset{\downarrow}{>C=O}}} \longrightarrow >C\overset{O-Pt}{\underset{H}{<}}$$

Among other studies on substituent effects in ketone hydrogenation are a comparison of cyclohexanone with its 4-non-alkyl-substituted derivatives (ref. 73), and a comparison of acetophenone with its substituted derivatives in various solvents (ref. 74).

3.6 CHARACTERIZATION OF ADSORBED SPECIES

3.6.1 IR spectroscopy

(i) *Preview.* IR spectral studies of the adsorption of simple carbonyl compounds on metal surfaces were initiated by Blyholder and co-workers (refs. 75-79), followed by Young and Sheppard (refs. 80-82) and also by Miyata and co-workers (refs. 83-86). All these groups failed to observe the O-H stretching and bending vibrations of the adsorbed species, suggesting the intermediacy of **5a** rather than **5b** in the ketone hydrogenation scheme of Fig. 3.3. The three groups also agree in that carbonyl compounds adsorb to form two distinct surface species: one exhibiting the characteristic absorptions of the carbonyl group and the other lacking these. However, they disagree in the assignment of these adsorbed species. According to the last two groups, the most likely carbonyl-containing species is the ketone or aldehyde itself π-coordinated to the surface metal (**7** in Fig. 3.4), and the dominant carbonyl-lacking species is the C,O-diadsorbed species

(**4k** in Fig. 3.3, corresponding to the associatively adsorbed olefin species). Blyholder and co-workers (refs. 75-79) considered instead that acyls and alkoxides were the significant carbonyl-containing species, although for acetone the acyl assignment in their earlier paper (ref. 77) was later reassigned as weakly adsorbed acetone (ref. 79).

Of interest here is Blyholder's special device for preparing metal adsorbent samples for spectral studies (ref. 87). He evaporated metals from a tungsten filament, and deposited them in a vacuum-pump oil film on the salt windows of an infrared cell. In this metal-oil mull the oil plays the role of the solvent. Thus, I believe that the spectral data for the oil-supported metals have relevance to liquid phase reactions, while those for ordinary SiO_2-supported metals are associated with gas phase reactions.

(ii) <u>Work of Blyholder and co-workers (refs. 75-79)</u>. The adsorption of about a dozen C_xH_yO compounds, including ketones, alcohols, aldehydes, and ethers, were examined on Fe, Co and Ni by Blyholder and co-worders (refs. 75-79). These studies are reviewed here without paying much attention to non-carbonyl compounds. The adsorption on SiO_2-supported and oil-supported metals exhibited the same broad features, and the differences were mainly quantitative rather than qualitative (refs. 76, 77). Most of the C_xH_yO compounds adsorb at room temperature on each metal to give the following dominant adsorption species: Fe, alkoxide (refs. 75, 77); Co, alkoxide, acyl, CO (refs. 77, 78); Ni, CO, acyl (refs. 76, 77). In brief, on Fe the alkoxide species enjoy a special stability with little tendency for further decomposition, on Ni the decomposition to CO is most significant and Co is intermediate between Fe and Ni. On Fe (ref. 75), as judged from the effect of CO addition upon adsorbate spectra, the order of adsorption strength of alcohols is primary > secondary > tertiary.

The subtle differences in adsorbate spectra between SiO_2- and oil-supported metal surfaces are of interest (ref. 77). When primary alcohols are adsorbed on Ni/oil the intensities of the resulting carbonyl and acyl bands are weaker than those on Ni/SiO_2. There is no IR evidence of 2-pronanol and acetone adsorbing on Ni/oil, whereas on Ni/SiO_2 they produce chemisorbed CO and a surface species with an acyl structure. The exposure of Co/oil to 1-propanol, 2-propanol or acetone results in only weakly held alkoxide structures, but on Co/SiO$_2$, bands due to acyl structures and CO as well as to alkoxide structures are observed. These results, together with other data, suggest that the adsorbate-metal interaction is stronger on metal/SiO$_2$ than on metal/oil, and that the oil inhibits dehydrogenation and alkyl migration. This distinction is reminiscent of the work of Friedman and Turkevich (ref. 21), who observed different reaction products and different extents of exchange for gas-phase and liquid-phase reductions of acetone with deuterium.

By a careful reinvestigation of acetone adsorption on Co and Ni, Blyholder and Shihabi (ref. 79) modified their earlier assignment to acyl structures for the species exhibiting absorption in the 1600-1700 cm^{-1} region. Their modified assignment is shown in Fig. 3.8; there are three species which are interconverted by hydrogen addition or elimination. On Ni, much less acetone adsorbs, and the main surface structure is the coordinated species (**19**).

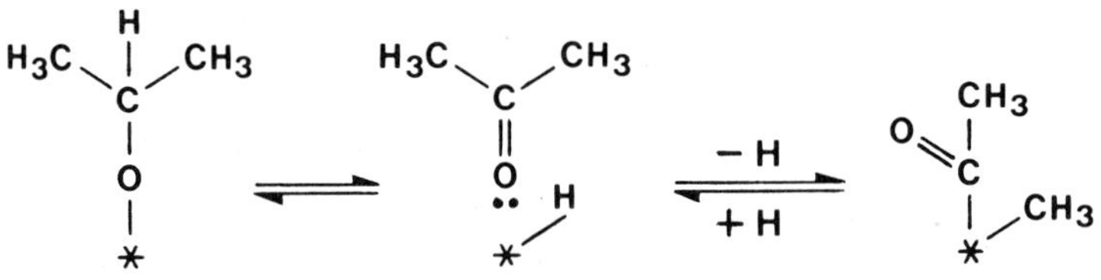

Fig. 3.8. Structures for chemisorbed acetone (ref. 79).

(iii) <u>Work of Young and Sheppard (refs. 80-82).</u> Young and Sheppard made careful studies of the adsorption of aliphatic ketones (refs. 80, 81) and acetaldehyde (refs. 80, 82) on Ni/SiO$_2$. Unlike Blyholder and co-workers (refs. 75-79) they did not observe the decompositon of acetone to CO at room temperature. Decarbonylation occurred only upon heating to 180°C. The order of ease of decarbonylation was acetaldehyde > acetone > diethyl ketone, probably reflecting steric effects of the bulky alkyl groups (ref. 81).

(iv) <u>Work of Miyata and co-workers (refs. 83-86) and other groups.</u> Miyata et al. (ref. 86) estimated the ratio of the associative to the coordinative adsorption of acetone on SiO_2-supported metals using two different techniques, i.e., IR spectroscopy and thermal desorption. The first method was based on a quantitative comparison of the intensities of the $\nu(C=O)$ and $\nu_{as}(CH_3)$ peaks, and the second method on the evaluation of the relative amounts of the acetone and hydrocarbons desorbed upon heating. The estimations from both methods were in close agreement, and are given in Table 3.3, together with the positions of $\nu(C=O)$.

TABLE 3.3
The percentages of surface species in acetone adsorption on metal/SiO_2 catalysts.

Metal	Surface species (%)		$\nu(CO)$ cm^{-1}
	Associative	Coordinative	
Pt	69	31	1690
Ni	62	38	1696
Pd	31	69	1700
Ru	26	74	—
Pd-Mo	5	95	1696

The proportion of the two adsorbed species varies with the amount of adsorption: associative adsorption is dominant at low coverage, and the proportion of coordinative adsorption increases with increasing coverage. Of particular interest is the observation (ref. 85) that, upon admission of hydrogen, the associatively adsorbed acetone remains intact, but the coordinatively adsorbed acetone is converted into 2-propanol on Ni/SiO_2, and into propane on Pt/SiO_2. This suggests that acetone hydrogenation proceeds through weak coordination to the surface rather than by strong associative adsorption.

Coordinative adsorption was also suggested by Szilagyi et al. (ref. 88) for cyclohexanone on Pt/SiO_2 at 313 K. By raising the temperature to 393 K, they observed a decrease in the intensity of the carbonyl band and concomitant formation of broad bands around 1610 and 1400 cm^{-1}, which are characteristic of a $C\!\!=\!\!C\!\!=\!\!O$ skeleton. This provides direct spectral evidence for the π-oxaallylic species inferred from kinetic data. Among other relevant systems studied by IR spectroscopy is that of acetylacetone adsorbed on evaporated films of Fe and Ni (ref. 89).

3.6.2   Thermal desorption

The thermal desorption technique used to distinguish and quantify the two different adsorption states of acetone, mentioned in Section 3.6.1, point iv, was also applied to other aliphatic ketones. The results are listed in Table 3.4 (ref. 90). It is

TABLE 3.4
The percentages of surface species in ketone adsorption on Pt/SiO$_2$ catalyst

Ketone	Surface species (%)	
	Associative	Coordinative
Acetone	79	21
Methyl ethyl ketone	59	41
2-Pentanone	36	64
Methyl isopropyl ketone	34	66

seen that the larger the ketone molecule, the greater is the percentage of coordinative adsorption, probably reflecting the decrease in attraction to the surface metal due to the bulky alkyl group. This trend is reminiscent of the molecular size effect upon decarbonylation of adsorbed carbonyl compounds (Section 3.6.1, point iii).

Alcohols, ethers and water commonly possess an out-of-plane oxygen lone-pair orbital. According to Rendulic and Sexton (ref. 91), each of these molecules adsorbs on Pt(111) in two distinct states at 100 K, a monolayer phase and a multilayer phase. The heat of adsorption, ΔH, for the monolayer phase is expressed as

$$\Delta H = 41.8 + 5.4n \quad (kJ/mol) \qquad (3.4)$$

where n is the carbon number as in $C_nH_{2n+1}OH$ and $(C_nH_{2n+1})_2O$ for normal alkyl groups ($n=2$ for isopropyl). This simple relationship suggests an adsorption model in which the contribution to the adsorption heat is from the oxygen lone pair (41.8 kJ/mol), with an additional weak van der Waals interaction between the alkyl chain and the platinum surface (5.4 kJ/mol per C atom). For ethers, however, only one of the side chains gives this additional contribution because only one alkyl chain is in parallel to the platinum surface and the other extends away from it.

3.6.3 Ultraviolet photoelectron spectroscopy

Chemisorption and decomposition of small oxygen-containing molecules including methanol, dimethyl ether, formaldehyde, acetaldehyde and acetone have been studied by ultraviolet photoelectron spectroscopy (UPS) on a polycrystalline palladium surface at 120 and 300 K (ref. 92). At 300 K these molecules decompose to form chemisorbed CO, but at 120 K they are chemisorbed, probably end-on to the surface via the oxygen atom through its lone-pair orbital. The chemisorption bond energies were estimated as 0.32 eV for acetaldehyde and as 0.35 eV for acetone, much lower than the heats of chemisorption for CO (1.21 eV) (ref. 93) and H_2 (0.9 eV) (ref. 94) on Pd.

3.6.4 Extrapolation from UHV-low temperature conditions

In order to use spectral data on adsorption for mechanistic studies of catalytic reactions, it is expedient to know how adsorbed states are altered upon going from ultrahigh vacuum (UHV)-low temperature conditions to the conditions of much higher pressures and temperatures. By thermodynamic considerations of the adsorption of diatomic molecules, Benziger (ref. 95) showed that molecular adsorption is preferred at low temperature and high pressure while dissociative adsorption is preferred at high temperature and low pressure. Figure 3.9 shows an example of his calculations — adsorption isotherms at 300 K for an arbitrary

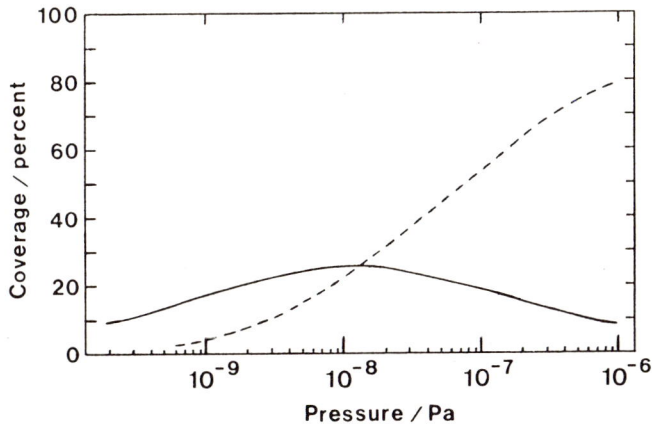

Fig. 3.9. Adsorption isotherms (at 300 K) for molecular (-----) and dissociative (———) adsorption of diatomic molecules, calculated for an arbitrary diatomic molecule whose heat of adsorption is 125 kJ/mol for both types of adsorption (ref. 95).

diatomic molecule whose adsorption energy is -125 kJ/mole for both molecular and dissociative adsorption.

The pressure effects observed by Davis and Somorjai (ref. 96) in the reaction of cyclohexene with hydrogen over Pt(223) seem to be in harmony with the above theory. The predominant reaction was dehydrogenation to benzene at low pressures (10^{-6}-10^0 Pa) and hydrogenation to cyclohexane at high pressures (10^2-10^3 Pa).

3.7 CONCLUSIONS

The adsorption states and hydrogenation mechanism of ketones and aldehydes on transition metals depend upon the environmental conditions under which the catalyst is operating, i.e., the temperature, pressure, reaction phase and solvent. In general, molecular adsorption and hydrogenation to the corresponding alcohol is favoured in the liquid phase and at low temperatures and high pressures, whereas dehydrogenation and decarbonylation are liable to occur in the gas phase and at high temperatures and low pressures.

Deuterium tracer studies have shown that on Ru, Os, Ir and Pt the hydrogenation to alcohols proceeds by simple hydrogen addition to the molecularly adsorbed ketones. Thus, one might extend the Horiuti-Polanyi mechanism to ketone hydrogenation on these metal catalysts. However, there is kinetic and spectral evidence that the active molecular species are either π-bonded or coordinated through the oxygen lone pairs rather than associatively adsorbed as with the olefins. Rhodium is characterized by the propagation of the α, β-process, and Pd by the formation of π-allylic species. The existence of π-oxaallyls is suggested for Pd and Pt. On Cu, possibly on Pd as well, more complex hydrogenation mechanisms may occur such as those involving the transfer of hydrogen from adsorbed alcohol to adsorbed ketone or the disproportionation of adsorbed ketone (ref. 26). In the hydrogenation of organic compounds, a variety of hydrogen sources has been proposed besides H(a), including H_2(a) (ref. 29), H_2^+(a) (ref. 97), solvent proton (ref. 46), H^-(a) (ref. 11), polarized hydrogen molecule (refs. 11, 98) and hydrogen transfer from adsorbed hydrocarbons (ref. 99).

Before the mechanism for the hydrogenation of carbonyl compounds can be established, much work has still to be done with emphasis on spectral studies to elucidate the adsorbed intermediate species.

REFERENCES

1. F. Zymalkowski, Katalytische Hydrierungen im Organische-Chemischen Laboratorium, Ferdinand Enke Verl., Stuttgart, 1965, Ch. 9, p. 91.
2. R. L. Augustine, Catalytic Hydrogenation, Marcel Dekker, New York, 1965.
3. P. N. Rylander, Catalytic Hydrogenation over Platinum Metals, Academic Press, New York, 1967, Ch.14, p. 238; Ch.15, p. 258.
4. M. Freifelder, Practical Catalytic Hydrogenation - Techniques and Applications, Wiley-Interscience, New York, 1971, Ch. 14, p. 282.
5. P. N. Rylander, Catalytic Hydrogenation in Organic Syntheses, Academic Press, New York, 1979, Ch.5, p.72; Ch.6, p.82.
6. W. Hückel, M. Maier, E. Jordan, and W. Seeger, Liebig Ann. Chem., 616 (1958) 46-81.
7. T. I. Taylor, in P. H. Emmett (Editor), Catalysis, Vol. V, Hydrogenation, Oxo-Synthesis, Hydrocracking, Hydrosulfurization, Hydrogen Isotope Exchange and Related Catalytic Reactions, Reinhold, New York, 1957, Ch.5, p.257.
8. G. C. Bond, Catalysis by Metals, Academic Press, London, 1962.
9. A. P. G. Kieboom and F. van Rantwijk, Hydrogenation and Hydrogenolysis in Synthetic Organic Chemistry, Delft University Press, Delft, 1977, pp.69-76.
10. M. Kraus, Adv. Catal., 17 (1967) 75-102.
11. M. Kraus, Adv. Catal., 29 (1980) 151-196.
12. S. Mitsui, in M. Imoto (Editor), Koza Yuki Hanno Kiko (Lectures on Organic Reaction Mechanism), Vol. 13, Sesshoku Kangen Hanno (Catalyzed Hydrogenation), Tokyo Kagaku Dojin, Tokyo, 1970, Ch. 12, 210.
13. K. Tanaka, Hyomen, 18 (1980) 249-259.
14. C. T. H. Stoddart and C. Kemball, J. Colloid Sci., 11 (1956) 532-542.
15. J. E. Kilpatrick, E. J. Prosen, K. S. Pitzer, and F. D. Rossini, J. Res. Natl. Bur. Stand, 36 (1946) 559-612.
16. K. K. Kearby, in P. H. Emmett (Editor), Catalysis, Vol. III, Reinhold, New York, 1955, p. 453.
17. S. W. Benson, Thermochemical Kinetics, Wiley, New York, 1968, Ch. 2, p.18.
18. S. W. Benson, F. R. Cruickshank, D. M. Golden, G. R. Hangen, H. E. O'Neil, A. S. Rodgers, R. Shaw, and R. Walsh, Chem. Rev., 69 (1969) 279-324.
19. T. Shimanouchi, Kagakuheiko Wa Donoyonishite Kimaru Ka (How is the Chemical Equilibrium Determined?), Tokyo Kagaku Dojin, Tokyo, 1962, p.31.
20. L. C. Anderson and N. W. MacNaughton, J. Am. Chem. Soc., 64 (1942) 1456-1459.
21. L. Friedman and J. Turkevich, J. Am. Chem. Soc., 74 (1952) 1669-1671.
22. C. Kemball and C. T. H. Stoddart, Proc. R. Soc. London, Ser. A, 241 (1957) 208-222.
23. J. R. Anderson and C. Kemball, Trans. Faraday Soc., 51 (1955) 966-973.
24. J. Newham and R. L. Burwell, Jr., J. Am. Chem. Soc., 86 (1964) 1179-1186.
25. W. R. Patterson and R. L. Burwell, Jr., J. Am. Chem. Soc., 93 (1971) 833-838.
26. W. R. Patterson, J. A. Roth, and R. L. Burwell, Jr., J. Am. Chem. Soc., 93 (1971) 839-846.
27. C. Kemball and C. T. H. Stoddart, Proc. R. Soc. London, Ser.

A, 246 (1958) 521-538.
28 D. Cornet and F. G. Gault, Bull. Soc. Chim. Fr., (1966) 2477-2480.
29 D. Cornet and F. G. Gault, J. Catal., 7 (1967) 140-151.
30 Y. Takagi, S. Teratani, and J. Uzawa, J. Chem. Soc. Chem. Commun., (1972) 280-281.
31 Y. Takagi, S. Teratani, and K. Tanaka, in J. W. Hightower (Editor), Proc. 5th Intern. Congr. Catalysis, Maiami Beach, FA, 1972, North-Holland, Amsterdam, 1973, pp. 757-770.
32 Y. Takagi, S. Teratani, and K. Tanaka, J. Catal., 27 (1972) 79-88.
33 S. Teratani, K. Tanaka, H. Ogawa, and K. Taya, J. Catal., 51 (1978) 372-379.
34 S. Teratani, K. Tanaka, H. Ogawa, and K. Taya, J. Catal., 70 (1981) 347-355.
35 J. K. A. Clarke and J. J. Rooney, Adv. Catal., 25 (1976) 125-183.
36 Y. Takagi, S. Yada, and K. Tanaka, J. Catal., 80 (1983) 469-471.
37 M. Bonnet, P. Geneste, Y. Lozano, C. R. Acad. Sci. Paris, Ser. C, 282 (1976) 1009-1012.
38 R. J. Wicker, J. Chem. Soc., (1956) 2165-2173.
39 W. G. Dauben, G. J. Fonken, and D. S. Noyce, J. Am. Chem. Soc., 78 (1956) 2579-2582.
40 E. L. Eliel and R. S. Ro, J. Am. Chem. Soc., 79 (1957) 5992-5994.
41 Y. Takagi, Sci. Papers Inst. Phys. Chem. Res. Jpn., 64 (1970) 39-61.
42 S. Mitsui, H. Saito, Y. Yamashita, M. Kaminaga, and Y. Senda, Tetrahedron, 29 (1973) 1531-1539.
43 K. Tanaka, Y. Takagi, O. Nomura, and I. Kobayashi, J. Catal., 35 (1974) 24-33.
44 E. G. Reppiatt and R. J. Wicker, Chem. Ind., (1955) 747-748.
45 R. P. Linstead, W. E. Doering, S. B. Davis, P. Levine, and R. R. Whetstone, J. Am. Chem. Soc., 64 (1942) 1985-1991.
46 J. H. Brewster, J. Am. Chem. Soc., 76 (1954) 6361-6363.
47 R. L. Augustine, D. C. Migliorini, R. E. Foscante, C. S. Sodano, and M. J. Sisbarro, J. Org. Chem., 34 (1969) 1075-1085.
48 M. Sugahara, S. Tsuchida, I. Anazawa, Y. Takagi, and S. Teratani, Chem. Lett., (1974) 1389-1392.
49 D. C. Wigfield, S. Feiner, and D. J. Phelps, J. Org. Chem., 40 (1975) 2533-2534.
50 K. L. Servis, D. J. Bowler, and C. Ishii, J. Am. Chem. Soc., 97 (1975) 73-80.
51 J. Simonikova, L. Hillaire, J. Panek, and K. Kochloefl, Z. Phys. Chem. (Neue Folge), 83 (1973) 287-304.
52 S. Kishida and S. Teranishi, J. Catal., 12 (1968) 90-96.
53 S. Kishida, Y. Murakami, T. Imanaka, and S. Teranishi, J. Catal., 12 (1968) 97-101.
54 I. Iwamoto, T. Yoshida, T. Aonuma, and T. Keii, Nippon Kagaku Zasshi, 91 (1970) 1050-1054.
55 I. Iwamoto, T. Yoshida, and T. Aonuma, Nippon Kagaku Zasshi, 92 (1971) 504-507.
56 I. Iwamoto and T. Yoshida, Nippon Kagaku Kaishi, (1973) 658-661.
57 S. L. Kiperman and G. I. Kaplan, Kinet. Katal., 5 (1964) 888-897.
58 P. B. Babkova, A. K. Avetisov, G. D. Ljubarskj and A. I. Gelbstein, Kinet. Katal., 10 (1969) 1086-1089; 11 (1970) 1451-1456.

59 O. A. Hougen and K. M. Watson, Chemical Process Principles, John Wiley, New York, 1947, Ch. 19, pp. 910-926.
60 J. Jenck and J.-E. Germain, J. Chim. Phys., 75 (1978) 810-814.
61 P. B. D. de la Mare, L. Fowden, E. D. Hughes, C. K. Ingold, and J. D. H. Mackie, J. Chem. Soc., (1955) 3200-3236.
62 C. D. Chalk, B. C. Hutley, J. McKenna, L. B. Sims, and I. H. Williams, J. Am. Chem. Soc., 103 (1981) 260-268.
63 T. Chihara and K. Tanaka, Bull. Chem. Soc. Jpn., 52 (1979) 507-511.
64 T. Chihara and K. Tanaka, Bull. Chem. Soc. Jpn., 52 (1979) 512-515.
65 T. Chihara, J. Catal., 89 (1984) 177-181.
66 S. Nishimura, M. Murai, and M. Shiota, Chem. Lett., (1980) 1239-1242.
67 T. Chihara and K. Tanaka, Chem. Lett., (1977) 843-846.
68 S. Choi and K. Tanaka, Bull. Chem. Soc. Jpn., 55 (1982) 2275-2276.
69 J. Jenck and J.-E. Germain, J. Catal., 65 (1980) 133-140.
70 J. Simonikova, A. Ralkova, and K. Kochloefl, J. Catal., 29 (1973) 412-420.
71 J. Jenck and J.-E. Germain, J. Catal., 65 (1980) 141-149.
72 P. Geneste, M. Bonnet, and M. Rodrigues, J. Catal., 57 (1979) 147-152.
73 T. Chihara and K. Tanaka, J. Catal., 80 (1983) 97-105.
74 M. Kajitani, N. Suzuki, T. Abe, Y. Kaneko, K. Kasuya, K. Takahashi, and A. Sugimori, Bull. Chem. Soc. Jpn., 52 (1979) 2343-2348.
75 G. Blyholder and L. D. Neff, J. Phys. Chem., 70 (1966) 893-900.
76 G. Blyholder and L. D. Neff, J. Phys. Chem., 70 (1966) 1738-1744.
77 G. Blyholder and W. V. Wyatt, J. Phys. Chem., 70 (1966) 1745-1750.
78 G. Blyholder and L. D. Neff, J. Phys. Chem., 70 (1969) 3494-3496
79 G. Blyholder and D. Shihabi, J. Catal., 46 (1977) 91-99.
80 R. P. Young and N. Sheppard, J. Catal., 7 (1967) 223-233.
81 R. P. Young and N. Sheppard, J. Catal., 20 (1971) 333-339.
82 R. P. Young and N. Sheppard, J. Catal., 20 (1971) 340-349.
83 M. Minobe, H. Miyata, and Y. Kubokawa, presented at 23rd Symposium on Catalysis, Sendai, October 3, 1968, Abstracts, pp. 181P-183P.
84 H. Miyata, M. Minobe, and Y. Kubokawa, presented at Annual Meeting of Catalysis Society, Fukuoka, October 18, 1969, Abstracts, p.55.
85 H. Miyata, K. Ohtani, and Y. Kubokawa, 27th Annual Meeting of The Chemical Society of Japan, Nagoya, October 12, 1972, Abstracts. pp. 363-364.
86 H. Miyata, Y. Nii, and Y. Kubokawa, presented at 29th Annual Meeting of The Chemical Society of Japan, Hiroshima, 1973, Abstracts, pp. 18-19.
87 G. Blyholder, J. Chem. Phys., 36 (1962) 2036-2039.
88 T. Szilagyi, A. Sarkany, J. Mink, and P. Tetenyi, J. Catal., 66 (1980) 191-199.
89 K. Kishi, S. Ikeda, and K. Hirota, J. Phys. Chem., 71 (1967) 4384-4389.
90 T. Matsubayashi, H. Miyata, and Y. Kubokawa, presented at 32nd Annual Meeting of The Chemical Society of Japan, Tokyo, April 2, 1975, Abstracts, p. 433.
91 K. D. Rendulic and B. A. Sexton, J. Catal., 78 (1982)

126-135.
92 H. Lüth, G. W. Rubloff, and W. D. Grobman, Surf. Sci., 63 (1977) 325-338.
93 J. C. Tracy and P. W. Palmbery, Surf. Sci., 14 (1969) 274-277.
94 H. Conrad, G. Ertl, and E. E. Latta, Surf. Sci., 41 (1974) 435-446.
95 J. B. Benziger, Applications Surf. Sci., 6 (1980) 105-121.
96 S. M. Davis and G. A. Somorjai, J. Catal., 65 (1980) 78-83.
97 T. Kwan, Shokubai , 1 (1946) 99-109.
98 Y. Izumi, Proc. Jpn. Acad., 53 (1977) 38-41.
99 S. J. Thomson and G. Webb, J. Chem. Soc. Chem. Commun., (1976) 526-527.

Chapter 4

HYDROGENATION OF NITRILES

JIŘÍ VOLF and JOSEF PAŠEK
Prague Institute of Chemical Technology, Suchbatarova 1905,
166 28 Prague 6 - Dejvice (Czechoslovakia)

4.1 INTRODUCTION

   Hydrogenation of nitriles is an important method for industrial preparation of diverse amines. It is usually carried out in the liquid phase at elevated hydrogen pressures in the presence of various metallic catalysts. The hydrogenation of nitriles of fatty acids leading to the corresponding primary, secondary or tertiary amines, and hydrogenation of adiponitrile to hexamethylenediamine, belong to the most important reactions of this kind. 1,3-Propylenediamine, dipropylenetriamine, N-alkyl-1,3-propylenediamines, 3-alkoxypropyleneamines, benzylamine, xylylenediamines, etc., are also prepared by hydrogenation of nitriles. In contrast to other hydrogenation or hydrogenolytic reactions, which usually proceed relatively simply, in the hydrogenation of nitriles a mixture of compounds is formed, consisting mostly of primary, secondary and tertiary amines. The reason for the lower selectivity of nitrile hydrogenation compared with other hydrogenation reactions is the formation of a reactive intermediate, aldimine, which can, in addition to hydrogenation to the primary amine, undergo condensation reactions and yield other compounds. The actual composition of the hydrogenation product depends on the properties of the starting nitrile, on the catalyst employed and also on the reaction conditions.
   The catalyst represents the most important factor determining the composition of the reaction product: a suitable choice of catalyst can enable one to carry out the hydrogenation with a high selectivity with respect to primary amines, but also to steer it towards a product where secondary or tertiary amines prevail.
   The majority of data on hydrogenation of nitriles have been published in the patent literature. The associated problems have not yet been systematically reviewed, and some original contributions to this field deal with partial aspects only. In

this contribution we try to summarize the existing knowledge on the hydrogenation of nitriles in the liquid phase, and supplement it by some, mostly unpublished results from our own research. Attention is focused on the chemism of nitrile hydrogenation, on the characterization of the main types of catalysts employed and, finally, on the effect of nitrile structure and of the reaction conditions on the course of the reaction. The diverse effects of some hydrogenation catalysts and reaction conditions, which often lead to contradictory results depending on the type of the catalyst and on the structure of the hydrogenated nitrile, are discussed from the point of view of the influence of these factors on the kinetics of reactions proceeding during hydrogenation of nitriles.

4.2 REACTION SCHEME IN THE FORMATION OF PRIMARY, SECONDARY AND TERTIARY AMINES

A mixture of primary, secondary and tertiary amines is usually formed when nitriles are hydrogenated in the presence of metallic catalysts. In the 1920´s, a number of investigators tried to explain the formation of secondary amines (refs. 1-3). In 1923, Braun et al. (ref. 4) published the results of liquid-phase hydrogenation of a number of nitriles catalysed by nickel. According to the reaction scheme proposed by these authors, secondary amines are formed as follows: in the first stage the nitrile is hydrogenated to aldimine, which yields the primary amine upon further hydrogenation:

$$RC \equiv N \xrightarrow{+H_2} RCH=NH \xrightarrow{+H_2} RCH_2NH_2 \quad (4.1)$$

An 1-aminodialkylamine is then formed by the reaction between the aldimine and the primary amine:

$$RCH=NH + RCH_2NH_2 \longrightarrow RCH\underset{NH_2}{-}NH-CH_2R \quad (4.2)$$

The 1-aminodialkylamine can lose ammonia to yield an alkylidenealkylamine which is in turn hydrogenated to a secondary amine:

$$RCH\underset{NH_2}{-}NH-CH_2R \xrightarrow{-NH_3} RCH=N-CH_2R \xrightarrow{+H_2} RCH_2NHCH_2R \quad (4.3)$$

The authors maintain that the latter can be also formed by hydrogenolysis of the 1-aminodialkylamine:

$$\underset{\underset{NH_2}{|}}{RCH-NH-CH_2R} \xrightarrow{+H_2} RCH_2NHCH_2R+NH_3 \qquad (4.4)$$

The experimental fact established by Braun et al. (ref. 4), that the content of secondary amines increases with the nitrile concentration in hydrogenations carried out in a solvent, is in accord with the scheme proposed for the formation of secondary amines.

Braun's scheme was later verified by a number of investigators (ref. 5-8). Greenfield (ref. 9) presented a similar scheme for the formation of tertiary amines, and maintained that they are formed analogously by the reaction between the aldimine and the secondary amine:

$$RCH=NH+(RCH_2)_2NH \longrightarrow \underset{\underset{NH_2}{|}}{RCH-N} \begin{matrix} CH_2R \\ CH_2R \end{matrix} \qquad (4.5)$$

Tertiary amines are, according to this author, formed by hydrogenolysis of substituted 1-aminotrialkylamines:

$$\underset{\underset{NH_2}{|}}{RCH-N} \begin{matrix} CH_2R \\ CH_2R \end{matrix} \xrightarrow{+H_2} (RCH_2)_3N+NH_3 \qquad (4.6)$$

Greenfield also noted that, according to Wheeler, tertiary amines could be formed via an intermediate enamine originating (similarly to imines) by elimination of ammonia from the 1-aminotrialkylamine

$$\underset{\underset{NH_2}{|}}{RCH-N} \begin{matrix} CH_2R \\ CH_2R \end{matrix} \xrightarrow{-NH_3} R\acute{}CH=CH-N \begin{matrix} CH_2R \\ CH_2R \end{matrix} \qquad (4.7)$$

and the tertiary amine could be formed by hydrogenation of the enamine:

$$R\acute{}CH=CH-N \begin{matrix} CH_2R \\ CH_2R \end{matrix} \xrightarrow{+H_2} (RCH_2)_3N \qquad (4.8)$$

It is known that enamines, which are very reactive compounds (ref. 10-12), are relatively easily formed by the reaction of aldimines or carbonyl compounds with secondary amines.

It follows that secondary and tertiary amines can be formed from diamino derivatives, either by elimination of ammonia (which leads to alkylidenealkylamines or enamines) or by hydrogenolysis of the C-N bond. The participation of alkylidenealkylamines in the formation of secondary amines has been confirmed experimentally (refs. 3, 8, 13); alkylidenealkylamines can even be produced by a partial hydrogenation of nitriles (ref. 14). The results of the hydrogenation of lauronitrile on a cobalt catalyst, presented in section 4.4.1, furnish evidence that secondary amines are not formed in the cobalt-catalysed hydrogenolysis of the 1-aminodialkylamine, since alkylidenealkylamine accumulates in the reaction mixture (as shown by chromatography and spectroscopy) and the secondary amine is in this instance formed only at the end of the reaction by hydrogenation of the alkylidenealkylamine.

Contrary to alkylidenealkylamines, the presence of enamines (as potential intermediates of tertiary amines) in the reaction mixture has not been confirmed. Some experimental results corroborating the view that enamines participate in the formation of tertiary amines will therefore be presented below.

An enamine is assumed to participate in the formation of triethylamine by disproportionation of diethylamine on a cobalt catalyst (ref. 15). The reaction scheme describing the disproportionation of diethylamine is very similar to that used to explain the formation of tertiary amines in the hydrogenation of nitriles. An enamine intermediate is also assumed in the reductive amination of carbonyl compounds by secondary amines, which gives rise to tertiary amines (ref. 16, 17). A tertiary amine is formed by hydrogenation of the enamine according to eqn. 4.8:

$$R_2NH + RCH_2CH=O \rightleftarrows RCH_2\underset{OH}{CH}-N{<}^R_R \qquad (4.9)$$

$$RCH_2\underset{OH}{CH}-N{<}^R_R \rightleftarrows RCH=CH-N{<}^R_R + H_2O \qquad (4.10)$$

1-Aminotrialkylamines and 1-hydroxytrialkylamines behave similarly in these reactions, leading to enamines. Owing to the similarity of the groups C=O and C=N-, the reactions 4.7, 4.8 and 4.9, 4.10 proceed analogously and represent typical reversible reactions.

The fact that benzonitrile (in contrast to aliphatic nitriles) does not yield a tertiary amine when hydrogenated on palladium or platinum (ref. 18) serves as an additional argument in favour of the enamine mechanism in the formation of tertiary amines; the latter should be formed by hydrogenation of benzonitrile as a result of the reaction between benzaldimine and dibenzylamine:

$$\text{Ph-CH=NH} + \text{Ph-CH}_2\text{NHCH}_2\text{-Ph} \rightleftharpoons \text{Ph-CH(NH}_2\text{)-N(CH}_2\text{Ph)}_2 \qquad (4.11)$$

However, an enamine cannot be formed from 1-aminotribenzylamine, see eqn. 4.11. It should be noted in this connection that the formation of a tertiary amine would in this instance also be affected by steric hindrance.

Neither is a tertiary amine formed when a mixture of benzonitrile and diethylamine is hydrogenated on palladium or platinum (ref. 18), while hydrogenation of a mixture of butyronitrile and diethylamine on a palladium catalyst does yield a tertiary amine, diethylbutylamine (ref. 19). The explanation lies in that the 1-aminotrialkylamine formed from benzaldimine and diethylamine cannot yield the corresponding enamine (and no tertiary amine is accordingly formed), in contrast to the analogous 1-aminotrialkylamine formed upon hydrogenation of butyronitrile and diethylamine mixture. This observation serves as indirect proof that enamines participate in the formation of tertiary amines.

All reactions involved in the gradual formation of tertiary amines in the hydrogenations of nitriles are reversible. Thus, e.g., a mixture of primary and secondary amines is formed when a tertiary amine reacts with ammonia in the presence of hydrogenation catalysts (ref. 20):

$$R_3N + NH_3 \rightleftharpoons R_2NH + RNH_2 \qquad (4.12)$$

Secondary amines react with ammonia and give rise to primary amines (ref. 21), and nitriles can be produced by dehydrogenation of primary amines (refs. 22, 23):

$$R_2NH + NH_3 \rightleftharpoons 2RNH_2 \qquad (4.13)$$

$$RCH_2NH_2 \rightleftharpoons RC\equiv N + 2H_2 \qquad (4.14)$$

Reactions 4.12 and 4.13 can easily be described by a scheme in which the starting amines are first dehydrogenated either to alkylidenealkylamines (in the case of secondary amines) or to enamines (tertiary amines) - the reverse of eqns. 4.3 and 4.8. In the presence of ammonia the latter two compounds then yield diamino derivatives that in turn give rise to primary or secondary amines by the reactions 4.3, 4.2, 4.1 or 4.7, 4.5, 4.1 proceeding in the opposite direction.

If trialkylamines were formed in the hydrogenation of nitriles by hydrogenolysis of 1-aminotrialkylamines, the same mechanism would apply also to the reverse reaction 4.12. However, the interaction of a tertiary amine with ammonia accompanied by a simultaneous elimination of hydrogen (the reaction which represents the reverse of the hydrogenolysis of 1-aminotrialkylamine) is highly improbable; the same explanation is also true of reaction 4.13.

One may conclude that secondary and tertiary amines are not formed by hydrogenolysis of the corresponding 1-aminodialkylamines and 1-aminotrialkylamines, respectively. In the hydrogenation of nitriles, alkylidenealkylamines and enamines represent the precursors of secondary and tertiary amines and their formation is described by the following comprehensive scheme:

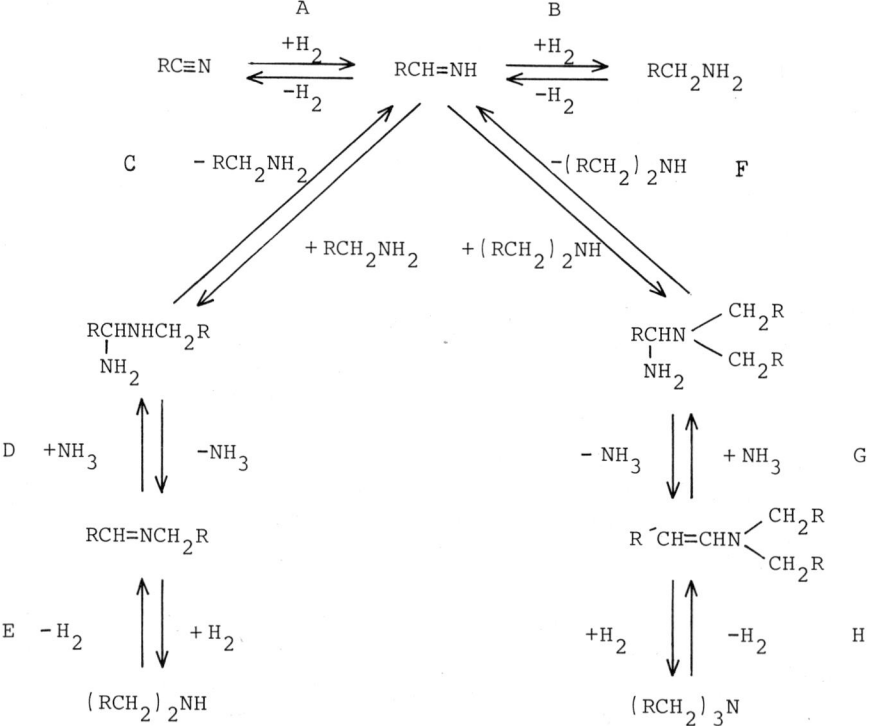

As shown, a system comprising competitive consecutive reactions of different characters is involved. A, B, E and H are typical hydrogenation reactions, while the condensation reactions which give rise to alkylidenealkylamine or enamines are acid-base ones (ref. 24). It is obvious that the composition of the hydrogenation products will depend on the reactivity of the starting nitrile, on the specific reaction conditions and on the ability of the employed catalyst to promote the individual reactions involved.

4.3 CATALYSTS FOR HYDROGENATION OF NITRILES

The catalysts employed is the most important factor determining the selectivity of nitrile hydrogenation. If we order metallic catalysts for hydrogenation of nitriles according to the increasing content of secondary and tertiary amines in the reaction product - beside the primary amines - we obtain the series (refs. 9, 25): Co, Ni, Ru, Cu, Rh, Pd, Pt.

For the production of primary amines, catalysts based on Co, Ni, Ru are mostly used. Copper and rhodium catalysts serve mainly for the preparation of secondary amines, while tertiary amines can sometimes be prepared with a high selectivity on Pt and Pd.

The selectivity of nitrile hydrogenation is of chief interest, particularly in the production of primary amines. In a number of cases the by-products cannot be utilized economically or, for example, in the production of primary amines derived from higher aliphatic acids it is difficult to separate primary and secondary amines owing to their high boiling points and the hydrogenation products containing 75 to 98% of primary amines are applied without any separation.

A great number of hydrogenation catalysts have been described in the literature (mostly in patents); they are modified by various admixtures with the aim of increasing their selectivity, activity, lifetime or of enhancing their mechanical strenght for applications in continuous hydrogenation. It is not the aim of this review to include all contributions to this area but rather to acquaint the reader with the basic types and modifications of metallic catalysts employed in the hydrogenation of nitriles in the liquid phase.

Nickel and cobalt catalysts rank among those most often described and used. Their properties are relatively similar and they are applied mostly in the production of primary amines derived from higher fatty acids by hydrogenation of the corresponding nitriles, of hexamethylenediamine, dipropylenetriamine, alkylpropylenediamines, xylylenediamines, etc. Raney nickel (refs. 26-31) and Raney cobalt (refs. 32-37) are probably the most often used catalysts. In most instances, various compounds are added in order to enhance their selectivity; various methods for leaching the original alloy have been described with the aim of improving its properties, and different modifications of the Raney alloy have been proposed. Raney cobalt is more selective with respect to primary amines than is Raney nickel (ref.8).

The pure metals - nickel and cobalt - without any support can also be used for hydrogenation; they can be prepared by a thermal decomposition of cobalt (II) or nickel (II) formates either in an hydrogen (nitrogen) atmosphere (ref. 25) or directly in the starting nitrile (ref. 38). A cobalt catalyst with a very high selectivity with respect to primary amines can be prepared by reducing Co_3O_4 (ref. 39) or a mixture of it with calcium oxide (refs. 40, 41). Cobalt catalysts modified by manganese and phosphoric acid admixtures are highly selective (refs. 42-44). Silver (ref. 45), zirconium (ref. 46) and lead (ref. 47) are recommended as promoters which enhance the selectivity of cobalt with respect to primary amines; palladium acts as a similar promoter for nickel (ref. 48). In addition, a number of catalysts have been described based on cobalt or nickel on various supports, which are allegedly suitable for selective hydrogenation. Let us mention cobalt (ref. 49) or nickel (ref. 50) on aluminium oxide, cobalt (ref. 51) or nickel (refs. 52, 53) on silica, cobalt (ref. 54) and nickel (ref. 55) on boron oxide, cobalt with magnesium oxide on kieselguhr (ref. 56), nickel on chromic oxide (ref. 57) and nickel with copper on kieselguhr (ref. 58). Under suitable reaction conditions (removal of NH_3), nickel catalysts serve also for the preparation of secondary amines (ref. 59).

A critical evaluation of the numerous patent dealing with cobalt and nickel catalysts is difficult, since the data presented in the individual patents are mostly not comparable: hydrogenations are performed on very diverse equipment, with different starting nitriles, and very often under disparate reaction conditions. The results of the hydrogenation of stearonitrile (ref. 60) on some commercially available catalysts or catalysts prepared on the basis of the patent literature are summarized for illustration in Table 4.1. It follows from these data that the type of metal is decisive for the selectivity, whereas the effect of the support is not significant. Manganese as a promoter of cobalt enhances the selectivity.

Copper catalysts are mentioned less often in the literature mainly because symmetric secondary amines, which are mostly produced on copper-based catalysts, are employed only to a limited extent in the preparation of some cation-active tensides (ref. 61). Adkins´copper (II)-chromium (III) catalyst (refs. 62-65) or copper on other supports (refs. 25, 66) are the most often described catalysts. The preparation of secondary amines in the

TABLE 4.1

Hydrogenation of stearonitrile on various commercial and in laboratory prepared nickel and cobalt catalysts

80 g stearonitrile, temperature 150 °C, pressure 6 MPa, autoclave volume 450 ml, total conversion of nitrile

Catalyst	Manufacturer	Product composition (mol%)		Reaction time (min)
		Primary amines	Secondary amines	
Ni-Al$_2$O$_3$ 1 g	Leuna-Werke No. 6524 (G.D.R.)	76.5	23.5	67
Ni-SiO$_2$ 1 g	Leuna-Werke No. 6500 (G.D.R.)	70.7	29.3	60
Raney-Ni 0.6 g	Kovohutě Mníšek (Č.S.S.R.) type W 4	81.1	18.9	37
Ni-Cr$_2$O$_3$ 1 g	U.S.S.R. 51-U-12	75.0	25.0	70
Co-SiO$_2$ 1 g	Ruhrchemie AG RCH-4520 TS Pulver	87.4	12.6	72
Co 3 g	Prepared by reduction of Co$_3$O$_4$	88.7	11.3	43
Co-Mn (5% Mn) 3 g	Prepared in laboratory (ref. 42)	95.4	4.6	23
Co-Pb 3 g	Prepared in laboratory (ref. 47)	90.4	90.6	50

presence of a copper (II) catalyst is often associated with the so-called desamination (disproportionation) of primary to secondary amines, where primary amines produced by hydrogenation are converted into secondary ones when ammonia is simultaneously removed from the reactor (refs. 62, 63). Copper possesses a low activity with respect to hydrogenation of the C=C bond and this is utilized for a selective hydrogenation of unsaturated nitriles to unsaturated secondary amines (ref. 62).

The hydrogenation of nitriles on catalysts based on platinum metals is not very widespread in industry because of the specific properties of these catalysts and also due to economic considerations. Most data comes from original papers where it is maintained that rhodium is very selective with respect to secondary amines (refs. 9, 18, 67, 68), while tertiary amines are formed in high yields on platinum and palladium catalysts (refs. 9, 18, 67). Hydrogenations on mixtures of platinum group metals have also been described (refs. 18, 69). Dibenzylamine is preferentially formed on platinum or palladium as the catalyst in the hydrogenation of benzonitrile (refs. 19, 70, 71). Rhodium serves as the catalyst for a selective hydrogenation of dinitriles to aminonitriles (ref. 72).

An interesting application is that of catalysts based on Zn in the preparation of primary amines (refs. 73, 74). In view of the their lower activity, the hydrogenation has to be carried out at high temperatures (250-300°C). Catalysts based on Fe have also been described in the patent literature; they are claimed to have suitable properties for the hydrogenation of adiponitrile to hexamethylenediamine (refs. 75-79).

4.4 PROPERTIES OF THE MAIN TYPES OF CATALYSTS FOR HYDROGENATION OF NITRILES

4.4.1 Nickel and cobalt catalysts

Nickel and cobalt have similar properties as catalysts for hydrogenation of nitriles; this is connected with their position in the Periodic Table. The first attempts at a detailed comparison of these catalysts date back to the 1970´s (refs. 8, 9, 80) and confirm a finding reported in the patent literature: the hydrogenations of palmitonitrile (ref. 80), adiponitrile (ref. 8) and butyronitrile (ref. 9) proceed more selectively (higher amounts of primary amines are formed) on the cobalt-based catalysts. These conclusions are valid both for the pure metallic catalysts (ref. 25) and for the Raney catalysts (ref. 8). The higher content of secondary amines in the hydrogenation product on nickel is explained by a stronger adsorption of the products on nickel in comparison with cobalt (refs. 8, 80). The method used to obtain the pure metal has been found to be essentially irrelevant as regards both the catalytic activity and the selectivity per unit surface area of cobalt or nickel in the hydrogenation of lauronitrile; cobalt and nickel prepared by reduction of the corresponding oxide and by decomposition of formates or oxalates were compared (ref. 25). On cobalt the mean initial reaction rate of lauronitrile hydrogenation, related to 1 m^2 of the metal surface, was 0.0102 mmol H_2 s^{-1} and the mean content of dilaurylamine in laurylamine was 11 mole per cent (160 °C, 4.4 MPa hydrogen pressure). The corresponding rate on nickel was 0.208 mmol H_2 s^{-1} and the mean content of dilaurylamine was 25 mole%. It was also ascertained that catalysts having different activities can be prepared by depositing nickel on various oxide supports, the values of the initial reaction rate per unit surface area of the metal being similar. This finding leads to the conclusion that the support affects only the degree of nickel dispersion; it does not substantially influence the selectivity. Although aluminium oxide and chromic oxide are considered to be supports having acidic character, on these supports nickel used as catalyst for the hydrogenation of nitriles does not lead to an increased content of secondary amines in comparison with other nickel catalysts.

The main difference between cobalt and nickel lies in the kinetics of the reactions. Fig. 4.1 shows the rate of lauronitrile hydrogenation as a function of the extent of conversion on pure cobalt and on a Ni-Al_2O_3 catalyst. While on nickel the reaction rate decreases monotonically with the extent of conversion, a distinct maximum located near the end of the reaction is seen in the case of pure cobalt.

In order to explain such differences, we also followed the composition of the reaction mixture as a function of time (ref. 81). The results obtained for the hydrogenation of lauronitrile on cobalt or nickel metals (always obtained by decomposition of formates) are plotted in Figs. 4.2 and 4.3. The main difference between the two catalysts is immediately apparent: while on nickel the didodecylamine is formed by hydrogenation of dodecylidenedodecylamine during the whole experiment, the latter compound accumulates in the reaction mixture in the case of cobalt catalysis and only at the end of the hydrogenation, when nitrile is no longer present, dodecylidenedodecylamine is rapidly hydrogenated to didodecylamine. This is obviously connected with the maximum reaction rate observed in cobalt-catalysed hydrogenation near the end of the reaction.

Completely analogous results have been obtained with the catalysts Ni-Al_2O_3 and cobalt-kieselguhr. The maximum was very

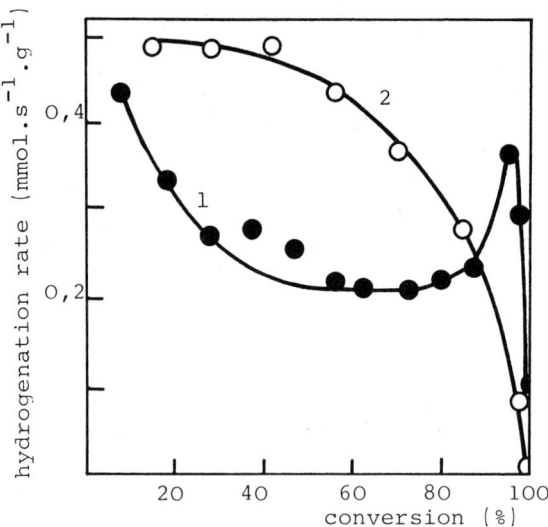

Fig. 4.1. Dependence of the hydrogenation rate on the lauronitrile conversion (ref. 25). (1) On cobalt catalyst, (2) on Ni-Al$_2$O$_3$ catalyst.

pronounced for the hydrogenation on the latter catalyst and amounted to ca. 2.5 times the minimum reaction rate; a higher amount of dilaurylamine was found in the product than was the case in the catalysis by pure cobalt. From the time dependences of the reaction rate and of the product composition, one can deduce that on cobalt the adsorption coefficient of the nitrile is substantially higher than that of the intermediate product, dodecylidenedodecylamine.

Dodecylidenedodecylamine is formed by condensation of the corresponding aldimine with the primary amine, and ammonia is eliminated from the 1-aminodidodecylamine formed primarily, yielding an imine by reactions 4.2 and 4.3 presented in Section 4.2. Since dodecylidenedodecylamine was detected by IR spectroscopy in the reaction mixture resulting from the hydrogenation of lauronitrile catalysed by cobalt, it is possible to conclude unambiguously that didodecylamine is formed on cobalt by hydrogenation of dodecylidenedodecylamine rather than by hydrogenolysis of 1-aminodidodecylamine as stated in several papers (refs. 4, 9, 69, 82).

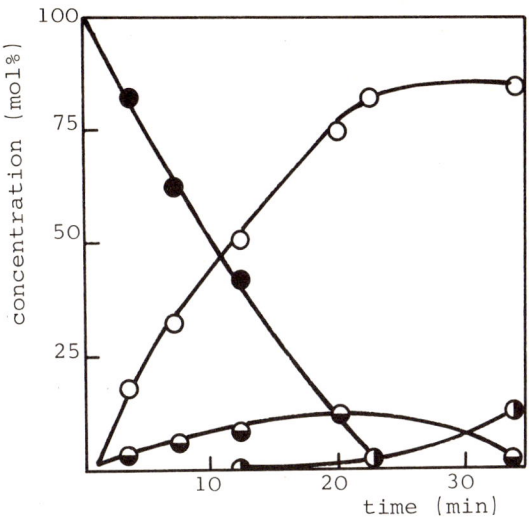

Fig. 4.2. The composition of the product of lauronitrile hydrogenation as a function of time. Cobalt catalyst: temperature 160 °C, pressure 5 MPa, initial amount of lauronitrile 0.45 mole, amount of catalyst 2 g (unsupported Co). ●, Lauronitrile; ○, dodecylamine; ◐, dodecylidenedodecylamine; ◑, didodecylamine.

One of several possible explanations for the higher selectivity observed in hydrogenations of nitriles on cobalt catalysts might be connected with the establishment of an equilibrium between the aldimine and primary amine on the one hand and between dodecylidenedodecylamine and ammonia on the other (reactions 4.2 and 4.3). Dodecylidenedodecylamine accumulates in the reaction mixture, its concentration becomes high and, if close to the equilibrium concentration, might result in a lowering of the rate of its formation. On nickel, where the concentration of dodecylidenedodecylamine is lower due to its continuous hydrogenation to didodecylamine, the larger distance from the equilibrium position might lead to a higher rate of dodecylidenedodecylamine formation, and thus also to a higher amount of didodecylamine. However, the explanation of higher cobalt selectivity, based on the equilibrium in reactions 4.2 and 4.3, does not seem to be probable, since, under otherwise identical conditions to those employed in the experiments depicted

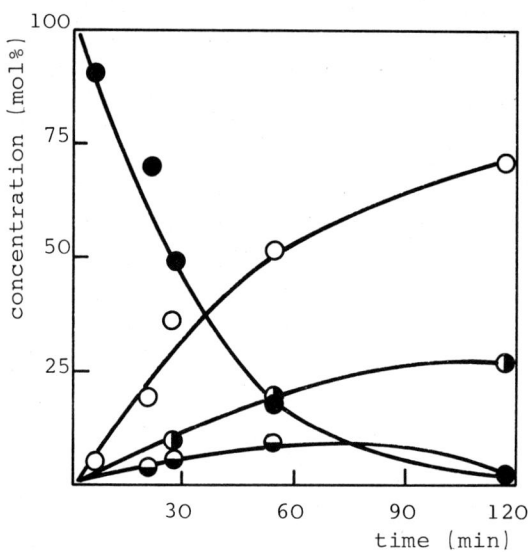

Fig. 4.3. The composition of the product of lauronitrile hydrogenation as a function of time. Nickel catalyst: temperature 160 °C, pressure 5 MPa, initial amount of lauronitrile 0.45 mole, amount of catalyst 1.25 g (unsupported Ni). ●, Lauronitrile; ○, dodecylamine; ◒, dodecylidenedodecylamine; ◐, didodecylamine.

in Fig. 4.2, with the catalyst Co-kieselguhr we obtained a final product with a higher content of didodecylamine, and also the concentration of dodecylidenedodecylamine observed in the course of the reaction was higher than that found in the catalysis by pure cobalt. Thus, the reason for the difference in selectivity between cobalt and nickel rests upon the difference in the relative rates of condensation and hydrogenation that lead to the formation of primary amines on these two metals.

In experiments with pure cobalt it has also been established that the selectivity is practically independent of the catalyst concentration in the reaction mixture (ref. 81). The selectivity also does not depend on the catalyst activity when nickel catalyst of different specific surface areas are used (ref. 25). If the catalyst were not involved in the condensation reactions that proceed in the homogeneous phase, the selectivity would be affected by the catalyst concentration and/or by its activity. Thus, the observation that the selectivity does not depend on these parameters furnishes evidence that the condensation reactions leading to secondary amines proceed on the surface of the metallic catalyst.

It also follows from the results plotted in Figs. 4.2 and 4.3 that the rates of the condensation reactions are comparable to the rate with which the nitrile is hydrogenated to the primary amine. Various changes in the reaction system, provided they affect the relationship between the hydrogenation and condensation reactions, might then significantly influence the composition of the product.

It has been established in the hydrogenation of acetonitrile and propionitrile (see Section 4.5) that tertiary amines are never formed on cobalt (in contrast to nickel). This can again be explained by the fact that, in hydrogenations catalysed by cobalt, secondary amines appear in the reaction mixture only when the concentration of the nitrile is already very low, as is the concentration of the aldimine which is necessary for the formation of tertiary amines.

4.4.2   Copper catalysts

Catalysts based on copper are used for the hydrogenation of nitriles whenever secondary amines, in particular those derived from higher fatty acids, are to be prepared in high yields. In comparison with nickel and cobalt, copper shows a considerably lower catalytic activity in the hydrogenation of nitriles. The ratio of the initial reaction rates (characterized by the rate of hydrogen consumption) per unit surface area of the metals is 20:10:1 for nickel, cobalt and copper, respectively (ref. 25).

In view of the low activity of copper, which is moreover difficult to prepare in pure and with a large surface area, we studied the hydrogenation of nitriles on supported catalysts. Table 4.2 summarizes the results obtained for stearonitrile on several commercial copper catalysts. A high content of secondary amines was always found in the reaction product, varying between 70 and 90 mole%. In contrast to the case of nickel or cobalt, the hydrogenation of nitriles on copper yields also tertiary amines, in some instances as high as 10%. The results collected in Table 4.2 again indicate that the properties of the support do not influence the selectivity to any appreciable extent. It then follows that the difference between nickel and cobalt on the one hand and copper on the other hand lies essentially in the properties of the metals themselves.

TABLE 4.2

Hydrogenation of stearonitrile on various copper catalysts

80 g stearonitrile, 5 g catalyst, temperature 210 °C, pressure 9 MPa, autoclave volume 250 ml.

Catalyst	Manufacturer	Composition of amine reaction products (mol%)			Reaction time[a] (min)
		Primary amine	Secondary amine	Tertiary amine	
$CuO-Cr_2O_3$	Harshaw No. 1107	7.8	88.1	4.1	250
	Harshaw No. 1134	11.8	78.1	10.1	220
	Harshaw No. 1808	10.7	85.1	4.2	163
	Leuna Werke (G.D.R.) No. 4492	10.3	71.2	8.5	89
$CuO-Cr_2O_3Fe_2O_3$	PO AZOT Kemerovo (U.S.S.R.)	18.5	75.2	6.3	195
$Cu-SiO_2$	BASF R 3-11	16.9	81.4	1.7	145
$Cu-Al_2O_3SiO_2$[b]	Prepared in laboratory (ref. 84)	10.5	85.4	4.1	183

[a] The time corresponding to the end of measurable hydrogen consumption
[b] Catalysts were reduced in hydrogen before experiments

Fig. 4.4 shows how the composition of the reaction product varies with time in the case of lauronitrile hydrogenated at 180 °C in the presence of the catalyst Cu 1107 marketed by Harshaw (ref. 83). In contrast to the corresponding dependences obtained with nickel or cobalt, with copper there is a very low content of dodecylamine and dodecylidenedodecylamine; even at low extents of conversion, didodecylamine is present as the main product. One can conclude from the time dependences of the concentrations of the intermediates that both the condensation of the primary amine with the aldimine and the hydrogenation of the alkylidenealkylamine proceed on copper considerably more rapidly than the hydrogenation

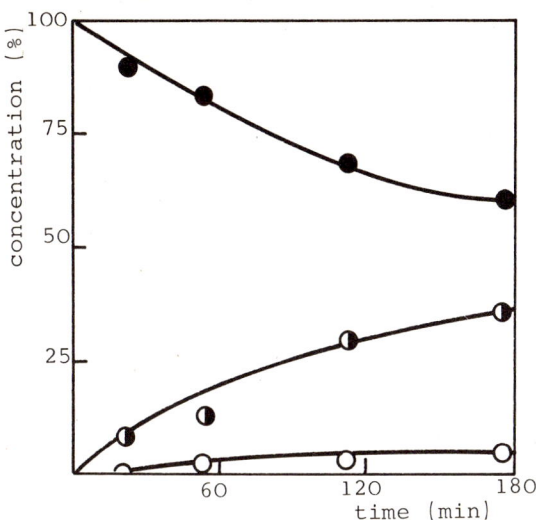

Fig. 4.4. The composition of the product of lauronitrile hydrogenation as a function of time. CuO-Cr$_2$O$_3$ catalyst: temperature 190 °C, pressure 5 MPa, initial amount of lauronitrile 0.45 mole, amount of catalyst 15 g (Harshaw Cu-1107 CuO-Cr$_2$O$_3$ catalyst). ●, Lauronitrile; ○, dodecylamine; ◐, didodecylamine.

of the nitrile to the primary amine.

The main reason why considerable amounts of secondary amines are formed on copper lies apparently in the low rate of nitrile hydrogenation at 160 °C (lower by a factor of about 20 than that observed with nickel (ref. 25)). The result then is a relatively higher rate of condensation in comparison to hydrogenation (see also Section 4.6.1) leading to a higher content of secondary amines.

The relatively low hydrogenation activity of copper-based catalysts and, on the other hand, their higher activity with respect to condensation reactions can be utilized with advantage in the preparation of tertiary amines by hydrogenation of nitriles in the presence of secondary amines. In the case of lauronitrile and dimethylamine the yield of dimethyldodecylamine can be as high as 95% (ref. 66).

4.4.3 Catalysts from the platinum group metals

Ruthenium, rhodium, palladium and platinum metals are the most often described catalysts from this group. The results of hydrogenation catalysed by these metals depend not only on the type of catalyst, but to some extent also on the structure of the hydrogenated nitrile (ref. 18). The ability of platinum-group metals to catalyse condensation of amines with aldimines to yield secondary and tertiary amines increases in the above order. While primary amines are formed in high yields under these conditions on ruthenium, catalysis by rhodium results in an almost quantitative yield of secondary amines. Platinum and palladium are the most interesting catalysts, which in hydrogenation of nitriles yield tertiary amines almost quantitatively; the content of secondary amines is usually very small and primary amines are as a rule completely absent (refs. 9, 18, 67, 68). In the case of platinum the reaction usually stops before a total conversion has been reached. This problem was studied by Koubek (ref. 86) who found that in the hydrogenation catalysed by platinum on charcoal the formal reaction order with respect to acetonitrile varied rapidly from the initial very high values (of about 10) at conversions below 10%, approaching 1 to 1.3 at the end. This is a substantially lower reaction order than that typical for cobalt or nickel (ref. 25). The rapid decrease in the reaction rate resulting finally in a complete interruption of hydrogenation cannot be ascribed to accumulation of ammonia in the autoclave: this effect was excluded experimentally, as was the possibility of catalyst poisoning by some intermediates or by poisons present in the starting compounds. The reason lies in the strong adsorption of tertiary amine on the catalyst surface.

It is also of interest that palladium on charcoal, when activated at 200 $^\circ$C in a stream of hydrogen, becomes completely inactive with respect to hydrogenation of nitriles, whereas its activity with respect to hydrogenation of cyclohexene is preserved (ref. 86).

Greenfield (ref. 9) also found that a total conversion could not be attained in the hydrogenation of butyronitrile in methanol on platinum or palladium. On the other hand, when water is used as the solvent, the reaction rate increases dramatically and the nitrile is totally hydrogenated (to about 90% of tertiary amine). This result is in accord with the retarding effect of tertiary amines on the hydrogenation of nitriles. The strong adsorption of the amine on the catalyst surface is obviously mediated by the free electron pair on nitrogen (ref. 15). In aqueous media a strong solvation of the amine by water molecules takes place via hydrogen bonding and the tertiary amine is also partially protonated. These interactions involve the free electron pair on nitrogen and compete with the strong adsorption of the amine. Solvation of the amine molecules thus increases the rate of hydrogenation.

Horyna (ref. 87) investigated the composition of the reaction mixture as a function of the extent of conversion in the hydrogenation of acetonitrile catalysed by palladium on charcoal (see Fig. 4.5), and ascertained that up to a conversion of about 20% the reaction mixture contains exclusively triethylamine with only traces of ethylamine and diethylamine. With increasing conversion the concentration of ethylamine remains approximately constant around 1%, while that of diethylamine rises gradually to 15% at 75% conversion, where the hydrogenation stops. One may reasonably assume, considering the chemism involved in the formation of secondary and tertiary amines, that palladium and also platinum shows a considerably higher activity with respect to condensation reactions than to nitrile hydrogenation. The rate of condensation of the aldimine with primary and secondary amines formed by hydrogenation is then very much higher; this leads directly to the conclusion that the addition of an other amine to the reaction mixture will result in the formation of mixed amines.

Rylander (ref. 18) studied the hydrogenation of benzonitrile and valeronitrile in the presence of various amines, using catalysts based on platinum, palladium and rhodium. The hydrogenation of benzonitrile was found to yield mostly dibenzylamine. When diethylamine was added to benzonitrile, diethylbenzylamine was never formed, but a total conversion to butylbenzylamine could be achieved in the presence of butylamine.

Horyna (ref. 87) established that diethylbutylamine is formed with a high selectivity when a mixture of acetonitrile and butylamine is hydrogenated in the presence of palladium on charcoal as the catalyst. It was also found, by following the composition as a function of the extent of conversion, that diethylbutylamine is formed from ethylbutylamine as an intermediate product; the reaction rate decreases with increasing initial concentration of butylamine in acetonitrile, in accord with the concept of a strong retarding effect of amines on hydrogenations catalysed by palladium. The formation of triethylamine commenced only after all the butylamine had been

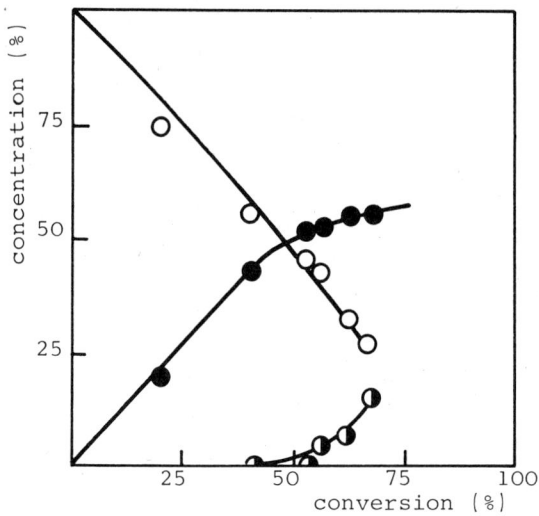

Fig. 4.5. The dependence of the acetonitrile hydrogenation product upon the extent of conversion. Catalysts: palladium supported on active carbon (ref. 87); temperature: 160 °C, pressure 4.4 MPa, 1.9 mole of acetonitrile, 13.2 g of CHEROX 4300 catalyst (5% Pd) marketed by CHZ ČSSP, Litvínov Czechoslovakia. ○ , Acetonitrile; ◐ , diethylamine; ● , triethylamine.

consumed, thus confirming the idea that the condensation of an aldimine with a primary (secondary) amine is much more rapid than the hydrogenation of acetonitrile. If dibutylamine is added to acetonitrile, considerable amounts of dibutylethylamine are formed (ref. 87). Similarly, when diethylamine is added to butyronitrile, diethylbutylamine, is formed by hydrogenation catalysed by palladium on charcoal (ref. 19).

These results for the hydrogenation of aliphatic nitriles in the presence of secondary amines, together with the observation that benzonitrile in the presence of diethylamine does not yield the mixed tertiary amine, are in accord with the concept that an enamine participates in the formation of tertiary amines (cf., Section 4.2).

The results obtained by Koubek (ref. 86), who followed the selectivity of various palladium-based catalysts (palladium black, palladium on different supports), in hydrogenation, confirm that regardless of the support these catalysts yield exclusively tertiary amines in the hydrogenation of aliphatic nitriles. This led to the conclusion that with palladium (as with other hydrogenation catalysts) the active centres in the catalysis of condensation reactions are localized on the metallic components, and the support contributes to a much lesser degree.

The high activity of palladium with respect to condensation reactions is probably associated with its acidic character, observed also in other reactions (ref. 88). The possibility to suppress the condensation (the formation of secondary amines) by the addition of various inorganic bases (refs. 9, 89) is another evidence in support of the acidic nature of the active centres in hydrogenation catalysts which enhance the condensation of amines with imines.

4.5 EFFECT OF NITRILE STRUCTURE ON HYDROGENATION

Investigation of the effect of the structure of nitriles on the kinetics and selectivity of their liquid-phase hydrogenation is complicated by several specific features of this reaction, mainly by the fact that ammonia is released when secondary amines are formed. Under the conditions of hydrogenation, ammonia is present partly in the vapour phase, where it lowers the partial pressure of hydrogen and affects the rate of hydrogenation. Its partitioning between the vapour and liquid phases is governed by the absorption coefficient, which depends on the structure of both the starting nitrile and the resulting amine. If the experiments are performed with the same initial molar amount in the autoclave, the starting mass concentrations of various nitriles are substantially different, and this affects the volume available for the vapour phase, which in turn influences the selectivity of hydrogenation (cf., Section 4.6.3). For example the molar volume of acetonitrile at 20 $^\circ$C is 52 cm^3 while that of lauronitrile is 202 cm^3. A study of the hydrogenation kinetics based on the measurement of the decreasing hydrogen pressure in the autoclave, commonly applicable to other types of hydrogenations, is in this case complicated by the gradual increase in ammonia partial pressure during hydrogenation, and by the partial pressures of both the starting compounds and the reaction products, which continuously change with time. The relatively large reaction heat (ΔH_r -120 kJ mol^{-1} for hydrogenation) also contributes to these problems. In order to suppress the unfavourable effect of the above factors in the measurement of reaction rate and selectivity

TABLE 4.3

The effect of nitrile structure on the selectivity and rate of hydrogenation on unsupported cobalt and nickel catalysts in the presence of methanol and cyclohexane

Temperature 180 °C, pressure 9.0 MPa, 0.3 mole of nitrile, the total volume of the starting reaction mixture was held constant (150 ml).
S=Ratio of the molar concentrations of primary and secondary amines. r_{10}=The relative initial rate of hydrogenation expressed as the ratio of amount of hydrogen consumed during the first 10 min of the experiment relative to butyronitrile. σ^*= Taft constants corresponding to the alkyl group in the amine produced.

Nitrile	Catalyst	CH_3OH		C_6H_{12}		σ^*
		S	r_{10}	S	r_{10}	
Benzonitrile	Co	43.6	2.3	39.3	1.5	0.215
Acetonitrile	Co	11.4	0.9	7.8	0.6	-0.100
Propionitrile	Co	9.6	1.2	5.0	0.6	-0.115
Butyronitrile	Co	9.0	1.0	7.4	1.0	-0.130
Lauronitrile	Co	5.8	0.7	2.6	0.8	-0.160[a]
Benzonitrile	Ni	86.4	2.2	22.4	1.2	0.215
Acetonitrile	Ni	-	0.9	-	0.8	-0.100
Propionitrile	Ni	5.7	0.7	3.6	1.0	-0.115
Butyronitrile	Ni	5.3	1.0	2.9	1.0	-0.130
Lauronitrile	Ni	4.0	0.7	3.2	0.7	-0.160[a]

[a] Approximation for hexyl

of nitrile hydrogenation, we followed the effect of nitrile structure using always the same molar amount of the nitrile (0.3 mole) and added a solvent to give a total volume of 150 ml (the volume of the autoclave was 250 ml). Even in this instance the conditions were not exactly comparable, since the molar ratio nitrile/solvent was not the same for different nitriles.

A preliminary study has shown that the polarity of the solvent influences the hydrogenation selectivity. The effect of nitrile structure was therefore followed in two solvents - methanol and cyclohexane - on cobalt and nickel metals prepared by thermal decomposition of formates (ref. 25); the results are summarized in

Table 4.3. In view of the above complications, we compared the individual nitriles on the basis of the relative initial reaction rate, r_{10}, expressed as the ratio of hydrogen which has reacted within the first ten minutes with the given nitrile and with butyronitrile as the standard, respectively. The selectivity, S, was then defined as the ratio of the amount of primary amine to that of secondary amine. No significant amount of tertiary amine was ever identified in the reaction mixture resulting from hydrogenations catalysed by cobalt. In nickel-catalysed hydrogenations the concentration of tertiary amine was several tenths of percent in most instances, while in the hydrogenation of acetonitrile the concentration of triethylamine reached tens of percent. Accordingly, for acetonitrile, on nickel the value of S is omitted in Table 4.3.

The results obtained clearly show the similarity of the trends in selectivity and in the inductive effect of the alkyl group in the resulting amine. Taking into consideration the reaction scheme for nitrile hydrogenation presented in Section 4.2, we may assume that the rate of formation of secondary amine by the reaction between the aldimine and the primary amine will be influenced by the electron density on the nitrogen atom in the primary amine; this conforms to the generally accepted views concerning this type of reaction. Results obtained by Rylander et al. (ref. 18), who studied the hydrogenation of nitriles on platinum and palladium catalysts, are also in agreement with the notion that the electron density, increasing with the inductive effect from benzylamine to dodecylamine, is important for the rate at which secondary amines are formed. In hydrogenations catalysed by palladium, valeronitrile yields 84% of tripentylamine and no primary amine. Under the same conditions, benzonitrile yields 63% of benzylamine and 34% of dibenzylamine (tribenzylamine cannot be formed, see Section 4.2). Benzylamine is less reactive with respect to benzaldimine (lower electron density on nitrogen), so that a considerable amount of primary amine is present in the reaction product. Valeronitrile, having similar properties to butyronitrile, yields upon hydrogenation pentylamine, which is more reactive with respect to condensation reactions leading to dipentylamine and tripentylamine.

4.6 EFFECT OF REACTION CONDITIONS ON HYDROGENATION OF NITRILES

4.6.1 Temperature

It is stated in all studies where the effect of temperature has been investigated that the content of secondary amines formed in the hydrogenation of nitriles on nickel or cobalt catalysts increases with temperature (refs. 8, 25, 90, 91). We have found (ref. 81) that in the hydrogenation of lauronitrile catalysed by cobalt the enhanced formation of secondary amine is due to the higher apparent activation energy (132 kJ/mol) of the condensation reaction between the amine and imine leading to the alkylidenealkylamine in comparison with the apparent activation energy of nitrile hydrogenation (52 kJ/mol). Similar conclusions can also be drawn for hydrogenations catalysed by nickel.

Surprising results were obtained in our study of the effect of temperature on the selectivity of hydrogenation of stearonitrile in the presence of a copper (II)-chromium (III) catalyst (Table 4.4), where the content of primary amine increased with temperature within the interval 210-300 °C. One possible explanation for this phenomenon might be the effect of a chemical

TABLE 4.4

The effect of temperature on hydrogenation of stearonitrile in the presence of $CuO-Cr_2O_3$ as catalyst (Cu 1107, marketed by Harshaw)

160 g stearonitrile, pressure 5 MPa, autoclave volume 450 ml, 10 g of catalyst.

Temperature (°C)	Reaction time[a] (min)	Conversion (%)	Composition of amine reaction products (mol%)		
			Primary amine	Secondary amine	Tertiary amine
180[b]	461	44.8	15.8	84.2	0
195	374	65.0	13.0	85.1	1.9
210	245	71.7	23.1	75.0	2.0
240	148	89.7	37.7	60.9	1.4
260	67	89.3	36.5	61.6	1.9
290	42	90.0	41.5	56.5	2.0

[a] The time corresponding to the end of measurable hydrogen consumption

[b] 15 g of catalyst

TABLE 4.5

The effect of temperature on the hydrogenation of stearonitrile in the presence of Ni-Al$_2$O$_3$ as catalyst (No. 6524 marketed by Leuna-Werke)

160 g stearonitrile, pressure 5 MPa, autoclave volume 450 ml, the total conversion of nitrile.

Temperature	Amount of catalyst	Reaction time	Composition of products (mol%)	
($^\circ$C)	(g)	(min)	Primary amine	Secondary amine
130	10	77	82.5	17.5
150	6	19	78.0	22.0
170	4	23	69.4	30.6
190	2	9	66.3	33.7
210	1	14	62.0	38.0
235	1	11	27.0	73.0
280	0.3	8	29.7	70.3

equilibrium on the composition of the reaction mixture at high temperatures, since the reversible reaction 4.13, which describes in a comprehensive manner the formation of secondary from primary amines, is slightly exothermic; the reaction heat is -15.3 kJ/mol (ref. 92). However, experiments performed with a nickel catalyst within the same temperature interval excluded this possibility, because in this instance the concentration of secondary amine increased monotonically (Table 4.5); the amount of secondary amine formed at higher temperatures on nickel is even higher than in the case of catalysis by copper. The explanation of the unusual behaviour of the copper catalyst must be sought in reaction kinetics. The activation energy for the hydrogenation of the nitrile to the primary amine is apparently higher on copper than the activation energy for the condensation between the imine and amine - see Fig. 4.4, which shows that the concentrations of the intermediates are very low when lauronitrile is hydrogenated at 180 $^\circ$C; the rate of nitrile hydrogenation is the lowest, and the remaining reaction steps are very rapid, so that mostly the secondary amine is present. The rate of hydrogenation increases much more rapidly with temperature than that of condensation, and consequently the reaction of the primary amine leading to the alkylidenealkylamine is relatively slower and the concentration of the former in the reaction product increases.

4.6.2 Hydrogen pressure

The reaction rate always increases (and the reaction time therefore decreases) with increasing hydrogen pressure in the autoclave. In view of the reaction mechanism presented in Section 4.2, one may state that increasing the pressure of hydrogen accelerates the hydrogenation reactions proper, while the condensation reactions leading to secondary or tertiary amines will be independent of hydrogen pressure; accordingly, the relative content of primary amines increases. Experimental results obtained for the hydrogenation of lauronitrile on metallic cobalt (ref. 81) were in accord with these concepts. Similarly, the content of primary amines increased with hydrogen pressure when valeronitrile was hydrogenated on a rhodium catalyst (ref. 18). However, some data presented below show that the effect of hydrogen pressure is rather complex and the overall result depends not only on the type of catalyst employed, but also on the structure of the hydrogenated nitrile.

Thus, in experiments where the effect of agitation on the reaction rate and selectivity was followed in the hydrogenation of stearonitrile (ref. 93), we found that a lowering of the agitation intensity such that the time of hydrogenation was prolonged by a factor of 6 did not affect the selectivity of a nickel-Al_2O_3 catalyst. Less intensive agitation of the reaction mixture decreases the concentration of hydrogen in the liquid phase and its effect thus parallels that of decreased hydrogen pressure.

Surprising results were obtained when the effect of hydrogen pressure on the hydrogenation of butyronitrile was investigated with the catalyst Ni-Al_2O_3, and also with metallic nickel and cobalt, obtained by decomposition of the respective formates (see Table 4.6). On pure nickel and on Ni-Al_2O_3 the content of primary amines increased with increasing pressure of hydrogen. On the other hand, under the same reaction conditions, the selectivity was practically independent of hydrogen pressure when cobalt was used as the catalyst, at variance with previous experience of the hydrogenation of lauronitrile. Data in Table 4.7 on the

TABLE 4.6

The effect of hydrogen pressure on the hydrogenation of butyronitrile

Temperature 160 °C, 0.8 mole of butyronitrile, autoclave volume 250 ml, the total conversion of nitrile.

Catalyst	Pressure (MPa)	Reaction time (min)	Composition of products (mass%)[a]	
			Monobutyl-amine	Dibutyl-amine
0.3 g Ni-Al$_2$O$_3$	9	48	66.5	31.9
	5	68	72.2	27.0
	3	100	74.5	24.8
2.5 g unsupported Ni	9	80	69.3	29.3
	5	130	75.8	23.4
	3	330	87.0	12.2
2.5 g unsupported Co	9	46	73.8	25.6
	5	80	75.4	23.7
	3	190	73.7	25.3

[a] The rest till 100% were tributylamine and the high boiling substances. In case of Co, no tributylamine was detected.

hydrogenation of lauronitrile on a cobalt catalyst prepared by decomposition of formate show quite unambiguously that the content of primary amine increases with the pressure of hydrogen.

The experimental results concerning the effect of hydrogen pressure on selectivity, obtained with various catalysts and different nitriles, indicate that complicated phenomena are involved, which are difficult to reconcile with the reaction scheme presented in Section 4.2. The fact that the selectivity with respect to primary amines increases with decreasing hydrogen pressure in the hydrogenation of butyronitrile on nickel-based catalysts is an important result. This phenomenon might be explained for example by a lower reaction order with respect to hydrogen in hydrogenation of the aldimine to the primary amine (in comparison with the order for hydrogenation of the nitrile to the aldimine). Another reason might be a change in the properties of the catalyst due to its interaction with hydrogen at elevated

TABLE 4.7

Effect of the hydrogen pressure on the hydrogenation of lauronitrile on unsupported cobalt as catalyst

Temperature 160 °C, 0.32 mole of lauronitrile, 5.1 g of cobalt catalyst prepared from cobalt formiate, autoclave volume 250 ml, the total conversion of nitrile.

Pressure (MPa)	Reaction time (min)	Composition of products (mol%)	
		Primary amine	Secondary amine
12	17	84.7	14.9
8	36	79.5	20.1
5	40	74.3	24.5
3	55	71.5	28.3

pressures. Such an interaction might have very diverse consequences, as evidenced by the fact (Section 4.4.3) that a palladium catalyst exposed to hydrogen at elevated temperature loses its activity for hydrogenation of nitriles, while its activity for hydrogenation of the C=C bond is preserved.

4.6.3 Ammonia

The formation of secondary amines in the hydrogenation of nitriles in the presence of nickel- or cobalt-based catalysts (as well with some others) is suppressed when ammonia is added to the autoclave. If it is desirable to prepare selectively a primary or secondary amine, ammonia is invariably added to the initial mixture, as described in almost all patents dealing with the preparation of primary amines. In addition to its effect on the formation of primary amines on nickel or cobalt catalysts, ammonia is known to influence favourably the formation of primary amines also with ruthenium on charcoal as the catalyst (ref. 9).

The higher content of primary amines observed in hydrogenations carried out in the presence of ammonia is explained by a scheme proposed by Schwoeleger and Adkins (ref. 94) and still accepted by other investigators (refs. 82, 91). Ammonia is deemed to favour the formation of primary amines due to the reactions:

$$\text{RCH=NH} + \text{NH}_3 \rightleftarrows \text{RCH}\begin{array}{c}\text{NH}_2\\[-2pt]<\\[-2pt]\text{NH}_2\end{array} \xrightarrow{+\text{H}_2} \text{RCH}_2\text{NH}_2 + \text{NH}_3 \qquad (4.15)$$

In our opinion this scheme does not adequately describe the action of ammonia. The aldimine does react with ammonia giving rise to a diamino derivative; but this reaction is reversible and leads to a lowering of the aldimine concentration in the reaction mixture, which in turn must result in a slowing down of the reaction between the aldimine and primary amine leading to the secondary amine. This is in accord with our results on the hydrogenation of lauronitrile on a cobalt catalyst (ref. 81). When ammonia was added before hydrogenation (by saturating the nitrile with gaseous ammonia at 20 $^\circ$C and a pressure of 0.7 MPa), the rate of dodecylidenedodecylamine formation was decreased by a factor of about five in comparison with an experiment without ammonia. The rate of hydrogenation was also affected by ammonia (ref. 81) and was lowered by a factor of about 2 in its presence. All in all, the relative rate of condensation (with respect to hydrogenation) reactions is therefore raised and results in enhanced hydrogenation selectivity.

The effect of ammonia on selectivity depends on the experimental arrangement (ref. 93): since in a batch arrangement ammonia accumulates in the autoclave, its increasing concentrations lowers the rate of the condensation reaction between the aldimine and primary amine. When the initial amount of reactants charged into the autoclave is high, the partial pressure of ammonia, released by the reactions leading to secondary amines, is large in the small volume of vapour phase available and consequently so is its concentration in the liquid phase. As a result of all these factors the formation of secondary amine is suppressed. The effect of the initial charge in the autoclave on selectivity is illustrated by the data collected in Table 4.8.

An analogous action of ammonia must be considered, e.g., when comparing the hydrogenation of nitriles that differ substantially in molar mass (cf., Section 4.5).

The formation of secondary amines is influenced decisively by the concentration of ammonia in the liquid phase, which can be controlled with advantage by the addition of compounds (such as water or some alcohols) capable of enhancing its solubility (refs. 85, 95, 96).

Ammonia shows an interesting effect on the composition of the

TABLE 4.8

Effect of the initial amount of nitrile in the autoclave on hydrogenation of stearonitrile

Pressure 5 MPa, catalyst Ni-Al_2O_3, autoclave volume 450 ml.

Starting amount of nitrile (g)	Reaction time[c] (min)	Composition of products (mol%)	
		Primary amine	Secondary amine
80[a]	64	73.0	27.0
120[a]	91	74.8	25.2
160[a]	120	77.7	22.3
240[a]	185	79.9	20.1
80[b]	30	66.3	33.7
120[b]	46	66.8	33.2
160[b]	56	71.4	28.6
240[b]	93	75.6	24.4

[a]Temperature 130 °C, 2 g of catalyst.
[b]Temperature 160 °C, 1 g of catalyst.
[c]The total conversion of stearonitrile.

hydrogenation product when copper is used as the catalyst. By comparing the results of stearonitrile hydrogenation carried out (i) in a reactor where agitation was achieved by bubbling hydrogen through the liquid layer and (ii) in a stirred batch autoclave (ref. 83), it was found that the hydrogenation in the bubble reactor, contrary to expectations, led to a product with a high content of primary amine (73 mol%) at a nitrile conversion of 71%, comparable to that obtained in experiments with a nickel--Al_2O_3 catalyst in a reactor of the same type. In the classical, stirred autoclave the content of primary amine in the product was 23.1% at a comparable extent of conversion; the remainder consisted mostly of secondary amine. In this connection it must be stressed that the design of the stirred autoclave, the intensity of agitation and other experimental conditions employed ensured that the hydrogenation proceeded without any effect of transport phenomena on the rate. On the other hand, the rate of hydrogenation was found to be several times higher in the bubble reactor - a similar extent of conversion was attained within a time shorter by a factor of about 8 and with a only half as much catalyst.

The main difference between these two experiments is due to ammonia: while its concentration in both phases was low in the bubble reactor because it was continuously removed by the stream of hydrogen, it accumulated in the autoclave, resulting in a lower partial pressure of hydrogen and a lower rate of hydrogenation. Ammonia decreased the rate probably also because it competed in the adsorption on the catalyst surface. In hydrogenations catalysed by copper the presence of ammonia in the reaction mixture also lowers the concentration of the aldimine, cf., reaction 4.15, and thus results in a lower rate of its condensation with the primary amine to yield secondary amine. However, this decrease in the rate of condensation is apparently smaller in the case of the copper catalyst than is the lowering of the hydrogenation rate brought about both by the smaller partial pressure of hydrogen and by the retarding effect of ammonia, so that the net result is a relative increase in the rate of condensation between the aldimine and amine in comparison with the rate of hydrogenation leading to primary amine, and the content of secondary amine in the autoclave is higher. In contrast, in the reactor with hydrogen bubbling through the reaction slurry, the concentration of ammonia was low and neither lowered the partial pressure of hydrogen nor retarded the hydrogenation, the substantially higher relative rate of hydrogenation resulted in a high content of primary amine, although ammonia was continuously removed.

In hydrogenations carried out in the presence of platinum or palladium on charcoal, the action of ammonia is also different from that in the case of catalysis by cobalt or nickel (ref. 9). In the presence of water the addition of ammonia leads to an increased content of tertiary amines, but the lack of detailed data prevents any discussion of possible reasons.

4.6.4   Water

It is known from the patent literature that the content of primary amines is raised, as is the rate of hydrogenation of nitriles, in the presence of nickel or cobalt catalysts, if water is added to the reaction mixture (refs. 95, 96). A combined action of water and ammonia is usually described, which is said to promote high yields of primary amines. In addition to this mode of nitrile hydrogenation in the presence of water, hydrogenations where aldehydes are formed upon addition of water are also known (ref. 97). Water added to hydrogenations of nitriles on platinum or palladium leads to a substantial increase in the reaction rate, thus enabling the reaction to proceed the total conversion (refs. 9, 71).

Our own results pertaining to the effect of water on the hydrogenation of stearonitrile and lauronitrile carried out in the presence of several nickel and cobalt catalysts are collected in Table 4.9 and show that the hydrogenation rate and the selectivity with respect to primary amines are positively influenced only on unsupported cobalt, while on catalysts deposited on supports the hydrogenation is slowed down and its selectivity is slightly lower.

We also followed (by gas chromatography) the effect of water on hydrogenations of propionitrile and butyronitrile in the presence of metallic cobalt and of various nickel catalysts (ref. 98). It was established that in the presence of water the hydrogenation proceeds more rapidly not only on pure cobalt, but also on pure nickel, both prepared by decomposition of the respective formate (ref. 25), while the hydrogenation rate is lowered by water when either nickel on aluminium oxide or Raney nickel are used as catalyst. The presence of water leads to an increased content of primary amines (to the detriment of secondary ones) in the hydrogenation of butyronitrile catalysed by pure cobalt or nickel; small amounts of butanol and of higher-molecular-weight compounds are also formed, by aldolization. Regardless of the type of catalyst, the hydrogenation of propionitrile in the presence of water leads to the formation of higher amounts of aldolization products; the content of 2-methylpentylamine was found to be the highest. However, the addition of water was never found to result in an increased content of primary amine. This is interesting, especially for the cobalt catalyst, since in the presence of water the formation of primary amines was always enhanced on this catalyst for all the other nitriles investigated. A possible positive effect of water on the selectivity of propionitrile hydrogenation catalysed by cobalt is apparently offset by the aldolization reactions whose extent is governed by the reactivity of the aldimines and alkylidenealkylamines formed.

The lower rate of nitrile hydrogenation observed in the presence of water can be explained by the hydration of the catalyst. Catalytic supports - silica and alumina - are strongly hydrophilic. Since in our experiments the amount of water added was at least comparable to the amount of the catalyst, the latter was mostly dispersed in the aqueous phase owing to the hydrophilic character of the support. It must be stressed that the hydrogenation - in particular in the initial stages, when only the water-insoluble nitrile is present - proceeds in an heterogeneous system containing two liquid, one solid and one vapour phase. This is then reflected in the kinetics, because the rate of hydrogenation depends on the concentration of hydrogen and of the nitrile in the aqueous phase.

The hydrophilic character of unsupported cobalt or nickel is apparently smaller than that of catalysts deposited on supports; accordingly, they are not preferentially dispersed in the aqueous phase. The positive influence of water on the rate of hydrogenation might be then ascribed to some unspecified influence on the sorption or on the catalytic properties of the metals. One cannot also exclude that this effect of water is associated with the solvation of the amines formed by water, leading to their weaker adsorption on the surface of the catalyst, which in turn might suppress their retarding effect. The behaviour of Raney nickel, which is not a supported catalyst, is not fully compatible with the above discussion. Its affinity towards water might be higher than that of pure metallic catalysts, since its surface after the leaching by sodium hydroxide may contain aluminium oxide.

In the presence of water, the increased content of primary amines, in the products of hydrogenation of nitriles catalysed by pure cobalt or nickel can be explained on the basis of the enhanced rate of hydrogenation - as compared with the condensation reactions - discussed above. The latter lead to the formation of alkylidenealkylamines. By following the concentration of the alkylidenealkylamine in the reaction mixture during the hydrogenation of lauronitrile catalysed by pure cobalt without water and in the presence of water (3%), we have ascertained (ref. 81), that the rate of hydrogenation was higher in the latter case and, on the other hand, the rate of the reactions leading to the alkylidenealkylamine was lower (Fig. 4.6). This effect can be explained by the reaction of water with the aldimine:

$$RCH=NH + H_2O \rightleftarrows RCH\begin{matrix}OH\\NH_2\end{matrix} \rightleftarrows RCH=O + NH_3 \qquad (4.16)$$

As a result the aldimine concentration decreases and consequently

TABLE 4.9

Hydrogenation of lauro- and stearonitrile in the presence of water on various nickel and cobalt catalyst

Catalyst	Amount of water added (g)	Composition of products (mol%)		Reaction time[e] (min)
		Primary amine	Secondary amine	
Stearonitrile[a]				
Ni-Al$_2$O$_3$	0	76.5	23.5	67
Ni-Al$_2$O$_3$	3	75.8	24.2	164
Co-SiO$_2$	0	87.4	12.6	72
Co-SiO$_2$	3	85.0	15.0	110
Co[c]	0	88.5	11.5	190
Co[c]	3	91.5	8.5	105
Lauronitrile[b]				
Co-Mn (5% Mn)	0	86.7	13.3	8
Co-Mn (5% Mn)	0[d]	91.5	8.5	11
Co-Mn (5% Mn)	3	94.6	5.4	4
Co-Mn (5% Mn)	3[d]	96.2	3.8	8

[a] 80 g Stearonitrile, temperature 150 °C, pressure 6.0 MPa, autoclave volume 450 ml, 1 g of catalyst.
[b] 150 g Lauronitrile, temperature 200 °C, pressure 6.0 MPa, autoclave volume 250 ml, 1.5 g of catalyst.
[c] Prepared by reduction of Cr$_3$O$_4$.
[d] Lauronitrile was saturated with ammonia in the autoclave at 0.8 MPa and 20 °C.
[e] The total conversion of nitrile.

the formation of the alkylidenealkylamine is slowed down, cf. eqns. 4.2 and 4.3. Provided the reaction between water and the aldimine proceeds to the stage where aldehyde is formed, the secondary amine can be alternatively formed by its reaction with the primary amine followed by hydrogenation. One cannot also exclude the possibility that the selectivity is influenced by solvation of the primary amine by water. The interaction between the amino group and water will take place by means of hydrogen bonding, which can be assumed to exist either between the electron pair on oxygen and the hydrogens of the amino group, or between

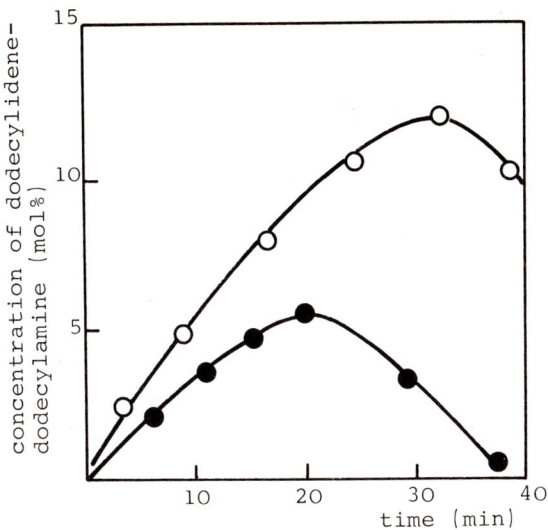

Fig. 4.6. The effect of water on the concentration of dodecylidenedodecylamine in the product of lauronitrile hydrogenation.
○, 1 g of unsupported Co-catalyst; ●, 1 g of unsupported Co-catalyst and 3 g of water.

the electron pair on nitrogen and hydrogens in the water molecule. The latter interaction might then lower the rate of condensation between the primary amine and the imine, since this competes with the reaction between the primary amine and the aldimine.

The problems connected with the effect of water on the hydrogenation of nitriles in the presence of palladium or platinum catalysts are somewhat different from those encountered in catalysis by cobalt or nickel. Trialkylamines, which in this case represent the major product of hydrogenation, significantly retard the reaction, because of their strong adsorption on the surface of the catalyst, stronger than that of the starting nitriles. The result is that the hydrogenation spontaneously stops before the complete conversion is attained (Section 4.4.3). Water, if present in a hydrogenation catalysed by palladium or platinum on charcoal, can hydrate these catalysts, which are then predominantly dispersed in the aqueous phase. The hydrogenation products -
- tertiary amines - are only partially soluble in water (as are

the starting nitriles). Since the concentration of the tertiary amine in the aqueous phase is low, and its molecules are solvated and partially protonated by water, the retarding effect of the tertiary amine on the hydrogenation is substantially reduced, hence, in the presence of water the hydrogenation of nitriles on palladium or platinum catalysts can proceed to total conversion.

Greenfield (ref. 71) found that, in the hydrogenation of benzonitrile in the presence of platinum on charcoal, with increasing content of water (up to the molar ratio 2:1 with respect to benzonitrile) the content of secondary amine in the reaction mixture raise to 97%, and no tribenzylamine was present. Tribenzylamine was found in the product of hydrogenation of benzonitrile in water (molar ratio 65:1). Considering the mechanism discussed in Section 4.2 as responsible for the formation of tertiary amines, we see that tertiary amines can be formed in exceptional cases by a mechanism other than that via enamine intermediates.

4.7 CONCLUSION

Hydrogenation of nitriles is a relatively complicated process, owing to the existence of reactive intermediates (aldimines and alkylidenealkylamines), comprising several consecutive and competitive reactions of different characters. Some of them are typical hydrogenation reaction promoted by metallic hydrogenation catalysts; others are condensation reactions of acid-base character. The active centres for the hydrogenation and condensation reactions are localized on the surface of the metallic component of the catalyst, and the type of metal used is decisive for the catalyst properties. The support has a much smaller influence on the final properties of the catalyst. The main differences between nickel and cobalt on the one hand (which mostly promote the formation of primary amines) and palladium and platinum on the other hand (where tertiary amines are predominantly formed) lie in the relative rates of the hydrogenation and condensation reactions catalysed by these metals. The rates are comparable on nickel and cobalt, while a substantially higher rate of condensation (as compared to hydrogenations proper) is typical in hydrogenations carried out in the presence of palladium or platinum, which sometimes result in an almost quantitative yield of tertiary amines. Copper-based catalysts, in comparison with nickel, cobalt, platinum and palladium, generally exhibit a low hydrogenation activity; the activity of copper with respect to condensation reactions is relatively higher, so that formation of secondary amines usually ensues.

By changing the reaction conditions (temperature, hydrogen pressure, addition of ammonia or water) one can influence the relationships between the rates of hydrogenation and condensation and thus vary the composition of the reaction product. The net result under the actual reaction conditions employed will depend on the type of catalyst and/or on the experimental arrangement (filling of the autoclave, intensity of agitation, etc.). The individual catalysts differ in their responses to a change in reaction conditions, and very often contradictory results are encountered. Thus, e.g., on nickel catalysts the content of secondary amines in the product increases with temperature, while the same is true for primary amines on copper catalysts. Addition of ammonia enhances the content of primary amines in hydrogenations catalysed by nickel or cobalt, promotes the formation of secondary amines on copper catalysts under certain conditions and leads to an increased yield of tertiary amines in hydrogenations catalysed by platinum or palladium. The majority of these results, but also of other interesting and often contradictory phenomena not mentioned here, can be discussed on the basis of the reaction scheme presented in Section 4.2, where it is assumed that aldimines, alkylidenealkylamines and enamines all participate in the formation of primary, secondary and tertiary amines.

REFERENCES

1   C. Paal and J. Gerum, Chem. Ber., 42 (1909) 1553-1560.
2   H. Rupe and R. Hodel, Helv. Chem. Acta., 6 (1923) 865-880.
3   G. Mignonac, C.R. Hebd. Seances Acad. Sci., 171 (1923) 114-119.
4   J. Braun, G. Blessing and F. Zobel, Chem. Ber., 36 (1923) 1988-2001.
5   Ch.F. Winans and H. Adkins, J. Am. Chem. Soc., 54 (1932) 306-312.
6   R. Juday and H. Adkins, J. Am. Chem. Soc., 77 (1955) 4559-4564.
7   P.N. Rylander, Catalytic Hydrogenation in Organic Syntheses, Academic Press, New York, 1979, pp. 138-140.
8   A.J. Lazaris, E.N. Zilberman, E.V. Lumitcheva, A.M. Vedin, Zh. Prikl. Khim., 38 (1965) 1097-1101.
9   H. Greenfield, Ind. Eng. Chem. Prod. Res. Develop., 6 (1967) 142-144.

10 J. Ernest, S. Heřmánek and M. Hudlický, Preparativní reakce v organické chemii IV, nakladatelství ČSAV, Praha, 1959, pp. 425-429.
11 A.G. Cook (Editor), Enamines: Syntheses, Structure and Reactions, Marcel Decker, New York, 1969.
12 R.L. Reeves in S. Patai (Editor), The Chemistry of Carbonyl Group, Interscience, London, New York, 1966, pp. 600-602.
13 L.D. Volkova, H.B. Kagarlickaya and G.D. Zakumbaeva, Izv. Akad. Nauk Kaz. S.S.R. Ser. Khim., 23 (1973) 70-73.
14 U. Schwenk and A. Becker, H. Mueler, Ger. Pat., 1 768 321 (1974).
15 J. Volf, J. Pašek and M. Duraj, Collect. Czech. Chem. Commun., 38 (1973) 1038-1047.
16 J. Volf, J. Pašek and B. Beneš, Chem. Prum., 52 (1967) 623-626.
17 Y. Kimura and D. Saika, Yukagaku, 53 (1979) 379-387.
18 P.N. Rylander and L. Hasbrouck, Engelhard. Ind. Tech. Bull., 11 (1970) 19-24.
19 J. Koubek, J. Volf and J. Pašek, in preparation.
20 T. Aonuma and T.K. Daigaku, Kogyo Kogaku Zasshi, 65 (1962) 1819-1824.
21 J. Pašek, P. Richter, L. Rusek, L. Jarkovský and V. Růžička, Czech. Pat., 131 353, 1969.
22 L.M. Saunders, A kinetic Study of the Reactions of Butylamine in the Vapor Phase over Raney Nickel, Dis. Abstr., 25 (1965), 6464.
23 Y. Ogino, Y. Saito and K. Okano, Jpn. Kokai Tokkyo Koho, 73 23, 725 (1973).
24 F. Hagedorn and H.P. Gelbke, in Ullmanns Enzyklopädie der technischen Chemie, Band 17, Verlag Chemie, Weinheim, 1979, pp. 333-337.
25 J. Pašek, N. Kostova and B. Dvořák, Collect. Czech. Chem. Commun., 46 (1981) 1011-1022.
26 J. Bolle, Fr. Pat., 2 063 378 (1971).
27 K. Hioki, K. Hirokawa, Y. Kuka and A. Shimade, Jpn. Kokai Tokkyo Koho, 75 47 909 (1975).
28 W. Jerzykiewicz, Z. Krasnodebski and G. Bekierz, Pol. Pat., 115 999 (1983).
29 Ch.E. Kuthens and L.M. Lanier, U.S. Pat., 4 429 159 (1984).
30 S. Taira, Jpn. Pat., 70 32, 410 (1970).
31 Toyo Rayon Co., Fr. Pat., 1 530 809 (1968).
32 R.A. Diffenbach, Fr. Pat., 2 149 987 (1973).
33 S. Taira, Jpn. Pat., 69 05, 844 (1969).
34 R.J. Alain, U.S. Pat., 4 375 003 (1976).
35 R. Uehara, T. Horii, T. Imai, Y. Tomita and K. Yamano, Jpn. Kokai Tokkyo Koho, 76 78, 795 (1976).
36 Societe des Usines Chimiques Rhone-Poulenc, Fr. Demande, 2 248 265 (1975).
37 A. Kuroda and S. Taira, Jpn. Pat., 72 31, 833 (1972).
38 J. Miwa, K. Ueno and N. Kiyoaki, Jpn. Pat., 78 (62) (1962).
39 M. Kalina, P. Przybylo, E. Pitřík and J. Pašek, Czech. Pat., 114 232 (1965).
40 D.S. Burleigh and B.D. Hawkins, Ger. Offen., 2 312 972 (1973).
41 W. Hansel and H. Hoffmann, Ger. Offen., 2 238 452 (1974).
42 K. Adam and E. Haarer, Fr. Pat., 1 483 300 (1967).
43 K. Baer, H. Hoffmann, D. Voges, A. Wegerich and S. Winderl, Ger. Offen., 2 301 139 (1974).
44 D. Voges, L. Hupfer, L. Winderl, K.W. Leonbard and H. Hoffmann, Ger. Offen., 2 312 591 (1974).
45 Z.S. Vanyushina, G.A. Chistyakova and P.N. Ovchinikov,

U.S.S.R. Pat., 265 113 (1972).
46 A.B. Stiles, U.S. Pat., 3 752 744 (1973).
47 S. Taira and A. Kuroda, Jpn. Pat., 71 06, 885 (1971).
48 Y. Kageyama and Y. Fukai, Jpn. Kokai Tokkyo Koho, 74 47, 304 (1974).
49 M.I. Yakushin, L.B. Gal´perin, U.S.S.R. Pat., 276 915 (1970).
50 M. Koštíř, P. Przybylo and J. Novotný, Czech. Pat., 190 253 (1981).
51 J.W. Reynolds, U.S. Pat., 3 728 284 (1973).
52 D. Szabo, Belg. Pat., 635 132 (1963).
53 W. Schroeder and W. Franzischka, U.S. Pat., 4 184 982 (1980).
54 Y.D. Waddan, Ger. Offen., 2 226 647 (1972).
55 D.E. Terry and L.J. Jacobsen, U.S. Pat., 2 784 232 (1957).
56 H. Noeske, F. Pfitzer and W. Rottig, Ger. Pat., 1 518 118 (1973).
57 A.A. Vedenskii, M.J. Yakushin, E.M. Bocharova and M.V. Blinova, U.S.S.R. Pat., 191 572 (1967).
58 Y. Kageyama and Y. Fukai, Jpn. Kokai Tokkyo Koho, 74 47, 3013 (1974).
59 S.H. Shapiro, U.S. Pat., 2 781 399 (1957).
60 J. Volf, J. Koubek and J. Pašek, in preparation.
61 V.M. Lenfield in E. Jungermann (Editor), Cationic Surfactans, Marcel Decker, New York, 1970, p. 29.
62 Armour and Co., Brit. Pat., 773 432 (1955).
63 Farbwerke Hoechst A.G., Ger. Offen., 1 941 290 (1971).
64 J. Mostecký, J. Zajíc, M. Bareš, M. Morák and R. Smrž, Czech. Pat., 183 342 (1980).
65 Farbwerke Hoechst A.G., Fr. Pat., 1 504 323 (1967).
66 M. Morák and F. Kršňák, Czech. Pat., 152 102 (1974).
67 P.N. Rylander and J.G. Kaplan, U.S. Pat., 3 117 162 (1964).
68 P.N. Rylander and D.R. Steele, Engelhard. Ind. Tech. Bull., 5 (1965) 113-120.
69 Ch.A. Drake, U.S. Pat., 3 898 286 (1975).
70 H. Greenfield and R.S. Sekelick, U.S. Pat., 3 923 891 (1975).
71 H. Greenfield, Ind. Eng. Chem. Prod. Res. Develop., 15 (1976) 156-158.
72 S.E. Diamond, F. Mares and A. Szalkiewicz, Eur. Pat. Appl., 79, 911 (1983).
73 Henkel und Cie GmbH, Brit. Pat., 1 153 919 (1969).
74 M.I. Yakushin and V.M. Evgrasin, U.S.S.R. Pat., 216 009 (1970).
75 K. Nishimura, K. Matsui, Y. Shiomi, S. Niida, T. Toshimitsu and S. Ono, Jpn. Kokai Tokkyo Koho, 76 48, 602 (1976).
76 T. Toshimitsu, S. Ono, K. Nishihira, K. Nishimura and K. Matsui, Jpn. Kokai Tokkyo Koho, 76 06, 907 (1976).
77 T.G. Dewdney, D.A. Dowden, B.D. Hawkins and W. Moris, Ger. Offen., 2 429 293 (1975).
78 D.B. Bivens, L.W. Paton and J.B. Wiggill, Ger. Offen., 2 304 278 (1973).
79 W.E. Thomas, Ger. Offen., 2 304 269 (1973).
80 M. Kalina and J. Pašek, Kinet. Katal., 10 (1969) 574-580.
81 J. Volf, J. Dlouhý and J. Pašek, in press.
82 D. Nowak and W. Jerzykiewicz, Przem. Chem., 49 (1970) 664-668.
83 J. Volf and J. Pašek, in preparation.
84 J. Volf and J. Pašek, Chem. Prum., 53 (1978) 464-467.
85 F. Kršňák, Czech. Pat., 168 074 (1975).
86 J. Koubek and J. Pašek, Chem. Prum., 56 (1981) 349-356.
87 J. Horyna, Diploma Thesis, Prague Institute of Chemical Technology, Prague, 1970.
88 L. Červený and V. Růžička, Sb. Vys. Sk. Chem. Technol. Praze,

Org. Chem. Technol., 27 (1981) 61-68.
89 Y. Takagi, S. Neshimura, K. Taya and K. Hirota, Sci. Pap. Inst. Phys. Chem. Res. Jpn., 61 (1967) 114-117.
90 Sh.A. Zelenaya, A.S. Basov, A.A. Pavlov, N.K. Petryakova and N.V. Gushin, Khim. Prom. Moscow, 46 (1970) 11-12.
91 M.I. Yakushin, M.V. Blinova and P.V. Bazyleva, Khim. Prom. Moscow, 44 (1968) 265-267.
92 J. Pašek, P. Kondelík and P. Richter, Ind. Eng. Chem. Prod. Res. Develop., 11 (1972) 333-337.
93 J. Volf and J. Pašek, Chem. Prum., 56 (1981) 590-593.
94 E.J. Schwoegler and H. Adkins, J. Am. Chem. Soc., 61 (1939) 3499-3502.
95 M. Morák, F. Kršňák and H. Hrdličková, Czech. Pat., 151 721 (1974).
96 N. Waddleton, Brit. Pat., 1 321 981 (1973).
97 P. Tinapp, Chem. Ber., 102 (1969) 2770-2776.
98 J. Volf and J. Pašek, in preparation.

Chapter 5

HYDROGENOLYSIS OF C-C BONDS ON PLATINUM-BASED BIMETALLIC CATALYSTS

F. GARIN, L. HILAIRE and G. MAIRE

Laboratoire de Catalyse et Chimie des Surfaces, U.A. 423 du CNRS, Université Louis Pasteur - Institut Le Bel, 4 Rue Blaise Pascal, 67070 STRASBOURG Cedex (FRANCE)

5.1 INTRODUCTION

In the late sixties it was realized that the addition of a second metal to platinum improved the properties of reforming catalysts. Since then a tremendous amount of effort has been devoted to the study of true alloys and other bimetallic systems. Nowadays, monometallic catalysts are no longer used and all industrial catalysts are a complicated mixture of at least two metals plus one or several dopants. This increasing complexity has had two major consequences. More sophisticated methods of preparation were required since the increasing number of elements comprising the catalyst caused a parallel increase in the parameters which, during the course of its preparation, are liable to affect the physical and chemical properties of the surface atoms and therefore the catalytic behaviour. A second consequence, directly linked to the preceding one, was the necessity of acquiring a better knowledge of the surface structure of such complex systems and thus more efficient tools of catalyst characterization. Fortunately, during this period, considerable progress was made in the development of physical methods, which became freely available about fifteen years ago : the best known, and most widely used, are Auger Electron Spectroscopy (AES) and Electron Spectroscopy for Chemical Analysis (ESCA), which are surface methods, and more recently Extended X-ray Absorption Fine Structure (EXAFS). These new tools, in addition to the improvement in classical methods such as chemisorption measurements and Electron Microscopy, enabled a considerable amount of information to be collected which was of great help in the design of new and more efficient catalysts.

This chapter will deal with the hydrogenolysis reactions of saturated hydrocarbons performed on platinum-based bimetallic catalysts. We will restrict ourselves to this metal since it is the most widely used in industrial plants, and attention will be focused on the modifications of its properties due to the nature of the second metal added. First a brief account will be given of the main parameters involved in such complex systems; in particular, the current ideas, often controversial, on the factors which are believed to govern the behaviour of these catalysts. A few results showing the activity and selectivity (and hence the kinetics and mechanisms) of platinum itself in hydrogenolysis reactions will then be given. Finally, there is a review of different bimetallic catalysts.

5.2 GENERAL PROPERTIES OF BIMETALLIC CATALYSTS

We do not intend to give an exhaustive study of the problems of bimetallic catalysts, only a general background to the results described later on. More detailed reviews have recently been published (see for example ref. 1).

One of the major problems with bimetallic catalysis is the determination of the atomic arrangements in the metallic crystallites. Many different situations may occur, from the formation of a true alloy to a completely heterogeneous surface, with a variety of intermediate configurations : bimetallic clusters, "cherry" model, etc. as described for example by Sachtler and Van Santen (ref. 2) in the particular case of Cu-Ni. The method of preparing the catalyst, the different possible treatments, calcination, reduction and the catalytic reaction itself will drastically modify the superficial arrangements. This demonstrates the importance of physical techniques for the characterization of the catalyst; e.g. the usefulness of EXAFS, the most recent one, has been demonstrated by Sinfelt's group (ref. 3, 4) with Os-Cu and Ru-Cu bimetallic catalysts.

When a transition metal A, say platinum, is alloyed (or associated) with a second metal B, two cases may be distinguished, depending on the nature of B :

- B may be another transition metal atom, possessing catalytic properties. Will the behaviour of A vary continuously as the concentration of B increases or will other phenomena (synergy, new reaction) occur?

- A may be diluted by an inactive metal, most often of the Group I. Will the activity and selectivity of A decrease regularly or will a different behaviour appear?

5.2.1 Geometric and electronic factors

Half a century ago Balandin (ref. 5) explained that the dissociation of a reacting molecule, which is a necessary step for any catalytic reaction, was possible only if there existed a good match between the interatomic distances in this molecule and the atomic arrangement of the superficial metal atoms; if the distances differed too greatly, the dissociation could not take place. This explanation was provided for monometallic catalysts. With bimetallics the situation is much more complicated. The question arises as to whether the introduction of a second metal merely changes the geometry of the system or the perturbation of the electronic structure is more important. Controversy has surrounded this matter for many years and the problem is not completely solved.

5.2.2 The rigid band model

From the nineteen fifties to the late sixties, the dominant model was the so-called "electronic theory of catalysis" (ref. 6) and the behaviour of bimetallic catalysts was explained in terms of the "rigid band theory". Alloying was supposed to result in the filling of a d band by the electrons of the second element, and as the holes in the d band were considered responsible for chemisorption, the catalytic behaviour disappeared when the d band was filled. As a consequence, charge transfers between elements were expected, and the individuality of each type of atoms was lost in the new band thus created.

Band-structure calculations (ref. 7) and the advent of photoemission led to the abandonment of this theory about fifteen years ago. Except in a few cases like Ni-Al (ref .8) or Pd-Zr (ref. 9), very small charge transfers were observed. Moreover X-ray Photoelectron Spectroscopy (XPS) and Ultra Violet Photoelectron Spectroscopy (UPS) spectra showed that the atoms in an alloy keep their individuality to a large extent : the resulting bands are formes by a convolution of the initial bands of the elements, weighted according to their respective concentrations in the alloy (ref. 10, 11). These experimental results led to the development of a new theoretical model, known as the coherent potential approximation (CPA) (ref. 12, 13).

5.2.3 The geometric effect

Boudart's concepts (ref. 14) on "demanding" reactions, which require several contiguous atomic sites, in contrast to "facile" reactions, were of great help is elucidating this complicated problem. In 1972 Sinfelt et al. (ref. 15) showed that the activity of Ni-Cu alloys for the dehydrogenation of cyclohexane was constant over a large range of concentrations, whereas a dramatic decrease was observed when a small amount of copper was added to nickel when the reaction was the hydrogenolysis of ethane. Since dehydrogenation reactions take place on a single atom while hydrogenolysis is known to be a "demanding" reaction, it is easy to understand why the addition of copper, which rapidly decreases the probability of finding several contiguous atoms of nickel, has such a drastic effect on the activity.

How large must be these patches of contiguous atoms ? A number of authors have tried to answer this question. Martin and co-workers (ref. 16, 17) in a study of Ni-Cu/SiO_2 catalysts by magnetic methods and kinetics measurements found that 12, 16 and 20 nickel atoms are required for the hydrogenolysis of ethane, propane and butane, respectively. According to Frennet et al. (ref. 18), the "landing site" of a methane molecule is composed of eight contiguous rhodium atoms. Christmann and Ertl (ref. 19) believe that an hydrogen atom needs five ruthenium atoms for adsorption to occur.

This use of the geometric effect (or ensemble effect) has had considerable success in the past few years since it provides a simple but satisfactory explanation for a large number of experimental findings.

5.2.4 The electronic factor (ligand effect)

One may find it hard to believe that electronic effects do not play an important part in bimetallic catalysis. The problem is that experimental evidence is difficult to obtain as we have seen above, from photoemission spectra or other techniques. However, it can be argued that the band modifications predicted by the rigid band model are long-range effects, whereas catalysis may be highly sensitive to short-range effects. It has been shown that, whereas the position of the band in a Ni-Cu alloy does not vary, its width decreases with increasing copper content (refs.10, 11). This means that the local densities of states are modified, due to the interaction between nickel and copper atoms. This perturbation of the electronic structure of nickel on adding copper might result in an inhibition of the reaction. The ensemble effect could therefore be viewed as a consequence of an electronic modification. This interpretation has been reviewed by Burch (ref. 20), and others (refs. 18, 21, 22).

At this stage the question arises as to whether these local electronic modifications are really large enough to influence the catalytic behaviour. Or, conversely, should only a geometric effect be considered ? This question is still open. We believe that geometric considerations provide a satisfactory explanation in many cases. However, there are results which can hardly be explained in this simple way, as we will see below.

5.2.5 Synergistic effects

One of the main difficulties of the geometric effect is the explanation of synergy. One of the main interest of bimetallic catalysis is to find systems more active than any of the components. A good example of synergy is provided by Boudart's experiments on the H_2-O_2 reactions on Pd-Au alloys (ref. 23). It was shown that the addition of 60% inactive gold to palladium resulted in an increase in the activity by nearly two orders of magnitude.

For many years it has been considered that the existence of synergy is good evidence for electronic effects and indeed it is difficult to believe that geometric considerations alone can provide a good explanation for this phenomenon.

The formation of carbonaceous deposits allows one to cope with this difficulty. Somorjai and co-workers (refs. 24, 25) have shown that kink sites, ledges and steps are the most reactive parts of a surface and that the development of carbonaceous deposits, which poison the surface, is inhibited in these areas. This "coking" effect takes place preferentially on flat surfaces. Similar results were found by Ponec and co-workers (ref. 26) who showed that small particles are less sensitive to coking effects than large ones.

If we consider alloying as the dispersal of for example nickel by inactive copper, it can be argued that the copper atoms will hinder the development of large carbonaceous deposits which poison the surface. A good example of this interpretation is provided by Sachtler for Pt-Re alloys (ref. 27). Sulphur, which is an usual impurity of the reactant, forms strong bonds with rhenium atoms, which inhibit the formation of coke, while on platinum itself poisoning can take place more easily since the Pt-S bond is much more labile. This interpretation has recently been confirmed by Coughlin et al.(ref.28) who showed by gravimetric measurements that coke deposition on Pt-Re catalysts is greatly retarded by sulphur and is closely associated with the rates of deactivation of these catalysts.

5.2.6 Surface segregation

It is hardly possible to talk about bimetallic catalysts without saying at least a few words about surface enrichment. This phenomenon has been known for many years but the advent of modern tools for surface studies was necessary to get experimental evidence for it, not only qualitative but also semiquantitative.

A number of theoretical considerations have been developed to predict surface segregation. Simple models consider the driving force for surface enrichment as due to the difference in heats of vaporization (ref. 29) of the elements and/or the differences in atom sizes (ref. 30). Good qualitative predictions can be made in this way (that is in good agreement with the experimental evidence), but it must be emphasized that platinum-based alloys, in particular Pt-Ni (ref. 31, 32) seem to behave in a different way (ref. 33). As far as quantitative predictions are concerned, the problem is much more complicated. Even sophisticated models do not really enable valuable quantitative predictions and much remains to be done in this direction.

The surface segregation may be completely modified in the presence of a gaseous atmosphere. The driving force is then the higher affinity of one of the elements towards the gas phase, together with the mobility of this element in the alloy, which requires a sufficiently high temperature. For example, decontaminated surfaces of Pd-Ni alloys are largely enriched in Pd (refs. 34, 36). At room temperature in the presence of air, the surface will be composed mainly of Pd, with a small amount of NiO; but at 300°C, NiO will be almost alone on the surface (ref. 36).

Chemisorption-induced segregations in the presence of O_2 and CO have been extensively studied. However little is known about the surface compositions in the presence of a hydrogen + hydrocarbon reaction mixture.

5.2.7 Small particles

A special mention must be made of small particles since they exhibit particular properties liable to influence the catalytic behaviour. From the electronic point of view, photoemission experiments have shown that in small particles the valence bands are shifted towards higher binding energies, and their width is decreased (ref. 37). This finding is in good agreement with experiments on stepped surfaces which show that atoms of low coordination possess a d-orbital population different from atoms of high coordination (ref. 38).

As regards surface segregation, a cubooctahedron of 1nm contains 55 atoms, of which 42 are at the surface (ref. 39). Obviously, surface enrichment can hardly occur in such particles. Even in 3 nm particles, where 40% of the atoms are on the surface, one may expect a strong perturbation of surface segregation.

5.3 HYDROGENOLYSIS REACTIONS

Hydrogenolysis of saturated hydrocarbons is defined as the rupture of C-C bonds in the presence of hydrogen. This reaction takes place mainly on Group 8 transition metals in the form of a film, foil, single crystal, metallic powder or supported metal.

The initial steps in such reactions are hydrogen dissociation and dissociation of C-H bonds before the C-C bond rupture occurs, followed by formation of carbon-hydrogen bonds. The C-C bond rupture may be considered as a highly irreversible process, whereas the C-H bond rupture is completely reversible. Extensive work has been done on exchange reactions and the mechanism, kinetic schemes and equations deduced.

The structure of the reactant is also important and archetypal hydrocarbon reactants have been used to study the mechanism of the degradation reactions.

As the aim of this chapter is to describe the particular behaviour of bimetallic catalysts in hydrocracking reactions, we will mention, first step, the most striking results obtained on monometallic catalysts.

5.3.1 Thermodynamic and kinetic data

(i) <u>Thermodynamic data</u>. To explain most of the features of the kinetics of these reactions, Guczi et al.(ref. 40) have assumed one type of sites on which a competitive adsorption of hydrogen and hydrocarbons takes place. It is known that, under the conditions of hydrogenolysis, hydrogen is on the surface in the atomic form.

The crucial point is whether the C-C bond rupture is preceded by the adsorption of hydrocarbons accompanied by a C-H bond rupture, or occurs immediately upon interaction of hydrocarbons with the surface.

Selwood (refs. 41, 42) has shown that, in the absence of hydrogen, ethane adsorption on Ni takes place at room temperature, whereas in the presence of hydrogen, adsorption occurs only at around 470 K which is also the temperature at which ethane hydrogenolysis occurs on nickel catalysts.

Frennet et al. (refs. 43, 44) have shown that, on other metals and with other hydrocarbons, there is no discontinuity in this inhibiting effect of hydrogen on the adsorption rate, under experimental conditions where only exchange occurs up to those where the main catalytic reaction is hydrogenolysis. However, the hydrocarbon deuterium-exchange reactions which are indicative of C-H bond rupture on the surface always have a threshold temperature lower than that for hydrogenolysis (ref. 45). The C-H bond rupture always proceeds more easily and thus before the C-C bond rupture;

in some cases, the reason may be the thermodynamics of the reactions : C-H + 2 S ⟶ C_{-s} + H_{-s} and C-C + 2S ⟶ $2C_{-s}$. However the kinetic aspects of adsorption must not be forgotten and in most practical cases this will most likely be the decisive factor. The C-H bonds are stronger than the C-C bonds, but they are immediately accessible to an interaction with the surface, whereas the C-C bonds are most likely not.

These considerations support the postulate of Guczi et al (ref. 40) that the hydrogenolysis of saturated hydrocarbons involving C-C bond rupture is preceded by adsorption of the reaction partners : the dissociative adsorption of the reaction components, including the H-H and the C-H bond ruptures in hydrogen and in hydrocarbon.

Characterizing the reaction rates by the "threshold temperature" T_t, where the rate reaches a low but well detectable value, a gap of 80-280 K between the values for ethane exchange and rupture was observed (ref. 45), except on Co which gave overlapping T_t values. The activation energy for hydrogenolysis was, as a rule, twice (three times for Pd and Pt) as high as that of H-D exchange in ethane.

TABLE 5.1
Energy of activation (KJ mol^{-1})

Cata.	Ref.	Exchange Reaction[a]			Hydrogenolysis reactions (taken from ref. 55)						
		CH_4	C_2H_6	C_2H_6 (43)	C_3H_8	n-C_4H_{10}	i-C_4H_{10}	n-C_5H_6	neo C_5H_{12}	C_7H_8[b]
Co	c	88	-		113	-	88	88	-	-	-
Ni	c	125	88		167	134	134	125	134 (142)[h]	151	130
Ru	c	97	59		119	-	84	-	-	150[e]	137
Rh	c	94	71		156-150[d]	-	123	-	117[d]	221[e]	135
Pd	c	112	73		222	-	155	161	151	224	162
Re	c				130	-	65[f]	78[f]	-	-	138
Os	e				-	-	-	-	-	134	106
Ir	d	111	82		163	-	100	-	106	209[e]	117
Ir/SiO_2	g				168-175	-	100-174	225-243	-	230-251	-
Pt	c	105	79		222	100	96	109	180[h]	113	141
Pt/Zeol.	i				-	-	-	-	-	134-154	-
Pt/SiO_2	i				-	-	-	-	180[j] 146-160[k]	104-146 (250e)	

[a] Activation energy of methane and ethane exchange calculated from the consumption of light hydrocarbons (ref. 45).
[b] Toluene hydrodemethylation (ref. 46) : catalyst 1-10% metals on Al_2O_3.
[c] Unsupported metals (ref. 47); [d](ref. 48); [e]SiO_2 supported metals (ref. 49); [f]Re black (ref. 50); [g](ref. 51); [h]unsupported metals (ref. 52) Ea for Pt was identical in refs. 47 and 52; [i](ref. 53); [j]catalyst Pt/Al_2O_3 internal C-C bond fission (ref. 54); [k](ref. 54) terminal C-C bond fission.

(ii) <u>Kinetics models</u>. A very good summary has been given by Guczi et al (ref. 40). When a power rate law is used to correlate the rate of hydrogenolysis with the reactant pressure, e.g., rate = $kP_{CH}^{n} P_{H_2}^{m}$, it is usually found that :
- n is positive and m is negative at pressures between 1.3 and 100 KPa ;
- m increases with increasing temperature and n approaches zero at a high temperature;
- a maximum in the rate may be found when the hydrogen pressure increases from very low to high values, but occurs less frequently with a varying hydrocarbon pressure;
- the maximum is shifted to a higher value of hydrogen pressure if
    a) the carbon number in the hydrocarbon increases,
    b) the temperature of the reaction increases,
    c) the metallic dispersion, D, of supported metal catalysts increases.
- normally the location of the maximum in the curves of rate versus hydrogen pressure is different for hydrogenolysis and isomerization ;
- the rate maximum occurring with varying hydrogen pressure is not found at the same hydrogen pressure when different metals are used.

From these experimental results it is clear that the elementary steps in the mechanisms of hydrogenolysis are not obvious on pure metals. The rate-determining step (RDS) may be different under various experimental conditions when using the same hydrocarbon and metal. Figure 5.1 shows the various processes possible, any of which can be the RDS.

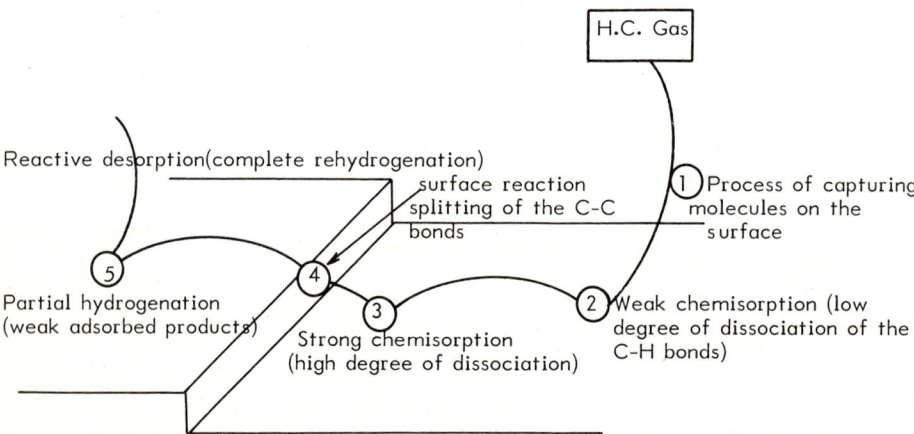

Fig. 5.1 . Elementary steps in the mechanism of hydrogenolysis.

The aim of this chapter is not to examine every kinetic equation proposed for hydrogenolysis, but to indicate that the dependence of the rate on the hydrocarbon and hydrogen pressures may be indicative for a certain mechanism, although is not sufficient to decide which steps is rate determining.

(iii) <u>Kinetic data</u>. A combination of kinetic and tracer studies (ref. 54) may help to solve the reaction mechanism. On the one hand, the use of labelled hydrocarbons allows one to isolate the various elementary steps, on the other hand, a knowledge of their apparent energies and orders, and particularly a comparison between them, may be very useful in order to determine the detailed mechanism.

TABLE 5.2

Kinetic data for isomerization and hydrogenolysis reactions on Pt/Al_2O_3 catalysts (ref. 72)
E_a = activation energy/KJ mol^{-1}
a : Kinetic order with respect to hydrogen.

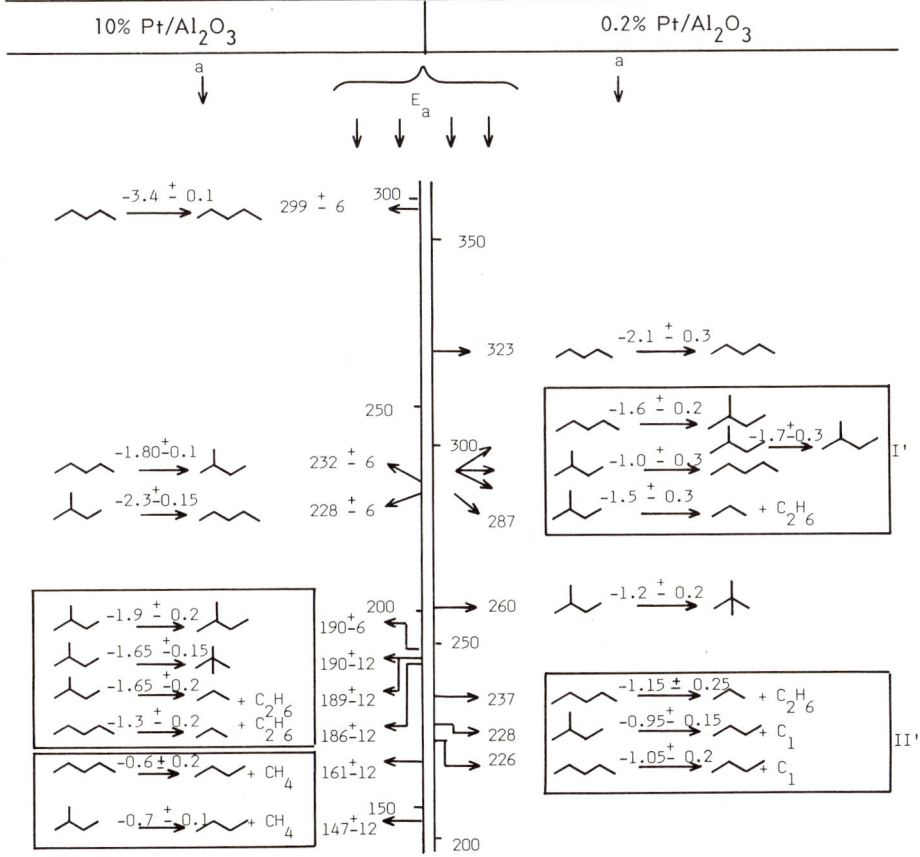

In Table 5.2 are collected the apparent activation energies and kinetic orders with respect to hydrogen for all the various skeletal reactions over 10 % and 0.2 % platinum catalysts. One of the most striking results is the very high negative orders with respect to hydrogen found for all the hydrogenolysis and isomerization reactions. From these results we can see that the cracking reactions can be classified into two groups (refs. 54, 56, 57); this means that two mechanisms are present on the Pt-Al_2O_3 catalysts with low and high metallic dispersion. On the one hand there are two different values of the apparent activation energies for the internal fission reaction and the demethylation reaction ; the latter has lower E_a values. On the other hand, on highly dispersed catalysts, there is a difference between the internal fission for linear and branched hydrocarbons, the latter having a higher E_a values.

5.3.2 Mechanisms of hydrogenolysis reactions on platinum catalysts

(i) <u>Low dispersion catalysts.</u> Since the isomerization of neopentane to isopentane takes place (Table 5.2) on platinum, and not on palladium (ref. 58), a metallacyclobutane species, well known in platinum coordination chemistry, has been proposed as a precursor for the reactions of Group I (ref. 54). This mechanism, shown in Fig. 2, involves a reversible dismutation of the metallacyclobutane into a metallacarbene and π-adsorbed olefin. Its accounts for the following phenomena. First the rotation of the adsorbed olefin results in the observed bond-shift reaction.

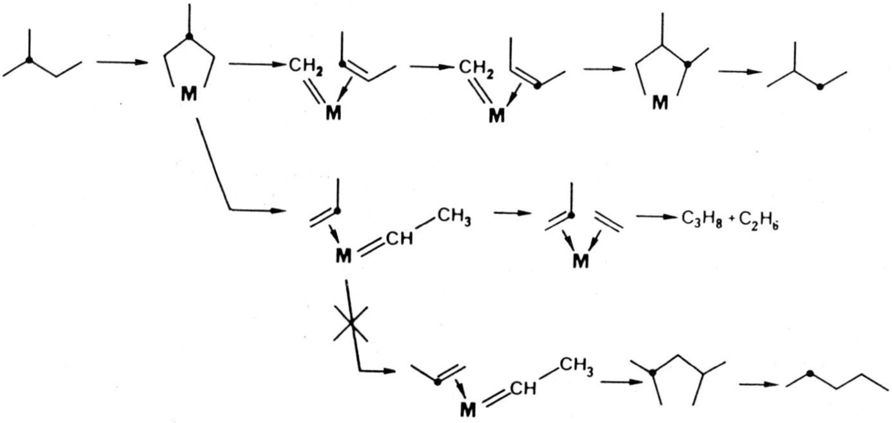

Fig. 5.2. Hydrogenolysis of the internal C-C bond.

Secondly, in the isomerization of isopentane to n-pentane, an adsorbed ethylidene is formed which is rapidly isomerized by hydrogen shift to an adsorbed olefin. The metallacyclobutane cannot then be reformed and isomerization is replaced by hydrogenolysis of the internal C-C bond.

Thus at the same time we can explain why internal fission has the same activation energy as methyl shift, and why chain lengthening does not take place by this mechanism but requires a more activated process. On the other hand, the demethylation reaction has the lowest activation energy.

(ii) <u>Highly dispersed catalysts</u>. The reactions of deethylation of n-pentane and isopentane have different activation energies, which shows that the internal fission reactions on highly dispersed catalysts, do not involve the same intermediate.

If only one metal is involved for the cracking reactions of Group II', the precursor species could be either αα diadsorbed or ααα triadsorbed. These species can be rejected because the first one is very labile and is only involved in exchange reactions as shown by Kemball and co-workers (refs. 59-61), while the second species does not explain the deethylation reaction of n-pentane. If the metal atom is bound, to two carbon atoms we can consider the following species : αβ diadsorbed (π olefinic), αγ diadsorbed (metallacyclobutane), ααβ triadsorbed (πΓ bound) and ααγ triadsorbed (Fig 3). Of these species, all but the last one (the ααγ species) can be rejected. The π olefinic species are associated with exchange reactions and the triadsorbed ααβ species exists only with the binuclear form (ref. 62).The αγ species can be used to explain the Group I reactions on 10 % Pt-Al$_2$O$_3$ catalysts.

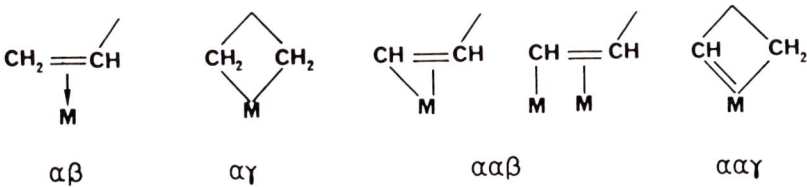

αβ αγ ααβ ααγ

Fig. 5.3. Adsorbed states of hydrocarbons on metal atoms.

We shall not consider the adsorption on two metal atoms, because the 0.2 % Pt-Al$_2$O$_3$ catalyst is mainly constituted of small aggregates and it is known that for the 1-2 diadsorbed intermediate in ethane hydrogenolysis, the activation energy is higher than for C$_3$ or C$_4$ alkanes which is close to that for neopentane for which the 1-2 adsorption mode is impossible (refs. 58, 63).

From these results it is necessary to invoke the presence of a 1-3 adsorbed intermediate for hydrocarbons higher than ethane. Anderson and Avery (ref. 64) have proposed an ααγ triadsorbed intermediate, responsible for bond-shift isomerization, which can lead to hydrogenolysis of the bond β to the carbon diadsorbed to the platinum (Fig. 5.4)

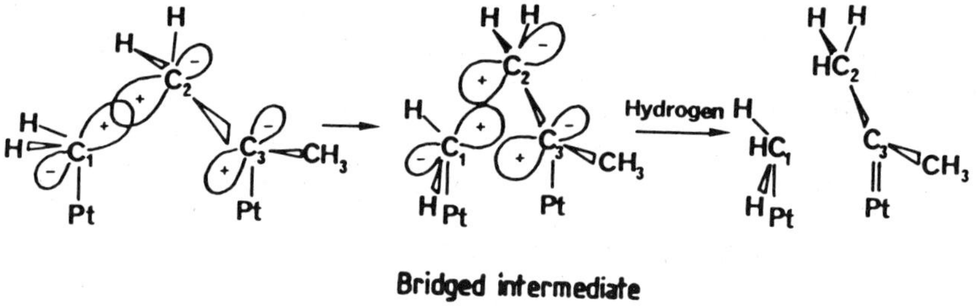

Bridged intermediate

Fig. 5.4. Anderson's cracking mechanism (ref. 65)

In contrast, Leclercq et al (ref. 66) pointed out, that the bond contiguous to the carbon doubly bonded to platinum is most likely to be broken. The 1-3 adsorbed species can be invoked in every case to explain the selectivity for hydrogenolysis of saturated hydrocarbons. We have suggested (ref. 56) that in such a species, the carbon σ-bonded to the platinum would be a tertiary or secondary carbon atom ; such carbon atoms have a partial positive charge (refs. 67, 68), and the intermediate will under go fission by rupture of the bond contiguous to the carbon doubly bonded to the platinum, leading to a π olefinic species and an adsorbed carbyne (Fig. 5.5). This ααγ triadsorbed intermediate accounts for the fact that demethylation and internal fission of n-pentane have the same activation energy as that of isopentane demethylation and explains why the internal fission of isopentane has a higher activation energy (Table 5.2).

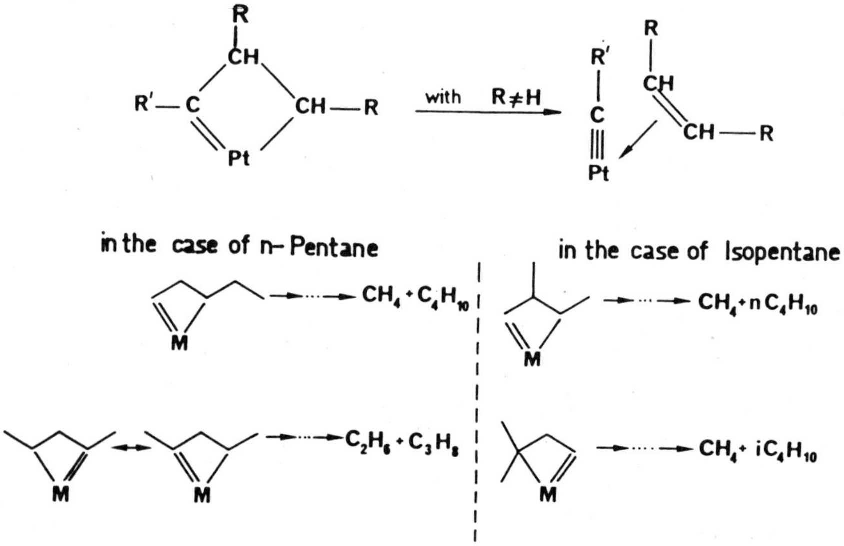

Fig. 5.5. Cracking mechanism involving an ααγ triadsorbed species.

This rapid overview of the mechanisms of hydrogenolysis on platinum catalysts has revealed :
- the influence of the dispersion of the metallic particles,
- the impossibility to separate hydrogenolysis from isomerization since analogous intermediate are found in both reactions.

5.3.3 Hydrogenolysis of C-C bonds on alloys and bimetallic systems

Before discussing the catalytic behaviour of alloys or bimetallic platinum based catalysts we will summarize the results obtained for the hydrogenolysis of different aliphatic hydrocarbons on platinum catalysts.

Maurel and co-workers (ref. 66, 69) have made an extensive study on the influence of the structural effects of saturated hydrocarbons on the rates and selectivities of cracking reactions. They found that the reactivity of a bond depends not only on the substitution of the two carbon atoms involved, but also on the substitution of the neighbouring atoms. Bonds in the β position to a tertiary carbon atom are extensively broken , with a quaternary carbon atom, the hydrogenolysis at the position α to this atom is faster than that at the β position. These very interesting results are summarized in Table 5.3. We have adopted, from ref. 66, the reactivity factor, ω , which is defined as the ratio of the actual rate of rupture of the bond under consideration to the statistical rate of rupture. The latter is obtained by multiplying the total hydrogenolysis rate by the number of bonds identical to those being considered, then dividing by the total number of bonds. It is the rate that would be expected if all the bonds exhibited the same reactivity.

These results cannot be rationalized by a single reaction mechanism, but support the view that many diadsorbed species can play a role in the reaction between hydrogen and hydrocarbons on platinum.

On the other hand there is the influence of the metallic dispersion. In the work done in Maurel's laboratory the metallic platinum particles of the catalyst were well dispersed (D=60%); on the contrary, in Maire's laboratory (ref. 70), the experiments were done on a platinum-black catalyst. The comparison made for 2-methylpentane shows that the deethylation reaction, which is always favoured, increases as the metallic dispersion decreases. In contrast, the demethylation reaction C_{III}-C_I (Number 1 Table 5.3) increases with increasing dispersion.

This very rapid overview of the influence of the hydrocarbon structure indicates that it will be very difficult to isolate the specificity of bimetallic compounds due to the fact that this involves a new parameter, the mutual interaction between the metallic "ensembles".

TABLE 5.3
Reactivity factor for hydrogenolysis of C-C bonds at 300°C. P_{H_2} = 760 torr; P_{Hc} = 5 torr ref. 70 on platinum black; P_{H_2} = 685 torr; P_{Hc} = 75 torr ref. 66 on 2% PtAl$_2$O$_3$.

Hydrocarbons	Brocken bonds	Reactivity ω factors Ref.66	Ref.70	Observations Ref. 70	Hydrocarbons	Brocken bonds	Reactivity ω factors Ref.66	Ref.70
C–C	1				(iso-C₄, br at 1,2)	1	1.15	0.6
C–C–C	1					2	0.45	2.6
(propane, br 1,2)	1	0.9			(iso-C₅, br 1,2)	1	0.7	
	2	1.2				2	1.65	
(butane, br 1,2)	1					1	0.95	0.6
	2					2	0.8	2.5
(hexane)	1	1.1	1.2		(br 3,4,5)	3	0.85	1.0
	2	0.6	0.6			4	1.75	1.3
	3	1.3	1.6			5	0.7	0.3
(iso-C₅, br 1,2,3)	1	0.8	0.6		(br 1,2,3)	1	1.1	
	2	0.5	1.2			2	0.5	
	3	1.9	1.6	5% 0.2% PtAl₂O₃		3	1.6	
(br 1,2,3,4)	1	0.85	0.2	0.55 0.8	(br 1,2,3,4,5)	1		0.5
	2	0.9	1.0	1.45 0.9		2		1.4
	3	1.2	1.8	1.75 1.45		3		2.0
	4	1.15	1.2	0.7 1		4		0.7
						5		0.5
						6		1.4
(br 1,2,3)	1	1.8	1.3		(br 1,2,3)	1		0.6
	2	0.4	0.8			2		0.5
	3	0.6	0.9			3		3.7
(br 1,2,3,4,5)	1		0.7		(neo, br 1,2,3)	1	0.8	
	2		1.0			2	2.2	
	3		1.7			3	0.35	
	4		0.5		(br 1,2,3,4)	1	0.25	
	5		1.4			2	3.7	
						3	0.09	
						4	1.45	
					(br 1,2,3,4)	1	0.35	
						2	3.8	
						3	0.09	
						4	1.05	
					(br 1,2,3)	1	0.7	
						2	2.7	
						3	0.6	
					(br 1,2)	1	0.45	
						2	4.3	

158

(i) Platinum-based catalysts alloyed with inactive components for cracking reactions.

The skeletal rearrangement of hydrocarbons on platinum and other metals has been reviewed (ref. 71, 72) with respect to reaction mechanisms and particle-size surface structure effects. Two basic mechanisms occur for the skeletal isomerization of hydrocarbons on metals :

- The bond-shift mechanism which corresponds to a simple carbon-carbon bond displacement, accounting for the isomerization of short-chain paraffins.

- The cyclic mechanism which involves dehydrocyclization to an adsorbed cyclopentane intermediate, followed by ring cleavage and desorption of the products, and is responsible for the isomerization of larger molecules on dispersed platinum-alumina catalysts.

Concurrently, these molecules may undergo hydrogenolysis, forming C_1 to C_5 products from hexane. An improvement in the knowledge of these destructive reactions can lead to a better selectivity by the addition of a second metallic component (by suppressing hydrogenolysis in favour of skeletal isomerizations).

(a) Pt-Group IB metals (Cu, Ag, Au); Platinum-Copper (ref. 73, 74).

Platinum is alloyed, i.e. diluted with an element (Cu) which itself is many times less active than Pt, but in contrast to Pt-Au (ref. 75), alloying does not lead to a decrease in hydrogenolysis activity, but rather an unexpected increase. The exact surface composition of the Pt-Cu alloys is not known but results obtained in Ponec's laboratory and by Tomanek et al (ref. 76) have shown that the alloy surfaces are only moderately enriched in Cu.

Fig. 5.6 demonstrates the variations of the selectivity with alloy composition. The selectivity of 2-methyl[2-^{13}C]pentane is plotted as a function of alloy composition, at 330°C and with similar extents of conversion (7-9 %).

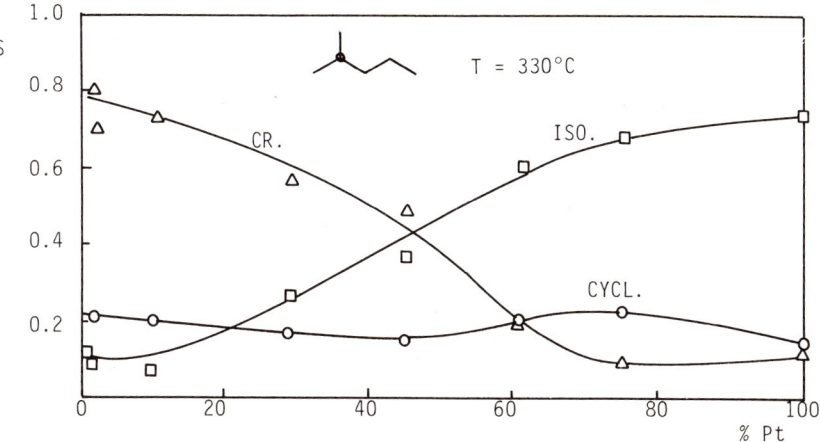

Fig. 5.6. Selectivity as a function of alloy composition (ref. 77) (ISO. isomerization; CR.; hydrogenolytic cracking; CYCL. : dehydrocyclization).

The increase in cracking activity of the Cu-rich alloys ($<$ 45 % Pt) requires an explanation. According to Ponec's argument (ref. 77), cracking seems to be a reaction which, of all the hydrocarbon reforming reactions, needs the largest ensemble of sites and should therefore be the first one to be suppressed (ref. 78). A similar effect has been found with very diluted alloys of Pt, Ir, Pd and Ni with Cu (ref. 78). All the data suggest that Cu plays a definite role in the formation of cracking products.

One can imagine two possible ways in which Cu could participate in the overall reaction : either intermediates are formed partially bound to the Group 8 metal and partially to Cu (mixed ensembles), or Pt (another Group 8 metal) produces a dehydrogenated species which is further converted (hydrogenolyzed) on Cu. The results were explained by an "ensemble size effect" which causes a shift from multisite mechanisms for isomerization (on Platinium-rich alloys) towards one site mechanisms for isomerization (on Copper rich alloys).

On the other hand, Anderson et al (ref. 79) have pointed out that the rates of hydrogenolysis, per unit area of platinum, decrease with increasing crystallite size.
This would suggest a geometric type explanation for the changes in selectivity, the reason being that due to the segregation of Cu at the surface, alloying Pt with Cu is very efficient in producing very small platinium ensembles.

- <u>Platinum -Silver (ref. 80)</u>. This combination was used to try to elucidate the reaction mechanisms and to identify active sites in hydrogenolysis.

The selectivity for the isomeric products is in one way or another related to the electronic structure of the metals and may also be influenced by the particle size of the metals (and alloys), since the isomeric products are all more or less influenced by the coordination of metal atoms. For example, the selectivity for bond-shift and cyclic mechanisms (ref. 71,72) is very sensitive to both alloying and particle size where 1-3 or 1-5 intermediates respectively are involved.

Ponec's laboratory (ref. 80) studied the influence of alloying on the preference of metals for a certain overall reaction (hydrogenolysis, isomerization) as well as for some intermediates ($\alpha\beta$ or $\alpha\gamma$). A suitable molecule to test the preference of metals and alloys for complexes involving either two ($\alpha\beta$) or three ($\alpha\gamma$) carbon atoms is neohexane (2,2-dimethylbutane). The products derived from complexes which are attached to the surfaces always through two carbon atoms (diadsorbed species), involve either two ($\alpha\beta$) or three ($\alpha\gamma$) carbon atoms and one or more metal atoms. Some representative results au collected in Table 5.4.

With regard to the effect of Ag (or Au) on Pt, the conclusion is straightforward addition of Ag (or Au) promotes isomerization and $\alpha\gamma$' adsorption and correspondingly suppresses other modes of adsorption and other reaction pathways.

TABLE 5.4

Typical product distributions for platinum and Pt-Ag and Pt-Au alloy catalysts (ref. 80)

Total metal Loading	Catalyst	T/K	Types of adsorption complexes (C_{2+}, C_{1+}) M			Types of adsorption complexes (C_{1+}) M			M'M
			$\alpha\gamma'$ isom.	$\alpha\gamma'$ hydro.	$\alpha\gamma'$ Tot.	$\alpha\gamma$ isom.	$\alpha\gamma$ hydro.	$\alpha\gamma$ Tot.	$\alpha\beta$
6 %	Pt/SiO$_2$	503	0.37	0.25	0.62	0.17	0.18	0.35	0.03
	Euro-Cat.	523	0.28	0.23	0.51	0.14	0.22	0.36	0.13
12%	Pt 82.3% Ag 17.7% SiO$_2$	665	0.94	0.01	0.95	0.04	0.01	0.05	
16%	Pt 4% Au 96% SiO$_2$	648	0.98	0.01	0.99	0.01	0.01	0.02	
Statistical rondom contribution by complexes					0.43			0.43	0.14

- <u>Platinum-Gold</u>. For this alloy, gold is the segregating element (ref. 76). Alloying with gold leads to a decrease in cracking activity (Table 5.4).

Reactions of hexanes labelled with ^{13}C on alumina-supported platinum-gold alloys have been studied by O'Cinneide and Gault (ref. 81). While the mechanism of hexane isomerization on platinum is mostly "Bond shift", on the 15 % Pt-Au alloy only the "cyclic" mechanism takes place. Similarly, a non-selective ring opening of methylcyclopentane occurs on this alloy, while on platinum only the CH_2-CH_2 secondary-secondary bonds are broken. The 15 % Pt-Au alloy is active only after pretreatment by air and is deactivated in the presence of hydrogen. Activation and deactivation are time-dependent, temperature dependent and reversible. No change in selectivity and mechanism is observed during the activation and deactivation processes.

It was concluded that a high dispersion of the platinum atoms and the presence of oxygen ions in their immediate vicinity are both required to induce the cyclic mechanism of hexane isomerization and the non-selective hydrogenolysis of methylcyclopentane.

The results are summarized in Table 5.5

TABLE 5.5
Reactions of ^{13}C Labelled methylpentanes on Pt and 15% Pt-Au alloy at 300°C (10% of metal on alumina). Relative abundance of the various isotopic species of the isomerized products (ref. 81).

Reactant	Catalyst	2-Methylpentanes			3-Methylpentanes		n-Hexanes		Total conversion %α
 (2-MP)	Pt	-2	43	59	-2	88		14	7.5
 (2-MP)	15 % Pt-Au	3	74	23	9	91		0	3.0
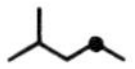 (3-MP)	15 % Pt-Au	-1	-2	103	0	49		51	4.2
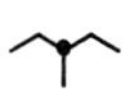 (3-MP 2-¹³C)	15 % Pt-Au	4	94	1		-1	99	2	1.8

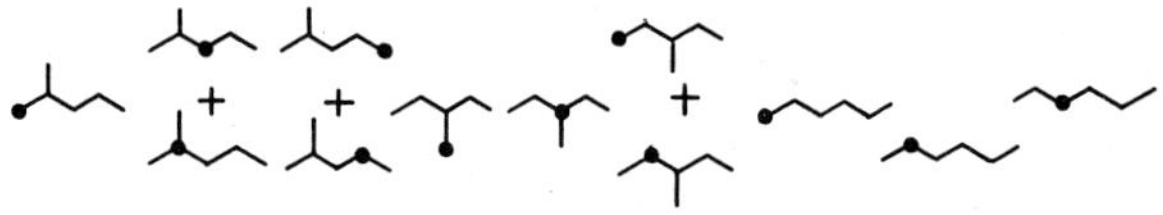

For the conversion of 2-methyl[2-^{13}C] pentane in to 3-methylpentane, the bond-shift mechanism should give 3-methyl[2-^{13}C] pentane, whereas the cyclic mechanism should give 3-methyl [3-^{13}C] pentane.

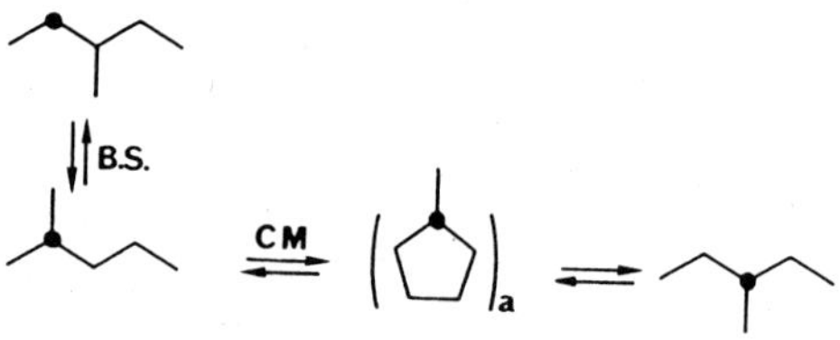

In the conversion of 2-methylpentane in to n-hexane the following reactions occur :

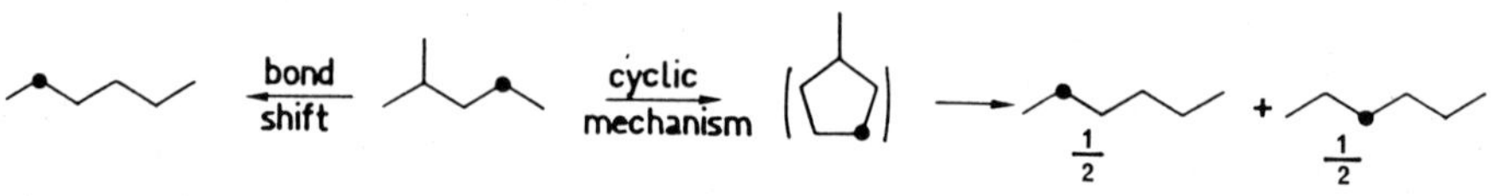

Results for the reactions of unlabelled hexanes are given in Table 5.6 ; it seems clear that the alloys are substantially less active than pure Pt.

TABLE 5.6

Comparison of product distributions for the reactions of 2-methylpentane, n-hexane and methylcyclopentane on Pt and on 15 % Pt-Au alloy (10 % of metal on alumina) (ref. 81).

Reactant	Catalyst	Temp (°C)	α(%)	C_1	C_2	C_3	iso-C_4	C_4	iso-C_5	C_5	cyclo-C_5	2,2-DMB	2,3-DMB	2MP	3MP	n-Hex	MCP	Be/CH	S(%)
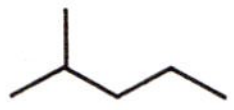 2-methylpentane	Pt	300	8.6	3.0	3.6	8.0	2.8	0.4	1.0	1.8	0.4	0.2	–	–	39.8	13.6	25.4	–	88
2-methylpentane	15%Pt-Au	300	2.65	5.1	4.5	1.7	4.2	–	1.4	3.6	–	–	–	–	14.7	24.8	39.9	–	80
 n-hexane	Pt	300	8.3	3.1	4.1	9.9	0.2	3.6	–	3.2	–	–	2.0	38.2	26.0	–	7.4	2.3	85
n-hexane	15%Pt-Au	300	6.95	2.4	3.3	6.8	–	2.7	–	1.9	0.4	–	1.0	24.0	18.5	–	24.8	14.2	91
 methylcyclopentane	Pt	170	2.2	–	–	–	–	–	–	–	–	–	–	71.3	24.3	4.4	–	–	100
methylcyclopentane	Pt	280	6.5	Tr[a]	–	–	–	–	–	–	Tr	–	–	72.9	23.9	3.3	–	–	100
methylcyclopentane	15%Pt-Au	280	5.55	Tr	–	–	–	–	–	–	Tr	–	–	44.2	19.1	36.6	–	–	100

Concentration (mole %)

a) Tr = Trace quantity.

On the other hand, in these experiments, the cracking reactions on Pt-Au are decreased only for n-hexane ; this is not the case with 2-methylpentane.

Sachtler and Somorjai (ref. 82) have studied the influence of ensemble size on the catalytic conversion of n-hexane by Au-Pt (III) bimetallic single-crystal surfaces. Epitaxial gold adlayers were formed by heating a Pt(III) crystal that was covered with a gold multilayer. Surface alloys were found to be more active than pure Pt (III). Large increases in the isomerization rate of n-hexane and simultaneous exponential decreases in the hydrogenolysis and aromatization rates with increasing gold concentration led to high selectivity for isomerization (Fig 5.7).

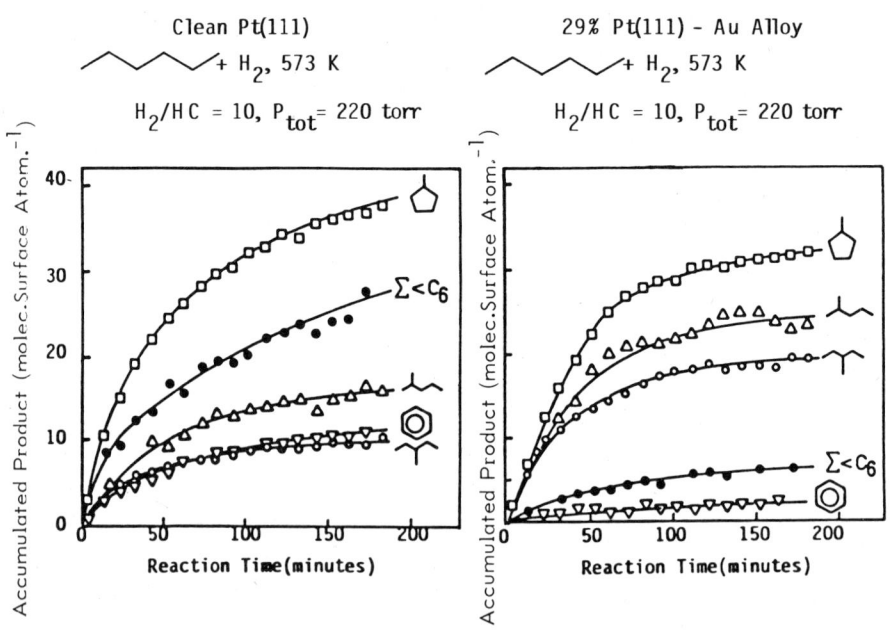

Figure 5.7
Product-accumulation curves for n-hexane reactions. (a) Clean Pt (III) surface. (b). Alloy surface with platinum surface concentration of 29 at. %

These results indicate that : cracking activity is decreased, which is in agreement with the previous works of Ponec and Gault (refs. 80, 81). There is, however, one major difference between the results of Sachtler and Somorjai (ref. 82) and those in (ref. 75). The Au-Pt alloys in the former work having small platinum clusters were always more active than surfaces where the same amount of Pt was arranged in larger ensembles. Van Schaik et al. (ref. 75) found that all reaction rates involving C-C bond rearrangements decreased with increasing concentration of gold.

Foger and Anderson (ref. 83) have studied skeletal reactions of neopentane over Pt-Au/SiO$_2$. The results are given in Table 5.7.

They observed that the selectivity for isomerization passes through a maximum at an intermediate composition. This behaviour is interpreted in terms of two different reaction mechanisms which require different types of active sites. The active sites are identified as different ensembles of surface platinum atoms. Reaction selectivity is influenced by both ensemble availability and the surface hydrogen concentration.

In the platinum-rich range, large platinum surface ensembles are dominant, and the reaction is dominated by the pathway requiring diadsorbed or triadsorbed neopentane. However, as the gold content increases, the concentration of adsorbed hydrogen decreases and the effect is to divert an increasing proportion of the reaction to isomerization, even though the nature of the dominant adsorbed reaction intermediate is unchanged. However, when the gold content becomes sufficiently high, the alternative pathway which has a high intrinsic hydrogenolysis selectivity becomes increasingly important : this becomes dominant at $X_{(s)Pt}$ = 0.17, the mole fraction of platinum in the surface atomic layer. Clearly, at an intermediate value of the surface composition, the hydrogenolysis selectivity will be minimized and the isomerization selectivity maximized, in agreement with Fig. 5.8

TABLE 5.7

Data for Neopentane Reaction (ref. 83)

Catalyst on Degussa 200 Aerosil (~0,9%wt metal)	Reaction temperature (K)	Reaction product ^{a}(%)^b						Isomerization Selectivity ^{c}
		M + E	P	i - B	n - B	i - P	n - P	
Pt 100. Au0	553	12.1	3.7	29.5	4.3	50.4	-	50.4
	593	16.6	4.4	29.6	4.6	42.0	2.8	44.8
Pt 98.Au2	553	8.8	2.0	23.4	1.6	64.2	-	64.2
	593	11.4	5.0	18.6	2.5	56.5	5.9	62.5
Pt 90.Au10	573	5.3	2.6	12.1	-	80.0	-	80.0
	613	9.3	3.5	13.6	-	63.9	9.7	73.5
Pt67.Au33	573	10.0	3.1	20.5	-	60.4	6.0	66.4
	613	15.1	7.4	21.4	-	50.7	5.4	56.1
Pt15.Au85	589	16.5	4.6	41.1	-	37.8	-	37.8
	620	24.2	9.8	44.0	-	22.0	-	22.0

a Primary products at low conversion (0.5-4%): M=methane: E=ethane: P=Propane i-B= isobutane: n-B=n-Butane: i-P=isopentane: n-P=n-pentane : b Expressed as mol % of parent converted in to indicated product. c mol % of parent reacting to C_5 products.

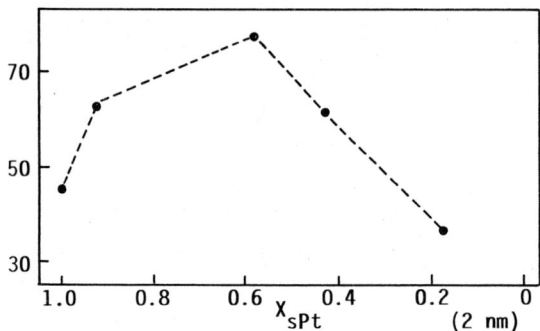

Fig. 5.8. Fraction of neopentane reacting by isomerization as a function of the mole fraction of the surface platinum, $(X_{(S)Pt})$, in a Pt-Au Aerosil catalyst. Reaction temperature 593 (ref. 83).

(b)<u>Platinum-tin</u>. The Pt-Sn system is complex below 1000 K. At equilibrium, as mentioned by Moss (ref. 73), it consists of the so-called α solution with up to ca. 8% Sn and stable intermetallic compounds, Pt_3 Sn, Pt Sn, etc. In the Pt Sn film catalysts of Karpinski and Clarke (ref. 84), with 9 and 31% Sn, X-ray diffraction revealed (α solution + Pt_3 Sn) and (Pt_3 Sn + Pt Sn) respectively. For reactions involving n-hexane and n-pentane in an excess of hydrogen, hydrogenolysis decreased with increasing tin content in the temperature range 593-673 K.

The isomerization of ^{13}C-labelled hexanes on 10% Pt-X% Sn catalysts supported on alumina has shown that, depending on the tin content, the catalytic results may be divided into three groups (ref. 85).
- At low tin content, up to an atomic ratio Pt/Sn of 7, the only apparent effect of tin is to increase the platinum dispersion. The reaction mechanisms are the same as the ones expected on monometallic platinum catalysts of similar dispersion.
- At Pt/Sn ratios ranging from 7 to 2.5 the dispersion remains constant although the H/Pt ratio decreases significantly (from 0.6 to 0.25). Similar effects have been found in the cases of Pt-Au (ref. 81) and Pt-Cu (ref. 77) and it was concluded that part of the tin is alloyed to platinum in the catalysts.
- Further addition of tin blocks the active sites. Fig. 5.9 shows the marked decrease in specific activity for the isomerization of 2MP at Pt-Sn ratios of around 3 for the series A (catalysts calcined at 600°C for 2 h in air after tin has impregnated the alumina and before deposition of platinum salt), and 7 for series B whereas the specific activity increased sharply at low Pt/Sn ratios.

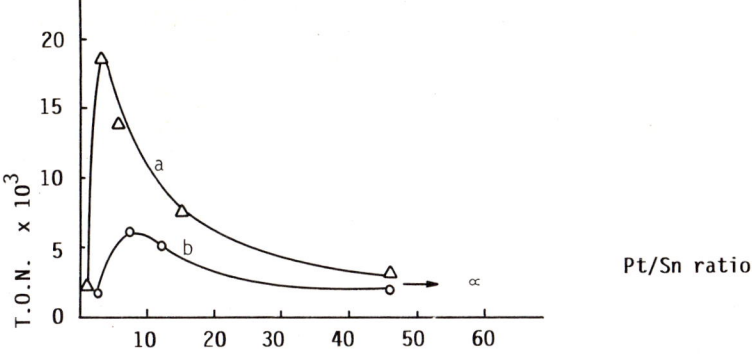

Figure 5.9
Specific activity isomerization at 2-Pt/Sn ratio, a) △ 10% Pt-X % Sn-A, b) ○ 10% Pt-X % Sn-B (ref. 85)

A study undertaken by Burch (ref. 86) on Pt-Sn reforming catalysts with 0.3 wt.% Pt and 0.3 wt.% tin on alumina led to the following conclusions about the oxidation state of tin and the interaction between platinum and tin :

- Sn on alumina is stabilized by the alumina, but some reduction can occur.
- Pt catalyses the reduction of Sn, but the lowest oxidation state is Sn(II).
- Pt adsorbs more hydrogen when Sn is present.
- No proper alloys of Pt and Sn are formed, so this cannot account for changes in catalytic properties.
- The special properties of Pt-Sn catalysts cannot be due to a geometric effect in which tin atoms divide the surface into very small clusters of Platinium atoms.
- The special properties of Pt-Sn catalysts must be due to a change in the electronic properties of small Platinium crystallites, either by interaction with a tin (II) ion stabilized on the alumina to give electron-deficient Pt, or by incorporation of a few percent metallic Sn as a solid solution in Pt to give electron-rich Pt.

Catalysts containing 1% Pt and 0.06-4% Sn on Al_2O_3 have been studied by Coq and Figueras (ref. 87) for the conversion of methylcyclopentane. The results obtained were similar to those shown in Fig. 5.9. In Fig. 5.10 are plotted the partial conversions measured at 673 K for different products, as a function of the tin content. Hydrogenolysis passes through a maximum at Pt/Sn = 20, aromatization at Pt/Sn = 5, expressed in weight ratios.

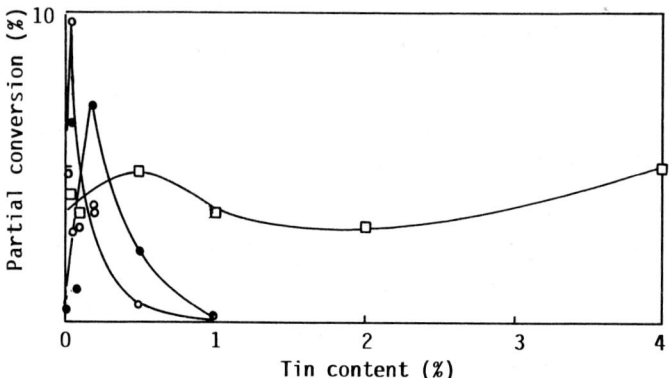

Fig.5.10. Partial conversion of MCP as a function of the catalyst tin content : ring opening (O), MCP^{2-} (□), and Bz (●) ; temp. 673 K. (ref. 87).

The authors concluded that the addition of tin to platinum decreases the hydrogenolysis. Coke and sulphur have similar effects. The main role of tin is to dilute the platinum surface. An enhancement of aromatization was observed, which passes through a maximum and then decreases to zero as a function of the tin content. Dehydrogenation of methylcyclopentane to methylcyclopentene remains unchanged. These changes in selectivity were attributed by the authors to the modification of the Pt-C bond strength induced by the changes in electronic density at platinum atoms.

c) <u>Platinum-Lead</u>. The conversion of n-heptane on Pt-Pb/Al$_2$O$_3$ has been studied by Völter et al. (ref. 88 see also Chapter 10) at a pressure of 0.1 M Pa. The influence of the second metal can be seen from the activity/time curves for a pure Platinium catalyst and for a Pt-Pb catalyst in Fig. 5.11.

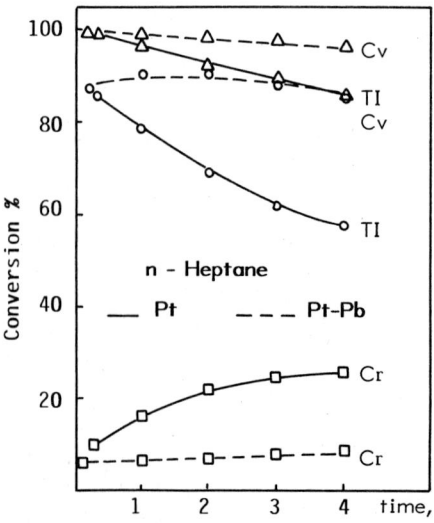

Fig. 5.11. Activity/time curves for the conversion of n-heptane at 0.1 MPa and 773 K on a bimetallic catalysts (0.35 wt.% Pt ; 0.37 wt. % Pb, broken lines) and on a platinum catalyst (0.35 wt. % Pt ; solid lines). Cv = conversion ; Tl = toluene ; Cr = cracking products. (ref. 84)

The bimetal effect can be characterized as follows. 1) The extent of conversion decreases less rapidly. The deactivation is retarded. 2) The content of toluene in the product is considerably higher. The total aromatization is increased by a higher selectivity and a higher extent of conversion. 3) The cracking is inhibited. On the pure platinum catalyst, many cracking products from C_1 to C_6 are formed and only small amounts of C_7 olefins and isomerization products. On the bimetallic catalyst the total amount of all by products is distinctly lower.

The authors concluded that additions of Pb or Sn have similar effects. It is generally agreed that in the interaction between Platinum and tin (ref. 86) or carbon (ref. 89) the electron donation occurs from the modifier to platinum. On Pt-Pb (ref.90), the infrared band of linearly adsorbed CO is significantly shifted (60 cm^{-1}) to lower frequencies, demonstrating the occurrence of electron donation from Pb to Pt.

ii) - Platinum-based catalysts alloyed with active components for cracking reactions

a) Pt - Group 6 metal ; - Platinum-Molybdenum The conversion of neopentane on Pt + Mo/SiO_2 supported catalysts obtained from organometallic compounds of Platinum and Molybdenum has been studied by Kuznetsov et al. (ref. 91). The turnover numbers for hydrogenolysis of neopentane were higher than those reported for Pt/SiO_2. This is true in spite of the fact that Mo/SiO_2 prepared in the same way as the bimetallic catalyst was found to be inactive with respect to neopentane.

Tri et al. (ref. 92) have studied bimetallic Pt-Mo catalysts supported on Y-zeolite.These catalysts are prepared in a different way to those studied by Kuznetsov et al (ref. 91). The molybdenum atoms deposited on the platinum act as adsorption sites for the hydrocarbon, whereas the platinum atoms dissociate hydrogen. These catalysts exhibit enhanced hydrogenolysis activity in n-butane conversion. Tri et al (ref. 92) have suggested that the atoms of platinum and molybdenum act primarily as adsorption sites for n-C_4 and hydrogen respectively. The metal atoms have different adsorption properties in the Pt-Mo aggregates. In n-butane conversion, the curve of activity versus Pt/Mo composition is volcano-shaped with a maximum near the 1 : 1 ratio (Fig. 5.12). The rate at the maximum is 7 and 34 times larger than on Pt Na HY and Pt/SiO_2 respectively.

Leclercq et al. (ref. 93) have studied the properties of platinum-molybdenum bimetallic catalysts deposited on silica. They observed that hydrogen chemisorption decreases when the molybdenum proportion increases. Figure 5.13 shows the variations in the activity of the catalysts. A strong synergistic effect is observed.

The authors proposed three hypotheses to explain these observations :
- an electronic modification of Pt by Mo resulting in an increased activity of this metal.
- a mixed site comprising both Pt and Mo could be more active than a site made from in atoms of only one metal.

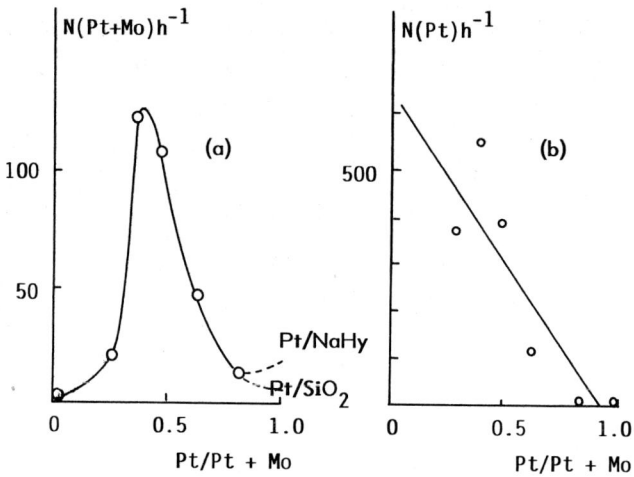

Fig. 5.12. Influence of molybdenum content on the total hydrogenolysis rate in n-butane conversion. (a) Turnover calculated from the total number of atoms : N(Pt + Mo) (b) Turnover calculated from the number of platinum atoms chemisorbing hydrogen : N(Pt). Data on Pt Na H y and Pt/SiO$_2$ are given for comparison. (ref. 92)

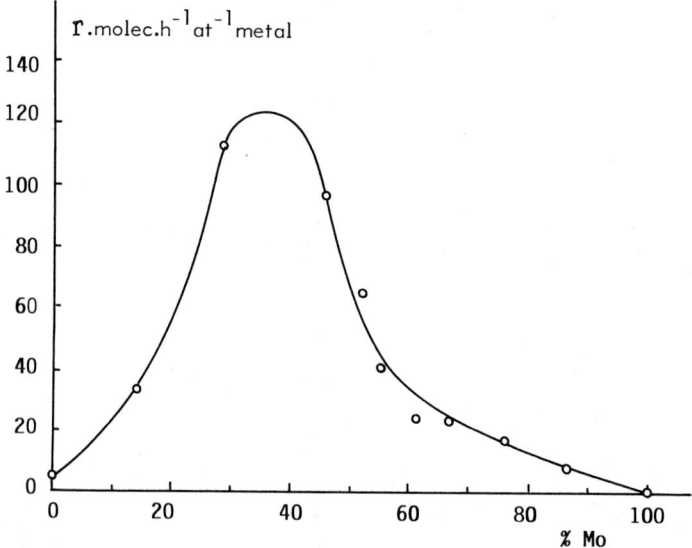

Fig. 5.13. Rate of butane hydrogenolysis vs. relative atomic percentage, of Mo. T = 300°C, $P_{C_4H_{10}}$ = 0.1 atm, P_{H_2} = 0.9 atm (ref. 93).

- as Mo is not totally reduced to Mo^0 it is possible that, in the presence of Pt at the surface, the extent of reduction of Mo would be increased leading to catalytically active Mo.

Similar results have been obtained by Yermakov et al. (ref. 94) for the hydrogenolysis of ethane on supported $(Mo + Pt)/SiO_2$ catalysts. These catalysts exhibit a much higher activity, a constant activity in a series of consecutive experiments and a lower activation energy as compared to hydrogenolysis on Pt/SiO_2 as catalyst. The results were interpreted in terms of a change in the electronic properties of platinum due to its interaction with molybdenum on the support surface.

Experiments done, in a flow system, on $Pt-Mo/SiO_2$ prepared by Leclercq et al (ref. 93) with 2-methylpentane and methylcyclopentane as starting hydrocarbons have shown the very high selectivity of 75 atom.% Mo and 50 atom. % Mo for cracking Table 5.8 (ref. 95).

TABLE 5.8

2-Methylpentane isomerization and hydrogenolysis of methylcyclopentane a 250°C (ref. 95).

Catalysts on SiO_2	2-Methylpentane isomerization					Methylcyclopentane hydrogenolysis		
Degussa 3wt.%metal	Selectivity for isomers	rate $\mu\ell S^{-1}$ V x 10^{-2}	⋎	$iC_{5/C5}$	$iC_{4/4C}$ cracked products	V x 10^2	⋏	⋎
75 % Mo	68.7	0.161	17.6	2.7	34.5	0.21	3.3	3.0
50 % Mo	45.0	0.237	17.3	9.5	85.5	0.430	15.5	4.4
14 % Mo	30.8	1.465	0.7	0.6	25.9	26.73	1.3	0.38
100 % Pt	61.4	1.910	0.6	0.5	18.2	9.01	1.1	0.39

- <u>Platinum-Tungsten</u>. Kuznetsov et al. (ref. 91) have studied the contact reaction of neopentane. They observed that the bimetallic sample exhibited a decreased selectivity towards isomerization but increased activity towards hydrogenolysis when compared with a sample containing only platinum. The catalytic results were quite similar to those obtained with $(Pt+Mo)/SiO_2$. The authors asked the question : why are the bimetallic catalysts (Pt+Mo and Pt+W) more active in hydrogenolysis than Pt/SiO_2 ? Their answer was as follows : for the hydrogenolysis of hydrocarbons on metals, the most likely rate-determining process is the rupture of a C-C bond in an adsorbed partially dehydrogenated hydrocarbon.

The formation of the latter is inhibited by hydrogen. Hence a reduction in the ability of the surface to adsorb hydrogen could give rise to an enhanced activity for hydrogenolysis. In their work, no inhibition by hydrogen was found for hydrogenolysis on the bimetallic catalysts at any value of the ratio P_{H_2}/P_{HC} for (Pt+W)/SiO$_2$ and at sufficiently large values of P_{H_2}/P_{HC} for (Pt+Mo)/SiO$_2$.

b) <u>Pt - Group 7 metal; Platinum-Rhenium</u>. This bimetallic reforming catalyst was first introduced by Chevron in February 1967 (ref. 96). The physical and chemical nature of the rhenium in the catalyst has been considered by various investigators (ref. 97-101). Whether or not a highly dispersed alloy or a bimetallic cluster of platinum and rhenium is present on the alumina is a matter of some interest.

Betizeau et al (ref. 99) observed that the curves of activity as a function of the composition of Pt-Re/Al$_2$O$_3$ catalysts exhibit maxima at 65 atom. % of Re in cyclopentane hydrogenolysis and at 85 atom.% Re in n-butane hydrogenolysis.

Tournayan et al. (ref. 102) have studied the conversion of n-heptane in hydrogen at 400°C, over six series of 2 wt.% (Pt-Re)/Al$_2$O$_3$ catalysts reduced at 500, 700 or 900°C. It appears that the state of these catalysts is sensitive to the conditions employed in their preparation, Fig. 5.14. This introduces considerable uncertainty in to projections about the nature of the catalyst under actual reforming conditions.

Approaches to the understanding of the enhanced activity of Pt-Re catalysts have been made by Sachtler (ref. 27) and Coughlin et al. (ref. 28). In the first paper, the enhanced activity of Pt-Re/Al$_2$O$_3$ catalysts relative to Pt/Al$_2$O$_3$ is shown to be caused not by a reduction in the rate of coke formation, but by changes in the nature of this coke. In the second paper, the authors mentioned that the metal crystallites are activated by hydrocracking at the kinked steps. The carbene and carbyne fragments formed by hydrogenolysis at the kinks combine to produce a disordered hydrocarbonaceous layer which grows outward from the edges and across the terraces. This layer deactivates the kinked cracking sites to a significant extent. The presence of Re at the kink sites increases their hydrogenolysis selectivity towards gaseous rather than layer-reforming products; thereby this Re somewhat retards the rate of formation of the carbonaceous deposits.

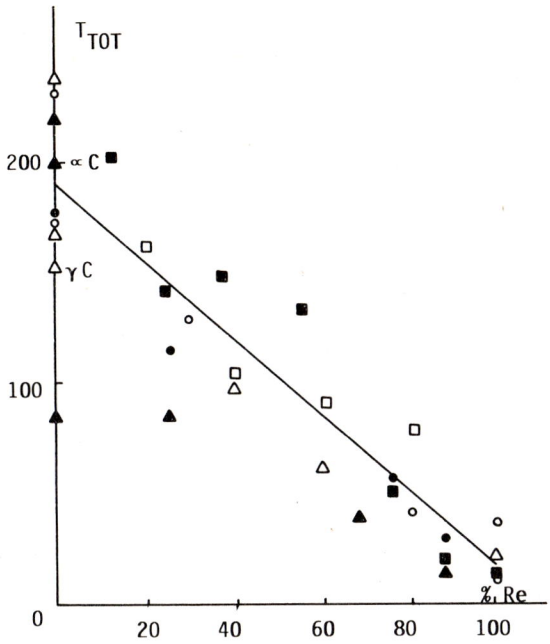

Fig. 5.14. Turnover Number per exposed metal atom (Pt_s + Re_s) in the overall conversion of n-heptane over (Pt + Re)/Al_2O_3. Ordinate : number of n-C_7H_{16} molecules converted per hour per metal atom exposed. Abscissa : relative percentage of Re versus (Pt + Re). Open symbols are for the γ-Al_2O_3-supported catalysts, closed symbols for the α-Al_2O_3-supported ones. The temperature of reduction was 500 (triangles), 700 (squares), or 900°C, (circles). γC and αC refer to 2% Pt/ γ -or α-Al_2O_3 catalysts calcined in air before reduction (ref. 102).

c) Pt-Group 8 metal ; Platinum-Ruthenium. No isomerization occurs on ruthenium supported on alumina in a flow system (ref. 103), Table 5.9. It is very significant that ruthenium does not affect the carbon-carbon bond where one carbon atom is tertiary ; when the starting hydrocarbon is 3-methylpentane it is easier to obtain isobutane formed by two consecutive C_{II}-C_I bond ruptures than to obtain n-butane from one C_{III}-C_{II} bond rupture. Such very high selectivity for cracking is also observed when the starting hydrocarbon is methylcyclopentane where a small amount of n-hexane is formed.

TABLE 5.9

Contact reaction of C_6 alkanes on 10 % and 0.2 % Ru and Alumina catalyst (Ref. 103).

Starting HC	%Ru on Al_2O_3	Red. temp. (°C) a)	Reac. temp. (°C) b)	Conver. % c)	Selecti. % d)	C_1	C_2	C_3	iC_4	nC_4	iC_5	nC_5	2-MP	3-MP	n-H	MCP	$\frac{iC_4}{nC_4}$	$\frac{iC_5}{nC_5}$	$\frac{3-MP}{n-H}$	$\frac{2-MP}{3-MP}$
2-MP	10	400	160	23	0.4	35	7.3	5.4	26.1	0.9	23	1.6	–	0.3	0.09	–	29	14.3		
3-MP	10	400	160	26	1.0	47	8.3	1.7	7.2	1.3	32.6	0.5	0.9	–	0.1	–	5.5	65.2		
N-H	10	400	160	46	0.3	50.7	22.7	9.8	0.03	8.7	0.05	8.2	0.09	0.2	–	–	–	–		
3-MP	0.2	400	160	12	1.2	40	5.2	1.6	10.2	1.6	39.7	0.5	1.2	–	–	–	6.4	79.4		
MCP	0.2	400	160	1.3	47	24	4.7	2.3	7.5	1.4	12	1.6	24	18	5	–	–	–	3.6	1.3

a) Reduction temperature °C
b) Reaction temperature °C
c) Conversion %
d) Selectivity %

If we use the nomenclature of Anderson (ref. 104) for the cracking reactions, one can define :

- The C_2 - unit mode reactions where only primary and secondary carbon atoms are involved. Bond types C_I-C_I, C_I-C_{II}, $C_{II}-C_{II}$.
- The Iso - unit mode reactions where tertiary and quaternary carbon atoms are involved. Bond types C_I-C_{III}, C_I-C_{IV}, $C_{II}-C_{III}$, $C_{II}-C_{IV}$.

On ruthenium, the ratio $P_1 = \frac{\text{iso-mode}}{C_2\text{-mode}}$ for 2-methylpentane and $P_2 = \frac{\text{iso-mode}}{C_2\text{-mode}}$ for 3-methylpentane at 160°C are very similar, $P_1 \sim P_2 \sim 0.05$, which means that it is 20 times more difficult to break a C-C bond where a tertiary carbon atom is involved.

What is the cracking pattern on the Pt-Ru catalysts ? For the cyclopentane hydrogenolysis, only a small increase was observed on Pt-Ru/Al_2O_3 catalysts compared to Pt or Ru; the turnover number is twice as large on the bimetallic catalysts (ref. 105). The authors interpreted this small change in activity in terms of the presence of a compensation effect due to a change in the adsorption strength of hydrogen upon alloying.

Blanchard et al. (ref. 106) have studied the conversion of n-heptane. Over Pt-Ru/SiO_2 (T= 380°C) and Pt-Ru/Al_2O_3 (T= 400°C), the activities and selectivities varied only marginally with time. The turnover number first decreases with increasing ruthenium content and then remains relatively constant. The 50 at.% Ru alloys are roughly half as active as Pt. The variations in the selectivites with the composition are shown in Fig. 5.15 for (Pt, Ru)/SiO_2 and in Fig. 5.16 for (Pt-Ru)/Al_2O_3.

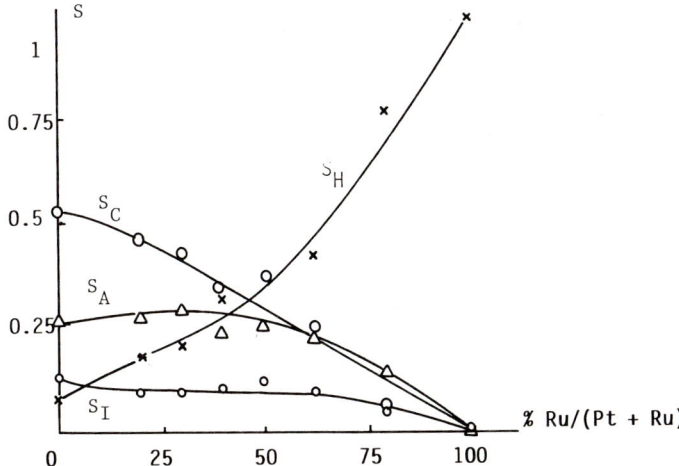

Fig. 5.15. Conversion of n-heptane on (Pt.Ru)/SiO_2 catalyst. Selectivities : A. aromatization (toluene); C. cyclization(ethyl. and dimethylcylopentane) ; 1. isomerization ; H. hydrogenolysis. (T = 380°C) (ref. 106).

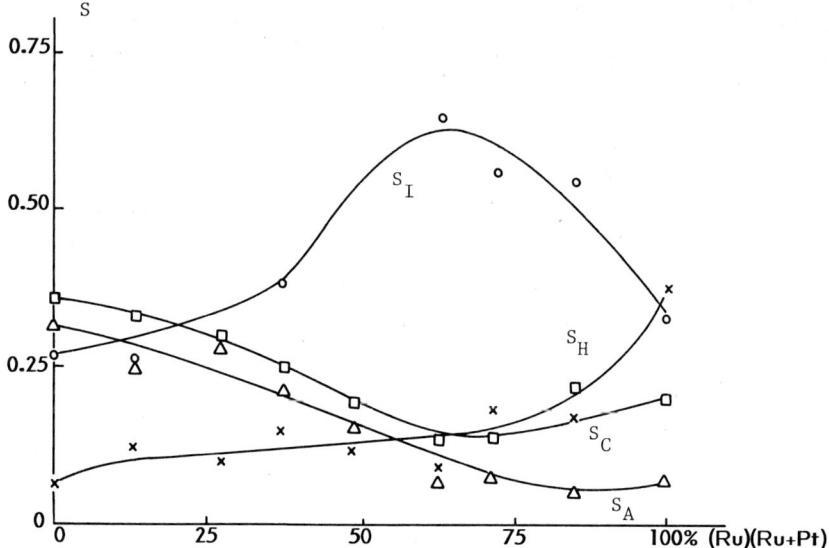

Fig. 5.16. Conversion of n-heptane on (Pt-Ru)/Al_2O_3 catalyst. Selectivities : A. aromatization (toluene) ; C. cyclisation (ethyl-and dimethylcyclopentane) ; I. isomerization ; H. hydrogenolysis (T = 400°C) (ref. 106).

The two catalysts Ru/SiO_2 and Ru/Al_2O_3 differ in their degree of reduction (Ru is converted into Ru° to the extents of 100% and 55% respectively), in their dispersion of Ru° (20 % and 80 % respectively) and in the acidic properties of the support.

The authors discussed their results in terms of the similarities between Pt-Ru and Pt-Re and invoked the possible influence of electron transfer, as suggested earlier between Pt and Re (ref. 107).

Contact reactions on 10 wt.% Pt 96 at.%-Ru 4 at.% /Al_2O_3 have been studied between 220 and 300°C (ref. 108). Skeletal rearrangements proceed from 220°C (where Pt is inactive for this type of reaction). The results of the hydrocracking reactions of 2-methylpentane, carried out at low extents of conversion to avoid repetitive processes, are given in Table 5.10. Several points can be emphasized :

- The selectivity increases with increasing reaction temperature.
- The cracking pattern seems to be very sensitive to the structural morphology of the metal crystallites in the catalyst. Deethylation is the main reaction until 300°C, in contrast to the case on pure platinum catalysts (ref. 72). It is seen that the amount of (C_2 + C_4) molecules is higher than that of (C_1 + C_5) up to 260°C, but at higher

TABLE 5.10

Hydrocracking and Isomerization of 2-Methylpentane on 0.02 of Pt-Ru/Al$_2$O$_3$ Catalyst as a Function of Temperature. (Low Conversion) (ref. 108).

Temp (°C)	α_T^a (%)	S^b (%)	Products (mole %)						iC$_4$/nC$_4$	iC$_5$/nC$_5$	3MP/n-H
			2C$_3$	C$_2$+C$_4$	C$_1$+C$_5$	3MP	n-H	MCP			
200	0.2	17.5	17.2	42.3	23	–	–	17.5	36	3.1	–
240	0.48	35	11.1	31.7	20.7	3.8	2.8	28.3	42	2.6	1.4
260	2.8	49.5	5.9	25	20.5	17.5	12.0	20	25	2.6	1.45
270	2.6	66	2.6	15.6	15.4	24.5	18.2	23.6	22	1.4	1.3
280	4.1	67	3.5	14.3	15.2	21.3	25.4	20.2	13	1.4	0.8
290	8.5	69	3.3	13.0	14.7	23.5	30.3	14.9	6	1.4	0.77
300	8.7	74	3.1	9.8	13.3	24.2	33.3	15.8	4.8	1.12	0.72

a) Total conversion
b) Selectivity

temperatures the statistical value, i.e. $3(C_2 + C_4) = (C_1 + C_5)$, is never reached. No repetitive processes are present, as confirmed by the high value of the ratio isobutane/n-butane.

- The ratio isopentane/n-pentane is always higher than the theoretical value of 0.5; it seems that the predominance of deethylation and the selective rupture of the C_{II}-C_I bond in the molecule are specific to the Pt-Ru/Al_2O_3 catalyst.

Experiments were also made at high extents of conversion (30%). The deethylation process was again predominant and a very high isopentane/n-pentane ratio was also observed.

Diaz (ref. 103) has also studied the contact reaction of 2-methylpentane over various Pt-Ru catalysts ; the results are summarized in Table 5.11. The selectivity for isomers is very low. Extensive cracking does not occur, and deethylation predominates. The very high values of the ratios iC_4/nC_4 show that there are no repetitive processes. Carbon - carbon bond rupture involving a tertiary carbon atom is always disfavoured, the ratio iC_5/nC_5 being higher than 3.

These results show the selectivity of these catalysts in hydrocracking reactions, whatever the atomic compositions.

- <u>Platinum-Cobalt</u>. Little work has been reported on this system. The catalytic behaviour of this bimetallic in hydrocracking reactions has been studied by Zyade et al. (107). Two series of 10 wt.% (Pt-Co)/Al_2O_3 catalysts were used, one reduced at 450°C for 24 h and one reduced at 700°C for 6 h.

First if we look at the behaviour of cobalt, the selectivity for isomers in the contact reaction of alkanes is nearly zero whatever the reduction temperature. Methane represents about 85% of all the cracked products.

In Table 5.12 are reported the results of the contact reactions of 2-methylpentane on Pt-Co/Al_2O_3 reduced at 700°C. First it was observed (ref. 110) that the bimetallic catalysts reduced at 700°C are more active than those reduced at 450°C, in contrast to the monometallic catalysts. Secondly, the demethylation reaction involving primary-secondary carbon bonds predominates, the ratio iC_5/nC_5 being 0.9.

These results can be generalized to the hydrocracking reactions of 3-methylpentane where the ratio iC_5/nC_5 is always higher than 2, and to the methylcyclopentane hydrogenolysis where the ratio 3MP/n-H is higher than 2, which means that tertiary carbon atoms are not affected in the cracking processes.

Taking into account the high selectivity for the demethylation and hydrogenolysis the authors have suggested the intervention of adsorbed alkynes as intermediates species (ref. 111).

TABLE 5.11

Contact reaction of 2-methylpentane at 220°C on various Pt-Ru on alumina catalysts

Pt/Ru atomic ratio	%Pt at.%	%Ru at.%	F/ω	αT	S	6C₁	3C₂	2C₃	C_2+C_4	C_1+C_5	M3P	n-H	MCP	$\frac{iC_4}{n-C_4}$	$\frac{iC_5}{nC_5}$	$\frac{\text{iso mode}}{C_2 \text{ mode}}$
25	9.8* (9 6)	0.2* (4)	80	0.2	17.5	-	-	17.2	42.3	23	ε	ε	17.5	36	3.1	0.38
4.6	9.0 (8 2)	1.0 (1 8)	275	13	5.4	-	-	7.1	62.1	25.3	2.4	1.9	1.04	22	3.1	0.17
2.1	8.0 (5 8)	2.0 (3 2)	385	33	3.5	0.6	-	7.4	66.8	21.6	2.3	0.97	0.3	20	6.5	0.12
0.7	6.0 (4 4)	4.0 (5 6)	770	26	3.1	8.5	3.0	7.7	57.3	20.2	1.9	1.03	0.2	15	6.5	0.14
0.34	4.0 (2 5)	6.0 (7 5)	900	50	1.2	8.8	0.6	9.6	63.4	16.3	0.8	0.4	0.05	18	7.8	0.15
		0.2 (100)	125	2.5	36.5	11	-	5.2	34	12.6	-	-	36.6	14	5	0.17

* Weight percent ; the atomic percentages are mentioned in brackets

TABLE 5.12

Contact reaction of 2-methylpentane at 300°C on Pt-Co/Al$_2$O$_3$ reduced at 700°C.

S : Selectivity in isomers

Activity expressed in $\mu\ell$ (Sec. g)$^{-1}$

iso-mode/C$_2$ mode as defined by Anderson (ref. 104).

Atomic Ratio Pt/Co	Pt At.%	Activity x 10^3	%S	6 C$_1$	3 C$_2$	2 C$_3$	C$_2$+C$_4$	C$_1$+C$_5$	3-MP	n-Hex	MCP	iC$_4$/nC$_4$	iC$_5$/nC$_5$	iso mode / C$_2$ mode
4	80	285	81	0	0	4.5	4.5	10	40.5	29.5	10.5	2.7	0.9	1.2
1.5	60	16	60	0	0	7.5	13	19.5	19	21.5	20	9.1	1	0.8
1*	50	20	63	1	1	4.5	14.5	15.5	18.5	12	33	12.1	1.2	0.5
	0	195	0	96.5	2.5	1								
	100	12	84	0	0	2.5	4	10	25.5	20	38	16	0.5	1.5

* Reaction temperature : 320°C

These species will undergo a cleavage reaction and produce surface carbynes, and can explain the unreactivity of the tertiary carbon atom.

From work done by De Kock et al. (ref. 112) on a model for the irreversible adsorption of alkyne in homogeneous catalysis, is was suggested that the basic driving force for the reaction between the metal surface and the alkyne should be the negative transfer to the hydrocarbon electron flow. As the transition metals used have lower electronegativities than carbon, this seems quite reasonable.

Finally, in 1980 Fritsch and Vollhardt (ref. 113) synthesized a bis (carbyne) cluster by the direct cleavage of alkynes with a cobalt cluster. These findings would support the above suggestion.

<u>Platinum-Rhodium</u>. Rhodium is predominantly a hydrogenolysis catalyst (ref. 65) with no detectable bond shift isomerization activity, although in a dual functional form it is said to have a good dehydrocyclization activity (ref. 114). It forms $\alpha\alpha$ intermediates very readily (ref. 115) but appears to lose this ability in rhodium-platinum alloys (ref. 116).

Karpinski and Clarke (ref. 84) have studied reactions of alkanes on Pt-Rh alloys to determine whether the catalytic character of the component elements is retained or not. Synergistic improvement of the activity in isomerization of n-pentane (ref. 117) has been reported for this alloy.

Patents have described the use of Rh-Pt in naphtha reforming because of its high paraffin aromatization activity (ref. 118).

TABLE 5.13

Product distribution in n-pentane hydrogenolysis on Pt-Rh films (ref. 84)

Film composition. at. % Rh	temp./K	methane (wt. %)	ethane (wt. %)	propane (wt. %)	n-butane (wt. %)
100	497	86.5	9	-	4.5
	533	86	10	-	4
91	497	42	30	11	17
	533	53	25	3	19
62	497	30.5	31	2	36.5
	533	32	31	0.5	36.5
20	497	23.5	33	20	23.5
	533	38	38	3	21
0	497	50	50	-	trace
	533	38	54	-	8

Karpinski and Clarke (ref. 84) found a supralinear decrease in the n-hexane hydrogenolysis selectivity of rhodium when it was diluted with platinum, which may be an indication that the reaction is associated with an ensemble of several contiguous rhodium sites. In contrast to platinum, no detectable methylcyclopentane was formed on rhodium films (up to 673 K), only benzene and products of hydrogenolysis. The reaction of n-pentane on 100% Rh gave only hydrogenolysis products, not cyclopentane or isopentane. Experiments on n-pentane conversion (Table 5.13) have been carried out at lower temperatures (497-533 K) because the rate of hydrogenolysis is much greater than that for n-hexane.

More recently, the reactions of n-butane and 2,2-dimethylpropane on silica-supported Rh-Pt bimetallic catalysts have been studied (ref. 119). The rates of reaction of 2,2-dimethylpropane at 538 K and n-butane at 465 K determined by the pulse system on the 100 series of catalysts (the precursor salt is $RhCl_3$) are shown in Fig. 5.17.

That of 2,2-dimethylpropane was not significantly reduced by increasing the Platinum content of the catalysts up to 40%, but thereafter fell by more than a factor of 400; with n-butane the rate declined by a factor of 10^4 from Rh to Pt and again the decrease was more marked at the platinum-rich end of the series.

Data on the selectivities of the reaction in the pulse reactor are shown in Fig.5.18. Some isomerization of 2,2-dimethylpropane to 2-methylbutane was observed on all catalysts, but the amounts rose dramatically on the catalysts containing more than 90%. Pt. With n-butane the main reaction on most of the catalysts was central-bond cleavage to form ethane, and the percentage of propane in the product resulting from terminal-bond cleavage rose slowly as the platinum content was increased, but again the major change occurred at more than 90% Pt.

The authors found no evidence for surface enrichment of the catalysts from the chimisorption of H_2 and CO, in agreement with results of Wang and Schmidt (ref. 120). They suggested that the greater variations in catalytic activity, in static or flow reactors for the n-butane hydrogenolysis, of 4×10^3 and 10^4 molecules $s^{-1}site^{-1}$, respectivily compared with the somewhat lower factors of 200 and 400, respectively for the reaction of 2,2-dimethylpropane, may be associated with different mechanisms for the two reactions.

Foger and Anderson (ref. 51) have distinguished two modes of hydrogenolysis on iridium catalysts, the C_2-unit and the iso-unit modes. The former has a lower activation energy but requires a larger ensemble of atoms to form the active site whereas the latter has a higher activation energy but may occur on a single iridium atom.

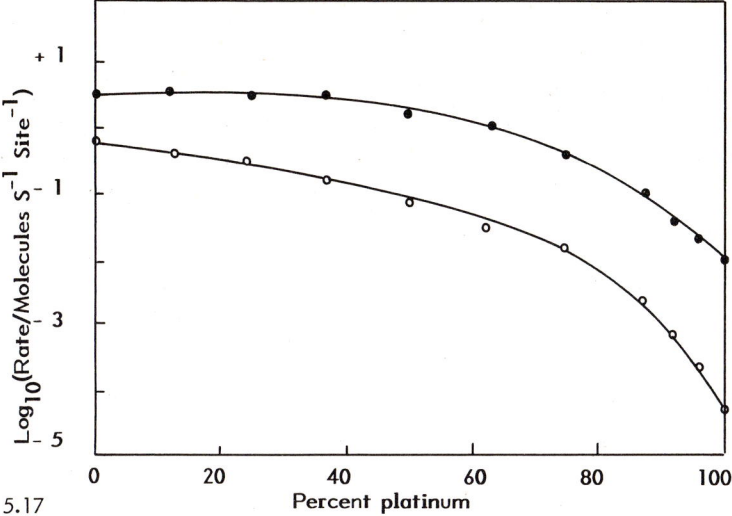

Fig. 5.17
Rates of reaction determined by the pulse reactor : ●, reaction of 2,2-dimethylpropane at 538 K ; ○, reaction fo n-butane at 465 K. (ref. 119).

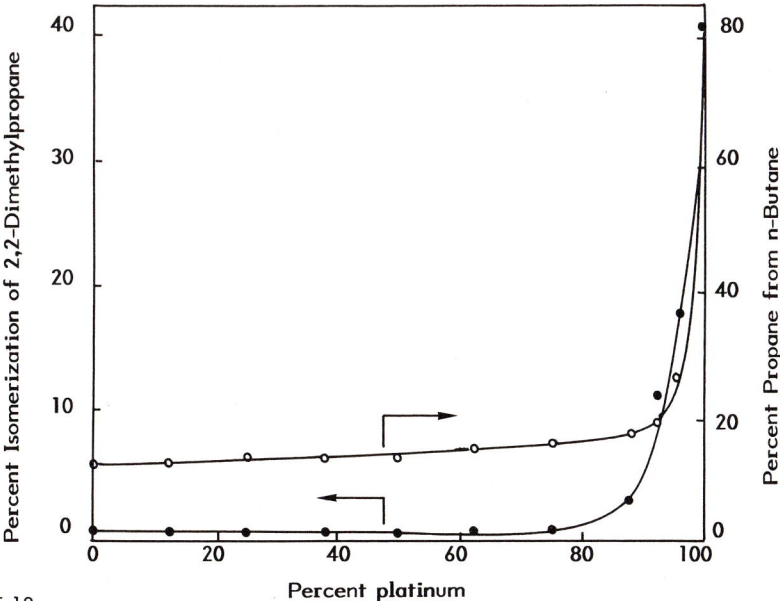

Fig. 5.18
Selectivities determined by the pulse reactor : ●, percentage isomerization of 2,2-dimethylpropane to 2-methylbutane at 538 K ; ○, percentage propane from n-butane at 465 K. (ref. 119).

The preferential cleavage of the central carbon-carbon bond in n-butane is evidence in support of a C_2-unit mechanism of hydrogenolysis for this reactant because the rate of this reaction is expected to be greater with a pair of secondary carbon atoms than with a primary and a secondary carbon atom. The marked decrease in activity, particularly with catalysts containing more than 85% Pt, suggest that Pt is not a very effective catalyst for the C_2-unit mechanism and also that an ensemble incorporating two or more rhodium atoms may be required. A curious feature of the selectivity results for n-butane is that preferential central-bond cleavage is not appreciably influenced by the platinum content until it exceeds 85- 90%. Thus the authors have evidence that over a composition range from pure Rh down to only 15 to 10% Rhodium the character of the reaction does not change, although the activity decreases by factors of more than 100. In general terms such results could be interpreted as the C_2-unit mechanism requiring an ensemble of, say, four metal atoms and ensembles consisting of only rhodium atoms having greater activity than mixed ensembles containing one or more platinum atoms. In this way the character of the reaction would remain rhodium-like over a considerable range of composition, while its rate substantially decreased.

Because of the nature of the molecule, the hydrogenolysis of 2,2-dimethylpropane cannot involve a C_2-unit mechanism but must occur by the iso-unit mode. Platinum is apparently a relatively more active catalyst for this type of mechanism than for the C_2-unit mechanism, although it is still much less active than Rh. The consequence is that the overall decrease in activity with catalyst composition is less dramatic with 2,2-dimethylpropane than with n-butane, and the decrease for the platinum-rich catalysts is less marked.

On the other hand, the main reaction on most of the catalysts is the cleavage of one carbon-carbon bond to give a molecule of 2-methylpropane and a molecule of methane. There is also some multiple carbon-carbon bond cleavage because the production of methane is always greater than that of the C_4 molecule. This type of reaction is of increased importance with catalysts containin > 50% Rh and is particularly marked for the pure rhodium catalysts. Put another way, the methane to 2-methylpropane ratio is 1.1 on the pure-platinum catalysts and rises sharply on the rhodium-rich catalysts to 2.1 on the pure rhodium catalysts in the 500 series (The precursor salt is $Rh(NO_3)_3$, $2H_2O$) as shown in Table 5.14.

The authors believe that a probable source of the ethane and propane is the cleavage of a carbon-carbon bond in an isomerized C_5 molecule on the catalyst surface.

TABLE 5.14

Selectivities for Reactions in the Static Reactor on the 500 Series of Catalysts (ref. 84).

Catalyst	(Rh/Pt)	Molecules product/molecule reactant used						
		2,2-Dimethylpropane at 463 K				n-Butane at 398 K		
		CH_4	C_2H_6	C_3H_8	C_4H_{10}[a]	CH_4	C_2H_6	C_3H_8
500	(100/0)	1.42	0.15	0.17	0.69	0.23	1.60	0.18
510	(90/10)	1.17	0.08	0.08	0.71	0.28	1.48	0.24
520	(80/20)	1.16	0.09	0.07	0.81	0.28	1.51	0.24
530	(60/40)	1.14	0.07	0.06	0.85	0.26	1.54	0.22
540	(40/60)	1.03	0.07	0.05	0.86	0.24	1.56	0.24
550	(20/80)	0.98	0.06	0.04	0.91	0.23	1.58	0.23
560	(15/85)	0.96	0.07	0.04	0.91	0.20	1.61	0.18
570	(10/90)	0.97	0.08	0.04	0.91			
580	(5/95)	0.92	0.10	0.05	0.88[b]			
590	(0/100)	0.85	0.02	0.04	0.82[d]	0.67[d]	0.67[d]	0.67[d]

a 2-Methylpropane
b Also 0.03 2-methylbutane
c Also 0.08 2-methylbutane and 0.05 n-butane
d Extrapolated from results at higher temperatures.

- Platinum-Nickel. Small amounts of platinum incorporated in alumina-supported nickel catalysts have been observed by Nowak and Koros (ref. 121), to enhance activation of the nickel by hydrogen. Hydrogenolysis reactions of n-heptane with hydrogen were used to characterize the catalytic activity. Platinum to nickel atomic ratios as low as 5×10^{-3} were sufficient to cause several-fold increases in the extent of conversion compared with the nickel component. These large effects appear to be due to precious metal catalyzed hydrogen reduction of supported nickel oxide.

The catalytic activity of alumina-supported platinum-nickel alloys has been studied by Renouprez et al. (ref. 122). A strong maximum in the turnover for the isomerization of neopentane to isopentane was observed in the composition range (Pt 80-90 - Ni 20-10). Similarly, the rate of hydrogenolysis of neopentane is increased by two orders of magnitude for the composition $Ni_{90}Pt_{10}$. This synergistic effect was attributed to modifications of the electronic properties of both metals at the two critical compositions.

Dominguez et al. (ref. 122) have studied the structure and the selectivity of graphite-supported Pt-Ni alloys using neopentane as a test molecule (Table 5.15). The distribution of the reaction products is quite different from one catalyst to an other, mainly due to the fact that cracking reactions occur on Ni-rich alloys. Dominguez (ref. 123) concluded that the structure of these platinum-rich alloys, (i.e., $Pt_{78}-Ni_{22}/C$) is comprised mainly of cubo-octahedral particles and of a smaller proportion of irregular, triangular and square shapes. Nickel-rich alloys, (i.e., $Pt_{12}-Ni_{88}/C$), exhibit many irregular shapes and a lower proportion of special and five fold decahedral and icosahedral shapes.

Skeletal rearrangements of ^{13}C-labelled hexanes on Pt-Ni bulk alloys have been studied by Aeiyach et al. (ref. 124). The three Pt-Ni alloys with 40, 60 and 70 at.% Ni bulk concentrations were prepared by high-temperature melting of mixtures of the elements. The bulk compositions and homogeneity were verified by X-ray diffraction and electron microprobe analysis. Studies on Pt-Ni alloys have shown (ref. 31) that clean surfaces are enriched in platinum by an amount which increases with increasing platinum concentration in the bulk : with 40, 60 and 70 at.% of Ni in the bulk, the surface nickel concentrations are 20, 50 and 70% respectively. The same trend has also been observed by Jugnet et al. (ref. 125) and Massardier et al. (ref. 126) with single crystals, although in these cases the platinum enrichment was even larger.

The experiments were performed in a Varian LEED chamber with four-grid optics. The Auger detector, used to monitor the surface composition, was retarding field analyzer (RFA). The catalytic reactions were carried out in an isolation cell housed within the main UHV chamber which remained under ultra-high vacuum during the reactions carried out up to atmospheric pressure as described in ref. 127.

TABLE 5.15

Neopentane hydrogenolysis on Pt-Ni catalysts. Distribution of the products obtained by Renouprez et al. (ref. 121), Dominguez et al (ref. 123).

Product	Pt (at. %) 0		10		23		38		50		70		84	95	100	
	a	c	a	c	a	c	a	c	a	c	a	c	a	a	a	c
C_1	96	50.5	97	45.3	66	72.8	54	52.1	51	40.2	46	41.1	43	38	-	43
C_2	-	23.7	-	16	21	10.7	22	16.3	23.5	18.9	14	19.3	17	15.5	-	13.5
C_3	2.4	14.6	1.1	16.4	10	6.6	16	20.1	14.5	23.3	4.4	5.9	12.4	12.3	-	12.3
C_4	1.8	10.9	1.6	22.1	3	9.83	7.5	11.2	11	17.4	35	33.5	28	35	-	31.1
Hydr.rate[b]	9	0.38	750	4.5	190	7	7	2.6	1.3	3.17	1.5	5.3	11	3	0	7.6
Isom.rate[b]	0	0	0	0	0	0	1	1.09	1.5	4.02	2.4	18.2	45	21	1	51.8
Selectivity % in isom.	0	0	0	0	0	0	13	29.8	54	55.9	62	77.4	80	88	100	87.2

a : Results obtained from ref. 122 at 350°C. The hydrogen to hydrocarbon ratio equals 10. The total conversion is lower than 3%; γ alumina (Degussa C. 180 m² g⁻¹) was coimpregnated with aqueous solutions of H_2PtCl_6 and $NiCl_2$. Homogeneous solid solutions were obtained after a reduction of 15 h at 1400 K in flowing hydrogen followed by homogenization of 6h at 1200 K under static conditions

b : Hydrogenolysis rate and isomerization rate expressed as 10^8 mol s⁻¹ m⁻² for the column a and for the column c specific activities in 10^8 mol s⁻¹ g⁻¹

c : Results obtained from ref. 123 at 300°C. H_2/neo. = 10. Supported platinum, nickel and platinum alloys were prepared by deposition from organic solutions ($H_2PtCl_6 \cdot 6H_2O$ and $NiCl_2 \cdot 6H_2O$). The substrate was a graphitized carbon LONZA-LTIO with 99.9% of carbon, a BET surface area of 18m² g⁻¹. Reduction at 850°C in pure H_2.

TABLE 5.16

Product distributions (in moles) from the isomerization of 2-methylpentane on Pt-Ni bulk alloys at 350°C 1 h (ref. 124).

Catalysts[a]	%α_T[b] (mol)	%S[c]	6C_1	3C_2	C_1+C_5	C_2+C_4	2C_3	3-MP	n-H	MCP	Bz	i-B/n-B	n-P/i-P
%Ni_v	%Ni_s												
0	0	1.6	69.4	0	6.3	13.6	10.7	14.8	10.8	42.7	1.1	4.4	1.5
40	20	3.5	62.5	0.8	0.4	8.9	15.9	14.9	12.1	33.4	2.1	3	1.4
60	50	0.7	57.7	12.2	1.8	10.2	10.4	7.7	8.4	5.3	40	4	2.4	1.3
70	70	2.2	38.9	17.0	5.4	11.2	14.2	13.3	6.3	3.5	17.4	11.7	1.2	1
100	100	26	0.9	89	7.4	0.2	0.8	1.7	0.4	0.2	0.3		3.3	1.2

a% Ni_v: at.% Ni bulk; % Ni_s: at.% Ni surface. b% $α_T$ moles: overall conversion per cm^2 of bulk alloys. c% S: selectivity defined as the percentage (in moles) of C_6 isomers in the reaction products. 3-MP: 3-methylpentane, n-H: n-hexane, MCP: methylcyclopentane, Bz: benzene, i-B: isobutane, n-B: n-butane, i-P: isopentane, n-P: n-pentane.

TABLE 5.17

Product distributions (in moles) from methylcyclopentane hydrogenolysis at 350°C, 1 h on bulk alloys (ref. 124).

Catalysts	%α_T (mol)	%S	% Σcrack	2-MP	3-MP	n-H	Bz	3-MP/n-H	2-MP/3-MP	r2-MP/rMCP [a]
% Ni_v	Ni_s									
0	0	2.9	93.1	6.9	50.9	23.4	18.4	0.4	1.3	2.2	0.5
40	20	2.1	86.2	13.8	43	22.4	18.8	2.0	1.2	1.9	1.6
60	50	0.4	74.1	25.9	40.7	18.2	11.3	3.9	1.6	2.2	1.7
70	70	1.2	70.8	29.2	26.3	11.7	9.7	23.1	1.2	2.2	1.8
100	100	5.1	15.9	84.1	7.6	3.6	2.7	2.0	1.3	2.1	5.1

a: r2-MP/rMCP: ratio between the activities for 2-methylpentane and methylcyclopentane.

Reactions of 2-methylpentane (2-MP : 5 torr) were studied at 350°C, 1 h, over the catalysts previously described, under a total pressure of 1 atm. (Table 5.16). The catalytic activity, as measured by the total conversion, α_T, of the Pt-Ni unsupported catalysts varied erratically as a function of the nickel concentration. The selectivity for isomers decreases when the amount of Ni increases. The product distribution shows that extensive cracking occurs only when 50 at.% nickel is present on the surface of the catalyst. From the isomer distribution, it is evident that methylcyclopentane is the major product. The high values of the ratio i-butane/n-butane indicates that no repetitive process occurs, except on the 70 at.% Ni surface where the ratio is equal to 1.2 and a large amount of benzene is present.

Hydrogenolysis of methylcyclopentane has been studied under the same experimental conditions as the 2-methylpentane reaction (Table. 5.17). The lower activity of this cyclic molecule as compared to the acyclic one is shown by the ratio of the rates, r 2-methylcyclopentane/r methylcyclopentane, which is larger than unity when Ni is present on the catalyst. Secondly, on the 70 at.% Ni surface, 23% benzene is formed. The selectivity ratio 3-methylpentane/n-hexane, determined at 350°C, was constant over all the catalysts studied.

TABLE 5.18.
Isomerization of 2-methyl[2^{13}C]pentane ; distribution of the various isotopic species at 350°C on Pt-Ni alloys in a static system. (ref. 124).

Catalysts						Mechanisms		
% Ni$_V$	% Ni$_S$					abnormal	cyclic	bond shift
0	0	0.8	48.3	50.9	4.8	47.5	47.7	
40	20	1.1	51.7	47.9	6.6	50.6	43.5	
60	50	1.7	64.1	34.2	10.2	62.4	27.4	
70	70	7.6	57.2	35.2	45.6	49.6	4.8	

In the isomerization of labelled hexanes (2-methylcyclopentane to 3-methylcyclopentane) (Table 5.18), abnormally labelled 3-methylpentane is formed. On the three catalysts with 0, 20 and 50 at.% surface Ni, the amount of 3-^{13}C methyl pentane is low and does not greatly affect the cyclic contribution, as measured by the amount of 3-methyl 3-^{13}C pentane. Howerer, on the 70 at. % nickel surface, the abnormal mechanism is as important as the cyclic mechanism. To confirm the presence of this repetitive reaction, the authors carried out further experiments with 3-methylpentane and n-hexane. They observed, whatever the starting molecule, that the catalysts with 70 at. % surface Ni stimulates the formation of benzene.

To understand why benzene formation is increased on this alloy Pt_{30}-Ni_{70} two remarks seem relevant :

- This catalyst has a large excess of nickel at the surface, and it is well known that this metal alone leads to successive C-C bond ruptures before desorbing CH_4 in large amounts.

- In the Pt-Ni alloys, these successive rearrangements, i.e., repetitive isomerization reactions, are suppressed for the superficial compositions Pt_{80}-Ni_{20} and Pt_{50}-Ni_{50}. The authors suggested that the platinum blocks the readsorption sites, either by decreasing the "Ni ensemble size" or by electronic modification. In this case they think that the desorption rate is faster than the successive rearrangement rate in the adsorbed phase.

They consider that an electronic effect may account for this phenomenon. Renouprez et al. (ref. 122) have proposed that in the range Pt_{15-25} the density of states at the Fermi level is considerably increased for platinum. Such modifications could favour a strong coupling at the adsorbates, as an important electron transfer from the 3d band of nickel to unoccupied 5d states in platinum occurs.

From these results we can easily understand why the catalytic results obtained by Renouprez (ref. 122) or Dominguez (ref. 123) are not very reproducible if a new reaction occurs.

(iii) <u>Platinum-based catalyst with active components for isomerization reaction ;</u> <u>Platinum - Palladium</u>(refs. 128.130.) The reactions of neopentane, n-pentane and n-hexane in the presence of an excess of hydrogen have been studied on evaporated films of Pt-Pd alloys as catalysts (ref. 128). The selectivity for neopentane isomerization on a pure palladium film was found to be non negligible, in contrast with previous work. A possible explanation, involving carbonization of the surface during the experiments and the effect of texture on the process was proposed. The neopentane isomerization versus alloy composition was interpreted in terms of an "inferred" surface composition. The results may easily be explained assuming a one-platinum-site mechanism for isomerization and a two-palladium-site mechanism for hydrogenolysis. The 1,5 -cyclization selectivity pattern in the n-pentane reaction was interpreted in terms of a change in the number of active sites. In the n-hexane conversion the ratio of the selectivity for 1,6 to that for 1,5-cyclization is higher on alloys than for pure metals.

Isomerization, cyclization and hydrogenolysis reactions of n-pentane and n-hexane have been studied on silica-supported Pt-Pd alloy catalysts (ref. 129). From the results of the kinetic studies, it can be inferred that the surface composition of highly dispersed Pt-Pd/SiO_2 catalysts is not substantially different from the bulk composition. The selectivity for n-pentane cyclization is higher for the Pt-Pd alloys than for the pure metals.

This synergistic effect was not seen for the Pt-Pd films, which is due to the different surface compositions of the samples examined. Self-poisoning of the Pt-Pd SiO_2 catalysts was observed, the cyclization and hydrogenolysis reactions being the most suppressed, whereas the isomerization process was relatively insensitive to this effect. For the n-hexane conversion the major reactions were isomerization and hydrogenolysis, whereas dehydrocyclization products were formed in minor amounts. It was suggested (without proof) that the n-hexane isomerization on Pt-Pd/SiO_2 catalysts occurs primarily by the cyclic mechanism.

The effects of hydrogen in the conversion of saturated C_6 hydrocarbons have been studied over Pt-Pd/SiO_2 catalysts (ref. 130). The activities and selectivities of a series of Pt-Pd/SiO_2 catalysts have been determined for the conversion of n-hexane, 3-methylpentane and methylcyclopentane . Saturated C_6 products predominate at high hydrogen pressures, whereas benzene and olefin prevail under hydrogen-deficient conditions. The overall activity exhibits a minimum for alloys with about 50 at.% Pd ; at the same time, the isomerization selectivity (at the expense of C_5 cyclization and hydrogenolysis) is maximal at similar compositions. On the basis of the hydrogen dependence of the yields as well as of the isomer ratios obtained from alkanes and methylcyclopentane, possible reaction mechanisms were suggested. These are in agreement with data published earlier for both pure metals and alloys. The contribution of the bond-shift mechanism is important especially for 3-methylpentane isomerization. This is supported also by the formation of 2,2-dimethylbutane over each catalyst. The bond shift as well as the C_5 cyclic mechanism of isomerization requires less dissociated surface species, as distinct from the dehydroisomerization producing benzene from 3-methylpentane and methylcyclopentane. Possible ensemble and hydrogen effects were discussed

Isomerization of ^{13}C-labelled hexanes is a very sensitive chemical probe in the study of mono- or multimetallic catalysts. It has been used to study Pt_x-Pd_{1-x} alloy catalysts (ref. 131). A series of six alumina supported catalysts with various percentages of Pd was prepared and carefully characterized by X-ray diffraction and transmission electron microscopy (T.E.M.). All these catalysts are true alloys comprising large aggregates (150Å $\leqslant$ d $\leqslant$ 500 Å). From the relative rates of the demethylation and internal fission reactions of methylpentanes at 300°C, a correlation was established with the oxidation mechanisms (ref. 132) and the electronic structure of the atoms forming the active sites. The active sites for the demethylation and internal fission reactions are different from those responsible for isomerization or ring opening.

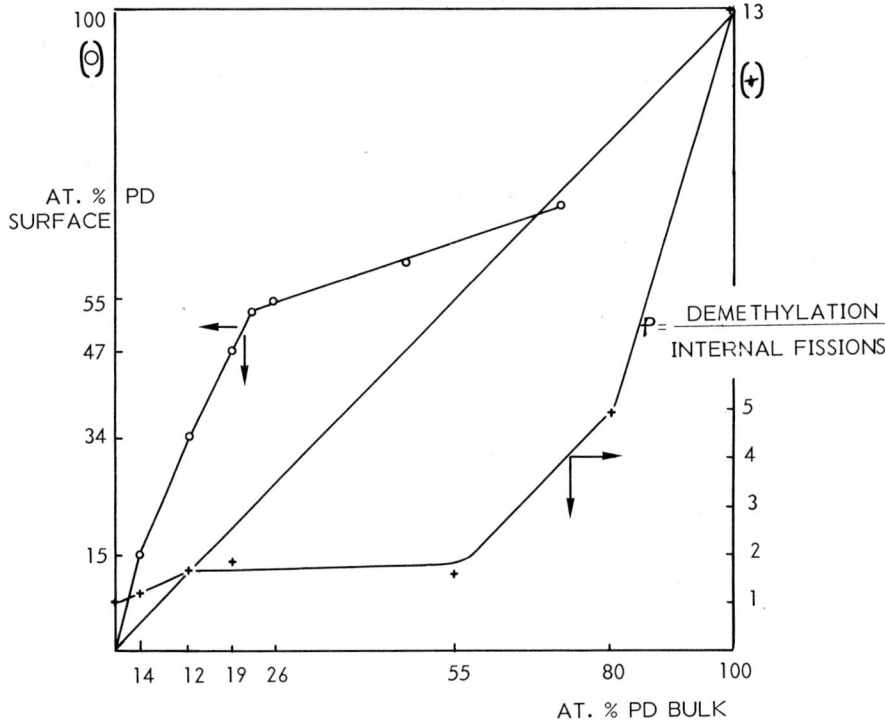

Fig. 5.18
Dependence of the ratio ρ = demethylation/internal fissions for hydrocracking of 2-MP and 3-MP on the 10% Pd bulk (+) and dependence of the % Pd surface on the % Pd bulk (O). Ref. 131

Fig. 5.18 from ref. 131 shows the percentage of surface palladium versus the percentage of palladium in the bulk, according to reference 132, for $Pt_x Pd_{1-x}$ bulk alloys. The average values of the ratio f = demethylation/internal fission obtained for the hydrocracking reactions of 2-and 3 - methylpentanes are also plotted versus the percentage of palladium as measured by X-Ray diffraction. At first sight the "character" of platinum remains unchanged up to bulk compositions higher than 60% of palladium for the hydrogenolysis of methylpentanes. On the other hand, a synergistic effect appeared for the cyclic mechanism on an alloy with 47 % surface Pd.

Finally, an inhibition effect of Pd on the activity could be related to the existence of ensembles of 7 ± 1 adjacent platinum atoms. Electronic and geometric factors govern parallel reactions (hydrocracking and isomerization) on different active sites.

- <u>Platinum-Iridium</u> The structural and catalytic properties of Pt-Al$_2$O$_3$ and Pt-Ir-Al$_2$O$_3$ have been compared by Ramaswamy et al. (ref. 133). For both catalysts, the mechanism and relative rates of sintering of the metal crystallites in hydrogen are similar. Pt-Ir-Al$_2$O$_3$ is less active in dehydrogenation reactions compared to Pt-Al$_2$O$_3$. In the presence of n-heptane, the deposition of "coke" is lower on the bimetallic catalyst under identical conditions. Pt-Ir-Al$_2$O$_3$ is more active in hydrogenolysis reactions. When the catalysts were presulphided at 523 K, a selective, permanent deactivation of the hydrogenolysis activity is observed on Pt-Ir-Al$_2$O$_3$. It was suggested that the "dilution" of Pt by an element with a relatively lower dehydrogenation activity, viz., Ir, leads to lower surface concentrations of "coke" precursors on Pt-Ir-Al$_2$O$_3$ and hence to reduced fouling rates. However, it possesses a higher hydrogenolysis activity, thereby necessitating presulphidation to poison selectively the sites responsible for that activity.

Rice and Lu (ref. 134) have studied the role of iridium in an alumina-supported platinum-iridium bimetallic catalyst by examining the activity/selectivity behaviour of platinum, iridium and platinum-iridium catalysts using n-heptane reforming as a test reaction at 135 and 790 KPa. The iridium imparted superior dehydrocyclization activity and deactivation resistance to the bimetallic catalyst, relative to platinum. There appeared to be considerable synergism between platinum and iridium which resulted in suppression of surface coke formation. However, the bimetallic, even when presulphided, exhibited an undesirable higher hydrogenolysis activity, particularly at higher pressures. Comparison of results for the bimetallic catalyst and a mechanical mixture of platinum and iridium catalysts provided indirect evidence for the existence of bimetallic clusters on the platinum-iridium catalyst.

The characterization of platinum-iridium reforming catalysts and their performance has been undertaken by Rasser et al (ref. 135). In view of the limited insight into the factors which play a part in improving the action of iridium in reforming obtained with platinum-iridium supported on α-alumina, a study was made of the texture and composition of a number of catalysts with various Ir/Pt ratios. Using Auger electron spectroscopy, an analysis was made of the surface composition of a number of platinum-iridium alloy powders, before and after contact with propane. n-Heptane and n-hexane reforming were studied as model reactions, and the influence of the Ir/Pt ratio on the isomerization, cyclization and aromatization selectivities was determined. The surface of the alloy powders was found to be strongly enriched in platinum ; in accordance with the platinum-like behaviour of the catalysts. The main function of iridium appears to be the suppression of surface carbiding.

CONCLUDING REMARKS.

In this short survey of hydrogenolysis on platinum-based alloys we have shown how the adjunction of a second metal may drastically change the catalytic behaviour of platinum, depending on the nature of the element added and on the method of preparation. Of all the parameters involved, the importance of surface segregation has been stressed. Another possible cause of unexpected results, not yet mentioned is the modification of the chemical reactivity due to alloying when oxidation-reduction treatments are involved in the preparation.

A photoemission technique has been used to study the surface composition of Pt-Pd, Pt-Ru and Pt-Ir alloys over a wide range of bulk compositions (ref. 132). The samples were exposed to oxygen in-situ at atmospheric pressure and various temperatures in the range 200 to 600°C. Oxygen-induced changes in surface composition were monitored. Special attention was paid to the various oxidation states of the metals characterized by shifts towards higher binding energies of the core levels. The oxidation properties of platinum and also of Pd, Ru and Ir were drastically changed by alloying. Under the same experimental conditions, pure platinum was inactive, that is no oxidation took place. The platinum-based alloys could be oxidized provided the impurity (i.e., Pd, Ru or Ir), surface concentration was higher than between 13 % and 20 % for all three systems. Above this limit, the impurity and the platinum atoms were both oxidized. Three different oxidation states of Pt were detected and identified with PtO, α-PtO$_2$ and β-PtO$_2$, the occurrence of which depended on the composition and the temperature of oxidation. At low impurity contents, neither the impurity nor platinum could be oxidized. Several hypotheses (kinetics of oxidation, oxygen spill-over, percolation effects, changes in local electronic states) were proposed to elucidate this phenomenon.

It is quite clear that, if one wishes to get a better understanding of the chemical behaviour of such catalytic systems, studies on the kinetics and mechanisms must be made in addition to the use of physical methods of characterization. Unfortunately, very little is known about the modification of the surface during the course of the reactions because there is a lack of physical tools able to solve this problem. In this respect, EXAFS in the dispersive mode seems to be a very promising technique (ref. 136).

REFERENCES

1. V. Ponec, Adv. Catal., 32 (1983) 149.
2. W.M.H. Sachtler and R.A. Van Santen, Adv. Catal., 26 (1977) 69.
3. J.H. Sinfelt, G.H. Via, F.W. Lytle and R.B. Greegor, J. Chem. Phys., 75 (1981) 5527.
4. J.H. Sinfelt, G.H. Via and F.W. Lytle, J. Chem. Phys., 72 (1980) 4832.
5. A.A. Balandin, Z. Phys. Chem., 132 (1929) 289.
6. D.A. Dowden, J. Chem. Phys., (1950) 242.
7. D.H. Seib and W.E. Spicer, Phys. Rev. B Condens. Matter, 2 (1970) 1676 and 1694.
8. K. Ichikawa, J. Phys. Soc. Jpn., 37 (1974) 377.
9. P. Steiner, M. Schmidt and S. Hüfner, Solid State Comm., 35 (1980) 493.
10. S. Hüfner, G.K. Wertheim, R.L. Cohen and J.H. Vernick, Phys. Rev. Lett., 28 (1972) 488.
11. S. Hüfner, G.K. Wertheim, R.L. Cohen and J.H. Vernick, Phys. Rev. B Condens. Matter, (1973) 4511.
12. P. Soven, Phys. Rev., 156 (1969) 809.
13. B. Velicky, S. Kirkpatrick and H. Ehrenreich, Phys. Rev., 175 (1968) 747.
14. M. Boudart, Adv. Catal., 20 (1969) 153.
15. J.H. Sinfelt, J.L. Carter and D.J.C. Yates, J. Catal., 24 (1972) 283
16. G.A. Martin and B. Imelik, Surface Sci., 42 (1974) 157
17. J.A. Dalmon and G.A. Martin, J. Catal. 66 (1980) 214
18. A. Frennet, G. Lienard, A. Crucq and L. Degols, J. Catal., 53 (1978) 150.
19. K. Christmann and G. Ertl, J. Mol. Catal., 25 (1984) 31.
20. R. Burch, Acc. Chem. Res., 15 (1982) 24.
21. M.V. Mathieu and M. Primet, Surface Sci., 58 (1976) 511.
22. J.A. Dalmon and G.A. Martin, in T. Seiyama and K. Tanabe (Editors), New Horizons in catalysis Proc. 7th Int. Congress Catalysis, Tokyo, June 30 - July 4, 1980, Elsevier, Amsterdam, 1981, Part. A p. 402.
23. Y.L. Lam, J. Criado and M. Boudart, Nouv. J. Chim., 1 (1977) 461.
24. D.W. Blakely and G.A. Somorjai, J. Catal., 42 (1976) 181.
25. G.A. Somorjai, in Proc. 8th Int. Congress Catalysis, Berlin, July, 1984, Vol. 1, Verlag Chemie, Weinheim, 1984, p 113.
26. P.P. Lankhorst, H.C. de Jongste and V. Ponec, Catalyst Deactivation, Elsevier, Amsterdam, 1980, p. 43.
27. W.M.H. Sachtler, J. Mol. Catal., 25 (1984) 1.
28. R.W. Coughlin, A. Hasan and K. Kawakami, J. Catal., 88 (1984) 163.
29. F.L. Williams and D. Nason, Surface Sci., 45 (1974) 377.
30. S.C. Fain and J.M. Mc David, Phys. Rev. B Condens. Matter 9 (1974) 5099.
31. J. Sedlacek, L. Hilaire, P. Légaré and G. Maire, Surface Sci., 115 (1982) 541
32. J.C. Bertolini, J. Massardier, P. Delichere, B. Tardy, B. Imelik, Y. Jugnet, Tran

Minh Duc, L. de Temmermans, C. Creemers, H. van Hove and A. Neyens, Surface Scie. 119 (1982) 95.
33 F.F. Abraham, N.H. Tsai and G.M. Pound, Surface Sci., 83 (1979) 406 .
34 C.T.H. Stoddart, R.L. Moss and D. Pope, Surface Sci., 53 (1975) 241.
35 D.A. Merwyn, R.J. Baird and P. Wynblatt, Surface Sci., 82 (1979) 79.
36 V. Mintsa-Eya, L. Hilaire, R. Touroude, F.G. Gault, B. Moraweck and A.J. Renouprez, J. Catal.,76 (1982) 169.
37 Y. Takasu, R.Unwin, B. Tesche, A.M. Bradshaw and M. Grunze, Surface Sci., 77 (1978) 219.
38 M. Mehta and C.S. Fadley, Phys. Rev. B Condens. Matter, 20 (1979) 2280.
39 R. Van Hardeveld and F. Hartog, Surface Sci., 15 (1969) 189.
40 L. Guczi, A. Frennet and V. Ponec, Acta. Chim. Acad. Sci. Hung.,112(1983) 127.
41 P.W. Selwood, J. Am. Chem. Soc. 19 (1957) 3346, 4673 .
42 P.W. Selwood in Adsorption and Collective paramagnetism, Academic Press. New York London, 1962.
43 A. Frennet, G. Liénard, A. Crucq, L. Degols, Surface Sci., $\underline{80}$, (1979) 412.
44 A. Frennet, Catal. Rev., 10 (1974) 37.
45 P. Tetenyi, L. Guczi and A. Sarkany, Acta. Chim.Acad. Sci. Hung., 97 (1978) 221.
46 D.C Grenoble, J. Catal., 56 (1979) 32, 40.
47 P. Tetenyi, L. Guczi, Z. Paal and A. Sarkany, Kèm. Közl., 47 (1977) 363.
48 A. Sarkany, K. Matusek and P. Tetenyi, J.Chem. Soc. Faraday Trans. 1,73 (1977) 1699.
49 M. Boudart and L.D. Ptak, J. Catal., 16 (1970) 90.
50 S. Palfi, A. Sarkany and P. Tetenyi, J. Chem. Soc. Faraday Trans. 1, 77 (1981) 177.
51 K. Foger and J.R. Anderson, J. Catal., 59 (1979) 325.
52 A. Sarkany, J. Gaal and L. Toth, in T. Seiyama and K. Tanabe (Editors), New Horizons in Catalysis, Proc. 7 th Int. Congress Catal., Tokyo, June 30 - July 4, 1980, Elsevier, Amsterdam 1981, Part A, p. 291.
53 K. Foger and J.R. Anderson, J. Catal., 54, (1978) 318.
54 F. Garin and F.G. Gault, J. Am. Chem. Soc., 97 (1975) 4466.
55 Z. Paal and P. Tetenyi, in G.C. Bond and G. Webb (Editors) Catalysis, Specialists Periodical Reports. The Royal Society of Chemistry, London, 1982, pp. 80-126.
56 F. Garin, F.G. Gault and G. Maire, Nouv. J. Chimie, 5 (1981) 553.
57 F. Garin, G. Maire and F.G. Gault, Nouv. J. Chimie, 5 (1981) 563.
58 J.R. Anderson, N.R. Avery, J. Catal., 5 (1966) 446.
59 J.R. Anderson and C. Kemball, Proc. R. Soc. London, Ser. A, 223, (1954) 361.
60 C. Kemball, Catal. Rev., 5 (1) (1971).
61 C. Kembal Proc. R. Soc. London, Ser. A, 217 (1953) 376.
62 R. Touroude and F.G. Gault J. Chem. Soc., Chem. Comm. 154 (1975).
63 J.R. Anderson and B.G. Baker, Proc. R. Soc. London, Ser. A. 271 (1963) 402.

64 J. R. Anderson and N.R. Avery, J. Catal., 7 (1967) 315.
65 J.R. Anderson, Adv. Catal., 23 (1973) 1.
66 G. Leclercq, L. Leclercq and R. Maurel, J. Catal., 57 (1977) 87.
67 J.A. Pople and M. Gordon, J. Am. Chem. Soc., 89 (1967) 4253.
68 L. Hall and L.B. Kier, Tetrahedron, 33 (1977) 1953.
69 Leclercq, L. Leclercq and R. Maurel,Bull. Soc. Chim. Belg., 88 (1978) 599.
70 F. Luck, Thesis, Strasbourg, 1983.
71 F.G. Gault, Adv. Catal. 30 (1981) 1.
72 G. Maire and F. Garin in J.R. Anderson and M. Boudart (Editors), Catalysis Science and Technology, Vol. 6 Spinger, Berlin, 1984 p. 161.
73 R.L. Moss, in C. Kemball (Editor) Catalysis, Specialist Periodical Reports The Royal Society of chemistry, London 1 (1977) 37.
74 H.C. de Jongste, V. Ponec in T. Seiyama and K. Tanabe (Editors), New Horizons in Catalysis, Proc. 7 th Intern. Congress, Catal., Tokyo, June 30-July 4, 1980, Elsevier, Amsterdam, 1981, Part A, p. 186.
75 J.R.H. Van Schaik, R.P. Dessing and V. Ponec, J. Catal., 38 (1975) 273.
76 D. Tomanek, S. Mukherjee, V. Kumar, J.H. Bennemann, Surface Sci., 114 (1982) 11.
77 H.C. deJongste, V. Ponec and F.G. Gault, J. Catal., 63 (1980) 395.
78 H.C. de Jongste and V. Ponec, J. Catal., 63 (1980) 389.
79 J.R. Anderson and Y Shimoyama, in J.W. Hightower (Editor), Proc 5th Int. Congress Catalysis Miami, 1972, Elsevier, Amsterdam, 1973 , p. 695.
80 M.W. Vogelzang, M.J.P. Botman and V. Ponec in Faraday Discussions of the Chemical Society, 72 (1981) 33.
81 A. O'Cinneide and F.G. Gault, J. Catal., 37 (1975) 311.
82 J.W.A. Sachtler and G.A. Somorjai, J. Catal., 8 (1983) 77.
83 K. Foger and J.R. Anderson, J. Catal., 61 (1980) 140.
84 Z. Karpinski and J.K.A. Clarke, J. Chem. Soc. Faraday Trans. 1,71 (1975) 893.
85 F. G. Gault, O. Zahraa, J.M. Dartigues, G. Maire, M. Peyrot, E. Weisang. P.A. Engelhard in T. Seiyama and K Tanabe (Editors), New Horizons in Catalysis, Proc. 7th Intern. Congress Catalysis Tokyo, 1980, Elsevier, Amsterdam, 1981, Part A, p. 199.
86 R. Burch, J. Catal., 71 (1981) 348.
87 B. Coq and F. Figueras, J. Catal., 85 (1984) 197.
88 J. Völter, G. Lietz, M. Uhlemann and M. Hermann, J. Catal., 68 (1981) 42.
89 R. Burch and L.C. Garla, J. Catal., 71 (1981) 360.
90 M.S. Kharson, G.B. Kadinov and A.N. Palazov, React. Kinet. Catal. Lett., 10 (1979) 267.
91 B.N. Kuznetsov, Y.I. Yermakov, M. Boudart, J.P. Collman,J. Mol. Catal., 4 (1978) 49.
92 T.M. Tri, J. Massardier, P. Gallezot and B. Imelik, J. Catal., 85 (1984) 244.

93 G. Leclercq, T. Romero, S. Pietrzyk, J. Grimblot and L. Leclercq, J. Mol. Catal., 25 (1984) 67.
94 Yu I. Yermakov, B.N. Kuznetsov and Yu. A. Ryndin, J. Catal., 42 (1976) 73.
95 C. Klein-Petit Specialist report, Diplôme d'Etudes Approfondies (D.E.A), Strasbourg 1985.
96 R.L. Jacobson, H.F. Kluksdahl, C.S. Mc Coy and R.W. Davis, Proc. Am. Petrol. Inst. Div. Refining 34th Midyear Meetg., Chicago, May 13, 1969.
97 C. Bolivar, H. Charcosset, R. Frety, M. Primet, L. Tournayan C. Betizeau, G. Leclercq and R. Maurel, J. Catal., 39 (1975) 249.
98 C. Bolivar, H. Charcosset, R. Frety, M. Primet, L. Tournayan C. Betizeau, G. Leclercq and R. Maurel, J. Catal., 45 (1976) 163
99 C. Betizeau, G. Leclercq, R. Maurel, C. Bolivar, M. Charcosset, P. Frety, L. Tournayan 45 (1976) 179.
100 H.Charcosset,. Platinum Met. Rev., 23 (1979) 18.
101 R. Burch, Platinum Met. Rev., 22 (1978) 57.
102 L. Tournayan, R. Bacaud, H. Charcosset and G. Leclercq J. Chem. Res. Synop. (1978) 290.
103 G. Diaz, Thesis, Strasbourg, 1982.
104 J.R. Anderson, Am. Chem. Soc., Div. Pet. Chem. Prepr., 26 (1981) 361.
105 R. Gomez, G. Corro, G. Diaz, A. Maubert and F. Figueras, Nouv. J. Chimie, 4 (1980) 677.
106 G. Blanchard, H. Charcosset, M. Guenin and L. Tournayan, Nouv. J. Chimie, 5 (1981) 85.
107 G. Leclercq, H. Charcosset, R. Maurel, C. Betizeau, C. Bolivar, R. Frety, D. Jaunay, H. Mendez and L. Tournayan, Bull. Soc. Chim. Belg., 88 (1979) 577.
108 G. Diaz, F. Garin and G. Maire, J. Catal, 82 (1983) 13.
109 S. Zyade, F. Garin, L. Hilaire, M.F. Ravet, G. Maire, Bul Soc. Chim. France, 3, (1985) 341.
110 S. Zyade, Thèse de specialité, Strasbourg, 1984.
111 S. Zyade, F. Garin, G. Maire, submitted for publication.
112 R.L. De Kock, T.D. Fehlner, C.E. Housecroft, T.V. Lubben and K. Wade, Inorg. Chem., 21 (1982) 25.
113 J.R. Fritsch and K.P.C. Vollhardt, Angew. Chem. Int. Ed. Engl., 7 (1980) 559.
114 F.G. Ciapetta, R.M. Dobres and R.W. Baker, in P.H. Emmett (Editors) Catalysis, Reinhold, New York, 1958, Vol. 6, p 495.
115 C. Kemball, Adv. Catalysis, 11 (1959) 223.
116 D.W. MC Kee and F.J. Norton, J. Catal., 4 (1965) 510.
117 T.J. Gray, N.G. Masse and H.G. Oswin, Actes 2e Congrès Int. Catalyse, Technip, Paris, 1961 Vol. 2, p. 1697.
118 J.H. Sinfelt, U.S. patent 3,684, 693, 1972 J.H. Sinfelt (to Esso Res. and Eng. C.O.) Ger. Offen. 2, 153, 475/1973.

119 T. Chen Wong, L.C. Chang, G.L. Haller, J.A. Oliver, N.R. Scaife and C. Kemball, J. Catal., 87 (1984) 389.
120 T. Wong and L.D. Schmidt, J. Catal., 71 (1981) 411.
121 E.J. Nowak and R.M. Koros, J. Catal., 7 (1967) 50.
122 A.J. Renouprez, B. Moraweck, B. Imelik, V. Perrichon, J.M. Dominguez - Esquivel and J. Jablonski, in T. Seiyama and R. Tanabe (Editors), New Horizons in Catalysis Proc. 7th Int. Congress Catalysis Tokyo, 1980, Elsevier, Amsterdam, 1981, Part. A, p. 173.
123 J.M. Dominguez A. Vazquez S., A.J. Renouprez and M.J. Yacaman, J. Catal., 75 (1982) 101.
124 S. Aeiyach, F. Garin, L. Hilaire, P. Légaré and G. Maire, J. Mol. Catal., 25 (1984) 183.
125 Y. Jugnet, J. Massardier, Tran Minh Duc, D.C. Bertolini, B. Tardy and J.C. Védrine, Surface Sci. 107 (1981) L320.
126 J. Massardier, B. Tardy, M. Abon and J.C. Bertolini, Surface Sci., 126 (1983) 154.
127 F. Garin, S. Aeiyach, P. Légaré and G. Maire, J. Catal., 77 (1982) 323.
128 Z. Karpinski and T. Koscielski, J. Catal., 56 (1979) 430.
129 Z. Karpinski and T. Koscielski, J. Catal., 63 (1980) 313.
130 T. Koscielski, Z. Karpinski and Z. Paal, J. Catal., 77 (1982) 539.
131 F. Garin, P. Girard, A. Chaqroune, F. Weisang and G. Maire, in Proc.8th Int. Congress catalysis, Berlin, July 1984, Vol. III, Verlag Chemie, Weinheim, 1984, p. 405.
132 L. Hilaire, G. Diaz Guerrero, P. Légaré, G. Maire and G. Krill, Surface Sci., 146 (1984) 569.
133 A.V. Ramaswamy, P. Ratnasamy and S. Sivasanker and A.J. Leonard., in G.C. Bond, P.B. and F.C. Tompkins (Editors), Proc. 6th Int. Congress Catalysis, London (1976) p.855.
134 R.W Rice and Kang Lu, J. Catal., 77 (1982) 104.
135 J. C. Rasser, WH. Beindorf and J.J.F. Scholten, J. Catal., 59 (1979) 211.
136 G. Maire, P. Bernhardt, F. Garin, P. Girard, J.L. Schmitt, E. Dartyge, P. Lagarde, A. Fontaine, H. Dexpert, Submitted for publication.

Chapter 6

HYDROGENATIVE DENITROGENATION OF MODEL COMPOUNDS AS RELATED TO THE REFINING OF LIQUID FUELS

HANS SCHULZ, MARCO SCHON and NURUM M. RAHMAN
Engler-Bunte-Institut, Universität Karlsruhe (FRG)

6.1 ORGANIC NITROGEN COMPOUNDS IN TARS, OILS FROM COAL, SHALE AND PETROLEUM

Organic nitrogen compounds are essential constituents of fossil fuels. Their impact during fuel processing to final products is generally that of undesirable and more or less harmful by-products which must be removed or converted in the most efficient and economic manner.

The first information to be obtained about the organic nitrogen

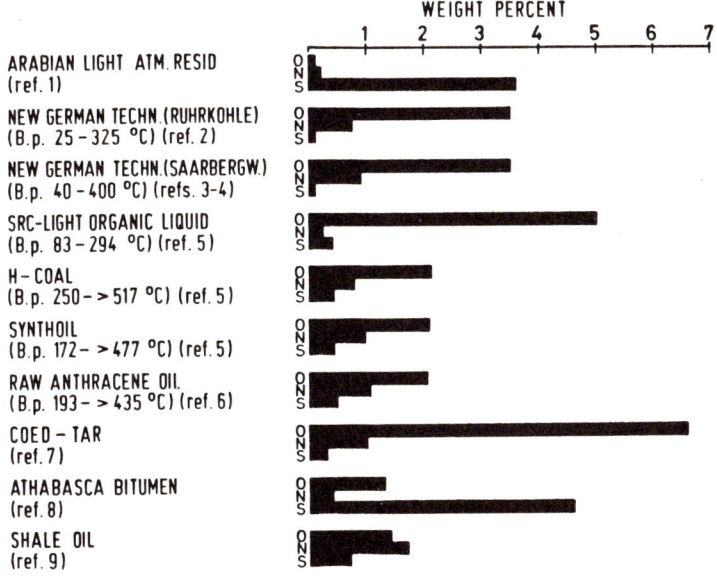

Figure 6.1. Content of oxygen, nitrogen and sulphur in selected liquid fuels (refs. 1-10).

compounds in a fuel will be its total nitrogen content. This is shown in Fig. 6.1 for a number of typical liquid fuels, together with their content of oxygen and sulphur (refs. 1-9). In petroleum the most abundant heteroatom is sulphur. The atmospheric residue of an Arabian light crude oil (No. 1 in Fig. 6.1) contains 3.6 wt-% sulphur, only 0.2 wt-% nitrogen and only 0.1 wt-% oxygen (ref. 1). Oils formed by coal hydrogenation are particularly rich in oxygen but also have substantially higher contents of nitrogen than does petroleum. Thus the oxygen content in such oils (Nos. 2 to 6 in the figure) ranges from 2 to 5 wt-% and the nitrogen content from 0.2 to about 1 wt-%. The fuel oils obtained via the new German technology for coal hydrogenation (Nos. 2 and 3 in the figure) contain 3.5 wt-% oxygen, 0.9 and 0.75 wt-% nitrogen respectively and only 0.1 wt-% sulphur. Coal tars obtained via high temperature carbonization (No. 7) or via low temperature fluid bed coking (No. 8) are rich in oxygen (2.0 and 6.6 wt-%) and in nitrogen (1.1 and 1.0 wt-%). The Athabasca bitumen (No. 9), an extract from Canadian tar sands, is related to petroleum because of its high sulphur content of 4.6 wt-%, however it is also rich in oxygen (1.3%) compared to petroleum, indicating the relatively young nature of this fossil fuel. The nitrogen content too is relatively high (0.4 wt-%). The highest nitrogen content reported in Fig. 6.1 is that of shale oil (1.7 wt-%) (ref. 5).

It may be concluded that when shifting from petroleum to coal-derived oils the nitrogen content in the fuel increases. However, the oils from tar sand and shale are also rich in nitrogen. Generally, the heavier the fuel the more stringent will be the need for removal of nitrogen in order to reduce NO_x emissions, to avoid interference with acidic catalysts and to meet the specifications for marketable products.

As regards the structure of the nitrogen compounds present in liquid fossil fuels, a great variety of substances has been identified (refs. 10-23). The type of nitrogen compounds found in fuels seldom reflects the structure of the original matter (as in the case of porphyrins which have survived in petroleum and shale oil (ref. 14)). In Table 6.1 relevant basic structures of organic nitrogen compounds are summarized.

Aliphatic amines (<u>1</u> and <u>2</u> in Table 6.1) are very reactive and therefore are not substantial constituents of original fuels or of products from high temperature thermal processes. They are formed however as intermediates during hydrodenitrogenation (HDN) of

TABLE 6.1

Basic structures of organic nitrogen compounds in liquid fuels.

Compound class	Basic structure	Remark
Aliphatic amines	**1** $\sim\sim NH_2$ **2** $\sim\sim NH\sim\sim$	Reactive intermediates of hydroconversion of cyclic N-compounds
Saturated monocyclic amines	**3** pyrrolidine **4** piperidine	HDN-intermediates
Aromatic monocyclic nitrogen compounds	**5** pyrrole **6** pyridine	Basic structures of monocyclic aromatic nitrogen compounds; occurrence primarily in coal tar
Anilines	**7** aniline	Occurrence primarily in coal tars and in oils from direct coal hydrogenation
Dicyclic and tricyclic derivatives of five-membered nitrogen compounds	**8** indole **9** carbazole **10** dihydroindole **11** tetrahydrocarbazole	Indoles (**8**) and carbazoles (**9**) are constituents of fuel oils of different types. Dihydroindole (**10**) and tetrahydrocarbazole (**11**) are important intermediates of hydrotreating reactions
Dicyclic and tricyclic derivatives of six-membered nitrogen compounds	**12** quinoline **13** acridine **14** 1,2,3,4-tetrahydroquinoline **15** 1,2,3,4-tetrahydroacridine **16** isoquinoline	Quinoline (**12**), acridine (**13**) occur in high-temperature tar and in coal hydrogenation products. 1,2,3,4-Tetrahydroquinoline (**14**) and 1,2,3,4-tetrahydroacridine (**15**) are intermediates of hydroconversion. Isoquinoline, occurrence in coal carbonization liquids
Porphyrins	**17** porphyrin	Occurrence in petroleum and shale oil

cyclic nitrogen compounds (refs. 25-27) and can therefore be observed in products from incomplete hydrogenative nitrogen removal. Saturated monocyclic amines (pyrrolidines $\underline{3}$ and piperidines $\underline{4}$) are mainly obtained as products from ring hydrogenation of the corresponding aromatic compounds (pyrroles and pyridines) (refs. 28-32) and are important intermediates in the HDN of fuels from high temperature processes like coke oven tar. Monocyclic aromatic nitrogen compounds ($\underline{5}$ and $\underline{6}$) are particularly stable at high temperature and mainly observed in various tars. Side chains (C_2+) on the aromatic ring have little thermal stability and methylated derivatives are the most common nitrogen-containing substances in these products. Anilines ($\underline{7}$) are the major nitrogen-containing substances in oils from coal hydrogenation (refs. 4,10-14). As shown below, they are formed as relatively stable intermediates from bi- or tricyclic nitrogen compounds (refs. 33-35). Anilines are also present in coal tars (refs. 15-18). Benzo-homologues of pyrrole, namely indole and carbazole (8 and 9) occur in many fuel oils. Their partially hydrogenated derivatives (10 and 11) are intermediates in hydrotreating processes. The same can be said of quinoline and acridine, the bicyclic and tricyclic benzo-homologues of pyridine (refs. 36-39). Isoquinoline (16) has been found in coal tars (refs. 15-18). Porphyrins (17) are important constituents of petroleum. They prove its fossil organic nature by their relation to the essential biosubstances chlorophyll and hemin. Their technical impact concerns the complex formation with some metals which makes nickel and vanadium very harmful constituents of heavy petroleum fractions.

Considering the multiplicity of nitrogen compounds in fossil fuels, the different modes of binding of nitrogen in the molecules and its usual integration into aromatic ring systems, it is evident that hydrodenitrogenation must be regarded as a multi-reaction conversion in which the individual C-N bonds show a broad range of reactivity and no simple and generally applicable reaction mechanism for HDN can be expected.

6.2 EVALUATION OF CATALYST SELECTIVITY FOR HDS, HDO AND HDN REACTIONS AND HYDROGENATION OF MULTIPLE BONDS

When Pier and co-workers (ref. 40) introduced the sulphur resistant hydrogenation catalysts $CoMoS/Al_2O_3$, $NiMoS/Al_2O_3$ and $NiWS/Al_2O_3$, the principal result desired - apart from obtaining maximum gasoline yields - was the removal of oxygen and nitrogen from oils produced via slurry phase coal hydrogenation. Later, the hydrotreatment of petroleum fractions was particularly concerned with hydrodesulphurization. Fossil fuels of the future will demand an increasing degree of purification and conversion of heavy and unconventional feedstocks and thereby hydrodeoxygenation and hydrodenitrogenation will be essential. The early work had shown that the removal of oxygen and nitrogen needed more severe reaction conditions than those for sulphur removal (ref. 41). $CoMoS/Al_2O_3$ appeared to be the most suitable catalyst for hydrodesulphurization and $NiMoS/Al_2O_3$ for hydrodenitrogenation (ref. 42).

6.2.1 Conversion of a four compound model mixture

The characterization of a catalyst in terms of its selectivity for individual hydrofining reactions requires measurements to be carried out with adequately chosen reactants. The selection of appropriate model compounds is decisive for any conclusions to be drawn about catalyst suitability and reaction networks during the conversion of technical feeds. In this section a method based on the conversion of a four-compound mixture and its application to the simultaneous determination of the HDS, HDO and HDN and aromatic ring hydrogenation activity of a hydrotreating catalyst are described (refs. 4,43).

The hydrodesulphurization activity is defined in terms of the conversion of 2-methylthiophene. Thiophene derivatives are the most common and least reactive sulphur compounds in liquid fuels. Hydrodeoxygenation activity is specified by reference to phenol as a model compound. Oxygen in coal or other fossil fuels commonly occurs as a bridging atom in multi-ring compounds (ref. 44). During hydroconversion such compounds are transformed into phenols which are relatively stable intermediates, so that the slow step of HDO is usually removal of oxygen from phenols. To specify HDN activity, o-toluidine has been selected as a model compound. As in the case of oxygen, nitrogen mainly occurs in multi-ring systems in untreated fuels. Anilines are particularly stable intermediates in hydro-

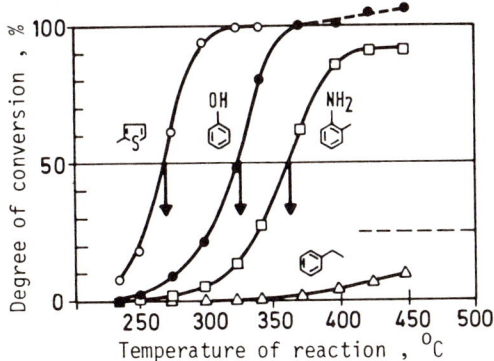

Fig. 6.2. Extent of conversion of the four compounds in the model mixture in HDS, HDO, HDN and aromatic ring hydrogenation as a function of reaction temperature (refs. 4,43). CoMoS/Al_2O_3-A, p_{H_2} = 96.7 bar, p_{MTh} = p_{Ph} = p_{o-T} = 0.67 bar, p_{Eb} = 1.3 bar, p_{H_2S} = 0.1 bar, τ_{eff} = 7 - 10 sec.

conversion. Finally, ethylbenzene is incorporated in the mixture as a measure of the catalyst hydrogenation activity for saturation of monocyclic aromatic rings.

Particular care has been taken to ensure that the hydrocarbon products from HDS, HDO, HDN and aromatic ring hydrogenation all have different structures. Individual degrees of conversion are easily calculated from the composition of the reactor effluent. The evaluation of catalyst selectivity is performed by means of a series of experiments at increasing reaction temperature. The individual degrees of conversion are plotted as a function of reaction temperature (Fig. 6.2). The reactivity of the compounds and the specific catalyst activity is quantified as the reaction temperature at which a 50% conversion is obtained, T_{50}, which can be read from Fig. 6.2. HDS (T_{50} = 268°C) is much easier than HDO (T_{50} = 323°C) and this is easier than HDN (T_{50} = 359°C). Hydrogenation of the monocyclic aromatic ring (T_{25} = 450°C) is much slower than HDN.

In Table 6.2 the results for the CoMoS/Al_2O_3-A, the NiMoS/Al_2O_3 and the NiWS/Al_2O_3 catalysts are compared. It is seen that the highest activity for hydrodesulphurization is obtained with CoMoS/Al_2O_3 at a temperature of about 270°C. However, for deoxygenation in the temperature range 300-320°C, this catalyst is less active than NiMoS/Al_2O_3 and NiWS/Al_2O_3, a 20°C higher temperature being necessary. For HDN, in the range 330-336°C, CoMoS/Al_2O_3-A needs a 11°C higher reaction temperature than NiMoS/Al_2O_3. For aromatic ring hydrogenation of ethylbenzene at higher reaction temperatures

TABLE 6.2
Reaction temperatures for 50% conversion (T_{50} in °C) of the four components of a model mixture (refs. 4,43). Conditions as indicated in Fig. 6.2.

Catalyst	Compounds to be converted			
	SH	OH	NH_2	[a]
CoMoS/Al_2O_3 -A	268	323	359	>500
NiMoS/Al_2O_3	273	303	334	424
NiWS/Al_2O_3	273	302	348	432

[a] For ethylbenzene, the T_{25} values are reported because reaction temperatures higher than 450°C were not applied.

the same sequence of activity is observed with NiMoS/Al_2O_3 being the most active and CoMoS/Al_2O_3 being the least active catalyst. The question then arises as to how these changes in sequence of catalyst activity can be explained; e.g., can the higher activity of CoMoS/Al_2O_3 for hydrodesulphurization be attributed to a higher activity for C-S bond hydrogenolysis. To answer this question it is necesary to discriminate between the individual reaction steps of the HDS, HDO and HDN.

6.2.2 Evaluation of specific HDS activity

There have been many investigations of the HDS mechanism (refs. 45-49) and several reviews have been published (refs. 50-52). Nevertheless a recent workshop (ref. 53) revealed considerable gaps in understanding. The HDS of 2-methylthiophene goes through the intermediates tetrahydro-2-methylthiophene and pentenes. No other compounds were observed among the products at 100 bar total pressure (refs. 4,43).

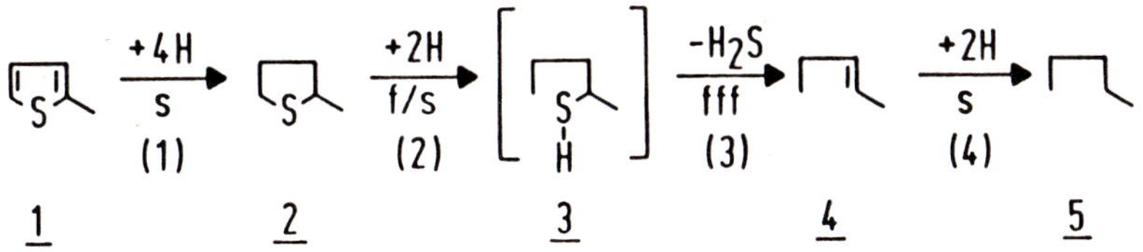

$$\underset{\underline{1}}{\boxed{S}} \xrightarrow[S]{+4H \atop (1)} \underset{\underline{2}}{\boxed{S}} \xrightarrow[f/s]{+2H \atop (2)} \left[\underset{\underline{3}}{\boxed{\underset{H}{S}}} \right] \xrightarrow[fff]{-H_2S \atop (3)} \underset{\underline{4}}{\boxed{}} \xrightarrow[S]{+2H \atop (4)} \underset{\underline{5}}{\boxed{}}$$

Scheme 6.1. Hydrodesulphurization of 2-methylthiophene.
s = Slow, f = fast, s/f, f/s = reaction steps with intermediate values of the relative rate constants; compounds in brackets have not been observed in the product and are regarded as very reactive intermediates.

Assuming first-order rate dependences with respect to the consumption of compounds $\underline{1}$ to $\underline{4}$ in Scheme 6.1 it follows from Fig. 6.3, which shows the product composition as a function of space time, that reactions 1 and 4 are slower than reaction 2. Thus it is concluded that splitting of a C-S bond in mercaptan is very fast; splitting of a C-S bond in tetrahydro-2-methylthiophene is much slower, however still faster than hydrogenation of the thiophene ring and faster than hydrogenation of pentene to yield pentane. The slow step of 2-methylthiophene desulphurization is therefore not the C-S bond splitting in tetrahydro-2-methylthiophene. High HDS activity for a thiophene ring is equivalent to high hydrogenation activity of the catalyst: hydrogenation and not C-S splitting is

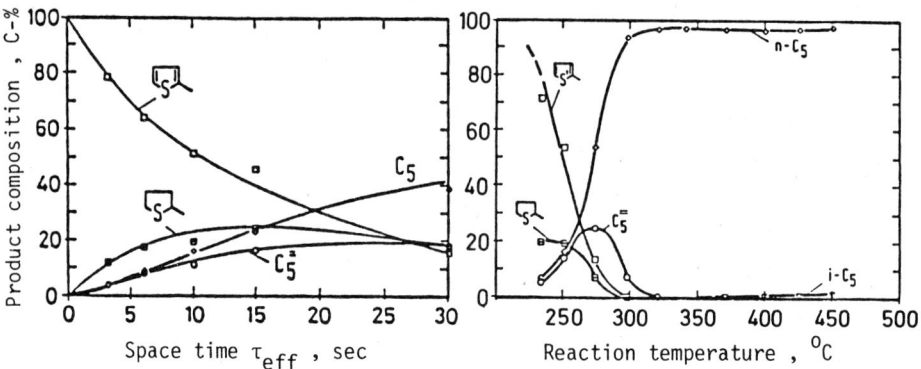

Fig. 6.3. Composition of the products (in C-%) of 2-methyl-thiophene hydrodesulphurization on CoMoS/Al$_2$O$_3$-B at 100 bar as a function of space time (left) and reaction temperature (right) (for other conditions see Fig. 6.2)(refs. 4,43).

limiting for HDS. In Fig. 6.3 (right) it is also seen that slow carbon skeleton hydroisomerization occurs only above 425°C. Hydroisomerization on CoMoS/Al$_2$O$_3$ as catalyst had been investigated earlier (ref.54).

A comparison of CoMoS/Al$_2$O$_3$-B, NiMoS/Al$_2$O$_3$ and NiWS/Al$_2$O$_3$ with respect to their specific activity for hydrogenation and C-S bond splitting during conversion of 2-methylthiophene is given in Table 6.3. CoMoS/Al$_2$O$_3$ gives more than twice the conversion (51%) during the same space time (10 sec) than the two other catalysts. It is twice as active for hydrogenation of the thiophene ring than the two other catalysts.

The C-S bond splitting activity is quantified by the ratio of the moles of C$_5$ hydrocarbons formed to the sum of the moles of C$_5$ hydrocarbons plus 2-methylthiophene. This is equal to the fraction of tetrahydro-2-methylthiophene converted via ring opening relative to the amount of tetrahydro-2-methylthiophene formed. The lowest values are obtained with CoMoS/Al$_2$O$_3$, and it is concluded that CoMoS/Al$_2$O$_3$ has the lowest activity for C-S bond splitting. This type of activity is highest with NiWS/Al$_2$O$_3$. The last column of Table 6.3 indicates that CoMoS/Al$_2$O$_3$ has the highest hydrogenation activity for olefin saturation.

Obviously CoMoS/Al$_2$O$_3$ is the most active hydrodesulphurization catalyst because of its superior hydrogenation activity. These results are quite contradictory to current mechanistic pictures (refs. 55-57), which assume butadiene to be a major primary product

TABLE 6.3

Composition of products from 2-methylthiophene obtained with CoMoS/Al_2O_3-B, NiMoS/Al_2O_3 and NiWS/Al_2O_3 under comparable reaction conditions (T = 250°C, p_{total} = 100 bar, further conditions see Fig. 6.2) (refs. 4,43).

Catalyst [a]	Conv. %	Space Time (sec)	Product composition (mol-%)			Molar ratios	
			[S]	Pentenes	Pentane	C_5/[S]+C_5	Pentane/ΣC_5
CoMoS/Al_2O_3-B	19	3	60	21	19	40	0.48
"	51	10	41	26	33	59	0.56
NiMoS/Al_2O_3	22	10	26	49	25	74	0.34
NiWS/Al_2O_3	20	10	14	53	33	86	0.38

a) Catalyst evaluation parameters:
(1) aromatic ring hydrogenation activity: conversion of 2-methylthiophene
(2) C-S bond breaking activity: molar ratio of C_5 hydrocarbons to 2-methylthiophene + C_5 hydrocarbons;
(3) olefin hydrogenation activity: molar ratio of pentane to sum of C_5 hydrocarbons

of thiophene hydrodesulphurization:

$$[S] \xrightarrow{+2H_2} [\,\,] + H_2S$$

The explanation is that butadiene can be observed only under quite unrealistic conditions of, e.g. only 1 atm of pressure which is much too low to provide sufficient activated hydrogen at the surface of the CoMoS/Al_2O_3 catalyst.

6.2.3 Evaluation of specific HDO activity

One result of the hydrodeoxygenation of phenol is given in Fig. 6.4 which shows the product composition as a function of space time obtained with a NiMoS/Al_2O_3 catalyst at 297°C and a total pressure of 100 bar. Only traces of benzene are obtained up to 60% conversion of the phenol: the C-O bond in phenol is very stable and is not split through hydrogenolysis.

The first intermediate observed is cyclohexene. Obviously the primary product cyclohexanol is very reactive at the temperature (>300°C) necessary for hydrogenation of the aromatic ring. The oxygen is removed through an elimination of water producing cyclohexene. The most important slow step of hydrodeoxygenation is again the hydrogenation of the aromatic ring and the HDO activity is equivalent to the hydrogenation activity (Scheme 6.2). Recent literature on HDO has been reviewed by Furimsky (ref. 57).

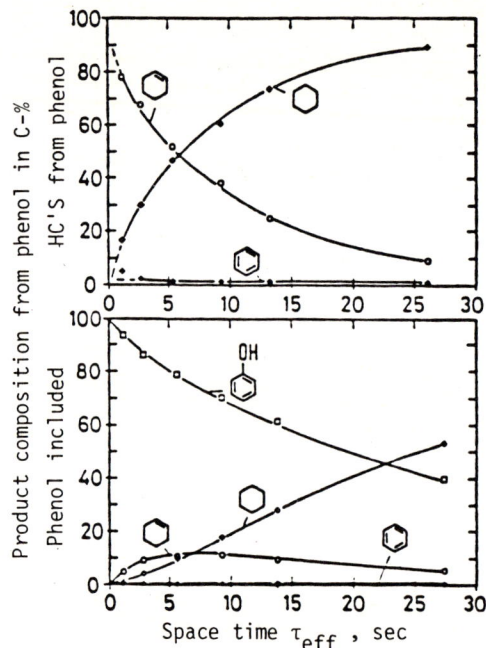

Fig. 6.4. Composition of products obtained during phenol deoxygenation as a function of space time (refs. 4,43). NiMoS/Al$_2$O$_3$, T = 297°C, p$_{total}$ = 100 bar, for other conditions see Fig. 6.2.

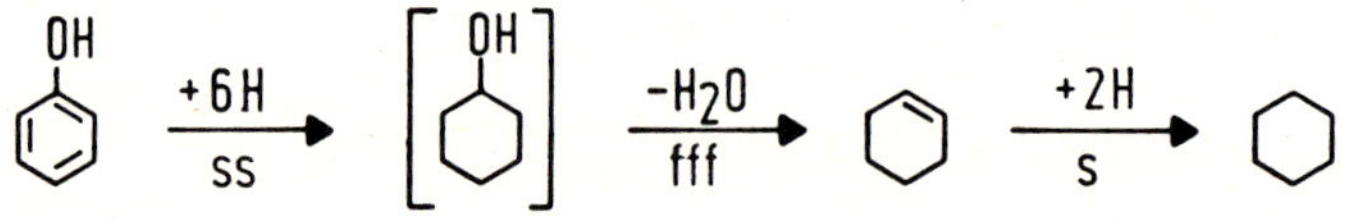

Scheme 6.2. Hydrodeoxygenation of phenol

6.2.4 Evaluation of specific HDN activity

Because nitrogen is present in liquid fuels predominantly in multi-ring compounds, the overall hydrodenitrogenation must proceed through a complex reaction network. Liquids formed from hydrogenation of coal contain the nitrogen as NH$_2$ groups attached to the aromatic ring. One can conclude (as shown below and from the conversion of multi-ring model compounds) that the denitrogenation of anilines is a particularly slow reaction and thus anilines are

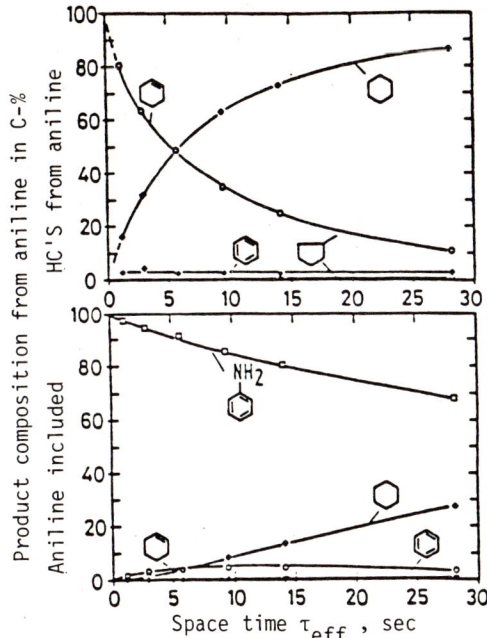

Fig. 6.5. Composition of products during aniline denitrogenation as a function of space time (refs. 4,43). NiMoS/Al$_2$O$_3$, T = 297°C, p_{H_2} = 96.7 bar, p_{BTh} = p_{o-Cr} = p_{An} = 0.67 bar, p_{o-Xy} = 1.3 bar, p_{H_2S} = 0.1 bar.

suitable model compounds for the testing of HDN catalyst activity (refs. 58,59). Fig. 6.5 shows the product composition as a function of space time for the hydrogenative conversion of aniline on NiMoS/Al$_2$O$_3$ at 297°C and 100 bar. The only substantial intermediate to be observed is cyclohexene (Scheme 6.3). Its selectivity of formation is higher than 80% at short space times.

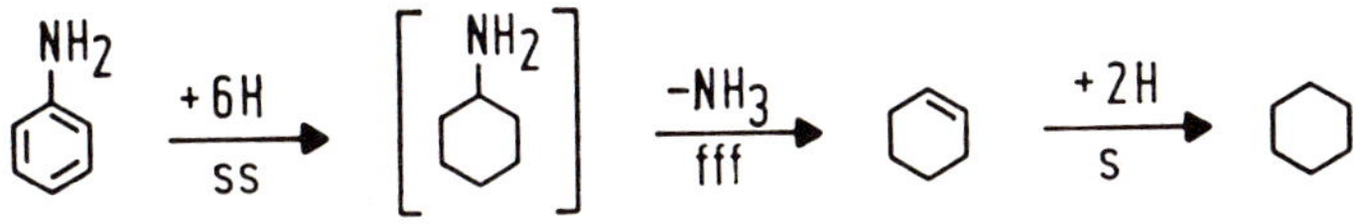

Scheme 6.3. Hydrodenitrogenation of aniline.

Aniline conversion has been investigated by several authors (refs. 60,61). The C-N bond in anilines is very stable and resists to hydrogenolysis on NiMoS/Al$_2$O$_3$ at 300°C and 100 bar. The aromatic ring is hydrogenated to cyclohexylamine, which is not stable at all under these conditions and not observed among the products. Generally two reaction pathways have to be considered for the conversion of cyclohexylamine: (1) elimination of NH_3 and simultaneous formation of cyclohexene; (2) substitution of NH_2 by hydrogen to yield NH_3 and cyclohexane. Extrapolation to zero conversion in Fig. 6.5 shows elimination to be the almost exclusive reaction under these conditions which are consistent with industrial processes.

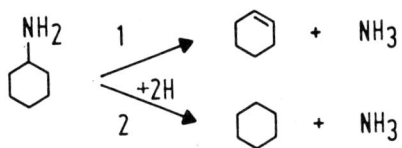

It can be concluded that a high specific activity of catalysts for HDN is identical with a high activity for aromatic ring hydrogenation.

6.2.5 Concluding evaluation of specific catalyst activity

From Table 6.3 it has been deduced that the HDS activity of $CoMoS/Al_2O_3$ is higher than that of $NiMoS/Al_2O_3$ and $NiWS/Al_2O_3$. This was correlated with a higher activity of the $CoMoS/Al_2O_3$ catalyst for the slow step of thiophene ring hydrogenation. Under these conditions at 250°C, $CoMoS/Al_2O_3$ shows the highest hydrogenation activity. In Table 6.2 and Figs. 6.4 and 6.5 it has been shown that $NiMoS/Al_2O_3$ and $NiWS/Al_2O_3$ are both more active hydrogenation catalysts than $CoMoS/Al_2O_3$ at the higher temperature (300°C and higher) necessary for HDO of phenol, HDN of aniline or hydrogenation of ethylbenzene.

The following explanation is proposed for this unusual change in the sequence of catalyst activity with reaction temperature. With $CoMoS/Al_2O_3$ the active hydrogenation sites are dynamically generated in the presence of H_2S and H_2 at a relatively low temperature, thus it operates at lower temperatures than do the others. Phenols and anilines can only be converted at higher reaction temperatures (>300°C). On $NiWS/Al_2O_3$ and $NiMoS/Al_2O_3$ the active sites are generated only at this temperature but are of higher activity than those obtained in the $CoMoS/Al_2O_3$ system, making $NiMoS/Al_2O_3$ and $NiWS/Al_2O_3$ the more active hydrogenation catalysts at the higher temperature. Low temperature activity sequence at <250°C:
$CoMoS/Al_2O_3 > NiMoS/Al_2O_3 = NiWS/Al_2O_3$.
High temperature activity sequence at >300°C:
$NiMoS/Al_2O_3 > NiWS/Al_2O_3 > CoMoS/Al_2O_3$.

6.2.6 Variation of the compounds of the mixture

To evaluate the specific catalyst activity, a second mixture comprising benzothiophene, o-cresol, aniline and o-xylene was chosen. The results (temperatures for a 50% conversion) are shown in Table 6.3. The conclusions which can be drawn are generally the same as those above. CoMoS/Al$_2$O$_3$ is more active for HDS at low temperature than NiMoS/Al$_2$O$_3$ and NiWS/Al$_2$O$_3$, and is less active for HDO, HDN and hydrogenation of o-xylene at higher temperatures.

TABLE 6.4
Reaction temperatures for a 50% conversion (T_{50} in °C) of the four components of a model mixture. Conditions as indicated in Fig. 6.2 (refs. 4,43).

Catalyst	Compounds to be converted			
	⌬S⌭	OH-CH$_3$ (phenol)	NH$_2$	CH$_3$/CH$_3$ [a]
CoMoS/Al$_2$O$_3$ -A	237	358	356	535
NiMoS/Al$_2$O$_3$	244	333	333	421
NiWS/Al$_2$O$_3$	241	339	338	418

a For o-xylene, the T_{25} values are reported because reaction temperatures higher than 450 °C were not applied.

As regards the HDS reaction, benzothiophene (T_{50} ca 240°C) is much more reactive than 2-methylthiophene (T_{50} ca 270°C). This is due to the fact that hydrogenation of the thiophene ring in benzothiophene is much faster than in 2-methylthiophene (ref. 43). Similar trends are observed for cyclic nitrogen and oxygen compounds.

When comparing phenol and o-cresol, it is found that o-cresol is considerably less reactive (ΔT = ca. 35°C) which could reflect steric hindrance due to the methyl group. On the other hand, comparison of the T_{50} values for aniline and o-methylaniline shows only slightly higher values for o-methylaniline.

6.2.7 Variation of the partial pressure of one of the compounds of the model mixture

Competitive chemisorption and reaction can considerably affect the rates of individual steps during hydrotreatment of a fuel and also influence the evaluation of catalyst selectivity with the four component model mixture as shown in Figs. 6.6 to 6.9. By

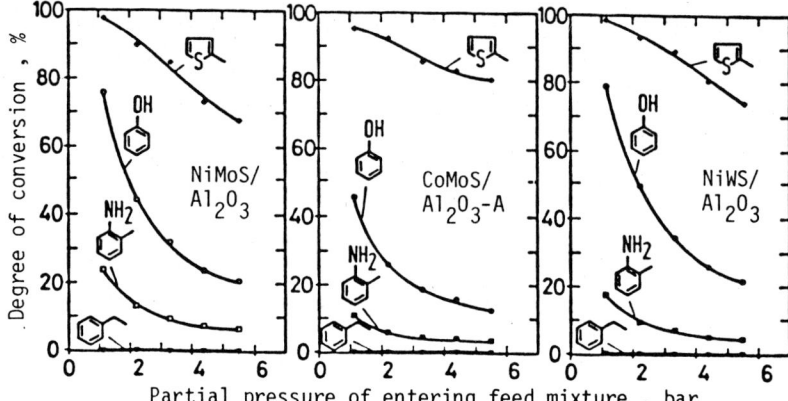

Fig. 6.6. Degree of conversion of the constituents of the four compound model mixture as a function of the total partial pressure of the feed compounds on three catalysts (refs. 4,43). T = 297°C, τ_{eff} = 9.1 sec, p_{total} = 100 bar, p_{H_2S} = 0.1 bar.

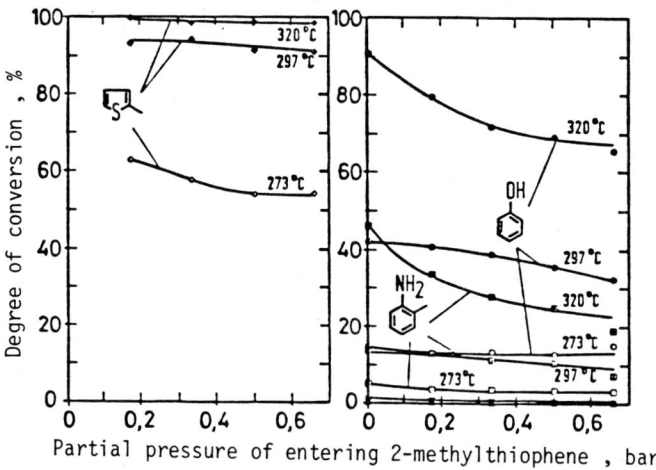

Fig. 6.7. Degree of conversion of 2-methylthiophene (left) and the other constituents of the four compound model mixture (right) as a function of 2-methylthiophene partial pressure in the feed at three temperatures (refs. 4,43). NiMoS/Al$_2$O$_3$, τ_{eff} = 9 sec, p_{H_2} = 96.7 bar, p_{Ph} = p_{o-T} = 0.67 bar, p_{Eb} = 2 bar - p_{MTh}, p_{H_2S} = 0.1 bar.

simultaneously increasing the partial pressures of all the four components of the test mixture at 297°C and p_{H_2} = 96,7 bar, as shown for three catalysts in Fig. 6.6, the degree of conversion of all the compounds decreases, indicating reaction orders of less

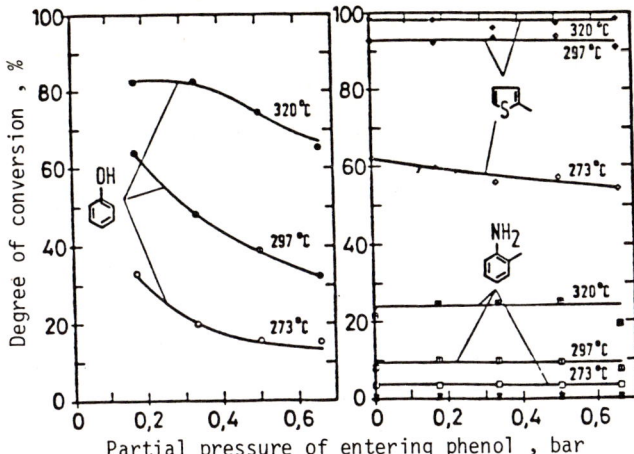

Fig. 6.8. Degree of conversion of phenol (left) and the other constituents of the four compound model mixture (right) as a function of phenol partial pressure in the feed at three temperatures (refs. 4,43). NiMoS/Al$_2$O$_3$, p$_{H_2}$ = 96.7 bar, τ_{eff} = 9 sec, p$_{MTh}$ = P$_{o-T}$ = 0.67 bar, p$_{Eb}$ = 2 bar - p$_{Ph}$, p$_{H_2S}$ = 0.1 bar.

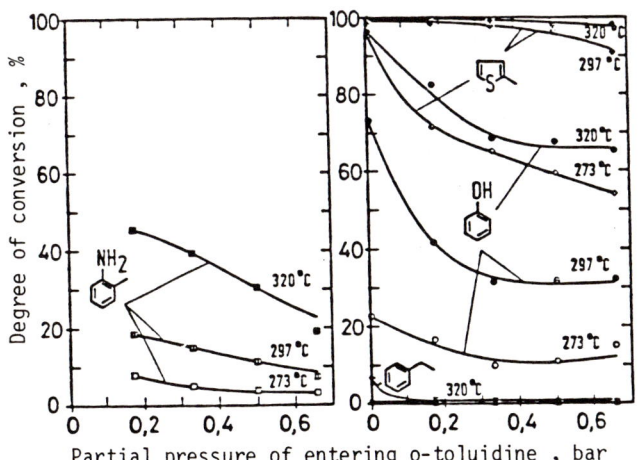

Fig. 6.9. Degree of conversion of o-toluidine (left) and the other constituents of the four compound model mixture (right) as a function of o-toluidine partial pressure in the feed at three temperatures (refs. 4,43). NiMoS/Al$_2$O$_3$, τ_{eff} = 9 sec, p$_{H_2}$ = 96.7 bar, p$_{MTh}$ = p$_{Ph}$ = 0.67 bar, p$_{Eb}$ = 2 bar - p$_{o-T}$, p$_{H_2S}$ = 0.1 bar

than one. However this decrease is slightly different on the three catalysts. When comparing the three catalysts it is seen that the diagrams for $NiWS/Al_2O_3$ and $NiMoS/Al_2O_3$ are very similar, however that for $CoMoS/Al_2O_3$-A is different.

With $CoMoS/Al_2O_3$ the degree of HDS conversion decreases less slowly with increasing partial pressure than with the other catalysts. Competitive chemisorption has least influence on HDS conversion with the $CoMoS/Al_2O_3$ catalyst. It is also seen in Fig. 6.6 that phenol conversion decreases the most with increasing total pressure of feed mixture.

As regards the influence of variation of the partial pressure of the individual components of the mixture on $NiWS/Al_2O_3$ as shown in Figs. 6.7-6.9, the following conclusions about the competitive action of the components can be drawn. Increasing the partial pressure of 2-methylthiophene from 0.2 to 0.7 bar results only in a small decrease in the conversion of 2-methylthiophene. Therefore the reaction order for 2-methylthiophene is very close to one (Fig. 6.7). The degrees of conversion of phenol and of toluidine show a declining trend indicating a competitive action and displacement of these reactants by 2-methylthiophene. Increasing the partial pressure of phenol (Fig. 6.8) causes a considerable decrease in phenol conversion, indicating a reaction order for phenol of substantially less than one. However, the degrees of conversion of 2-methylthiophene and 2-methylaniline are not noticeably affected. It can be concluded that the active sites are of a partially acidic nature so that the adsorption of methylaniline and of methylthiophene is favoured. Adsorption of the more acidic phenol cannot displace anilines or thiophenes. Increasing the partial pressure of 2-methylaniline (Fig. 6.9) leads to a decrease in methylaniline conversion and also to a decrease in 2-methylthiophene and phenol conversion. This shows that methylaniline is the most strongly adsorbed of the compounds of the mixture and is again indicative of a partially acidic type of active catalyst site.

6.3 SIMULTANEOUSLY PROCEEDING HYDROGENATION REACTIONS DURING REFINING OF OILS FROM "SUMPFPHASE" COAL HYDROGENATION

The model mixture containing 2-methylthiophene, phenol, o-toluidine and ethylbenzene was mixed in a mass ratio 1:3 with an oil produced from coal hydrogenation according to new German technology and hydrogenated with a $NiMoS/Al_2O_3$ catalyst at 100 bar (Table 6.5) (ref. 4). A small extent of catalyst deactivation was

TABLE 6.5

Composition with respect to elements and selected compounds (in C-%) of the 1:3 mass ratio mixture of the model substances 2-methylthiophene, phenol, o-toluidine and ethylbenzene with an oil produced by coal hydrogenation via new German technology (ref.4). $NiMoS/Al_2O_3$, p_{total} = 100 bar, τ_{eff} = 8 sec, p_{H_2} = 96.7 bar, p_{H_2S} = 0.1 bar, composition of the four-compound test mixture as indicated in Fig. 6.2.

	Feed	Products obtained at			
		325 °C	375 °C	425 °C	450 °C
Elemental composition (mass-%)					
C	83.9	84.2	86.7		88.5
H	9.3	9.7	10.8		11.2
O	3.6	3.3	1.5		0.1
N	1.5	1.4	1.0		0.1
S	1.7	0.8	0.14		0.08
Selected compounds (C-%)					
2-Methylthiophene	4.1	1.5	0.0	0.0	0.0
Phenol	8.7	8.7	3.1	0.0	0.0
o-Toluidine	6.0	5.8	4.8	1.3	0.5
Ethylbenzene	15.3	15.5	15.1	13.5	14.2
Methylphenols	4.6	3.8	1.4	0.0	0.0
Cyclohexene	0.2	0.3	1.1	0.0	0.03
Methylcyclohexene	0.2	0.3	1.0	0.2	0.1
$C_6 - C_9$ Aromatics	6.8	7.4	8.7	11.3	14.1
$C_6 - C_9$ Naphthenes	5.6	6.0	13.8	21.2	21.0
Naphthalin	1.9	1.7	1.1	1.1	1.6
Tetralin	2.6	2.7	3.6	3.5	2.7
$C_1 - C_4$ HC'S	0.2	0.7	0.9	1.2	1.7

observed during hydrotreatment of the oil from coal. To avoid this, a higher pressure would be required. A few results of these experiments are reported in the table. As regards the degree of conversion of the four compounds, the same trends as those observed with the mixture in the absence of the oil were found with the sequence of reactivity being:

2-methylthiophene > phenol > o-toluidine >> ethylbenzene

Of particular interest are the concentrations of cyclohexene and methylcyclohexene. These olefins are important for the stability of the oil and are also an indicator of coke formation. They are obtained as intermediates of the HDO and HDN of phenols and anilines and mainly observed when the conversion is not complete (see Scheme 6.2 and 6.3). As regards the concentrations of the C_6 to C_9 aromatic and alicyclic hydrocarbons it is evident that mainly naphthenes and not aromatics are obtained in HDO and HDN. The molar ratio of naphthalin/tetralin is also of great practical interest and is characteristic of the catalytic system. It has been shown that at about 400°C this ratio shifts from kinetic to thermodynamic control. The temperature at which this shift occurs can be taken as a measure of the severity of hydrogenation.

6.4 REACTION STEPS IN DENITROGENATION

Denitrogenation of organic nitrogen compounds generally involves a number of individual reactions, or reaction steps, to form a complex reaction network. Each of the reaction steps must basically be regarded as having a different dependence on the reaction conditions and catalyst properties. In addition, the rate for each type of reaction depends on the nature of the substituents at those atoms which participate in bond splitting and bond formation. The underlying approach in this article is to discriminate and generalize the principal types of reaction steps taking part in denitrogenation and thus to learn their impact in distinct reaction networks. This goal is complicated by the fact that a number of reaction steps can proceed consecutively on the catalyst surface with no desorption of the intermediates which are then not observable as products and so the kinetics of these reaction steps are not individually accessible.

6.4.1 Nitrogen in aliphatic amines

Aliphatic amines are generally not constituents of fossil fuels, however they are principal intermediates in ring-opening reactions of cyclic nitrogen compounds. Their reactivity is relatively high, and almost quantitative conversion is possible below 300°C (Table 6.6) (refs. 25,36), a temperature much below that necessary for HDN of, e.g. anilines. The compounds observed in the product during conversion of pentylamine are dipentylamine, pentane, pentene-1, trans- and cis-pentene-2. The data in Table 6.6 are consistent with the mechanism in Scheme 6.4.

Generally, the question arises as to whether the rupture of a C-N bond is of the hydrogenolytic type occuring on hydrogenation sites of the catalyst or of an heterolytic nature affected through

TABLE 6.6
Conversion of pentylamine on NiMoS/Al$_2$O$_3$ (ref. 26). Product composition in C-%, p_{H_2} = 39.5 bar, p_{H_2S} = 0.08 bar, p_{Pa} = 0.65 bar.

	Space time (sec) at reaction temp. (°C)						
	250 °C [a]			262 °C [a]	300 °C [a]	380 °C [a]	262 °C [b]
Compounds (C-%)	5 sec	10 sec	20 sec	10 sec	10 sec	10 sec	10 sec
Pentylamine	95.1	88.5	78.0	82.9	2.0	1.5	84.2
Dipentylamine	1.9	0.9	0.0	0.8	0.15	0.0	0.9
Pentane	2.1	7.4	16.4	11.7	92.6	98.5	11.2
Pentene-1	0.41	0.98	1.2	0.7	0.96	0.0	0.6
trans-Pentene-2	0.36	1.47	2.9	2.9	2.9	0.0	2.1
cis-Pentene-2	0.21	0.81	1.5	1.0	1.4	0.0	0.9

[a] p_{H_2} = 39.5 bar
[b] p_{H_2} = 80 bar

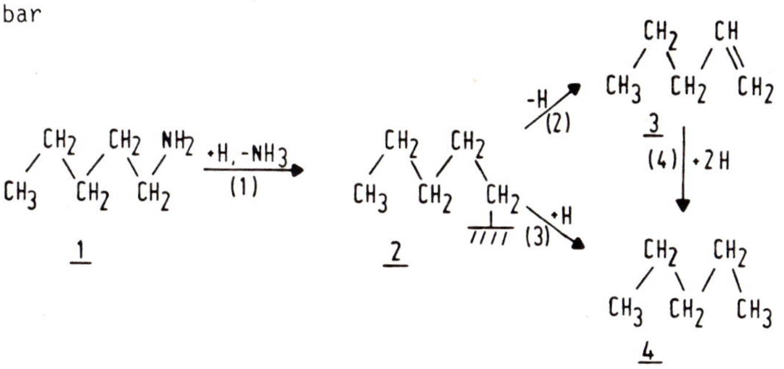

Scheme 6.4. Hydrodenitrogenation of pentylamine

electrophilic attack by acidic sites. The heterolytic rupture would lead to ammonia elimination and an olefin as the primary hydrocarbon product, which could then be isomerized and hydrogenated. As regards the influence of space time on the hydrocarbon composition in Table 6.6, it is concluded that both pentane and pentene-1 are primary products. The primary formation of pentane however has to be related to hydrogenolytic bond splitting. Primary formation of an olefin is not a proof of ammonia elimination on acidic sites. It has been shown that in systems with limited hydrogenation activity the hydrocarbon product can be obtained as an olefin even from hydrogenation sites, e.g., the hydrocarbons obtained through hydrogenation of carbon monoxide on iron catalysts (ref. 62) are mainly olefins.

During partial hydrogenation of α-olefins the double bond shift is faster than the double bond saturation, which shows that the addition of one hydrogen atom to the olefin (ref. 62) (reactions 1, 2, and 3 in Scheme 6.5) is fast and reversible and that the associative desorption of alkyl plus one hydrogen to yield the paraffin (reactions 4 and 5) is slow and irreversible. Scheme 6.5 explains the formation of pentenes-2 through the double bond shift on hydrogenation sites.

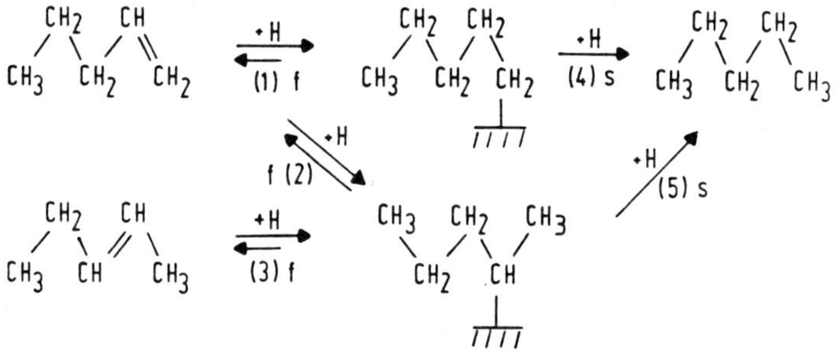

Scheme 6.5. Double bond shift during partial olefin hydrogenation.

In addition, amine disproportionation - formation of dipentylamine plus NH_3 from two molecules of pentylamine - takes place:

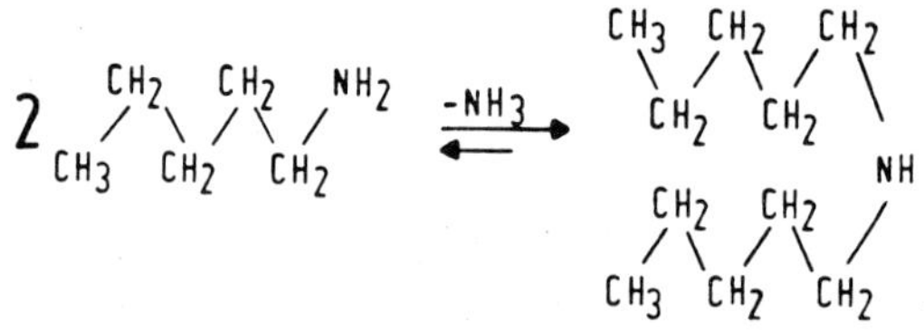

This reaction is favoured by low temperature (Table 6.6) and proceeds on weak acidic sites (ref. 63) as in the formation of ethers from alcohols (ref. 64). The dipentylamine formed is more reactive as regards C-N bond cleavage than is the primary amine. Thus the maximum dipentylamine concentration is low (Fig. 6.10). The conversion of dipentylamine generally confirms the conclusions derived above for pentylamine. Figs. 6.10 and 6.11 show the product composition as a function of reaction temperature and space time.

Additional reaction products observed in small concentrations are tripentylamine (formed according to: 2 dipentylamine ⟶ 1 tripentylamine + 1 pentylamine) and C_{10} hydrocarbons.

The main course of the reaction is the primary hydrogenolytic cleavage of a C-N bond yielding pentylamine, pentane and pentene-1, followed by secondary hydrogenation and double bond shift of the pentene-1. Pentylamine - being less reactive than dipentylamine -

attains a relatively high intermediate concentration (Fig. 6.10).

An increase in the hydrogen partial pressure from 40 to 60 bar did not noticeably affect the rate of dipentylamine conversion. It may be concluded that the slow "hydrogenolytic" reaction step (that is splitting on hydrogenation sites) of C-N bond rupture does not involve hydrogen (ref. 26).

With respect to industrial hydrodenitrogenation, it can be stated that the reactivity of aliphatic nitrogen compounds is high compared to that of cyclic and aromatic nitrogen compounds. Denitrogenation of aliphatic nitrogen containing intermediates is not a slow step in complex reaction networks. With increasing

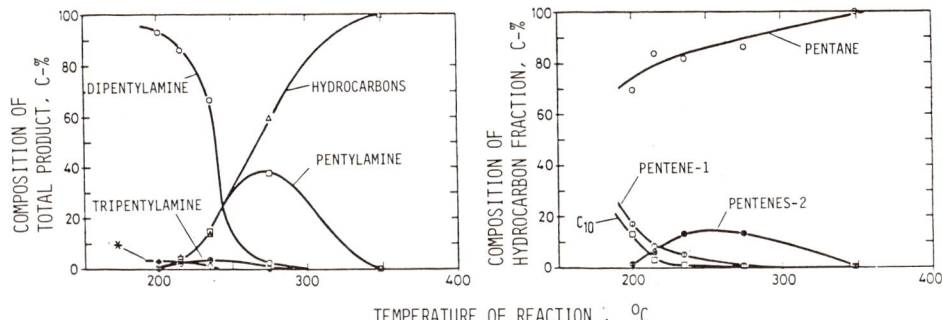

Fig. 6.10. Composition of the products from dipentylamine conversion as a function of reaction temperature (ref. 26). NiMoS/Al$_2$O$_3$, τ_{eff} = 10 sec, p_{DPa} = 0.6 bar, p_{H_2} = 40 bar, p_{H_2S} = 0.08 bar.

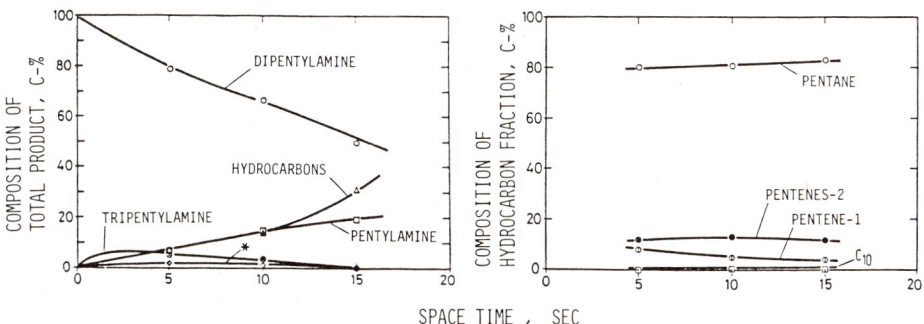

Fig. 6.11. Composition of the products from dipentylamine conversion as a function of space time (ref. 26). NiMoS/Al$_2$O$_3$, T = 235°C, p_{DPa} = 0.6 bar, p_{H_2} = 40 bar, p_{H_2S} = 0.08 bar.

acidity of the catalyst support, the mechanism of this step shifts to an ionic type of C-N bond rupture, however without any benefit for the overall reaction rate.

6.4.2 <u>Nitrogen in amino groups attached to aromatic rings: aniline and homologues</u>

The reactions of aniline and of 2-methylaniline have already been mentioned in section 6.2 in connection with the catalyst evaluation. The results of a more detailed investigation (ref. 26) are discussed below.

The degree of conversion and the product composition for aniline on NiMoS/Al_2O_3 at 40 bar as a function of reaction temperature,

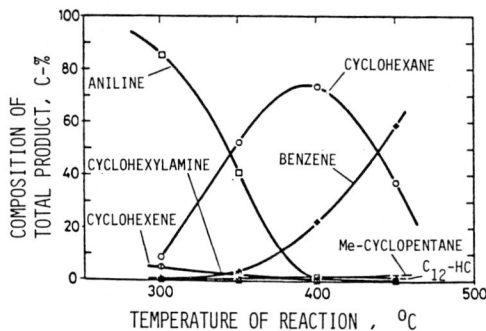

Fig. 6.12. Composition of the products from aniline conversion as a function of reaction temperature (ref. 26). NiMoS/Al_2O_3, τ_{eff} = 20 sec, p_{An} = 1.9 bar, p_{H_2} = 38 bar, p_{H_2S} = 0.08 bar.

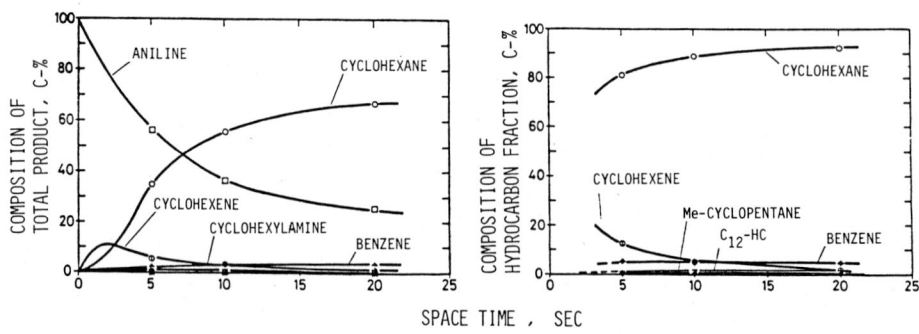

Fig. 6.13. Composition of products from aniline conversion as a function of space time (ref. 26). NiMoS/Al_2O_3, T = 350°C, p_{An} = 1.9 bar, p_{H_2} = 38 bar, p_{H_2S} = 0.08 bar.

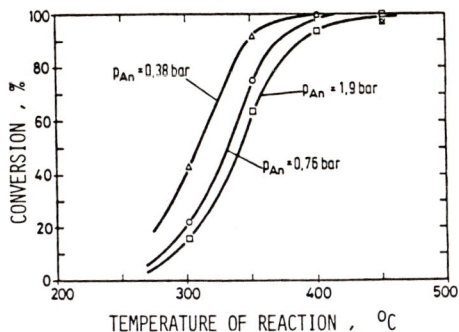

Fig. 6.14. Degree of aniline conversion as a function of reaction temperature at three partial pressures of aniline (ref. 26). NiMoS/Al$_2$O$_3$, τ_{eff} = 20 sec, p_{H_2} = 38 bar, p_{H_2S} = 0.08 bar.

space time and aniline partial pressure is shown in Figs. 6.12-6.14. A very small intermediate concentration of cyclohexylamine was found ($\approx$ 0.1 C-%). The main hydrocarbon product is cyclohexane under all reaction conditions. At high temperatures (>400°C) the formation of methylcyclopentane is observed. At this temperature also some opening of the carbon ring occurs (ca. 2 C-% at 450°C). Additional by-products are phenylcyclohexane and biphenyl (maximum yield ca. 1 C-%).

The reaction order for aniline, as estimated from Fig. 6.14 is close to zero at 300°C and increases with temperature. Strong adsorption of the aniline can therefore be assumed. In terms of the 50% conversion temperature, T_{50}, as a measure of the reactivity of the aniline, it is seen that T_{50} increases by about 40°C upon increasing the aniline partial pressure from 0.38 to 1.9 bar. Mechanistic conclusions are more safely drawn from experiments on simultaneous conversion of a two compound mixture of 78 moles aniline and 22 moles 4-methylcyclohexene (Fig. 6.15) and the comparison of experimental and equilibrium molar ratios for aromatics, olefins and saturated hydrocarbons. When comparing the curves in Fig. 6.15 (left) with those of Fig. 6.15 (right), it is seen that the olefin fraction at, e.g. 300°C in C$_7$ is even lower than in C$_6$. The cyclohexene is obtained from aniline. The methylcyclohexene (mixture of isomers) is that remaining from the 4-methylcyclohexene which was initially employed together with the aniline. Evidently the cyclohexene obtained in the product is only a minor part of the amount being formed and cyclohexene must be regarded as the main hydrocarbon formed initially from aniline. The

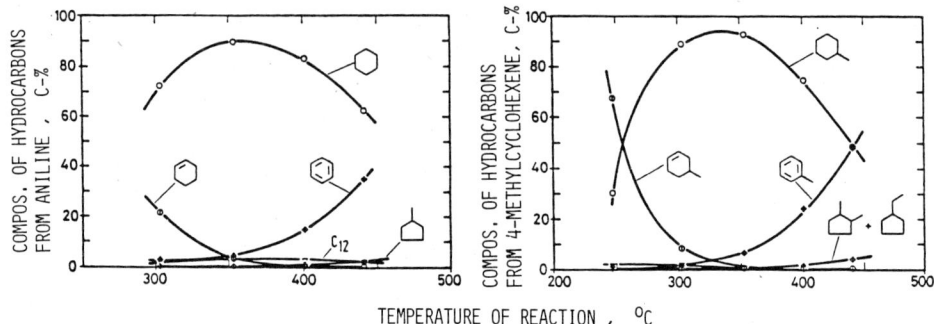

Fig. 6.15. Simultaneous conversion of aniline and 4-methylcyclohexene (molar ratio 78:22). Composition of the hydrocarbons from aniline (left) and of those from 4-methylcyclohexene (right) as a function of reaction temperature (ref. 26). NiMoS/Al$_2$O$_3$, τ_{eff} = 5 sec, p_{An} = 0.6 bar, p_{MCh} = 0.17 bar, p_{H_2} = 39.2 bar, p_{H_2S} = 0.09 bar.

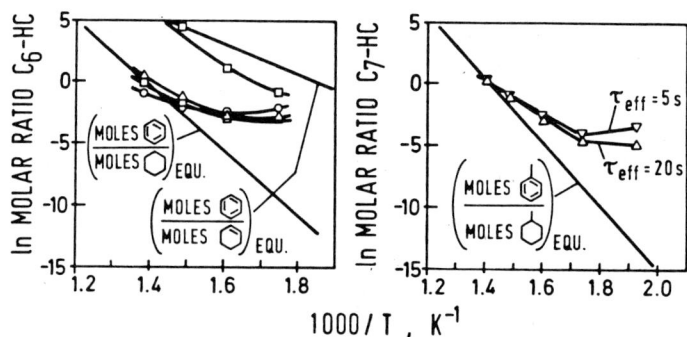

Fig. 6.16. Experimental and equilibrium molar ratios for benzene, cyclohexene and cyclohexane (left, o CoMoS/Al$_2$O$_3$-A, △ NiWS/Al$_2$O$_3$, □ NiMoS/Al$_2$O$_3$) and for toluene and methylcyclohexane (right, NiMoS/Al$_2$O$_3$) (ref. 26). Experimental data as in Fig. 6.15.

formation of benzene and of toluene is very similar.

In Fig. 6.16 a comparison of experimental and thermodynamic values is given. The concentrations of benzene (toluene) are always higher than those in thermodynamic equilibrium with cyclohexane (methylcyclohexane). Formation of the aromatics via dehydrogenation of the saturated six-membered-ring is therefore excluded (except at high temperatures >400°C). Formation of benzene from aniline via hydrogenolysis is possible on a thermodynamic basis. However, the formation of toluene starting from 4-methylcyclohexene proceeds

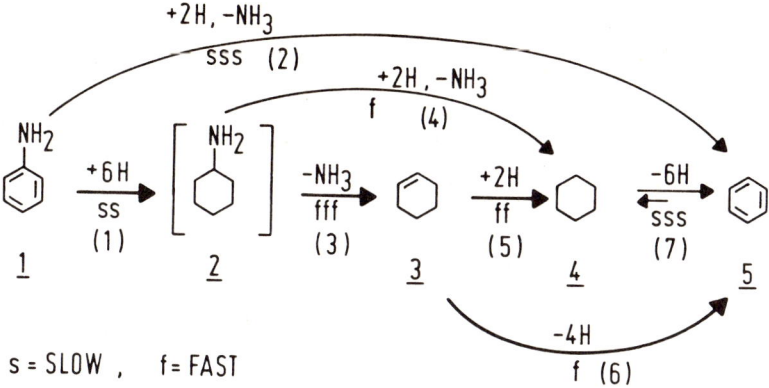

Scheme 6.6. Hydrodenitrogenation of aniline

similarly to that of benzene from aniline at temperatures above 300°C. Assuming the same mechanism for both reactions, this will be the formation of toluene and benzene via dehydrogenation of methylcyclohexene and cyclohexene respectively. Both reactions are thermodynamically feasible. The mechanism of denitrogenation of aniline can now be written as shown in Scheme 6.6.

6.4.3 <u>Nitrogen in aromatic six-membered heterocycles: Pyridine, quinoline, isoquinoline, acridine</u>

 (i) <u>Pyridine denitrogenation</u>. Pyridine and methylpyridines are prominent constituents of high temperature tars. HDN of pyridine is therefore of commercial interest in connection with tar

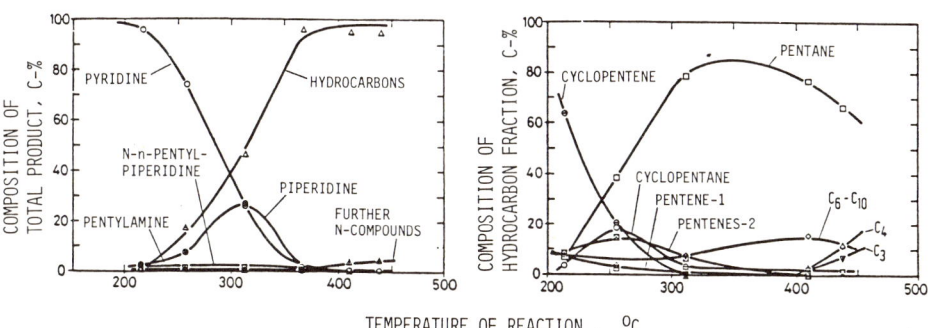

Fig. 6.17. Composition of products from pyridine conversion as a function of reaction temperature (ref. 26). NiMoS/Al_2O_3, τ_{eff} = 15 sec, $p_{pyridine}$ = 0.64 bar, p_{H_2} = 40 bar, p_{H_2S} = 0.08 bar.

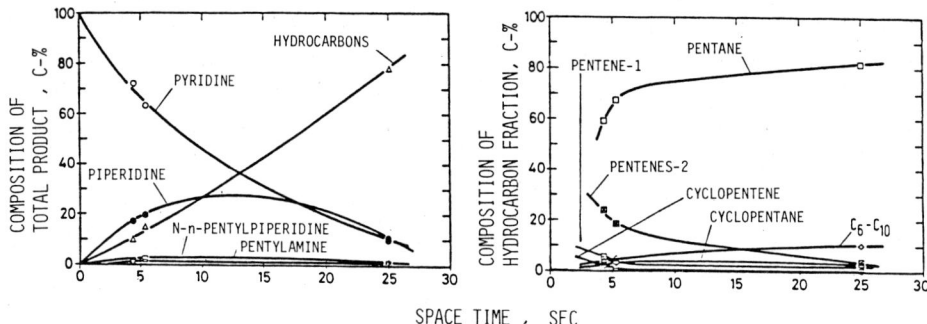

Fig. 6.18. Composition of products from pyridine conversion as a function of space time (ref. 26). NiMoS/Al$_2$O$_3$, T = 312°C, $p_{pyridine}$ = 0.64 bar, p_{H_2} = 40 bar, p_{H_2S} = 0.08 bar.

hydrotreatment and has been studied by several authors (refs. 31,56,65,66).

The product composition as a function of reaction temperature and space time is shown in Figs. 6.17 and 6.18 (refs. 4,26). In addition to the main expected reaction products piperidine, pentylamine, pentenes and pentane, the compounds N-n-pentylpiperidine (maximum yield of 4.3 C-% obtained at 256°C, τ_{eff} = 74 sec and 75% conversion of pyridine), N-cyclopentylpiperidine (maximum yield of 0.9 C-% obtained at 214°C, τ_{eff} = 33 sec and only 12% conversion of pyridine), pentylpyridines (maximum yield of 3.7 C-% obtained at 438°C, τ_{eff} = 11 sec and 99.9% conversion of pyridine), cyclopentene and cyclopentane (at temperatures below 300°C), 4-methylnonane and cyclopentylcyclopentane have been observed.

The main reaction mechanism is as shown in Scheme 6.7. The first reaction step 1, hydrogenation of pyridine to piperidine, has often

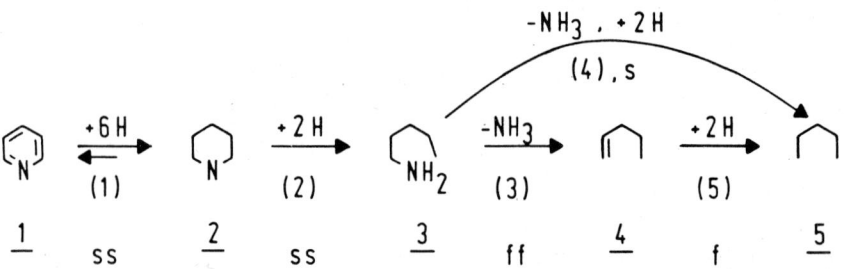

Scheme 6.7. Hydrodenitrogenation of pyridine

been regarded as a fast reversible reaction which leads to an equilibrium concentration ratio of pyridine and piperidine (refs. 56,66). The experimental values of the molar ratio of piperidine to pyridine in Table 6.7, however are generally much lower than the equilibrium ratio. Equilibrium values are approached only at relatively high temperature (ca. 350°C) and high degrees of conversion. Under most reaction conditions the first step (reaction 1) as well as the second step (formation of pentylamine from piperidine) are both similarly slow.

TABLE 6.7

Molar ratios piperidine/pyridine in the products of pyridine conversion on $NiMoS/Al_2O_3$, $NiWS/Al_2O_3$ and $CoMoS/Al_2O_3$-A and of equilibrium (ref.26). $p_{pyridine}$ = 0.64 bar, p_{H_2} = 40 bar, p_{H_2S} = 0.08 bar.

T (°C)	$NiMoS/Al_2O_3$			$NiWS/Al_2O_3$	$CoMoS/Al_2O_3$-A	Equilibrium
	5 sec	10 sec	20 sec	20 sec	20 sec	
214	0.006	0.014	0.034	0.02	0.04	$1 \cdot 10^6$
256	0.075	0.09	0.10	0.2	0.23	$4.6 \cdot 10^4$
312	0.26	0.45	0.82	0.7	0.9	40
365	-	-	-	-	-	1.8
400	-	-	-	-	-	0.26

In Table 6.8 the molar ratios of piperidine/pyridine are reported as obtained under the conditions of maximum piperidine intermediate concentration in the product as a function of the reaction temperature. Assuming first order kinetics and also that at this maximum the rates of formation and of consumption of piperidine are equal, then this ratio is equal to the reciprocal of the ratio of the corresponding rate constants:
$c_{piperidine}/c_{pyridine} = k_1/k_2$.

It is concluded, that at 310°C with $NiMoS/Al_2O_3$ the rate constant for pyridine hydrogenation, k_1, is equal to that for piperidine ring opening, k_2, and with $NiWS/Al_2O_3$ at 305°C and $CoMoS/Al_2O_3$ at 310°C, k_1 is only about half k_2. The T_{50} values required (temperatures for a 50% conversion of pyridine) show a considerable difference in catalyst activity for the conversion of pyridine (pyridine hydrogenation):
$NiMoS/Al_2O_3$ > $NiWS/Al_2O_3$ > $CoMoS/Al_2O_3$

The maximum concentration of pentylamine 3 found in the products was only 2.9 C-% ($NiMoS/Al_2O_3$, 256°C, τ_{eff} = 74 sec) showing that

TABLE 6.8
Molar ratios of piperidine/pyridine in the products of reaction of pyridine at the maximum piperidine yield, as a function of reaction temperature (ref. 26). τ_{eff} = 15 sec, $p_{pyridine}$ = 0.64 bar p_{H_2} = 40 bar, p_{H_2S} = 0.08 bar, T_{50} = Temperature for 50% conversion of pyridine, $T_{pip\ max}$ = Temperature of maximum concentration of piperidine.

	T_{50}, (°C)	$T_{pip\ max}$, (°C)	C_{pip}/C_{pyr}
NiMoS/Al$_2$O$_3$	280	310	1.0
NiWS/Al$_2$O$_3$	315	305	0.45
CoMoS/Al$_2$O$_3$	350	310	0.45

this compound is very reactive and that step 3 is very fast. Pentylamine hydrogenation has already been discussed above.

Pyridine conversion increases almost linearly with increase in hydrogen partial pressure from ca. 30 to 100 bar (ref. 26), which has to be associated with an increase in rate of step 1 (pyridine hydrogenation).

Formation of the by-products during pyridine conversion is explained as follows:

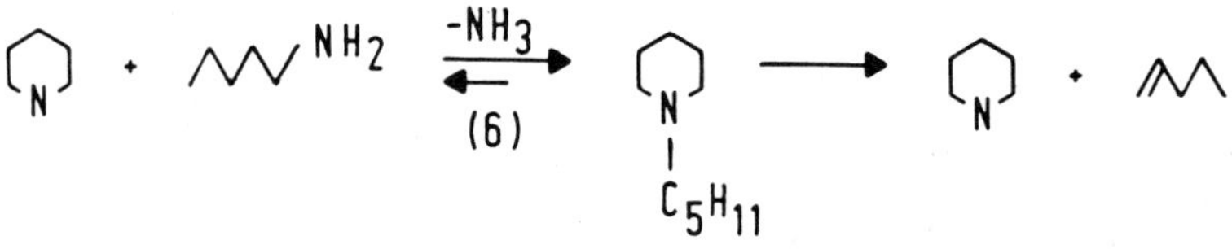

The formation of N-n-pentylpiperidine (6) is related to the well known reaction of amine disproportionation as discussed above for the pentylamine conversion. It has been described by Sonnemans and van den Berg (ref. 31). The reaction occurs at low temperatures and the product is very reactive towards denitrogenation.

A new type of reaction is the highly selective formation (7) of cyclopentene <u>7</u> at low temperature:

It is remarkable that C-C bond formation occurs during elimination of NH_3. This reaction necessarily includes several rearrangement steps and could be visualized as follows:

The formation of cyclopentylpiperidine 8 is easily explained in terms of addition of the amine piperidine to the olefin cyclopentene.

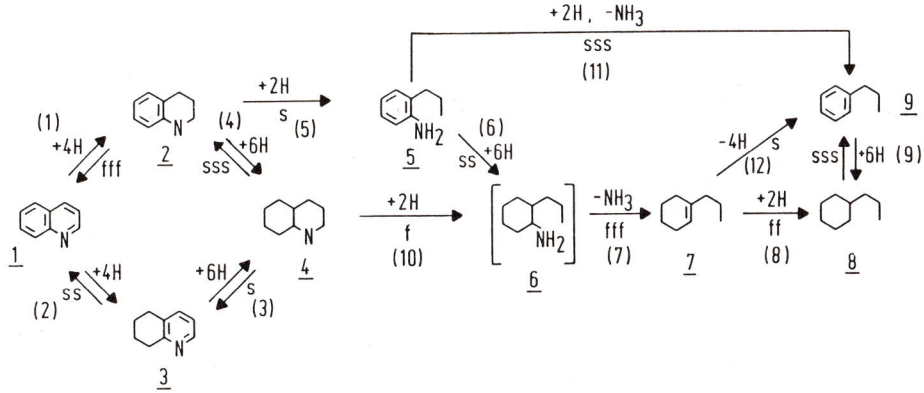

8

(ii) <u>Quinoline denitrogenation.</u> In spite of quinoline being a benzo-homologue of pyridine, the reaction mechanism for its denitrogenation is substantially different. Mainly by the work of Satterfield and Cocchetto (refs. 67,68), Katzer et al. (ref. 35) and Schulz and Eichhorn (refs. 69,70), the reaction network in Scheme 6.8 has been substantiated:

Scheme 6.8. The hydrodenitrogenation of quinoline

The hydrogenation of quinoline 1 to yield 1,2,3,4-tetrahydroquinoline 2 (1) is a very fast reaction. It proceeds almost quantitatively up to the equilibrium concentration at temperatures even below 200°C. The aromatic nitrogen-containing six-membered ring in quinoline is obviously much more reactive than the corresponding ring of pyridine. Above 300°C the equilibrium concentration of quinoline increases substantially, and its concentration in the product increases.

The hydrogenation of quinoline 1 to 5,6,7,8-tetrahydroquinoline 3 (2) on the other hand is a very slow reaction. It only starts at about 350°C. This indicates a very much lower reactivity for the carbocyclic ring of quinoline. At this temperature also the increasing concentration of quinoline 1 in equilibrium with 1,2,3,4-tetrahydroquinoline 2 favours the formation of 5,6,7,8-tetrahydroquinoline 3.

Cleavage of the C-N bond of the saturated heterocyclic six-membered ring in 1,2,3,4-tetrahydroquinoline 2 (5) is a very slow irreversible reaction, proceeding only above 350°C. The reactivity of the hydrogenated heterocyclic six-membered ring in 1,2,3,4-tetrahydroquinoline 2 towards ring opening is much lower than that of piperidine probably due to steric restrictions of the transition state.

Comparing reactions (3) and (4) - hydrogenation of the heterocyclic and of the aromatic ring in 3 and 2 respectively - both are very slow, however, reaction (3) starting at ca. 400°C is less slow. Thus at T >400°C HDN proceeds mainly via hydrogenation of the 5,6,7,8-tetrahydroquinoline 3 (3) with decahydroquinoline 4 being an important reactive intermediate. Ring opening in 1,2,3,4-tetrahydroquinoline 2 (5) to yield o-propylaniline 5 is the significant pathway at lower reaction temperature.

Conversion of o-propylaniline 5 proceeds almost exclusively through hydrogenation of the aromatic ring in a slow reaction (6) again needing about 400°C to give o-propylcyclohexylamine 6 which is highly unstable at this temperature (compare pentylamine above) and eliminates ammonia in a very fast reaction (7) to produce the olefin propylcyclohexene 7 as a reactive intermediate. Primary formation of propylbenzene 9 from o-propylaniline 5 via hydrogenolysis (11) is almost excluded as shown by its low concentration at low temperature. It is possible via propylcyclohexene dehydrogenation (12), whereas at high temperature (>450°C) the equilibrium (9) between propylcyclohexane 8 and propylbenzene 9 is established.

The reaction network of quinoline HDN outlined above (Scheme 6.8) is substantiated with the help of the following figure. Fig. 6.19 shows the product composition as a function of reaction temperature. From these data additional by-product formation is

evident:
(1) Formation of alkyltetrahydroquinolines and alkylquinolines with a maximum yield at 400 - 450°C.
(2) Formation of aniline, o-methylaniline and o-ethylaniline via C-C bond rupture of 1,2,3,4-tetrahydroquinoline instead of C-N bond rupture.
(3) Formation of the hydrocarbons C_6, C_7, C_8 and C_1-C_3 resulting from HDN of the anilines.
(4) Hydrocarbons C_{10}, C_{11}, C_{12} obtained via HDN of the alkyl-quinolines and alkyltetrahydroquinolines.
(5) Indanes and indenes formed in a similar reaction to that producing cyclopentene from piperidine discussed above.

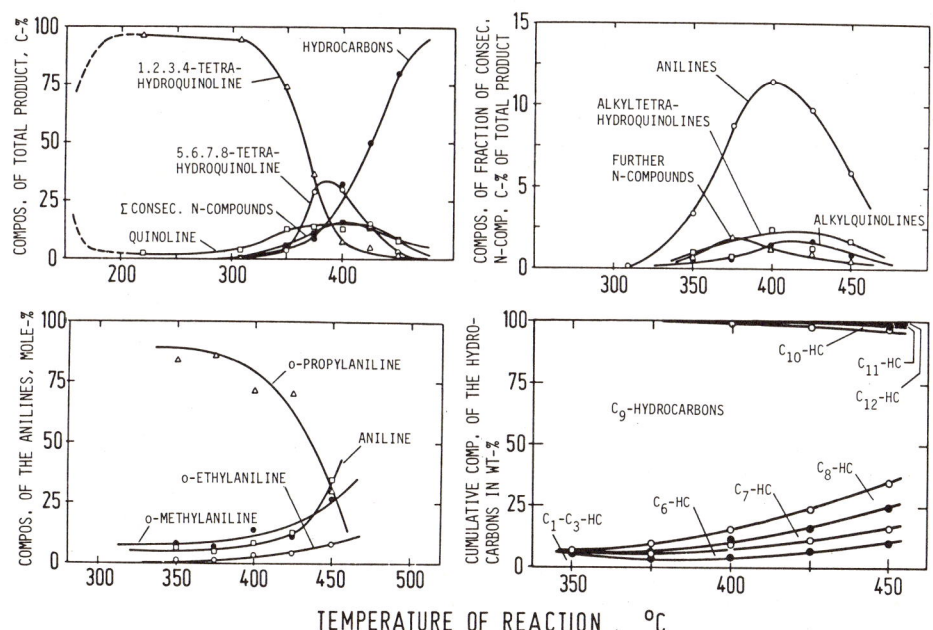

Fig. 6.19. Composition of the products from quinoline conversion as a function of reaction temperature (ref. 70). NiWS/Al2O3, p_{Qu} = 0.46 bar, p_{H_2S} = 0.127 bar, p_{total} = 46 bar, τ_{eff} = 10 sec.

Quinoline conversion has also been performed at different space times (ref. 70). The mechanistic conclusions are as stated above.
The attainment of the thermodynamic equilibria for reactions (1), (2) and (9) (Scheme 6.8) during quinoline HDN as a function of

temperature is illustrated with the diagrams of Fig. 6.20. Because of the great stoichiometric surplus of hydrogen, its relative variation with the degree of conversion can be neglected and the linear plot of the experimental logarithmic concentration ratios against 1/T indicate attainment of equilibrium. Reaction (1) is at equilibrium over the entire temperature range (300-450°C) on both catalysts. Reaction (2) is at equilibrium on $NiWS/Al_2O_3$ above 400°C and on $NiWO/Al_2O_3$ only above 450°C. It follows that nickel-tungsten-alumina is a much more active hydrogenation catalyst in the sulphide than in the oxide state. With $NiWS/Al_2O_3$ above 400°C, reaction (9) is also at equilibrium.

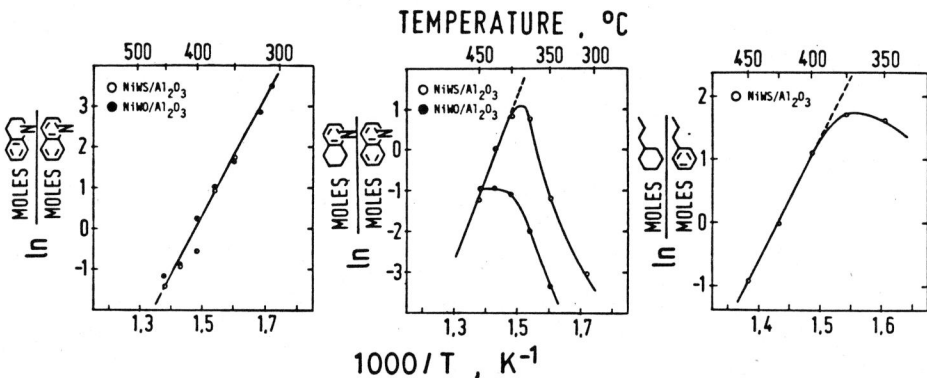

Fig. 6.20. Logarithmic molar concentration ratios as a function of 1/T for the evaluation of equilibria attained during quinoline HDN (ref. 70). $NiWS/Al_2O_3$ and other reaction conditions as indicated in Fig. 6.19, $NiWO/Al_2O_3$ same catalyst, but with no H_2S pretreatment and no H_2S in the reaction mixture.

In order to discriminate the relative reactivity of 5,6,7,8-tetrahydroquinoline and of o-propylaniline which would be indicative of the two alternative HDN routes of quinoline (see Scheme 6.8) a 1:1 molar mixture of 5,6,7,8-tetrahydroquinoline and o-isopropylaniline was converted on $NiWS/Al_2O_3$ and the reaction temperature varied (ref. 4). The results are presented in Fig. 6.21. The reactivity of 5,6,7,8-tetrahydroquinoline characterized by a 50% conversion temperature, T_{50}, of 305°C is much higher than that of o-isopropylaniline (T_{50} = 380°C). This indicates a preference for the quinoline HDN route via 5,6,7,8-tetrahydroquinoline and decahydroquinoline as intermediates as soon as

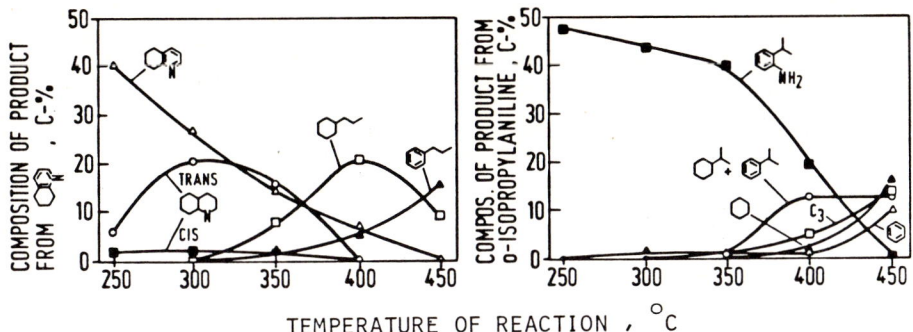

Fig. 6.21. Conversion of a 1:1 molar mixture of 5,6,7,8-tetrahydroquinoline with o-isopropylaniline (ref. 4). NiWS/Al$_2$O$_3$, τ_{eff} = 10 sec, p_{Mix} = 0.5 bar, p_{total} = 46 bar, p_{H_2S} = 0.1 bar.

5,6,7,8-tetrahydroquinoline is being formed sufficiently rapidly. At low temperature (<320°C) the 5,6,7,8-tetrahydroquinoline is converted very selectively to trans-decahydroquinoline. Noticeable decomposition of decahydroquinoline is observed above 310°C.

The system of quinoline HDN is kinetically characterized by:
- decreasing conversion with increasing p_{Qu} (0.1-0.9 bar)
- increasing conversion with increasing p_{H_2} (10-60 bar)
- increasing conversion with increasing p_{H_2S} to 0.12 bar (ref. 70).

(iii) <u>Isoquinoline denitrogenation.</u> Denitrogenation of isoquinoline proceeds in a quite different manner to that of

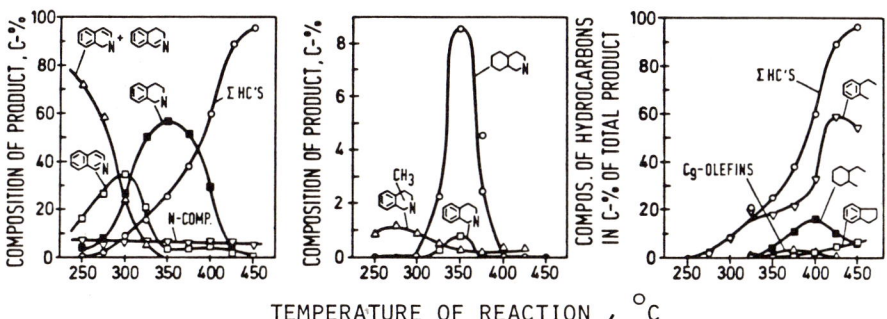

Fig. 6.22. Composition of the products from isoquinoline conversion as a function of reaction temperature (ref. 71). NiWS/Al$_2$O$_3$, τ_{eff} = 10 sec, p_{total} = 69 bar, p_{Iq} = 0.5 bar, p_{H_2S} = 0.12 bar.

*C_9 olefins (isomers)

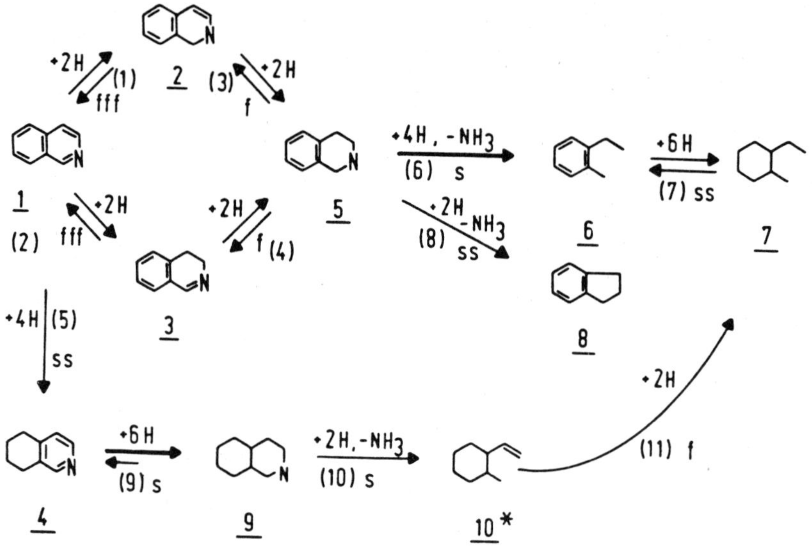

Scheme 6.9. Hydrodenitrogenation of isoquinoline

quinoline, due to the change in position of nitrogen in the heterocyclic ring. The reaction pathway has been derived from kinetic investigations with $NiWS/Al_2O_3$ (refs. 4,71). Results showing the temperature influence are presented in Fig. 6.22.

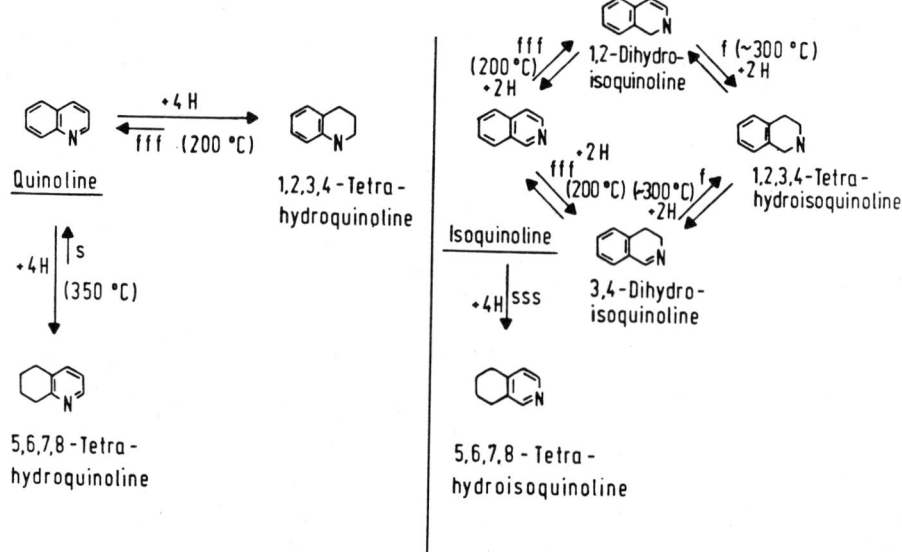

Scheme 6.10. Comparison of the first hydrogenation steps of quinoline and isoquinoline.

At low temperatures (200-275°C) the main reaction products are 1,2-dihydroisoquinoline 2 and 3,4-dihydroisoquinoline 3 (reactions (1) and (2)). 5,6,7,8-Tetrahydroisoquinoline 4 is formed very slowly at temperatures above 300°C. In HDN of quinoline no dihydro compounds were observed, whereas 5,6,7,8-tetrahydroquinoline is a main reaction intermediate (Scheme 6.10). How can these essential differences be explained? Presumably they have to be related to differences in the chemisorption of the compounds and transition states of the reaction. With isoquinoline, an edgeways chemisorption of the heterocyclic ring seems to prevail which favours individual hydrogenation of the double bonds of this ring, the half hydrogenated state being orientated almost perpendicularly to the surface (Scheme 6.11).

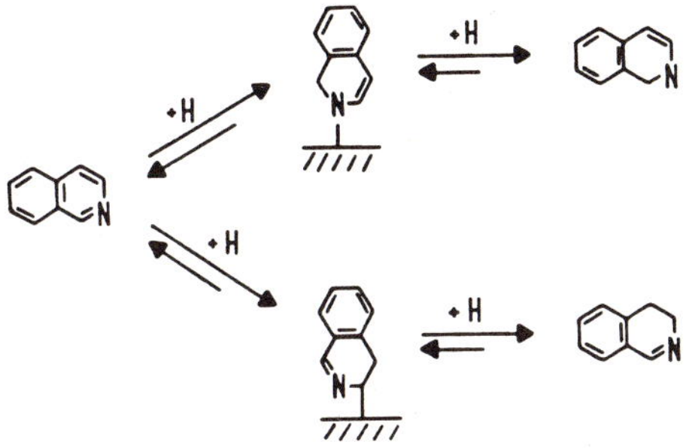

Scheme 6.11. Proposed initial steps of isoquinoline hydrogenation.

With quinoline, facial chemisorption of the aromatic heterocyclic ring is dominant, leading to transfer of four hydrogen atoms and saturation of the ring during one chemisorption step and no intermediates in this hydrogenation sequence are obtained (Scheme 6.12).

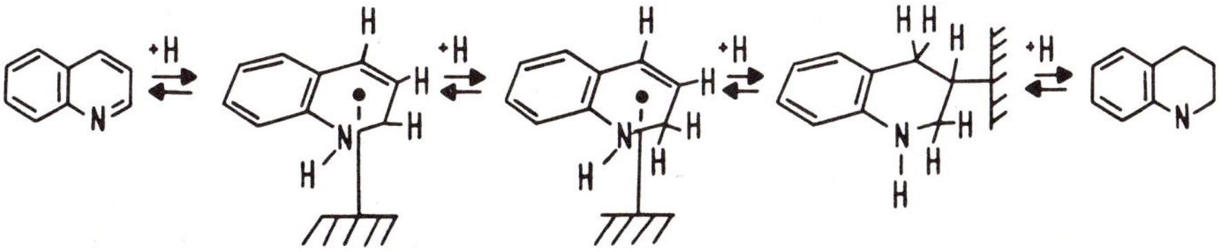

Scheme 6.12. Proposed initial steps of quinoline hydrogenation.

The lower reactivity of the carbocyclic ring of isoquinoline can be attributed to its relationship to the benzene structure rather than to the aniline structure (as with quinoline), and accordingly aniline has been shown above to be more reactive than, e.g. o-xylene or ethylbenzene.

The equilibrium between isoquinoline and the two dihydroisoquinolines is attained below 250°C. With increase of temperature (250 to 300°C) the concentration of the reactant isoquinoline increases again due to the shift in this equilibrium with reaction temperature.

Formation of 1,2,3,4-tetrahydroisoquinoline 5 in Scheme 6.9 occurs only at about 300°C, whereas formation of 1,2,3,4-tetrahydroquinoline 2 in Scheme 6.8 is possible below 200°C. 1,2,3,4-Tetrahydroisoquinoline is formed very selectively and is an important intermediate in isoquinoline HDN, whereas quinoline HDN, particularly at high temperature, proceeds to a high degree via 5,6,7,8-tetrahydroquinoline and decahydroquinoline.

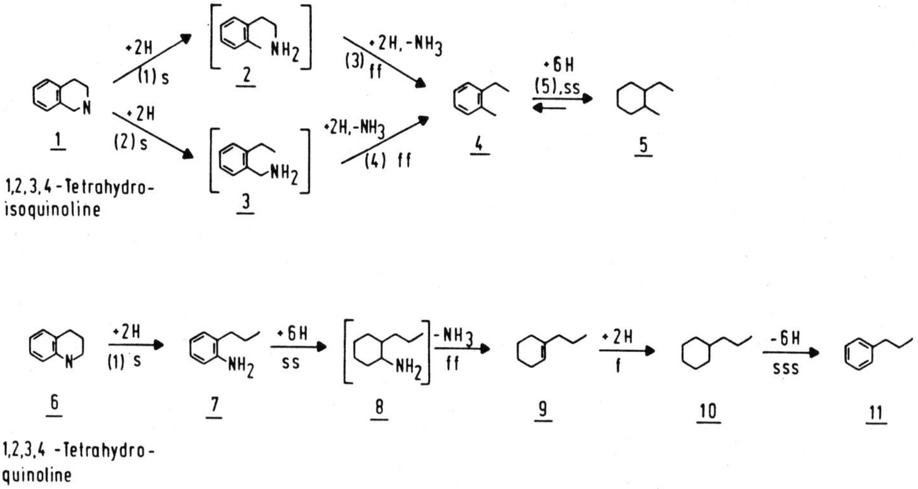

Scheme 6.13. Comparison of HDN of 1,2,3,4-tetrahydroisoquinoline and 1,2,3,4-tetrahydroquinoline.

Like 1,2,3,4-tetrahydroquinoline, 1,2,3,4-tetrahydroisoquinoline is rather stable up to 320°C. The rate of C-N bond splitting is not very different for both the isomers. However the products from both reactions behave very differently (Scheme 6.13).

The amines 2 and 3 obtained from compound 1 are saturated, very reactive, not observed as products and transformed in very fast reactions into the aromatic hydrocarbon o-methylethylbenzene 4. Methylethylcyclohexane 5 is formed only as a secondary product. The amine 7 obtained from compound 6 is of the alkylaniline type, very stable and its carbon-nitrogen bond only splits after hydrogenation of the aromatic ring. The saturated amine 8 yields mainly propyl-cyclohexene 9 as the primary hydrocarbon product in a very fast reaction.

As with cyclopentane from piperidine and indane from quinoline, some indane is formed from isoquinoline.

(iv) Acridine denitrogenation. Acridine denitrogenation on NiWS/Al$_2$O$_3$, at 46 bar and 375-475°C, proceeds via the reaction network in Scheme 6.14 (ref. 73).

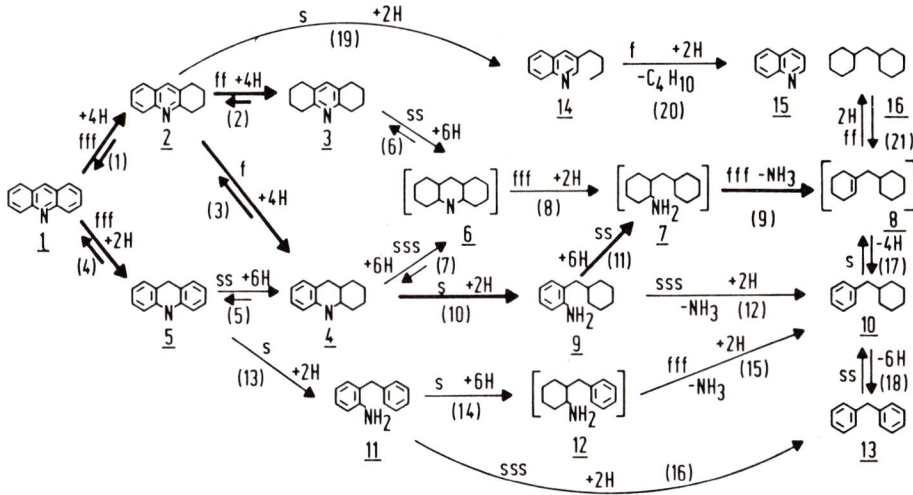

Scheme 6.14. Hydrodenitrogenation of acridine

Because of the greater complexity of the reactant molecule, as compared to those discussed above, the number of alternative and interconnected reaction routes is larger, although the nature of the reaction steps is mainly that elucidated for the simpler molecules. Experimental problems arise from the low volatility and solubility of acridine. Earlier work was performed by Katzer and co-workers (ref. 72).

Initial hydrogenation of one of the three condensed aromatic acridine rings is very fast (reactions (1) and (4)). Hydrogenation of a second ring is fast, as long as it is condensed to another aromatic ring (reactions (2) and (3)) and equilibrium between the respective compounds - acridine 1, 1,2,3,4-tetrahydroacridine 2, 9,10-dihydroacridine 5, 1,2,3,4,5,6,7,8-octahydroacridine 3 and the octahydroacridine 4 - is established at, e.g. 400°C.

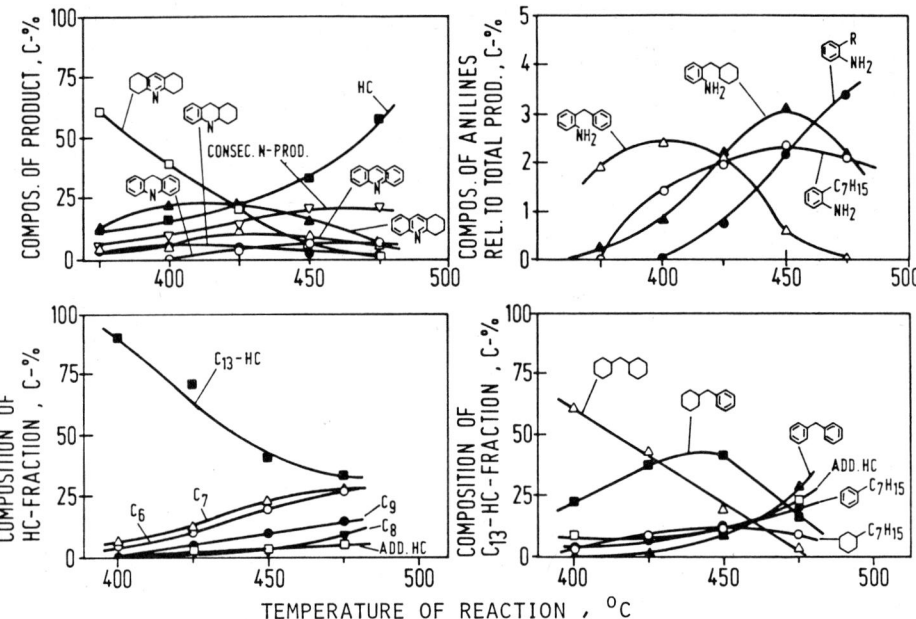

Fig. 6.23. Composition of the products of HDN of acridine as a function of reaction temperature (ref. 73). NiWS/Al$_2$O$_3$, τ_{eff} = 5 sec, p_{H_2} = 45.9 bar, p_{Ac} = 0.09 bar, p_{H_2S} = 0.09 bar.

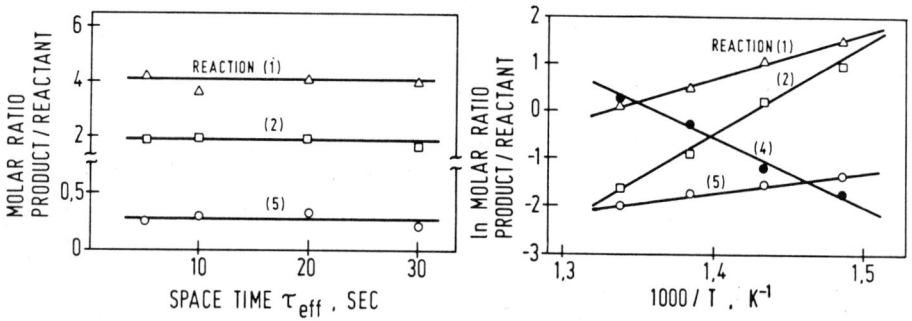

Fig. 6.24. Proof of equilibrium attainment during HDN of acridine (ref. 73). Conditions as in Fig. 6.26. Left hand: T = 400°C, right hand: τ_{eff} = 5 sec. (reactions from Scheme 6.14)

C-N Bond cleavage in 9,10-dihydroacridine <u>5</u> is facilitated by weakening of these aniline type bonds by the second aromatic ring. Thus o-benzylaniline <u>11</u> can be formed at relatively low temperature However, the reaction rate remains low because of the low equilibrium concentration of 9,10-dihydroacridine <u>4</u> whose C-N single bond in β-position to the aromatic system is weak and reacts to form o-methylcyclohexylaniline 9 a relatively stable intermediate. The maximum concentration of this intermediate (ca. 3.3 C-%) is attained only at the high temperature of 450°C. It is then hydrogenated to the very unstable saturated primary amine 7, which forms cyclohexenylcyclohexylmethane 8, dicyclohexylmethane 16, cyclohexylphenylmethane 10 and diphenylmethane 13.

A non-negligible side reaction concerns the formation of quinoline 15 according to reactions (19) and (20). At 400°C and τ_{eff} = 30 sec its yield amounted to 12 C-%.

6.4.4 Nitrogen in aromatic five-membered heterocycles: pyrrole, indole, carbazole

(i) *Pyrrole denitrogenation.* The reaction steps in the main route of pyrrole hydrodenitrogenation (Scheme 6.15) involve slow hydrogenation of the aromatic five-membered ring (1), cleavage of a C-N bond in pyrrolidine (2) and the further conversion of butylamine as outlined in section 6.4.1.

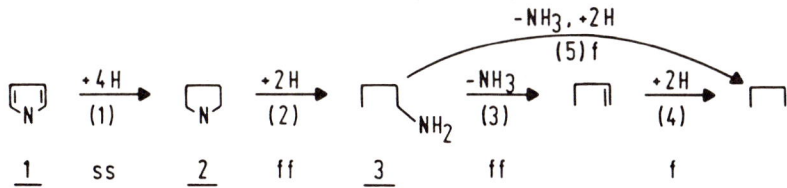

Scheme 6.15. Hydrodenitrogenation of pyrrole.

The high 50% conversion temperature for pyrrole (T_{50} = 350°C, NiWS/Al2O3, p_{total} = 46 bar, τ_{eff} = 10 sec) characterizes a lower reaction rate than observed with pyridine (compare section 6.4.3). In the temperature range 300 to 400°C the equilibrium constant, K_p for reaction (1) is very small ($K_{p\,300°C}$ = 1.5 x 10^{-2} and $K_{p\,350°C}$ = 1.9 x 10^{-3}). Thus the equilibrium yields of pyrrolidine are restricted to values of less than 1 %. However the experimental ratios of moles pyrrolidine to moles pyrrole are still much smaller than the equilibrium concentration ratio. At 350°C and a 50% conversion of pyrrole a yield of pyrrolidine of only 0.22 C-% has been obtained. Thus the low rate of pyrrolidine formation is due not only to the unfavourable thermodynamic equilibrium but also to a low value of the rate constant for the forward reaction. The concentration of the next intermediate n-butylamine 3 remains small because of its high reactivity at this high temperature. The maximum butylamine concentrations observed are as low as 0.04 C-% of the total product. Fig. 6.25 shows the composition of the products from pyrrole conversion as a function of the reaction temperature (ref.74).

The final maximum yield of n-butane - the expected hydrocarbon from complete HDN of pyrrole - is only about 50 C-%. This illustrates the appreciable importance of side reactions in the system, as can be deduced from Fig. 6.25. The corresponding products are n-octane $\underline{4}$, ethylcyclohexane $\underline{5}$, 3-methylheptane $\underline{6}$, dibutylamine $\underline{7}$, tributylamine $\underline{8}$, 2-butylpyrrole $\underline{9}$, 3-butylpyrrole $\underline{10}$ and N-butylpyrrole $\underline{11}$.

Alkylation of the pyrrole with butylamine is a very important reaction because it accounts for 31.5 C-% of the C_8 hydrocarbon products. Alkylation in the 2-position of pyrrole (yielding 16 C-% n-octane) is much favoured compared to alkylation in the 3-position (yielding 3.5 C-% 3-methylheptane). A very interesting reaction is also the formation of ethylcyclohexane (12 % yield) probably

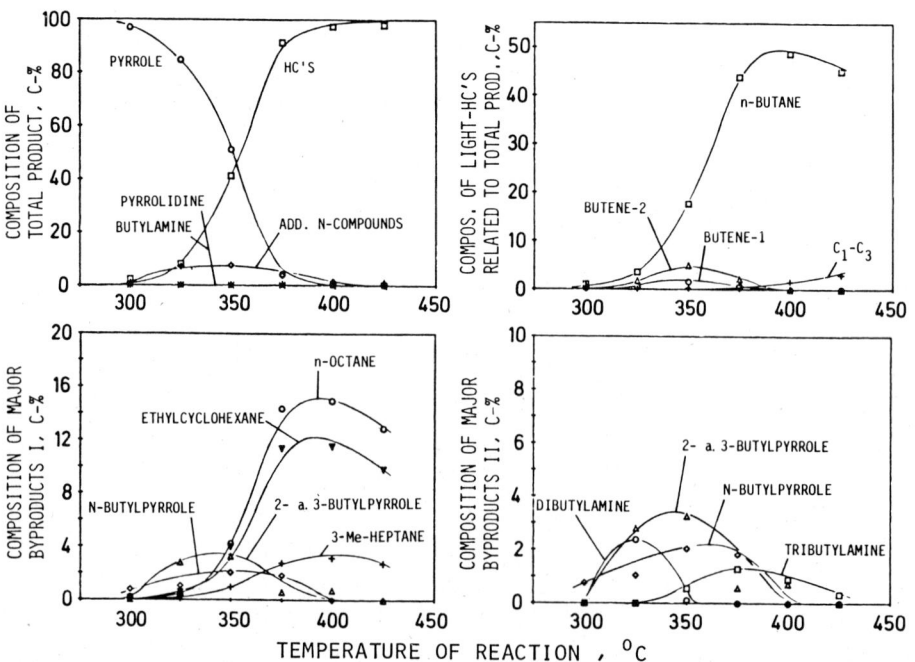

Fig. 6.25. Product composition as a function of reaction temperature for HDN of pyrrole (ref. 74). NiWS/Al_2O_3, p_{H_2} = 45 bar, $p_{pyrrole}$ = 0.5 bar, p_{H_2S} = 0.09 bar, τ_{eff} = 10 sec.

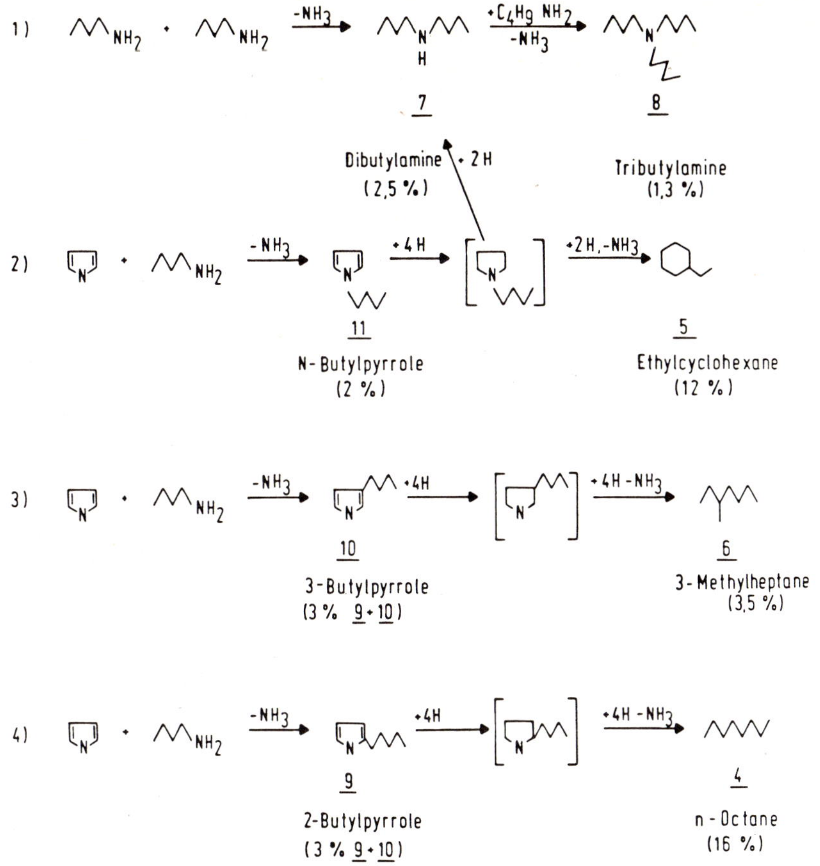

Scheme 6.16. By-product formation during HDN of pyrrole.

through rearrangement and NH_3-elimination from N-butylpyrrole.

(ii) <u>Indole denitrogenation.</u> Indole denitrogenation follows mainly the route shown in Scheme 6.17. The product composition as a function of space time and reaction temperature is shown in Figs. 6.26 and 6.27 (ref. 74).

A plot of the molar ratio of 2,3-dihydroindole <u>2</u> to indole <u>1</u> over space time at 250°C was linear, showing the hydrogenation (1) to be in equilibrium. Thus in Scheme 6.17 reaction (1) is very fast and similar to the first step of hydrogenation of quinoline. However, thermodynamic equilibrium restricts the maximum obtainable 2,3-dihydroindole <u>2</u> yield to ca. 10-15 C-% at 350°C. C-N bond rupture in dihydroindole starts above 300°C as a relatively slow reaction (2) to form the o-ethylaniline <u>3</u>. This is

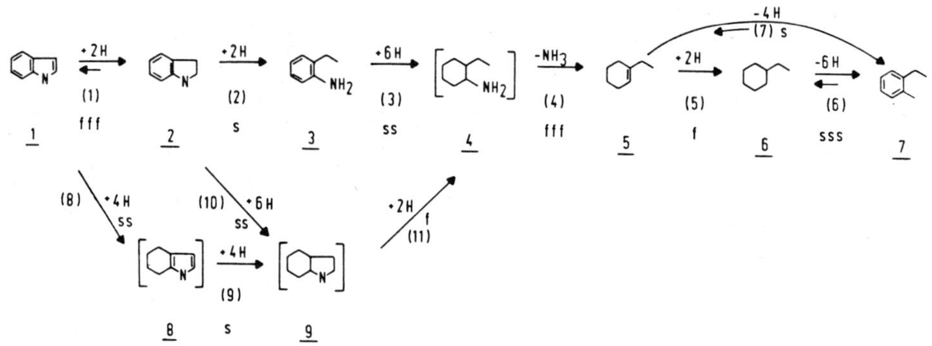

Scheme 6.17. Basic route of indole denitrogenation.

very stable and hydrogenated only above 350°C (3) to give the very reactive saturated amine 4, which undergoes fast NH_3-elimination (4)

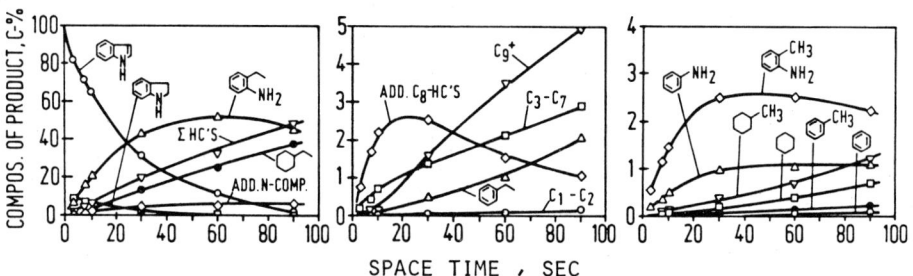

Fig. 6.26. Product composition in indole denitrogenation as a function of space time (ref. 74). NiWS/Al_2O_3, T = 350°C, p_{indole} = 0.5 bar, p_{H_2S} = 0.09 bar, p_{H_2} = 45.4 bar.

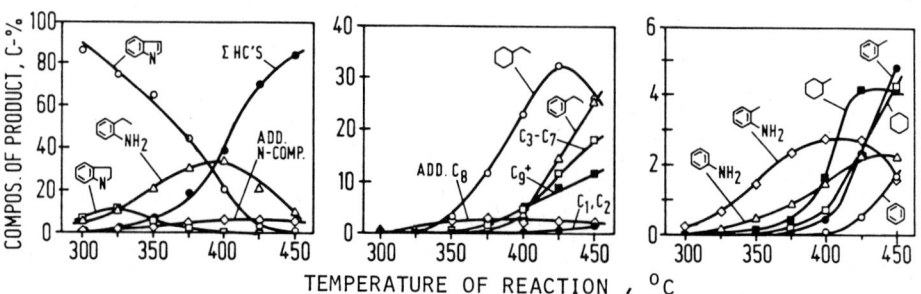

Fig. 6.27. Product composition in indole denitrogenation as a function of temperature (ref. 74). NiWS/Al_2O_3, τ_{eff} = 10 sec, conditions as in Fig. 6.26.

to ethylcyclohexene 5. At the relatively high reaction temperature (400°C) necessary for the aniline conversion (3), the olefin 5 is easily hydrogenated to yield ethylcyclohexane 6 as the main hydrocarbon product. Thus the particularly slow step of indole denitrogenation is the hydrogenation of the substituted aniline 3 and therefore high HDN activity of a catalyst for indole is equivalent to high hydrogenation activity for aromatic rings.

By-product formation, particularly of aniline, toluidine, cyclohexane, methylcyclohexane, benzene and toluene, results from C-C bond rupture in 2,3-dihydroindole (reactions (11) and (12)):

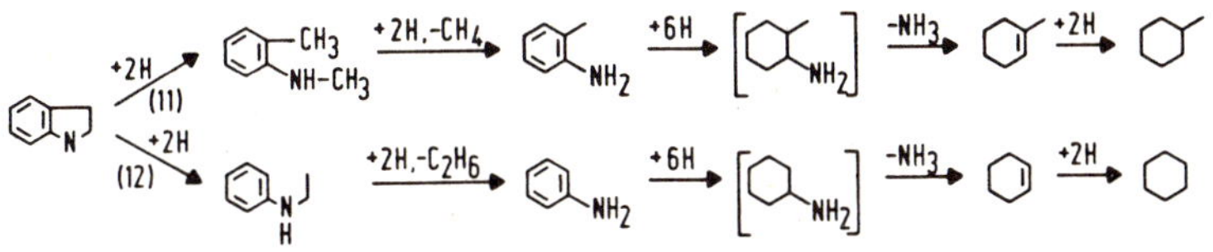

(iii) <u>Carbazole denitrogenation.</u> A preliminary investigation (ref. 75) indicated the main route of HDN of carbazole to be as in Scheme 6.18, and being consistent with the results of product composition in Fig. 6.28.

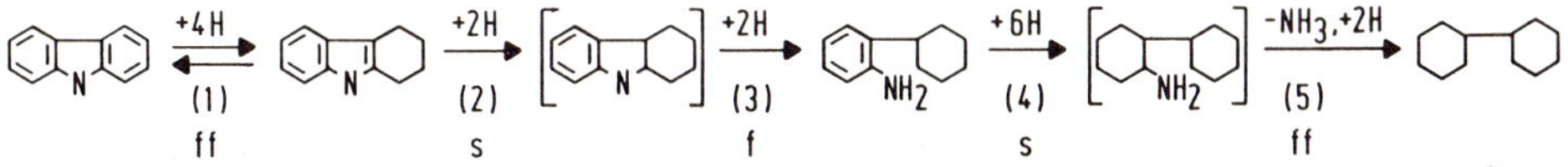

Scheme 6.18. Basic route of carbazole denitrogenation.

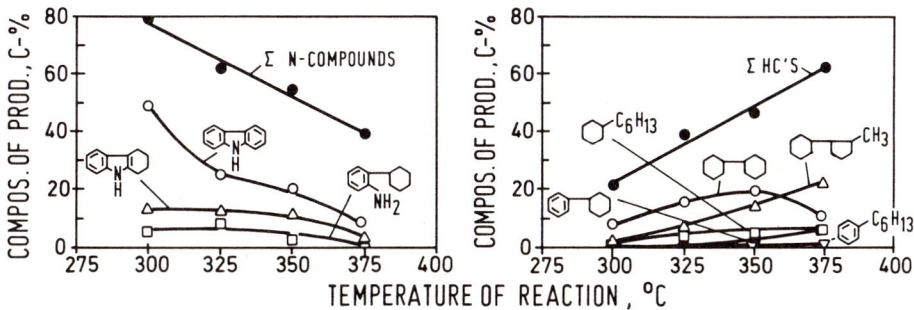

Fig. 6.28. Composition of the products from carbazole as a function of reaction temperature (ref. 75). NiWS/Al$_2$O$_3$, p_{H_2} = 45.8 bar, τ_{eff} = 10 sec, p_{Ca} = 0.09 bar, p_{H_2S} = 0.1 bar.

6.4.5 Nitrogen in saturated monocyclic five- and six-membered rings: Pyrrolidine and piperidine.

(i) <u>Pyrrolidine denitrogenation.</u> The denitrogenation of pyrrolidine is of interest because it is the first observable intermediate in pyrrole conversion. The product composition as a function of reaction temperature is shown in Fig. 6.29 (ref. 27).

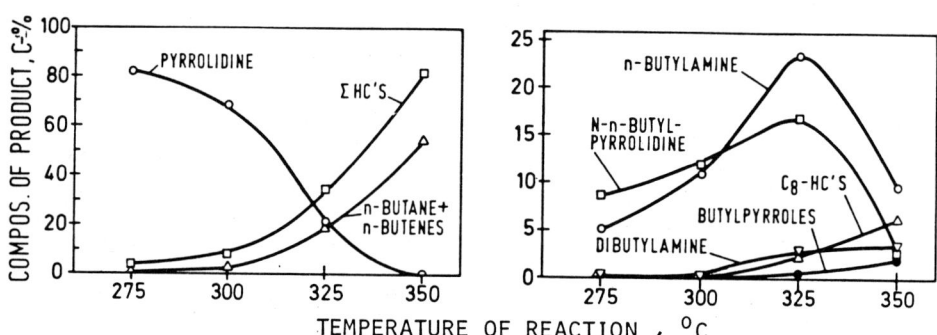

Fig. 6.29. Composition of the products from pyrrolidine as a function of reaction temperature (ref. 27). NiWS/Al$_2$O$_3$, τ_{eff} = 10 sec, $p_{pyrrolidine}$ = 1 bar, p_{H_2S} = 0.09 bar, p_{H_2} = 44.9 bar.

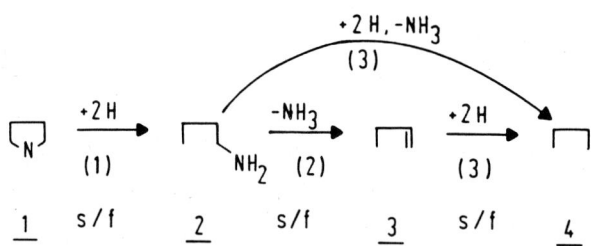

Scheme 6.19. Hydrodenitrogenation of pyrrolidine.

The main HDN route is shown in Scheme 6.19. Cleavage of a C-N bond in pyrrolidine occurs above 280°C. The reaction products attain their maximum yield of ca. 23 C-% at 225°C indicating a comparatively high stability in this reaction sequence. Butylamine conversion occurs as outlined above (section 6.4.1).

When comparing the relative rate of pyrrolidine ring cleavage to that of pyrrolidine dehydrogenation yielding pyrrole, the latter is found to be substantially slower:

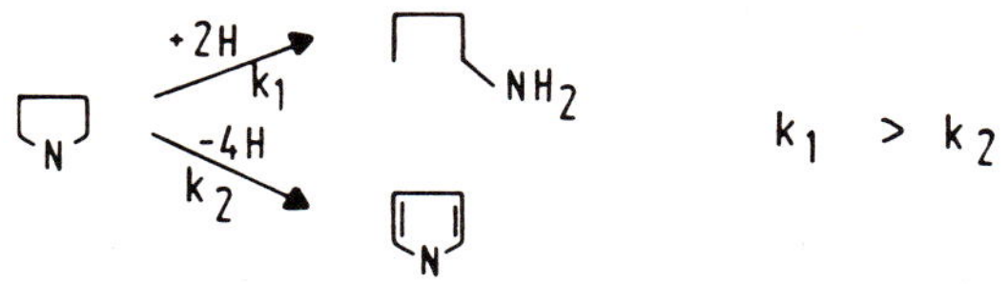

$$\text{pyrrolidine} \xrightarrow[k_2]{\overset{+2H}{\underset{-4H}{k_1}}} \text{aminobutene / pyrrole} \qquad k_1 > k_2$$

The yield of pyrrole remains far below 1 C-% despite the fact that pyrrole is thermodynamically favoured at equilibrium (in the temperature range in question). Correspondingly, the hydrogenation of pyrrole was found to be a slow step in its denitrogenation (section 6.4.4).

By-product formation from pyrrolidine is much lower than from pyrrole. This is due to the much lower pyrrole concentration in the reacting mixture, because most of the by-products (n-octane, 3-methylheptane) are produced through C-alkylation of the pyrrole. The main intermediate by-product being formed is butylpyrrolidine (with a maximum yield of 17 C-% at 325°C):

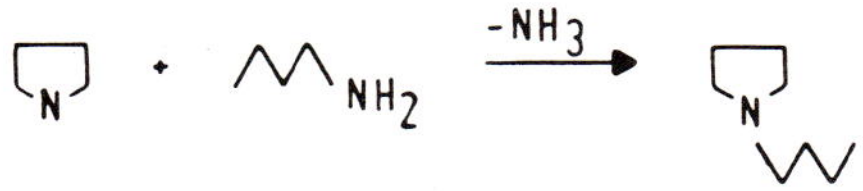

$$\text{pyrrolidine} + \text{butylamine} \xrightarrow{-NH_3} \text{butylpyrrolidine}$$

It may be assumed that this is also the precursor of the main by-product hydrocarbon, the 4-methylheptane, which is not formed in noticeable amount from pyrrole.

  (ii) <u>Piperidine denitrogenation.</u> Piperidine denitrogenation mainly follows reaction Scheme 6.20 (see also Fig. 6.30).

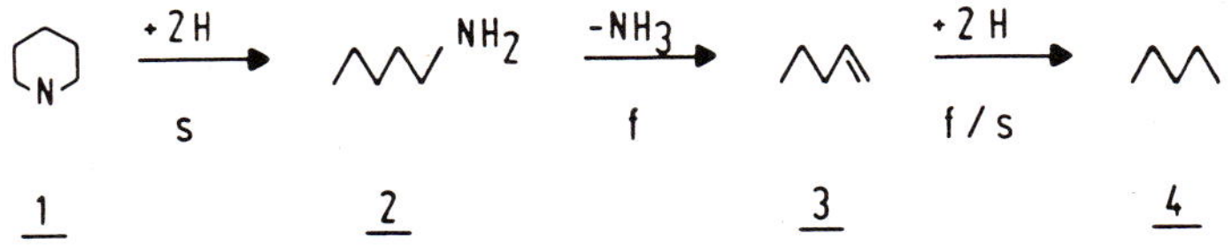

$$\underset{\underset{\underline{1}}{s}}{\text{piperidine}} \xrightarrow{+2H} \underset{\underset{\underline{2}}{}}{\text{pentylamine}} \xrightarrow{-NH_3} \underset{\underset{\underline{3}}{f}}{\text{pentene}} \xrightarrow[f/s]{+2H} \underset{\underline{4}}{\text{pentane}}$$

Scheme 6.20. Hydrodenitrogenation of piperidine.

In this sequence the C-N bond rupture is the slow step. Piperidine is substantially less reactive than pyrrolidine. At higher temperatures the intermediate aliphatic amine reacts more rapidly and its concentration is lower. The amount of pyridine formed via dehydrogenation remains low, the maximum yield being ca. 3 C-% at 350°C. The main intermediate by-product formed is N-n-pentyl-piperidine, the maximum yield being ca. 9 C-% at ca. 320°C:

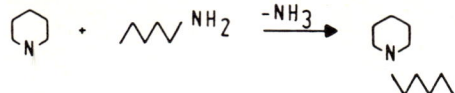

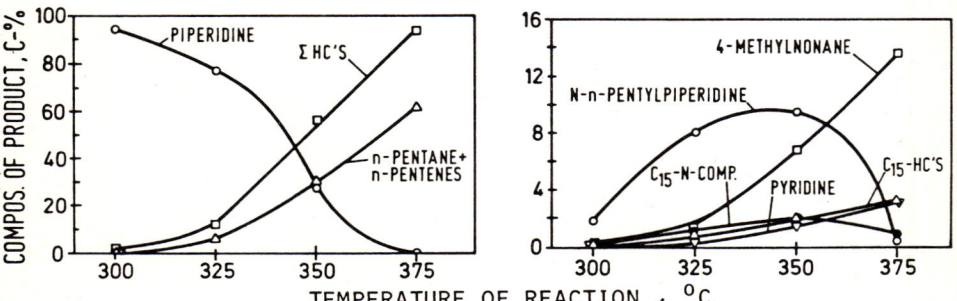

Fig. 6.30. Composition of products from piperidine as a function of reaction temperature (ref. 27). NiWS/Al$_2$O$_3$, τ_{eff} = 10 sec, $p_{piperidine}$ = 1 bar, p_{H_2S} = 0.09 bar, p_{H_2} = 44.9 bar.

A particularly high yield of 4-methylnonane (ca. 14 C-%) is obtained at high temperature (350°C). The following reaction could account for its formation:

6.5 GENERALIZATION FOR FAST AND SLOW STEPS IN HDN REACTION NETWORKS

The reactivity of reactants and intermediates in HDN networks covers a wide range. Power law rate constants appear not to be an appropriate measure of reactivity for several reasons. They are substantially dependent on the concentrations of the other compounds in the mixture and the range of reactivity is so broad that one reference temperature would not be suitable for all the reactions. Thus a temperature scale, the 50% conversion temperature, is used in this paper as a coarser measure of reactivity and the T_{50} values determined cover the range from 200 to 450°C.

Principally two types of reaction have to be distinguished: hydrogenation of unsaturated systems and cleavage of C-N bonds. However, the range of reaction rates for both types of reactions is very broad, due to differences in the molecular structures. In Table 6.9 the reactivities for 24 unsaturated compounds in 29 reactions are presented as T_{50} values.

TABLE 6.9
Reactivity of aromatic nitrogen compounds for hydrogenation reactions as characterized by approximate T_{50} values (temperature for 50% conversion), NiMoS/Al$_2$O$_3$ (NiWS/Al$_2$O$_3$), p_{H_2} 40 - 50 bar, τ_{eff} ca. 10 sec.

No.	Reaction	T_{50} (°C)	Ref.
One-ring compounds			
1	pyrrole +4H → pyrrolidine	350 *	74
2	pyridine +6H → piperidine	280	26
3	aniline +6H → cyclohexylamine	350	43
4	2-ethylaniline +6H → 2-ethylcyclohexylamine	380	74
Hydrocarbons for comparison			
5	o-xylene +6H → dimethylcyclohexane	>450	43
6	propene +2H → propane	~250	26
Two-ring compounds, first step			
7	indole +2H → indoline	330	74
8	+4H →	**	74
9	quinoline +4H → 1,2,3,4-tetrahydroquinoline	<200	70
10	+4H → 5,6,7,8-tetrahydroquinoline	370	70
11	isoquinoline +2H →	<200	71
12	+2H →	<200	71
13	isoquinoline +2H →	300	71
14	+2H →	300	71
For comparison			
15	naphthalene +4H → tetralin	>300	76

TABLE 6.9 (continued)

No.	Reaction			T_{50} (°C)	Ref.
Two-ring compounds, second step					
16	[pyrrole-benzene fused]	+6H →	[pyrrolidine-cyclohexane fused]	**	75
17	[dihydroquinoline]	+6H →	[decahydroquinoline]	~400	70
18	[dihydroisoquinoline isomer]	+6H →		>400	4
19	[dihydroisoquinoline]	+6H →	[decahydroisoquinoline]	380	71
20	[tetrahydroisoquinoline]	+6H →		>380	71
For comparison					
21	[tetralin]	+6H →	[decalin]	>350	76
Three-ring compounds, first step					
22	[carbazole]	+4H →	[tetrahydrocarbazole]	250	75
23	[acridine]	+2H →	[dihydroacridine]	250	73
24		+4H →	[tetrahydroacridine]	250	73
For comparison					
25	[anthracene]	+2H →	[dihydroanthracene]	~200	77
Three-ring compounds, second step					
26	[tetrahydrocarbazole]	+2H →	[hexahydrocarbazole]	**	75
27	[acridine]	+6H →		~400	73
28	[dihydroacridine]	+4H →		~400	73
29		+4H →	[octahydroacridine]	>350	73

* Conversion restricted due to low equilibrium concentration of pyrrolidine.
** Not observed

Comparing isolated rings having aromatic character, the following sequence of reactivity is observed:

pyridine > pyrrole ≈ aniline > toluene
280 °C   350 °C   350 °C   >450 °C

Reactivity is strongly increased by the presence of a condensed aromatic ring (ΔT to ca. 80°C). In condensed ring systems the N-containing ring is much more reactive than the carbocyclic ring.

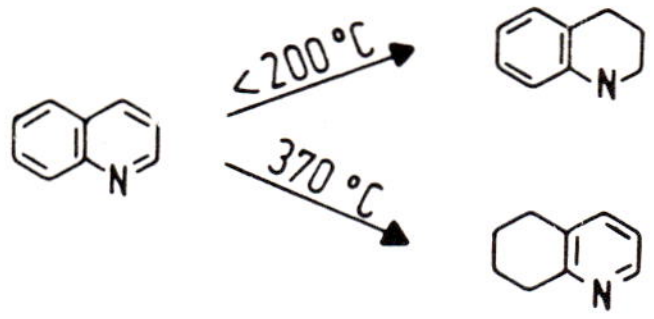

Hydrogenation of anilines is less difficult than that of benzene homologues.

For tricyclic systems it is generally true that hydrogenations are fast for condensed and slow for isolated aromatic rings and hydrogenation of the nitrogen containing ring is preferred to that of the carbocyclic ring.

TABLE 6.10
Reactivity of organic nitrogen compounds for C-N bond splitting as characterized by approximate T_{50} values, $NiMoS/Al_2O_3$ ($NiWS/Al_2O_3$), p_{H_2} 40 - 50 bar, τ_{eff} ca. 10 sec.

(1) Saturated amines, substituents exchange	Ref.

(1a)

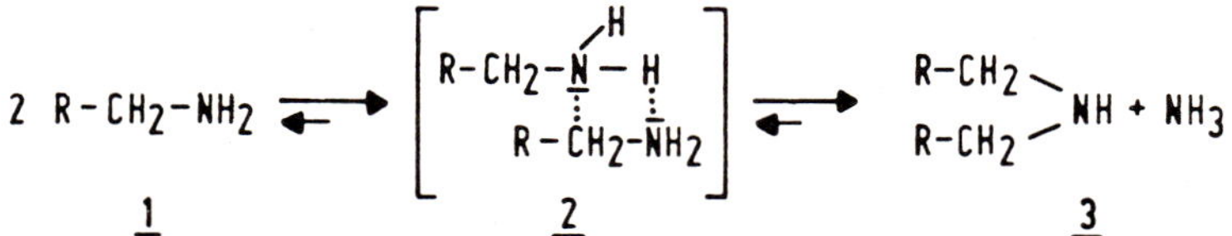

$2\ R-CH_2-NH_2 \rightleftarrows$ [intermediate $\underline{2}$] $\rightleftarrows$ $(R-CH_2)_2NH + NH_3$

$\underline{1}$ $\underline{2}$ $\underline{3}$

2 molecules primary amines $\longrightarrow$ 1 molecule sec. amine + 1 NH_3

(1b) $R-NH_2$ + [pyrrolidine-NH] $\longrightarrow$ [pyrrolidine-NR] + NH_3

$\underline{4}$

(1c) $R-NH_2$ + [piperidine-NH] $\longrightarrow$ [piperidine-NR] + NH_3

$\underline{5}$

TABLE 6.10 (continued)

C-N bond splitting		Ref.

(2) Aliphatic amines
(2a) Primary amines

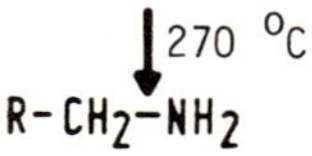

$\downarrow$ 270 °C
R–CH$_2$–NH$_2$
6

Ref. 26

(2b) Secondary amines

230 °C $\downarrow$
R–CH$_2$ \
⟩NH
R–CH$_2$ /
7

220 °C, n.o.
$\downarrow$ $\downarrow$
⟨C$_6$H$_5$⟩–CH$_2$–NH–C$_2$H$_5$
8

26

71

(3) Saturated amines with cyclic substituents

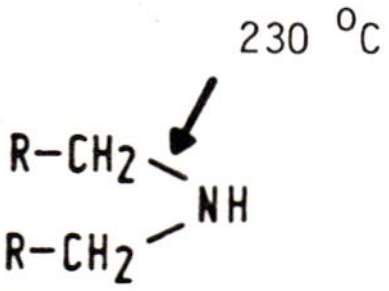

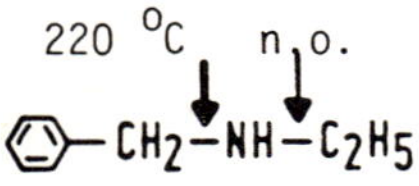

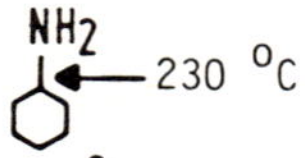

NH$_2$
|
⟨C$_6$H$_{11}$⟩ ← 230 °C
9

26

(4) Anilines

NH$_2$
|
⟨Ph⟩ → >450 °C
10

NH$_2$
|
⟨Ph⟩–CH$_3$ → >450 °C
11

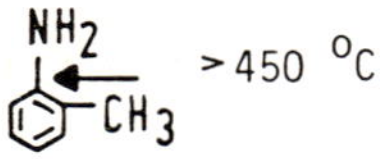

43

(5) Saturated cyclic amines

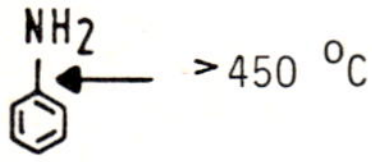

(5a) One ring

⟨N⟩ 300 °C   ⟨N⟩ 340 °C   ⟨N⟩ 275 °C
12   **13**   **14**

27

(5b) Saturated five-membered ring in bicyclic systems

325 °C ↗⟨N-fused⟩↘ LESS REACTIVE
15

↗⟨N-fused⟩ REACTIVE UNDER CONDITIONS OF FORMATION FROM **15**
16

74

(5c) Five-membered rings in tricyclic systems

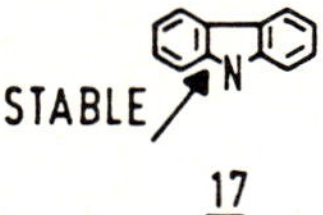

STABLE ↗⟨carbazole⟩
17

>400 °C ↗⟨tricyclic⟩↘ 400 °C
18

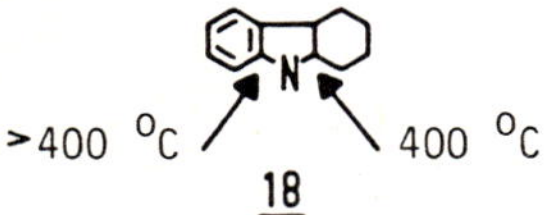

75

(5d) Saturated six-membered ring in bicyclic systems

360 °C ↗⟨N-fused⟩↘ REL. STABLE
19

330 °C ↗⟨decahydroquinoline⟩↘ 330 °C
20

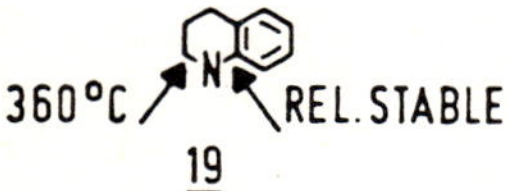

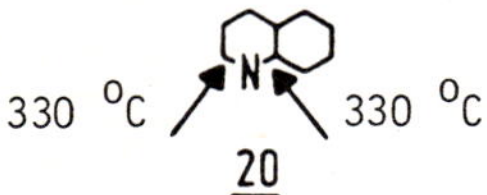

70

TABLE 6.10 (continued)

	Ref.
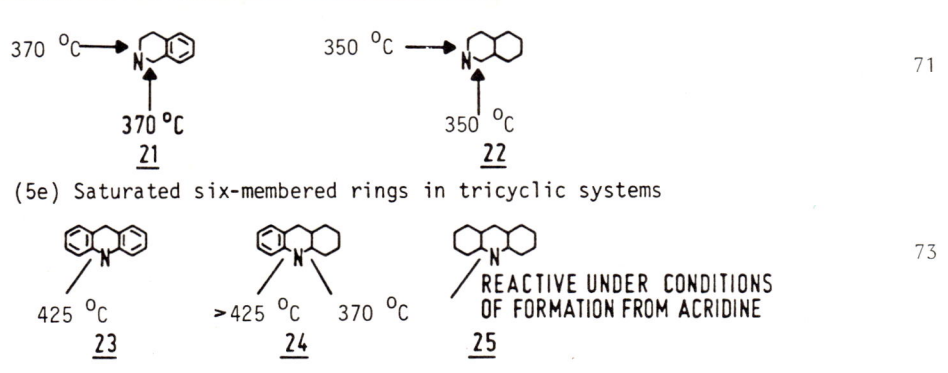	71
(5e) Saturated six-membered rings in tricyclic systems	73

The reactivity patterns in Table 6.10 for cleavage of C-N bonds may be summarized as follows. At low temperature (<200°C) the exchange of amine substituents prevails, interconverting primary, secondary and tertiary amines. In this reaction NH_3 may be liberated: e.g., 2 primary amines = 1 secondary amine plus NH_3. However the number of C-N bonds in the mixture remains the same, so that this NH_3 release is not of use for a complete removal of nitrogen.

C-N bond splitting in saturated aliphatic amines is generally easy (e.g., T_{50} = 270°C). Aliphatic amines as intermediates in HDN networks are commonly denitrogenated in fast reactions or reaction step sequences. When only relatively fast hydrogenation reactions are encountered, as in the case of pyridine conversion (T_{50} = 280°C), then the rate of denitrogenation of the aliphatic amines can also contribute noticeably to the overall reaction rate. The reaction rate is increased when the C-N bond is weakened by a phenyl group in β-position as in N-ethylbenzylamine (T_{50} = 220°C). In contrast, the C-N bond in anilines - being in α-position to the aromatic ring - is remarkably strengthened (T_{50} = 400°C) and breakage of this bond usually requires preceding hydrogenation of the aromatic ring. This behaviour is of particular importance for technical hydrodenitrogenation.

C-N bond cleavage in saturated rings is a complex subject. As regards five- and six-membered rings, a considerably higher reactivity of the former (300°C) as compared to the latter (340°C) is noticed. Seven-membered rings are more reactive than five- and six-membered rings.

C-N bond breaking in the benzo-homologues of pyrrolidine and piperidine, e.g., dihydroindole 15 and 1,2,3,4-tetrahydroquinoline 19, requires a ca. 20°C higher reaction temperature (325 and 360°C respectively). However, only that C-N bond of the hydrogenated ring is split which is in β-position to the aromatic system whereas that in α-position is strengthened and comparatively stable. The C-N bonds in decahydroquinoline 16 are very reactive (T_{50} ca. 320°C) under conditions of its formation from quinoline via 5,6,7,8-tetrahydroquinoline (≈380°C). 1,2,3,4-Tetrahydro-isoquinoline 21 (370°C) is approximately as stable as 1,2,3,4-tetrahydroquinoline 19 (360°C). Both compounds have an activated C-N bond in β-position to the aromatic system, which will be split first. For 1,2,3,4-tetrahydroquinoline the following relative reactivities of the bonds in the saturated ring towards splitting have been derived from the composition of the alkylanilines at 400°C (ref. 70). The C-C bond in β-position to the aromatic ring: 10; the corresponding C-N bond: 100; γ-C-C bond: 5; α-C-C bond: 12.

Appendix

All the experiments reported in this chapter have been performed with the reactants being in the vapour phase. The experimental set-up consisted essentially of a high pressure metering pump and a fixed bed flow reactor. The reaction products were analyzed by capillary gas chromatography. Space time (τ_{eff}) is defined as the ratio of catalyst volume/total gas flow (NTP). Product concentrations are given in terms of mass percent of carbon related to the mass of carbon in the feed (C-%). Catalyst volumes of 1 to 1.5 cm³ were used. The catalyst was slowly heated in flowing pure hydrogen and prereduced for eight hours at 450°C and p_{H_2} = 46 bar. The same procedure was used for sulphiding with 1 vol.-% of H_2S in H_2. The sulphided catalysts are designated by CoMoS, NiMoS and NiWS. The following catalysts were used:

Catalyst	Id.-No	Metal content	Spec. surf. area, (m^2/g)	Manufacturer
CoMo/Al_2O_3-A	M 8-10	3.9% Co 9.0% Mo	220	BASF
CoMo/Al_2O_3-B	S 444	3.2% Co 9.6% Mo	220	Shell
NiW/Al_2O_3	M 7-11	3.0% Ni 20% W	250	BASF
NiMo/Al_2O_3	M 8-21	2.4% Ni 10% Mo	150	BASF

List of abreviations.

An = aniline, Ac = acridine, BTh = benzothiophene, Ca = carbazole, Dpa = dipentylamine, Iq = isoquinoline, MCh = 4-methylcyclohexene, MTh = 2-methylthiophene, mix = mixture, o-Cr = o-cresol, o-T = o-toluidine, o-Xy = o-xylene, Pa = pentylamine, Ph = phenol, Qu = quinoline

References
1   L.D. Rollmann, J. Catal., 46 (1977) 243.
2   A. Jankowski, W. Döhler and U. Graeser, Fuel, 61 (1982) 1032.
3   E. Gallei and T. Jacobsen, Erdöl, Kohle, Erdgas, Petrochem., 34 (1982) 447.
4   H. Schulz, Dac Vong Do, W.Köhler, M. Schon and Van Hung Nguyen, Forschungsbericht T83-175, Bundesministerium für Forschung und Technologie, Fachinformationszentrum Karlsruhe, 1983.
5   R.B. Callen, S.G. Bendoraitis, C.A. Simpson and S.E. Voltz, Ind. Eng. Chem. Prod. Res. Dev., 15 (1976) 222.
6   B.L. Crynes and R. Sivasubramanian, Ind. Eng. Chem. Prod. Res. Dev., 18 (1979) 179.
7   J.F. Jones, M.R. Schmid, M.E. Sacks, Y.-C. Chen, C.A. Gray and R.T. Eddinger, Office of Coal Research Technical Report, NTIS PB-173916, Oct., 1965.
8   C.W. Bowman, in 7th World Petrol. Congr. Proc., 3, 1967, p.583.
9   Refining Synthetic Liquids from Coal and Shale, Final Report of the Panel on R&D needs in Refining, Nat. Acad. Sci. Press, Washington, DC, 1980, cited in ref. 57.
10 D.M. Parees and A.Z. Kamzelski, J. Chromatogr. Sci., 20 (1982) 441.
11 S.E. Schiller, Prepr. Div. Pet. Chem., Amer. Chem. Soc., 20 (1977) 638.
12 D.W. Later, M.L. Lee and B.W. Wilson, Anal. Chem., 54 (1982) 117.
13 M. Novotny, R. Kump, F. Merdi and L.S. Todd, Anal. Chem., 52 (1980) 401.
14 M.V. Buchanan, Anal. Chem., 54 (1982) 570.
15 M. Novotny, D. Wiesler and F. Merdi, Chromatographia, 15 (1982) 374.
16 S.Macak, V.M. Nabivach, P. Buryan and J. S. Berlizov, J. Chromatogr., 209 (1981) 472.
17 H.D. Sauerland and M. Zander, Erdöl, Kohle, Erdgas, Petrochem., 25 (1972) 526.
18 S. Vymetal, Erdöl, Kohle, Erdgas, Petrochem., 25 (1972) 537.
19 L.R. Snyder, Prepr. Div. Pet. Chem. Amer. Chem. Soc., 15 (1970) C-44.
20 J.F. McKay, J.H. Weber and D.R. Lattiam, Anal. Chem., 48 (1976) 891.
21 D. Brown, D.G. Earnshaw, F.R. McDonald and H.B. Jensen, Anal.Chem., 42 (1970) 146.
22 G.U. Dinneen, G.L. Cook and H.B. Jensen, Anal. Chem., 30 (1958) 2026.
23 F.F. Shue and T.F. Yen, Anal. Chem., 53 (1981) 2081.
24 G.Eglinton and M.T.S. Murphy, Editors, Organic Geochemistry,Springer, Berlin, Heidelberg, New York, 1969.

25 J. Sonnemans and P. Mars, J. Catal., 34 (1974) 215.
26 W. Köhler, Dissertation, Universität Karlsruhe, 1982.
27 H. Schulz, E. Faber and M. Schon, 1983, unpublished results.
28 A.K. Aboul-Gheit and I.K. Abdou, J. Inst. Pet., 59 (1979) 188.
29 E.W. Stern, J. Catal., 57 (1979) 390.
30 K.E. Cox and L. Berg, Chem. Eng. Progr., 58(12) (1962) 54.
31 J. Sonnemans, G.H. van den Berg and P. Mars, J. Catal., 31 (1973) 220.
32 H.G. McIlvried, Ind. Eng. Chem. Proc. Des. Dev., 10 (1971) 125.
33 A.K. Aboul-Gheit, Appl. Catal., 16 (1985) 39.
34 C.N. Satterfield, M. Model, R.A. Hites and C.J. Declerck, Ind. Eng. Chem. Proc. Des. Dev., 17 (1978) 141.
35 J.R. Katzer, S.S. Shih, H. Kwart and A.B. Stiles, Prepr. Div. Pet. Chem. Amer. Chem. Soc., 22 (1977) 919.
36 J. Doelman and J.C. Vlugter, Proc. 6th World Petrol. Congr., Section III, Paper 12 PD 7, 1963.
37 S.S. Shih, E. Reiff, R. Zawadski and J.R. Katzer, Prepr. Div. Fuel Chem. Amer. Chem. Soc., 23 (1978) 99.
38 O. Weisser and S. Landa, Sulfide Catalysts, Their Properties and Applications, Pergamon Press, Oxford, 1973.
39 B.C. Gates, J.R. Katzer, T.H. Ohlsen, H. Kwart and A.B. Stiles, Fe-2028-6, Dep. of Chem. Eng. and Chemistry, University of Delaware, Quarterly Report, June-Sept., 1976.
40 M. Pier, Z. Elektrochem., 53 (1949) 297.
41 Ullmanns Encyklopädie der Technischen Chemie, Band 10, Urban & Schwarzenberg, München-Berlin, 1958, pp 483-570.
42 C.M. Cowley, Proc. 3rd World Petrol. Congr., Sect. IV (1950), p. 294.
43 Dac Vong Do, Dissertation, Universität Karlsruhe, 1982, Fortschr. Ber. VDI Z., Reihe 3 Nr. 69,
44 R.G. Ruberto and D.C. Cronauer, in Organic Chemistry of Coal, J.W. Larsen, (Editor), ACS Symp. Ser., Vol. 71, Amer. Chem. Soc., Washington, DC, 1978.
45 C.H. Amberg and P. Desikan, Can. J. Chem., 41 (1963) 1966 and 42 (1964) 843.
46 M. Zdrazil, React. Kinet. Catal. Lett., 6 (1977) 479.
47 H. Kwart, G.C.A. Schuit and B.C. Gates, J. Catal., 61 (1980) 128.
48 C.N. Satterfield and G.W. Roberts, AIChE J., 14 (1968) 159.
49 J.F. Le Page, (Editor), in Catalyse de Contact, Editions Technip, Paris, 1978, Ch. E4.
50 M.L. Vrinat, Appl. Catal., 6 (1983) 137.
51 M. Zdrazil, Appl. Catal., 4 (1982) 107.
52 B.C. Gates, G.C.A. Schuit and J.R. Katzer, (Editors) in Chemistry of Catalytic Processes, McGraw-Hill, New York, 1979, Ch. 5.
53 Second Workshop Meeting on Hydrotreating Catalysts, Louvain, Proc. published in Bull. Soc. Chim. Belg., 93 (8) and 93 (9), 1984.
54 H.O. Reitemeyer, Dissertation, Universität Karlsruhe, 1967.
55 P.S. Owens and C.H. Amberg, Adv. Chem. Ser., 32 (1961) 182.
56 C.N. Satterfield, M. Model and S.A. Wilkens, Ind. Eng. Chem. Proc. Des. Dev., 19 (1980) 154.
57 E. Furimsky, Catal. Rev. Sci. Eng., 25 (1983) 421.
58 Y.T. Shah, (Editor), G.S. Stiegel, S. Krishnamurthy and S.V. Panvelker, in Reaction Engineering in Direct Coal Liquefaction, Addison-Wesley, Reading, Mass., 1981, Ch. 6.
59 J.R. Katzer and R. Sivasubramanian, Catal. Rev. Sci. Eng., 20 (2) (1979) 155.
60 R.A. Flinn, O.A. Larson and H. Beuther, Hydrocarbon Process. Pet. Refiner, 42 (1963) 129.
61 W. Stengler, J. Welker and E. Leibnitz, Freiberger Forschungsh., 329 A (1964) 51.
62 H. Schulz, C1 Mol. Chem., 1 (1985) 231.
63 J. Pasek, J. Tyrpekl and M. Machova, Collect. Czechoslov. Chem. Commun., 31 (1966) 4108.

64 C. Triadis, Dissertation, Universität Karlsruhe, 1981.
65 S.A. Anabtawi, R.S. Mann and K.C. Khulbe, J. Catal. 63 (1980) 456.
66 F. Goudriaan, H. Gierman and J.C. Vlugter, Inst. Petrol. Tech. Pap. IP, 59 (565) (1973) 41.
67 C.N. Satterfield and J.C. Cocchetto, Ind. Eng. Chem. Proc. Des. Dev., 20 (1981) 53.
68 J.C. Cocchetto and C.N. Satterfield, Ind. Eng. Chem. Proc. Des. Dev., 20 (1981) 49.
69 H. Schulz and H. D. Eichhorn, Preprints 7th Int. Congr. Catal. Tokyo, 1980, E 7.
70 H. D. Eichhorn, Dissertation, Universität Karlsruhe, 1979.
71 M. Schon, Dissertation, Universität Karlsruhe, 1986.
72 S. Shih, E. Reif, R. Zawadzki and J. R. Katzer, Prepr. Div. Pet. Chem. Amer. Chem. Soc., 23 (1978) 99.
73 H. Schulz, A. El Fayoumi and M. Schon, 1985, unpublished results.
74 H. Schulz, D. Blank and M. Schon, 1984, unpublished results.
75 H. Schulz, G. Kasturi and M. Schon, 1984, unpublished results.
76 J.F. Patzer II, R.J. Farrauto and A.A. Montagna, Ind. Eng. Chem. Prod. Des. Dev., 18 (1979) 625.
77 W.H. Wiser, S. Singh, A. Qader and G.R. Hill, Ind. Eng. Chem. Prod. Res. Dev., 9 (1970) 350

Chapter 7

EFFECT OF CATALYST COMPOSITION ON REACTION NETWORKS IN HYDRODESULPHURIZATION

M. ZDRAŽIL and M. KRAUS
Institute of Chemical Process Fundamentals, Czechoslovak Academy of Sciences, Rozvojová 135, 165 02 Prague 6 - Suchdol (Czechoslovakia)

7.1 INTRODUCTION

Composite sulphides, like Co-Mo, Ni-Mo and Ni-W, are more active for hydrorefining of petroleum fractions than simple sulphides of molybdenum or tungsten. Cobalt and nickel are usually described as "promotors" of molybdenum or tungsten in spite of the fact that cobalt and nickel sulphides possess their own activities. Upon their mixing with Group 6 metal sulphides the resulting activity is more than additive. Therefore, it is more appropriate to use the term "synergism" instead "promotion" (cf. refs. 1-4).

Considerable effort is being made to explain the synergic effect, especially for the most common system Co-Mo. The cause is seen as the formation of special surface structures by the interaction of molybdenum and cobalt precursors and evidence for this is being sought by modern physical methods of structure and surface research (for reviews of the results and theories see refs. 5-10).

However, little attention has been paid to the influence of synergism on the complex reaction system of hydrorefining, which has been treated mostly as an aside in papers dealing with general effects on activity. Almost all the information on the chemistry of refining concerns hydrodesulphurization.

In the present chapter, we summarize the literature data on the influence of the catalyst composition on the product distribution in the hydrodesulphurization of model sulphur compounds, i.e., thiophene, benzothiophene and dibenzothiophene. Most data concern the Co-Mo catalytic system, however, some information can also be found on the Ni-Mo and Ni-W catalysts.

7.2 THE CHEMISTRY OF HYDRODESULPHURIZATION

The basic features of hydrodesulphurization with respect to the chemical nature of the starting compounds, intermediates and products have been discussed by Zdražil (ref. 11). It was concluded that the opening of the thiophene ring must be preceded by at least its partial saturation which destroys the aromatic system. In thiophene and its derivatives, the sulphur atom is part of the conjugated system of π-electrons and the molecule behaves like a hydrocarbon. The adsorption on the catalyst surface must be realized through the π-electrons and is therefore weak. By addition of one or more hydrogen atoms to the ring, the free electron pairs of the sulphur atom allow a coordinative bond to the surface, which results in weakening of the two C-S bonds. These two C-S bonds are split in two separate elementary steps. As intermediates, species with a single C-S bond must be considered, corresponding in the case of thiophene to butanethiol or butenethiol. The types of substances appearing subsequently in hydrodesulphurization of heteroaromatic sulphur compounds are the hydrocarbons.

Thus, we have four distinct groups of compounds in the reaction scheme for hydrodesulphurization, which will be denoted A - D. First are the starting compounds containing sulphur in the aromatic ring (A: thiophene, benzothiophene and dibenzothiophene), secondly, the cyclic, non-aromatic intermediates with two C-S bonds (B: dihydrothiophenes and tetrahydrothiophene in the case of thiophene), thirdly, the intermediates with one C-S bond (C) and fourthly, the hydrocarbons plus hydrogen sulphide (D). It is evident that the progress of the reaction towards the final products is achieved by a number of hydrogenation and hydrogenolytic steps; moreover, the splitting of the C-S bond by acid-catalyzed elimination of hydrogen sulphide has to be considered.

The situation for thiophene, benzothiophene and dibenzothiophene is depicted in Schemes 7.1-7.3. The compounds enclosed in boxes are regarded as adsorbed, the compounds above them are in the bulk phase. The species within each segment are related by their principal structural features and their interconversion by hydrogenation-dehydrogenation is possible; however, an equilibrium between them is, in general, not established. Whereas all hydrogenation and adsorption steps are reversible, the splitting of the C-S bonds is considered irreversible under the conditions of the hydrodesulphurization process. The transfer of reactants from segment B to segment C and from segment C to segment D, respectively, is possible in any point from left to right, i.e. in any state of reactant unsaturation. The position of the predominating transfers determines the main reaction pathway through the scheme. In order to keep the Schemes 7.1 - 7.3 easy to survey, all reaction steps (transfers) between two segments are represented by only one arrow-head.

Each species in segments B and C has three possibilities of transformation, by hydrogenation-dehydrogenation, C-S bond splitting and adsorption-desorption steps, respectively. The relative possibility of each transformation depends not only on the structure of the species but also on the reaction conditions, that is on the temperature and on the partial pressures of hydrogen and other reaction components. We will show here that it depends to a great extent also on the composition of the catalyst, that is on its hydrogenation, hydrogenolytic and possibly elimination activities, respectively.

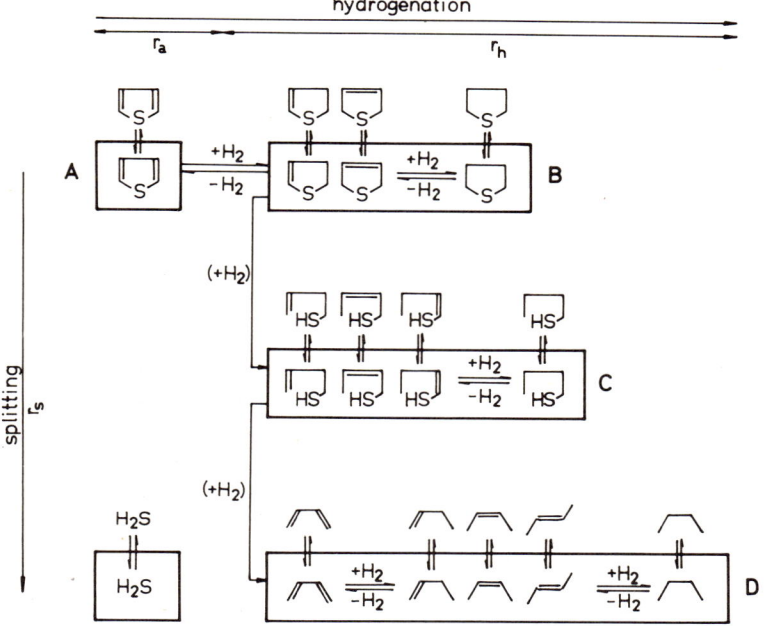

Scheme 7.1

A similar approach to a multistep catalytic reaction has been used by several other authors. For example, Mills et al. (ref. 12) explained the reforming of hydrocarbons on dual function catalysts by the concept of a surface network of hydrogenation-dehydrogenation and isomerization reactions in mutual competition, and also with adsorption-desorption steps, for each surface intermediate. Kemball and co-workers (ref. 13) described the hydrodesulphurization of diethyl sulphide by a multistep scheme in which each surface intermediate can either be desorbed or react further.

7.3 KINETIC CONSEQUENCES

The reaction systems in Schemes 7.1-7.3 are rather complex. Even when the splitting steps are assumed to be irreversible we need almost thirty rate constants for the formal kinetic description of the thiophene reaction. This number can be reduced to about half if we consider all adsorption-desorption steps to be in equilibrium. This is still too high for a complete treatment.

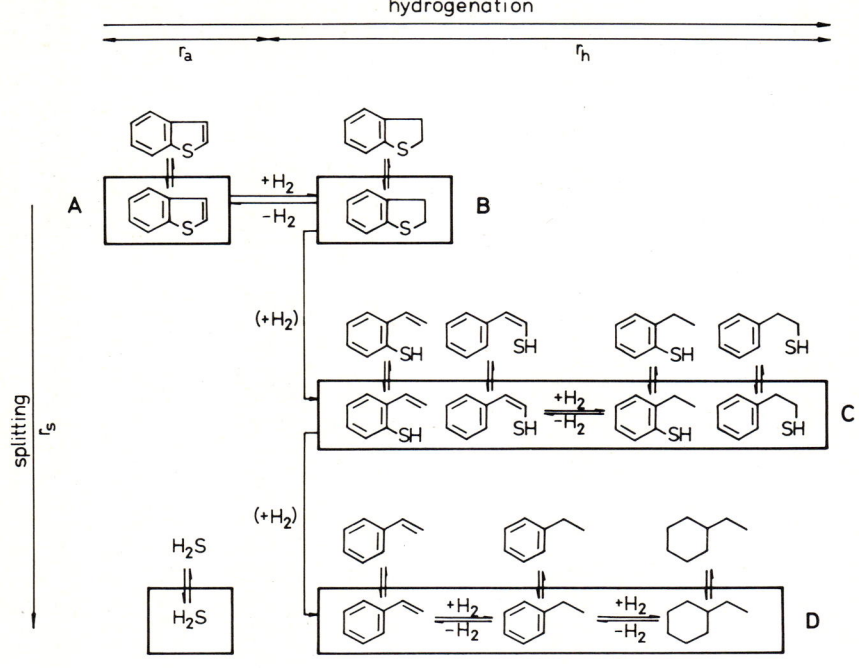

Scheme 7.2

However, satisfactory kinetic descriptions of the hydrodesulphurization have been achieved by a number of authors (for a review see ref. 14, further, see refs. 15,16) on the basis of simplified reaction schemes and with acceptable numbers of adjustable constants. The derived equations are valid only for the catalyst used and for the narrow range of experimental conditions. With other catalyst compositions and at different conditions, other reaction steps from Schemes 7.1-7.3 would predominate and other reaction steps could be neglected.

This can be illustrated by a disagreement in the literature on the composition of the C_4 hydrocarbons obtained from the hydrodesulphurization of thiophene on Co-Mo/alumina as catalyst. At low hydrogen pressure, 0.003-0.3 MPa, butenes predominate in the C_4 products (refs. 16-22) and some authors have also found 1-3 % of butadiene (refs. 19,23-26). At high pressures of hydrogen, 3-10 MPa, the amounts of olefins and alkanes are comparable and dienes were not found (refs. 16,27).

It should be noted that even Schemes 7.1-7.3 are simplifications because we

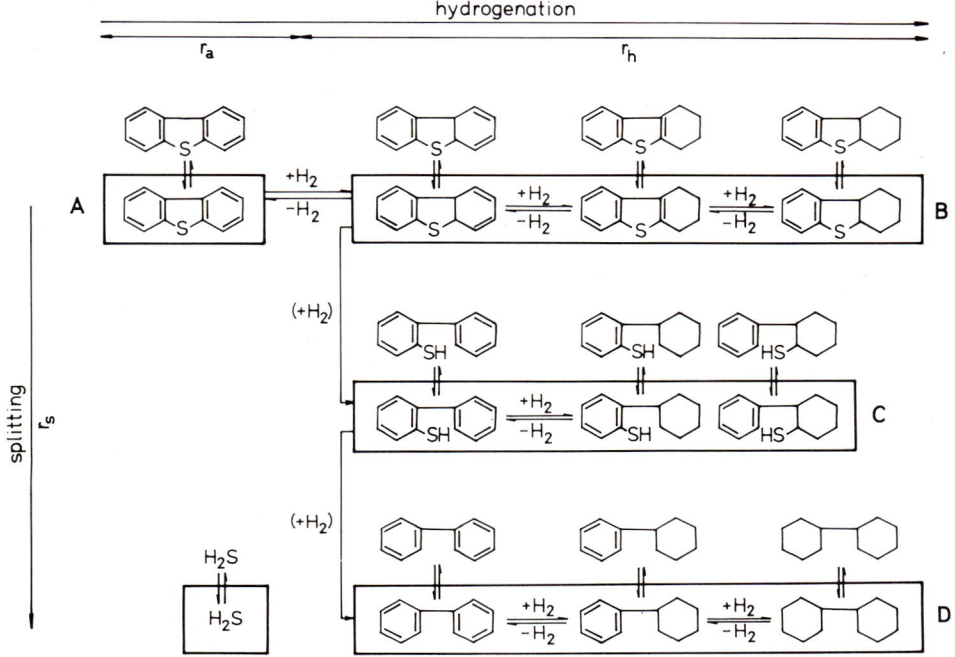

Scheme 7.3

assume always that two hydrogen atoms take part in each reaction step. The stepwise addition of hydrogen atoms to the C=C double bond seems to have been demonstrated (cf. refs. 28,29), with different rate constants for each elementary step.

For the purpose of the present discussion we are not interested in the overall kinetics, but in the relative rates of the individual steps. For simplification, we will distinguish only three rates: the rate of addition of the first hydrogen(s) to the starting fully aromatic compound, r_a, the rate of saturation of any C=C double bond in any molecule, r_h, and the rate of splitting of any C-S bond, r_s. The rate of hydrodesulphurization, r_{HDS}, is related in a complicated way with all three rates r_a, r_h and r_s.

In general, r_a is always less than r_h, but the ratios r_a/r_s and r_h/r_s change in parallel with the reaction conditions. The dependence of these ratios on temperature, hydrogen pressure and synergism indicates that hydrogenation and splitting have different temperature coefficients, different orders with respect to hydrogen and that synergism affects the quality of the active centres.

7.4 EFFECTS OF STRUCTURE, TEMPERATURE AND HYDROGEN PRESSURE

At every point in the networks in Schemes 7.1-7.3, corresponding to individual adsorbed species, there are several possible directions which the reaction can take. Beside desorption and dehydrogenation, which are of minor interest for our discussion, hydrogenation and splitting determine the main reaction pathway and the composition of the products at various times from the beginning of the transformation. When the rates r_h are larger than the rates r_s, the principal path is via right-hand sides of the schemes; the opposite is true when r_s is larger than r_h.

Therefore, some of the compounds in Schemes 7.1-7.3 have not been found in the bulk phase or in negligible concentrations. In comparison with thiophene, the benzene rings in benzothiophene and dibenzothiophene not only make the networks more complex but also influence the final (equilibrium) composition. For thiophene, the most stable products, even at low hydrogen pressure, are butane and hydrogen sulphide, whereas with dibenzothiophene, the equilibrium mixture even at relatively high pressure contains biphenyl and cyclohexylbenzene, besides dicyclohexane.

At higher hydrogen pressures, the increase in r_h is greater than that in r_s. The prevailing reaction pathway includes more saturated compounds which are also desorbed and are then found in the bulk phase. There is ample experimental evidence for this. With increasing hydrogen pressure, the yield of tetrahydrothiophene increases and that of butenes decreases in the reaction of thiophene, the yield of dihydrobenzothiophene increases in the reaction of benzothiophene and the yields of tetrahydro- and hexahydrodibenzothiophene, of phenylcyclohexane and dicyclohexane increase in the reaction of dibenzothiophene (Tables 7.1-7.5).

The influence of temperature on the r_h/r_s ratio is opposite to that of pressure. With increasing temperature, the reaction path is shifted to the left-hand sides of the reaction schemes. Again, a number of examples of this behaviour have been reported (Tables 7.1 - 7.5).

7.5 SYNERGIC EFFECTS ON THE DISTRIBUTION OF INTERMEDIATES
7.5.1 Non-aromatic cyclic sulphur compounds

The yields of non-aromatic cyclic sulphur compounds from aromatic precursors are governed by the ratio of the rates, r_h, in segments B and rates, r_s, between segments B and C. At high r_h/r_s ratios, the concentrations of substances in segments B are increased and more of them can be desorbed into the bulk phase.

TABLE 7.1

Effect of pressure and temperature on the yield of tetrahydrothiophene (THT) in thiophene hydrodesulphurization on Co-Mo/Al_2O_3 catalysts

Pressure (MPa)	Temperature (°C)	Conversion (%) Overall	Conversion (%) to THT	Ref.
0.1	400	40	0.03	30
0.1	350	20	0.1	30
0.1	300	22	0.4	30
0.1	260-360	3-10	not found	31
0.1	250-310	0-10	max. 3	32
0.9	220	0-100	max. 9	33
1.75	220	0-100	max. 12	33
1.5	250	17	10	30
2.1	260	0-100	max. 4	34
2.8	290	0-100	max. 10	20
9.7	250	19	11.5[a]	27

[a] 2-Methyltetrahydrothiophene from the reaction of 2-methylthiophene.

TABLE 7.2

Effect of pressure and temperature on the yield of butenes in thiophene hydrodesulphurization on Co-Mo/Al_2O_3 catalysts

Pressure (MPa)	Temperature (°C)	$100 \frac{\Sigma \text{ mol butenes}}{\Sigma \text{ mol hydrocarbons}}$	Ref.
At 20 % thiophene conversion			
0.1	350	95	18
0.1-0.3	320-430	93	19
0.1	350	90	17
0.1	400	88	21
2.8	290	50	20
9.7	250	53[a]	27
At 40 % thiophene conversion			
0.1	350	93	22
0.2	305	87	16
0.8	282	75	16
1.8	282	66	16

[a] Pentenes from the reaction of 2-methylthiophene.

TABLE 7.3

Effect of pressure and temperature on the yield of dihydrobenzothiophene (DHBT) in benzothiophene hydrodesulphurization on Co-Mo/Al_2O_3 catalysts

Pressure (MPa)	Temperature (°C)	Conversion (%)		Ref.
		Overall	to DHBT	
0.1	350-400	0-20	not found	35
0.1	300-400	40-100	not found	36
0.1	350-400	0-50	negligible	37
0.1	350-400	10-80	negligible	38
0.1	400	18	2	39
0.1	300	20	9	39
0.1	270	0-100	max. 4	40
0.2	250	0-100	max. 25	33
2.1	270	0-100	max. 17	41
2.1	270	0-100	max. 19	34
2.1	330	0-100	max. 9	34
5.0	250	0-100	max. 20	42
8.8	250	0-70	max. 55	43
9.7	230	40	20	27
9.7	250	70	15	27

TABLE 7.4

Effect of pressure on the yields of tetrahydrodibenzothiophene (THDBT) and hexahydrodibenzothiophene (HHDBT) in dibenzothiophene hydrodesulphurization on Co-Mo/Al_2O_3 catalysts

Pressure (MPa)	Temperature (°C)	Conversion (%)		Ref.
		Overall	to (THDBT+HHDBT)	
0.1	350-400	40-80	0	36
1.5-4.0	200-250	0-15	0	44
3.1	310	55	0.4	45
7.1	300	low	low	46
8.5[b]	300	15	3	47

[b]Estimated value.

TABLE 7.5

Effect of pressure and temperature on the composition of the hydrocarbon fraction in dibenzothiophene hydrodesulphurization on Co-Mo/Al$_2$O$_3$ catalysts

Pressure (MPa)	Temperature (°C)	$100 \dfrac{\text{mol biphenyl}}{\Sigma \text{ mol C}_{12} \text{ hydrocarbons}}$ [a]	Ref.
0.1	350	100	36
1.0	240	94	44
2.4[b]	300	95	46
2.6	310	92[c]	48
4.0	240	86	44
4.0	200	82	44
8.5[b]	300	67	47

[a] At 15 % conversion of dibenzothiophene.
[b] Estimated values.
[c] By extrapolation.

(i) <u>Thiophene-tetrahydrothiophene</u>. The influence of cobalt added to molybdenum on the tetrahydrothiophene content during hydrodesulphurization of thiophene is demonstrated in Fig. 7.1. The yield of tetrahydrothiophene, i.e., the r_h/r_s ratio, is decreased. Because the overall activity is increased by cobalt, its effect on selectivity must be due to a greater increase in r_s than in r_h.

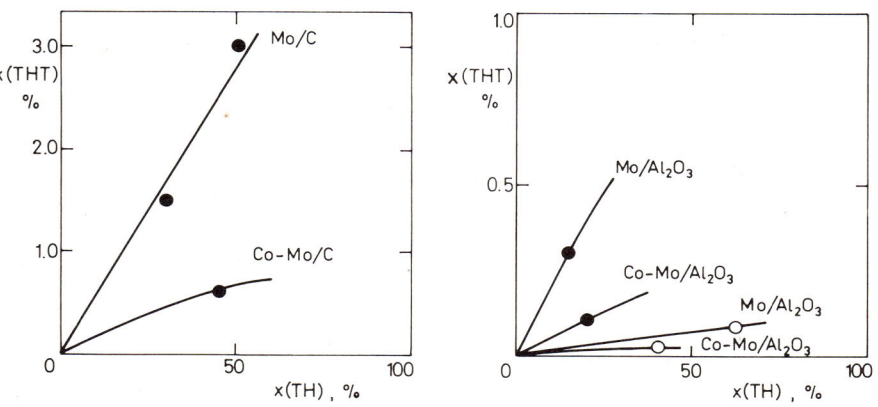

Fig. 7.1. Synergic effect on tetrahydrothiophene (THT) selectivity in thiophene (TH) hydrodesulphurization. Atmospheric pressure, temperature: ● 350°C, ○ 400°C; x(TH)=per cent overall conversion of thiophene, x(THT)=per cent conversion to tetrahydrothiophene (ref. 30).

Analogous conclusions can be drawn from experiments with tetrahydrothiophene (Table 7.6). On the molybdenum catalyst, the production of thiophene by dehydrogenation (r_d) is high and addition of Co increases the overall rate, but decreases the r_d/r_s ratio by increasing r_s. (A proportionality exists between r_d and r_h under any conditions, so that the ratios r_d/r_s and r_h/r_s change in parallel with the catalyst composition).

TABLE 7.6

Ratio of the rate of dehydrogenation, r_d, to the rate of desulphurization, r_s, in the reaction of tetrahydrothiophene over Mo and Co-Mo catalysts; conditions of low conversion (under 10 %)

Temperature (°C)	Pressure (MPa)	Medium	r_d/r_s Mo	r_d/r_s Co-Mo	Ref.
250	subatmospheric	He	3.3	0.82	26
300	atmospheric	H_2	0.5	0.03	49

The nature of the catalyst components affects the selectivity. In Table 7.7, data for Co-Mo, Ni-Mo and Ni-W catalysts are compared for the reaction of 2-methylthiophene; a decrease in the ratio r_h/r_s is manifested not only in a decrease in the ratio of methyltetrahydrothiophene/hydrocarbons but also in a simultaneous increase in the ratio of pentenes/pentane (cf., Section 7.5.2(i)).

TABLE 7.7

Selectivities of various hydrorefining catalysts in hydrodesulphurization of 2-methylthiophene at 250°C, 9.7 MPa total pressure and about 20 % conversion (ref. 27)

Catalyst	Product composition (mole %)		
	Methyltetrahydrothiophene	Pentenes	Pentane
Co-Mo/Al_2O_3	60	21	19
Ni-Mo/Al_2O_3	26	49	25
Ni-W/Al_2O_3	14	53	33

(ii) <u>Benzothiophene-dihydrobenzothiophene</u>. The yield of dihydrobenzothiophene is always higher than of tetrahydrothiophene under comparable conditions (ref. 11). This is caused by the smaller difference in electronic structure between dihydrobenzothiophene and benzothiophene than between tetrahydrothiophene and thiophene.

The selectivity of dihydrobenzothiophene formation was studied systematically with Co, Ni, Mo and W and their combinations Co-Mo, Ni-Mo, Co-W and Ni-W. Fig. 7.2 shows the results plotted, for convenience, without experimental points as triangular diagrams which include also the equilibrium line. The ratio r_h/r_s decreases with decreasing height of the lines over the baseline. For Mo and W as catalysts (upper lines in Fig. 7.2), the value of r_h is so high that the equilibrium between dihydrobenzothiophene and benzothiophene is established in the bulk phase. The addition of Co or Ni decreases the ratio r_h/r_s by increasing r_s and the dihydrobenzothiophene content is lower. The molecules in segment B are split rapidly and equilibrium is not reached.

Fig. 7.2 also contains a line for a Co/C catalyst; the corresponding catalyst Co/Al$_2$O$_3$ had low activity. This line demonstrates the inherent high hydrogenation activity of cobalt and the synergetic effect on selectivity is clearly demonstrated by this diagram. The lines for Mo/C and Co/C lie much higher than that for the mixed catalyst which therefore must yield much lower r_h/r_s ratios.

Also when the reaction network was entered through dihydrobenzothiophene as the starting compound the influence of Co on the selectivity was clearly seen.

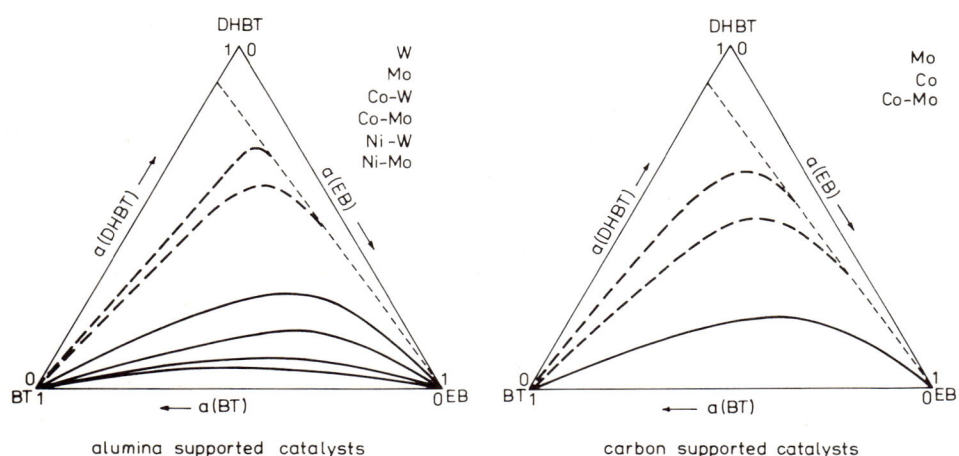

Fig. 7.2. Synergic effect on dihydrobenzothiophene (DHBT) selectivity in benzothiophene (BT) hydrodesulphurization. Pressure 2.1 MPa, temperature 270°C, a(i) = molar fraction, EB = ethylbenzene; ----- equilibrium line BT + H$_2$ ⇌ DHBT, kinetic lines: – – – monometallic catalysts, ——— bimetallic catalysts; the order of curves corresponds to the order of catalysts (ref. 41).

On Mo/Al$_2$O$_3$, the equilibrium between dihydrobenzothiophene and benzothiophene is rapidly established by dehydrogenation. The addition of cobalt leads to an increase in r_s, split competing with dehydrogenation and the equilibrium is not reached (ref. 40).

(iii) <u>Dibenzothiophene-hydrodibenzothiophenes</u>. Information on the formation of hydrogenated dibenzothiophenes on monocomponent and bicomponent catalysts is scattered but comparison is possible.

The conversions to hydrodibenzothiophenes are more extensive at high hydrogen pressures and in the liquid phase. However, no work has been done on monometallic and bimetallic sulphide catalysts simultaneously under such conditions and so Fig. 7.3 compares the data from different authors.

Vrinat and Mourgues (ref. 44) studied the reaction both on Mo/alumina and on Co-Mo/alumina as catalysts, but at lower hydrogen pressures and in the gas phase where the yields of hydrodibenzothiophenes are small. They reported low yields of hydrodibenzothiophenes on the molybdenum catalyst while on Co-Mo these products were not identified.

In spite of the limited data available, the trend is clear: synergism diminishes the ratio r_h/r_s which benefits the desorption of more saturated species from the surface.

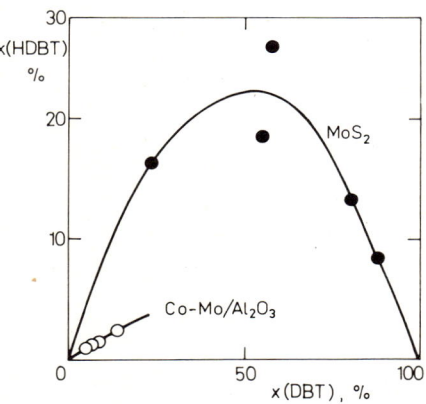

Fig. 7.3. Formation of hydrodibenzothiophenes in dibenzothiophene hydrodesulphurization at 300°C and in the liquid phase. Pressures for MoS$_2$, 5.0 MPa (ref. 50); for Co-Mo/alumina, 8.5 MPa (estimated value) (ref. 47); x(DBT) = conversion of dibenzothiophene, x(HDBT) = the sum of conversions to tetrahydro-, hexahydro- and perhydrodibenzothiophene.

7.5.2 Unsaturated hydrocarbons

(i) <u>Hydrodesulphurization of thiophene</u>. The literature reveals that, on bicomponent catalysts, more butenes are formed than on monocomponent molybdenum and tungsten catalysts. This is in agreement with the principal reaction pathway proceeding through the left-hand side of Scheme 7.1.

The evidence is based on two types of data. The integral dependences of the product composition on the contact time or overall conversion have been published (ref. 17,51) for a broad range of thiophene conversion. Fig. 7.4 presents them in the form of a triangular diagram for the system thiophene-butenes-butane. Other data (see refs. 18,21,22) cover a narrower range of conversion but the effect is the same as in Fig. 7.4. Other authors have demonstrated the influence of cobalt addition to molybdenum on the rate of formation of butenes and butane at a constant overall conversion of 50 % (ref. 52) or after extrapolation to zero conversion (ref. 25). The ratio of r(butane)/r(butenes) was much higher on the monocomponent catalyst than on the bicomponent one.

The second type of data are the ratios of rate constants, k(HYD)/k(HDS), calculated from one composition of the reaction mixture under the assumption that the consecutive reactions thiophene → butenes → butane are first order. This is a rough approximation which causes large deviations in k(HYD)/k(HDS) at

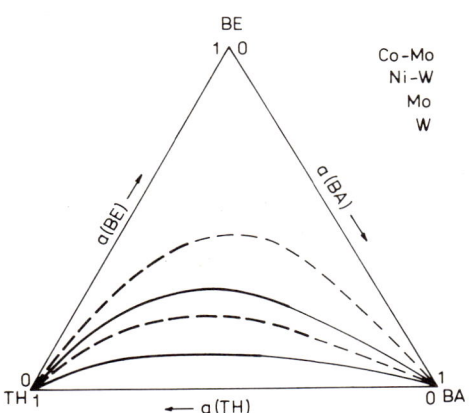

Fig. 7.4. Synergic effect on butene (BE)/butane (BA) selectivity in thiophene (TH) hydrodesulphurization at 300°C. W and Ni-W catalysts supported on silica, hydrogen pressure 0.2 MPa (ref. 51), Mo and Co-Mo catalysts supported on alumina, atmospheric pressure (ref. 17); a(i)=molar fraction, thick line measured, thin line extrapolated, the order of the curves corresponds to the order of the catalysts.

low degrees of conversion. The first-order consecutive kinetics requires r(butane) to fall to zero at x(thiophene) = 0, but this is not an experimental fact (ref. 15,25,53). The reason is evident from Scheme 7.1; the unsaturated intermediates in segment D are transformed to butane in the adsorbed state. The consequence is that the value of k(HYD)/k(HDS) calculated from integral data at low degrees of conversion increases strongly with decreasing conversion. The strong dependence of k(HYD)/k(HDS) on metal content in the Mo/alumina catalyst (ref. 54) is probably an artifact which need not be taken into account for comparison of the selectivity of monocomponent and bicomponent catalysts. Table 7.8 contains selected data calculated from the product composition at higher degrees of conversion where the approximation of the kinetics causes only slight distortion. The data show that, under comparable conditions, the selectivity of butenes formation is higher on bicomponent catalysts than on monocomponent ones.

TABLE 7.8
Selectivity of mono- and bicomponent sulphide catalysts in thiophene hydrodesulphurization at atmospheric pressure and 400°C calculated under the assumption that the consecutive reactions thiophene → butenes → butane are first order

Catalyst	Carrier	k(HYD)/k(HDS)	Ref.
Mo Co-Mo	C	1.8 0.5	4
Mo Co-Mo	C	4.0 1.2	55
Mo Co-Mo	Al_2O_3	4.5 2.2	55
W Ni-W	C	3.3 0.6	4
Mo Ni-Mo	C	1.8 0.6	4

The explanation for this effect is as follows: in segment D the rate of hydrogenation of surface intermediates is much higher than the rate of their dehydrogenation even at low hydrogen pressure. This is connected with the equilibrium in which butane predominates under the conditions of the hydrodesulphurization. Thus the influence of dehydrogenation on the concentrations in segment D is negligible. On the monocomponent catalyst, the ratio r_h/r_s is high and the majority of molecules enter segment D on the right-hand side, which is in the state of higher saturation. The rate of their further hydrogenation in segment D is not very different from the rate at which they enter this segment. In this situation, the surface concentration of butenes is lower than that of butane and this ratio of concentrations is preserved also in the gas phase. On the bicomponent catalysts, the r_h/r_s ratio is lower and molecules enter segment D in a less saturated state. Furthermore the rate of their further hydrogenation in this segment is lower than the rate of their formation (only relatively because the rate of formation of butane increases). Both effects lead to an increase in the surface, and consequently also the bulk, concentration of butenes.

It is of interest that hydrogen sulphide influences the selectivity by increasing butane formation (ref. 32); because the overall rate is diminished, hydrogen sulphide must retard r_s more than r_h.

(ii) <u>Hydrodesulphurization of benzothiophene</u>. In analogy with butenes/butane selectivity in the reaction of thiophene, the ratio styrene/ethylbenzene in benzothiophene hydrodesulphurization should be higher on bicomponent than on monocomponent catalysts. Several authors have reported styrene in the products on Mo (ref. 56) and Co-Mo (refs. 35-37,43) as catalysts but the data are not suitable for comparison.

The hydrogenation of ethylbenzene to ethylcyclohexane is very difficult and starts only when all the benzothiophene has been converted; thus this selectivity cannot be measured simply.

(iii) <u>Hydrodesulphurization of dibenzothiophene</u>. The selectivity of formation of biphenyl, phenylcyclohexane and bicyclohexane on monocomponent and bicomponent catalysts was compared by Vrinat et al. (refs. 44,57), who defined the hydrogenation selectivity, S_h, as the ratio of the sum of bicyclohexane + + phenylcyclohexane to total hydrocarbons. They found that this hydrogenation selectivity depends only slightly on the overall extent of dibenzothiophene conversion in the range of 2 to 15 %.

With unsupported catalysts at 260°C and 2.3 MPa, MoS_x yielded S_h = 0.65 to 0.75 and Co-Mo catalysts S_h = 0.17 to 0.35. Bicomponent catalysts prepared by the comaceration method exhibited weak synergism in activity and higher values of S_h of 0.27 to 0.35, while those prepared by homogeneous precipitation exhibited strong synergism and low values of S_h of 0.17 (ref. 57). With alumina supported catalysts, over a broad range of conditions, S_h was always higher on Mo than on Co-Mo; e.g., at 220°C and 4 MPa the values were 0.35 and 0.14, respectively (ref. 44).

The explanation is based again on the change in the ratio r_h/r_s with the addition of cobalt to molybdenum. At a high r_h/r_s ratio characteristic of the monocomponent catalyst, the principal pathway is through the right-hand side of Scheme 7.3. The low r_h/r_s ratio found on the bicomponent catalysts shifts the pathway to the left.

Further, S_h increased with increasing hydrogen sulphide concentration (ref. 44). Since the overall rate was diminished, hydrogen sulphide must retard r_s more than r_h. This is in accord with the influence of hydrogen sulphide on the selectivity for butenes/butane in thiophene hydrodesulphurization.

7.6 SYNERGIC EFFECTS ON INDIVIDUAL REACTIONS

In the previous Section, hydrodesulphurization was considered to comprise hydrogenation (r_h) and splitting (r_s) steps taking place on the catalyst surface in a complex parallel consecutive reaction network. The comparison of the ratio r_h/r_s on mono- and bicomponent catalysts was then based on the distribution of hydrodesulphurization intermediates. However, such a comparison can also be made using the rates of two separate reactions, hydrodesulphurization, r(HDS), and hydrogenation of an olefin, r(HY). Some examples are shown in Tables 7.9-7.11. In this case, the rate r(HDS) depends on both r_h and r_s, but r(HY) only on r_h. The synergic effect increases r_s more than r_h and this leads to a decrease in both the ratios r_h/r_s (as discussed in Section 7.5) and r(HY)/r(HDS) (as shown in Tables 7.9-7.11). Thus, the effect of synergism on the relative rates of individual reactions is the same as in complex reaction systems.

TABLE 7.9

Effect of catalyst composition on the ratio of the rate constants for cyclohexene hydrogenation, k(HY), and thiophene hydrodesulphurization, k(HDS), at 350°C and atmospheric pressure (ref. 58)

Molar ratio Co/Mo [a]	k(HY)/k(HDS)
0.0	3.84
0.2	0.79
0.5	0.25
1.0	0.17

[a] Alumina-supported catalysts, molar ratio (Co+Mo)/Al_2O_3 was always 0.1.

TABLE 7.10

Effect of catalyst composition on the ratio of the rate constants for cyclohexene hydrogenation, k(HY), and thiophene hydrodesulphurization, k(HDS), at 400°C and atmospheric pressure (ref. 59)

Catalyst carrier [a]	k(HY)/k(HDS)	
	Mo	Co-Mo
Al_2O_3	11.7	2.0
10 % SiO_2 + Al_2O_3	17.3	3.1
25 % SiO_2 + Al_2O_3	25.1	3.0
75 % SiO_2 + Al_2O_3	21.3	1.7
SiO_2	23.2	0.7
73 % SiO_2 + MgO	25.8	2.5
TiO_2	14.8	2.7

[a] Molybdenum catalysts contained 8 % Mo, Co-Mo catalysts contained 8 % Mo and 3 % Co.

TABLE 7.11

Effect of catalyst composition on the ratio of the initial rates of propene hydrogenation, r(HY), and thiophene hydrodesulphurization, r(HDS), at 420°C and atmospheric pressure (ref. 60)

Catalyst composition (mass %) [a]		r(HY)/r(HDS)
MoO_3	NiO	
7	–	1.33
16	–	1.73
7	2	0.82
16	4.3	0.53
14	3.5 [b]	0.55

[a] Alumina-supported catalysts.
[b] Industrial catalyst.

7.7 CONCLUSIONS

The results presented show clearly that, beside the synergic effect on activity, the selectivity is also strongly influenced by adding cobalt or nickel to molybdenum or tungsten as catalyst. A change in selectivity is observed both in the distribution of intermediates and products of complex hydrodesulphurization reactions and in the relative rates of individual hydrodesulphurization and hydrogenation reactions.

The composition of hydrodesulphurization intermediates on monocomponent molybdenum or tungsten catalysts is characterized by higher contents of hydrogenated sulphur-containing cyclic compounds. The synergism in bicomponent catalysts results in a decrease in the relative concentration of these intermediates. At the same time, more saturated hydrocarbons are found with monocomponent catalysts than with bicomponent ones. These observations are generally valid irrespective of the structure of the starting aromatic heterocyclic compound.

In order to explain the seemingly contrasting simultaneous presence of unstable tetrahydrothiophene (and its analogues) and the thermodynamically most stable saturated hydrocarbons, like butane, in one case and of butenes in the absence of tetrahydrothiophene in another case, we have introduced the concept of the "principal reaction pathway" in Schemes 7.1-7.3 which is shifted to the left by synergism. We believe that all precursors of the species depicted in Schemes 7.1-7.3 exist on the surface of the catalysts, but only those which, under given conditions of catalyst composition, temperature and hydrogen pressure, are present in higher concentrations are desorbed into the bulk phase and detected as products.

In kinetic terms, the synergic effect is based on changes in the relative activities of the catalysts for surface hydrogenation of the C=C double bond and for surface splitting of the C-S bond. Both activities are increased by synergism, the rate of splitting much more so than that of hydrogenation. While surface hydrogenation of the C=C double bond can be mechanistically described as addition, for surface splitting of the C-S bond several different, experimentally not easily distinguishable possibilities exist. This elementary step might be realized as a substitution on sulphur or carbon involving a hydrogen species, as a β-elimination without participation of hydrogen, producing a C=C double bond, or as homolytic or heterolytic splitting by two active centres with participation of hydrogen in the consecutive step. The results summarized in Section 7.4 show that, in comparison with elementary C=C hydrogenation, the elementary C-S splitting is accelerated less (or not at all) by hydrogen and has a higher activation energy.

Several microscopic theories of the synergic effect have been suggested in the literature (for reviews see refs. 5-10). They concentrate on the structure of the catalyst and the conclusions of the present work do not contribute to a discussion of them. However, the hypothesis of the formation of "pseudobinary" metal sulphides (ref. 3) is of interest. According to it, sulphur is bonded too strongly in sulphides of Mo and W and too weakly in sulphides of Ni and Co. The synergism is then related to the formation of a mixed or "pseudobinary" sulphide with an optimum sulphur bond strength.

The conclusions of the present work show that the overall rate of hydrodesulphurization is increased mainly by increasing the surface rates, r_s. The latter are determined in a dynamic way by two consecutive processes: (i) transfer of the sulphur from non-aromatic organic sulphides to the catalyst and (ii) transfer of sulphur from the catalyst to the bulk phase in the form of hydrogen sulphide. We suggest that on sulphides of Mo and W the process (i) is fast but the rate r_s is slow because process (ii) is slow. On sulphides of Co and Ni the process (ii) is fast but the rate r_s is again slow because process (i) is slow. On a bicomponent catalyst the rate r_s is fast because the processes (i) and (ii) are balanced.

On the other hand, the results summarized here show the relatively small effect of the addition of cobalt or nickel on the hydrogenation activity, r_h, of molybdenum or tungsten. This is understandable as the reaction does not depend on sulphur mobility. The observed small synergic effect may be explained by increased reduction of the molybdenum and tungsten sulphides and formation of more active centres (coordinatively unsaturated surface atoms).

REFERENCES

1 J.P.R. Vissers, T.J. Lensing, F.P.M. Mercx, V.H.J. de Beer and R. Prins, Comm. Eur. Communities Rep. EUR, (1983) EUR 8651.
2 J. Bachelier, J.C. Duchet and D. Cornet, J. Catal., 87 (1984) 283-291.
3 R.R. Chianelli, T.A. Pecoraro, T.R. Halbert, W.-H. Pan and E.I. Stiefel, J. Catal., 86 (1984) 226-230.
4 J.C. Duchet, E.M. van Oers, V.H.J. de Beer, and R. Prins, J. Catal., 80 (1983) 386-402.
5 F.E. Massoth, Adv. Catal., 27 (1978) 265-310.
6 B. Delmon, in H.F. Barry and P.C.H. Mitchell (Editors), Proc. Climax 3rd Int. Conf. Chemistry and Uses of Molybdenum, Ann Arbor, August 19-23, 1979, Climax Molybdenum Co., London, 1980, pp. 73-85.
7 P. Grange, Cat. Rev. Sci. Eng., 21 (1980) 135-181.
8 P.C.H. Mitchell, Catalysis (Chem. Soc. Spec. Periodic Rep., London 1981), 4 (1981) 175-209.
9 H. Topsøe in J.P. Bonnelle, B. Delmon and E. Derouane (Editors), Surface Properties and Catalysis by Non-metals: Oxides, Sulfides and other Transition Metal Compounds, D. Reidel, Dordrecht, 1983, p. 329.
10 H. Topsøe, R. Candia, N.Y. Topsøe and B.S. Clausen, Bull. Soc. Chim. Belg., 93 (1984) 783-806.
11 M. Zdražil, Appl. Catal., 4 (1982) 107-125.
12 G.A. Mills, H. Heinemann, T.H. Bulliken and A.G. Oblad, Ind. Eng. Chem., 45 (1953) 134-137.
13 J.G. Williamson, C.S. John and C. Kemball, J. Chem. Soc., Faraday Trans., 1, 76 (1980) 1366-1379.
14 M.L. Vrinat, Appl. Catal., 6 (1983) 137-158.
15 P. Fott and P. Schneider, Collect. Czech. Chem. Commun., 45 (1980) 2728-2741.
16 I. van Parys, G.F. Froment and B. Delmon, Bull. Soc. Chim. Belg., 93 (1984) 823-829.
17 K. Wakabayashi and Y. Orito, Kogyo Kagaku Zasshi, 74 (1971) 1317-1320.
18 F.E. Massoth and K.S. Chung, in T. Seiyama and K. Tanabe (Editors), Studies in Surface Science and Catalysis, Vol. 7A, New Horizons in Catalysis (Proc.

7th Int. Congr. on Catalysis, Tokyo, 30 June - 4 July, 1980), Kodansha, Tokyo and Elsevier, Amsterdam, 1981, p. 629.
19 M.R. Blake, M. Eyre, R.B. Moyes and P.B. Wells, in T. Seiyama and K. Tanabe (Editors), Studies in Surface Science and Catalysis, Vol. 7A, New Horizonts in Catalysis (Proc. 7th Int. Congr. on Catalysis, Tokyo, 30 June - 4 July, 1980), Kodansha, Tokyo and Elsevier, Amsterdam, 1981, p. 591.
20 J.M. Pazos and P. Andréu, Can. J. Chem., 58 (1980) 479-484.
21 M. Bladuri and P.C.H. Mitchell, J. Catal., 77 (1982) 132-140.
22 P.C.H. Mitchell and C.E. Scott, Bull. Soc. Chim. Belg., 93 (1984) 619-625.
23 S. Kolboe and C.H. Amberg, Can. J. Chem., 44 (1966) 2623-2630.
24 S. Kolboe, Can. J. Chem., 47 (1969) 352-355.
25 A.E. Hargreaves and J.R.H. Ross, in G.C. Bond, P.B. Wells and F.C. Tompkins (Editors), Proc. 6th Inter. Congr. on Catalysis, London, July 12-16, 1976, The Chemical Society, London, 1977, pp. 937-950.
26 A.E. Hargreaves and J.R.H. Ross, J. Catal., 56 (1979) 363-376.
27 H. Schulz and Dac-Vong Do, Bull. Soc. Chim. Belg., 93 (1984) 645-651.
28 J. Uchytil, E. Jakubíčková and M. Kraus, J. Catal., 64 (1980) 143-149.
29 J. Uchytil, E. Kočová and M. Kraus, Collect. Czech. Chem. Commun., 46 (1981) 2076-2082.
30 J. Kraus and M. Zdražil, React. Kinet. Catal. Lett., 6 (1977) 475-480.
31 S. Morooka and C.E. Hamrin, Chem. Eng. Sci., 32 (1977) 125-133.
32 H.C. Lee and J.B. Butt, J. Catal., 49 (1977) 320-331.
33 J. Devanneaux and J. Maurin, J. Catal., 69 (1981) 202-205.
34 P. Pokorný and M. Zdražil, Collect. Czech. Chem. Commun., 46 (1981) 2185-2196.
35 E. Furimsky and C.H. Amberg, Can. J. Chem., 54 (1976) 1507-1511.
36 R. Bartsch and C. Tanielian, J. Catal., 35 (1974) 353-358.
37 D.R. Kilanowski, H. Teeuwe, V.H.J. de Beer, B.C. Gates, G.C.A. Schuit and H. Kwart, J. Catal., 55 (1978) 129-137.
38 S. Morooka and C.E. Hamrin, Chem. Eng. Sci., 34 (1979) 521-525.
39 E.N. Givens and P.B. Venuto, Amer. Chem. Soc. Div. Fuel Chem., 14 (1970) 135-164.
40 R. Peter, V. Matějec and M. Zdražil, Collect. Czech. Chem. Commun., submitted for publication.
41 R. Peter and M. Zdražil, Collect. Czech. Chem. Commun., 51 (1986) 327-339.
42 P. Geneste, P. Amblard, M. Bonnet and P. Graffin, J. Catal., 61 (1980) 115-127.
43 F.P. Daly, J. Catal., 51 (1978) 221-228.
44 M.L. Vrinat and L. Mourgues, J. Chim. Phys., 79 (1982) 45-52.
45 G.H. Singhal, R.L. Espino and J.E. Sobel, J. Catal., 67 (1981) 446-456.
46 M. Houalla, N.K. Nag, A.V. Sapre, D.H. Broderick and B.C. Gates, AIChE J., 24 (1978) 1015-1021.
47 D.H. Broderick and B.C. Gates, AIChE J., 27 (1981) 663-673.
48 G.H. Singhal, R.L. Espino, J.E. Sobel and G.A. Huff, J. Catal., 67 (1981) 457-468.
49 V. Morávek and M. Kraus, Collect. Czech. Chem. Commun., 50 (1985) 2159-2169.
50 H. Urimoto and N. Sakikawa, Sekiyu Gakkai Shi, 15 (1972) 926-931.
51 Yu.I. Yermakov, A.N. Startsev and V.A. Burmistrov, Appl. Cat., 11 (1984) 1-13.
52 V.I. Yerofeyev and I.V. Kaletchits, J. Catal., 86 (1984) 55-66.
53 K.E. Givens and J.G. Dillard, J. Catal., 86 (1984) 108-110.
54 R. Thomas, E.M. van Oers, V.H.J. de Beer, J. Medema and J.A. Moulijn, J. Catal., 76 (1982) 241-253.
55 C.K. Groot, M. Stolarski, W.S. Niedzwiedz, V.H.J. de Beer and R. Prins, Ind. Eng. Chem., Process Des. Develop., submitted for publication.
56 S.W. Cowley and F.E. Massoth, J. Catal., 51 (1978) 291-292.
57 M. Vrinat, M. Breysse and R. Frety, Appl. Cat., 12 (1984) 151-163.
58 V. Vyskočil and M. Kraus, Collect. Czech. Chem. Commun., 44 (1979) 3676-3687.
59 G. Muralidbar, F.E. Massoth and J. Shabtai, J. Catal., 85 (1984) 44-56.
60 J. Bachelier, J.C. Duchet and D. Cornet, J. Catal., 87 (1984) 283-291.

Chapter 8

CARRIER EFFECT ON HYDROGENATION PROPERTIES OF METALS

G.M.PAJONK and S.J. TEICHNER
Laboratoire de Thermodynamique et Cinétique Chimiques, L.A. 231 du CNRS, University Claude Bernard LYON I, 43 Boulevard du 11 novembre 1918, 69622 Villeurbanne Cedex (France).

8.1 INTRODUCTION

In heterogeneous catalysis hydrogenation reactions are usually performed on metals supported on a carrier, except for catalysts like Raney Ni or Co which are used without a support. There are many advantages of using a supported metal instead of a pure one, e.g., a higher dispersion of the active phase (metal crystallites, clusters or even individual atoms), a higher resistance to sintering and therefore a longer life. In particular, for precious metals the use of a carrier is an economic necessity which leads to a better turnover. Other favourable characteristics are bifunctionality, resistance to poisons, geometric and/or electronic interaction between the metal and the carrier, a strong metal-support interaction (SMSI) and finally the spillover onto the carrier of species initially adsorbed on the metal. The property of bifunctionality has often been described in the literature, whereas the spillover has attracted attention only very recently (ref. 1). These two phenomena will not be included in the present chapter, except for one example given of spillover.

The hydrogenation reactions discussed here will be restricted to the addition of hydrogen to multiple bonds in linear, aromatic or cyclic organic molecules and to the reduction of simple oxygen containing molecules like CO, CO_2 and NO. The very large variety of reactions of hydrogenolysis, hydrodesulphuration (HDS) or hydrodenitrogenation (HDN) are not included.

If a carrier affects the catalytic properties (activity and selectivity) of a supported metal it is because the dispersion of the metal by the carrier influences these properties for structurally sensitive reaction (ref. 2) or because the interaction between the carrier and the metal contributes to the creation of a new type of active centres. All the following comments concern the effects which do not arise from the dispersion (metal crystallites size) of the active phase. In other words, it is supposed that the reactions described are either structure-insensitive or are performed on dispersed metal crystallites exhibiting a size above the critical one, where the structure sensitivity is no longer observed.

In recent years increasing attention has been paid to the role played by promoters. However general rules of the effect of such substances (like K_2O) on the catalytic activity of a metal are not yet well formulated. Also, it is more convenient, for the moment, to combine the effects of the promoter and support and to consider the support as a promoter present in a particularly high proportion (by weight) with respect to the metal. Alternatively, the promoter may be considered as a carrier present in a low proportion. The behaviour of the doubly promoted (K_2O and Al_2O_3) iron catalyst for ammonia synthesis is well understood, but the effect of these promoters in the same iron catalyst used in the Fischer-Tropsch synthesis is much less obvious. Structural and/or textural promotion may be brought about by the carrier itself and improved by some additives (ref. 3). It is therefore difficult to differentiate between the effects of the carrier and promoter.

Although, a carrier usually does not exhibit hydrogenation properties (activity and selectivity) evidence has recently been obtained which shows that carriers like silica and alumina may acquire very unusual hydrogenating properties after their activation by hydrogen spillover (ref. 1). These properties may be screened by those of the metal and therefore may remain undetected unless the metal is removed prior to the reaction.

In the present review we consider first some early data on the hydrogenation of ethylenic and acetlyenic bonds in linear hydrocarbons, of aromatics or cycloolefins and also of other functions. These data do not allow one to formulate any clear correlations between the nature of the support and the catalytic activity. Recent studies take into account the metal-support interaction, which is now well documented, and the interpretation of the role of the support becomes easier. Finally some enantioselective hydrogenations will be reviewed. For the sake of convenience the following sections are organized according to the type of the hydrogenation reaction and to the nature of the supported metal.

8.2 SOME PREVIOUS DATA

A thorough review of catalytic hydrogenation reactions was given by Rylander in 1967 (ref. 4). The effect of the carrier was not the main point of interest in this review but a few data are pertinent to this. It has been reported, for instance, that reduction by hydrogen of halogen-containing nitro-aromatics, without dehydrohalogenation, is difficult. Results concerning the reduction of p-chloronitrobenzene, in ethanol solution, in the presence of the precious metals, Pd, Pt and Rh supported on various carriers, are given in Table 8.1. (ref. 5). The best metal with respect to the absence of dehydrochlorination was Rh, but its activity, with one exception ($Rh/BaCO_3$), was rather low.

TABLE 8.1

Hydrogenation of p-chloronitrobenzene : effect of catalyst support and metal (after ref. 5) All catalysts contained 5 % by weight of metal. Hydrogenations were conducted at atmospheric pressure and room temperature with 200 mg catalyst 1.5 g substrate and 50 ml ethanol.

Support	Palladium		Platinum		Rhodium	
	Dehydrohalogenation (%)	Time (sec)	Dehydrohalogenation (%)	Time (sec)	Dehydrohalogenation (%)	Time (sec)
Carbon	53	340	23	550	2	2070
Calcium carbonate	34	745	21	570	-	-
Strontium carbonate	-	-	14	1200	1	3900
Barium carbonate	26	780	6	1260	1	300
Alumina	48	490	22	770	4	4800
Barium sulphate	35	510	21	1500	-	-
Kieselguhr	50	385	15	1360	-	-
Magnesium silicate	35	445	-	-	-	-

Platinum and palladium were more active but again their efficiency depended on the nature of the carrier. In thise case $BaCO_3$ was the best carrier. It is of interest that rather acidic carriers like kieselguhr and alumina are not suitable, except for Pd. However a general rule cannot be formulated from these data. Metals deposited on carbon were active but dehydrochlorination was observed to the greatest extent. When a secondary undesired reaction is not encountered, because of the structure of the reactant, a carbon support generally gives a more active catalyst with Pd or Pt than any other support. This is the case in the hydrogenation of aromatic aldehydes where alumina, zinc carbonate or calcium carbonate as supports give less active catalysts. In the same way, in the reduction of acetone and methyl-ethyl-ketone, ruthenium supported on carbon was the most active catalyst, but for the reduction of heptanal, cyclohexanone and levulose (ref. 6), calcium carbonate was a better support than carbon. Any general rule concerning the activity of these supported catalysts is again difficult to formulate.

The effect of the support is better perceived with respect to the selectivity than to the activity. Crotonaldehyde may be hydrogenated at 25°C to 2-buten-1-ol or to butyraldehyde (ref. 7). Table 8.2. shows data for the hydrogenation with platinum on various supports. The presence of Fe and Zn did not obscure the results.

TABLE 8.2
Catalytic hydrogenation of crotonaldehyde (after ref. 7).
a) Iron was added as ferrous chloride, zinc as zinc acetate, and silver as silver nitrate.

Experiment	Catalyst	Amount of catalyst	Atoms of metal[a] per atom of Pt	Substrate (0.1 mole)	Solvent (50 ml)	Product
1	5% Pt/C	2000	0.4 Fe 0.06 Zn	Crotonaldehyde	Ethanol	2-buten-1-ol
2	5% Pt/CaCO$_3$	1000	3.75 Fe 0.12 Zn	Crotonaldehyde	Ethanol	2-buten-1-ol
3	5% Pt/C	2000	0.4 Fe 0.06 Ag	Crotonaldehyde	Ethanol	2-buten-1-ol
4	5% Pt/BaSO$_4$	2000	0.4 Fe 0.06 Zn	Crotonaldehyde	Ethanol	Butyraldehyde
5	5% Pt/Al$_2$O$_3$	1000	0.4 Fe 0.06 Zn	Crotonaldehyde	Ethanol	Butyraldehyde
6	10% Pt/C	1000	0.4 Fe 0.6 Zn	Crotonaldehyde	Ethanol	2-buten-1-ol
7	30% Pt/C	300	0.4 Fe 0.06 Zn	Crotonaldehyde	Ethanol	2-buten-1-ol

Only two carriers, Al$_2$O$_3$ and BaSO$_4$, of rather acidic character, direct the selectivity towards butyraldehyde, whereas carriers like carbon and CaCO$_3$ (non-acidic) yield 2-buten-1-ol.

When the selectivity is no longer the main parameter and when the solvent is, in addition, changed, the carrier effect is very much obscured. This is the case in the reduction of cinnamaldehyde to hydrocinnamaldehyde on supported palladium, as shown in Table 8.3 (ref. 7).

TABLE 8.3
Effect of support on hydrogenation of cinnamaldehyde (after ref. 7).
200 mg 5 % palladium-on-support ; 2.00 ml cinnamaldehyde, 50 ml solvent ; room temperature, atmospheric pressure,% HC = Percent hydrocinnamaldehyde in product when reduction stopped spontaneously.

Support	Methanol		Ethanol		Acetic acid	
	% HC	Rate (ml H$_2$/min)	% HC	Rate (ml H$_2$/min)	% HC	Rate (ml H$_2$/min)
Carbon	55	18	50	21	73	29
Barium sulphate	96	17	91	6	60	5
Barium carbonate	54	12	_a	_a	_a	_a
Calcium carbonate	72	16	95	4	54	11
Alumina	94	17	100	6	77	24
Kieselguhr	93	14	_a	_a	99	7
Magnesium carbonate	56	12	84	6	35	15

a) Substantial poisoning.

A limited carrier effect on the selectivity for geometric stereoisomerization was found in the hydrogenation of xylenes to cis and trans dimethylcyclohexanes in the presence of supported rhodium or ruthenium (Table 8.4.)(ref. 8).

TABLE 8.4
Effect of catalyst carrier in hydrogenation of xylenes (after ref. 8).
All experiments carried out at 50 psig initial pressure and room temperature.

Catalyst	Percent trans isomer in dimethylcyclohexane		
	1,2-dimethyl	1,3-dimethyl	1,4-dimethyl
5% Rh/C	6.5	15.6	22.7
5% Ru/C	3.0	9.3	23.8
5% Rh/kieselguhr	7.1	19.7	31.9
5% Ru/kieselguhr	6.0	-	27.0
5% Rh/SrCO$_3$	8.0	23.8	25.0
5% Ru/SrCO$_3$	4.3	11.8	31.8
5% Rh/BaCO$_3$	15.8	30.3	30.3
5% Ru/BaCO$_3$	3.6	15.7	27.6
5% Rh/BaSO$_4$	14.2	18.2	34.5
5% Ru/BaSO$_4$	5.2	14.4	-
5% Rh/Al$_2$O$_3$	8.0	27.8	29.5
5% Ru/Al$_2$O$_3$	9.7	14.7	29.4

In almost all cases Rh is more selective than Ru towards the formation of trans isomers. However the effect of the support is not well perceived except for Rh where BaCO$_3$ seems to be the most efficient carrier for trans steroisomerism.

Asymmetric hydrogenation in the presence of palladium or platinum giving chiral products requires an asymmetric support like d or l quartz or natural silk fibroin. Optically active phenylalanine is obtained by hydrogenation of ethyl-α-acetoximino-β-phenylpyruvate or of 4-benzylidene-2-methyloxazole-5 on the latter support (ref. 9, 10). The effect of a chiral carrier on the synthesis of chiral products by hydrogenation has since been found to be quite general.

The carrier effect was better perceived when the selectivity was the main parameter observed in the hydrogenation of complex molecules, as in the case of chiral precursors. The reduction of trans bifuranedione (I) has been studied in the presence of palladium supported on carbon or alumina. The reaction gives two isomers (ref. 11) :

$$\text{(I)} \longrightarrow \text{(II)} + \text{(III)}$$

(where (III) contains a $(CH_2)_3\,COOH$ group)

Palladium on carbon gives mainly the isomer III (69 %), whereas Pd on alumina is very selective for the isomer II (83 %).

One of the first clear observations of the carrier effect on the rate of the reaction of 1-butene (hydrogenation and isomerization) was described by Brownlie et al. (ref. 12) for Pd on graphite. When graphite is decorated, i.e., when metal-support interaction is achieved, the rates of isomerization and of hydrogenation of 1-butene are significantly increased in comparison with non-decorated Pd/graphite catalysts. When the metal-support interaction prevails (as in the case of decoration) the support modifies the number of electrons in the bonding orbitals of the metal, thus accounting for the variation in the selectivity.

A pecular metal-support interaction was described by Maurel et al. (ref. 13) in the hydrogenation of benzene on Pt/Al_2O_3 catalysts. Commercial Al_2O_3 may contain sulphate ions. However the poisoning of catalysts containing this intrinsic sulphur is different from the extrinsic poisoning by a H_2S/SO_2 mixture prior to the reaction.

In the hydrogenation of benzene in the presence of supported nickel a very definite support effect on the activity was found by Taylor and Steiffin (ref. 14), but only at low metal loadings. When the rate of hydrogenation is expressed per unit surface area of the metal, a silica-alumina support leads to lower rates than does pure alumina. However for supported Pd the opposite effect was found by Romero and Figueras (ref. 15). With various supports like Al_2O_3, MgO, SiO_2 and $SiO_2-Al_2O_3$ the activity per unit surface area of the metal was always higher on $SiO_2-Al_2O_3$ than on other supports. Among various $SiO_2-Al_2O_3$ supports tested, the most active palladium catalyst was that supported on the most acidic mixed oxide. The acidity of the catalyst as a whole does not influence the activity as the addition of an acidic zeolite to palladium supported on SiO_2, MgO or Al_2O_3 is without any effect. Therefore Pd must interact directly with the acidic $SiO_2-Al_2O_3$ support and it is difficult to invoke, any bifunctionality.

The effect of the nature of the metal on the interaction with the support, already mentioned above, must be important, as shown by Vannice (ref. 16, 17) for methanation ($CO + H_2$) on Pd, Pt and Ni supported on various inorganic oxides like Al_2O_3, SiO_2 H-Y zeolite and also graphite. No support effect (determined as the turnover number in methanation) was detected for Pt on these various supports. Only a particle size effect was evidenced for supported or unsupported Pt. For Pd and Ni, the particle size effect plays a secondary role in comparison with the nature of the support whose acidity is correlated with the turnover number. This behaviour was explained by an electron transfer from the metal to the acidic support (electron acceptor). It is also in accord with the observation that unsupported metals always exhibited smaller turnover numbers than the same supported metals. The absence of the effect with Pt was not explained.

The $CO + H_2$ reactions are reviewed in more detail below, in connection with the metal-support interaction and, in particular, strong metal-support interaction (SMSI) effects.

8.3 RECENT VIEWS ON THE METAL-SUPPORT EFFECTS

A paper, published in 1978 by Schwab (ref. 18), attracted attention to the electronic effect in supported catalysts. A general definition of the influence of a carrier on the activity of a multiphasic catalyst has been provided : the carrier increases the surface area of the supported phase and/or it interacts energetically with the active phase which results in the formation of a new kind of active contact. Almost simultaneously Tauster et al. (ref. 19) reported a new effect of the support on the metal, called by them SMSI (strong metal-support interaction). Since then a number of studies have been published on this subject. The reactions most thoroughly studied in this respect are CO/H_2 reactions and the hydrogenation of benzene. A brief description of the SMSI effect is given below.

This effect occurs mainly with supports like TiO_2, Nb_2O_5 and other which are reducible (at least partially) in H_2 at 500°C. However, at higher temperatures of reduction the SMSI effect was observed on more refractory oxides like SiO_2, Al_2O_3 or MgO. The SMSI effect is generally perceived by a loss of chemisorptive properties towards H_2 or CO by a metal (Ni, Pt, Rh,...) supported on TiO_2 and prereduced at $\geqslant$ 500°C. This loss is not due to the sintering of metal particles as shown by electron microscopy and/or X-ray line-broadening techniques. Despite the inability of a metal in the SMSI state to sorb CO and H_2, its catalytic activity in the $CO + H_2$ reactions (and even a particular selectivity in methanation) is abnormally enhanced with respect to the same catalysts not in the SMSI state. In contrast, the catalytic activity for the hydrogenation of benzene or the hydrogenolysis of ethane and n-hexane is decreased in the SMSI state.

In almost all the explanations offered to account for the SMSI state a partial reduction of the support is involved. This reduction proceeds by a hydrogen spillover mechanism (ref. 1). Electronic effects (electron donation by the reduced support to empty d-orbitals of group 8 metals, formation of alloys) or geometric effects (migration of the reduced suboxide onto the metal particle) have been invoked, with many variations, to explain the loss of sorbing properties of the metal for H_2 and CO. These properties are restored by reoxidation of the catalyst at $\ll 500°C$ followed by a mild reduction (below 500°C). Simultaneously, "normal" catalytic activity and selectivity in the methanation or hydrogenation of benzene are recovered.

A somewhat refined picture of the SMSI state was given by Burch and Flambard (ref. 20) in the form of the IFMSI (interfacial metal-support interaction) state which is encountered upon progressive transition from the "normal" to the SMSI state. The IFMSI state is supposed to occur by the interaction between the partially reduced oxide support and the metal particles, at their interfaces (see below). It results in a moderate decrease (or even not at all) in the chemisorption of CO and H_2 and in an increase in the methanation activity. However, it has recently (ref. 21) been reported that even for TiO_2 as a support an increase in the methanation activity may be observed well below 500°C, i.e., without the need for the SMSI or IFMSI state. As the chemisorptive properties, toward CO and H_2 are not modified, the catalyst is not in the SMSI state. This behaviour is explained by a limited migration of the titanium suboxide onto the metal, creating a new type of active centres.

From the abundant literature on carrier effects, published since 1978, it appears that these effects depend on the nature of the metal/support combination, its previous history in the reduction (low or high temperature reduction : LTR or HTR), the metal loadings and the test reaction chosen. This means that besides the criteria of structure-sensitive and -insensitive reactions proposed by Boudart (ref. 2), there are also reactions which are sensitive to the metal-support interaction, like methanation. On the contrary, these reactions are not influenced by the particle dimensions in the Boudart sense.

8.4 HYDROGENATION OF CARBON OXIDES

The metals which are reviewed in this section are Ni, Pt, Pd, Fe, Rh, Co and Ru, and the supports involved are SiO_2, Al_2O_3, TiO_2, ZrO_2, Nb_2O_5 and graphite.

8.4.1 Nickel-based catalysts

From a survey of the literature data it is possible to draw a thorough comparison, in terms of specific activity (turnover number or frequency) in CO conversion and selectivity (toward CH_4), between bulk nickel and supported nickel. It appears that the unsupported nickel is always : (i) less active in the overall hydrogenation of CO, (ii) more selective in the methanation, than nickel supported on TiO_2, Al_2O_3 (η or α), SiO_2 or graphite (refs. 20, 22-24).

This does not depend on the mode of preparation of the catalyst (impregnation, coprecipitation), on the nickel loading (high or low) or on the achievement or not of the SMSI state. For instance, Burch and co-workers (refs. 20, 25, 26) studied the hydrogenation of CO over Ni supported on TiO_2, SiO_2 or SiO_2-Al_2O_3. TiO_2, which is more easily reducible than SiO_2 or SiO_2-Al_2O_3, may be expected to give birth to the SMSI state under conditions in which the other two supports would not. However, even when Ni/TiO_2 is not in the SMSI state (as defined classically, see above) it is more active and selective (for C_{2+} hydrocarbons) than Ni supported on SiO_2 or SiO_2-Al_2O_3. To discriminate between these two classes of catalysts the previous authors proposed the IFMSI state (refs. 20, 25) mentioned above. Table 8.5. shows typical results for the hydrogenation of CO at 275°C with Ni on various supports.

TABLE 8.5
Comparison of the activities and product selectivities for nickel catalysts (after refs. 26-28).

Catalyst	Methanation activity as CO converted by Ni atoms	Selectivity (%)		
		CH_4	C_2H_6	CO_2
10 % Ni/TiO_2	0.37	85.3	10.3	4.4
10 % Ni/SiO_2	0,016	79	0	21
5 % Ni/SiO_2	1.2 [a]			
7 % Ni/SiO_2-Al_2O_3	0.5 [a]			
5 % Ni/TiO_2	40.56 [a]			

[a] Steady state activity, defined as the percentage of CO converted into each hydrocarbon product.

The beneficial influence of the IFMSI state on the weakening of the CO bond is explained by Fig. 8.1a which describes the situation for a Ni/TiO_2 catalyst after a low temperature reduction (LTR). The presence of the partially reduced titania (Ti^{3+}) in the vicinity of the metal particle creates a dual adsorption site for CO. This type of site is also obtained in the SMSI state (Fig. 8.1b) after the migration of the suboxide onto the metal, upon high temperature reduction (HTR). However, a serious drawback to the IFMSI concept is revealed by the following observation : the increase in the activity (per Ni atom) occurs at low dispersion of Ni on TiO_2 and not at high dispersion of the metal (refs. 29, 30). Now, the interface between Ni and TiO_2 (Fig. 8.1a) increases when the dispersion of Ni increases and therefore the IFMSI state is more liable to be observed at high dispersion. Finally, the advantage of the SMSI state over the IFMSI state lies in the stability of the catalytic activity of Ni/TiO_2 with time in the SMSI state, showing in the opposite case (IFMSI) a rapid decline in activity.

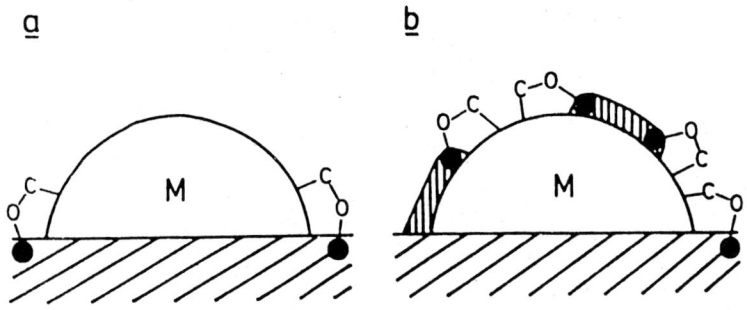

Fig. 8.1. Interfacial metal-support interactions in the CO/H_2 reaction. M, metal particles ; ●, Ti^{3+} ion ; ▨, TiO_2 ; ▥, TiO_x.

According to Vannice and Garten (ref. 21), Ni/TiO_2 is deactivated much less than is Ni/Al_2O_3. Also its methanation activity is much higher and is practically independent of the reactant pressure. Another difference between the two supports, TiO_2 and Al_2O_3, is found in the influence of the addition of sulphur to the stream of reactants. In both cases a shift in the selectivity towards C_{2+} hydrocarbons is observed, but a severe decrease in the activity is found for the Ni/Al_2O_3 catalyst (ref. 31). This peculiar behaviour of the Ni/TiO_2 catalyst was attributed to the SMSI state. Also the formation of $Ni(CO)_4$ is inhibited with Ni/TiO_2, in contrast to Ni/Al_2O_3 (ref. 22). Finally, carbon and SiO_2 used as supports for Ni do not show any effect (or a very small one) because the activity of these catalysts is very similar to that of unsupported nickel. The behaviour of silica is in agreement with its low reducibility which is detrimental to the observation of the SMSI state (ref. 32, 33).

Vance and Bartholomew (refs. 23, 24, 34) observed a very strong influence of the preparation method on the metal-support interaction for nickel on SiO_2, Al_2O_3 and TiO_2, when used in the hydrogenation of CO. Catalysts prepared by precipitation produced unsual yields of C_{2+} products (Fig. 8.2), in contrast to catalysts prepared by impregnation. In all cases, however, the beneficial influence of the support on the production of C_{2+} hydrocarbons follows the sequence $SiO_2 < Al_2O_3 < TiO_2$, which is also the sequence of the increased interaction between the metal and the support. The same sequence is also observed in the methanation of CO_2 as regards the activity and selectivity for CH_4 with Ni catalysts on SiO_2, Al_2O_3 and TiO_2 (ref. 34). It is of interest that the metal-support interaction (increasing in the sequence SiO_2, Al_2O_3, TiO_2) exerts an opposite effect concerning the selectivity for CH_4 in the hydrogenation of CO and CO_2. Whereas in the first case (CO) the production of CH_4 is reduced (and that of C_{2+} hydrocarbons increased) when the interaction increases (Fig. 8.2), the opposite is observed in the second case (CO_2).

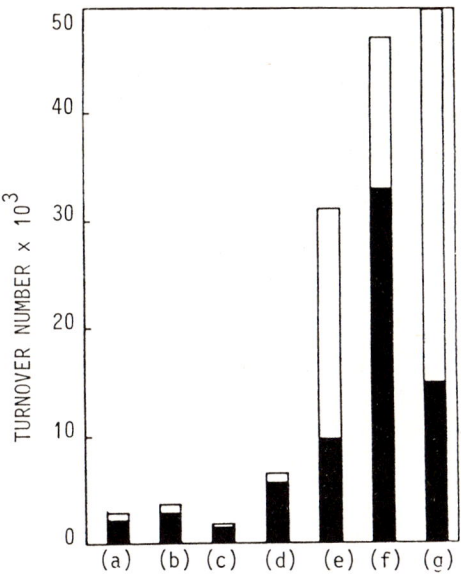

(a) 100 % Ni
(b) 2.7 % Ni/SiO_2 impregnated
(c) 3.6 % Ni/SiO_2 precipitated
(d) 3 % Ni/Al_2O_3 impregnated
(e) 2.9 % Ni/Al_2O_3 precipitated
(f) 2.8 % Ni/TiO_2 impregnated
(g) 2.8 % Ni/TiO_2 precipitated

Fig. 8.2. Effects of support and its method of preparation on the methane turnover number at 525 K for Ni : ■, CH_4 turnover No. ; ☐, C_{2+} hydrocarbon turnover No. ; total bar length is CO turnover No.

More recently Ko et al. (ref. 29, 32, 33) investigated the influence of the SMSI state on the olefin/paraffin ratio in the hydrogenation of CO with Ni supported on SiO_2 (not in the SMSI state) or on Nb_2O_5 (in the SMSI state). Table 8.6. shows that, besides the classical shift in product distribution towards C_{2+} a significant increase in the olefin/paraffin ratio is favoured by the SMSI state.

TABLE 8.6
Activity and olefin/paraffin ratios of nickel catalysts in CO hydrogenation. (after ref. 32, 33).

Catalyst	CH_4 formation % at 200°C	C_{2+}/CH_4	Olefin/paraffin		
			C_2	C_3	C_4
Ni/SiO$_2$ 30 wt %	0.7	0.42	0	0	0
Ni/Nb$_2$O$_5$ 10 wt %	0.73	0.96	0.02	0.55	1.45

The nature of the interaction between the support (Nb_2O_5) and the metals was discussed by these authors by comparing the results with those obtained on a classical nickel catalyst (on SiO_2) promoted by K. For this catalyst this promotion also increases the selectivity towards C_{2+} hydrocarbons but simultaneously decreases its activity, which is opposite to the behaviour of Ni in the SMSI state (on Nb_2O_5). The SMSI state of Ni cannot be explained by a simple electron transfer from Nb_2O_5 to the metal, which is also invoked for potassium-promoted nickel, as both catalysts behave differently.

An entirely different conclusion was drawn by Turlier et al. (ref. 35) who studied Ni supported on SiO_2 (not in the SMSI state) and on ZrO_2 and TiO_2 (in the SMSI state). They claimed that the SMSI state does not result in improved activity and selectivity for C_{2+} hydrocarbons. The explanation they offered is based on the more or less facile obtention, by reduction, of zerovalent nickel on these supports. It is the zerovalent nickel which is the active site for hydrogenation and therefore the activity pattern may follow the extent of reduction of the catalyst.

Ozdogan et al. (ref. 36) used, in the hydrogenation of CO, nickel catalysts supported on SiO_2, Al_2O_3, SiO_2-Al_2O_3 and TiO_2. The temperature-programmed reaction (TPR) technique showed that the activity pattern followed the generally accepted sequence Ni/Al_2O_3-SiO_2 < Ni/SiO_2 < Ni/Al_2O_3 < Ni/TiO_2. This sequence is in good agreement with that published by Burch and Flambard (refs 26, 27) where seemingly the most acidic support (SiO_2-Al_2O_3) gives the less active catalyst, but it is also in agreement with the facility of development of the SMSI state. Contrary to these findings concerning the hydrogenation of CO, the activity in the methanation of carbon, as determined by the temperature-programmed surface reaction (TPSR) technique (ref. 36) does not depend on the nature of the support, for the same catalysts. This suggests that in the methanation of CO the rate of hydrogenation of carbon (formed as an intermediate) is not the rate-determining step.

Recently, a clear picture of the phenomena occurring during the reduction of Ni/TiO$_2$ catalysts has been given by Chung et al. (ref. 37) who used XPS and Auger spectroscopies. The sequence of events is as follows : partial reduction of TiO$_2$ to TiO$_x$, diffusion of TiO$_x$ (presumably as Ti$_4$O$_7$) onto nickel particles, formation of Ni-Ti bonds (intermetallic) which decreases the sorptive properties towards CO and H$_2$ and finally spreading of Ni on Ti$_4$O$_7$ (raft or pill-box-like metal particles) but maintaining its former dispersion. By oxidation, which separates Ni (as NiO) and TiO$_2$, and a further LTR, the initial sorption properties of the catalyst towards CO and H$_2$ are restored. The SMSI state is destroyed by oxidation and LTR.

8.4.2 Platinum-based catalysts

Vannice and co-workers (refs. 21, 38, 39) extensively studied platinum catalysts supported on SiO$_2$, Al$_2$O$_3$, SiO$_2$-Al$_2$O$_3$ and TiO$_2$. They found that for a given support the methanation of CO was a structure-insensitive reaction according to Boudart's classification (ref. 2). The size of the platinum crystallites had no effect on the turnover for methane production (ref. 39), but the nature of the support has an effect on this turnover and, as previously found for supported Ni, the reaction had to be classified as a support-sensitive one. Only Pt/TiO$_2$ was regarded as being in the SMSI state and a cooperative effect of the metal and the support was proposed (ref. 21). The activity (for CH$_4$) sequence was established as follows : Pt/SiO$_2$ < Pt/SiO$_2$-Al$_2$O$_3$ ≈ Pt/η-Al$_2$O$_3$ < LTR Pt/TiO$_2$ < HTR Pt/TiO$_2$ (SMSI). For this series of catalysts the lowest coverage by CO was found, by IR measurements, on Pt/TiO$_2$ (SMSI). Simultaneously, the vibration frequency of adsorbed CO indicated a seriously weakened C-O bond (ref. 38). During the CO + H$_2$ reaction on this catalyst no IR-detectable CO bond was found. This was explained as due to a small fraction of platinum sites active in the SMSI state. Another possibility is that the reactive CO species is IR inactive. The above sequence of activities also suggests that the acidity of the support (in the absence of the SMSI state) has an effect on the efficiency) of the platinum. The authors proposed a model, similar to that of Burch et al. (ref. 26)(see above), in which a cooperative effect in the adsorption of CO is exerted by Pt and by the coordinatively unsaturated surface (CUS) ions Ti^{3+} (for LTR) or Al^{3+} (refs 21, 25).

In the hydrogenation of CO at high pressure (30 atm), Meriaudeau et al. (ref. 40) reported the selective formation of CH$_3$OH at 280°C on Pt supported on Al$_2$O$_3$, SiO$_2$, MgO, La$_2$O$_3$, CeO$_2$, ThO$_2$ and TiO$_2$. The pure supports La$_2$O$_3$, CeO$_2$ and ThO$_2$ are also active in the methanol synthesis, whereas SiO$_2$, Al$_2$O$_3$, TiO$_2$ and MgO are inactive. Taking into account this activity it was found that for mechanical mixtures with some extra carrier the activity of supported platinum catalysts in the formation of methanol was strongly dependent on the nature of the support. These results were explained by variations in the ease of hydrogen spillover on the supports, which depended on the nature of the support.

The synthesis of methanol was also studied at atmospheric pressure by Szymanski et al. (ref. 41) for platinum catalysts supported on ZrO_2, Al_2O_3, SiO_2 and carbon. The activity sequence for methanol formation (based on the turnover) was : $Pt/C < Pt/SiO_2 < Pt/Al_2O_3 < Pt/ZrO_2$ and is the same as in methanation. However the selectivity for CH_3OH follows a rather different order : $Pt/SiO_2 \simeq Pt/ZrO_2 < Pt/C < Pt/Al_2O_3$. Finally, alloying Pt with Zr increased the selectivity for CH_3OH to the same extent as for carbon or ZrO_2 as support.

8.4.3 Palladium-based catalysts

For Pd, Vannice and co-workers (ref. 42, 43) used the same supports as for Pt (see above). Many similarities between the two classes of catalysts (based on Pd and Pt) are observed for the methanation reaction. The most active catalyst is Pd/TiO_2 in the SMSI state, whereas Pt/SiO_2 exhibits the same activity as that of unsupported Pd. Again no IR band due to CO was found in the reaction on the Pd/TiO_2 (SMSI) catalyst. The reaction was also structure insensitive with respect to the size of the palladium crystallites (in the range 3-30 nm) on a given support (ref. 39). The activity sequence follows that established previously for platinum-supported catalysts (see above) : $Pd/SiO_2-Al_2O_3 < Pd/\eta-Al_2O_3 <$ LTR $Pd/TiO_2 <$ HTR Pd/TiO_2 (SMSI). Also, the same considerations for Pt apply for Pd.

More recently Bracey and Burch (ref. 44) described the methanation with Pd supported on TiO_2 or on SiO_2. The much higher activity of Pd/TiO_2 was explained in terms of the IFMSI state, as shown in Fig.8.3. Poels et al. (ref.45) hypothesized an intermediate formation of CH_xO species. According to Bracey and Burch (ref. 44) this species may be formed either on Pd or on the cation of the support (M^{n+}), e.g., Ti^{3+}, Si^{4+}, Zr^{4+}, Al^{3+} (Fig. 8.3a). In the absence of CH_xO as intermediate (Fig. 8.3b) the cation assists in the activated adsorption of CO.

The role of the acidity of the support was examined by Saha and Wolf (ref. 46) for Pd supported on a series of Y-zeolites and ZSM5 zeolites. The activity sequence for methanation was as follows ; $Pd/SiO_2 < Pd/Na-Y \ll Pd/Na-ZSM5 < Pd/H-ZMS5 < Pd/H-Y$. Both the higher structural stability and acidity of ZMS5 were expected to yield an active and a stable catalyst, even in the case of the Pd/Na-ZMS5 system.

As regards the formation of methanol, Poels et al (ref. 45) and Fajula et al. (ref. 47) showed that the selectivity of supported Pd strongly depends on the nature of the support, as pointed out simultaneously by Ichikawa and Shikakura (ref. 48), whereas the selectivity for methanation depends essentially on the <u>acidic</u> properties of the support.

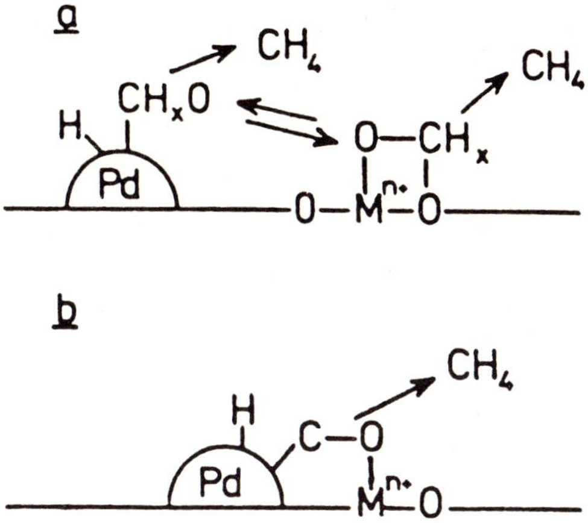

Fig. 8.3. Proposed reaction scheme for the formation of methane over supported palladium catalysts.

More recently, Pd and Rh supported on TiO_2 or on Nb_2O_5 in the SMSI state were examined by Kunimori et al. (ref. 49). In contradiction with the general rule concerning methanation catalysts in the SMSI state (see above), these authors found that Pd and Rh supported on Nb_2O_5 are much less active in the SMSI state (two orders of magnitude) than when they are not in this state. However for TiO_2 as support no differences in activities were found. Also, the product selectivity was the same with both states. It can therefore be concluded that there are as many different SMSI states as there are metal-support combinations.

8.4.4 Ruthenium-based catalysts

Although Ru is known to be active for the hydrogenation of CO to CH_4 and higher hydrocarbons, studies concerning ruthenium-based catalysts are rather scarce. Morris et al. (ref. 50) compared the catalytic properties of supported Ru on SiO_2, Al_2O_3, MgO, TiO_2, SiO_2-Al_2O_3, TiO_2-SiO_2 and 13X zeolite. The product distribution was sensitive to the nature of the support. Essentially the same distribution (CH_4 as the major product and higher hydrocarbons) was found for unsupported ruthenium powder as for Ru/Al_2O_3 and Ru/SiO_2. Ru/TiO_2 gave much higher yields of C_{2+} hydrocarbons and alkenes as well as the highest activity. The activity of Ru/MgO was even lower than that of unsupported Ru and the propene/propane ratio was decreased. Ru/TiO_2 catalysts gave the same results irrespective of the presence or not of the SMSI state. The peculiar influence of MgO can be correlated with the basicity of this oxide because promotion of

Ru by K gave the same effect.

Recently Leith (refs. 51, 52) proposed an interpretation of the correlation between the selectivity for olefins and the acidity of Ru/Y-zeolites. The increase in the basic character of the zeolite obtained by a partial exchange with K^+ or Cs^+ increased the fraction of C_2 and C_3 olefins (in C_2 and C_3 paraffins), whereas an increase in the acidic character increased the selectivity for branched paraffins (mainly isobutane). The selectivity in the hydrogenation of CO seems therefore to be determined by the direction of the electron shift between the supported metal and the support and depends on the electron donor-acceptor character of the zeolite.

Similar findings were presented by Yang and Goodwin (ref. 53) for potassium promoted Ru supported on SiO_2 and TiO_2 (not in SMSI state). For both supports, promotion by K increased the selectivity for olefins and, for the SiO_2 support only, it increased the chain growth probability. When the Ru/TiO_2 catalyst is in the SMSI state its high methanation activity and its selectivity for olefins are not modified by potassium promotion.

In the same way, the chain growth probability (or the selectivity for CH_4) was found to be altered by alkali-metal ions exchanged into various zeolites, whereas the fraction of olefins in C_2-C_4 hydrocarbons remained unchanged (ref. 54).

8.4.5 Rhodium-based catalysts

Few results are available concerning this type of catalysts. Meriaudeau et al. (ref. 55) studied Rh supported on SiO_2, Al_2O_3, MgO and SiO_2 as catalysts for the hydrogenation of CO. Rh/TiO_2 in the SMSI state or in the "normal" state was again the most selective for olefins and long chain hydrocarbons. The selectivity for CH_4 (low chain growth probability) followed the sequence $Rh/Al_2O_3 < Rh/MgO < Rh/SiO_2$ which is not the sequence of increasing acidity (MgO $<$ SiO_2 $<$ Al_2O_3). Clearly, the acidity does not seem to be the relevant parameter as regards this selectivity as, for instance, Rh/TiO_2 (LTR) has a comparable acidity to that of Rh/Al_2O_3. However, the first catalyst is much more selective for C_{2+} hydrocarbons as well as for olefins.

Kunimori et al. (ref. 56) examined rhodium catalysts supported on Nb_2O_5 in the SMSI state or in a "normal" state. Whereas the selectivities for various products were little affected by LTR ("normal" state) or HTR (SMSI state), the activities in the SMSI state were drastically lowered. It was suggested that the nature of the SMSI in the Nb_2O_5-supported systems may be different from that in the TiO_2-supported systems.

A study concerning Rh supported on ThO_2 which was pure or doped by 1°/oo weight of Na, K, Ca, Ce and Eu prior to the deposition of Rh was reported by Bardet et al. (ref. 57). Important variations in the activity and selectivity towards CH_3OH and C_2H_5OH as well as CH_4 were observed, depending on the nature of the dopant. The highest yield in alcohols was observed for Europium-doped catalysts. In all cases, doping of ThO_2 by an alkali decreases the selectivity for CH_4 and the ethylene/ethane ratio. Finally all dopants other than K favoured the formation of ethanol over that of methanol.

Solymosi et al. (ref. 58) studied the hydrogenation of CO_2 to CH_4 over Rh supported on Al_2O_3, SiO_2, MgO and TiO_2. For the hydrogenation of CO, the activity sequence was as follows : Rh/MgO < Rh/SiO_2 < Rh/Al_2O_3 < Rh/TiO_2.

8.4.6 Cobalt-based catalysts

For CO hydrogenation, Reuel and Bartholomew (ref. 59) reported, a similar sequence of activities for supported Co as for other metals (see above) : Co/MgO < Co/Carbon < Co/SiO_2 < Co/Al_2O_3 < Co/TiO_2. The method of preparing the catalysts (by impregnation or by precipitation) also influenced the activity and the selectivity. Table 8.7. gives the selectivities of these catalysts.

Blanchard and Vanhove (ref. 60) reported the conversion of CO in the liquid phase in the presence of unsupported cobalt powder or Co/Al_2O_3 (slurried catalysts). The olefin/paraffin ratio was higher on the Al_2O_3-supported catalyst than on pure metal, as was the parameter α (chain growth probability) of the Schulz-Flory distribution of products. This parameter was shown to depend on the porosity of the alumina support at low cobalt loadings (ref. 61).

Peuckert and Linden (ref. 62) studied cobalt catalysts supported on various zeolites like ZSM5, silicalite and mordenite. The activity in the hydrogenation of CO was not sensitive to the nature of the zeolite, in contrast to the selectivities for C_2, C_3 and C_4 olefins. The most selective catalyst for olefins was Co/silicalite and it was also the most efficient in the suppression of CH_4. The product distribution was explained by the basicity of the zeolite supports.

8.4.7 Iron-based catalysts

Vannice et al (ref. 63) compared the activities and selectivities of iron supported on various forms of carbon (glassy carbon, carbon blacks and graphite) and on alumina in the hydrogenation of CO. The carbon-supported iron catalysts were four times more active than Fe/Al_2O_3, and the olefin/paraffin ratio was ten times higher. They also observed that carbons interact with Fe in the same way as alkali-metal promoters, increasing both the activity and the production of olefins.

Pure iron and iron supported on SiO_2, SiO_2-Al_2O_3 and MnO were investigated by Egiebor et al (ref. 64). Silica-supported catalysts with extra silica added exhibited increased selectivity towards cis/trans internal olefins in non-equilibrium ratio whatever the carbon number. This indicates that these internal olefins were produced by secondary reactions from α-olefins due to the presence of SiO_2 in the catalysts. The two other catalysts (SiO_2-Al_2O_3 and MnO supports) also exhibited an abnormal cis/trans ratio for internal olefins.

TABLE 8.7
Hydrocarbon and carbon dioxide/water product distribution for supported and unsupported cobalt catalysts (after ref. 59).

Catalyst	Temp. (°C)	Weight % CO_2 selectivity[a,b]	Weight percentage hydrocarbon group selectivities [a,c]					Average carbon number [d]
			C_1	C_2-C_4	C_5-C_{12}	C_{13+}	Alcohols	
100 % Co	225	0	29	42	28	0	1.1	3.4
Co/SiO_2 3 %	225	41	47	34	15	0	3.9	2.5
10	225	0	29	27	42	0.2	1.3	4.0
3[e]	325	33	99	1.1	0	0	0	1.0
Co/Al_2O_3 1 %	-[g]	-[g]	-[g]	-[g]	-[g]	-[g]	-[g]	-[g]
3	225	71	41	36	18	0	4.0	3.1
10	225	18	32	31	35	0.7	1.3	3.8
15	215	0	3.8	5.5	86	4.4	0	9.5
3[e]	225	18	27	31	39	0	2.7	4.0
Co/TiO_2 3 %	225	0	31	35	15	0	3.5	3.5
10	225	0	16	30	52	1.7	1.1	5.0
3[e]	225	0	22	39	37	0	2.1	4.2
Co/MgO 3 %	-[g]	-[g]	-[g]	-[g]	-[g]	-[g]	-[g]	-[g]
10	300	36	55	39	6.2	0	2.1	1.9
Co/C (type UU)								
3%[f]	275	84	30	54	16	0	0	2.7
10[f]	225	44	53	31	16	0	0	2.3
Co/C (Spheron)								
3%[f]	250	24	85	8.1	7.0	0	0	1.5
10[f]	225	8	66	23	22	0	0	2.1

[a] Measured at temperature shown above at H_2/CO = 2 and 1 atm.
[b] Weight percentage of CO_2 in oxygen-containing, non hydrocarbon products: $CO_2(10^2)/(CO_2 + H_2O)$.
[c] Weight percentage of hydrocarbon group based on total hydrocarbons in the product.
[d] Weight-averaged carbon number
[e] Controlled-pH precipitation
[f] Evaporative deposition
[g] Inactive up to 400°C.

Tau and Bennett (ref. 65) used Mössbauer spectroscopy to follow the hydrogenation of CO on Fe/TiO$_2$ catalysts as well as on iron supported by carriers not giving the SMSI state. After reduction by H$_2$ at 500°C, titania-supported iron was found to be in the zerovalent state whereas on other supports a mixture of Fe°-Fe^{2+} was always present. Increasing the reduction temperature to 500°C promoted an electron transfer from the metal to the titania support (which is not a behaviour commonly observed with other metals), whereas the bulk electron density remained unchanged. The decrease in the activity of this catalyst in the CO + H$_2$ reaction and a shift in the selectivity towards higher hydrocarbons were explained by the IFMSI state including iron decoration by TiO$_2$ during reduction, leading finally to well dispersed TiO$_x$ species which facilitate the electron transfer to iron.

In contrast to the above results, a conventional structure sensitivity for hydrogenation of CO on Al$_2$O$_3$- and SiO$_2$-supported bimetallic Ru-Fe catalysts was advanced by Boszomrenyi et al (ref. 66). Ru-Fe/Al$_2$O$_3$ showed a higher activity than Ru-Fe/SiO$_2$. However Al$_2$O$_3$ allowed a better dispersion of the metallic clusters due to a pronounced interaction with their initial (before reduction) oxidized form. On SiO$_2$ such an immobilization effect of the oxidized metals did not occur and larger clusters resulted from the migration of the oxidized precursor.

8.4.8 Molybdenum-based catalysts

This metal does not belong to group 8. Its behaviour in the hydrogenation of CO is of interest as pointed out by Concha et al. (ref. 67). The activity sequence for supported catalysts was found as follows : Mo/CeO$_2$ < Mo/Carbon < Mo/Al$_2$O$_3$ < Mo/SiO$_2$. This is entirely different from that found for Group 8 metals. High resistance to poisoning by H$_2$S was also observed for the less active catalysts (Mo/CeO$_2$ and Mo/Al$_2$O$_3$), while for Mo/SiO$_2$ a severe decrease in the activity was observed. Also the selectivity for C$_{2+}$ hydrocarbons was greater for Mo/SiO$_2$ than for other catalysts.

At least for the group 8 metals supported on titania, which is the support leading most easily to the SMSI state, it can be concluded (ref. 68) that the specific activity in the hydrogenation of CO is enhanced for metals having more completely filled d bands (Ni, Pd,Pt,Rh,Ir); it is almost without effect on Ru and Co and it inhibits Fe.

8.5 HYDROGENATION OF BENZENE AND OTHER AROMATICS

For the sake of clarity this section is subdivided according to the nature of the supports : those giving the SMSI state (like TiO$_2$), other inorganic oxides and finally organic carriers like polyamides.

8.5.1 Metals supported on TiO_2 catalysts

It is generally found that platinum on TiO_2 in the SMSI state exhibits a marked decrease in activity in the hydrogenation of benzene in contrast with the results described above for the hydrogenation of CO.

Meriaudeau et al. (ref. 69) studied the hydrogenation of benzene and styrene at 15°C in the presence of Pt, Ir and Rh deposited on TiO_2, either in the SMSI state HTR or not LTR. The catalytic activity of these metals when not in the SMSI state was very comparable to that of the same metals supported by SiO_2 or Al_2O_3. In the SMSI state, all metals supported on TiO_2 exhibited a marked decrease in activity. Similar behaviour was also observed with these metals supported by CeO_2 (ref. 70) and subjected either to LTR or HTR. The formation of an intermetallic phase, Pt_5Ce, under HTR conditions was hypothesized by these authors. Comparable intermetallics are liable to be formed for other metals supported by CeO_2 or TiO_2. This peculiar behaviour in the SMSI state seems to be restricted to only a few metals in group 8 and is not found for all. Indeed, Burch and Flambard (refs. 26, 27) found that Ni supported on TiO_2 shows almost the same activity in the hydrogenation of benzene as on Ni/SiO_2 catalysts, irrespective of LTR or HTR. It is Ni/TiO_2 which exhibits a larger turnover afterHTR rather than Ni/SiO_2.

Another metal which does not follow the previous trend is Pd, as described recently by Vannice and Chou (ref. 71). The turnover in the hydrogenation of benzene in the presence of Pd/TiO_2 (HTR) compares very closely with that of $Pd/SiO_2-Al_2O_3$ which is not in the SMSI state. It has been shown above that HTR of Pd/TiO_2 catalysts leads to the SMSI state, which, in particular, is favourable for the hydrogenation of CO, as is Pt/TiO_2 in the SMSI state. The latter catalyst is not active in the hydrogenation of benzene, whereas Pd/TiO_2 indeed is active in this reaction. This difference in the behaviour of the two metals supported on TiO_2 has not yet been satisfactorily explained. Finally, the reverse reaction, i.e., dehydrogenation of cyclohexane to benzene, was studied by Meriaudeau et al (ref. 69) for Pt, Ir and Rh supported on TiO_2 catalysts. The same trend as for the hydrogenation of benzene was observed. The activities of these catalysts after HTR (SMSI state) were strongly decreased in contrast to those after LTR (not in SMSI state). However, Resasco and Haller (ref. 72) did not confirm this for Rh/TiO_2 catalysts as the decay in the dehydrogenation activity in the SMSI state was only moderate. However a parallel could be traced by these authors between the catalytic behaviour of Ni-Cu alloys of variable composition (progressive filling of the d-band of Ni) and of Rh/TiO_2 catalysts reduced at progressively increasing temperatures from 250° (LTR) to 500°C (HTR)(transition towards SMSI state).

8.5.2 Metals supported on other inorganic supports

It has been shown previously that supports like SiO_2, Al_2O_3, carbon, MgO and various zeolites do not lead to the SMSI state. Burch and Flambard (ref. 26, 27) confirmed that Ni/Al_2O_3 and Ni/SiO_2-Al_2O_3 catalysts exhibit the same behaviour in the hydrogenation of benzene after LTR or HTR. However, Del Angel et al. (ref. 73) reported that Rh supported on SiO_2, Al_2O_3 or carbon differs in the activity and deactivation by thiophene during hydrogenation of benzene. In particular, for similar good dispersions of Rh the silica catalyst was deactivated much faster than the alumina-supported catalyst. However for low dispersions of Rh all three supports behave similarly. It follows therefore that a phenomenon like poisoning may depend on the rhodium particle size and simultaneously on the nature of the support. At low dispersions the nature of the support is without influence, whereas at high dispersions some supports must undergo interaction with the metal which thus affects its poisoning.

Rh supported on various zeolites like NaY, NH_4Y, Na Ω, NH_4 Ω, NH_4 mordenite, H-erionite and H-offretite was recently examined by Del Angel et al. (ref. 74) who compared the activity of those catalysts in the hydrogenation of benzene with that of a conventional Rh/Al_2O_3 catalyst. The turnovers were similar for all catalysts containing large pore zeolites and for Rh/Al_2O_3. Only small pore zeolites (mordenite, Ω and offretite) yielded low turnovers, presumably because of the inaccessibility of benzene to the active (metal) sites. This explanation was based on the fact that, on omega-supported Rh, 1-hexene was hydrogenated, in contrast to benzene. These results are detailed in Table 8.8. and demonstrate the requirement of considering the pore structure of the support.

Competitive hydrogenation of benzene (B) and toluene (T) was studied by Tri et al. (ref. 75) on unsupported platinum sponge and over Pt supported on SiO_2 and Y-zeolite. These authors showed that the dependence of the ratio of the adsorption coefficients, $K_{T/B} = b_T/b_B$, on the nature of the support is in accord with the following sequence : Pt sponge < Pt/Na-Y $\approx$ Pt/SiO_2 < Pt-Y. This sequence is similar to that obtained by considering the modification of the electrophilic character of the metal which decreases upon adsorption of NH_3 (electron-donor molecule) but increases upon adsorption of H_2S (electron-acceptor molecule). For a given catalyst the $K_{T/B}$ value decreased upon adsorption of NH_3 and increased upon adsorption of H_2S. For the most acidic support, Pt/Y catalyst, $K_{T/B}$ was also decreased by the neutralization of the acidity with NaOH.

TABLE 8.8

Catalytic activities of rhodium-supported catalysts (after ref. 74)

Support	% Rh	% D	Turnover for		
			Benzene at 80°C	1-hexene at 25°C	1-pentene at 25°C
Al_2O_3	1	53	234		
Al_2O_3	2.35	68	380		
Al_2O_3	0.44	87	726		
Al_2O_3	1.85	22	540		
NaY	1	100	200-231		
NH_4Y	1	70	312		
$NH_4 \Omega$	1.36	80	0	14	33
$NH_4 \Omega$	1.36	60	0		
Na Ω	1.5	74	2.2		
NH_4-mordenite	1.0	78	3	18	41.4
NH_4-mordenite	4.0	49	60-77		
NH_4-mordenite	4.0	50	40	389	825
NH_4-mordenite	2.0	21	49		
H-erionite	1.0	19	1760		
H-offretite	1.0	47	50.5		

It has been shown previously that the hydrogenation of CO and of benzene do not show a parallel trend on catalysts in the SMSI state. This behaviour is also observed for catalysts which are not in the SMSI state, as shown by Martin and Dalmon (ref. 76) for Ni supported on SiO_2 reduced at temperatures in the range 627° to 847°C. Despite this treatment, the SMSI state is not observed for SiO_2 as the carrier. The turnover in the hydrogenation of benzene decreased when the reduction temperature increased. Such a behaviour could be attributed to a special requirement concerning the number of adjacent active sites and their geometry which may vary when the sintering increases. Many examples of such a trend are known. However, the partial restoration of the activity of the calcined Ni/SiO_2 catalyst after treatment with oxygen at 527°C bears some resemblance to the SMSI state of metals supported on TiO_2. At the same reduction temperatures, the turnover in the hydrogenation of CO was not modified. This emphasizes the fact that the nature of the reaction which allows some type of classification of catalysts is of prime importance in this classification.

Recently Goldwasser et al. (ref. 77) claimed that Pt supported on MgO catalysts can be in the SMSI state after reduction at high temperature (between 427°C and 527°C), but the activity in the hydrogenation of benzene decreased only slightly after these treatments. This is in contrast with the behaviour of Pt/TiO$_2$ catalysts which are shown to be in the SMSI state and for which the activity in the hydrogenation of benzene is almost suppressed, as shown above. The explanation of this difference may probably be found in the direction of the electron transfer for Pt/MgO on the one hand and for Pt/TiO$_2$ on the other hand. The previous authors proposed an electron transfer from Pt to MgO as monitored by the shift in the IR absorption band of adsorbed CO. For Pt/TiO$_2$ catalysts, it is the Pt which is enriched in electrons after HTR (ref. 78, 79).

8.5.3 Metals supported on organic carriers

Organic polymers like various polamides (Nylon, Nomex, Kevlar) have recently attracted attention as supports for Group 8 metals. In contrast to the same metals deposited on conventional supports like SiO$_2$ or Al$_2$O$_3$, a definite selectivity for <u>partial</u> hydrogenation is observed. Teichner et al. (ref. 80) reported a selectivity of the order of 80 % for hydrogenation of benzene to cyclohexene on Pt/nylon catalysts. With styrene the selectivity for ethylbenzene was 99 %. Pt supported on inorganic oxides gives respectively cyclohexane and ethylcyclohexane. The nature of the polyamide used as support is of prime importance. Whereas various aliphatic nylons with Pt give active and selective catalysts, aromatic nylon (Nomex) leads to an inactive catalyst. The activity and selectivity sequence for Group 8 metals on Nylon-66 was as follows : Pd < Pt < Rh. PTFE which does not contain amide groups was entirely non-selective. The explanation for the selectivity of the various nylons was based on the interactions between the amide functions, which are electron donors, and the Group 8 metal. Transition-metal carbides like WC, TiC, TaC and NbC are metallic type conductors. However, platinum catalysts supported by these carbides did not exhibit any selectivity for the partial hydrogenation of benzene (ref. 80). It is probable that only very partial filling of the d-orbitals of Pt leads to selective hydrogenation catalysts.

8.6 HYDROGENATION OF MONO- AND DIOLEFINS

Meriaudeau et al. (ref. 69) reported that the hydrogenation of ethylene was also sensitive to the metal-support interaction for Pt, Ir and Rh supported on TiO$_2$ subjected to LTR or HTR. The activity is decreased in the SMSI state. Briggs et al. (ref. 81) observed that Pt supported on SiO$_2$, Al$_2$O$_3$ and SiO$_2$-Al$_2$O$_3$ or pure Pt yielded the same product distribution in the deuteration of ethylene, which shows that the SMSI state is not realized for these supports in agreement with previous observations. The SMSI state for Pt/TiO$_2$ as catalyst leads to a different product distribution, which was attributed to a decrease in the bond strength of ethylene on Pt. The behaviour of Pt/MgO was very similar to that of Pt/TiO$_2$ despite the former's difficulty in achieving the SMSI state. In contrast to TiO_2, a partial reduction of MgO is difficult, even under HTR conditions. Also, the explanation provided by the previous authors (ref. 81) invoked a geometrical effect for Pt/MgO as catalyst, with the formation of adsorption sites possessing high coordinating ability.

Taghavi et al. (refs. 82-84) used Cu supported on δ-Al_2O_3 and on MgO and SiO_2 aerogels (ref. 85) for the hydrogenation of ethylene. With Cu/MgO the turnover was at least 20 times higher than on the other supports, demonstrating some type of interaction between Cu and MgO. This enhanced activity of Cu/MgO may be detrimental to selective hydrogenation. Indeed, for the competitive hydrogenation of a mixture of acetylene and ethylene the best support for minimizing the formation of ethane, during hydrogenation of acetylene to ethylene, was silica (0.02 % of ethane formed). For Cu/MgO, which is more active in the hydrogenation of ethylene as pointed out above, 7 % ethane was formed during hydrogenation of acetylene.

The selective hydrogenation of cyclopentadiene to cyclopentene (100 %) on the same catalysts was also studied by these authors (refs. 83, 86). Figure 8.4 shows the catalytic activity per gram of Cu supported on various carriers as a function of the surface area per gram of Cu. The activity does not increase in proportion to the surface developed by the metal, as is normally observed for structure-insensitive reactions. Instead, it remains constant for each type of support (each family of catalysts). The position in this diagram of unsupported Cu shows that some interaction between the metal and each type of support must occur, leading to higher activity. It was proposed that the active sites are produced by interaction with the support of peripheral surface atoms in each copper crystal in contact with the support. The interaction is somewhat similar to that described later by Burch and Flambard (ref. 20) and called IFMSI. However, in this case it is the support which determines the number of active sites and not the total number of surface atoms in the copper crystals. For the family of Cu/MgO catalysts a conventional structure insensitivity was found as the activity increased linearly with the surface area of Cu.

Stadler et al. (ref. 87) reported that the selectivity for 1-hexene, in the hydrogenation of unconjugated alkadienes like 1,5-hexadiene, in the liquid phase is much higher for Pt/TiO_2 than for Pd/Al_2O_3 as catalyst. This behaviour was explained by an interaction of Ti^{3+} with the diene.

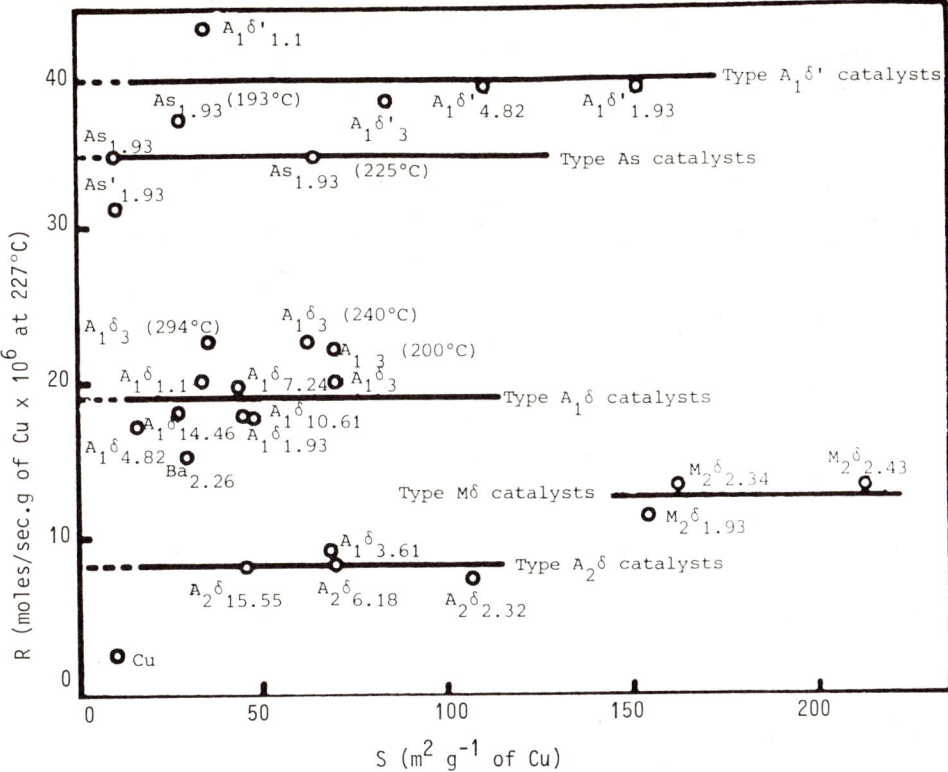

Fig. 8.4. Catalytic activities, R, versus metallic surface areas.

8.7 MISCELLANEOUS HYDROGENATIONS

8.7.1 Organic reactants

Titania is the support for Group 8 metals which most easily exhibits the SMSI state after HTR. Copper, which is not a Group 8 metal, when supported on TiO_2 behaves in a very similar way to Pt when it is subjected to LTR or HTR before the hydrogenation of 2-methylbutanal (ref. 88). HTR results in a much decreased activity compared to that resulting from LTR. In contrast, the activity of Cu/SiO_2 is not modified by these two treatments. Reduced Cu is a very poor adsorbent for H_2 or CO ; also the criterion of the SMSI state, i.e., a reduced chemisorption of these gases, cannot be applied and other tests should be used.

The hydrogenation of formaldehyde to methanol on Group 8 metals supported on SiO_2, ZnO and Cs_2O was described by Aika et al. (ref. 89). Ru, Rh and Pt were more selective for CH_3OH when they were supported on ZnO rather than SiO_2 or used without a support.

Srinivasan et al. (ref. 90) carried out the hydrogenation of an unsaturated fatty oil in the presence of Ni supported on Kieselguhr, silica, γ-alumina, silica-alumina and mordenite. The reaction was found to be structure insensitive in the classification of Boudart (ref. 2). The interaction between the metal and support was evidenced by the dispersion of nickel. Indeed this dispersion was not directly related to the BET area of the support. For instance, low surface areas of metal were measured for high-surface-area supports like γ-Al_2O_3, SiO_2-Al_2O_3 and mordenite. The opposite was observed for low-surface-area supports like kieselguhr. The interaction between the metal or its precursor and the support leading to the control of metal particle size had been observed previously (ref. 91).

The dehydrogenation of formic acid on Rh supported on TiO_2, Al_2O_3, SiO_2 and MgO was studied by Solymosi and Erdohelyl (ref. 92), who found that the highest turnover for the production of CO_2 is observed on Rh/TiO_2.

The effect of the support was also studied by Nitta et al (ref. 93) in the enantioselective hydrogenation of methyl acetoacetate (MAA) to methyl hydroxybutyrate in the presence of Ni supported on SiO_2 samples of various mean pore diameters. The optical yield increased with increasing mean pore diameter. For Ni/Al_2O_3 as catalyst a low optical yield was observed and attributed to the interaction with the support. The same reaction was also studied by Hoek and Sachtler (ref. 94) and the results are summarized in Table 8.9. Unsupported nickel powder and Ni/SiO_2 led to a similar enantioselectivity, whereas Ni/Al_2O_3 was inactive. It is difficult at the present time to identify the nature of the interaction (if any) between the metal and the support which suppresses the enantioselectivity.

TABLE 8.9
Influence of the carrier on the enantioselectivity of nickel catalysts, modified with tartaric acid at 293 K, in the hydrogenation of MAA at 343 K (after ref. 94).

Catalyst	ES (%) [a]
Raney Ni	+ 13 to + 14
Ni/Al_2O_3	0
Ni/SiO_2	+ 8 to + 10
Ni powder	+ 9 to + 15

[a] numbers give the range of data from different experiments.

The solid-state hydrogenation of thymol and 4-tert-butylphenol in the presence of Pd, Pt, Rh and Ru supported on carbon or on alumina was carried out by Lamartine and Perrin (ref. 95). Data presented in Tables 8.10 for thymol and 8.11 for 4-tert-butylphenol show a strong effect of the support on the stereoselectivity.

TABLE 8.10

Hydrogenation of solid thymol at 20°C. Effect of support (after ref. 95).

Catalyst (metal %)	Conversion ratio (%)	Isomer ratio (%) [a]					
		A	B	C	D	E	F
Pd/C(5)	50	14	86	0	0	0	0
Pd/Al$_2$O$_3$(5)	26	17	83	0	0	0	0
Rh/C(5)	77	10	38	9	42	0	1
Rh/Al$_2$O$_3$(5)	81	2	17	9	54	11	7
Pt/C(5)	99	1	8	12	42	29	8
Pt/Al$_2$O$_3$(5)	87	1	5	11	66	12	5
Ru/C(5)	20	2	16	3	61	0	18
Ru/Al$_2$O$_3$(5)	0	0	0	0	0	0	0
Adams PtO$_2$	21	25	75	0	0	0	0
Raney Ni	20	0	1	6	75	9	9

[a] A : menthone ; B : isomenthone ; C : neomenthol ; D : neoisomenthol ; E : menthol ; F : isomenthol.

TABLE 8.11

Hydrogenation of solid 4-tert-butylphenol at 20°C. Effect of support (after ref. 95).

Catalyst	Conversion ratio (%)	Products (%) [a]		
		1	2	
			cis	trans
Rh/C	57	5	45	7
Rh/Al$_2$O$_3$	67	5	28	34

[a] 1 : 4-tert-butylcyclohexanone ; 2 : 4-tert-butylcyclohexanol.

These results are very unusual considering the fact the solid reactant is mechanically admixed with the solid catalyst. As diffusion of the reactant activated on the catalyst is unlikely in the solid state at 20°C, it is probable that hydrogen is activated on the catalyst and spills over onto the reactant. However, it is difficult to explain how this spilled-over hydrogen remembers the catalyst (the metal as well as the support) to induce different stereoselective hydrogenations.

8.7.2. Inorganic reactants

One of the most spectacular effects of the support was demonstrated by Galvagno and Parravano (ref. 96) in the reduction of NO by hydrogen in the presence of gold. Pure Au is inactive in the catalytic hydrogenation, but it acquires activity when supported on SiO_2, Al_2O_3 and MgO. The nature of the interaction between the two inert components after reduction at 420°C is still open to question. Figure 8.5 shows the dependence of the selectivity on the nature of the carrier. A geometric effect related to the dispersion of gold might have been expected. The authors however suggested a back donation of electrons from Au to adsorbed NO species, which favours the formation of N_2 over that of NH_3. Nevertheless this explanation may be correlated with the dispersion state of Au, higher dispersions leading to the loss of metallic character.

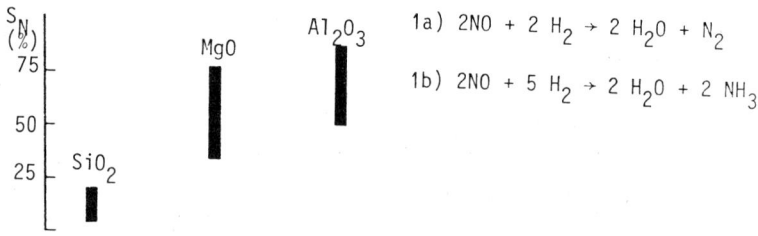

Fig. 8.5. Selectivity for N_2, S_N, for reactions (1a) and (1b) over supported gold catalysts. 350°C ; $3 < p_{H_2}/p_{NO} < 5$; total pressure, 1 atm.

A metal-support interaction was envisaged by Nielsen (ref. 92) in the synthesis of NH_3 on Fe supported on Al_2O_3, SiO_2 and carbon. This interaction was assumed to stabilize iron in the zerovalent state on the surface of the support. However it should not be too strong otherwise the iron would be ionized. The following sequence of interaction was suggested : carbon < SiO_2 < Al_2O_3. For the same reaction, Topsoe et al. (ref. 98) compared the turnover for N_2 based on the chemisorption of H_2, CO and N_2 on unsupported iron catalysts (labelled KMI and KMH) and on Fe/MgO. Turnovers based on chemisorption of N_2 were the same on the three catalysts having various iron dispersions, as shown in Fig. 8.6., but differed from those based on chemisorption of CO or H_2. These results emphasize the importance of collecting data on correctly choosen bases in order to avoid misinterpretations.

Shiflett and Dumesic (ref. 99) also tested the synthesis of NH_3 on Re and bimetallic RePt supported on SiO_2 and Al_2O_3. They found a small degree of interaction between Pt and Re supported on SiO_2 but no interaction for the alumina-supported bimetallic catalyst.

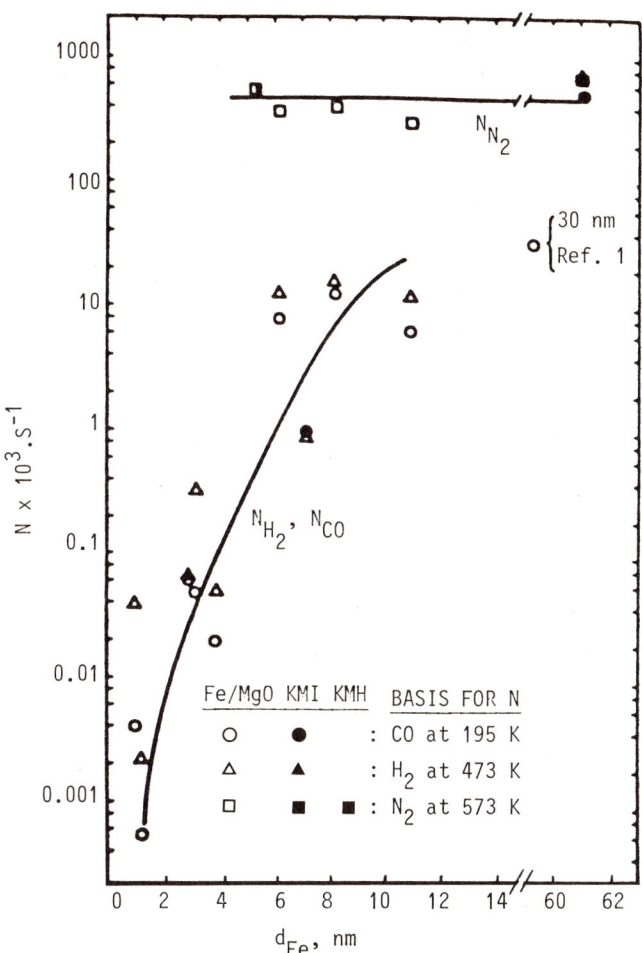

Fig. 8.6. Turnover for ammonia synthesis.

Finally the metal-support interaction in coal conversion was discussed by Cusumano et al. (ref. 100) with respect to the active phase stabilization, the ease of recovery of the catalytic activity and the resistance towards poisons like sulphur compounds.

8.8 INFLUENCE OF THE METAL-SUPPORT INTERACTION ON THE DEACTIVATION OF SUPPORTED CATALYSTS

The deactivation of supported catalysts may be due to various physical factors like sintering, changes in their structure or to chemical factors like poisons in the feed or coking. Two typical examples are given in this section, concerning the second type of factors, which show that the interaction between the metal and the support may increase the resistance of the catalyst to deactivation.

Bartholomew and Katzer (ref. 101) studied methanation on Ni supported on Al_2O_3 and ZrO_2. The resistance to sulphur with time of both catalysts is shown in Fig. 8.7. Ni/ZrO_2 performed much better in the presence H_2S than did Ni/Al_2O_3. The nature of the interaction which improves the resistance of Ni/ZrO_2 to sulphur is not known.

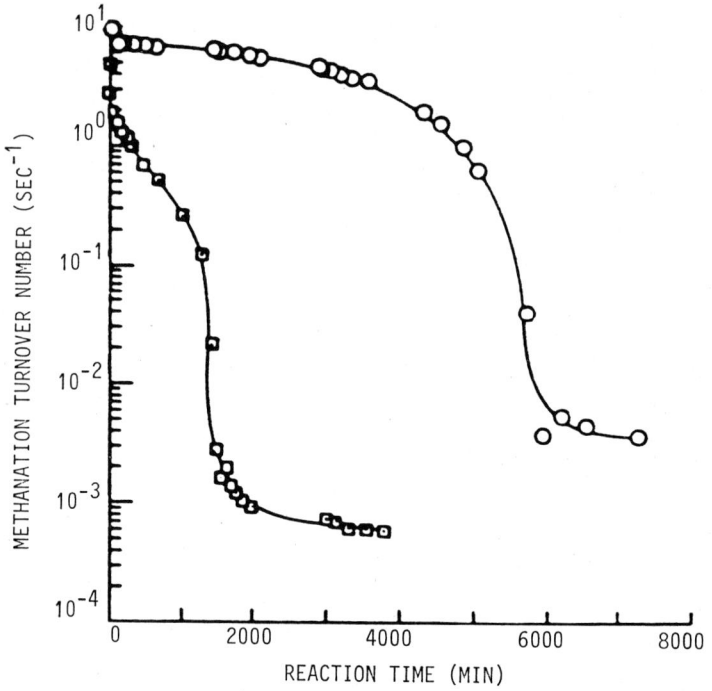

Fig. 8.7. Sulphur deactivation of □ Ni/Al_2O_3 (pellets) and ○ Ni/ZrO_2 (pellets) catalysts in the presence of low concentrations of H_2S; t= 388°C, 4% CO, 96 % H_2. P_{H_2S} = 55 ppb at steady state.

The second example concerns the coking of Pt/Al_2O_3 catalysts described by Parera et al. (ref. 102). Alumina of various chlorine contents (0.34 - 1.09 % Cl) was used as the support and the reaction studied was the reforming of n-heptane. The extent of coke formation on the catalyst, the decrease in activity (measured as the octane number) after the deactivation period and the slope of deactivation were minimal for the catalyst containing 0.89 % Cl. The D.T.A. curves also showed that for this catalyst the coke deposit is the smallest and the least polymerized. This behaviour was attributed to a maximum in hydrogen spillover from Pt to Al_2O_3, which keeps the alumina surface relatively free from unsaturated coke precursors. An optimum concentration of chlorine results from the simultaneous need for hydroxyls (chlorine replaces the hydroxyls) which are also important for the migration (spillover) of the solvated proton.

8.9 CONCLUSIONS

Differences between the hydrogenating properties of metals on various supports can be caused by a number of factors like the particle size effect. However if this effect is neglected for reaction which are not structure sensitive the SMSI effect seems to be the one which accounts for these differences.

General correlations between the enhancement (or the reduction) of catalytic activity and/or selectivity and the nature of the phenomena occurring in the SMSI state cannot at present be formulated. The nature of the hydrogenation reaction and of the metal-support combination introduce unpredictable consequences for the catalytic activity and selectivity in the SMSI state. Electronic and geometric effects and their combination are involved in this state and need to be better understood before a clear picture of the effect of the support on the hydrogenating properties of the metal can be given.

REFERENCES

1   W.C. Conner, G.M. Pajonk and S.J. Teichner, Adv. Catal., (1986) in press.
2   M. Boudart, Adv. Catal., 20 (1969) 156.
3   B. Imelik, C. Naccache, G. Coudurier, H. Praliaud, P. Meriaudeau, P. Gallezot, G.A. Martin and J.C. Vedrine (Editors), Metal-Support and Metal-Additive Effects in Catalysis, Elsevier, Amsterdam, 1982.
4   P.N. Rylander, Catalytic hydrogenation over platinum metals, Academic Press, New-York, 1967.
5   P.N. Rylander, N. Kilroy and V. Coven, Engelhard Ind. Techn. Bull., 6 (1965) 11.
6   J.H. Koch, US Patent 3,055,840, Sept. 1966.
7   P.N. Rylander, N. Himelstein and N. Kilroy, Engelhard Ind. Techn. Bull., 4 (1963) 49.
8   P.N. Rylander and D.R. Steele, Engelhard Ind. Techn. Bull., 5 (1965) 113.
9   S. Akabori, S. Sakurai, S.Y. Izumi and Y. Fugii, Nature, 178 (1956) 323.
10 S. Akabori, S. Sakurai, S.Y. Izumi and F. Fugii, Biokhimiya, 22 (1957) 154.
11 J.C. Sauer, R.D. Cramer, V.A. Engelhard, T.A. Ford, H.E. Holmquist and B.W. Howk, J. Am. Chem. Soc., 81 (1959) 3677.
12 I.C. Brownlie, J.R. Fryer and G. Webb, J. Catal., 14 (1969) 263.
13 R. Maurel, G. Leclercq and J. Barbier, J. Catal., 37 (1975) 324.
14 W.F. Taylor and H.K. Steiffin, Trans. Faraday Soc., 63 (1967) 2309.
15 R.G. Romero and F. Figueras, C.R. Acad. Sci., Ser.C, 275 (1972) 769.

16 M.A. Vannice, J. Catal., 40 (1975) 129.
17 M.A. Vannice, J. Catal., 44 (1976) 152.
18 G.M. Schwab, Adv. Catal., 27 (1978) 1.
19 S.J. Tauster, S.C. Fung and R.L. Garten, J. Am. Chem. Soc., 100 (1978) 170.
20 R. Burch and A.R. Flambard, J. Catal., 78 (1982) 389.
21 M.A. Vannice and C. Sudhakar, J. Phys. Chem., 88 (1984) 2429.
22 M.A. Vannice and R.L. Garten, J. Catal., 86 (1979) 236.
23 C.H. Bartholomew, R.B. Pannell and J.L. Butler, J. Catal., 65 (1980) 335.
24 C.H. Bartholomew, R.B. Pannell, J.L. Butler and D.G. Mustard, Ind. Eng. Chem. Prod. Res. Dev., 20 (1981) 296.
25 J.B.F. Anderson, J.D. Bracey, R. Burch and A.R. Flambard, Proc. 8th Int. Congr. Catalysis, Vol. V Dechema, Berlin, 1984, p. 111.
26 R. Burch and A.R. Flambard, J. Catal., 85 (1984) 16.
27 R. Burch and A.R. Flambard in B. Imelik, C. Naccache, G. Coudurier, H. Praliaud, P. Meriaudeau, P. Gallezot, G.A. Martin and J.C. Vedrine (Editors), Metal-Support and Metal-Additive Effects in Catalysis, Elsevier, Amsterdam, 1982, p. 193.
28 R. Burch and A.R. Flambard, React. Kinet. Catal. Lett., 17 (1981) 23.
29 J.A. Dalmon and G.A. Martin in T. Seiyama and K. Tanabe (Editors), New Horizons in Catalysis, Proc. 7th Int. Congr. Catal., Tokyo, Part. A, Elsevier, Amsterdam 1981, p. 402.
30 R.Z.C. Van Meerten in T. Seiyama and K. Tanabe (Editors), New Horizons in Catalysis, Proc. 7th. Congr. Catal., Tokyo, Part. B, Elsevier, Amsterdam 1981, p. 1440.
31 R.A. Dalla Betta, A.G. Piken and M. Shelef, J. Catal., 40 (1975) 173.
32 E.I. Ko, J.M. Hupp and N.J. Wagner, J. Catal., 86 (1984) 315.
33 E.I. Ko, J.M. Hupp and N.J. Wagner, Chem. Commun, (1983) 94.
34 C.K. Vance and C.H. Bartholomew, Appl. Catal., 7 (1983) 169.
35 P. Turlier, J.A. Dalmon and G.A. Martin in B. Imelik, C. Naccache, G. Coudurier, H. Praliaud, P. Meriaudeau, P. Gallezot, G.A. Martin and J.C. Vedrine (Editors), Metal-Support and Metal-Additive Effects in Catalysis, Elsevier, Amsterdam, 1982, p. 203.
36 S.Z. Ozdogan, P.D. Gochis and J.L. Falconer, J. Catal., 83 (1983) 257.
37 Y.W. Chung, G. Xiong and C.L. Kao, J. Catal., 85 (1984) 237.
38 M.A. Vannice, C.C. Twu and S.H. Moon, J. Catal., 79 (1983) 70.
39 M.A. Vannice and C.C. Twu, J. Catal., 82 (1983) 213.
40 P. Meriaudeau, M. Dufaux and C. Naccache, Proc. 8th. Int. Congr. Catalysis, Vol. II, Dechema, Berlin, 1984, p. 185.
41 R. Szymanski, H. Charcosset and V. Perrichon, Proc. 8th. Int. Congr. Catalysis, Vol. II, Dechema, Berlin, 1984, p. 151.
42 M.A. Vannice, S.Y. Wang and S.H. Moon, J. Catal., 71 (1981) 152.

43 S.Y. Wang, S.H. Moon and M.A. Vannice, J. Catal., 71 (1981) 167.
44 J.D. Bracey and R. Burch, J. Catal., 86 (1984) 384.
45 E.K. Poels, E.H. Van Broekhoven, W.A.A. Van Barneveld and V. Ponec, React. Kinet. Catal. Lett., 18 (1981) 223.
46 N.C. Saha and E.E. Wolf, Appl. Catal., 13 (1984) 101.
47 F. Fajula, R.G. Anthony and J.H. Lunsford, J. Catal., 73 (1982) 237.
48 M. Ichikawa and K. Shikakura in B. Imelik, C. Naccache, G. Coudurier, H. Praliaud, P. Meriaudeau, P. Gallezot, G.A. Martin and J.C. Vedrine (Editors), Metal-Support and Metal-Additive Effects in Catalysis, Elsevier, Amsterdam, 1981, p. 925.
49 K. Kunimori, H. Abe, E. Yamaguchi, S. Matsui and T. Uchijima, Proc. 8th Int. Congr. Catal., Vol. V, Dechema, Berlin, 1984, p. 251.
50 S.R. Morris, R.B. Moyes, P.B. Wells and R. Whyman in B. Imelik, C. Naccache, G. Coudurier, H. Praliaud, P. Meriaudeau, P. Gallezot, G.A. Martin and J.C. Vedrine (Editors), Metal-Support and Metal-Additive Effects in Catalysis, Elsevier, Amsterdam, 1982, p. 247.
51 I.R. Leith, Chem. Soc., J. Chem. Commun., 1984, p. 93.
52 I.R. Leith, J. Catal., 91 (1985) 283.
53 C.H. Yang and J.G. Goodwin, Proc. 8th Int. Congr. Catalysis, Vol. V, Dechema, Berlin 1984, p. 263.
54 Y.W. Chen, H.T. Wang and J. Goodwin, J. Catal., 85 (1984) 499.
55 P. Meriaudeau, H. Ellestad and C. Naccache, J. Mol. Catal., 17 (1982) 219.
56 K. Kunimori, H. Abe and T. Uchijima, Chem. Lett., (1983), 1619.
57 R. Bardet, J. Thivolle-Cazat and Y. Trambouze in B. Imelik, C. Naccache, G. Coudurier, H. Praliaud, P. Meriaudeau, P. Gallezot, G.A. Martin and J.C. Vedrine (Editors), Metal-Support and Metal-Additive Effects in Catalysis, Elsevier, Amsterdam, 1982, p. 241.
58 F. Solymosi, A. Erdöhelyi and T. Bansagi, J. Catal., 68 (1981) 371.
59 R.C. Reuel and C.H. Bartholomew, J. Catal., 85 (1984) 78.
60 M. Blanchard and D. Vanhove in B. Imelik, C. Naccache, G. Coudurier, J. Praliaud, P. Meriaudeau, P. Gallezot, G.A. Martin and J.C. Vedrine (Editors), Metal-Support and Metal-Additive Effects in Catalysis, Elsevier Amsterdam, 1982, p. 219.
61 D. Vanhove, P. Makambo and M. Blanchard, J. Chem. Soc., Chem. Comm., (1979) 605.
62 M. Peuckert and G. Linden, Proc. 8th Int. Congr. Catalysis, Vol. II, Dechema, Berlin, 1984, p. 135.
63 M.A. Vannice, P.L. Walker, H.J. Jung, C. Moreno-Castilla and O.P. Mahajan in T. Seiyama and K. Tanabe (Editors), New Horizons in Catalysis, Proc. 7th Int. Congr. Catal., Tokyo, Part. A, Elsevier, Amsterdam 1980, p. 460.

64 N.O. Egiebor, W.C. Cooper, J. Cameron and B. Farrand in S. Kaliaguine and A. Mahay (Editors), Catalysis on the Energy Scene, Elsevier, Amsterdam 1984, p. 481.
65 L.M. Tau and C.O. Bennett, J. Catal., 89 (1984) 285.
66 I.Boszormenyi, S. Dobos, L. Guczi, L. Marko, K. Lazar, W.M. Reiff, Z. Schay, L. Takacs and A. Vizi-Orosz, Proc. 8th Int. Congr. Catalysis, Vol. V, Dechema, Berlin 1984, p. 183.
67 B.E. Concha, G.L. Bartholomew and C.H. Bartholomew, J. Catal., 89 (1984) 536.
68 M.A. Vannice, J. Catal., 74 (1982) 199.
69 P. Meriaudeau, O.H. Ellestad, M. Dufaux and C. Naccache, J. Catal., 75 (1982) 243.
70 P. Meriaudeau, J.F. Dutel, M. Dufaux and C. Naccache in B. Imelik, C. Naccache, G. Coudurier, H. Praliaud, P. Meriaudeau, P. Gallezot, G.A. Martin and J.C. Vedrine (Editors), Metal-Support and Metal-Additive Effects in Catalysis, Elsevier, Amsterdam, 1982, p. 95.
71 M.A. Vannice and P. Chou, Proc. 8th. Int. Congr. Catalysis, Vol. V, Dechema, Berlin 1984, p. 99.
72 D.E. Resasco and G.L. Haller, J. Catal., 82 (1983) 279.
73 G. Del Angel, B. Coq and F. Figueras in B. Imelik, C. Naccache, G. Coudurier, H. Praliaud, P. Meriaudeau, P. Gallezot, G.A. Martin and J.C. Vedrine (Editors), Metal-Support and Metal-Additive Effects in Catalysis, Elsevier, Amsterdam, 1982, p. 85.
74 G. Del Angel, B. Coq, R. Dutartre, F. Fajula, F. Figueras and C. Leclercq in G.M.Pajonk, S.J. Teichner and J.E. Germain (Editors), Spillover of Adsorbed Species, Elsevier, Amsterdam, 1983, p. 301.
75 T.M. Tri, J. Massardier, P. Gallezot and B. Imelik in B. Imelik, C. Naccache, G. Coudurier, H. Praliaud, P. Meriaudeau, P. Gallezot, G.A. Martin and J.C. Vedrine (Editors), Metal-Support and Metal-Additive Effects in Catalysis, Elsevier, Amsterdam, 1982, p. 141.
76 G.A. Martin and J.A. Dalmon, React. Kinet. Catal. Lett., 16 (1981) 325.
77 J. Goldwasser, C. Bolivar, C.R. Ruiz, B. Arenas, S. Wanke, H. Royo, R. Barrios and J. Giron, Proc. 8th Int. Congr. Catalysis, Vol. V, Dechema, Berlin 1984, p. 195.
78 J.M. Herrmann and P. Pichat, J. Catal., 78 (1982) 425.
79 J.A. Horsley, J. Am. Chem. Soc., 101 (1979) 2870.
80 S.J. Teichner, C. Hoang-Van and M. Astier in B. Imelik, C. Naccache, G. Coudurier, H. Praliaud, P. Meriaudeau, P. Gallezot, G.A. Martin and J.C. Vedrine (Editors), Metal-Support and Metal-Additive Effects in Catalysis, Elsevier, Amsterdam, 1982, p. 121.
81 D. Briggs, J. Dewing, A.G. Burden, R.B. Moyes and P.B. Wells, J. Catal., 65 (1980) 31.

82 M.B. Taghavi, G. Pajonk and S.J. Teichner, Bull. Soc. Chim. France (1978) 180.
83 M.B. Taghavi, G.M. Pajonk and S.J. Teichner, J. Colloïd Interface Sci., 71 (1979) 451.
84 M.B. Taghavi, G. Pajonk and S.J. Teichner, Bull. Soc. Chim. France (1978) 302.
85 M. Astier, A. Bertrand, D. Bianchi, A. Chenard, G.E.E. Gardes, G. Pajonk, M.B. Taghavi, S.J. Teichner and B.L. Villemin in B. Delmon, P.A. Jacobs and G. Poncelet (Editors), Preparation of Catalysts I, Elsevier, Amsterdam, 1976, p. 315.
86 M.B. Taghavi, G. Pajonk and S.J. Teichner, Bull. Soc. Chim. France (1978) 285.
87 K.H. Stadler, M. Schneider and K. Kochloefl, Proc. 8th Int. Congr. Catalysis, Vol. V, Dechema, Berlin, 1984, p. 229.
88 F.S. Delk and A. Vavere, J. Catal., 85 (1984) 380.
89 K. Aika, H. Sekiya and A. Ozaki, Chem. Lett. (1983) 301.
90 G. Srinivasan, R.S. Murthy, K.M. Vijaya Kumar and P.V. Kamat, J. Chem. Tech. Biotechnol., 30 (1980) 217.
91 J. Cosyns, P. Courty, E. Freund, J.P. Franck, Y. Jacquin, B. Juguin, C. Marcilly, G. Martino, J. Miquel, R. Montarnal, A. Sugier and H. Van Landeghem in J.F. Le Page et al. (Editors), Catalyse de Contact, Technip, Paris 1978, p. 250.
92 F. Solymosi and A. Erdohelyi, J. Catal., 91 (1985) 327.
93 T. Nitta, O. Yamanishi, T. Sekine, T. Imanaka and S. Teranishi, J. Catal., 79 (1983) 475.
94 A. Hoek and W.M.H. Sachtler, J. Catal., 58 (1979) 276.
95 R. Lamartine and R. Perrin in G.M. Pajonk, S.J. Teichner and J.E. Germain (Editors), Spillover of Adsorbed Species, Elsevier, Amsterdam, 1983, p. 251.
96 S. Galvagno and G. Parravano, J. Catal., 55 (1978) 178.
97 A. Nielsen, Catal. Rev. Sci. Eng., 23 (1981) 17.
98 H. Topsoe, N. Topsoe, H. Bohlboro and J.A. Dumesic in T. Seiyama and K. Tanabe (Editors), New Horizons in Catalysis, Proc. 7th Int. Congr. Catal., Tokyo, Part. A, Elsevier, Amsterdam 1980, p. 247.
99 W.K. Shiflett and J.A. Dumesic, J. Catal., 77 (1982) 57.
100 J.A. Cusumano, R.A. Dalla Betta and R.B. Levy in "Catalysis in Coal Conversion". Academic Press, New-York, 1978.
101 C.H. Bartholomew and J.R. Katzer in B. Delmon and G.F. Froment, (Editors), Catalyst Deactivation, Elsevier, Amsterdam, 1980, p. 375.
102 J.M. Parera, N.S. Figoli, E.L. Jablonski, M.R. Sad and J.N. Beltramini in B. Delmon and G.F. Froment (Editors), Catalyst Deactivation, Elsevier, Amsterdam, 1980, p. 571.

Chapter 9

ROLE OF BIMETALLIC CATALYSTS IN CATALYTIC HYDROGENATION AND HYDROGENOLYSIS

L. GUCZI AND Z. SCHAY
Institute of Isotopes of the Hungarian Academy of Sciences,
H-1525 Budapest, P.O.Box 77, Hungary

9.1 INTRODUCTION

In the 1950s there was a strong interest in heterogeneous catalysis by alloys, due to the alleged validity of the Rigid Band Model (RBM). According to this model, upon alloying, the holes in the d-band of transition metals could be filled and as a consequence the chemisorption and adsorption would be considerably influenced. This interest ceased as results contradicting the RBM model were accumulated.

Since the late 1960s there has been a renaissance of interest in alloy catalysis due to the discovery of the superior properties of the Pt-Re/Al_2O_3 catalyst. At the same time, considerable progress was achieved both in the characterization of alloy surfaces and in the quantum theory of alloys. The catalysis by alloys has been reviewed by Moss and Whally (ref. 1), Bond and Allison (ref. 2), Clarke (ref. 3), Sachtler (ref. 4), Sachtler and Van Santen (ref. 5), Khulbe and Mann (ref. 6) and Ponec (refs. 7,8).

The present status of our knowledge about catalysis by alloys can briefly be summarized as follows:

(a) The composition of the surface of equilibrated alloys can differ substantially from that of the bulk, and chemisorption or the catalytic reaction itself can induce additional changes in the surface composition.

(b) The selectivity patterns are mainly determined by the ensemble size effect, but a ligand effect may also occur simultaneously. It is now generally accepted that the individual surface atoms retain their character in alloys, and their catalytic properties are mainly determined by their nearest neighbours.

(c) Alloys often exhibit higher catalytic activity than either of their constituents because of a lower degree of poisoning. This can result in higher catalyst stability and in a change in selectivity for a given reaction. Typically, hydrogenolysis is suppressed by alloying, while hydrogenation and isomerization are not significantly affected.

If metal atoms are supported on an "inert" carrier some additional factors become operative. The particle size is normally diminished, thus the ensemble size is generally decreased (ref. 9). As a consequence, the small particles easily interact with the coordinatively unsaturated sites (CUS) resulting in an incomplete reduction of the metals. Furthermore, the texture of the metal particles also changes since the number of kinks and step sites increases as compared to the terraces. Finally, in a highly dispersed system, the d-band structure also changes due to the small number of metal atoms in the particles.

In a bimetallic system, in addition to the dispersion effect, there is a further narrowing of the d-band structure as shown by Van Santen (ref. 10). Another important factor is that the ensemble size can be further diminished at a given particle geometry and, finally, dispersion of a bimetallic catalyst can be achieved nearly on the atomic scale if one component is "grafted" onto the support (ref. 11).

In the context of supported alloys, the following factors should be considered:
   (i) ensemble size variation
   (ii) particle size effect, dispersion
   (iii) matrix effect
   (iv) change in hydrogen coverage and in the structure of adsorbed hydrogen
   (v) separation of bimetallics into components
   (vi) metal-support interaction (MSI)
   (vii) suppression of self-poisoning reactions.

In this chapter we will consider how the above factors control the behaviour of bimetallic catalysts in some hydrocarbon transformations, and it will be shown that the geometric effects, including the ensemble size effect, together with changes in the hydrogen supply are the most important factors in hydrogenation and hydrogenolysis. Some examples will be given of supported bimetallic catalysts involving Pd, Pt and Ru with other transition and Group 1B metals. The most widely studied alloy, Ni-Cu, will be reviewed only briefly. For full details of the alloys and their reactions the reader should consult the excellent review articles already mentioned.

9.2 ENSEMBLE SIZE EFFECT AND SURFACE SEGREGATION

As mentioned previously, the main effects in a supported bimetallic system result from the ensemble size variation. When an active metal is alloyed with Group 1B metals the size of the active metal ensemble decreases as atoms of the second metal are inserted into the lattice of the first one and therefore the effect of the ensemble size variation is entirely different in monometallic and bimetallic systems. It has been demonstrated that the rate of hydrogenolysis of cyclohexane to form n-hexane increases with kink density while the rate of cyclohexane dehydrogenation to form benzene is not affected by the kink density on a stepped platinum surface (ref. 12). As the number of kinks may increase with dispersion, a similar effect is also expected for highly dispersed platinum. Indeed, the turnover frequency for hydrogenolysis of n-butane increases with dispersion of platinum (ref. 13), whereas ethane-deuterium exchange and cyclohexane--deuterium exchange are not influenced by dispersion (ref. 14).

On the contrary, on unsupported Cu-Ni and Ru-Cu alloys the rate of hydrogenolysis of ethane markedly decreases as copper is added either to Ni (ref. 15) or to Ru, indicating that as the nickel and ruthenium ensemble size decreases in a large crystallite containing mainly terraces, the population of the multiple bonded species of chemisorbed ethane decreases. On the other hand, reactions such as cyclohexane dehydrogenation are not influenced since here the reaction is controlled by single nickel atoms and not by nickel ensembles with a given geometry. However, if ruthenium-copper is supported in the form of very small particles, the decrease in hydrogenolysis rate at the same Cu/Ru ratio is much less than for unsupported alloys. This is due to the much higher ratio of the surface to the bulk, which in turn is a result of the high dispersion.

Ni-Cu, Pt-Au and Ru-Cu formed the subjects of other studies in which the dispersion effect could be clearly demonstrated.

9.2.1 Ni-Cu alloys

(i) <u>Unsupported Ni-Cu alloys.</u> The Ni-Cu system is one of the most extensively studied alloys, and plays an important role in testing alloy catalysis. The electronic structure of Ni-Cu has been reviewed by Ponec (ref. 16). These alloys form only one phase above 500 K, therefore phase separation should not be a problem. The composition of the surface of the alloy can strongly deviate from that of the bulk. Up to about 10% Cu addition to the bulk, the surface is highly enriched in Cu. When the copper concentration in the bulk is 10-80% the surface composition of Ni is nearly constant at 5-15% Ni, and decreases further on increasing the copper concentration. No electron transfer was observed from Cu to Ni, in agreement with predictions of the coherent potential approximation.

Data for about 40 reactions have been summarized by Ponec (ref. 17) and can be split into two groups (see Fig. 9.1):

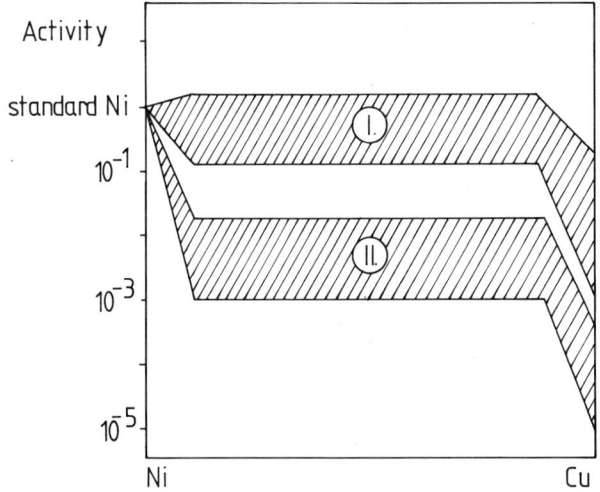

Fig. 9.1. Summary of the literature data on various reactions (≈40) studied on Ni-Cu alloys. Relative activity as a function of the bulk alloy composition. All data can be subdivided into two groups of reactions: sensitive(II) or insensitive(I) to alloying. The surface nickel concentration follows approximately the lower limit of the band for reactions of group I (From ref. 8).

Group I: Insensitive to alloying by Cu. Alloying leads to an increase or a slight decrease in activity the latter being less than the decrease in surface nickel concentration. HC/D_2 exchange reactions and hydrogenation, dehydrogenation reactions belong to this group. It is believed that the active sites for these reactions are isolated nickel atoms. The increase in activity is explained by the suppression of side reactions of the type in Group II which lead to poisoning of the surface by carbonaceous deposits.

Group II: Sensitive to alloying by Cu. The activity decreases by two or more orders of magnitude. Hydrogenolysis, isomerization and dehydrocyclization of alkanes and the CO + H_2 reaction belong to this group. These reactions require multiple sites.

At extremely low nickel concentrations (below 1%), isolated nickel atoms are present at the surface (ref. 18) and as a consequence no hydrogenation of ethylene is observed, but stepwise exchange to C_2H_3D and HD occurs readily. Cu must be treated as more than a diluent as copper atoms and the isolated nickel atoms form an active site for the formation of a half-hydrogenated species.

We may conclude that for the unsupported Ni-Cu alloys the geometric effects are the most important, and in some cases poisoning is also suppressed.

(ii) <u>Supported Ni-Cu catalysts.</u> It is well known that the activity of supported catalysts depends on their method of preparation. The most important factors are the type of support, nature of starting metal compounds, metal loading and activation procedures. Unfortunately, no systematic studies of supported Ni-Cu catalysts have been reported, therefore no definite conclusions can be drawn as regards the somewhat erratic activity and selectivity patterns reported in the literature. In general, no simple relationship between catalytic activity and composition (or any other property) of Ni-Cu catalysts has been found. A reason for this is the complicated structure of the catalyst, which strongly depends on the experimental conditions.

Sachtler (ref. 19) proposed the cherry model to describe the structure of these alloys at temperatures below the miscibility gap. In general, four ranges of concentration can be distinguished. At very high copper concentrations only one phase exists, the outer layer being enriched in copper. At medium concentrations there is not enough copper present to form a continuous skin and some patches of alloy will cover the nickel kernel. At over 95% Ni a homogeneous alloy is formed, the surface of which is enriched in copper.

Obviously, depending on the method of preparation, different types of crystallites can be present. During a hydrogenation reaction, the surface composition will change, as hydrogen will force some additional nickel into the surface layer. Another important factor is the self-poisoning, which again strongly depends on the experimental conditions. Van Barneveld and Ponec (ref. 20) demonstrated this for benzene hydrogenation on Ni, 5% and 10% Cu catalysts. At low temperatures (below 420 K) the specific activity of the alloys is lower than that of pure nickel, while above 500 K the opposite is true. At these temperatures, hydrogenolysis occurs with nickel, while copper seems to inhibit it and protects the catalyst at high temperatures.

9.2.2 Pt-Au alloys

The change of ensemble size on addition of a second metal and its effect on the catalytic reaction can be illustrated by use of the Pt-Au system.

This system has a wide miscibility gap and according to the cherry model (ref. 19) at equilibrium a kernel of platinum-rich phase is enveloped by a gold-rich phase. The constant surface composition has been confirmed by hydrogen chemisorption and a constant turnover in H-D exchange (ref. 21). In contrast to platinum where at least two types of chemisorbed hydrogen atoms can participate in the exchange, on the Pt-Au alloys the weakly bound hydrogen governs the exchange. Temperature-programmed desorption measurements have ruled out any electronic effect as essentially the same adsorption states (bond strengths) have been observed and only the population of the states changes upon alloying. The results can be explained by ensemble size effects (ref. 22). Addition of gold controls the ensemble size, as shown in the dehydrogenation of cyclohexane (ref. 23). With small ensembles, cyclohexane appears as an intermediate as a consequence of the stepwise dehydrogenation mechanism. However, as a side effect, the self-poisoning reactions are retarded on small platinum ensembles and as a consequence the rate of isomerization of n-hexane increases. On small ensembles of platinum the main product of the deuterium exchange in cyclopentane shifts from cyclopentane-d_{10} to cyclopentane-d_1 (ref. 24), as the dissociation of more than one C-H bond requires a larger ensemble size.

In the dehydrogenation of propane a diminution of the surface concentration of hydrogen has been observed on alloys (ref. 25), resulting in a change in the reaction order with respect to hydrogen.

By supporting the Pt-Au system (16 wt% metal), no drastic

change can be seen compared to alloy film or powder (ref. 26). Diluted platinum (1-4 at.%) catalyses the isomerization of n-hexane and n-pentane by a one-site mechanism, i.e., the activity of Pt is not influenced by the surrounding gold atoms. 10 at.% Pt favours dehydrocyclization due to the larger ensemble size, whereas pure platinum isomerizes n-hexane mainly via a bond-shift mechanism. Apparently no segregation occurs, unlike the highly dispersed silica-supported Pt-Au catalyst with low metal loading (0.7 wt.%) where separate gold particles are present along with Pt-Au bimetallics (ref. 27). The ensemble size-effect principle is also valid here because, in the hydrogen-transfer reaction between benzene and cyclohexane, gold atoms as diluent change the ensemble size of platinum. However, at higher metal loadings the reaction rate is increased with increasing addition of gold, contrary to the explanation given by the authors, this effect may be due to inhibition of the poisoning reaction.

Surface enrichment of gold after thermal treatment seems to be a general phenomenon. For a series of bimetallic Pt-Au systems supported on different carriers with 1-2 wt.% metal loading, the temperature programmed desorption (TPD) profile of adsorbed hydrogen was seen to be independent of the gold content (ref. 28) as a consequence of the equilibrium gold coverage.

9.2.3 Ru-Cu alloys

These two metals do not alloy with each other in the bulk (ref. 29), therefore it is of particular interest whether or not they form bimetallic mixed clusters. Sinfelt and co-workers (refs. 30-32) found that in silica-supported Ru-Cu clusters copper is present only on the ruthenium surface. Extended X-ray absorption fine structure (EXAFS) (ref. 33) has shown, that the copper atoms form an adsorbed layer on the ruthenium kernel.

Christmann and co-workers (refs. 34-36) studied the behaviour of a thin copper layer on a single crystal of ruthenium as a model for a Ru-Cu bimetallic cluster. Slight electron transfer from Ru to Cu was found and interpreted by the formation of a covalent bond between the two metals. The deposition of Cu, however, tremendously decreases the adsorption capacity for hydrogen, with very little influence on the heat of adsorption. Similar results were obtained for adsorption of CO. This suggests a virtual absence of the ligand effect, and all the following catalytic results can be attributed to the change in the ensemble size.

Bond and Yide (ref. 37) characterized some 1 wt.% Ru-Cu/SiO$_2$ catalysts by temperature programmed reduction (TPR), and X-ray photoelectron spectroscopy (XPS). Their main conclusion was that reduction of co-impregnated RuCl$_3$ and Cu(NO$_3$)$_2$ in hydrogen at 623 K leads to the formation of bimetallic particles of about 1 nm in diameter. Oxidation at 623 K destroys the clusters, and subsequent reduction at 430 K results in separate ruthenium and copper particles. Additional reduction at 623 K partially regenerates somewhat larger bimetallic particles. The formation of bimetallics rather than separate particles of Ru and Cu in the first reduction is explained by the easy reduction of RuCl$_3$ and hydrogen spillover from ruthenium particles to Cu(NO$_3$)$_2$. The resulting copper atoms migrate to the ruthenium particles and are trapped on their surface.

Because of the absence of the ligand effect and the layer type structure of the bimetallic particles, the Ru-Cu/SiO$_2$ system is an attractive model for studying ruthenium ensemble size effects in hydrocarbon reactions. The catalytic activity in the hydrogenolysis of ethane and cyclohexane clearly indicated that the ensemble size variation of Ru markedly diminished the rate of hydrogenolysis (which requires a large ensemble size), whereas the dehydrogenation of cyclohexane or the hydrogenation of benzene is hardly affected (ref. 30). The hydrogenolysis of n-butane (ref. 37) is also considerably suppressed on addition of copper. The formation of clusters was also evidenced catalytically. When catalysts of low dispersion were applied (D = 1%), that is most of the ruthenium atoms are in the bulk phase, the same activity decrease was observed at a much smaller Cu/Ru ratio than in the cluster case (ref. 31). Obviously, if most of the ruthenium atoms are located at the surface, a large Cu/Ru ratio is required for the same effect.

The hydrogenation of CO has also been studied on Ru-Cu bimetallic catalysts (refs. 38,39). In this reaction the addition of copper probably has an even more dramatic effect than in the hydrocarbon reactions. A pure ensemble size effect has been found, and an ensemble consisting of four ruthenium atoms seems to be necessary for activity in the hydrogenation of CO.

9.2.4 Pt-Pd alloys

The alloying of two Group 8 metals, both of which are active in hydrogenation and hydrogenolysis, results in mutual changes in the catalytic properties. Platinum and palladium are completely miscible over the full range of composition, but segregation and palladium enrichment on the surface have been found by Auger electron spectroscopy (AES) (ref. 40) and by catalytic methods (ref. 41). Likewise, in a supported system, segregation occurs in a reactive atmosphere, in particular when oxidation takes place between 300 and 700°C (ref. 42).

The most intriguing problem is whether there is perturbation of the electronic structure with these two metals, or whether the resulting catalytic behaviour can be explained by ensemble size variation. An early paper (ref. 43) denied the formation of bimetallics, on the basis of oxygen chemisorption experiments. However, the formation of Pt-Pd bimetallic clusters was also assumed, based upon the results of benzene hydrogenation. No change was observed in the turnover number (TON) of benzene down to 20 at% Pt. However, a pronounced change can be expected only at extreme dilutions. Cyclohexane dehydrogenation was also investigated (refs. 44,45); a five-fold activity decrease was observed for the Pt-Pd bimetallic catalyst as compared to pure platinum. The authors assumed that, on dissociation of the C-H bond, hydrogen is attached to Pt as H^-; consequently, if the "softness" of Pt decreases due to the presence of Pd, the rate of the C-H dissociation decreases. However, this observation can be explained by the ensemble size effect. According to Gonzales and co-workers (refs. 46,47), adsorption of CO in the bridged form markedly decreases with increasing platinum content of silica--supported Pt-Pd catalysts, i.e., on increasing the platinum content, the number of adjacent palladium atoms decreases. The opposite is also true, i.e., on increasing the number of palladium atoms the amount of contiguous surface platinum atoms decreases. Consequently, the rate of dehydrogenation, which requires large platinum ensembles, decreases. Moreover, with mild oxygen treatment, which was the case here, surface enrichment in Pd occurs which further diminishes the large platinum ensembles on the surface (ref. 47).

Surface enrichment in Pd is a determining factor in the reactions of C_5 and C_6 alkanes on Pt-Pd films. Karpinski and co--workers (ref. 48) explain the selectivities for neopentane reactions by a one-platinum site for isomerization and a two--palladium site for hydrogenolysis. However, as for the Pt-Au catalyst, we are inclined to speculate on a simple ensemble size effect, i.e., we assume that on large platinum ensembles isomerization takes place, whereas on small ensembles a slow one-site hydrogenolysis occurs. Evidence in support of this is provided by the fact that the selectivity for isomerization of C_5-C_6 hydrocarbons is markedly lowered, which could be due to the abrupt enrichment in Pd at low palladium concentrations. On supported Pt-Pd catalysts (refs. 48-50) the activity and selectivity for isomerization decreases, hydrogenolysis increases and cyclization passes through a maximum as the amount of Pd increases. Because of the high dispersion a cyclic mechanism is suggested, but that leaves the maximum in cyclization without explanation. In our opinion, instead of the synergistic effect suggested by the authors, the change in surface hydrogen coverage can be taken into account, which at the medium range of composition suppresses the formation of carbonaceous deposits.

It seems straightforward to suggest that for the Pt-Pd system, although alloy formation is possible over the entire composition range, the perturbations in electronic structure have a negligible influence upon the H-M and C-M bond strengths and the catalytic behaviour can be well explained by an ensemble size effect.

9.3 ENSEMBLE SIZE EFFECT ACCOMPANIED BY SECONDARY EFFECTS

Only in very rare cases is the ensemble size effect the only one operative. The structure of the surface intermediates of a reaction, especially the number of surface bonds per molecule, is altered as the hydrogen supply is affected upon alloying. When a non-noble transition metal is alloyed with a noble metal, which activates hydrogen, the ensemble size effect is accompanied by a matrix effect. The Fe-Pt system is a good example to illustrate this complex effect.

9.3.1 Matrix effect

(i) Iron-platinum. Iron-platinum catalysts have been thoroughly characterized by Mössbauer spectroscopy and as a result the formation of Fe-Pt bimetallic clusters was established. At high Fe/Pt ratios, part of the iron remains in an unreduced state, whereas Fe-Pt bimetallic particles are formed at the surface (refs. 51,52).

Considering the activity of these catalysts in the hydrogenolysis of ethane (refs. 53-55) and n-butane (refs. 55,56) as well as in the dimerization of propene (ref. 57), the following mechanism was suggested. At low iron loadings the dispersion of the silica-supported Pt-Fe catalyst increased, resulting in an increase in the catalytic activity (dispersion effect), whereas at higher Fe-Pt ratios the catalytic action of the platinum--modified Fe surface was manifested (matrix effect). In this latter case, hydrogen is mainly supplied by Pt inserted into the iron matrix (hydrogen effect). On platinum-rich catalysts there is no doubt about the catalytic action of Pt, because the adsorption data reveal a parallelism between dispersion and the rate of reaction (ref. 55). Moreover, selective splitting in the middle of the carbon chain is also preferred at higher dispersions. For iron-rich systems there are additional data to support the participation of the iron surface in the reaction: (i) the reaction order with respect to hydrogen is positive; (ii) preferred splitting at the end of the carbon chain (ref. 56) and preferential adsorption of ethane in a highly dissociated form (ref. 58).

A kinetic analysis of the hydrogenolysis of ethane on Pt/SiO_2 and $Fe-Pt/SiO_2$ catalysts by Gudkov et al. (ref. 58) clearly demonstrated a matrix effect. On both catalysts the dissociation of the C_2H_5 or C_2H_2 intermediates is the rate-determining step, depending on the hydrogen and ethane pressures in the gas phase. It was found that addition of iron promotes ethane adsorption in the form of C_2H_2. When the ratio of the gas-phase pressures of ethane to hydrogen is about 4 and 2 on Pt/SiO_2 and $Fe-Pt/SiO_2$ catalysts, respectively, a switch is found from C_2H_5 to C_2H_2 as intermediate in the rate-determining step. This means that on $Fe-Pt/SiO_2$ the ethane is adsorbed in a considerably dehydrogenated form at lower C_2H_2/H_2 ratios.

The ensemble size effect may be a possible explanation for the different reactions of propene on Pt and Fe-Pt catalysts. Although this explanation is in contradiction with that given originally (ref. 57), the speculation on the ensemble size effect seems to be justified. Namely, on platinum where large ensembles are present, multi-site hydrogenolysis is the main reaction path. This is because a large ensemble is required, not only for the extensive dissociation of propene, but also for the accommodation of a sufficient amount of hydrogen and the hydrogenolysis products. At the Fe-Pt surface, when atoms or small ensembles of Pt are inserted into the iron matrix, hydrogenation and dimerization are the main routes because both reactions require smaller platinum ensembles. In addition, Fe itself may also be active in hydrogenation.

In the $CO + H_2$ reaction Hughes et al. (ref. 59) observed a significant increase in the activity of catalysts derived from $Fe_3(CO)_{12}$ supported on γ-Al_2O_3 when Pt was added. At the same time the high selectivity for C_3-C_6 alkenes was retained. Rather surprisingly, an increase in activity has also been observed when platinum is added on separate Al_2O_3 particles. The relative positions of the 0.8 wt.% Pt/Al_2O_3 and the 1.3 wt.% Fe, $Fe_3(CO)_{12}/Al_2O_3$ catalysts within the reactor have a considerable effect on the product distribution. Data are given in Table 9.1 together with the iron analysis. The authors explain the increase in iron content of Pt/Al_2O_3 in the downstream configuration by vaporization of $Fe_3(CO)_{12}$ or some subcarbonyl species during thermal activation. The subsequent decomposition on the platinum surface results in bimetallic Pt-Fe particles, whereby the platinum surface is partially covered with iron.

TABLE 9.1
Selectivities, activities and iron analyses for various configurations of equal weights of $Fe_3(CO)_{12}/Al_2O_3$ and Pt/Al_2O_3 catalysts. Conditions: temperature = 473 K, P = 10 bar 2:1 H_2:CO v/v, ~10 g total catalyst weight, ~2 h on stream; $Fe_3(CO)_{12}/Al_2O_3$ = 1.3 wt.% Fe, Pt/Al_2O_3 = 0.8 wt.% Pt. NA = not applicable. (From ref. 59).

Product	Configuration		
	Fully mixed	Pt/Al_2O_3 upstream	Pt/Al_2O_3 downstream
CH_4	3.3	90.4	29.8
C_3H_6	84.3	4.1	31.9
1-C_4H_8	6.6	trace	8.5
CO_2	5.8	5.5	29.8
% CO conversion	0.18	0.22	0.14
Fe analysis, wt.% Fe:			
Fe layer	NA	1.10	1.10
Pt layer	NA	0.03	0.23

Platinum might increase the concentration of hydrogen atoms on the iron surface via spillover and thus might increase the rate of hydrogenation of the alkenic intermediates.

In summary, it can be shown that in reactions catalyzed by a combination of Pt with a more electropositive metal, such as Fe, the ensemble size effect is a controlling factor, and the catalytic activity of Fe is considerably improved. Depending on the reaction, either the ensemble size or the hydrogen supply is dominant.

9.3.2 Hydrogen effect

As already indicated in the Fe-Pt system, the availability of hydrogen for the reaction plays a decisive role in the matrix effect. However hydrogen can also be depleted as a second metal is added to platinum.

Both the ensemble size and hydrogen effects can be demonstrated in the conversion of neopentane on 1 wt.% Pt-Au alloys (ref. 60). As shown in Fig. 9.2, the isomerization passes through a maximum as the platinum content decreases.

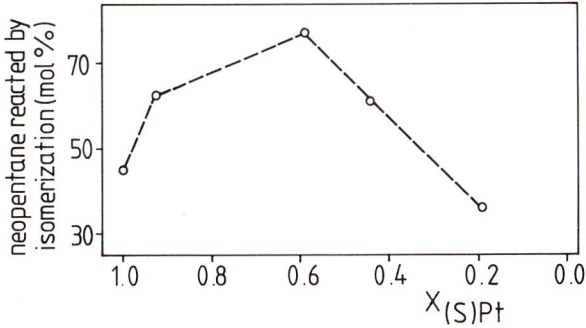

Fig. 9.2. Selectivity for neopentane isomerization as a function of the number of surface Pt atoms in the Pt-Au system (From ref. 28).

A possible explanation is as follows: on a large platinum ensemble the bond-shift mechanism operates via di- and triadsorbed species; on an extremely diluted platinum catalyst, a single-site mechanism of hydrogenolysis is preferred. However, over the medium range of composition, as the gold content increases the surface hydrogen is depleted (as found for propane dehydrogenation) and thus, even with the same surface intermediate, a larger proportion of the intermediate is converted into the isomeric product. With diluted platinum, the single-site hydrogenolysis mechanism becomes predominant again, resulting in a decrease in isomerization. Similar results were found for silica-supported Pt-Au catalysts (ref. 61).

(i) <u>Selective hydrogenation by palladium-based bimetallic catalysts.</u> Hydrogen depletion is a possible reason for the effect of the addition of copper and tin to palladium catalysts in the selective hydrogenation of acetylene.

Palladium itself is one of the most important metals in hydrogenation. Its unique ability selectively to hydrogenate alkynes and alkadienes to olefins has formed the subject of many in-depth investigations (refs. 62-74).

Palladium hydride is one of the key intermediates which controls the selectivity of acetylene hydrogenation (refs. 72,73). The rate of reaction increases in parallel with the formation of the β-hydride, whereas the selectivity defined as the ratio of the ethylene formed to the acetylene consumed is markedly decreased. This supports the mechanism proposed on unsupported palladium (ref. 71) according to which ethane is the initial product of the reaction and is directly formed from acetylene. The same mechanism is applicable on Pd/Al_2O_3, and it was established that three types of intermediates were sufficient to describe the mechanism (Fig. 9.3.).

Laboratory scale investigations using a large excess of ethylene have shown that the initial selectivity for ethylene is always high, even in the absence of gas-phase CO, but it decreases with catalyst aging time, resulting in net ethylene consumption (ref. 75). It has been shown (ref. 76) that the build-up of strongly bound carbonaceous material on the support is responsible for the change in selectivity. Recently, it has also been shown (ref. 74) that the intrinsic selectivity (the reaction routes of acetylene) does not change during aging, but the rate of ethylene hydrogenation from the gas phase increases as deposits are formed.

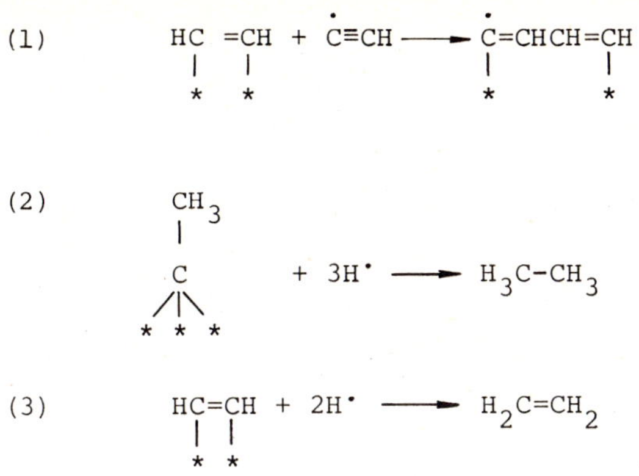

(1)   HC=CH + Ċ≡CH ⟶ Ċ=CHCH=CH
       | | | |
       * *   *   *

(2)   CH$_3$
       |
       C   + 3H· ⟶ H$_3$C-CH$_3$
      /|\
      * * *

(3)   HC=CH + 2H· ⟶ H$_2$C=CH$_2$
       | |
       * *

Fig. 9.3. Surface intermediates in acetylene hydrogenation
(1) dissociative adsorption of acetylene to form polymer species,
(2) reactive adsorption of acetylene to form ethane directly,
(3) associative adsorption of acetylene to form ethylene.

The addition of trace amounts of CO (max. 100 ppm) hardly reduces the reaction rate of acetylene, but drastically suppresses ethylene hydrogenation (ref. 77). At higher concentrations of CO a decrease in the overall reaction rate of acetylene has been found. $^{14}C_2H_2$ labelling (ref. 78) has demonstrated that the intrinsic selectivity is only marginally influenced by aging or by addition of CO which affect only ethylene hydrogenation from the gas phase. Obviously there are at least two different surface sites for acetylene and ethylene hydrogenation, the latter being most probably associated with hydrogen-deficient carbonaceous deposits on the support (ref. 79). Hydrogen migrating to the support seems to participate in the unselective process, in the hydrogenation of ethylene (ref. 80). The effect of added CO is to reduce the rate of ethylene hydrogenation by blocking sites for hydrogen dissociation on the metal (ref. 79).

In order to improve the selectivity of the Pd/Al$_2$O$_3$ catalyst, Cu, Sn and Au have been added in the belief that they might:
- change the intrinsic product selectivities in the reaction of acetylene;
- change the relative number of sites active for acetylene and ethylene hydrogenation;
- increase and stabilize the dispersion of Pd and thereby inhibit the formation of β-phase palladium hydride.

Results obtained by an American-Hungarian group (refs. 74-80) have clearly demonstrated that the addition of copper to palladium causes a significant decrease in the overall rate of ethane formation (see Fig. 9.4.).

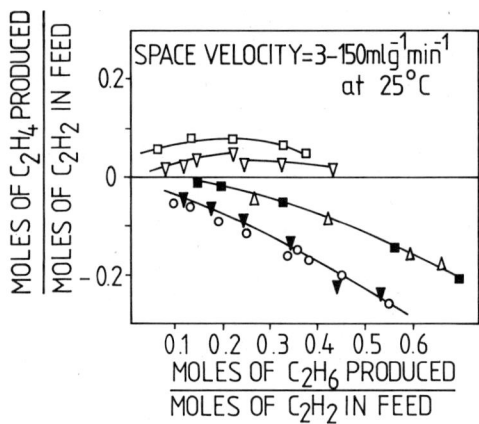

Fig. 9.4. Reaction paths near steady state showing the selectivity improvement of 0.04 wt.% Pd/Al_2O_3 catalysts impregnated with Cu (From ref. 78).

At the same time there is a decrease in the catalyst activity as well as a marginal decrease in oligomer selectivity. $^{14}C_2H_2$ and $^{14}C_2H_4$ labelling show that the intrinsic selectivity does not change upon addition of copper. The products formed from acetylene are about 40-50% ethylene, 40-50% C_4+ and 1-5% ethane.

In contrast to the Pd/Al_2O_3 catalyst, the addition of gas-phase CO in trace amounts has practically no effect on the selectivity with copper-containing catalysts. However, when large amounts of CO have been added, the overall reaction rate decreases. Fig. 9.5 shows the effect of addition of CO to copper-free and copper-containing catalysts. It should be noted that the effects of CO are reversible.

It has also be shown that the rate of deposition of oligomers on the support is also reduced on addition of copper (ref. 79).

Replacing hydrogen with deuterium in the feed has no effect on the selectivity of copper-containing catalysts (ref. 78), while it has a pronounced effect on that of the Pd/Al_2O_3 catalyst (ref. 81).

Based upon these results the following picture emerges. The addition of copper and of CO have similar effects on acetylene hydrogenation. Both reduce the unselective hydrogenation of gas-

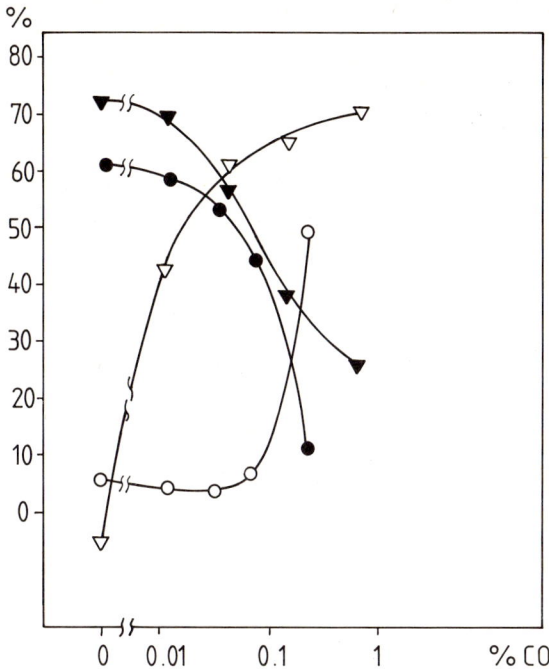

Fig. 9.5. Effect of CO on acetylene conversion and ethylene selectivities on Pd/Al$_2$O$_3$ and Pd - Cu/Al$_2$O$_3$ catalysts. Open symbols: ethylene selectivity (ethylene produced/acetylene consumed); closed symbols: acetylene conversion; (▽) 0.04 wt.% Pd/Al$_2$O$_3$; (o) 0.04 wt.% Pd/Al$_2$O$_3$ modified by 80 at.% Cu.

-phase ethylene, most probably by reducing the hydrogen supply to the deposits on the support. CO blocks adsorption sites for adsorbed hydrogen. Part of the copper covers palladium (as does CO) resulting in a decrease in activity and in a suppression of acetylene dissociation. It stabilizes the palladium dispersion and in this way no β-phase palladium hydride can be formed. The remainder of the copper partially covers the support and hinders the formation of deposits, resulting in a decrease in ethylene hydrogenation and in a better overall selectivity. Obviously, with Pd-Cu/Al$_2$O$_3$ catalysts, the geometric effects and the hydrogen supply are the most important factors.

Similar conclusions can be drawn from results obtained on tin-modified Pd/Al$_2$O$_3$ catalysts (ref. 80). Even the replacement of

Al_2O_3 by SiO_2 and a simultaneous increase in the palladium load to 5 wt.% do not significantly affect the intrinsic selectivity. Alloying of Pd with Au in the Pd/SiO_2 catalyst has a similar effect to that of addition of copper to the Pd/Al_2O_3 catalyst. The data in Table 9.2. clearly show that the intrinsic selectivity remains unchanged and there is a decrease in the hydrogenation of gas-phase ethylene resulting in an improved overall selectivity.

TABLE 9.2
The effect of the addition of Cu and Au to Pd/Al_2O_3 and Pd/SiO_2 catalysts on the overall and intrinsic selectivities for hydrogenation of trace amounts of acetylene in ethylene

Catalyst	Overall selectivity (%)			Intrinsic selectivity[a] (%)		
	$S_{C_2H_6}$	$S_{C_2H_4}$	$S_{C_4^+}$	$S'_{C_2H_6}$	$S'_{C_2H_4}$	$S'_{C_4^+}$
0.04 wt.% Pd/Al_2O_3	90	-20^b	30	6	67	27
0.04 wt.% Pd + 0.4 wt.% Cu/Al_2O_3	58	15	27	5	66	29
5 wt.% Pd/SiO_2	87	-23^b	36	2	54	44
95 at.% Au + 5 at.% Pd/SiO_2 total metal load 5 wt.%	35	21	44	2	43	55

[a] Measured by ^{14}C labelling.
[b] C_2H_4 is consumed.

In this system the geometric effects and the hydrogen supply are again the most important factors.

Acetylene hydrogenation on unsupported Pd-Au alloys has been studied by Visser et al. (ref. 82) in a pulse reactor. Besides the ensemble effect, a pronounced increase in the apparent activation energy was observed when d-band holes in palladium were filled (40% Au). Changes in the bond strengths and ligand effects were suggested to be responsible for this increase, as well as the elimination of side reactions leading to self-poisoning.

9.3.3 Metal-support interaction and segregation

(i) <u>Platinum-gold.</u> $Pt-Au/SiO_2$ catalysts have been discussed in paragraph 9.2.2. If the silica support is replaced by alumina, the metal support interaction (MSI) state may not be ruled out (ref. 83). Al_2O_3- and SiO_2-supported Pt-Au catalysts exhibit different behaviours after two oxidation-reduction cycles (see Fig. 9.6). As shown the activity and selectivity of SiO_2-supported Pt--Au catalysts do not change, whereas those of the Al_2O_3-supported bimetallic catalyst do. A possible explanation, given by De Jongste and Ponec (ref. 82), is that upon oxidation, segregation occurs on the alumina-supported catalyst and the bimetallic particle disintegrates to small particles of Pt and large crystallites of Au. Consequently, the extent of cyclization, which requires a single site, increases compared to isomerization.

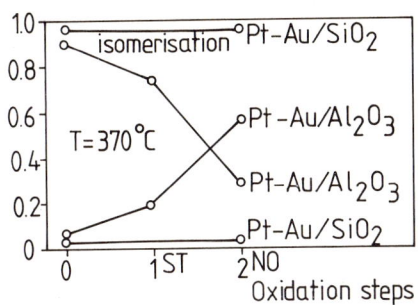

Fig. 9.6. Effect of oxidation on the isomerization and cyclization selectivity of n-hexane in the Pt-Au system (From ref. 83).

(ii) <u>Ruthenium-gold.</u> The addition of Au to Ru, mainly studied by the Ann Arbor group (refs. 84-88), is a rather interesting attempt to influence the catalytic activity of Ru. According to these studies, the support exerts a stabilization effect on the bimetallic cluster. That is, on a silica support, a bimodal particle-size distribution was detected because Au tended to aggregate in the form of large particles, whereas Ru remained in a highly dispersed state (whether the latter contains Au or not is still and open question). On the other hand, on MgO, evidence was found for the existence of bimetallic Ru-Au particles with a diameter of 10 nm.

Cyclopropane conversion has been investigated on a Ru-Au/MgO catalyst; upon addition of Au the selectivity for hydrogenolysis to methane decreased as compared to hydrogenation and hydrogenolysis toward methane + ethane, the latter being low-temperature reactions. One can again attempt to explain this selectivity change by the ensemble size effect, because multiple bond rupture definitely requires larger ensembles than does hydrogenation and single C-C bond-splitting reactions.

It seems more difficult to interpret other catalytic behaviour of the SiO_2 and MgO supported Ru-Au catalysts. On Ru-Au/SiO_2, the addition of Au decreases the catalytic activity in ethane hydrogenolysis and in the CO + H_2 reaction, while MgO-supported catalysts show a maximum in activity with addition of Au, as shown in Fig. 9.7. According to the authors this might be due to an electronic effect of alloying the two metals.

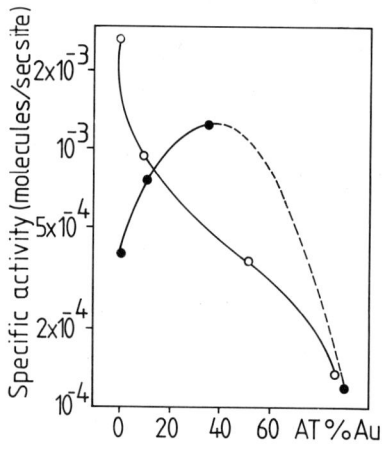

Fig. 9.7. MgO (●) and SiO_2 (o) supported Ru-Au system activity in hydrogenolysis of ethane (From ref. 84).

However, with the experimental data available we can also speculate that the effect of Au is simply an ensemble size effect in the self-poisoning reaction. As was shown with silica, small particles of Ru are present and are subjected to carbon deposition, which partially deactivates the active ruthenium component and the catalytic activity thus decreases. On the other hand, the existence of bimetallic particles has been demonstrated on MgO-supported catalysts. Thus the addition of Au probably decreases the ruthenium ensemble size and consequently also the self-poisoning reactions leading to deactivation. At higher levels of Au, too much of the total catalyst surface becomes covered by Au, leading to a decrease in the overall activity. This interpretation is supported by the fact that the behaviour of the catalyst in both reactions is the same, although their mechanisms are completely different. The data presented seem to be explained satisfactorily by the effects of the support, the ensemble size and the suppression of side reactions.

9.4 CONCLUSIONS

Hydrogenation and hydrogenolysis on supported bimetallic catalysts are primarily controlled by the ensemble size effect. In most cases this effect does not operate in isolation but is accompanied by other effects, among which the hydrogen supply and matrix effects are the most important ones. Other effects are generally marginal or strongly overlap with the ensemble size effect. On alloying a Group 8 metal with one in Group 1B, Au and Ag are usually inactive diluents but Cu must be treated as an active component which changes the hydrogen supply and eliminates side reactions leading to self-poisoning.

REFERENCES

1. R.L. Moss and L. Whalley, Adv. Catal., 22 (1972) 115.
2. G.C. Bond and E.G. Allison, Catal. Rev., 7 (1973) 233.
3. J.K.A. Clarke, Chem. Rev., 75 (1975) 291.
4. W.M.H. Sachtler, Catal. Rev. Sci. Eng., 14 (1976) 193.
5. W.M.H. Sachtler and R.A. Van Santen, Adv. Catal., 26 (1977) 69.
6. K.C. Khulbe and R.S. Mann, Catal. Rev. Sci. Eng., 24 (1982) 311.
7. V. Ponec, Catal. Rev. Sci. Eng., 11 (1975) 41.
8. V. Ponec, Adv. Catal., 32 (1983) 149.
9. L. Guczi, J. Mol. Catal., 25 (1984) 13.
10. R.A. Van Santen, Recl. Trav. Chim. Pays-Bas. Belg., 101 (1982) 127.
11. Yu.I. Yermakov, B.N. Kuznetsov and V.A. Zakharov, Catalysis by Supported Complexes, Elsevier, Amsterdam, 1981.
12. G.A. Somorjai and W. Blakely, Nature, 258 (1975) 580.
13. B.S. Gudkov and L. Guczi, React. Kinet. Catal. Lett., 9 (1978) 343.
14. L. Guczi and J. Sárkány, J. Catal., 68 (1981) 190.
15. J.H. Sinfelt, J.L. Garten and D.J.C. Yates, J. Catal., 24 (1972) 283.
16. V. Ponec, in E.G. Derouane and A.A. Lucas (Editors) Electronic Structure and Reactivity of Solid Surfaces, Plenum, New York, 1976, p. 537.
17. V. Ponec, J. Quantum Chem., 12 (1977) 1.
18. Z. Schay and P. Tétényi, J. Chem. Soc., Faraday Trans. 1, 75 (1979) 1001.

19 W.M.H. Sachtler, Vide Couches Minces, 164 (1973) 67.
20 W.A.A. Van Barneveld and V. Ponec, Recl. Trav. Chim. Pays-Bas. Belg., 93 (1974) 243.
21 F.J. Kuijers, R.P. Dessing and W.M.H. Sachtler, J. Catal., 33 (1974) 316.
22 J.J. Stefan and V. Ponec, J. Catal., 42 (1976) 1.
23 J.W.A. Sachtler and G.A. Somorjai, Prepr. Div. Pet. Chem. Am. Chem. Soc., 28 (1983) 491.
24 R.P. Dessing and V. Ponec, J. Catal., 44 (1976) 494.
25 P. Biloen, F.M. Dautzenberg and W.M.H. Sachtler, J. Catal., 50 (1977) 77.
26 J.R.H. Van Schaik, R.P. Dessing and V. Ponec, J. Catal., 38 (1975) 273.
27 S. Galvagno and G. Parravano, J. Catal., 57 (1979) 272.
28 J.R. Anderson, K. Foger and R.J. Breakspere, J. Catal., 57 (1979) 458.
29 M. Hansen, Constitution of Binary Alloys, Mc Graw-Hill, New York, 2nd ed., 1958, p. 607.
30 J.H. Sinfelt, J. Catal., 29 (1973) 308.
31 J.H. Sinfelt, Y.L. Lamm, J.A. Cusumano and A.E. Barnett, J. Catal., 42 (1976) 227.
32 J.H. Sinfelt, Acc. Chem. Res., 10 (1977) 15.
33 J.H. Sinfelt, G.H. Via and F.W. Lytle, Catal. Rev. Sci. Eng., 26 (1984) 81.
34 K. Christmann, G. Ertl and H. Shimizu, J. Catal., 61 (1980) 397.
35 H. Shimizu, K. Christmann and G. Ertl, J. Catal., 61 (1980) 412.
36 J.C. Vickerman, K. Christmann and G. Ertl, J. Catal., 71 (1981) 175.
37 G.C. Bond and X. Yide, J. Mol. Catal., 25 (1984) 141.
38 L.J.M. Luyten, M. Van Eck, J. Van Grondelle and J.H.C. Van Hooff, J. Phys. Chem., 82 (1978) 2000.
39 G.C. Bond and B.D. Turnham, J. Catal., 45 (1976) 128.
40 F.J. Kuijers, B.M. Tieman and V. Ponec, Surf. Sci., 75 (1980) 657.
41 L. Guczi and Z. Karpinski, J. Catal., 56 (1979) 438.
42 M. Chen and L.D. Schmidt, J. Catal., 56 (1979) 198.
43 R. Gomez, S. Fuentes, F.J. Fernandez del Valle, A. Campero and J.M. Ferreira, J. Catal., 38 (1975) 47.
44 J. Haro, R. Gomez and J.M. Ferreira, J. Catal., 45 (1976) 326.
45 M.E. Ruiz-Vizcya, O. Novaro, J.M. Ferreira and R. Gomez, J. Catal., 51 (1978) 108.
46 C.M. Grill and R.D. Gonzales, J. Catal., 64 (1980) 487.
47 C.M. Grill, M.L. McLaughlin, J.M. Stevenson and R.D. Gonzales, J. Catal., 69 (1981) 454.
48 Z. Karpinski and T. Koscielski, J. Catal., 56 (1979) 430.
49 Z. Karpinski and T. Koscielski, J. Catal., 63 (1980) 313.
50 T. Koscielski, Z. Karpinski and Z. Paál, J. Catal., 77 (1982) 539.
51 M.A. Vannice and R.L. Garten, J. Mol. Catal., 1 (1975) 201.
52 I. Dézsi, D.L. Nagy, M. Eszterle and L. Guczi, React. Kinet. Catal. Lett., 8 (1978) 301.
53 S. Engels, W. Mörke and J. Siedler, Z. Anorg. Allg. Chemie, 431 (1977) 181.
54 S. Engels, W. Mörke and J. Siedler, Z. Anorg. Allg. Chemie, 431 (1977) 191.
55 L. Guczi, K. Matusek and M. Eszterle, J. Catal., 60 (1979) 121.
56 L. Guczi, K. Matusek, A. Sárkány and P. Tétényi, Bull. Soc. Chim. Belg., 88 (1979) 497.
57 L. Guczi, G. Kemény, K. Matusek, J. Mink, S. Engels and W. Mörke, J. Chem. Soc., Faraday Trans. 1, 76 (1980) 782.

58 B.S. Gudkov, L. Guczi and P. Tétényi, J. Catal., 74 (1982) 207.
59 I.S.C. Hughes, J.O.H. Newman and G.C. Bond, J. Mol. Catal., 25 (1984) 219.
60 K. Foger and J.R. Anderson, J. Catal., 61 (1980) 140.
61 J.K.A. Clarke, A.F. Kane and T. Baird, J. Catal., 64 (1980) 200.
62 G.C. Bond, Catalysis by Metals, Academic Press, New York, 1962.
63 G.C. Bond and P.B. Wells, Adv. Catal., 15 (1964) 92.
64 W. McGown, C. Kemball, D. Whan and M. Scurrel, J. Chem. Soc., Faraday Trans. 1, 73 (1977) 632.
65 W. McGown, C. Kemball and D. Whan, J. Catal., 51 (1978) 173.
66 A.S. Al-Ammar and G. Webb, J. Chem. Soc., Faraday Trans. 1, 74 (1978) 195.
67 A.S. Al-Ammar and G. Webb, J. Chem. Soc., Faraday Trans. 1, 74 (1978) 657.
68 A.S. Al-Ammar and G. Webb, J. Chem. Soc., Faraday Trans. 1, 75 (1979) 1900.
69 A.H. Weiss, B. Gambhir, R. LaPierre and W. Well, Ind. Eng. Chem. Proc. Des. Dev., 16 (1977) 352.
70 L. Guczi, R. LaPierre, A.H. Weiss and E. Biron, J. Catal., 60 (1979) 83.
71 J. Margitfalvi, L. Guczi and A.H. Weiss, J. Catal., 72 (1981) 185.
72 J. Moses, M.S. Thesis, Worcester Polytechnic Institute, Worcester, MA, 1980.
73 A. Borodzinski, R. Dus, R. Franko, A. Janko and W. Palczewska, Proceedings of the Sixth International Congress on Catalysis, Vol. 1, Chemical Society, London, 1977, p. 150.
74 A.H. Weiss, S. LeViness, V. Nair, L. Guczi, A. Sárkány and Z. Schay, 8th International Congress on Catalysis, Vol. 5, Verlag Chemie, Weinheim, 1984, p. 591.
75 Z. Schay, A. Sárkány, L. Guczi, A.H. Weiss and V. Nair, Proc. V. Intern. Symp. on Heterogeneous Catalysis, Part I, Varna, 1983, p. 315.
76 A. Sárkány, L. Guczi and A.H. Weiss, Appl. Catal., 10 (1984) 369.
77 S. LeViness, A.H. Weiss, Z. Schay and L. Guczi, J. Mol. Catal., 25 (1984) 131.
78 L. Guczi, Z. Schay, A.H. Weiss, V. Nair and S. LeViness, React. Kinet. Catal. Lett., 27 (1985) 147.
79 A. Sárkány, L. Guczi and A.H. Weiss, Appl. Catal., 10 (1984) 369.
80 S. LeViness, Thesis, Worcester Polytechnic Institute, Worcester, MA, 1984.
81 J.M. Moses, A.H. Weiss, K. Matusek and L. Guczi, J. Catal., 86 (1984) 417.
82 C. Visser, J.G.P. Zuidwijk and V. Ponec, J. Catal., 35 (1974) 407.
83 H.C. De Jongste and V. Ponec, J. Catal., 64 (1980) 228.
84 J. Schwank, G. Parravano and H.L. Gonber, J. Catal., 61 (1980) 19.
85 S. Galvagno, J. Schwank and G. Parravano, J. Catal., 61 (1980) 223.
86 I.W. Bassi, F. Garbassi, G. Vlak, A. Marzi, G.R. Tauszik, G. Cocco, S. Galvagno and G. Parravano, J. Catal., 64 (1980) 405.
87 S. Galvagno, J. Schwank, G. Parravano, F. Garbassi, A. Marzi and G.R. Tauszik, J. Catal., 68 (1981) 283.
88 A.K. Datye and J. Schwank, 8th International Congress on Catalysis, Vol. IV, Verlag Chemie, Weinheim, 1984, p. 587.

Chapter 10

SUPPORTED MONO- AND BIMETALLIC CATALYSTS IN HYDROCARBON CONVERSIONS

J. VÖLTER
Central Institute of Physical Chemistry, Academy of Sciences of the G.D.R., Rudower Chaussee 5, DDR-1199 Berlin (G.D.R.)

10.1 INTRODUCTION

The platinum-group metals are still predominant among catalysts employed for the conversion of hydrocarbons. But now they are often combined with other metals, ranking on much lower levels in the hierarchy of the catalytic activity. Kluksdahl's discovery in 1968 that addition of Re improves the unique Pt/Al_2O_3 catalyst for petroleum reforming (ref. 1) has had a great impact. Only a few years later, other bimetallic systems for catalytic upgrading of naphtha fractions were patented, the most successful being Pt-Sn (refs. 2, 3). This has stimulated much research on the role of the second component, but many questions remain unanswered (ref. 4).

Several modifications of the Pt/Al_2O_3 system by a second metal can be anticipated: (i) an alloy is formed, (ii) an ion of the second metal interacts with Pt, (iii) an ion interacts with the support and (iv) the second metal interacts with the dual system Pt-C, the active platinum catalyst being partly covered with carbon (refs. 5, 6). Support for each of these possibilities is to be found in the literature and a corresponding variety of mechanisms have been suggested (refs. 7 - 17). In spite of these competing or even controversial proposals, a new attempt will be made in the present article to determine the relationships between the structures of the bimetallic catalysts and their specific catalytic properties. The literature has been surveyed beginning with the preparation of the catalyst and ending with its self-poisoning. The number of active metals will be confined to Pt, Pd, Rh and Ni, with predominance of Pt. As second metals, only those are included which display no intrinsic catalytic activity, especially the Group 4a metals.

The structure of these catalysts is considered in connection with recent progress in the preparation of mono- and bimetallic carrier catalysts. The catalytic bimetal effects are illustrated by the reactions of hydrocarbons with hydrogen, i.e., hydrogenation, dehydrogenation, dehydroisomerization, dehydrocyclization and hydrogenolysis. Coke, deposited during the reactions, must be regarded as a modifier of Pt, too. Therefore, the interactions of coke with the second metal are included. In the discussion, some general aspects of the relationships between the preparation, structures and catalytic effects of bimetal catalysts in hydrocarbon conversions are outlined, especially the relevance of ensemble and ligand effects.

10.2 PREPARATION AND STRUCTURE

The structures of supported catalysts are largely determined by the methods and conditions employed for their preparation. Nevertheless, knowledge of these processes has remained empirical. Major scientific progress has been achieved only recently.

10.2.1 Monometallic systems (Pt/Al_2O_3)

Alumina-supported Pt is one of the most frequently used catalysts in reforming, hydrogenation and dehydrogenation processes. The individual steps of its preparation might be used as an example for the synthesis of other supported catalysts, too.

(i) <u>Impregnation.</u> Industrial Pt/Al_2O_3 catalysts are almost exclusively prepared by impregnation of gamma or eta alumina with an aqueous solution of H_2PtCl_6. Provided there is sufficient dilution, the impregnation proceeds as a true adsorption process. $[PtCl_6]^{2-}$ is adsorbed on acidic sites of the alumina until an equilibrium is reached (ref. 18). The number of adsorbed anions is approximately 10^{14} per cm^2 of the alumina. This number is of the same order of magnitude as the number of hydroxyl groups per cm^2 (ref. 19). This indicates a direct chemical interaction with functional surface groups, probably via an ion exchange.

The active role of alumina can be seen by its adsorption of Cl^- from aqueous solutions containing either HCl or HCl and H_2PtCl_6 (ref. 20). This is demonstrated in Fig. 10.1.

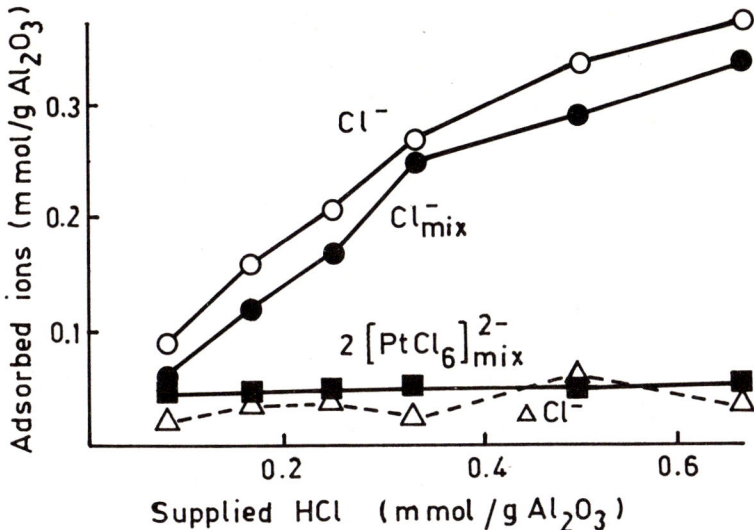

Fig. 10.1. Impregnation of γ-Al_2O_3 with H_2PtCl_6 and HCl. Adsorbed anions as a function of chloride supplied from solutions containing only HCl (Cl^-), or H_2PtCl_6 and HCl (Cl^-_{mix}); ΔCl^- = difference in adsorbed chloride between these solutions. $2\,[PtCl_6]^{2-}_{mix}$ twice number of adsorbed platinate anions.

The amount of HCl in the solution before the adsorption is denominated supplied HCl. The amounts of adsorbed chloride are increasing with increasing supply of HCl, but these amounts are always higher from solutions containing only HCl (curve Cl^-) than from mixed solutions containing HCl and a constant amount of H_2PtCl_6 (curve Cl^-_{mix}). The difference $\Delta Cl^- = Cl^- - Cl^-_{mix}$ is nearly constant. Moreover this difference, the number of nonadsorbed chloride from mixtures, is roughly the same as twice of the number of the adsorbed divalent chloroplatinate (curve $2x[PtCl_6]_{mix}$).

This result clearly indicates a competitive adsorption between Cl^- and $[PtCl_6]^{2-}$. The platinate is always completely adsorbed and an equivalent number of Cl^- remains unadsorbed.

This means that the total number of adsorption sites is constant at a given HCl concentration, but slowly increases with increasing HCl concentration and approaches a saturation value. The competitive adsorption, demonstrated under equilibrium conditions, is also responsible for kinetic effects. H_2PtCl_6 is adsorbed very quickly from aqueous solutions, resulting in a layer on the outer surface of an alumina pellet. The consequences are relatively large crystals of Pt outside and nearly no Pt inside the alumina pellet.

This disadvantage can be overcome by the addition of HCl to the impregnation solution (ref. 21). $[PtCl_6]^{2-}$ is adsorbed more slowly from acidic solutions, resulting in more highly dispersed Pt, distributed homogeneously throughout the alumina pellet. The mechanism of this controlled adsorption process could be as follows: Cl^- and $[PtCl_6]^{2-}$ compete for the adsorption sites. The Cl^- is in excess and, have a smaller size, diffuses more rapidly. Hence the chloride is adsorbed first and temporarily blocks the adsorption sites. Meanwhile $[PtCl_6]^{2-}$ diffuses throughout the whole pellet until it is finally adsorbed by ion exchange with Cl^-. Thus an homogeneous distribution is achieved.

A theoretical model of competitive adsorption as a means of controlling the adsorption profile has been proposed by Kulkarni et al. (ref. 22). A competition between adsorbing $[PtCl_6]^{2-}$, $[ReO_4]^-$ and Cl^- has been observed by de Miguel et al. (ref. 23).

(ii) <u>Calcination.</u> After impregnation the samples are usually dried and then calcined at temperatures of about 500°C. Several authors (refs. 24 - 28) have confirmed that the oxidation state of Pt is or remains +4 after treatment in air or oxygen. A detailed identification and description of different surface species has been given by Lietz and co-workers (refs. 19, 29). They proposed the following reaction scheme

$$[PtCl_6]^{2-}_s \xrightarrow[300°C]{+H_2O} [Pt^{IV}(OH)_xCl_y]_s \xrightarrow[500°C]{-H_2O} [Pt^{IV}O_xCl_y]_s \quad (1)$$

where the index s denotes the formation of a surface complex with the alumina. The proposed reactions are a hydrolytic ligand exchange at medium temperatures and a dehydration at 500°C. The complex formed at high temperature could have the formula $\left[Pt^{IV}O_2Cl_2\right]_s^{2-}$, being two-coordinated by alumina. Evidence in support of this is provided by extended X-ray absorption fine structure (EXAFS) experiments, which indicated an oxide surface species, bound to the oxygen of the alumina, but still containing two halogen atoms (ref. 30).

In chloride-free samples, $\left[PtO_2\right]_s$ is formed (ref. 29). The surface interaction stabilizes the oxide with respect to thermal decomposition and reduction. This stabilization is much stronger on Al_2O_3 than on SiO_2 or TiO_2 (ref. 31).

Calcination of supported Pd results in easily reducible PdO (ref. 32). Treatment of Ni/Al_2O_3 with oxygen gives two oxidized species, one similar to bulk NiO, the other interacting strongly with the alumina (refs. 33,34). The calcination of Pt/Al_2O_3 is related to the dehydration of the alumina. This dehydration is very beneficial for the catalyst preparation. Water in the alumina has a detrimental influence on the dispersion of Pt, as is seen from the data in Table 10.1. In order to get the highest platinum dispersion, it is necessary to dehydrate the Pt/Al_2O_3 by calcination at 500°C directly before the reduction. A rehydration of about 10 mass % occurs within 2 days from the moisture in the air.

TABLE 10.1
Dispersion of Pt as a function of the water content of alumina in Pt/Al_2O_3 catalysts (ref. 35)

Pt dispersion (%)	Water in Al_2O_3 (mass %)
100	0
96	0.8
83	3.5
58	10.0

(iii) <u>Reduction state of the active metal.</u> The different platinum (IV) surface complexes formed with alumina are easily reduced by hydrogen below 300°C. Highly dispersed metallic Pt can be formed with clusters of about 1 nm in diameter. The state of Pt in these clusters has been extensively investigated. From XPS spectra it was concluded that Pt (refs. 36, 37), Pd and Ni (ref. 36) are present in an electron-deficient state and that alumina causes a higher deficiency than silica (ref. 36). X-ray absorption spectra confirm this, indicating deficiency of 0.06 electrons for Pt/SiO_2 and 0.09 for Pt/Al_2O_3 (ref. 38). EXAFS experiments indicate that clusters of about 13 platinum atoms are derived with some links to the alumina support (ref. 30). Even UV-Vis spectra reveal a certain interaction with the alumina (ref. 19).

Such an interaction should not be confused with the so-called soluble platinum, proposed by McHenry et al. (ref. 39), which should be an irreducible ionic form of Pt. It has been shown that all the highly dispersed Pt becomes soluble on exposure to air prior to extraction (ref. 19). However, under inert conditions, small-particle catalysts can be prepared virtually free from soluble Pt (ref. 40).

Strong interactions with the support result upon hydrogen treatment above 550°C, known as the strong metal support interaction (SMSI) effect (refs. 41, 42). This effect is not beneficial for the catalytic reactions discussed here and therefore will not be considered further.

(iv) <u>Reoxidation, sintering, regeneration.</u> A fundamental drawback of Pt and other metal catalysts in hydrocarbon conversions is the long-term instability of the activity. Deactivation occurs mainly for two reasons: (i) blocking of the surface by the deposition of coke and (ii) sintering of the dispersed Pt. A well-known regeneration procedure is burn off the coke by treatment with oxygen. However, sintering may also occur during this treatment.

The chemical reactions of Pt occuring during this oxygen treatment have been identified only recently (refs. 19,29,43).

$$Pt_{disp} \xrightarrow{300°C} [PtO_2]_s \xrightarrow{450°C} [Pt^{IV}O_xCl_y]_s \xrightarrow{700°C} Pt_{cryst} \quad (2)$$

Dispersed Pt is first oxidized to a chloride-free $[PtO_2]_s$. At higher temperatures $[Pt^{IV}O_xCl_y]_s$ is formed with chloride from the alumina. This species seems to be a key product, a two-dimensional phase involved in sintering as well as in redispersion of Pt. Its stability is restricted to the temperature range between 450 and 700°C. At higher temperatures it decomposes. The platinum atoms released during the decomposition are obviously very mobile. They instantly agglomerate and form very large crystals. The product is a heavily sintered crystalline Pt. This is a new type of sintering mechanism, denominated decomposition-coalescence mechanism. It is initiated by a chemical reaction, a decomposition, and results in a distinctly bimodal particle-size distribution (ref. 43).

On the other hand, the reverse process, the formation of the two-dimensional $[PtO_xCl_y]_s$, is directly related to a redispersion of crystalline Pt. The following mechanism has been proposed: crystallites of Pt are oxidized by a combined attack of O_2 and Cl^- and the oxidation product $[Pt^{IV}O_xCl_y]_s$ is spread over the alumina. Thus, platinum atoms are removed from the crystallites and dispersed on the carrier. Later results are in accordance with this mechanism (ref. 44). Treatment with oxygen is thus not only a means of burning off the coke, but also a means to restore or to control the dispersion of Pt.

Analogous mechanisms of sintering and redispersion have been observed in the Pd/Al_2O_3 system, occuring via PdO. Redispersion is not related to the presence of Cl^- (refs. 32, 45).

10.2.2 Bimetallic systems

(i) <u>Preparation.</u> Bimetallic systems can be prepared in many different ways, most of them involving impregnations, but there are also possibilities to anchor (ref. 17) or to decompose (ref. 46) a metal complex. The impregnation with both metals can be performed simultaneously. The preparation of active bimetallic catalysts by co-impregnation with mixed solutions of H_2PtCl_6 and nitrates of the following metals has been reported: Pb (refs. 9, 47), Cd (ref. 47), Ga (refs. 47, 48) and Bi (ref. 48).

The Pt-Sn system exhibits some peculiarities. Sn can form the complexes $[Pt(SnCl_3)_2Cl_2]^{2-}$, $[SnCl_3]^-$ and $[SnCl_6]^{2-}$. Impregnation with $[Pt(SnCl_3)_2Cl_2]^{2-}$ has been described several times (refs. 8, 49-51). However, the Sn/Pt ratio is fixed and rather high. Moreover, in aqueous solution this complex is found predominantly on the outer surface of the alumina. The complex formation can be avoided by using Sn^{4+} instead of Sn^{2+}. However, catalysts prepared with Sn^{4+} display a much lower interaction between Pt and Sn, as shown by Mössbauer spectroscopy (ref. 8) and consequently a less strong effect on hydrogen adsorption and catalysis (ref. 52). Similarly diminished effects were reported in a recently published paper, moreover tin is enriched on the outer surface (ref. 14).

Effective bimetal systems can be prepared by consecutive impregnation, using Sn^{2+} in the first step (refs. 53, 54). Gault et al. (ref. 54) and Becaud et al. (ref. 55) used the same procedure without mentioning the oxidation state of tin. Surprisingly, Dautzenberg et al. (ref. 7) found no bimetal effect with catalysts impregnated first with tin. The reason is probably that they used neutralized alumina, thus blocking common adsorption sites. Pt-Sn catalysts first impregnated with H_2PtCl_6 have been used by Dautzenberg et al. (ref. 7), Davis et al. (ref. 49) and Burch (ref. 56).

A new controlled preparation of bimetallic clusters has been found by Yermakov (ref. 17). Mononuclear and polynuclear organic complexes are anchored by reaction with surface hydroxyl groups. Very specific procedures have been developed, but they are complicated and time consuming. Another method has been proposed by Nuzzo et al. (ref. 46). Volatile organometallic compounds can be thermally decomposed on dispersed Ni, resulting in the formation of intermetallic compounds.

(ii) <u>Structure and properties of the system Pt-Sn; state of tin in Pt-Sn/Al_2O_3</u>. Much research has been done on the system Pt-Sn, probably because of its industrial application. A fundamental question concerns the state of tin. Is it metallic and alloyed with Pt, or is it in the form of an oxide interacting with the alumina or with Pt? Many methods have been used to investigate this problem. A very sensitive method is Sn Mössbauer spectroscopy, but the spectra are very complicated. The signals of a variety of compounds of Sn^0, Sn^{II} and Sn^{IV} with different ligands (Pt, Cl, O) overlap and complicate the interpretation. Nevertheless, nearly all of the investigations (refs. 8, 11, 57, 58) revealed the existence of at least three species: Pt-Sn alloys, Sn^{II} and Sn^{IV}.

Various temperature programmed reduction (TPR) experiments have been reported (refs. 7, 14, 53, 56). One result is unambiguous. The TPR profile of Pt is significantly changed by Sn and vice versa. Hence it has been concluded that platinum and tin species are interacting: they are not spatially separated. A common conclusion is that the majority of tin is in the Sn^{II} state.

The formation of Pt-Sn alloys is still a subject of discussion. Burch (refs. 12, 56) excluded Sn^0 and alloy formation, except in minor amounts. Sexton et al. (ref. 14) only indicate that 80% of the tin is not in the zerovalent state. Dautzenberg et al. (ref. 7) explicitly excerted alloy formation. Lieske and Völter (ref. 53) gave a semiquantitative estimation: Sn^0 and the corresponding Pt-Sn alloys are formed, comprising 10-30% tin depending on the total tin content. The residual tin is in the form of Sn^{II} on the surface of the alumina.

Recent investigations were performed with X-ray photoelectron spectroscopy (XPS). Adkins and Davis (ref. 13) stated that the majority of tin is present in a valence state higher than Sn^0. Sexton et al. (ref. 14) estimated 30% alloying and 70% oxide on SiO_2, only Sn^{II} on Al_2O_3, but they did not exclude the formation of a dilute alloy. A quantitative evaluation of the XPS spectra obtained would be very difficult.

From oxygen and hydrogen adsorption experiments, Lieske and Völter (ref. 53) calculated an alloying in good accordance with TPR results, but Muller et al. (ref. 51) denied the formation of alloys. The main difference in the interpretation of the experiments is that the latter assumed a hydrogen adsorption on SnO which seems to be very improbable.

IR spectroscopy of CO, exclusively adsorbed on Pt, is a powerful tool in the examination of bimetallic clusters. With supported Pt-Pb (refs. 59 - 62) and Pt-Sn (refs. 62, 63) samples, a significant downward shift of the main absorption band (2090 cm^{-1}) has been observed. The unanimous conclusion is that the shift is caused by alloying. In one case, the presence of alloys was confirmed by X-ray diffraction (ref.

59). Moreover, an extinction of the small band (1850 cm^{-1}) assigned to bridged CO adsorption has been observed (refs. 60, 63). This bridged form needs two adjacent adsorption sites. Obviously they are blocked by dilution of Pt with Pb or Sn respectively. Hence this is a further hint as regards alloy formation. One question remains controversial: whether the alloying is accompanied by an electronic interaction or not. Bastein et al. (ref. 62) stated that the frequency shift with Pt-Sn can be (almost) fully explained by a dilution effect, whereas with Pt-Pb an additional effect might be related to the electronic structure. On the other hand, Palazov et al. concluded that the alloying of Pt-Pb (refs. 60, 61) and of Pt-Sn (ref. 63) is accompanied by a ligand effect, an electron transfer towards Pt. Such an interaction is consistent with Mössbauer results for Pt-Sn (refs. 8, 11).

Taken together, the numerous investigations reveal the following:

(1) Metallic Pt interacts with tin species; they are not spatially separated.

(2) The majority of tin is in the divalent state and is localized on the support.

(3) A minority of tin is in the zerovalent state and is alloyed with Pt. Alloying depends on the ratio Sn/Pt, on the carrier properties and on the methods of preparation and pretreatment. As observed with bulk alloys (ref. 64), the surface of the clusters could be enriched with tin.

(iii) <u>Structure and properties of the system Pt-Sn; chemisorption of hydrogen and oxygen on Pt-Sn.</u> The chemisorption of hydrogen and of oxygen on Pt is influenced by tin in a very characteristic manner, as is seen in Fig. 10.2. The adsorption of hydrogen is suppressed, but the adsorption of oxygen is promoted. The increase in oxygen adsorption has been observed by several authors (refs. 51, 53, 65, 66). The decrease in hydrogen adsorption, also confirmed (refs. 8, 53, 54, 65, 66), is in accordance with the observation that deuterium adsorption on bulk alloys is strongly inhibited by Sn (ref. 64). An increase in hydrogen adsorption has been reported only by Burch, but his experiments were performed under less well-defined conditions (ref. 56).

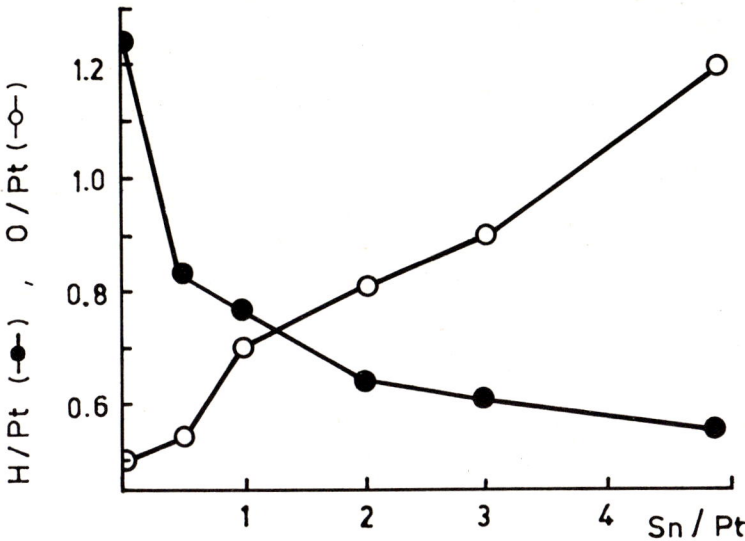

Fig. 10.2. Adsorbed oxygen (O_{ad}) and hydrogen (H_{ad}) per Pt as a function of the tin content in Pt-Sn/Al_2O_3 catalysts. (Sn/Pt) = atomic ratio. Data from ref. 71.

The reason for the increased oxygen adsorption on Pt-Sn obviously lies in the adsorption on metallic tin in the alloy (ref. 53). The decrease in hydrogen adsorption is explained by an electronic interaction between Pt and Sn or Pb respectively (ref. 65). Verbeek and Sachtler (ref. 64) also took into account the possibility of an ensemble effect, inhibiting dissociative adsorption of hydrogen.

Finally it should be stressed that the opposite effects of tin on the adsorption of hydrogen and oxygen, respectively, excludes the possibility of determining the dispersion of Pt-Sn samples by chemisorption of these gases.

(iv) <u>Structure and properties of the system Pt-Sn; dispersion of supported Pt-Sn.</u> Until now, transmission electron microscopy has been the best method for the determination of the particle size and hence of the dispersion. Gault et al. (ref. 54) observed a very remarkable effect. Upon addition of tin the particle size decreased from 9 to 3 nm. Zaikovskii et al. (ref. 67) found the same trend in anchored complexes. Qiao et al. (ref. 68) studied technical Pt-Sn catalysts and concluded that, at low tin contents, solid solutions are formed whereas at higher tin contents separate tin particles appear. On the whole, a beneficial effect of added tin has been observed, the dispersion being increased.

10.2.3 Further bimetallic systems of Pt, Pd, Rh and Ni

(i) Pt-(Pb, Ge, Sb). Supported Pt-Pb samples are quite similar to Pt-Sn. TPR (ref. 69), IR (refs. 59 - 62) and XPS (ref. 70) studies revealed alloyed Pb together with Pb^{II}. Hydrogen adsorption is suppressed, and oxygen adsorption increased (refs. 65, 66). Such adsorption properties are also observed with Pt-Pb/Al_2O_3 (ref. 71). Alloying was observed in Pt-Ge/Al_2O_3 samples by XPS (ref. 37).

(ii) Pd-(Pb, Sb). TPR profiles indicate a bimetallic interaction between Pd and Pb (ref. 72). In the Lindlar catalyst the Pd_3Pb phase has been demonstrated directly by XPS and X-ray diffraction (ref. 73). The formation of Pd_3Sb has been detected in an silica-alumina-supported hydrogenation catalyst (ref. 74).

(iii) (Rh, Ni)-Sn. Supported Rh-Sn catalysts display the typical bimetallic chemisorption properties, increasing oxygen and decreasing hydrogen adsorption (ref. 75). In silica-supported Ni-Sn catalysts the main part of tin is in the metallic state (ref. 76).

10.3 CATALYSIS WITH BIMETALLIC SYSTEMS
10.3.1 Hydrogenation

Monometallic hydrogenation catalysts are strongly modified by the addition of a second metal. Supported Pt is severely poisoned by added tin in the following reactions: hydrogenation of benzene (refs. 55, 63, 77, 78), of ethylene (refs. 63, 77) and of hexene (refs. 63, 78). An increase in benzene hydrogenation with Pt-Sn has been reported only in one case (ref. 50), but this may be a result of the different platinum dispersions. An inhibiting effect of tin was also observed with the systems Pd-Sn/SiO_2 and Ni-Sn/SiO_2 in the hydrogenation of ethylene and benzene (ref. 79) and with the system Rh-Sn/SiO_2 in benzene hydrogenation (ref. 80).

One can state that the activity of all the well-known hydrogenation catalysts Pt, Pd, Rh and Ni is decreased by added tin. The degree of poisoning by the same amount of bimetal strongly depends on the type of hydrocarbon. Benzene is especially sensitive. This will be demonstrated and discussed in Section 4.3.

Pd has a unique position among the hydrogenation catalysts. It is the best metal for the semihydrogenation of the carbon-carbon triple bond. Still more selective is the bimetallic Lindlar catalyst, containing Pd_3Pb supported on $CaCO_3$. Recent investigations (ref. 73) revealed the following. The activity is generally suppressed by lead. The selectivity is increased by a catalyst, containing Pd, partly covered by a deposit of lead oxide. However, the highest selectivity was observed with a sample containing the intermetallic phase Pd_3Pb. This is clear evidence that the bimetal effect is connected with a distinct metal-metal interaction. The authors assumed that a fixed geometry of the lead atoms on the palladium surface is responsible for the increased selectivity.

10.3.2 Dehydrogenation

(i) <u>Platinum catalysts.</u> At first sight the influence of a second metal on the dehydrogenation activity seems to be very contradictary. With the same catalyst and the same reaction, a promotion, a positive bimetal effect, as well as an inhibition, a negative bimetal effect, can be achieved. The difference is caused by the reaction conditions. This has been found in the dehydrogenation of cyclohexane on supported Pt-Pb and Pt-Sn systems (ref. 9) and will be demonstrated here with the new example of Pt-Sb (ref. 71). Typical curves for the cyclohexane dehydrogenation are shown in Fig. 10.3. A negative bimetal effect is observed under mild, non-destructive reaction conditions, at about $300°C$. The process is selective without side reactions and without self-poisoning. Bimetallic samples are distinctly less active than monometallic ones (Fig. 10.3 a). With increasing content of Sb the extent of conversion decreases (Fig. 10.3 c).

A positive bimetal effect develops under destructive reaction conditions, causing C-C bond splitting and severe deactivation by coke. The process becomes non-selective. Such conditions prevail at temperatures of about $450°C$ and higher. Depending on the reaction time, the monometallic catalyst suffers severe deactivation. The extent of conversion as well as the selectivity for benzene decrease. Surprisingly, the bimetallic catalyst is stable and the selectivity for benzene remains nearly 100% (Fig. 10.3 b). Depending on the amount of Sb, one gets a typical maximum for the conversion as well as for the benzene formation (Fig. 10.3 d). Small amounts of the bimetal have a distinct promotion effect; large amounts block the active sites.

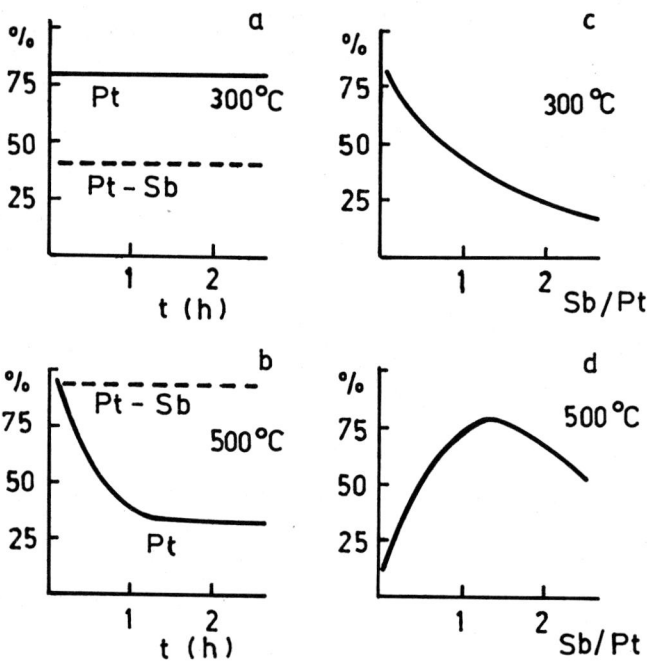

Fig. 10.3. Conversion of cyclohexane (%) on Pt/ γ-Al$_2$O$_3$ and Pt-Sb/ γ-Al$_2$O$_3$ as functions of time on stream (t), reaction temperature (300°C, 500°C) and content of Sb (atomic ratio). Pt-Sb/Al$_2$O$_3$ in (a) and (b) contain 1 Sb per Pt. Data from ref. 71.

Severe or mild reaction conditions obviously determine whether the second metal promotes or inhibits. The validity of this conclusion is confirmed by other results. Savostin et al. (ref. 81) and Lin et al. (ref. 77) reported a poisoning of the cyclohexane dehydrogenation by Pt-Sn, at 300°C, i.e., mild conditions. Burch and Garla (ref. 12) found a promotion of the dehydrogenation at 477°C, i.e., under severe conditions. The same is true for the dehydrogenation of butane at 570°C on Pt-Sn/ZnAl$_2$O$_4$, examined by Pakhomov et al. (ref. 82). They found that tin-lean alloys promote, tin-rich ones inhibit the selective dehydrogenation.

The results of Bursian et al. (ref. 15) are very interesting. They modified Pt/Al$_2$O$_3$ by a large variety of metals: Cu, Ag; Zn, Cd; Ga, In, Tl; Ge, Sn, Pb and Bi. Surprisingly, all these second metals resulted in the same catalytic effect: the selective dehydrogenation of dodecane was increased, and the lifetime of the catalysts prolonged.

(ii) <u>Palladium and nickel catalysts.</u> Pd as well as Ni have been modified by inert metals, especially by tin. Results, corresponding to a positive bimetal effect, have been reported by Masai et al. (ref 79) for the dehydrogenation of cyclohexanone and of cyclohexylamine on Pd-Sn/SiO$_2$.

A new interesting method for the preparation of bimetallic nickel catalysts has been proposed by Nuzzo et al. (ref. 46). They decomposed volatile, oxygen-free compounds of Ge, Si and B on highly dispersed Ni, thus ensuring the zerovalent state of the second metal. These catalysts display the typical bimetal effect in the high-temperature dehydrogenation of cyclohexane: inhibited cracking and promoted aromatization. This is unambiguous evidence that the bimetal effect with Ni is caused by a zerovalent and not by an oxidized second metal.

To summarize, a large variety of bimetallic systems, i.e., Pt modified with Cu, Ag, Zn, Cd, Ga, In, Tl, Ge, Sn, Pb, Bi and Sb, Ni modified with B, Si, Ge and Sn, and Pd modified with Sn, exhibit very similar catalytic effects, positive ones under destructive, negative ones under non-destructive reaction conditions.

10.3.3 Dehydrocyclization (reforming)

Catalytic reforming is a process designed to increase the octane number of naphtha fractions. It involves several types of reactions, dehydrogenation of cyclohexanes, dehydro-isomerization of cyclopentanes and dehydrocyclizations of paraffins. The catalysts of practical interest are bifunctional, containing a metallic and an acidic function. The most important reactions are those producing aromatics, especially the dehydrocyclization of paraffins. Therefore this reaction is often used as a model of reforming.

Besides Pt-Re catalysts, the bimetallic system Pt-Sn has attracted most interest. Pt-Sn/Al_2O_3 catalysts have been studied in the conversion of n-hexane (refs. 7, 12, 47, 83), of n-heptane (refs. 8, 9, 55) and of n-octane (ref. 49). With one exception (ref. 47), the results confirm the beneficial effects of tin: improved aromatization, decreased cracking and prolonged lifetime of the catalyst. Moreover, improved naphtha reforming by Pt-Sn catalysts forms the subject of many patents.

Little has been published about the pressure dependence of the bimetal effect, although pressures of 1-4 MPa are used in technical plants. An increase of pressure from 0.1 to 1.0 MPa reduces the beneficial tin effect (ref. 9). Nevertheless at 2 MPa a distinct promotion of aromatization by tin has been reported (ref. 84). Surprisingly, for catalysts containing only small amounts of tin, this was accompanied by an increase in cracking.

Several other elements from various groups have been used to modify Pt/Al_2O_3: Group 5, Sb (refs. 47, 48), Bi (ref. 48); Group 4, Pb (refs. 9, 47), Ge (ref. 47); Group 3, Ga (refs. 47, 85); Group 2, Cd (ref. 47). The results for the conversions of n-hexane and n-heptane are qualitatively the same as with the above Pt-Sn catalysts. Finally, it should be stressed that the positive bimetal effects are of the same type as in the non-selective dehydrogenation.

10.3.4 Dehydroisomerization

The dehydroisomerization of methylcyclopentane (MCP), forming benzene, is only one reaction of MCP. MCP is a strong coke producer, leading to severe deactivation of the catalyst. Hence an increased selectivity for dehydroisomerization is of twofold advantage.

The conversion of MCP on Pt-Sn/Al_2O_3 has recently been studied nearly simultaneously by three groups (refs. 10, 12, 16). All of them confirmed the typical bimetal effects: increased benzene formation, decreased ring splitting and increased stability. Moreover, two of them (refs. 10, 16) concluded that coke deposits cause similar effects to tin additions. The reason for the bimetal effect is controversial. Völter and Kürschner (ref. 16) as well as Coq and Figueras (ref. 10) assumed metallic Pt-Sn clusters, but Burch and Garla (ref. 12) proposed that modification of the acidic sites of the support by SnO was predominant, resulting in higher extent of isomerization and a lower extent of cracking.

10.3.5 Hydrogenolysis

Hydrogenolysis on bimetals has been studied not only as a side reaction of aromatization, as mentioned above, but also as the main reaction. The splitting of butane is inhibited by Pt-Sn and Pt-Cd catalysts (ref. 15). Pt-Sn and Pt-Pb samples suppress the hydrogenolysis of propane (ref. 86) and a similar inhibition was observed in the hydrogenolysis of ethylene (ref. 77).

10.4 COKING

10.4.1 Initial coking

Hydrocarbon conversions at higher temperatures are always accompanied by formation and deposition of coke on the catalyst. This is generally regarded as a harmful process, but this is an oversimplification. The effects of carbon or coke depositions during the lifetime of the catalyst may be explained by the activity-time plots in Fig. 10.4 (ref. 87).

The initial stage becomes observable with a feed containing very low concentrations of hydrocarbon (HC). Under such conditions a fresh Pt/Al_2O_3 catalyst displays an extreme crack activity, only methane being formed from heptane (Fig. 10.4 a). With a higher heptane charge (H_2 : HC = 75, Fig. 10.4 b), one finds decreasing methane production and increasing aromatization. Finally, with a ratio of H_2 : HC = 11, which is similar to reforming conditions, the normal behaviour of decreasing aromatization and cracking is observed (Fig. 10.4 c). One must conclude that a certain initial deposition of coke is necessary in order to transform the cracking catalyst into an aromatization catalyst. Hence, for reforming catalysts, one should distinguish between beneficial initial and detrimental long-term coking.

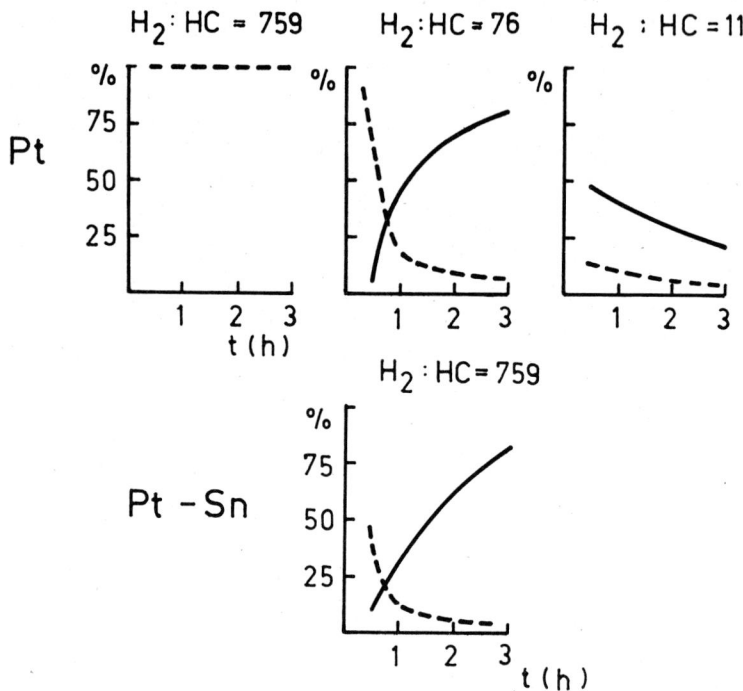

Fig. 10.4. Products of n-heptane conversion (%) as a function of time on stream (t) with feeds containing different H_2/hydrocarbon ratios on Pt/Al_2O_3 and Pt-Sn/Al_2O_3 (0.55 wt.% Pt, 0.20 wt.% Sn). Reaction temperature 500°C. Products: (———) toluene, (- - -) methane, only in the case of H_2 : HC = 11 mixed with C_2-C_5 products of cracking. Data from ref. 87.

These could be a relationship between two types of coke, that from methane according to a C_1 route and that from higher hydrocarbons according a polyene route (refs. 87, 89). Tentatively one could correlate the beneficial initial coke with the C_1 route and the detrimental long-term coke with the polyene route. The C_1 route initially must produce more or less dehydrogenated carbon atoms on the platinum surface. This should inhibit the hydrogenolysis and shift the selectivity according to an ensemble effect (Section 10.5.2). An inhibition of the hydrogenolysis is also observed with bimetallic catalysts (Section 10.3.5). This is a hint that Pt can be modified in the same way by C as by second metals.

Further evidence will be given below.

The polyene route should quickly produce large carbon deposits by polymerization of C_{2+} species. The consequence is a deactivation of the catalyst, but not a selectivity shift as with the C_1 route.

In general the presence of any carbon on the platinum surface during the reaction is well established and a model of a stationary working catalyst with coexisting free and carbon-covered metal atoms has recently been proposed (refs. 5, 6).

Details of the suggested interaction between carbon and a second metal can be derived from the n-heptane conversion on Pt/Al_2O_3 and $Pt-Sn/Al_2O_3$ shown in Fig. 10.4 a and d. On Pt only methane, but on Pt-Sn, toluene is formed under the same conditions. This indicates that the selectivity shift from hydrogenolysis to dehydrocyclization, caused only by carbon on the platinum catalyst, is partly caused by tin in the bimetallic catalyst. This shows that carbon and tin exert the same moderating action on Pt. They can substitute each other to a certain degree. Such a relationship, at first assumed by Biloen et al. (ref. 90), has since been confirmed in several experiments (refs. 10, 16, 87). However, this is only one of the possible relationships between tin and carbon, connected with the selectivity. Another relationship, discussed in the next section, is connected with the transportation and location of the coke.

10.4.2 Long-term coking

Long-term coking is a crucial problem in the maintenance of the activity of catalysts, the lifetime being definitely limited by coke deposition. The lifetime is improved by bimetallic catalysts. Hence, there must be a second relationship between bimetals and carbon. The most simple correlation would be that the coke production is inhibited by the bimetals, but this is ruled out by contrary results (refs. 9, 16, 91). A more detailed discussion must take account of several questions. Where is the coke production initiated, where is the coke deposited and what is the mode of action of the second metal?

The final location of the majority of the coke is on the

alumina. This can be derived from the quantity of the coke deposits. Catalysts still active can contain 500 or more C atoms per Pt atom (ref. 16). Recently Cabrol and Oberlin (ref. 92) have directly observed by transmission electron microscopy (TEM) stacks of aromatic rings on alumina, forming porous carbon particles. One could assume that the coke is directly formed on the acidic sites of the alumina and the deactivation by coke would occur by blocking of the acidic sites (ref. 93). However, Shum et al. (ref. 94) showed that the poisoning of the metallic function by carbonaceous residues controls the long-term deactivation of reforming catalysts.

Hence the whole coking process can be divided into three stages: (a) initiation of coke precursors on the Pt, (b) transportation to the alumina, (c) final coke formation and deposition on the alumina.

The question is where or how do the bimetals interfere. Sachtler (ref. 91) recently proposed a topographical factor in the deactivation of bimetallic catalysts. The growth of harmful graphitic entities, being relatively easy on pure Pt, is hindered by Re-S barriers on the surface of Pt-Re catalysts. This model relies only on a reorganization of carbonaceous fragments on the metal; it neglects the transportation to the alumina. Considering that nearly all the fragments must be transported to the alumina, the transportation might be the dominant problem. This aspect is stressed in the drain-off model (ref. 89). Essentially, it is assumed that the adsorption of coke precursors, which are olefinic or aromatic species, is weakened on bimetallic catalysts. They become more mobile and are drained off from the metal to the alumina where they finally polymerize and are deposited as coke. The reason for the weakened adsorption could be an ensemble effect, assuming that the unsaturated coke precursors require relatively large adsorption sites.

This model is supported by the following results: (a) the adsorption of the olefins 1-hexene (ref. 89) and ethylene (ref. 64) is weakened on Pt-Sn in comparison with Pt, (b) adsorbed olefins flow from Pt onto the alumina (ref. 89) and (c) a larger portion of the Pt remains free, not covered by coke, during the reaction with bimetallic than with

monometallic catalysts. This remarkable freeing effect of
of the second metal has recently been observed in several
bimetallic systems (refs. 15, 89). Finally, it should be
mentioned that the idea that a weakened adsorption is the
reason for the observed stability of Ni-Sn and Pd-Sn systems
had already been put forward by Masai et al. (ref. 79).

Tin has a twofold beneficial effect on Pt; it accelerates
the transformation of Pt from a cracking to an aromatizing
catalyst during the initial coking, and it stabilizes the
free platinum surface area during long-term coking by the
drain-off effect.

10.5 GENERAL DISCUSSION
10.5.1 Preparation and structure

The described structures and catalytic properties always
indicate a distinct interaction between the two metals in
the bimetallic catalysts. The components must not be spatially
separated. Considering that in many bimetallic systems only
about 1% of the surface is covered with the metals, this is
an astonishing result. Special forces must be responsible
for the combined fixation of the two metals. These are
relatively well-known chemical bonds formed in the preparation
methods using anchoring (ref. 17) or decomposition of
of complexes (ref. 46), but the relationships and forces in
the most frequently used impregnations are less clear. Hence
this will be discussed briefly in terms of how and when do
the two components find each other. Using the data in Section
10.2, a somewhat tentative generalization can be proposed.
Two conditions must be fulfilled: (a) at least one component
must be mobile and (b) a special trapping mechanism must fix
both components.

These conditions are fulfilled during the impregnation.
The components are mobile in the solution and are trapped
by adsorption sites of the carrier via an ion exchange.
Complex formation in the solution may be useful, but is not
a necessary condition. A further mobile stage is related to
the formation of $\left[PtO_xCl_y\right]_s$ during the calcination. At least
this and perhaps also other chlorine-containing species can
migrate on the surface. Finally there is a stage of enhanced
mobility during the reduction process. The platinum atoms

in statu nascendi migrate and coalesce, forming small clusters. An immobilized ion of the second metal could act as a nucleation centre, thus trapping the platinum atoms. Finally this ionic second metal, being in direct contact with a hydrogen-activating metal, is itself reduced. Altogether there are three stages in the impregnation and pretreatment procedures, where the two components can "find" each other on the surface of the carrier. This is in accordance with the practical experience that bimetallic clusters are relatively easily obtainable when the steps of impregnation, calcination and reduction are carefully controlled.

10.5.2 Classification of reactions

As outlined above, many hydrocarbon conversions are influenced by bimetallic catalysts and the effects can be positive as well as negative. These reactions can be divided into two groups, selective and non-selective processes. The essentials are summarized in Table 10.2.

TABLE 10.2
Classification of reactions

	Group of selective processes	Group of non-selective processes
Bonds involved	C-H	C-H and C-C
Reactions	Hydrogenation, selective dehydrogenation	Hydrogenolysis, dehydrocyclization, dehydroisomerization, non-selective dehydrogenation
Ensemble requirement	Small ensembles	Larger ensembles
Catalysts	Monofunctional	Mono- and bifunctional
Bimetal effect	Negative	Positive, except hydrogenolysis

The reactions are classified on the basis of the bonds involved in the reactions. In the selective processes only C-H bonds are split or formed, the C-C skeleton remaining unchanged. The corresponding reactions are hydrogenations and selective dehydrogenations without C-C bond ruptures. The ensembles, necessary for these reactions, are small. The bimetal effect is inhibiting. In the non-selective processes, formation and rupture of C-H and C-C bonds are involved. These reactions correspond to dehydrocyclization, dehydroisomerization, hydrogenolysis and non-selective dehydrogenations. The necessary ensembles are relatively large, and the catalysts are often bifunctional, containing metallic and acidic sites. The bimetal effects are positive, except in hydrogenolysis.

The reason for the bimetal effects will be considered in the following chapters mainly in terms of concept of alloy formation in the supported catalysts. Therefore, known theoretical aspects of the catalysis by bulk alloys will first be briefly outlined.

The activity and selectivity of alloys can be attributed to two sets of physical phenomena: the ensemble effect and the ligand effect (refs. 91, 95). The concept of the ensemble effect is based on the assumption that many adsorption complexes require more than one surface atom. The following sequence of ensemble size requirements is often observed (ref. 95):

dehydrogenation < dehydrocyclization < hydrogenolysis

So-called facile reactions have a low, demanding reactions a high ensemble size requirement. The mechanism of the ensemble effect consists in the selective blocking of ensembles. Upon dilution of an active metal by an inactive one, the largest ensembles are blocked first. This causes a selectivity shift towards the reaction with a smaller ensemble requirement. The concept of the ligand effect is based on the assumption that the nature and strength of the bond between the adsorbate and the surface atom are influenced by an electronic interaction with neighbouring atoms.

The ensemble requirement is included in Table 10.2, indicating that the group of selective reactions requires the smallest ensembles. A detailed discussion of ensemble and ligand effects is given in the following chapters.

10.5.3 Bimetal effect in selective conversions

Hydrogenations as well as selective dehydrogenations are strongly suppressed by added second metals. The reason for this may be an influence on the hydrogen adsorption and or on the hydrocarbon adsorption. The hydrogen adsorption is diminished on bimetallic catalysts (Section 10.2.2), probably due to a ligand effect, i.e., an electronic modification of Pt. In other words, it is decreased less than is the dehydrogenation rate of cyclohexane on Pt-Sn and Pt-Pb systems. However, a good correlation has been achieved only by considering the weakly adsorbed portion of the adsorbed hydrogen (ref. 65). As is seen in Fig. 10.5, in two bimetallic systems, Pt-Pb/Al_2O_3 and Pt-Sn/Al_2O_3, the weak adsorption and the activity exhibit a nearly linear relationship. This means the second metals block the sites for weak hydrogen adsorption, with the consequence of a corresponding blocking of the dehydrogenation.

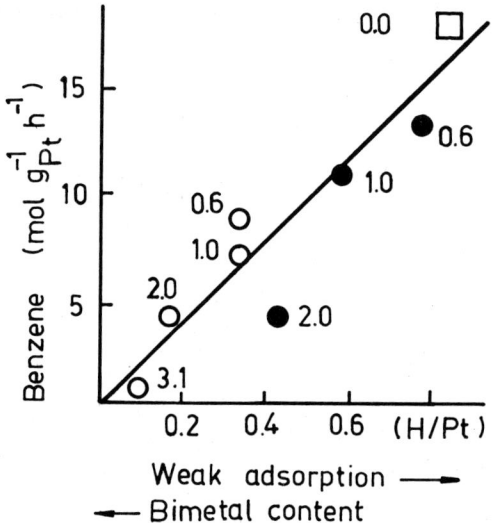

Fig. 10.5. Benzene formation from cyclohexane as a function of the weak adsorption of hydrogen on Pt-Pb/Al_2O_3 (●), Pt-Sn/Al_2O_3 (○) and Pt/Al_2O_3 (□) catalysts. The numbers indicate the atomic ratios Pb/Pt and Pt/Sn, respectively. Reaction temperature 315°C. Data from ref. 65.

However, this cannot be the only reason for the negative bimetal effect. The diminished hydrogen adsorption should be a similar handicap in all hydrogenations and dehydrogenations. Nevertheless, significant differences exist between the reactions. In Fig. 10.6 the activities with the same bimetallic catalyst are shown for various reactions taking the activity of the monometallic Pt/Al_2O_3 as 100% (refs. 20, 63, 65). Three trends are obvious: (a) the olefin hydrogenations are only weakly inhibited, (b) the benzene hydrogenation is strongly suppressed and (c) the reverse reaction, the dehydrogenation of cyclohexane, is only weakly inhibited.

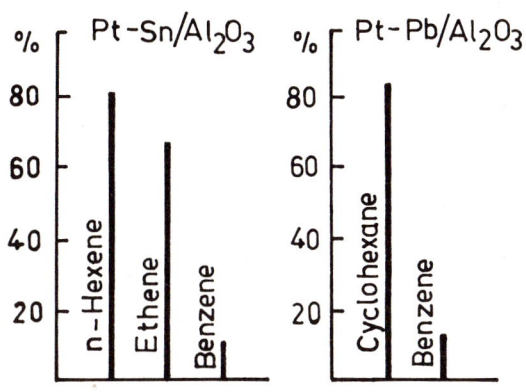

Fig. 10.6. Comparison of bimetal effects in different reactions, taking the extent of conversion on the monometallic Pt/Al_2O_3 sample as 100% in each case. Hydrogenations of 1-hexene, ethene and benzene, determined at 30°C, 60°C and 90°C respectively, dehydrogenation of cyclohexane determined at 315°C. The bimetallic catalysts contained 17 at.% of the second metal. Data for $Pt-Sn/Al_2O_3$ from ref. 63, for $Pt-Pb/Al_2O_3$ from refs. 9, 20.

This can be explained as follows. Olefin hydrogenations are facile reactions, requiring very small ensembles (ref. 95). The differences between the ethylene and 1-hexene hydrogenation are unlikely to be due to the ensemble effect and hence are probably caused by an electronic modification of Pt, i.e., by a ligand effect. The similar, only slight inhibition of the cyclohexane dehydrogenation probably has the same cause. The hydrogenation of benzene is reported to be more demanding than the olefinic hydrogenation (ref. 96). Hence one can assume that the inhibition by a ligand effect is strengthened by an ensemble effect. Evidence in support of an ensemble effect is provided by adsorption measurements for benzene and 1-hexene on $Pt-Sn/Al_2O_3$ (refs. 89, 63). It was concluded that these molecules, being adsorbed horizontally on Pt, are forced into a tilted position on bimetallic clusters. It seems conceivable that the latter effect would be more detrimental for the hydrogenation of benzene than of 1-hexene. A special geometric arrangement has also been assumed to be the reason for the selective hydrogenation with Pd_3Pb catalysts (ref. 73).

The negative bimetal effect in selective reactions can essentially be explained by a ligand effect in metal clusters, causing decreased hydrogen adsorption and modified hydrocarbon adsorption. A ligand effect has also been proposed on the basis of IR and Mössbauer experiments, which indicate an electronic interaction (refs. 8, 11, 60, 61, 63). In addition, an ensemble effect would cause extreme suppression of the benzene hydrogenation and suppression of the C = C bond hydrogenation in acetylene..

10.5.4 Bimetal effect in non-selective conversions

This group of reactions includes dehydrocyclization, dehydroisomerization, non-selective dehydrogenation and hydrogenolysis. Here the typical bimetal effects are increased aromatization, prolonged lifetime and decreased cracking.

Many bimetallic catalysts, especially the reforming catalysts, are bifunctional having metallic and acidic sites, both of which can be active in rearrangements of the C-C skeleton. Hence the question arises as to whether the metallic or the acidic sites are modified by the second metal. The answer can be derived from heptane-conversion experiments on supported and unsupported Pt and Pt-Sn catalysts, shown in Fig. 10.7 (ref. 97). The unsupported Pt-Sn alloy exhibits the same bimetal effects as the alumina-supported, acidic Pt-Sn catalysts. This clearly indicates that the active sites are Pt-Sn alloys or Pt-Sn clusters, respectively, present in the unsupported as well as in the supported samples. This is in accordance with temperature-programmed conversion profiles of n-heptane on unsupported Pt-Sn alloys, indicating increased desorption of benzene and decreased desorption of cracking products (ref. 7). An inhibition of the hydrogenolysis of hexane has been observed with another unsupported system with Pt-Sn films (ref. 98).

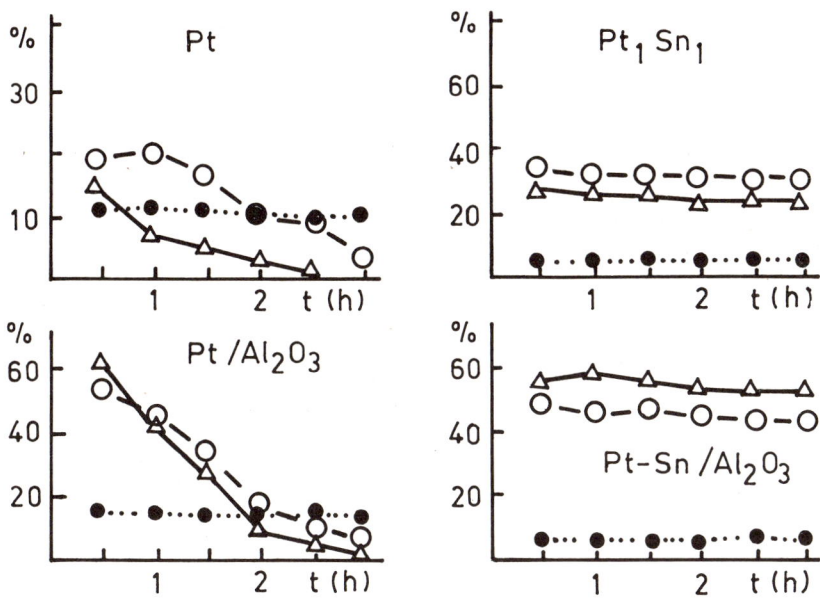

Fig. 10.7. Conversion of n-hexane on supported and unsupported Pt and Pt-Sn catalysts as a function of time on stream (t). (o - - - o) conversion, (△——△) selectivity for benzene, (o·····o) selectivity for C_1 - C_3 cracking products. From ref. 97.

To summarize, the quoted experiments unambiguosly demonstrate that alloying causes the bimetal effect. Alloys or bimetallic clusters are also active sites for this group of reactions. This statement raises other questions. What are the mechanisms of the bimetal effects, relating the structure of bimetallic clusters with the inhibition or promotion of specific reaction steps? This concerns the mechanisms of the three effects: (a) inhibition of hydrogenolysis, (b) promotion of aromatization and (c) prolonged lifetime.

Since Silfelt (ref. 99) observed the dramatic inhibition of C-C bond splitting, direct relationships between alloying and hydrogenolysis have become well-known. It has been suggested that the necessary intermediates in the hydrogenolysis of C_{3+} alkanes are α, α, γ triadsorbed species (ref. 5). Large ensembles are required for adsorption, and upon alloying with an active metal such ensembles are easily blocked. Consequently the hydrogenolysis is severely inhibited. Thus the observed bimetal effect in hydrogenolysis can be fully explained by the well-known ensemble effect in bulk alloys.

The same mechanism can help to explain the second bimetal effect, i.e., the improved aromatization. Cracking or hydrogenolysis is one of the main competitive reactions besides dehydrocyclization, especially at high temperatures of about $500°C$. A suppression of the demanding hydrogenolysis would shift the selectivity towards the more facile aromatization. Hence an increased formation of aromatics is an indirect consequence of the ensemble effect on the hydrogenolysis.

The third bimetal effect is the prolonged lifetime. This is probably connected with the drain-off effect of bimetals, proposed in Section 10.4. The coke, deposited on Pt, limits the lifetime. But the portion of Pt surface area, remaining free of blocking coke, is larger in bimetallic than in monometallic catalysts. This is explainable by a promoted drain-off of coke precursors, due to weakened adsorption on bimetallic clusters. Again an ensemble effect could cause the weakening.

On the whole, all the three bimetal effects can be explained by an ensemble effect. If this is the true explanation, then a further condition of the ensemble effect should be fulfilled. The effect should be independent of the chemical nature of the second metal, provided only that the diluting material is inactive. Hence a large variety of second metals should cause similar bimetal effects. This is indeed observed. Bursian et al. (ref. 15) found that the selectivity and stability of platinum catalysts is improved by an astonishing number of elements: Cu, Ag; Zn, Cd; Ga, In, Tl; Ge, Sn, Pb and Bi. A more detailed picture is shown in Fig. 10.8. The formation of benzene from cyclohexane under non-selective conditions is plotted as a function of the content of second metals. A maximum is typically obtained with additives of Group 4, Sn and Pb, as well as with additives of Group 5, Sb and Bi. Atomic ratios from 1:1 to 2:1 prove to be the best. In detail, the maxima, obtained with the tin- and lead-modified catalysts are somewhat broader than those with the antimony- and bismuth-modified ones. The reasons could be that the dominant but non-specific ensemble effect is slightly modified by element specific ligand effects.

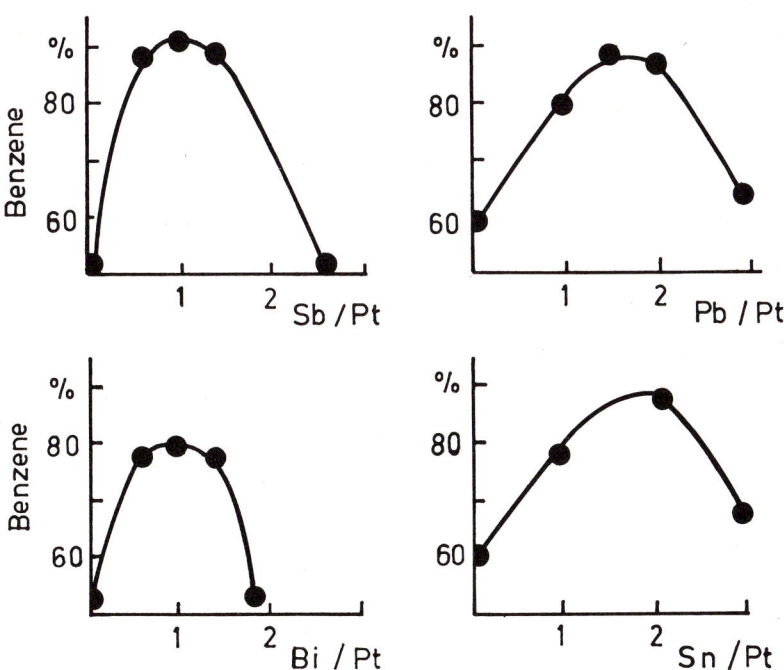

Fig. 10.8. Dehydrogenation of cyclohexane under non-selective conditions. Benzene formation as a function of bimetal content (atomic ratio) in the supported systems Pt-Sb, Pt-Bi, Pt-Sn and Pt-Pb. Data from refs. 48, 9.

The uniformity of the effects, caused by so many different second metals, is a further argument in favour of the concept of bimetallic clusters, exhibiting dominantly ensemble effects.

Finally this mechanism can be compared with other mechanisms proposed in the literature. The model is in good accord with the conclusions of Dautzenberg et al. (ref. 7). They proposed an ensemble effect by alloys, and their speculative deactivation mechanism now has been substantiated. Coq and Figueras (ref. 10) stated that the main role of tin is the dilution of Pt. Contrary concepts have been put forward by Burch and Garla (ref. 12), Adkins and Davis (ref. 13) and by Sexton et al. (ref. 14). They assume an electronic interaction between Pt and Sn^{II} to be the main reason for the bimetal effects. However, no explanation as to how this electronic interaction controls the different bimetal effects was proposed. Their main objection against the alloy concept is that the majority of tin is in the Sn^{II} and not in the Sn^0 state.

Hence the crucial question is which of the two tin species, Sn^0 or Sn^{II}, modifies the active Pt. In order to answer this the main evidence in support of the alloy concept is briefly summarized:

(a) Analogous catalytic effects are observed with bulk alloys and with supported bimetallic samples

(b) Very small amounts of the second metal in an alloy are sufficient to exhibit a strong ensemble effect, especially in hydrogenolysis (ref. 99)

(c) The neighbourhood of hydrogen-activating noble metals, distinctly promotes the reduction of oxides, as found in solid-state chemistry

These arguments strongly support the concept that platinum alloys are the active sites and not metallic Pt, electronically modified by an oxide of the second metal. This does not exclude that Sn^{II} modifies the acidic sites of the alumina, but such a modification obviously does not play the dominant role in the reactions discussed here.

10.6 CONCLUSIONS

10.6.1 Preparation and structure

Substantial progress has been achieved in understanding the different steps in the preparation and pretreatment, especially of Pt/Al_2O_3 catalysts. The impregnation of gamma Al_2O_3 with aqueous solutions containing $[PtCl_6]^{2-}$ and other anions is determined by competitive adsorption and ion exchange with surface hydroxyl groups. Calcination leads to the formation of a migrating surface complex, $[Pt^{IV}O_xCl_y]_s$. After reduction, Pt and other highly dispersed metals are present in a slightly electron-deficient state. Treatment with oxygen can cause a redispersion of Pt by formation of $[Pt^{IV}O_xCl_y]_s$ and, at higher temperatures, a sintering of Pt by thermal decomposition of this complex.

Various successful methods have been developed for the preparation of bimetallic catalysts by consecutive or simultaneous impregnation, by anchoring or by decomposition of complexes. The resulting active catalysts always contain both the metals in close proximity. In many cases, there is evidence for alloy formation. However, as observed with $Pt-Sn/Al_2O_3$, the majority of the second metal can remain in an oxidized state.

The adsorption properties are changed by the second metals in a consistent manner, the hydrogen adsorption being decreased and the oxygen adsorption increased. This is in accord with the concept of alloy formation. The metal dispersion is often higher in bimetallic than in monometallic catalysts, at least in $Pt-Sn/Al_2O_3$ samples.

10.6.2 Coking

Hydrocarbon conversions at higher temperatures are accompanied by severe self-poisoning, caused by coke deposition. It was shown recently that a fresh platinum catalyst initially exclusively exhibits total fragmentation to methane. Thereupon the selectivity shifts towards aromatization, obviously due to a modification of Pt by deposited carbon. A similar shift is observable when Pt is modified by tin. This indicates that Sn can substitute initially deposited carbon. In both cases the shift is explainable as an ensemble effect. The long-term coking

mainly causes a deactivation of the catalyst. This type
of coking displays a further interaction with second metals.
During hydrocarbon conversions a larger portion of the Pt
remains free, not covered by coke. This can be explained by
a mechanism in which the coke precursors are more readily
drained-off from bimetallic clusters to the alumina than
from platinum clusters.

10.6.3 Catalysis

The catalytic effects of very many second metals exhibit
a surprising uniformity; negative under mild, positive under
severe reaction conditions. These bimetal effects can be
explained in terms of bimetallic clusters as the active
sites. This concept is in accord with structural
investigations indicating alloys in many supported systems,
and with catalytic investigations of unsupported alloys.

Selective conversions, i.e., hydrogenations and
dehydrogenations without accompanying C-C bond scissions,
are only inhibited by second metals. This could be due to
inhibition of the hydrogen adsorption, probably caused by
a ligand effect. A correlation between the amount of weakly
adsorbed hydrogen and the rate of cyclohexane dehydrogenation
has been found. However, the same bimetallic catalysts
inhibit benzene hydrogenation much more strongly than olefin
hydrogenation. This is explained by an additional steric
effect of the second metals, in which the adsorbed molecules
are tilted on bimetallic clusters but horizontal on
monometallic clusters.

Non-selective conversions, i.e., dehydrocyclization,
dehydroisomerization and non-selective dehydrogenations,
can be classified by the accompanying reactions of C-C bond
rupture and of self-poisoning coke deposition. The typical
bimetal effects are inhibition or hydrogenolysis, promotion
of aromatization and maintenance of activity. The role of
the second metals in these effects is proposed to be as
follows. They block large ensembles, necessary for
hydrogenolytic splitting. Thus, not only is hydrogenolysis
inhibited, but also the aromatization is increased, due to
a shift of selectivity. The prolonged lifetime is probably
due to a freeing of the platinum surface in bimetallic

catalysts, caused by an easier drain-off of coke precursors from Pt to Al_2O_3.

Supported noble metal catalysts can generally be improved by the addition of inert metals. The bimetal effects displayed are consistent with the concept of alloy formation, causing predominantly ensemble effects. This could provide guidelines for the synthesis of new effective bimetallic catalysts.

REFERENCES

1. H.E. Kluksdahl, U.S. Pat. 3,415,737 (1968).
2. F.M. Dautzenberg, Ger. Offen. 2,121,765 (1971) and 2,153,891 (1972).
3. J.E. Weisang and P. Engelhard, U.S. Pat. 3,700,588 (1972).
4. G.C. Bond, Platinum Metals Rev., 29 (1985) 28.
5. S.M. Davis, F. Zaera and G.A. Somorjai, J. Catal., 85 (1984) 206.
6. S.M. Davis, F. Zaera and G.A. Somorjai, J. Catal., 77(1982) 439.
7. F.M. Dautzenberg, J.N. Helle, P. Biloen and W.M.H. Sachtler, J. Catal., 63 (1980) 119.
8. H. Berndt, H. Mehner, J. Völter and W. Meisel, Z. Anorg. Allg. Chem., 429 (1977) 47.
9. J. Völter, G. Lietz, M. Uhlemann and M. Hermann, J. Catal., 68 (1981) 42.
10. B. Coq and F. Figueras, J. Catal., 85 (1984) 197.
11. R. Bacaud, P. Bussiére anf F. Figueras, J. Catal., 69 (1981) 399.
12. R. Burch and L.C. Garla, J. Catal., 71 (1981) 360.
13. S.R. Adkins and B.H. Davis, J. Catal., 89 (1984) 371.
14. B.A. Sexton, A.E. Hughes and K. Foger, J. Catal., 88 (1984) 466.
15. N.R. Bursian, B.B. Zharkev, S.B. Kogan, G.A. Lastovkin and M.M. Potkletnova, Proc. 8th Congr. Catal., Berlin 1984, Vol. II, Verlag Chemie, Weinheim, 1984, p. 481.
16. J. Völter and U. Kürschner, Appl. Catal., 8 (1983) 167.
17. Yu. I. Yermakov, in T. Seiyama and T. Tanabe (Eds.), New Horizons in Catalysis, Proc. 7th Int. Congr. Catal., Tokyo 1980, Elsevier, Amsterdam, 1981, Part A, p. 57.
18. E. Santacesaria, S. Carra and I. Adami, Ind. Eng. Chem. Prod. Res. Dev., 16 (1977) 41.
19. G. Lietz, H. Lieske, H. Spindler, W. Hanke and J. Völter, J. Catal., 81 (1983) 17.
20. G. Lietz and J. Völter, unpublished results.
21. J. Völter et al., DDR Pat. 125758 (1976).
22. S.S. Kulkarni, G.R. Mauze and J.A. Schwarz, J. Catal., 69 (1980) 445.
23. S.R. de Miguel, O.A. Scelza, A.A. Castro and J.M. Parera, Appl. Catal., 9 (1984) 309.
24. N. Wagstaff and R. Prins, J. Catal., 59 (1979) 434 and 67 (1981) 255.
25. B.D. McNicol, J. Catal., 46 (1977) 438.
26. H.C. Yao, M. Sieg and H.K. Plummer, J. Catal., 59 (1979) 365.

27   R.W. Joyner, J. Chem. Soc. Faraday Trans. I, 76 (1980) 357.
28   P. Birke, S. Engels, K. Becker and H.-D. Neubauer, Chem. Tech., 31 (1979) 473.
29   H. Lieske, G. Lietz, H. Spindler and J. Völter, J. Catal., 81 (1983) 8.
30   P. Lagarde, T. Murata and G. Vlaic, J. Catal., 84 (1983) 333.
31   T. Huizinga, J. van Grondelle and R. Prins, Appl. Catal., 10 (1983) 199.
32   H. Lieske, G. Lietz, W. Hanke and J. Völter, Z. Anorg. Allg. Chem., 527 (1985) 135.
33   J. Zielinski, J. Catal., 76 (1982) 157.
34   Hoang Dang Lanh, Ho Si Thoang and J. Völter, in preparation.
35   J. Völter, H. Lieske, M. Uhlemann and G. Lietz, Z. Anorg. Allg. Chem., 452 (1979) 77.
36   G.V. Antoshin, E.S. Shpiro, O.P. Tkachenko, S.B. Nikishenko, M.A. Ryashentseva, V.I. Avaev and Kh.M. Minachev, in T. Seiyama and T. Tanabe (Eds.), New Horizons in Catalysis, Proc. 7th Int. Congr. Catal., Tokyo 1980, Elsevier, Amsterdam, 1981, Part A, p. 302.
37   R. Bouwman and P. Biloen, Proc. 7th Int. Vac. Congr., Vienna, 1977, p. 1129.
38   A.N. Mansour, J.W. Cook Jr., D.A. Sayers and J.R. Katzer, J. Catal., 89 (1984) 462.
39   K.W. McHenry, R.J. Bertolacini, H.M. Brennan, J.L. Wilson and H.S. Seelig, Act. Deux. Congr. Int. Catal., Paris, 1960, Technip, Paris, 1961, Vol. II, p. 2293.
40   M.J.P. Botman, Li-Qin She, Jia-Yu Zhang, W.L. Driessen and V. Ponec, submitted for publication.
41   M.A. Vannice and C. Sudhakar, J. Phys. Chem., 88 (1984) 2429.
42   D.E. Resasco and G.L. Haller, J. Catal., 82 (1983) 279.
43   H. Lieske and J. Völter, Z. Anorg. Allg. Chem., 512 (1984) 65.
44   W.I. Callender and J.J. Miller, Proc. 8th Int. Congr. Catal., Berlin, 1984, Verlag Chemie, Weinheim, 1984, Vol. II, p. 491.
45   H. Lieske and J. Völter, J. Phys. Chem., 89 (1985) 1841.
46   R.G. Nuzzo, L.H. Dubois, N.E. Bowles and M.A. Trecoske, J. Catal., 85 (1984) 267.
47   Yu.N. Usov, L.G. Zubanova and N.I. Kushinova, Neftechima, 17 (1977) 69.
48   Ho Si Thoang, Hoang Dang Lanh and J. Völter, Proc. 8th Int. Congr. Catal., Berlin, 1984, Verlag Chemie, Weinheim, 1984, Vol. II, p. 509.
49   B.H. Davis, G.A. Westfall, J. Watkins and J. Pezzanite Jr., J. Catal., 42 (1976) 247.
50   A. Campero, M. Ruiz and R. Gómez, React. Kinet. Catal. Lett., 5 (1976) 177.
51   A.C. Muller, P.A. Engelhard and J.E. Weisang, J. Catal., 56 (1979) 65.
52   J. Völter and H. Berndt, unpublished additional results to ref. 8.
53   H. Lieske and J. Völter, J. Catal., 90 (1984) 96.
54   F.G. Gault, O. Zahraa, J.M. Datigues, G. Maire, M. Peyrot, J.E. Weisang and P.A. Engelhardt, in T. Seiyama and T. Tanabe (Eds.), New Horizons in Catalysis, Proc. 7th

Int. Congr. Catal., Tokyo, 1980, Elsevier, Amsterdam, 1981, Part A, p. 199.
55 R. Bacaud, P. Bussière, F. Figueras and J.P. Mathieu, in B. Delmon, P. Jacobs and G. Poncelet (Eds.), Preparation of Catalysts I, Elsevier, Amsterdam, 1976, p. 509.
56 R. Burch, J. Catal., 71 (1981) 348.
57 V.I. Kuznetsov, E.N. Yurchenko, A.S. Belyi, E.V. Zatolokina, M.A. Smolikov and V.K. Duplyakin, Reak. Kinet. Catal. Lett., 21 (1982) 419.
58 N.A. Pakhomov, R.A. Buynov, E.N. Jurchenko, A.P. Chernyshev, G.R. Kotelnikov, E.M. Moroz, N.A. Zaitseva, V.A. Patanov, Kinet. Katal., 22 (1981) 488.
59 M.S. Kharson, G.B. Kadinov and A.N. Palazov, React. Kinet. Catal. Lett., 10 (1979) 267.
60 A. Palazov, Ch. Bonev, G. Kadinov, D. Shopov, G. Lietz and J. Völter, J. Catal., 71 (1981) 1.
61 A. Palazov, Ch. Bonev, G. Kadinov and D. Shopov, J. Catal., 83 (1983) 253.
62 A.G.T.M. Bastein, F.J.C.M. Toolenaar and V. Ponec, J. Catal., 90 (1984) 88.
63 A. Palazov, Ch. Bonev, G. Lietz and J. Völter, submitted for publication.
64 H. Verbeek and W.M.H. Sachtler, J. Catal., 42 (1976) 257.
65 J. Völter, H. Lieske and G. Lietz, React. Kinet. Catal. Lett., 16 (1981) 87.
66 B.I. Zaikovskii, Yu.A. Ryndin, V.I. Kovalohuk, L.M. Plyasova, B.N. Kuznetsov and Yu.I. Yermakov, Kinet. Katal., 22 (1981) 443.
67 V.I. Zaikovskii, V.I. Kovalchuk, Yu.A. Ryndin, L.M. Plyasova, B.N. Kuznetzov and Yu.I. Yermakov, React. Kin. Catal. Lett., 14 (1980) 99.
68 G.W. Qiao, J. Zhou and K.H. Kuo, Proc. 8th Int. Congr. Catal., Berlin, 1984, Verlag Chemie, Weinheim, 1984, Vol. III, p. 93.
69 H. Lieske and J. Völter, React. Kinet. Catal. Lett., 23 (1983) 403.
70 S.V. Bogdonov, A.P. Shepelin, V.I. Kovalcuk, E.M. Moroz, P.A. Zhdan, Yu.A. Ryndin, B.N. Kuznetsov and Yu.I. Yermakov, React. Kinet. Catal. Lett., 15 (1980) 233.
71 Hoang Dang Lanh, Ho Si Thoang, H. Lieske and J. Völter, Appl. Catal., 11 (1984) 195.
72 St. Karski and T. Paryjszak, React. Kinet. Catal. Lett., 15 (1980) 419.
73 W. Palczewska, A. Jablonski and Z. Kaszkur, J. Mol. Catal., 25 (1984) 307.
74 H. Shinohara, Appl. Catal., 10 (1984) 27.
75 V.I. Zaikovskii, E.M. Chalganov and Yu.E. Yermakov, React. Kinet. Catal. Lett., 16 (1981) 43.
76 Yu.A. Ryndin, P.A. Chemyshev, V.I. Zaikovskii, E.N. Yurchenko and Yu.I. Yermakov, React. Kinet. Catal. Lett., 21 (1982) 125.
77 L. Lin, J. Tsang, J. Wu and P. Chiang, in T. Seiyama and T. Tanabe (Eds.), New Horizons in Catalysis, Proc. 7th Int. Congr. Catal., Tokyo, 1980, Elsevier, Amsterdam, 1981, Part B, p. 1466.
78 A.S. Belyj, V.K. Dupljakin, Yu.A. Ryndin, W.S. Alfeev and E.G. Kryvzova, Proc. All-Unions Conf. Catal., Novosibirsk, 1978, Institute of Catalysis, Novosibirsk,

1978, p. 187.
79 M. Masai, K. Honda, A. Kubota, S. Ohnaka, Y. Nishikawa, K. Nakahara, K. Kishi and S. Ikeda, J. Catal., 50 (1977) 419.
80 Yu.I. Yermakov, B.N. Kuznetsov and E.M. Chalganov, React. Kinet. Catal. Lett., 14 (1980) 37.
81 Yu.A. Savostin, V.V. Chents, N.G. Kozeknikova, N.M. Zaidman, Chim. Technol. Topliv Masel, 1 (1979) 18.
82 N.A. Pakhomov, R.A. Buyanov, E.M. Moroz, E.N. Yurchenko, A.P. Chernyshev, N.A. Zaitseva and G.R. Kotelnikov, React. Kinet. Catal. Lett., 14 (1980) 329.
83 J.K.A. Clarke, I. Manninger and T. Baird, J. Catal., 54 (1978) 230.
84 H.J. Leuchs, Thesis, Technische Hochschule Aachen, 1978.
85 Yu.A. Ryndin, S. Göböls, V.I. Zaikovskii, J. Margitfalvi and Yu.I. Yermakov, React. Kinet. Catal. Lett., 21 (1982) 91.
86 U. Kürschner and J. Völter, in preparation.
87 G. Lietz, J. Völter, M. Dobrovolszky and Z. Paál, Appl. Catal., 13 (1984) 77.
88 A. Sárkány, H. Lieske, T. Szilagyi and L. Toth, Proc. 8th Int. Congr. Catal., Berlin, 1984, Verlag Chemie, Weinheim, 1984, Vol. II, p. 613.
89 H. Lieske, A. Sárkány and J. Völter, submitted for publication in Appl. Catal.
90 P. Biloen, J.N. Helle, V. Verbeek, F.M. Dautzenberg and W.M.H. Sachtler, J. Catal., 63 (1980) 112.
91 W.M.H. Sachtler, J. Mol. Catal., 25 (1984) 1.
92 R.A. Cabrol and A. Oberlin, J. Catal., 89 (1984) 256.
93 B.C. Gates, J.R. Katzer and G.C.A. Schuit, Chemistry of Catalytic Processes, McGraw-Hill, New York, 1979.
94 V.K. Shum. J.B. Butt and W.M.H. Sachtler, Appl. Catal., 11 (1984) 151.
95 V. Ponec, Adv. Catal., 32 (1983) 149.
96 S. Puddu and V. Ponec, Rec. Trav. Chim., 95 (1976) 255.
97 Hoang Dang Lanh, G. Lietz, Ho Si Thoang and J. Völter, React. Kinet. Catal. Lett., 21 (1982) 429.
98 Z. Karpinski and K.A. Clarke, J. Chem. Soc. Faraday Trans. I, 1975, 893.
99 J.H. Sinfelt, Acc. Chem. Res. 10 (1977) 15.

Chapter 11

SUPPORTED BIMETALLIC CATALYSTS PREPARED BY CONTROLLED SURFACE REACTIONS

J. MARGITFALVI, S. SZABÓ and F. NAGY
Central Research Institute for Chemistry of the Hungarian Academy of Sciences, P.O.Box 17, 1525-Budapest (Hungary)

11.1 INTRODUCTION

Since the introduction of the alumina-supported rhenium-platinum reforming catalyst in 1969 (ref. 1) bimetallic catalysts have attracted great scientific and industrial interest. In addition to Chevron's Re-Pt/Al_2O_3 catalysts other bimetallic systems such as Ir-Pt/Al_2O_3 (ref. 2) and Sn-Pt/Al_2O_3 (ref. 3) became of importance in reforming and aromatization technologies. The advantage of these bimetallic catalysts is their continued high activity and the suppression of undesired side reactions.

With respect to the properties of supported bimetallic catalysts, the most disputed questions are as follows: (i) is there an intimate contact between the two metals or are do they form separate metallic phases, (ii) does the second metal exist in metallic or in ionic forms.

In the preparation of supported bimetallic catalysts one of the most crucial problems is how to create an intimate contact between the two metals and how to control surface reactions responsible for the formation of bimetallic surface entities. For this type of catalysts, the techniques applied for the preparation of unsupported bimetallic systems or alloys cannot generally be used. Co- or subsequent impregnation techniques are often used, followed by high-temperature decomposition of the precursor compounds in oxidative or reductive atmospheres. During these high-temperature processes, the formation of bimetallic surface species takes place only by chance. Due to the lack of knowledge about the chemistry of surface reactions involved in the formation of bimetallic surface entities, it is not known how to control surface reactions in order to obtain such species. The surface reactions responsible for the formation of bimetallic species are very complex, and their control requires different approaches, new techniques and even a new mode of thinking.

New chemical technologies need new catalysts with "tailored" catalytic properties. However, this means new scientific approaches to the preparation of supported mono- and bimetallic catalysts, based on the accumulated knowledge of surface reactions involved in catalyst preparation.

In our approach emphasis is placed on the mode of controlling the reactions mentioned above. The chemical nature of the alumina--supported platinum catalyst is used as a driving force to obtain surface entities with direct metal-metal interactions. This approach is based on the reactivity of hydrogen preadsorbed on metal surfaces. In parallel with the creation of an intimate contact between two metals, we have tried to find ways of introducing the second metal without the formation of a bimetallic species by anchoring the second metal to the support via a surface reaction in which either OH or other reactive surface groups are involved.

The aim of this paper is to demonstrate that the principle of controlled surface reactions (CSRs) is a very powerful approach, and supported bimetallic catalysts prepared by these methods possess unique catalytic properties, in hydrocarbon conversion and in hydrogenation of different organic compounds.

11.2 THE MODE OF CONTROL OF SURFACE REACTIONS INVOLVED IN THE PREPARATION OF SUPPORTED BIMETALLIC CATALYSTS

11.2.1 General aspects

In the course of preparation of supported bimetallic catalysts by conventional methods the formation of bimetallic surface species can take place during the high-temperature decomposition of the precursor compounds, adsorbed on the inorganic supports. In general, little is known about the surface reactions involved. This means that surface reactions responsible for the formation of bimetallic surface species cannot be considered to be controlled.

One of the challenges facing us is the elucidation of the elementary processes involved, both in the preparation of catalysts and in the course of catalytic reactions, at the molecular level. The next step would be to learn how to control these processes.

The use of different organometallic compounds in the preparation of different mono- and bimetallic catalysts is of great scientific and practical importance. These methods of catalyst preparation have been reviewed (refs. 4-6). For the preparation of supported bimetallic catalysts two approaches appeared to be very promising:

(i) via anchored ionic surface complexes (refs. 4,5),
(ii) the use of bimetallic complexes or clusters (including carbonyl compounds) (ref. 6).

11.2.2 Preparation via anchored ionic surface complexes

The key step in this approach is based on the reaction between the surface hydroxyl groups of the support and an organometallic compound containing highly reactive ligands (allyl, alkoxy, acetate, alkyl, etc.) (reaction 11.1), followed by treatment of the primary surface species with hydrogen (reaction 11.2) (ref. 4)

$$\underset{-OH}{\overset{-OH}{\rule{0pt}{0pt}}} + MR_x \longrightarrow (-O)_n MR_{x-n} + nRH \qquad (11.1)$$

$$(-O)_n MR_{x-n} \xrightarrow[H_2]{\Delta} (-O)_n M + (x-n)RH \qquad (11.2)$$

(I)

where M = Re, Mo, W. The surface complexes (I) are used to anchor a second metal (M') (usually platinum from corresponding π-allyl complexes). After subsequent treatment with hydrogen, the formation of surface species (II) is assumed (ref. 7)

$$\underset{-O}{\overset{-O}{\rule{0pt}{0pt}}}\!\!> M\, Pt_n \qquad (II)$$

although the mechanism of its formation was not elucidated. In these bimetallic catalysts one of the metals (M) is in its higher oxidation state.

With respect to the surface reactions involved, it can be suggested that only reactions 11.1 and 11.2 can be considered to be controlled, i.e., the chemistry of the key step, namely the formation of the bimetallic surface entities (surface species II) is still unknown.

11.2.3 The use of bimetallic complexes or clusters

In this approach different bimetallic carbonyl clusters are adsorbed on or caused to react with dehydroxylated silica or alumina (ref. 6), followed by the thermal decomposition of the primary formed subcarbonyls in reductive or inert atmospheres. A study of the surface species formed from these carbonyls revealed that a process of cluster decomposition, i.e., formation of separate metallic phases, occurred after thermal treatment (ref. 8).

A variety of tin-platinum bimetallic complexes such as $H_2Pt_3Sn_8Cl_{20}$, $[PtCl_2/SnCl_3/_2][N/C_2H_5/_4]_2$, have also been used to prepare supported Sn-Pt catalysts (refs. 9,10). In this case, at high temperatures in an hydrogen atmosphere, the formation of monometallic surface species, i.e., of metallic platinum and mostly Sn^{2+} bound to the support, was observed. This fact as well as the lack of formation of a Sn-Pt alloy (refs. 3,11) indicated decomposition of the parent bimetallic compounds.

11.3 NEW APPROACHES

In the light of these experimental data it can be concluded that in order to obtain direct metal-metal interaction in a supported bimetallic catalyst new approaches should be used. The main aim is to find the driving force which quarantees the control of surface reactions involved in the formation of bimetallic surface entities.

In our approach a supported monometallic catalyst is used to prepare bimetallic catalysts and hydrogen preadsorbed on the metals provides the driving force for the primary interaction, i.e., for the exclusive formation of bimetallic surface entities, with metal--metal interaction. Two different types of reactions were applied for the control of surface reactions resulting in species with direct metal-metal interactions. Both are based on the reactivity of hydrogen adsorbed on metals.

11.3.1   Electrochemical approach

(i) Adsorption of metals on metals. The basic principle of this approach has been described previously (refs. 12-20). Electrochemical adsorption of a metal on another metal surface takes place at a more positive potential than the Nernst potential of the adsorbing metal. This potential difference is a measure of the difference in free enthalpy, (ΔG^o), of the metal atom adsorbed on a foreign metal surface and in its own crystal lattice. This difference varies from a few joules to about 300 kJ as shown in Table 11.1.

TABLE 11.1
Standard free enthalpy change, ΔG^o, for adsorption of metals on platinum and the site requirement, S, of these adsorbed metal atoms

	Adsorbed metal								
	Cu	Ag	Pd	Bi	Sn	Pb	Re	Cd	Au
$-\Delta G^o$ kJ/mol	65; 82	35; 55	~50	188	330	166; 198	~100	216	1.5
S	1.0; 1.2	0.75; 1.12	1.5	2.0; 3.0	2.0; 2.2	1.9; 2.3	2.3	–	1.2; 2.5
References	21, 32, 33	22, 24, 30	23	24, 34, 35	25, 26	26, 31	27	28	29

It has been demonstrated that metal atoms adsorb on platinum at the same active sites as does hydrogen, consequently they may inhibit hydrogen adsorption. The number of hydrogen adsorption sites occupied by one adsorbed metal atom can be determined (Table 11.1).

In the adsorption of metals on other metal surfaces the adsorbed layers are mostly formed in aqueous solution, consequently the discharge of metal ions precedes the adsorption. It has been shown that the adsorbed metal layers prepared by this technique comprise not only discharged metal atoms, but also metal ions (ref. 30). Consequently, the adsorbed layer should also contain amounts of anions (ref. 36). In suitable circumstances the adsorbed layer containing partially discharged metal atoms can be reduced further (ref. 32).

In summary, the electrochemical process of metal adsorption can be written in the following general form

$$Me^{Z+} + \lambda e \longrightarrow Me_{ads}^{Z-\lambda} \qquad (11.3)$$

where Me^{Z+} is the depositing ion and λ is the mean partial charge-transfer coefficient (ref. 14). λ indicates the extent to which the average adsorbed metal atom is reduced. In the case of total reduction, λ is equal to the ionic charge.

(ii) <u>Metal adsorption via ionization of adsorbed hydrogen.</u> As it has already mentioned that the electrochemical adsorption of a metal is practically a charge-transfer process. An adsorbed monolayer of metal can be formed only in the presence of some source of electrons. In electrochemistry, as usual, the source of electrons is a system of polarization.

Naturally, the source of electrons can be any reducing agent. For instance, in the case of metals, (e.g., platinum), which adsorb hydrogen readily the adsorbed hydrogen can be considered as a source of electrons, according to the following equation:

$$H_{ads} \longrightarrow H^+ + e^- \qquad (11.4)$$

Combining reactions 11.3 and 11.4 and assuming the complete discharge of the metal ion, for the sake of simplicity, the spontaneous metal adsorption via ionization of preadsorbed hydrogen can be written as:

$$Me^{Z+} + z\, H_{ads} \longrightarrow H_{ads} + z\, H^+ \qquad (11.5)$$

Experimental results have led to the conclusion that metal adsorption via the ionization of preadsorbed hydrogen results in an adsorbed metal layer of the same character as that obtained by electrochemical methods (ref. 34).

In addition to the formation of an adsorbed metal layer, bulk deposition can also be expected if the Nernst potential of the depositing ions (in reaction 11.3) is more positive than the reversible hydrogen potential in the same solution. Experimental results have verified this prediction, adsorption of Cu, Ag, Pd, Bi, Au onto Pt via the ionization of preadsorbed hydrogen takes place in two steps (refs. 21,23,24,29):

First step $\quad H_{ads} \longrightarrow H^+ + e \qquad (11.6)$

$\qquad\qquad\quad Me^{Z+} + ze \longrightarrow Me_{bulk} \qquad (11.7)$

Second step $\quad Me_{bulk} \longrightarrow Me^{Z+} + ze \qquad (11.8)$

$\qquad\qquad\quad Me^{Z+} + ze \longrightarrow Me_{ads} \qquad (11.9)$

In the case of metals which have several oxidation states, (e.g. Cu) some part of the adsorbed hydrogen may take part in a side reaction that does not result in adsorbed metal atoms (ref. 33). Experimental results indicate that, under non-equilibrium conditions, bulk deposition should always be considered if it is thermodynamically possible (ref. 23).

In the presence of metal ions under open circuit conditions, when the potential rises to a value higher than the Nernst potential the metal adsorption can be considered to take place as shown in Scheme 11.1 (ref. 23)

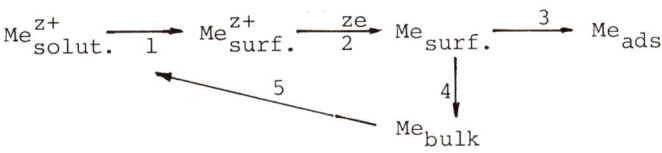

Scheme 11.1

where in step 1 the metal ion diffuses to the surface, in 2 a charge transfer occurs, in 3 metal adsorption, in 4 bulk deposition and in 5 resolution of the bulk metal. If step 3 is very slow, then the product of metal adsorption can only be bulk phase metal and not an adsorbed metal overlayer.

(iii) <u>Application of electrochemical metal adsorption to the preparation of supported bimetallic surface species.</u>
The electrochemical method is not suitable to prepare an adsorbed layer on the surface of small metal particles randomly distributed at an insulator surface, since electrical contact cannot be made with each metal particle. However, the ionization of preadsorbed hydrogen makes possible the formation of adsorbed layers of finely dispersed supported metal particles under suitable experimental conditions. This will result in the formation of supported bimetallic surface species with direct metal-metal interaction. Details on the preparation technique and its use to prepare different supported bimetallic catalysts from alumina--supported platinum catalysts will be given in Section 11.4.1.

11.3.2 Organometallic approach

As demonstrated in Section 11.2, different types of organometallic compounds can be used to prepare supported mono- and bimetallic catalysts. However, in this method the reactivity of the surface hydroxyl groups rather than that of the hydrogen adsorbed on the metals is used to control the surface reactions.

According to our knowledge the reaction of adsorbed hydrogen with organometallic compounds is a new reaction (ref. 37). We have observed that $Pb(C_2H_5)_4$ and $Sn(C_2H_5)_4$ react with hydrogen adsorbed on nickel or platinum with the formation of the corresponding alkane:

$$PtH_a + Sn(C_2H_5)_4 \longrightarrow Pt\,Sn(C_2H_5)_3 \text{ (IIIa)} + C_2H_6 \quad (11.10a)$$

$$NiH_a + Pb(C_2H_5)_4 \longrightarrow Ni\text{-}Pb(C_2H_5)_3 \text{ (IIIb)} + C_2H_6 \quad (11.10b)$$

Complexes (IIIa), (IIIb) decompose in hydrogen at higher temperatures with the formation of ethane (reaction 11.11):

$$(III) \xrightarrow[H_2]{\Delta} Pt\text{-}Sn(Ni\text{-}Pb) + 3C_2H_6 \quad (11.11)$$

In parallel to these reactions in the absence of preadsorbed hydrogen, lead or tin alkyl compounds can decompose catalytically on the parent metals with the formation of equivalent amounts of ethane and ethylene:

$$Sn(C_2H_5)_4 \xrightarrow{Me} Sn^0 + 2C_2H_6 + 2C_2H_4 \quad (11.12)$$

Reactions involving surface hydroxyls (see reaction 11.1) can also take place, i.e., special care should be taken in order to assure the predominance of reaction 11.10.

More details on these processes will be given in Section 11.5.

11.4 PREPARATION OF SUPPORTED BIMETALLIC CATALYSTS VIA METAL ADSORPTION

11.4.1 Technical aspects of the preparation

In the preparation of supported catalysts modified by adsorbed metals, one of the most important requirements is the use of oxygen-free inert gases, hydrogen and solutions. A second important requirement is a multifunctional reactor in which the following procedures can be carried out with the exclusion of air: hydrogen adsorption, metal adsorption, intensive washing and drying of the catalysts prepared. The reactor is described in ref. 38.

When preparing supported platinum catalysts modified by adsorbed metals, a base-supported platinum catalyst should be used in which the platinum is already reduced.

Since the formation of the adsorbed metal layer takes place in aqueous solution, the first procedure is to transfer the catalyst to the required electrolyte. As Pt/Al_2O_3 catalysts always contain chlorine it is advisable to use HCl solution as the electrolyte to avoid contamination of the support with additional anions. Prior to the introduction of the HCl solution to the reactor containing the base platinum catalyst, it is necessary to destroy the chemisorbed oxygen overlayer on the platinum sites by addition of hydrogen to the reactor. This is followed by addition of an HCl solution saturated with hydrogen. The treatment with hydrogen should be continued until the surface of the catalyst is completely saturated.

Having finished the saturation with hydrogen, the reactor is flushed with inert gas in order to remove the hydrogen dissolved in the electrolyte. This is followed by the addition of a solution of the chloride of the adsorbing metal.

After completion of the processes of metal adsorption the catalyst should be washed free from metal ions, first with deoxygenated electrolyte, then with distilled water and finally it is dried under oxygen-free conditions. The drying of the bimetallic catalyst is a very important step as in the simultaneous presence of water and oxygen the dissolution, i.e., reoxidation of the adsorbed metal, can take place:

$$Me_{ads} + \frac{z}{4} O_2 + \frac{z}{2} H_2O \longrightarrow Me^{z+} + z/OH^- \qquad (11.13)$$

This may result in the formation of ionic surface species on the support, and loss of the second metal especially from the outer sphere of the catalyst pellet.

11.4.2 Preparation of bimetallic $Pd-Pt/Al_2O_3$ catalysts

Catalysts with different Pd/Pt ratios have been prepared using the same parent base Pt/Al_2O_3 catalyst (refs. 39,40). Different approaches were used resulting in changes not only in the Pd/Pt ratio but also in the adsorption properties and in the catalytic behaviour of the $Pd-Pt/Al_2O_3$ catalysts. In these experiments a 3.39×10^{-2} mol solution of $PdCl_2$ in 0.2 mol HCl was used for the preparation, carried out at room temperature. The base Pt/Al_2O_3 catalyst containing 0.5% Pt was prepared by impregnation with H_2PtCl_6 in HCl solution, followed by drying and reduction in hydrogen at $400°C$.

(i) Regulation of the metal content; (a) The influence of the duration of palladium adsorption. When the duration of the palladium adsorption was 5 min, 24 h and 72 h, the palladium content of the catalysts was 0.088, 0.092 and 0.087% (mass), respectively (ref. 40). From these results it can be concluded that the processes involved in the palladium adsorption are very fast, i.e., the duration of adsorption cannot be used to control the palladium content.

(b) Successive metal adsorption (ref. 39), (Catalysts Type A, Series 1). Another simple way to control the amount of palladium adsorbed is the use of successive metal adsorption and saturation with hydrogen after each palladium adsorption step. Typical results are shown in Table 11.2.

TABLE 11.2

Characteristic data for Pd-Pt/Al_2O_3 catalysts prepared via successive palladium adsorption (Series 1)

Catalysts	Precursor	Palladium adsorption step	Palladium content	Pd/Pt ratio	Relative increase in Pd/Pt	Hydrogen uptake[a]		Chlorine[d] content (%)
						0,500[b]	400,500[c]	
	Pt/Al_2O_3	-	-	-	-	0.08	0.41	1.37
1 Pd-Pt	Pt/Al_2O_3	1	0.087	0.33	0.33	0.14	0.47	1.07
2 Pd-Pt	1 Pd-Pt	2	0.164	0.59	0.24	0.15	0.52	1.14
3 Pd-Pt	2 Pd-Pt	3	0.215	0.80	0.21	-	0.48	1.28

[a] In H/(Pt+Pd) measured by pulse method at 20°C.
[b] Without oxygen treatment, reduction at 500°C.
[c] Oxygen and hydrogen treatments at 400 and 500°C, respectively.
[d] Measured after palladium adsorption.

In three subsequent palladium adsorption steps the absolute amount increased to 0.215%, however, the relative increases expressed as Pd/Pt ratios were 0.33, 0.24 and 0.21, respectively. This means that after formation of the first adsorbed layer of palladium the amount of palladium deposited in the next step is less. These experimental data are in good agreement with those obtained for the adsorption of palladium on platinized platinum electrodes, i.e., after palladium adsorption the amount of hydrogen adsorbed on the platinum covered by palladium appeared to be lower than that prior to the palladium adsorption.

(c) <u>The influence of different experimental parameters,</u> (Catalysts Type A, Series 2). An attempt to change the amount of hydrogen chemisorbed on the parent Pt/Al_2O_3 catalyst was made by applying different thermal treatments. The amount of hydrogen adsorbed on Pt was measured by pulse hydrogen chemisorption and by standard hydrogen temperature programmed desorption (TPD). The experimental data are given in Table 11.3.

TABLE 11.3

The influence of the pretreatment of the Pt/Al_2O_3 catalyst on the palladium adsorption (Series 2)

Catalysts	Precursors and pretreatments[a]	Pd (%)	Pd/Pt ratio	Hydrogen uptake H/ Pt+Pd[c] O_2 treatment[d]		H/(Pt+Pd) by TPD	Cl (%)
				No	Yes		
-	Pt(0,400)	-	-	0.08	-	-	1.57
1 Pd-Pt	Pt(0,400)	0.087	0.33	0.14	0.47	-	1.07
-	Pt(0,500)	-	-	0.08	-	0.93	1.31
(Pd-Pt)$_A$	Pt(0,500)	0.084	0.31	0.10	0.46	-	1.00
-	Pt(400,500)	-	-	-	0.41	1.78	1.06
(Pd-Pt)$_B$	Pt(400,500)	0.139	0.51	0.18	0.61	-	1.05
-	Pt(400,575)	-	-	-	0.20	1.42	0.86
(Pd-Pt)$_C$	Pt(400,575)	0.151	0.55	-	0.58	-	1.27
-	Pt(400,575)[b]	-	-	-	0.20	0.10	0.86
(Pd-Pt)$_D$	Pt(400,575)[b]	0.152	0.55	-	0.57	-	1.12

[a] (0,500) no oxygen treatment, hydrogen treatment at 500°C; (400,500), oxygen and hydrogen treatment at 400 and 500°C, respectively; catalysts were cooled in H_2.
[b] After the pretreatment the catalyst was cooled in an atmosphere of nitrogen.
[c] By pulse adsorption at 20°C measured after H_2 treatment at 500°C.
[d] At 400°C.

From these data it can be concluded that neither the temperature of hydrogen treatment (catalysts 1 Pd-Pt vs. (Pd-Pt)$_A$) nor the atmosphere used for cooling (catalysts (Pd-Pt)$_C$ vs. (Pd-Pt)$_D$) appeared to be important in controlling the palladium content. No correlation was found between the H/Pt ratios of the parent catalysts, determined from the hydrogen uptake either by the pulse method or by TPD, and the amount of palladium adsorbed. An increase in the H/Pt ratio upon treatment with oxygen before final reduction resulted in a significant increase in the palladium content (catalysts 1 Pd-Pt vs. (Pd-Pt)$_B$), although the amount of palladium adsorbed was not sensitive to the slight changes in the H/Pt values of the parent catalysts (catalysts (Pd-Pt)$_B$ vs. (Pd-Pt)$_C$).

The fact that identical palladium contents were found for (Pd-Pt)$_C$ and (Pd-Pt)$_D$ catalysts clearly indicated that it is not the hydrogen adsorbed at high temperature from the gas phase but that adsorbed at room temperature in an aqueous solution which controls the amount of palladium adsorbed.

(d) <u>The influence of oxygen-hydrogen treatment cycles on the palladium adsorption</u> (Catalysts Type B) (ref. 39). The strong influence of the oxygen treatment on the hydrogen uptake of both the Pt/Al$_2$O$_3$ and Pd-Pt/Al$_2$O$_3$ catalysts (see Tables 11.2 and 11.3) directed us to study the influence of oxygen and hydrogen treatment cycles between each adsorption step on the palladium content. In these experiments, prior to the palladium adsorption, the catalysts were treated with oxygen at 400°C followed by reduction at 500°C. The results are given in Table 11.4. It is seen that upon applying oxygen-hydrogen treatment cycles between each palladium adsorption step the total amount of palladium introduced appeared to be twice as much as that obtained by successive palladium adsorption (see Table 11.2). The Pd/Pt ratios were 0.51, 0.55 and 0.66, respectively after each adsorption step. The treatment of the Pd-Pt/Al$_2$O$_3$ catalysts with oxygen changed markedly the values of hydrogen uptake measured as H/(Pd+Pt) ratios. Without such treatment, the H/(Pd+Pt) ratios were relatively low for all Pd-Pt/Al$_2$O$_3$ catalysts.

We believe that by treatment with oxygen we may alter the primary formed adsorbed palladium overlayer. The formation of a mixed Pd-Pt surface oxide phase is suggested. Upon subsequent reduction in hydrogen, this may result in a mixed alloy-type

TABLE 11.4

The influence of oxygen-hydrogen cycles on the palladium adsorption (catalysts Type B)

No. of adsorption steps	Catalysts	Precursors[a]	Pd (%)	Pd/Pt	Relative changes in Pd/Pt	Hydrogen uptake H/(Pd-Pt)[b] O_2 treatment[c] No	Yes	Cl[a] (%)
-	-	Pt(400,500)	-	-	-	-	0.41	1.06
1	(Pd-Pt)$_B$	Pt(400,500)	0.13	0.51	0.51	0.18	0.61	1.05
2	Pd(Pd-Pt)$_B$	(Pd-Pt)$_B$	0.23	1.06	0.55	0.21	0.47	1.04
3	2Pd(Pd-Pt)$_B$	Pd(Pd-Pt)$_B$	0.47	1.72	0.66	0.18	0.43	1.07

[a] After treatment with oxygen and hydrogen at 400 and 500°C, respectively.
[b] By pulse method at 20°C and measured after treatment with hydrogen at 500°C.
[c] Treatment with oxygen at 400°C.

bimetallic surface species which adsorbs hydrogen more readily than the parent monometallic platinum catalyst. The increased amount of hydrogen may be responsible for the higher amount of palladium adsorbed in each adsorption step.

11.4.3 Preparation of Pt/Al$_2$O$_3$ catalysts modified by adsorbed Re and Sn

The theoretical background for rhenium and tin adsorption onto platinum has been studied by standard electrochemical methods using a platinized platinum electrode (refs. 26,27). The base Pt/Al$_2$O$_3$ catalyst used for the preparation was the same as that used earlier to prepare different Pd-Pt/Al$_2$O$_3$ catalysts. Characteristic data for the catalysts prepared are given in Table 11.5.

11.5 PREPARATION OF SUPPORTED BIMETALLIC CATALYSTS BY USING ORGANOMETALLIC COMPOUNDS

11.5.1 The experimental technique and procedures used

The use of organometallic compounds requires special conditions for all steps of catalyst preparation. All gases and solvents used must be carefully deoxygenated and dried, and all manipulations carried out under an inert gas atmosphere. Special care should be

Table 11.5

Characteristic data for Pt/Al_2O_3 catalysts modified by adsorbed Re and Sn (refs. 41,42)

No.	Adsorbed metal	Precursor compounds [a]	Amount of metal adsorbed (%)	M/Pt ratio	Hydrogen uptake H/(Pt+M) [b]
1.	Re	$(NH_4)_2ReCl_6$	0.56	1.2	0.085
2.	Re	$(NH_4)_2ReCl_6$	0.54	1.1	0.080
3.	Sn	$SnCl_4$	0.21	0.7	0.090
4.	Sn	$SnCl_4$	0.27	0.9	0.076
5.	Sn	$SnCl_4$	0.33	1.1	0.047

[a] Duration of metal adsorption: $Re-Pt/Al_2O_3$, 60 hours; $Sn-Pt/Al_2O_3$, 40 h.
[b] By pulse method at 20°C, after treatment with hydrogen at 500°C.

taken to reduce the solvent content of the carrier after its preparation.

11.5.2 Preparation of supported Sn-Pt catalysts

(i) Catalysts with direct Sn-Pt interaction (refs. 42,43). Controlled surface reactions involved in the preparation of this type of $Sn-Pt/Al_2O_3$ catalysts have been described in Section 11.3.2. 0.5% Pt/Al_2O_3 catalysts (see Section 11.4.2.1) were used. Benzene solutions of $Sn(C_2H_5)_4$ were used for the reaction with hydrogen preadsorbed on the platinum sites. Prior to this reaction the supported platinum catalysts were treated with hydrogen at 500°C and then cooled down in a hydrogen atmosphere.

The two basic reactions involved in the preparation are:

$$Pt-H + Sn(C_2H_5)_4 \longrightarrow Pt-Sn(C_2H_5)_{4-x} + xC_2H_6 \qquad (11.14)$$
$$(IV)$$

$$(IV) \xrightarrow[H_2]{\Delta} Pt-Sn + (4-x)C_2H_6 \qquad (11.15)$$

Reaction 11.14 was monitored by gas volumetric determination of the gases evolved and by their gas chromatographic (GC) analysis. Special care was taken to avoid side reactions in which protons of the support may be involved. In this context the following surface reaction has been suggested upon adsorption of $Sn(C_2H_5)_4$ on $\gamma\text{-}Al_2O_3$ (ref. 44):

$$\}\!\!-\!\!OH + (Sn\text{-}C_2H_5)_4 \longrightarrow \}\!\!-\!\!O\!-\!\!\underset{\underset{C_2H_5}{|}}{\overset{\overset{C_2H_5}{|}}{Sn}}\!-\!C_2H_5 + C_2H_6 \qquad (11.16)$$

(V)

In order to avoid side reactions, involving the formation of surface species such as (V), the concentration of tin alkyl compounds in the benzene solution used for the preparation was relatively low.

In the presence of adsorbed hydrogen and if the initial concentration of the tin alkyl in the reaction mixture was not higher than 0.0025 mol/dm^3, only ethane formation was observed in reaction (11.14). This reaction appeared to be slow at 35°C, but measurable rates were obtained at 50°C.

The lack of formation of ethylene indicated that under these experimental conditions the catalytic decomposition of $Sn(C_2H_5)_4$ (reaction 11.12) is negligible, however, when the initial Pt-Sn atomic ratio in the reaction mixture was higher than 1:2 or in the absence of preadsorbed hydrogen, besides the formation of ethane, formation of ethylene was also observed. The rate of the catalytic decomposition (reaction 11.12) appeared to be much lower than the rate of reaction (11.14) (ref. 37).

On pure alumina under analogous experimental conditions at 50°C, the formation of ethane and ethylene was observed only in trace amounts. The temperature programmed reactions (TPR) technique was used to study the decomposition of surface complexes (IV, V) formed in the course of the interaction of different amounts of $Sn(C_2H_5)_4$ with Pt/Al_2O_3 containing preadsorbed hydrogen or pure alumina. Typical TPR curves are shown in Fig. 11.1. In the case of direct Sn-Pt interaction, the main reaction product was ethane, but ethylene and methane were also detected in very small amounts. Both the methane formation and the ethane peak at around 240-260°C were attributed to the decomposition, i.e., hydrogenolysis, of the traces of solvent used in the course of the preparation. We did not observe this peak with pure Al_2O_3 impregnated with $Sn(C_2H_5)_4$ (see Fig. 11.2) (ref. 43).

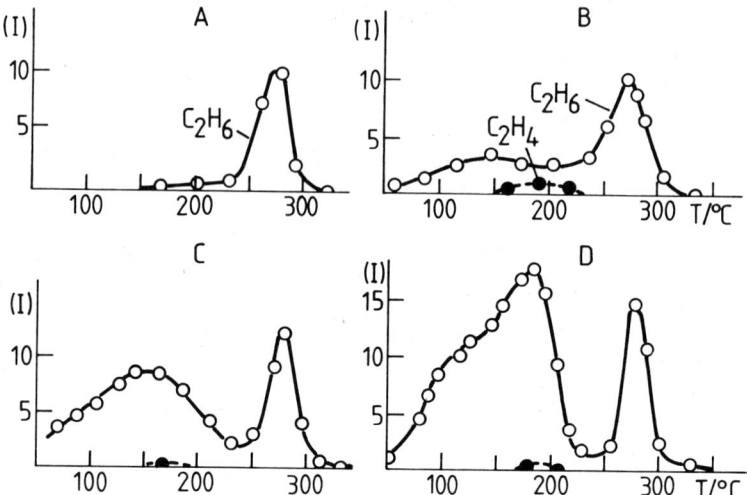

Fig. 11.1. Catalysts: Sn-Pt/Al$_2$O$_3$. TPR curves obtained during the preparation according to reaction 11.15 (catalysts of type B). Heating rate: 5°C/min.

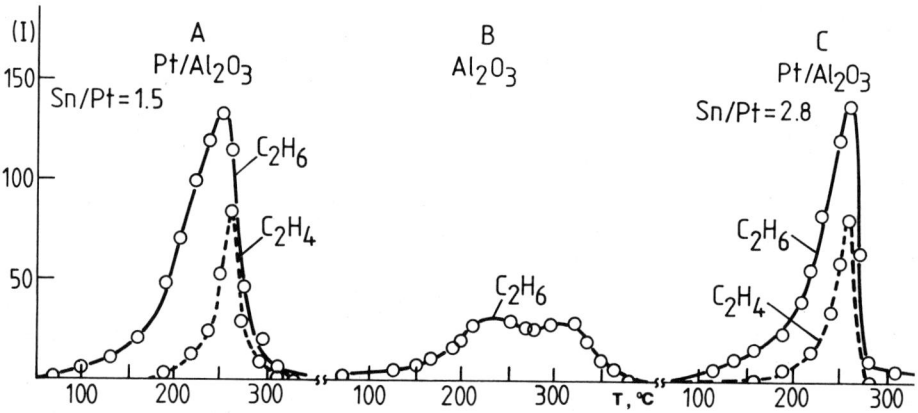

Fig. 11.2. Catalysts: Sn-Pt/Al$_2$O$_3$. TPR curves obtained during the preparation (catalysts of type D and Sn(Et)$_4$/Al$_2$O$_3$). Heating rate: 5°C/min.

Good agreement was obtained between the integrated peak areas and the tin content of the catalysts measured before and after reaction 11.15. No loss of tin was observed during the TPR experiment. The value of x in reactions 11.14 and 11.15 was calculated to be 1.0 $\pm$ 0.12 (ref. 42). From these data it was concluded that, under appropriate experimental conditions, the surface reactions involved in the preparation of this type of

Sn-Pt/Al_2O_3 catalysts are controlled. These Sn-Pt/Al_2O_3 catalysts will be distinguished from other catalysts as catalysts of type B (see Section 11.6).

It was also of interest to determine the kind of surface reactions which take place if control of the surface reactions involved in the preparation cannot be achieved. Two samples of the base Pt/Al_2O_3 catalysts were treated with excesses of $Sn(C_2H_5)_4$ and $SnCl_2(C_2H_5)_2$ (Catalysts type D, samples D1 and D2). The TPR curves of these catalyst samples are shown in Fig. 11.2. Due to the increased amount of the precursor, the surface reaction 11.14 was not controlled, i.e., part of the tin precursors reacted with the alumina. This side reaction resulted in significant changes in the TPR peak profile and in the peak positions. The low temperature shoulder of the ethane peak is related to the formation and decomposition of surface complex (IV), with direct Sn-Pt interaction, while the high temperature ethane and ethylene peaks are attributed to surface species (V), which were formed by an interaction of tin alkyls with the alumina.

(ii) <u>Preparation of Sn-Pt/Al_2O_3 catalysts with tin-alumina interaction</u> (refs. 42,43,44). In this approach special care has to be taken to prevent the formation of surface species with Sn-Pt interaction. Although a different kind of Sn-Al_2O_3 interaction has been discussed in the previous section, these uncontrolled surface reactions could not be used to obtain tin-alumina surface entities exclusively.

In this case the following surface reactions are used:

$$\}\text{-OH} + C_4H_9Li \longrightarrow \}\text{-O-Li} + C_4H_{10} \quad (11.17)$$

$$n \}\text{-O-Li} + SnCl_4 \longrightarrow \}\text{-O-)}_n SnCl_{4-n} + n\ LiCl \quad (11.18)$$
$$(VI)$$

$$(VI) \xrightarrow[\Delta]{H_2} \}\text{-O-)}_n Sn + (4-n)HCl \quad (11.19)$$
$$(VII)$$

Prior to reaction (11.17) the Pt/Al_2O_3 catalyst is treated with an inert gas atmosphere at 150-300°C to obtain partially dehydroxylated alumina surfaces.

In these catalysts (C type Sn-Pt/Al_2O_3 catalysts) the tin is strongly bound to the support via Al-O-Sn bridges. In the surface complexes (VII) the tin remains in its higher oxidation state even after reduction in hydrogen at 500°C (ref. 46).

11.5.3 Preparation of lead-modified alumina-supported nickel catalysts

A semiindustrial Ni/Al_2O_3 catalyst with 40% nickel content has been used for the preparation of $Pb-Ni/Al_2O_3$ catalysts (refs. 43,47). The general principles of the organometallic approach were not altered, i.e., the procedures used for the lead introduction were the same as applied for the preparation of $Sn-Pt/Al_2O_3$ catalysts (Catalysts type B). The only significant difference is in the amount of the second metal to be introduced. Due to the relatively high nickel content of this catalyst, the amount of lead required for the modification is about two orders in magnitude higher than that of the tin used in the case of Pt/Al_2O_3. Prior to the modification, the Ni/Al_2O_3 catalyst is treated with hydrogen at 150°C for 6 h.

The first step of the modification, i.e., the reaction of the preadsorbed hydrogen with $Pb(C_2H_5)_4$ (reaction 11.10b), was carried out between 25 and 50°C. At 25°C, a good correlation was obtained between the amount of ethane formed and the amount of hydrogen detected by TPD. Under these conditions, besides the formation of ethane, methane and ethylene were also formed. The methane formation is attributed to the hydrogenolysis of the C_2H_5 groups on the nickel sites. Ethylene formation indicates the catalytic decomposition of the lead alkyl by nickel:

$$Pb(C_2H_5)_4 \xrightarrow{Ni} Pb^0 + 2C_2H_6 + C_2H_4 \qquad (11.20)$$

Reaction 11.20 became of great importance at 50°C. Additionally, decomposition of $Pb(C_2H_5)_4$ was observed on the alumina at 50°C and higher lead alkyl concentrations. Under these experimental conditions the formation of the surface complex (VIII) was suggested:

$$\rangle\text{-OH} + Pb(C_2H_5)_4 \longrightarrow \rangle\text{-O-Pb}(C_2H_5)_{4-x} + x\, C_2H_6 \qquad (11.21)$$
$$(VIII)$$

The decomposition of the primary $Ni-Pb(C_2H_5)_x$ surface species has been studied by temperature-programmed reaction in an hydrogen atmosphere. The formation of ethane was followed by GC analysis.

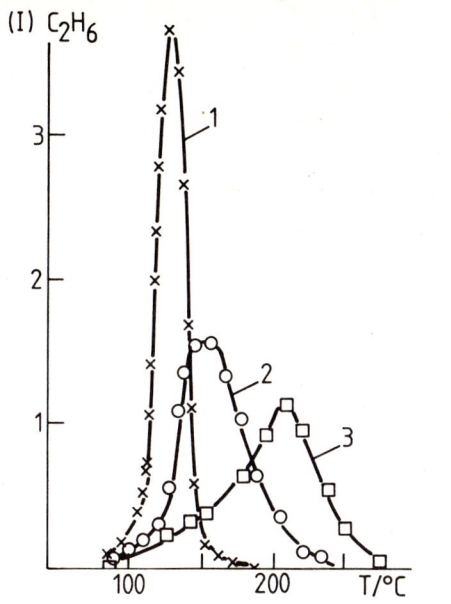

Fig. 11.3. Catalysts: Pb-Ni/Al$_2$O$_3$. TPR curves obtained during the preparation. Heating rate: 10°C/min. (1) Catalyst of type C, (2) Catalyst of type B, (3) Catalyst of type A.

In order to demonstrate the importance of the first step of the preparation (reaction 11.10b) three catalyst samples were prepared as follows:

Catalyst type A: adsorption of Pb(C$_2$H$_5$)$_4$ at 25°C on Ni/Al$_2$O$_3$ without thermal treatment;

Catalyst type B: adsorption of Pb(C$_2$H$_5$)$_4$ at 25°C on Ni/Al$_2$O$_3$ treated with nitrogen at 150°C;

Catalyst type C: reaction of Pb(C$_2$H$_5$)$_4$ at 25°C with hydrogen adsorbed on Ni/Al$_2$O$_3$ treated with hydrogen at 150°C.

Typical TPR curves of these catalysts are shown in Fig. 11.3 (ref. 47). The amount of ethane formed increased in the following order: catalyst B < catalyst A < < catalyst C. The position of the peak maxima shifted to the opposite direction. The low temperature peak of catalyst C can be considered as indirect evidence for the formation of Ni-Pb surface entities, while the high temperature peak of the catalysts A and B can be attributed to the interaction between the alumina and lead alkyl.

11.6 CATALYTIC PROPERTIES OF SUPPORTED BIMETALLIC CATALYSTS PREPARED VIA CONTROLLED SURFACE REACTIONS

11.6.1 General aspects

The properties of supported bimetallic catalysts prepared via CSRs will depend on:
(1) The type of the base catalyst used for the preparation.
(2) The kind and amount of the second metal.
(3) The nature of controlled surface reactions.
(4) The thermal treatment applied before or after the preparation.

The role played by these factors has already been mentioned in the previous sections. Here again we should like to emphasize the importance of the thermal treatment. Having prepared a supported catalyst, the only possibility to change its catalytic properties is by using different thermal treatments.

The thermal treatment of bimetallic catalysts prepared via CSRs induces complex changes in their properties as a variety of surface interactions can take place during treatment with oxygen or hydrogen at high temperatures. During these processes the primary bimetallic surface species can be altered significantly. For this reason, in our approach, special care was taken to use different thermal treatments in order to find additional ways of controlling the properties of different bimetallic catalysts.

In this section, we shall try to demonstrate the variety of the applications of supported bimetallic catalysts prepared via CSRs. Most of these applications are closely related to industrial processes, the aim being to improve the properties, of different industrial catalysts. The only exception that of benzene hydrogenation, where an attempt was made to illustrate the effect of thermal treatment on the catalytic properties of different $Pd-Pt/Al_2O_3$ catalysts.

11.6.2   Benzene hydrogenation

In these experiments $Pd-Pt/Al_2O_3$ catalysts of Type A have been used (see Table 11.2). Before the reaction, the bimetallic catalysts were treated with hydrogen at different temperatures and then cooled in an hydrogen atmosphere to the temperature of the reaction (ref. 48). The results are shown in Fig. 11.4. The parent Pt/Al_2O_3 catalysts had high activity in this reaction. As is seen in Fig. 11.4a, after treatment with hydrogen at 500°C the activity of the Pd-Pt catalysts is much lower than that of the parent Pt/Al_2O_3.

The initial activity of the bimetallic catalysts decreased in the order: 1 Pd-Pt > 2Pd-Pt ≥ 3Pd-Pt, i.e., upon increasing the palladium content the activity decreased due to the site-blocking effect of the deposited palladium. These results are in a good agreement with an earlier observation in which the electrochemical hydrogenation of acetone on a palladium-covered platinum electrode did not proceed at all (ref. 49). It should be noted that in the electrochemical hydrogenation there was no possibility of treating the modified platinum electrodes with a second metal at high temperatures.

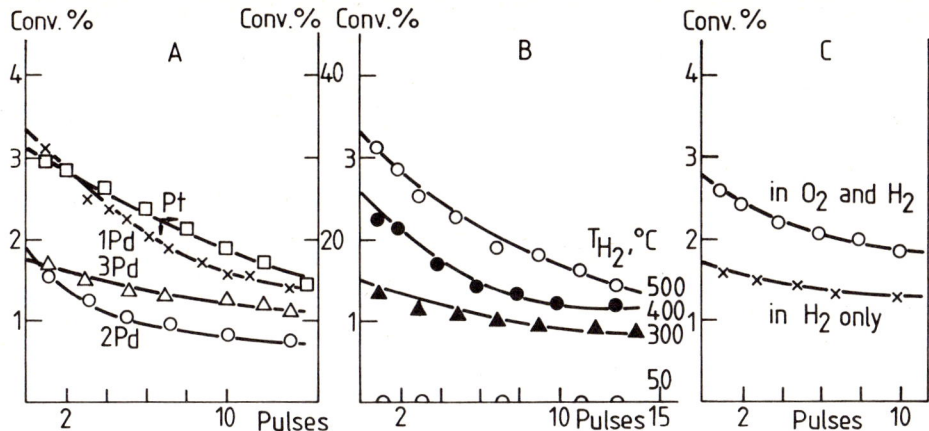

Fig. 11.4. Benzene hydrogenation over Pd-Pt/Al$_2$O$_3$ catalysts. Reaction temperature: 50°C; Partial pressure of benzene: 0.104 kPa; Flow-rate: 1.5 ml × sec^{-1}. (A) Influence of content of Pd, (B) Influence of the temperature of hydrogen treatment, (C) Influence of oxygen treatment.

If prior to the reaction, the treatment of the catalysts with hydrogen was carried out at 50°C, all of the bimetallic catalysts appeared to be entirely inactive towards benzene hydrogenation. In addition, no hydrogen uptake was observed on these catalysts at 20°C after such a low-temperature hydrogen treatment.

From these experiments it can be concluded that the high-temperature hydrogen treatment results in certain changes in the primary form of the adsorbed palladium overlayer. The higher the temperature of hydrogen treatment the higher is the extent of these changes, resulting in higher catalytic activity. This can be deduced from the series of experiments carried out on catalysts 1Pd-Pt and 2Pd-Pt in which the temperature of hydrogen treatment applied before the reaction was varied (Fig. 11.4b and 11.4c). More pronounced changes were obtained on catalyst 1Pd-Pt.

Additional increases in the catalytic reactivity were obtained for all of these catalysts if they were treated with oxygen at 400°C prior to the final reduction at 500°C.

All of these data are in good agreement with the phenomena discussed in Section 11.4.4., i.e., the primary palladium overlayer can be altered during the high-temperature treatment.

11.6.3   n-Hexane conversion

The conversion of n-hexane over different supported mono- and bimetallic catalysts has been extensively studied, since it can be used as a model of a very important industrial process, namely naphtha reforming.

In all of our studies attention was focused on the formation of the following reaction products: hydrogenolysis products, i.e., light hydrocarbons including pentanes; benzene; C_6-isomers (2-methylpentane and 3-methylpentane); methylcyclopentane and toluene. In the kinetic measurements the formation of olefins, i.e., 1-hexene and 2-hexenes, was also monitored. Two reactor set-ups were used: a microcatalytic continuous-flow reactor operated in slug-pulse mode (ref. 50) and a conventional continuous-flow reactor. Details of the analysis of the reaction products can be found elsewhere (ref. 51). All of these experiments were carried out at atmospheric pressure.

(i) The catalytic properties of Pd-Pt/Al_2O_3 catalysts prepared via metal adsorption (refs. 39,52,53)

(a) Pulse experiments. Catalysts of Type A (Series 1.). These catalysts were tested in two different ways; by treating them with hydrogen at 500°C without oxygen treatment and by treating them with oxygen at 400°C followed by hydrogen treatment at 500°C. The extents of conversion and selectivity data obtained are given in Table 11.6. Upon comparing the results given in Table 11.6 it is seen that the properties of the Pd-Pt/Al_2O_3 catalysts are strongly affected by the oxygen treatment.

Without oxygen treatment the activity of the Pd-Pt/Al_2O_3 catalysts was lower than that of their monometallic precursor. On these Pd-Pt catalysts the conversion of n-hexane was almost independent of the palladium content, however, the selectivities for benzene and isohexanes appeared to be very sensitive to variation of the amount of palladium. Upon increasing the palladium content the selectivity for benzene decreased with a simultaneous increase in the selectivity for isohexane. The selectivities for other reaction products were influenced little by the palladium content. However, it is interesting that on the palladium-containing catalysts almost twice as much toluene was formed than on the base Pt/Al_2O_3 catalysts.

Upon treatment with oxygen the activity of these Pd-Pt catalysts increased significantly. The extents of conversion achieved in the case of 2Pd-Pt even exceeded that with the parent Pt/Al_2O_3 catalyst. A slight increase in the selectivity for benzene was observed on the 2Pd-Pt catalysts, attributed to the higher extent of conversion. Data given in Table 11.6 indicate that, after treatment with oxygen catalysts of type A strongly resemble the base Pt/Al_2O_3 catalysts.

TABLE 11.6

Catalytic activity and selectivity of catalysts Type A (Series 1)

Catalysts	Pd (%)	Conversion (%)	Selectivities (%)				
			B	I	Hy	MCP	T
Pt/Al_2O_3[a]	-	59	67	8	8	6	8
1Pd-Pt[a]	0.087	38	52	11	10	8	15
2Pd-Pt[a]	0.164	39	46	18	9	8	13
3Pd-Pt[a]	0.215	37	37	22	11	8	14
Pt/Al_2O_3[b]	-	76	62	14	8	4	9
1Pd-Pt[b]	0.087	79	65	15	6	3	8
2Pd-Pt[b]	0.164	89	71	9	6	2	9
3Pd-Pt[b]	0.215	77	61	15	8	4	10

[a]Catalysts without oxygen treatment.
[b]Catalysts treated with oxygen at 400°C. Final reduction at 500°C.
Reaction temperature 480°C, amount of catalysts: 0.10 g; flow-rate: 0.5×10^{-3} sec^{-1}; reaction mixture: H/CH = 5:1, results obtained in first pulses; (B = benzene; I = isohexanes, Hy = hydrogenolysis products, MCP = methylcyclopentane, T = toluene).

As discussed in Section 11.4, the oxygen treatment may result in a significant alteration of the primary adsorbed palladium overlayer. This process should be very fast as variation of the duration of the oxygen treatment did not result in significant changes in the activity and selectivity data (ref. 53).

Catalysts of type A (Series 2). In this series of catalysts the same parent Pt/Al_2O_3 catalyst was used but it was treated in a different way before palladium adsorption (see Section 11.4, Table 11.3). These catalysts were treated with oxygen at 400°C and then reduced at 500°C. The results obtained are given in Table 11.7. As is seen from Table 11.7, the activity of these catalysts was independent of the palladium content. Upon increasing the palladium content, only slight changes were observed in the selectivity for benzene and isohexanes.

This series of experiments can also be used to demonstrate the excellent reproducibility of our method, both as regards the amount of metal adsorbed and the properties of the catalysts obtained.

Catalysts of type B. These catalysts were tested also in two different ways in order to demonstrate the effect of oxygen treatment. The results are given in Table 11.8. The data in Table 11.8 strongly resemble those obtained on catalysts of type A (see Table 11.6).

TABLE 11.7

Catalytic activity and selectivity of catalysts of type A (Series 2)[a]

Catalysts	Pd (%)	Conversion (%)	Selectivities (%)				
			B	I	Hy	MCP	T
1Pd-Pt	0.087	79	66	15	6	3	8
(Pd-Pt)$_A$	0.084	74	62	17	6	4	9
(Pd-Pt)$_B$	0.139	72	56	17	8	4	11
(Pd-Pt)$_C$	0.151	77	58	16	8	4	10
(Pd-Pt)$_D$	0.152	79	63	14	7	4	8

[a] Reaction conditions and abbreviations: see Table 11.6. Catalysts treated with oxygen at 400°C.

TABLE 11.8

Catalytic activity and selectivity of catalysts of type B[a]

Catalysts	Pd (%)	Conversion (%)	Selectivities (%)				
			B	I	Hy	MCP	T
Pt/Al$_2$O$_3$[b]	–	81	61	13	11	3	8
(Pd-Pt)$_B$[c]	0.139	34	41	18	10	10	13
Pd(Pd-Pt)$_B^c$	0.293	41	42	24	10	8	11
2Pd(Pd-Pt)$_B^c$	0.470	42	48	14	11	7	16
(Pd-Pt)$_B^b$	0.139	72	56	17	8	4	11
Pd(Pd-Pt)$_B^b$	0.293	90	68	7	8	1	13
2Pd(Pd-Pt)$_B^b$	0.470	81	61	12	8	3	13

[a] Reaction conditions and abbreviations see Table 11.6.
[b] Catalysts treated with oxygen and hydrogen at 400°C and 500°C, respectively;
[c] Catalysts without oxygen treatment.

In the first palladium adsorption step a relatively large amount of palladium was adsorbed [Catalyst (Pd-Pt)$_B$]. This resulted in a very sharp decrease in the extent of conversion and in the selectivity for benzene. However, contrary to catalysts of type A, a slight increase in the catalytic activity was observed after the second palladium adsorption step. The higher activity of catalysts Pd(Pd-Pt)$_B$ and 2Pd(Pd-Pt)$_B$ prepared from catalyst (Pd-Pt)$_B$ and the lower activity of latter can be related to the differences between their precursors. For catalyst (Pd-Pt)$_B$ the sites for hydrogen adsorption were strongly altered by the high-temperature oxygen-hydrogen treatment cycle carried out before successive palladium adsorption. The formation of a mixed platinum-palladium alloy-type surface cluster was mentioned earlier (see Section 11.4.).

It is suggested that, after successive palladium adsorption and when omitting the oxygen treatment before the hydrocarbon reaction, only the palladium-free platinum sites are involved in the hydrocarbon reaction. These sites contain only platinum in catalysts (Pd-Pt)$_B$, however, in catalysts Pd(Pd-Pt)$_B$ and 2Pd(Pd-Pt)$_B$ these sites may resemble a mixed platinum-palladium alloy-type surface cluster with different catalytic properties.

(b) <u>Continuous-flow experiments.</u> In the continuous-flow reactor only oxygen-treated catalysts were tested. The same pretreatment processes were applied as used in the pulse experiments. The reaction was carried out at 500°C. Samples used in these experiments were Pt/Al$_2$O$_3$ catalysts of type A: 1Pd-Pt, (Pd-Pt)$_B$, 2Pd-Pt; catalysts of type B: Pd(Pd-Pt)$_B$ and 2Pd(Pd-Pt)$_B$ (ref. 39).

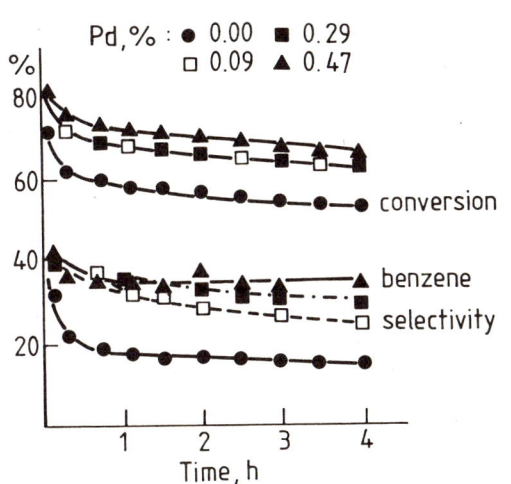

Fig. 11.5. n-Hexane conversion, continuous-flow experiments, Pd-Pt/Al$_2$O$_3$. Reaction temperature: 500°C; catalyst: 0.2 g; flow-rate: 0.5 ml x sec^{-1}; CH/H = 1:5.

All of these bimetallic catalysts appeared to be more active than the base Pt/Al$_2$O$_3$ catalyst. Typical results are shown in Fig. 11.5 (ref. 39). As is seen, the bimetallic catalysts retain their activity. Their selectivity for benzene formation was much higher than that of the base catalyst. The catalyst 2Pd(Pd-Pt)$_B$ with the highest palladium content had the highest selectivity for benzene formation and what is even more significant there was no decrease in the benzene selectivity during the ageing process.

Experimental results obtained after four hours of reaction are compared in Fig. 11.6 (ref. 39). The differences between catalysts

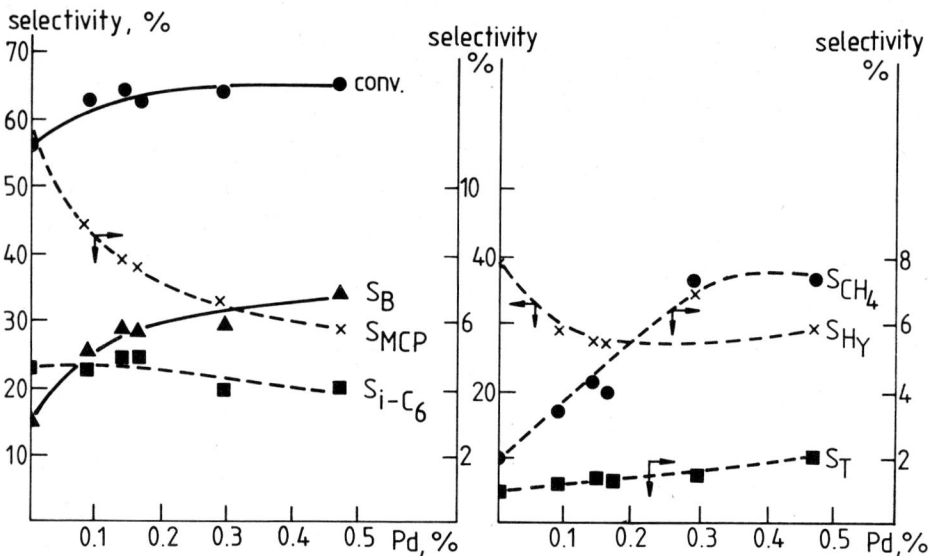

Fig. 11.6. Conversion and selectivity data (after 4 h). Reaction conditions: see Fig. 11.5. Abbreviations see Table II.6.

of types A and B are slightly blurred and, with the exception of the selectivity for hydrogenolysis products, a good correlation was found between the conversion or selectivity data and the palladium content. Upon increasing the palladium content, the activity increased slightly, but a more pronounced increase was observed in the selectivity for benzene. The selectivity for isomerization was almost independent of the palladium content, those for methane and toluene increased, and that for methylcyclopentane decreased with increasing palladium content.

(ii) <u>Catalytic properties of Re-Pt/Al_2O_3 catalysts prepared via metal adsorption (ref. 41). Pulse experiments</u>

The results obtained on Re-Pt/Al_2O_3 catalysts prepared via metal adsorption are summarized in Table 11.9. These catalysts were first tested for reproducibility. In the course of these experiments, the stability was studied and two different charges were compared (see experiments 1-7).

The freshly prepared catalysts did not have a constant activity.

TABLE 11.9

Activity and selectivity data for Re-Pt/Al$_2$O$_3$ catalysts in n-hexane dehydrocyclization (reaction temperature 480°C, H$_2$: n-hexane = 5:1, amount of catalysts: 0.2 g (ref. 41))

No.	Catalyst sample[a]	Temperature of pretreatment (°C)		Conversion (%)	Selectivity data (%)				
		O$_2$	H$_2$		B	I	Hy	MCP	T
1	Re-Pt (IA)	400	400	81	75.0	10.0	7.6	2.1	3.7
2	Re-Pt (IB)	400	400	66	70.5	6.2	8.4	4.0	6.7
3	Re-Pt (IIB)	400	400	67	66.0	11.0	10.0	4.6	5.7
4	Re-Pt (IA)	400	575	18	61.0	7.5	12.2	5.6	12.0
5	Re-Pt (IB)	400	575	80	71.2	14.5	6.4	2.4	2.5
6	Re-Pt (IB)	400	500	75	72.5	11.0	7.0	2.6	4.2
7	Re-Pt (IIB)	400	500	78	74.0	10.0	7.0	2.4	4.4
8	Pt	0[b]	500	39	60.0	6.1	19.0	6.9	6.6
9	Re-Pt (IB)	0[b]	500	23	52.5	12.0	9.2	9.6	12.0
10	Pt	0[b]	575	30	56.0	8.9	19.0	9.3	3.5
11	Re-Pt (IB)	0[b]	575	25	47.3	10.0	16.0	14.0	9.4
12	Pt	400	500	61	69.0	8.4	7.7	2.7	9.8
13	Re-Pt (IB)	400	500	75	72.5	11.0	7.0	2.6	4.2
14	Pt	500	500	78	70.0	9.6	11.0	2.4	4.4
15	Re-Pt (IB)	500	500	80	74.0	11.0	5.1	2.3	4.4

[a] (I) and (II) are different samples of Re-Pt/Al$_2$O$_3$ catalysts. A = tested within three days after contact with air, B = tested following one week after contact with air.
[b] Catalysts without oxygen treatment. Abbreviations see Table 11.6.

This can be attributed either to the sensitivity of the adsorbed rhenium species toward oxygen and moisture, or to the formation of partially reduced rhenium which reoxidizes relatively slowly. Catalysts tested within a few days of contact with air had different activities and selectivities than those in contact with air for a longer period (experiments 1 vs. 2 and 4 vs. 5). However, the stabilized catalyst showed constant activity and selectivity. Catalysts prepared as different charges had good reproducibilities after stabilization (experiment 2 vs. 3 and 6 vs. 7).

In additional experiments the role of oxygen treatment was studied and activity and selectivity measured on Re-Pt/Al$_2$O$_3$

catalysts were compared with that of obtained on Pt/Al_2O_3 (experiments 8-15).

Without oxygen treatment the catalytic activity of the $Re-Pt/Al_2O_3$ catalyst is lower, but after calcination at 400°C it is higher, than that of the Pt/Al_2O_3 catalyst. After calcination at 500°C each catalyst showed the same activity. The $Re-Pt/Al_2O_3$ catalyst possessed lower selectivities for hydrogenolysis products and higher selectivities for benzene and isohexanes, independently of the oxygen treatment.

As for $Pd-Pt/Al_2O_3$ catalysts, the $Re-Pt/Al_2O_3$ catalyst without oxygen treatment had a lower catalytic activity than the base Pt/Al_2O_3 catalyst, and its selectivity was rather different from that of Pt/Al_2O_3. The selectivities for hydrogenolysis products and benzene were lower, those for isohexanes, methylcyclopentane and toluene were higher. Increasing the temperature of hydrogen treatment resulted in a slight increase in the selectivities for hydrogenolysis and methylcyclopentane and a slight decrease in the selectivity for benzene.

The oxygen treatment has changed markedly the catalytic properties of the $Re-Pt/Al_2O_3$ catalyst. For a catalyst treated with oxygen at a lower temperature, upon increasing the temperature of hydrogen treatment the activity increases considerably, but only minor changes are obtained in the selectivities for isohexanes and toluene.

It should be noted that $Re-Pt/Al_2O_3$ catalysts prepared by conventional techniques usually possess relatively high selectivity for hydrogenolysis (ref. 54) and sulphur poisoning is used to suppress this undesired side reactions (ref. 55).

The lower activity of the $Re-Pt/Al_2O_3$ catalyst treated only with hydrogen may be attributed to the site-blocking effect of rhenium species adsorbed on the platinum crystallites. The lower selectivity for cracking may indicate the blocking of terrace sites of platinum by rhenium. After oxygen treatment a mixed Re-Pt oxide phase may be formed, which decomposes during the subsequent hydrogen treatment with the formation of Re-Pt bimetallic clusters.

(iii) <u>Catalytic properties of $Sn-Pt/Al_2O_3$ catalysts prepared in different ways</u> (refs. 42,43,45,56). Tin-modified Pt/Al_2O_3 catalysts have been prepared in different ways as described in Section 11.4 and 11.5. <u>Catalysts of type A</u> have been obtained via tin adsorption. <u>Catalysts of type B</u> were prepared to obtain direct tin-platinum interaction via the reaction of adsorbed hydrogen with $Sn(C_2H_5)_4$. In <u>catalysts of type C,</u> tin was introduced in order to obtain strong tin-alumina interaction. In <u>catalysts of type D,</u> tin may interact both with the platinum sites and the alumina support due to the lack or loss of control of surface reactions.

(a) <u>Kinetic studies.</u> In these experiments the extent of conversion of n-hexane and the yields of the main reaction products were determined after various contact times, and the initial rates of reaction were calculated. The details of this approach can be found elsewhere (ref. 50). Results are given in Table 11.10 (ref. 42).

<u>Catalysts of type A.</u> The introduction of tin in this form resulted in a positive bimetallic effect. The rate of n-hexane conversion had a maximum at Sn:Pt = 0.7. The Sn:Pt ratio had only a minor influence on the rate of hydrogenolysis, contrary to those of the aromatization and isomerization reactions, which were strongly affected. There was a permanent decrease in the rate of benzene formation upon increasing the Sn/Pt ratio from 0.7 to 1.1.

<u>Catalysts of type B.</u> Over these catalysts, upon introduction of tin, the rate of n-hexane conversion decreased only slightly. No significant decrease in the initial rate of benzene formation was observed. The activity and selectivity data for catalysts B1-B4 and D1 are given in Fig. 11.7/A.

It is of interest that upon introduction of tin the selectivity for benzene calculated from the initial rates increased slightly with a parallel decrease in the selectivity for isohexanes. The selectivity for hydrogenolysis showed a very complex dependence on the Sn:Pt ratio. For catalyst D1 the observed selectivity changes for benzene and hydrogenolysis products can be considered as additional indirect evidence of the loss of control of the surface reactions used during the catalyst preparation.

<u>Catalyst of type C.</u> The introduction of tin into the alumina support resulted in a very strong alteration in the selectivity data. The formation of isohexanes was almost entirely suppressed. This catalyst also possessed a lower hydrogenolysis activity. The rate of benzene formation was more than halved. This catalyst had a very unusual high selectivity (more than 60%) for n-hexenes.

TABLE 11.10

Initial rates of n-hexane conversion obtained on different Pt/Al_2O_3 and $Sn-Pt/Al_2O_3$ catalysts (ref. 42)

Catalyst	Precursor	Sn/Pt (atoms)	Initial rates[b] [mol g(cat.)$^{-1}$ sec^{-1} x 10^5]			
			$r_{C_6H_{14}}$	r_B	r_{Hy}	r_I
Base 1	-	-	1.20	0.27	0.38	0.25
AI	$SnCl_4$	0.7	2.09	0.97	0.18	0.39
A2	$SnCl_4$	0.9	1.69	0.45	0.24	0.60
A3	$SnCl_4$	1.1	1.31	0.33	0.20	0.35
Base 2[c]	-	-	0.93	0.28	0.28	0.20
BI	$Sn(C_2H_5)_4$	0.07	0.89	0.26	0.20	0.20
B2	$Sn(C_2H_5)_4$	0.22	0.83	0.25	0.16	0.19
B3	$Sn(C_2H_5)_4$	0.31	0.73	0.25	0.19	0.14
B4	$Sn(C_2H_5)_4$	0.42	0.75	0.25	0.21	0.12
CI[d]	$SnCl_4$	0.40	0.57	0.11	0.03	0.01
DI[e]	$SnCl_4$	0.40	0.64	0.14	0.28	0.07
D2	$Sn(C_2H_5)_4$	1.50	0.51	0.04	0.29	0.07
D3[d]	$SnCl_2(C_2H_5)_2$	2.80	0.19	0.02	0.11	0.03
D4	$SnCl_2(C_2H_5)_2$	3.80	0.10	0.01	0.06	0.01

[a] Reaction temperature = 520°C, temperature of hydrogen treatment 550°C, amount of catalyst: 0.05 g; B = benzene, Hy = hydrogenolysis products, I = isohexanes.
[b] Expressed in moles of n-hexane converted.
[c] Prepared as base 1, but on this blank catalyst the preparation procedures in organic solvents were simulated.
[d] Lithiated alumina was used for the preparation.
[e] Prepared by impregnation from acetone solution (ref. 57).

Catalysts of type D. Although these catalysts were prepared by using different types of tin precursors, very good correlations were obtained between the rates of n-hexane reaction or the reaction selectivity data and the Sn:Pt ratios as shown in Fig. 11.7/B. Unusually high selectivities for hydrogenolysis products were observed on all of these catalysts.

The effect of oxygen treatment. In order to study the stability of the Sn-Pt or Sn-alumina surface species formed in the course of

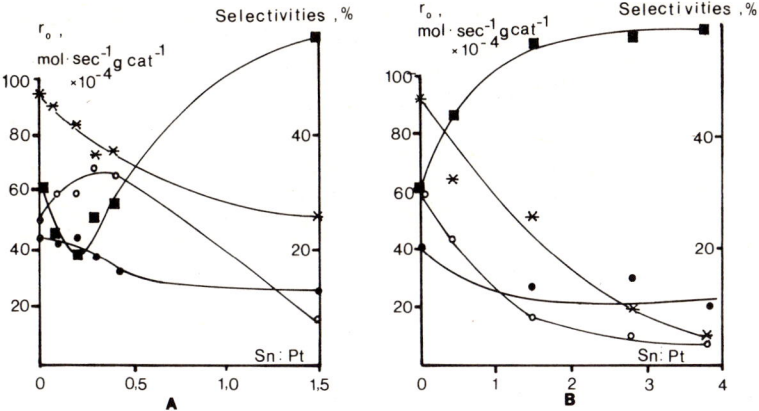

Fig. 11.7. Catalysts: Sn-Pt/Al$_2$O$_3$. Reaction rates and selectivity data as a function of tin content. A - Catalysts; B1-B4 and D$_2$. B - Catalysts D1-D4. Reaction conditions: see Table 11.10. ✕ - rate of n-hexane conversion; selectivities: o - benzene; ● - -isohexanes, ■ - hydrogenolysis.

preparation of these catalysts, an oxygen treatment was used before final reduction. This resulted in an increase in the activity of all catalysts. Experimental data are given in Table 11.11. In order to compare the intrinsic properties of the catalysts after oxygen treatment, selectivity data were measured at the same extent of conversion.

All of the catalysts with high tin contents were strongly activated by the oxygen treatment, their relative activity changes being much more pronounced than that of the base catalyst. Catalysts of Type A were only slightly influenced by oxygen treatment. The unusually high selectivities for hydrogenolysis of catalysts D1 and D4 (see Fig. 11.7) were lowered to that of the base catalyst. The oxygen treatment did not alter the unusual properties of the catalyst of type C.

(b) <u>Continuous-flow experiments.</u> The Sn-Pt/Al$_2$O$_3$ catalysts were also tested in a continuous-flow reactor, as was done for the Pd-Pt/Al$_2$O$_3$ catalysts. In this way the long-term activity and the stability of these catalysts could be compared. Typical results are shown in Fig. 11.8 (ref. 42). Selectivity data and the coke content of the catalysts measured after four hours are given in Table 11.12.

TABLE 11.11

Activity changes and selectivity data upon applying oxygen treatment (ref. 42)

Catalysts	Relative activity changes[a]	Selectivities (%)[b]			
		Benzene	Isohexanes	Hydrogenolysis	Olefin
base 1	1.28	30.1	24.7	18.0	15.0
A2	1.08	25.7	27.1	8.7	33.0
B4	1.52	33.3	26.6	12.0	17.0
D1	1.98	38.0	22.7	12.8	17.0
D4	16.10	25.0	16.7	11.6	35.0
C	2.96	16.3	1.4	3.6	75.0

[a]The ratio of r_o values obtained after oxygen treatment and without it.
[b]Measured at 4.5 ± 0.1% conversion.

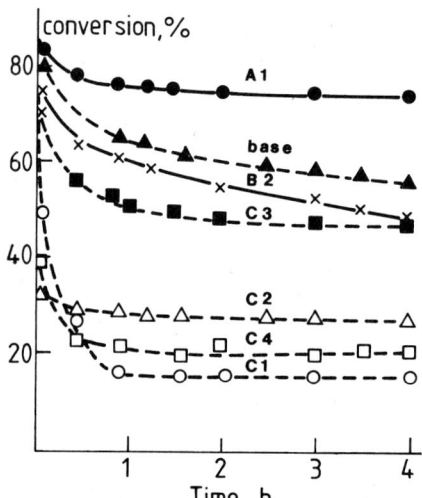

Fig. 11.8. Catalysts: Sn-Pt/Al$_2$O$_3$ conversion versus time dependence for n-hexane reaction. Reaction conditions: see Table 11.12.

One the characteristics of the Sn-Pt/Al$_2$O$_3$ catalysts is their relatively high stability. The changes in activity between one hour and four hours of experiment, with the exception of those for the base and B4 catalysts, are less than 2-3 conversion percents. Catalyst A2 had the highest activity. The coke formation was the lowest on this catalyst. Catalyst B4 strongly resembled the base catalysts, however, its selectivity for benzene formation was much lower than that of the base catalyst. Catalyst C1 strongly retained its unusuall properties under continuous-flow conditions. For catalysts of type D a good correlation was found between the tin content and the overall activity or the decreased selectivity for benzene.

TABLE 11.12

Results obtained in continuous-flow experiments (ref. 42)

Catalysts	Sn/Pt	Conversion (%)	Selectivities (%)					Coke content (mass %)
			B	Hy	$I-C_6$	MCP	Ol	
basel	-	55,4	16.8	27.8	29.5	11.4	12.4	1.50
A2	0.90	73.6	18.2	21.9	47.8	5.9	4.4	0.40
B4	0.42	48.2	9.6	29.2	29.7	12.9	13.4	0.63
C1	0.40	23.0	6.7	4.9	12.6	9.3	57.1	0.78
D1	0.40	48.2	11.4	40.9	20.6	13.9	10.9	-
D3	2.80	26.4	3.2	33.2	26.8	18.1	16.4	0.78
D4	3.80	15.1	2.9	39.7	23.0	19.1	13.3	0.77

Reaction temperature: $500°C$; catalysts treated in H_2 at $500°C$, amount of catalysts: 0.2 g, flow-rate: 0.5×10^{-3} l sec^{-1}, CH/H = = 5:1. (Abbreviations see Table 11.6; Ol = C_6-unsaturated).

11.6.4 Hydrogenation of acrylonitrile

The lead-modified catalysts prepared as described in Section 11.5.3. were studied in the selective hydrogenation of acrylonitrile to propionitrile (refs. 43,58). A conventional trickle bed reactor operated in the pressure range 10-20 bar was used.

On a semi-industrial Ni/Al_2O_3 catalyst (with 40% Ni), parallel to reaction 11.22:

$$CH_2=CH-CN \xrightarrow{H_2} CH_3-CH_2-CN \qquad (11.22)$$

the formation of propylamine, di- and tri-n-propylamines and n-propylimine were observed in the temperature range 80-135°C. Typical experimental data are shown in Table 11.13.

Three types of $Pb-Ni/Al_2O_3$ catalysts were prepared in order to demonstrate the effect of controlled surface reactions on the properties of these catalysts (see Section 11.5.3). Catalysts of Type A were prepared with 4.0 and 6.0% lead content. On catalyst A4, although 100% conversion was obtained, the selectivity for propionitrile appeared to be below 50%, i.e., the introduction of the lead did not resulted in an improvement in the PN selectivity. Upon increasing the lead concentration to 6% a significant improvement was obtained in the selectivity for PN with a parallel decrease in the activity as the extent of conversion dropped up to 72.8%. Four different catalysts of type B were prepared. Upon increasing the tin content the selectivity for propionitrile was

TABLE 11.13

Hydrogenation of acrylonitrile on Ni/Al_2O_3 catalyst[a]

Temperature (°C)	Conversion (%)	Selectivities (%)[b]				
		NPA	NPI	DNPA	TNPA	PN
81	97.6	8.8	5.9	1.4	-	83.9
100	100.0	13.1	10.0	3.9	-	73.0
104	99.6	23.9	10.2	9.9	0.3	53.7
135	100.0	51.2	-	41.4	6.1	1.3

[a] Liquid flow-rate: 0.3 kg kg^{-1} h^{-1} acrylonitrile (1.92 mol - n - n-hexane solution): gas flow-rate: 6.6 l h^{-1}; P = 15 bar.
[b] NPA = n-propylamine, NPI = n-propylimine, DNPA = di-n-propylamine, TNPA = tri-n-propylamine, PN = propionitrile.

TABLE 11.14

Activity and selectivity of Pb-Ni/Al_2O_3 catalysts of types A, B and C[a]

Catalysts	Lead content (%)	Temperature (°C)	Conversion (%)	Selectivities (%)				
				NPA	NPI	DNPA	TNPA	PN
A4	4.0	106	100.0	25.5	15.5	15.5	0.9	41.7
A5	6.0	104	72.8	-	1.1	-	-	98.9
B4	3.1	102	99.5	24.4	14.1	13.2	0.4	47.9
		108	100,0	29.8	14.1	26.4	4.2	25.4
B5	4.3	105	98.1	5.7	7.2	2.7	-	84.4
B6	6.1	106	98.9	2.5	4.0	1.3	0.3	91.9
B7	9.9	102	81.0	-	-	-	-	100.0
		111	98.9	6.4	7.2	11.0	1.3	74.1
C1	5.1	100	100.0	1.4	3.6	1.1	-	93.9

[a] Reaction conditions and abbreviations are given in Table 11.13.

increased to 100%, although high propionitrile selectivity and high extents of conversion could not be obtained simultaneously. The best results were obtained on a catalyst of type C prepared by controlled surface reactions. This catalyst had high activity and resulted in strong suppression of the most undesirable side reactions, i.e., the formation of di-n-propylamines and tri-n-propylamines. The results obtained clearly demonstrate that the best selectivity control is achieved when a direct lead-nickel interaction occurs. Upon introduction of lead into the nickel by CSR, the most reactive sites of the nickel responsible for the formation of di-n-propylamines, and tri-n-propylamines can be selectively blocked. Over these types of Pb-Ni/Al_2O_3 catalysts, acrylonitrile can be hydrogenated selectively to propionitrile.

11.7 CONCLUSIONS

An attempt has been made to demonstrate that different supported bimetallic catalysts can be prepared by means of CSRs. At least two types of catalysts could be prepared: (i) catalysts with direct metal-metal interaction, (ii) catalysts in which the second metal is strongly bound to the support. The properties of these catalysts strongly depended on (i) the type of surface reactions used during the preparation and (ii) the thermal treatment applied before the reaction.

Catalysts prepared in this way possess unique catalytic properties and it has been clearly demonstrated that the principle of CSRs is a powerful tool in the design of bimetallic catalysts with "tailored" selectivities.

REFERENCES

1. Chevron, U.S. Pat. 3,415,737 (19 May, 1967).
2. Exxon, U.S. Pat. 3,953,368 (27 April, 1976).
3. C.F.R., Fr. Pat. 2,031,984 (14 February, 1969).
4. Yu.I. Yermakov and B.N. Kuznetsov, J. Mol. Catal., 9(1980)13.
5. Yu.I. Yermakov, B.N. Kuznetsov and V.A. Zakharov (Editors), Catalysis by Supported Complexes, Elsevier, Amsterdam, 1981.
6. D.C. Bailey and S.H. Langer, Chem. Rev., 81(1981)110.
7. Yu.I. Yermakov and B.N. Kuznetsov, Kinet. Katal., 18(1977)1167.
8. I. Böszörményi, S. Dobos, L. Guczi, L. Markó, K. Lázár, W.M. Reiff, Z. Schay, L. Takács, A. Vizi-Orosz, 8th International Congress on Catalysis, Proceedings, Vol. V, Verlag Chemie, Weinheim, 1984, p. 183.
9. W. Keim, H. Leuchs and B. Engler, Forschungsber. Landes Nordrhein-Westfalen, 2838(1979)1.
10. B.N. Kuznetsov, in Yu.I. Yermakov (Editor), Supported Metallic Catalysts for Hydrocarbon Reactions, New Approaches to their Preparation and Study (in Russian), Vol. 2, Institute of Catalysis, Novosibirsk, 1978, p. 43-66.
11. R. Bacaud, P. Bussiere and F. Fugueras, J. Catal., 69(1981)399.
12. M.D. Kolb, M. Przasnyski and H. Gerischer, J. Electroanal. Chem., 54(1974)25.
13. D.M. Kolb, in H. Gerischer and C. Tobias (Editors) Advances in Electrochemistry and Electrochemical Engineering, Vol. II. Wiley Interscience, New York, 1978, p. 195.
14. J.W. Schultze and K.J. Vetter, J. Electroanal. Chem., 44(1973) 63.

15 J.W. Schultze and F.D. Koppitz, Electrochim. Acta, 21(1976)327.
16 F.D. Koppitz and J.W. Schultze, Electrochim. Acta, 21(1976)337.
17 B.E. Conway and S. Marshall, Electrochim. Acta, 28(1983)1003.
18 S. Swathirajan and S. Bruckenstein, Electrochim. Acta, 28(1983) 865.
19 K.J. Vetter and J.W. Schultze, J. Electroanal. Chem., 53(1974) 67.
20 K.J. Vetter and J.W. Schultze, Ber. Bunsenges. Phys. Chem., 76(1972)920.
21 N. Furuya and S. Motoo, J. Electroanal. Chem., 72(1976)165.
22 R.G. Barradas, S. Fletcher and S. Szabó, Can. J. Chem., 56(1978)2029.
23 S. Szabó and F. Nagy, Isr. J. Chem., 18(1979)162.
24 S. Szabó and F. Nagy, J. Electroanal. Chem., 70(1976)357.
25 S. Szabó, J. Electroanal. Chem., 172(1984)359.
26 N. Furuya and S. Motoo, J. Electroanal. Chem., 98(1979)195.
27 M. Ifandi, S. Szabó and F. Nagy, Elektrokhimiya, 18(1982)1205.
28 F. Mikuni and T. Takamura, Denki Kagaku, 38(1970)113.
29 S. Szabó and F. Nagy, J. Electroanal. Chem., 85(1977)339.
30 N. Furuya and S. Motoo, J. Electroanal. Chem., 98(1979)189.
31 S. Szabó and F. Nagy, J. Electroanal. Chem., 160(1984)299.
32 B.J. Bowles, Electrochim. Acta, 15(1970)589.
33 S. Szabó and F. Nagy, J. Electroanal. Chem., 84(1977)93.
34 B.J. Bowles, Electrochim. Acta, 15(1970)737.
35 S. Szabó and F. Nagy, J. Electroanal. Chem., 88(1978)259.
36 N. Furuya and S. Motto, J. Electroanal. Chem., 107(1980)159.
37 J. Margitfalvi, E. Kern-Tálas and S. Göbölös, Abstracts, XIIth International Conference on Organometallic Chemistry, Vienna, September 1985, p. 518.
38 S. Szabó and F. Nagy, Appl. Catal., 16(1985)209.
39 J. Margitfalvi, S. Szabó, F. Nagy, S. Göbölös and M. Hegedüs, in G. Poncelet, P. Grange and P.A. Jacobs (Editors), Preparation of Catalysts III, Elsevier, Amsterdam, 1983, p. 473.
40 J. Margitfalvi, M. Hegedüs, S. Göbölös, S. Szabó and F. Nagy, submitted for publication.
41 J. Margitfalvi, M. Hegedüs, S. Szabó and F. Nagy, React. Kinet. Catal. Lett., 18(1981)175-179.
42 J. Margitfalvi, M. Hegedüs, S. Göbölös, E. Kern-Tálas, P. Szedlacsek, S. Szabó and F. Nagy, 8th International Congress on Catalysis, Proceedings, Vol. IV, Verlag Chemie, Weinheim, 1984, p. 903.
43 J. Margitfalvi, E. Kern-Tálas, S. Göbölös and M. Hegedüs, Abstracts 9th North American Meeting of the Catalysis Society, Houston, March 1985, p. C-18.
44 A.N. Karasov, L.S. Polak and E.B. Shlikhter, Probl. Kinet. Katal., Akad. Nauk SSSR, 12(1968)297-301 (in Russian)
45 J. Margitfalvi, M. Hegedüs, P. Szedlacsek, E. Kern-Tálas and F. Nagy, Abstracts 9th North American Meeting of the Catalysis Society, Houston, March 1985, p. P-24.
46 Z. Schay, J. Margitfalvi, E. Kern-Tálas and L. Guczi, in preparation.
47 J. Margitfalvi, S. Göbölös and E. Kern-Tálas, submitted for publication.
48 J. Margitfalvi, M. Hegedüs, S. Szabó and F. Nagy, in preparation.
49 S. Szabó and F. Nagy, Magy. Kém. Folyóirat, 86(1980)371.
50 J. Margitfalvi, P. Szedlacsek, M. Hegedüs and F. Nagy, Appl. Catal., 15(1985)69.
51 J. Margitfalvi, M. Hegedüs, S. Göbölös and F. Nagy, React. Kinet. Catal. Lett., 18(1981)73-78.

52 J. Margitfalvi, S. Göbölös, E. Kwaysser, S. Szabó, F. Nagy and
    L. Koltai, React. Kinet. Catal. Lett., 18(1981)
53 J. Margitfalvi, M. Hegedüs, S. Göbölös, S. Szabó and F. Nagy,
    submitted for publication.
54 L. Guczi, K. Matusek, A. Sárkány and P. Tétényi, Bull. Soc.
    Chim. Belg., 88(1979)497.
55 P. Biloen, I.N. Helle, H. Verbeek, F.M. Dautzenberg and W.M.H.
    Sachtler, J. of Catal., 63(1980)112-118.
56 J. Margitfalvi, E. Kern-Tálas, M. Hegedüs and P. Szedlacsek,
    in preparation.
57 R. Burch, J. Catal.,71(1981)348.
58 J. Margitfalvi and S. Göbölös, submitted for publication.

Chapter 12

NEW SUPPORTED METALLIC NICKEL SYSTEMS

J. M. MARINAS, J. M. CAMPELO and D. LUNA
Department of Organic Chemistry, Faculty of Sciences, Córdoba University, E-14004 Córdoba (Spain)

12.1 INTRODUCTION

In recent years, there has been increasing interest in the application of nickel metal, or several nickel metal compounds supported on various carriers, as heterogeneous catalysts in organic reduction processes on the laboratory scale as well as in the petrochemical, chemical and automotive industries. Evidence has been found that in many systems the carrier exerts a marked influence on the properties of metal particles supported on it. Furthermore, the incorporation of additives, usually known as promoters, and the employment of supported bimetallic systems allow the choice of the most suitable catalyst for a particular process.

The advantages of supported catalysts, in preference to other metals or unsupported nickel such as Raney nickel, lie in their relatively low cost, remarkable thermal stability and resistance to poisoning. Furthermore, the support facilitates the formation of very small metal particles having a very high metal surface area.

The most common applications of supported nickel catalysts are in the hydrogenation of a large number of organic functional groups, in hydrogenation-dehydrogenation, cracking and hydrogenolysis of hydrocarbons and in the methanation of coal synthesis gas.

According to recent literature, the study of supported nickel catalysts is conducted basically in four wide areas:

(1) Development of new methods for preparing highly dispersed nickel metal catalysts deposited on classical supports, such as Al_2O_3, SiO_2, SiO_2-Al_2O_3, Kieselguhr, other metal oxides or carbon.

(2) Development of supported bi- or multimetallic nickel systems or modification of the habitual nickel systems by adsorption of several metals.

(3) Development of heterogeneous supported nickel systems for asymmetric hydrogenation by the utilization of modifiers of different structures and different modifying conditions.

(4) Development of new carriers in order to obtain new supported nickel

catalysts exhibiting higher activity and/or selectivity than conventional ones.

The proliferation of the literature means that the present review can only highlight the achievement on the last point. However, as a consequence of the interrelationships between the topics and techniques, overlaps are inevitable.

12.2 PREPARATION AND CHARACTERIZATION OF NICKEL CATALYSTS
12.2.1 Introduction

According to Delmon and Houalla (ref. 1), the two major factors which affect the reduction of supported oxides are the dispersion of the deposited oxide and its interaction with the carrier. We think that both parameters determine the catalytic behaviour of the supported metal systems. These factors are in turn affected by two fundamental parameters: the method of preparation and the nature of the carrier. Thus, it is well known that small variations in the experimental procedure may lead to profound changes in the properties of the prepared catalyst. The pore structure, porosity, surface area, adsortive properties, etc., of the supports can have an important influence on the catalyst performance, on the activity and/or selectivity as well as on secondary support effects, viz., adsorption of poisons, variation in metal dispersion or pore size and pore-size distribution. Thus, it is conceivable that large organic molecules will be prevented from entering cavities of small size and, therefore, only the active sites found within large pores contribute significantly to the reaction.

Accordingly, new preparation methods and new materials employed as supports are treated in separate sections. The catalyst characterization is indicated in every case with emphasis on the measurement of metal dispersion, although the acidity of the supports is sometimes studied.

12.2.2 New preparation methods

In most of the works dealing with the preparation of supported nickel catalysts, some of the following three methods are employed: wet impregnation, deposition/ coprecipitation or ion exchange. However, an increasing number of papers deal with new synthetic procedures to enhance the catalytic activity of the classical supported systems, frequently causing a supported nickel system to be classified as "new" since a classical support is subjected to some treatment or obtained by a new synthetic procedure.

For example, Teichner and co-workers (refs. 2,3) described a new nickel catalyst supported on alumina, obtained by thermal decomposition of nickel hydroaluminate, which is very active in the hydrogenolysis of alkylbenzenes. Klabunde et al. (ref. 4) used a metal atom-vapour method for the deposition of highly dispersed nickel metal catalysts [solvated metal atom dispersed (SMAD) catalysts] supported on Al_2O_3 (refs. 4.5) or MgO (ref. 6). This method is an extension of the work of Yermakov (ref. 7) and Ichikawa (ref. 8). The catalytic activities of the SMAD catalysts greatly outstrip those of the conventional systems in the hydrogenation-dehydrogenation, hydrogenolysis and methanation reactions.

The development of techniques to control metal particle size in supported catalysts is currently of interest, not only for increasing the metal surface area, bult also in order to study the effect of particle size on the activity and/or selectivity, since it has been reported that the physical properties of a metal particle should change when it is < 100 Å (ref. 9). Thus, several authors have obtained excellent results by using a nickel salt as precursor. For example, Martin and co-workers (refs. 10-12) synthesised Ni/SiO_2 systems by making the support react with a solution of nickel nitrate hexamine. The degree of reduction as measured by saturation magnetization was found to be nearly equal to unity and the nickel particle diameters span the range 5-15 nm.

Zagli and Falconer (ref. 13) described a method involving precipitation of nickel dimethylglyoxime onto silica. Such systems may be reduced at low temperature (498 K) and present higher dispersion and metal reduction than catalysts prepared by nickel nitrate impregnation and reduction under similar conditions.

On the other hand, Ueno and co-workers (refs. 14,15) described a new method to control nickel metal particle size in a silica-supported catalyst in which nickel particles are immobilized by forming chemical bonds with oxygen atoms in the support. The dispersion and particle-size distribution of the catalysts were determined by transmission electron microscopy (TEM) and small-angle X-ray scattering (SAXS), the state of bonding of the nickel ions or metallic particles to the carrier by infrared (IR) spectroscopy and the fraction of nickel reduced to metal by magnetization measurements. The catalytic activity/selectivity was determined for the hydrogenation of propionaldehyde.

Gut et al. (ref. 16) obtained excellent results for silica-supported nickel systems prepared by reduction of nickel formate in sunflower oil at 500 K under nitrogen (wet reduction).

A new and original technique for obtaining nickel catalysts on Al_2O_3, characterized by small metal crystallites in spite of the quite drastic reduction conditions (3 h at 773 K) was reported by Barcicki et al. (ref. 17). The support was previously impregnated with a solution of disodium EDTA at 343 K. After filtration, the solid was dried at 393 K for 2 h and then impregnated with a solution of nickel nitrate.

Umani-Ronchi and co-workers (refs. 18,19) reported the preparation of highly active nickel catalysts dispersed on a potassium-graphite surface by addition of a solution of bis(dimethoxyethane)dibromonickel in tetrahydrofuran-hexamethyl phosphoric triamide (THF-HMPTA) to a slurry of graphite in THF at room temperature under an argon atmosphere. The reduction of the nickel salt was fast and the catalysts thus obtained were highly active for liquid-phase semihydrogenation of alkynes to alkenes in the presence of ethylenediamine as catalyst modifier. However, exposure to air modifies the catalyst and markedly reduces its activity.

On the other hand, catalysts with oriented nickel faces (ref. 20) have been prepared starting from a synthetic clay with lamellar structure, such as nickel antigorite $Ni_3(OH)_4Si_2O_5$ (ref. 21), which upon heating and reduction is transformed into Ni/SiO_2 exhibiting a (111) or (110) orientation of the nickel faces (depending upon the conditions of reduction). The system thus obtained shows a weak nickel-support interaction (refs. 22,23).

Other nickel catalysts have been prepared by the homogeneous precipitation of slowly decomposing urea, as described by Richardson and Dubus (ref. 24). This method generates narrow and reproducible crystallite-size distributions, the time of precipitation being the most important factor in the preparation method (ref. 25).

The design and development of a new range of methanation catalysts has been described by Trimm and co-workers (refs. 26,27). Of over 120 supported catalysts of nickel and other metals (Ni-Rh, Ni-Co, Ni-Zr, etc), they chose the most convenient catalysts to accomplish the methanantion process under standard conditions. The authors follow the methodology proposed by Trimm in "Design of Industrial Catalysts" (ref. 28). Finally, Geus (ref. 29), on consideration of the deficiencies of the presently used procedures, arrived at the conclusion that the precipitation of an active precursor onto a separately prepared carrier can provide the best results.

Nevertheless, we think that the development of new supported nickel systems, based on new synthesis procedures is a field of growing interest whose study will proceed in parallel with that of the development of new carriers.

12.2.3 New materials

An increasing number of papers deal with the applicability of very different materials as supports for nickel, some being modifications of classic carriers. For example, graphite cannot be directly intercalated with transition metals, but intercalation is claimed to be possible by the reduction of the intercalated salts of these metals. A wide range of metallic intercalates is available from the Ventron Corporation under the general name "Graphimets". A detailed description of their preparation is included in a U.S. patent by Lalancette (ref. 30). After reduction, this process is said to produce the metal substantially in atomic dispersion between the sheets of the graphite. In view of the higher catalytic activity of these materials (ref. 31), Smith et al. (ref. 32) carried out a TEM study on "Graphimets" of several metals, including nickel, confirming that the graphite occurred as overlapping flakes with the metal clusters present on the surface.

Likewise, the use of some naturally occurring materials as supports has recently been studied. Thus, mica is an excellent support material for several reasons (ref. 33): first, gases such as CO and CO_2 are not adsorbed on mica; secondly, clean surfaces are easily obtained; and finally, specimens are easily prepared for subsequent TEM analysis. The mica-supported nickel catalyst is obtained by vapour deposition using electron-beam evaporation of a high-purity nickel source on the mica films. In these systems the metal-support interaction increases with diminishing particle size.

Granquist (ref. 34) described a novel clay-like silicate that consists of silica-alumina-silica layers (2:1 layers) with the following unit-cell composition

$$(Al)_4^{octa} (Al_xSi_{8-x})^{tetra} O_{20} (OH,F_4)^{x-} xNH_4^+, H_2O$$

where x is about 1.5. Since these systems contain both mica and montmorillonite-like layers, they have been called synthetic mica-montmorillonite (SMM) (ref. 35). Much attention has been given to the synthesis of nickel-substituted SMM and several catalytic applications have been reported. These included hydroisomerization, hydrocracking (refs. 36,37) and hydrotreating (ref. 38). Heinemann et al. (ref. 39) found that an important pretreatment step of the catalyst is its reduction in hydrogen. Palladium promotes the formation of zero valent nickel. Removal of metallic nickel from reduced Ni-SMM (not containing Pd) by carbon monoxide destroys the pentane hydroisomerization activity, while leaving the number of acidic sites unchanged. Apparently, the acidic sites alone are not capable of isomerizing pentene, the presence of a metal fraction being necessary.

Covert et al. (ref. 40) described a Ni/pumice catalyst obtained by impregnation with nickel nitrate and further precipitation of nickel carbonate at room temperature. After filtration and drying overnight, the solid is reduced at 725 K. The catalyst thus obtained is active in alkene hydrogenations, as shown by Gault and co-workers (ref. 41). The reactivity and structure of nickel-exchanged Prayssac vermiculite have also been studied (ref. 42). The authors determined the mechanism of the hydrogen reduction of exchangeable Ni^{2+} in a trioctahedral vermiculite. The very weak reactivity of nickel metal particles towards ethylene hydrogenation is in acord with the assumption that most of nickel atoms are encaged in the ditrigonal cavities.

The honeycomb ceramic supports are currently of considerable technical interest in pollution control. They are also studied as a nickel support (refs. 43,44). These catalysts are active in the hydrogenation of olefins, selective hydrogenation of acrylonitrile and benzene hydrogenation (ref. 44). On the other hand, we have obtained promising results in the liquid-phase hydrogenation of 1-hexene over a catalyst supported on a natural sepiolite from Vallecas-Madrid, Spain (ref. 45).

However it is in the field of inorganic supports where the majority of new systems are to be found. Of the new nickel inorganic supports, four kinds of compound are especially important: zeolites, titania, magnesia and aluminium orthophosphate and related systems $AlPO_4$-Al_2O_3 and $AlPO_4$-SiO_2. They will be discussed later in more detail.

Notwithstanding, other inorganic supports have been used for nickel. Thus, Mardashev and co-workers (refs. 46,47) developed new nickel catalysts supported on ionic carriers, NaF and CaF_2, prepared by saturating the carrier with an aqueous solution of nickel nitrate and treating the product with UV ligh and hydrogen. These catalysts when treated with different organic modifiers (acetone, benzene, isopropanol, etc) showed appreciable changes in the nickel coordination state, according to X-ray photoelectron spectroscopy (XPS) and diffuse reflectance (DR)studies, and higher specificity towards the substrate in which the carrier was treated, i.e., "memory". The specificity increases with the amount of modifier adsorbed on the carrier. Thus, large differences in the catalytic activity of the sample have been noted in acetone hydrogenation (ref. 47).

The use of magnesium silicate as a support for nickel (ref. 48) leads to a catalyst which exhibits a turnover number about forty times smaller than that of Ni/SiO_2 in methanation. This behaviour was associated with a metal-support interaction, according to the metal dispersion determined by nitrogen and hydrogen adsorption, TEM, XPS and mercury porosimetry measurements.

Several metal oxides have been studied as supports for nickel and the metal-supported system has been prepared either by the deposition of metal on the metal oxide or by thermal treatment, reduction or reaction of appropriate compounds. Table 12.1 summarizes the various methods employed.

TABLE 12.1
Methods of preparation of some metal oxides used as supports for nickel

Metal oxide	Preparative components	Refs.
CaO	$Ca(OH)_2$, thermal decomposition in air	53
Cr_2O_3	$Cr(NO_3)_3$, NH_4OH	49,50
La_2O_3	$La(NO_3)_3$, NH_4OH	53
ThO_2	$Th(NO_3)_4$, thermal decomposition in air	53
ZnO	$Zn(NO_3)_2$, NH_4OH	54
ZrO_2	$ZrCl_4$, NH_4OH	52
ZrO_2	$ZrOCl_2$, NH_4OH	53

Thus, Sokolskii and co-workers (refs. 49,50) prepared a Ni/Cr_2O_3 catalyst of high nickel content which is active in the liquid-phase hydrogenation of α,β-unsaturated carbonyl compounds. Zirconium oxide has been used as the support for nickel by several authors (refs. 51-53). Nickel salts are easily reduced to nickel metal, although above 700 K the reduction results in a decrease in hydrogen adsorption and catalytic activities in ethane hydrogenolysis and benzene hydrogenation (ref. 52). These properties are restored by treatment with oxygen followed by a subsequent reduction at moderate temperatures. The loss of chemisorption is of the same order of magnitude for large and small nickel particles and suggests that the whole metal surface is modified.

On the other hand, Tanabe and co-workers (ref. 53) studied the hydrogenation of 1-butene and N,N-dimethyl-2-propenylamine over nickel catalysts supported on various kinds of metal oxides MgO, CaO, La_2O_3, ThO_2, ZrO_2, Al_2O_3, TiO_2, SiO_2 and SiO_2-Al_2O_3, to examine the influence of the support on the catalytic activity. The activities did not correlate with the amounts of hydrogen adsorbed nor the amounts of the metallic nickel titrated with iodine, although the activity was roughly proportional to the number of nickel atoms exposed on the surface (ref. 54). Nickel supported on solid base catalysts, MgO, CaO, La_2O_3 and ThO_2, exhibited high activity as compared to nickel on solid acidic catalysts, and the reaction proceeded via double bond migration to N,N-dimethyl-1-propenylamine. Other supported nickel systems have been described: Ni/WO_3-Al_2O_3 (ref. 55), Ni/SiO_2-ZnO (ref. 56), Ni/Al_2O_3-ZnO_2 (refs. 57-59), Ni/Al_2O_3-MgO (ref. 60), and nickel on small area supports, such as Ni/silicon carbide (refs. 61-63), which have greater thermal conductivity and are more inert. The latter are mainly used in the hydrogenolysis of alkanes.

A new class of supported catalysts Ni/MO_2, where M= Si, Th, Ce, U or Zr, obtained from the intermetallic compounds MNi_5, have recently been studied in relation to the conversion of CO/H_2 mixtures into hydrocarbons (refs. 64-77). Auger spectroscopy (ref. 64) and electron spectroscopy for chemical analysis (ESCA) studies (ref. 73) demonstrated the active catalyst to be the system Ni/MO_2 rather than the original intermetallic compound. The catalysts thus prepared exhibit exceptional activity compared to the supported catalysts obtained by conventional impregnation.

Similarly, Siegmann et al. (ref. 78) have recently found that the surface of the closely related material $LaNi_5$ consists of La_2O_3, $La(OH)_3$ and metallic nickel; $LaNi_5$ differs from $ThNi_5$ in that the former absorbs hydrogen (ref. 79) whereas the latter does not (ref. 80). Barrault et al. (ref. 81) reported that some intermetallic compounds in the form RE_xNi_y, where RE represents lanthanum or a rare earth mischmetal alloy, exhibit an appreciable activity, particularly in carbon monoxide hydrogenation, toluene dealkylation and ethane hydrogenolysis. X-ray powder analysis of the Ni-LaH samples used in ethane hydrogenolysis showed a transformation of the intermetallic compounds during the reaction. It is clear that the intermetallic compounds provide a new way of preparing supported metals. Indeed, through this method, the use of metallic salts, "anion effect", and the impurities that supports often contain, are avoided. This provides new possibilities for solving some important problems in conventional catalyst preparation.

Kudlacek and co-workers (refs. 82-84) reported the use of Ni/ZnO catalysts, prepared by reduction of NiO-ZnO mixed oxides, which exhibit activity in the liquid-phase hydrogenation of maleic and fumaric acids. The catalytic systems obtained have a low degree of reduction, except when starting with a high content of NiO.

On the other hand, rather more sophisticated supports for nickel have been described. A new type of heterogeneous nickel catalyst with a high durability and hydrogenation activity was prepared (refs. 85, 86) by embedding Raney nickel catalysts, at room temperature, in a vulcanized silicone rubber having high permeability to hydrogen gas and a moderate permeability to reactants. When this catalyst is modified with a mixture of NaBr and tartaric acid it shows enantioselectivity and no detectable decrease in activity due to storage in air at room temperature. Other hydrogenation studies using polymer-bound nickel catalysts have been reported. A catalyst may be prepared by treating a supported nickel complex with alkylaluminium (ref. 87), or Ni^{2+} may be used without any activating agent by anchoring $NiCl_2$ to calcined polyacrylonitrile and then heating the system to 575 K in the presence of hydrogen, with the formation of nickel metal (ref. 88). In both cases the hydrogenation studies were limited to alkenes and alkynes.

Finally, it should be noted that the preparation of new supported nickel systems, in order to develop active catalysts for hydrogenation processes, is continually been reported.

(i) <u>Molecular sieve zeolites</u>. Zeolites are peculiar supports incorporating large amounts of acidic sites of different natures. Moreover, transition metals at a high degree of dispersion (approaching the atomic one) may be inmmobilized on zeolites via ion exchange (ref. 89) or impregnation (ref. 90). Furthermore the catalytic activity can be enhanced by the addition of suitable amounts of catalyst-promoting compounds to the reaction mixture (ref. 91). Accordingly, we expect the catalytic properties of metallic clusters on zeolites to show some peculiarities, and there is a growing interest in their applicability to many different catalytic processes (ref. 92).

However, the zeolitic matrix is especially suited for the preparation of finely divided metals, with a narrow particle-size range (ref. 93). The different catalytic behaviour is primarily determined by the difference in the state of the nickel (ref. 94), because it has been shown that zeolites are a suitable matrix to stabilize unusual oxidation states (although the stabilization of a metal of the first transition row is more difficult to obtain than that of most of the other noble metals in Group 8 (refs. 95-99). Thus, the reduction by molecular hydrogen of nickel in faujasite-type zeolites is complete only at 873 K. Large particles are formed, so that systematic studies have been undertaken to analyze the factors governing the reducibility of Ni^{2+} in faujasite-type zeolites, and the optimun conditions for obtaining well dispersed nickel have been established (refs. 100-108). It has been shown that the reducibility of nickel ions and the formation of very small nickel particles in Ni-X-zeolites are favoured by low support acidity, the presence of reduced noble metals (Pd or Pt) or rare-earth metal cations, Ce^{3+}, and reduction under flowing hydrogen. In this context, Jeanjean et al. (ref. 103) determined through X-ray diffraction analysis the crystal structure of $Ni_{17}Pt_{0.5}Na_{40}H_{11}$-X-zeolite before and after hydrogen reduction, finding that when the reduction is conducted under flowing hydrogen the presence of platinum greatly increases the degree of reduction and the stabilizing effect is maintained since an homogeneous dispersion of particles of about 25 Å is obtained. Klyneva et al. (ref. 90) reported that the introduction of chromium into NiCaY increased the nickel dispersity and promoted a decrease in the partial agglomeration of the metal. Furthermore, the difference between the impregnated and exchanged samples was determined. In the first case, the metal particles are located mainly on the outer surface of the zeolite crystals, while in the second the nickel particles are located mainly inside zeolite cavities. Exner et al. (ref. 109) found that a monodispersed nickel phase with a narrow aggregate size distribution (6-9 nm) remains stable with respect to sintering and catalytic activity in the hydrogenation of carbon monoxide up to 625 K, if fajausite is loaded simultaneously with nickel and calcium ions in the ion-exchange step of the preparation. The introduced calcium ions exhibit an unfavourable influence on the catalytic activity, as seem from Table 12.2. All samples were analyzed by X-ray diffraction and transmission electron microscopy and were prepared as described elsewhere (ref. 92).

Besides XRD and TEM, many physical, chemical and physico-chemical techniques have been used to obtain information on the surface area and particle-size distribution of supported nickel catalysts. Thus, a nickel-exchanged X-zeolite was studied by electron spinning resonance (ESR) (ref. 110) and by magnetic and UV-visible spectroscopic methods (ref. 111). Briese-Gülban et al. (ref. 112) obtained the reduction degree of a reduced monodispersed nickel faujasite by comparison of particle-size distributions determined by transmission electron microscopy and dynamic oxygen chemisorption. Olivier et al. (ref. 99), by the combined use of electron paramagnetic resonance (EPR), electron nuclear double resonance (ENDOR) and reflectance spectroscopy, showed that the reduction of NiCa-X zeolites by hydrogen in the range 473-623 K leads to the formation of nickel (I) species located in the sodalite units and identified as hydrocomplexes $Ni(H_2)^+$. Evidence was provided to show that the genesis and the decomposition of this species is governed by the reaction:

$$Ni^0 + Ni^{+2} + 2H_2 \longrightarrow 2Ni(H_2)^+ \qquad (12.1)$$

TABLE 12.2
Composition and catalytic properties of the nickel faujasites analyzed by transmission electron microscopy according to ref. 109.

Abbreviation	Composition	Catalytic activities[a]					
		250°C		300°C		350°C	
		1h	6h	1h	6h	1h	6h
NiCaX 5[b]	$Ni_{9.6}Ca_{10.6}Na_{45.5}X$	14	27	209	214	499	520
NiCaX 1.1[c]	$Ni_{10.1}Ca_{10.7}Na_{44.5}X$	4	4	82	103	164	70
NiX 17	$Ni_{7.6}Na_{70.8}X$	53	46	242	242	786	544

[a] Hydrocarbon yields, mg(hydrocarbon)/g(nickel).h, in the carbon monoxide hydrogenation reaction: 0.1 MPa; CO/H_2 = 3/7; 1 and 6 h time on stream; fluidized bed reactor.

[b] Simultaneous ion exchange of calcium and nickel.

[c] Consecutive ion exchange of calcium and nickel.

Romanowski (ref. 113) determined the mean sizes of nickel particles of different populations, formed by the hydrogen reduction of Ni^{2+} contained in zeolites A, X and Y, from experimental magnetization curves. The smallest particles were formed in large cages of the zeolite structure, whereas the others are formed on the surface of zeolite grains by diffusion and coalescense, proceeding simultaneously with the reduction process. Ferromagnetic resonance measurements of small nickel particles on catalysts can provide information on the particle sizes and shapes, degree of reduction, surface anisotropy and support -metal interaction. Thus, Sauvion et al. (ref. 114) showed that the hydrogen chemisorption and butane hydrogenolysis on well dispersed nickel metal on CeX zeolites can be considerably modified by the support acidity. The dependence of the catalyst activity on the ratio of acidic to metallic components has been confirmed for nickel zeolite catalysts (refs. 115-118), the rate of hydrocracking of n-octane increasing sharply with increasing nickel content in the zeolite.

Ione and co-workers (refs. 117,118) studied the peculiar catalyst action of nickel in zeolites of type Y in hydrogenation reactions. They found that the specific catalytic activity of the metal in the hydrogenation of 1-hexene, 1-octene, cyclohexene, styrene, 4-phenyl-1-butene and benzene does not depend on its dispersion and only differs slightly from that for the Ni/SiO_2 catalysts.

Besides, a decrease in the oxygen-specific chemisorption and in the specific activity in the hydrogenation of the aromatic ring of benzene was observed on increasing the nickel dispersion on zeolites. The latter was related to the specificity of atomic states in nickel clusters located in zeolite cavities.

On the other hand, Briend-Faure et al. (ref. 119) concluded that, at first sight, the activity of Ni/X-zeolite in benzene hydrogenation may be considered a measuremem of the superficial metallic nickel since the activity is a function of the extent of reduction. Davidova et al. (120) studied some aspects of the bifunctionality of nickel-zeolite catalysts in the disproportionation of toluene. They concluded that such catalysts containing only metallic nickel are not active in this reaction.

In conclusion, in the light of recent data on zeolite supports for nickel, their catalytic dual functionality may be considered as their most important feature. This dual functionality is due to the presence of large amounts of acidic sites of different natures in zeolites, although the problem of incomplete reduction to metallic nickel always exists.

(ii) <u>Titanium dioxide</u>. In the last few years titanium dioxide has been widely used as a support for several metals, especially nickel, in order to study the strong metal-support interaction (SMSI), following the observation by the Exxon group that metals supported on titania could exhibit unusual properties (refs. 121-125). Thus, particularly in the case of Ni/TiO_2 catalysts, it has been shown (refs. 126-134) that the ability to adsorb hydrogen and carbon monoxide is greatly supressed. At the same time, an exceptionally high activity in the hydrogenation of CO to methane was obtained, together with very similar activities and selectivities for hydrogenolysis to those shown by Ni/SiO_2 (refs. 127,128, 132), or drastically reduced (refs. 135,136). Actually, titania-supported nickel shows turnovers one to two orders of magnitude higher than other nickel catalysts and, perhaps more intriguinly, has the capacity to produce higher-molecular-weight paraffins (refs. 126,137). Such alterations in the adsorption and catalytic properties of the Ni/TiO_2 catalysts have been associated with a strong metal-support interaction which will be discussed in a later section.

The system Ni/TiO_2 is generally prepared by wet impregnation, of titania either purchased commercially or obtained by hydrolysis of $TiCl_4$ (refs. 23,53, 126,128,132). However, several other methods have been used, such as ion exchange, precipitation with alkali or urea (ref. 138) or reduction of a mull of the support and nickel carbonate (ref. 60). The method of preparation can affect the structure, reducibility, dispersion and even the morphology of the catalyst (ref. 138), with the possible formation of titanate species. The acid-base or redox character of the support could also have an effect (refs. 53,139,140). Menon and Froment (ref. 141), from temperature-programmed desorption (TPD) experiments, using 5% hydrogen in argon, showed that the TiO_2 carrier is reduced at higher temperatures, about 770 K, but the hydrogen consumption is so low that only a partial reduction of the surface layer alone may occur with the formation of TiO_x species (ref. 142).

The mean crystallite size and size distribution of nickel dispersed on TiO_2 have been determined by using several techniques such as gas chemisorption (H_2, CO and O_2), XRD and TEM (refs. 130,143). Thus, Vannice and co-workers (ref. 143) calculated nickel crystallite sizes from chemisorption of H_2, CO and O_2 compared them with those determined by XRD and TEM. As seen in Table 12.3, close agreement

TABLE 12.3
Chemisorption and X-ray results for nickel catalysts (ref. 143)

Catalyst	Percentage reduction[a]	Adsorption(μmol g^{-1})[b]			$\bar{d}_v$ from XRD (nm)	Sample No.
		H_2	CO	O_2		
1.4% Ni/TiO_2	86	5.0	34.9	25.3	ND[c]	1
		5.5	33.1	26.3	ND	2
7.0% Ni/TiO_2	89	20.7	248.0[d]	76.3	12	1
		20.5	53.7	76.5	9	2
		20.6	—	—	—	2
		23.0	58.8	80.0	8	3
		22.5	54.2	—	—	4
8.6% Ni/TiO_2	94	30.5	84.4	119.0	8	1
		31.2	87.5	120.0	12	2
12.3% Ni/TiO_2	89	45.2	91.4	645.0	9	1
		46.9	77.0	202.0	8	2
		48.0	—	186.0	—	3
		44.0	—	178.0	—	3
6.5% Ni/TiO_2	83	80.0	165.0	186.0	ND	1
		—	—	191.0	—	1
		80.0	—	211.0	8	2
		—	—	247.0	—	3
6.8% Ni/SiO_2	100	98.2	184.0	196.0	10	1
		105.0	—	—	—	2
Pure TiO_4		0	0	7.1		
Pure γ-Al_2O_3		0	1.0	0.5		
Pure SiO_2		0	0.7	0.3		

[a] Corrected for 12.1 mol g^{-1} O_2 uptake on pure TiO_2 at 698 K for Ni/TiO_2.
[b] Total uptake before correction for irreversible adsorption on the support.
[c] ND, not detectable.

was obtained with all techniques for typical Ni/SiO$_2$ and Ni/Al$_2$O$_3$ catalysts, whereas the sizes calculated from chemisorption of H$_2$ and CO are far too large for Ni/TiO$_2$ catalysts. In contrast, oxygen chemisorption provides values in excellent agreement with those from XRD and TEM. Low-contrast nickel crystallites on titania were observed in the TEM micrographs, due to the lower electron density of nickel and the opacity of the crystalline titania particles, indicating that nickel may have a raft-like morphology on this support.

Bartholomew and Pannell (ref. 130) obtained similar trends when comparing the metal crystallite size distribution in Ni/SiO$_2$, Ni/Al$_2$O$_3$ and Ni/TiO$_2$ catalysts obtained by hydrogen chemisorption, XRD and TEM as shown in Table 12.4. The

TABLE 12.4
Accuracy of crystallite size measurements (ref. 130)

Catalyst	H$_2$ ads.	TEM	XRD
Ni/SiO$_2$	Very good (±10%)	Very good	Fair (±30-50%)
Ni/Al$_2$O$_3$	Very good	Fair-good	Fair
Ni/TiO$_2$	Poor (high by 50-100%)	Good (±20%)	Fair-good

results were found to agree over wide ranges of metal dispersion and loading for Ni/SiO$_2$ and Ni/Al$_2$O$_3$. On the other hand, poor agreement was evident for Ni/TiO$_2$ system, suggesting that hydrogen adsorption was greatly suppressed.

On the basis of these results it is possible to suggest that X-ray diffraction as well as oxygen chemisorption at room temperature, can be utilized as accurate methods for measuring the nickel surface area of titania-supported nickel catalysts. On the other hand, titania surface area itself decreases during heat treatment and hence sintering of the support may be superimposed over the SMSI and any other effects taking place in the presence of hydrogen at high temperatures.

(iii) <u>Magnesium oxide</u>. Magnesium oxide has frequently been used as a support for metals, particularly for nickel, since it is one of the strongest solid bases (refs. 144-146). The oxide supported nickel catalyst exhibits some peculiarities (refs. 23,53,147,148). Magnesium oxide has electron-donor properties (refs. 139,146) associated with hydroxyl groups, with weakly coordinated O^{2-} and with surface hydroxyl groups having acidic character (ref. 149).

The Ni/MgO catalysts are prepared by different methods: impregnation of MgO with an aqueous nickel nitrate solution (refs. 53,150), an aqueous ammonia solution of nickel nitrate (refs. 23,151) or nickel hydroxide dissolved in ethylene glycol (ref. 53); reaction of nickel hydroxide with a nickel hexamine nitrate solution, where the transformation of Mg(OH)$_2$ into MgO occurs during the hydrogen activation and carbonate formation is avoided (ref. 148); thermal decomposition of corresponding mixed oxalates in vacuo (ref. 152); superficial reduction of nickel and magnesium mixed oxides (refs. 154-156); metal atom vapour deposition (ref. 6), etc. In all cases, the resulting catalyst exhibits a very low degree of reduction, although in some cases the metal dispersion is high, as confirmed by hydrogen adsorption, thermogravimetric, magnetic, XRD and TEM measurements. Previous irradiation with gamma-rays or fast neutrons (refs. 154-156) has a positive effect on the reduction of the catalyst and changes the sorption centres of hydrogen.

Since the two oxides, NiO and MgO, form solid solutions over the whole composition range (ref. 154), the degree of reduction is appreciably affected by the presence of unreduced MgO so that higher extents of reduction, up to 95%, are obtained at higher NiO contents (refs. 154,155). Such behaviour is explained not in terms of definite surface compounds but rather as due to the modification of the electronic properties of NiO which loses its individuality as a Mott insulator (i.e., with a narrow Ni 3d band) and undergoes a much larger delocalization of unpaired spin toward the support, resulting in a larger interaction NiO-MgO (ref. 23). Also no bulk Ni-Mg alloy formation is detected (ref. 148) when the reduction is carried out at high temperatures, 1203 K. On the other hand, Marcelin and Vogel (ref. 60) described a Ni/Al_2O_3-MgO catalyst prepared by mulling Al_2O_3-MgO (obtained by coprecipitation from nitrate salts using a mixture of ammonium hydroxide and ammonium bicarbonate) and nickel carbonate, with water. The catalyst exhibits a high degree of reduction and no supression of hydrogen adsorption when it is reduced at 773 K.

(iv) <u>Aluminium phosphate and related systems</u>. The applicability of metal phosphates as heterogeneous catalysts in a wide variety of reactions catalysed by acidic sites, viz., dehydration, isomerization, polymerization, alkylation, etc., has been the subject of several studies in the last decade. The results relative to the surface properties and catalytic activity have been reviewed by Moffat (refs. 157,158). However, with the exception of aluminium phosphate, these compounds have not been employed as catalytic supports. Even the use of aluminium phosphate has received little attention. Thus, apart from the work of Marinas and co-workers, who studied not only the surface and catalytic properties of aluminium phosphate and alumina-aluminium phosphate and silica-aluminium phosphate systems (refs. 159-191), but also their applicability as supports for platinum (refs. 192-194), palladium (refs. 195-198), rhodium (refs. 199-204) and nickel (refs. 205-216), only a number of U.S. patents (refs. 216-220) and some of Marcelin's very recent papers (refs. 60,221) have reported the use of aluminium orthophosphate or some mixed system as metal supports.

However, there is a growing interest in such compounds, considering the relatively high number of papers concerned with their preparation (refs. 60,157,

222-231) or in some cases, their applicability as catalysts (refs. 157,227,229, 232-234). It is believed that these inorganic solids will be useful in most of the applications now using zeolites (refs. 223,225), since they are completely isostructural with silica (ref. 235) and, furthermore, can readily be prepared as stable solids with high surface areas (refs. 236-238).

- <u>Aluminium phosphate.</u> In 1962 Kearby (ref. 237) described the synthesis and catalytic activity of a new form of aluminium phosphate (the authors call it "clear transparent") prepared by precipitation with ethylene oxide from aqueous solutions of aluminium chloride and phosphoric acid:

$$AlCl_3 \cdot 6H_2O + H_3PO_4 + 3CH_2\underset{O}{-}CH_2 \longrightarrow AlPO_4 + 3ClCH_2-CH_2OH \qquad (12.2)$$

Its surface area and acidity seem to be higher than the most popular catalysts used in cracking processes, Al_2O_3 and $SiO_2-Al_2O_3$. An interesting discussion is taking place on the variation in the surface properties of these aluminium phosphates when prepared from different precipitation agents and at varying pH values (refs. 60, 221,226-231). Thus, Marinas and co-workers (refs. 159,160,186, 239) studied the influence of the precipitation agent and washing solvent on the textural and acidic properties of $AlPO_4$. They obtained the best results when ethylene (refs. 159,160) or propylene (refs. 186,239) oxides were used as bases and when isopropanol was the washing solvent. On the other hand, Peri (ref. 235) concluded that $AlPO_4$ is a mixed oxide of Al and P, completely isostructural with SiO_2, and exhibits a number of Broensted and Lewis acid sites depending on the activation temperature (ref. 240). The textural and acidic properties of some selected $AlPO_4$ samples collected in Table 12.5 illustrate such conclusions.

All the aluminium phosphates were washed with isopropanol and then calcined at 920 K for 3 h. In the preparation of sample A only, 65% of the theoretical amount of ethylene oxide according to Kearby (ref. 237) was employed. Sample B was prepared exactly according to Kearby; C is a commercial sample from PROBUS; D, F and P were obtained like B by using dioxane, ammonium hydroxide and propylene oxide respectively as precipitation agents instead of ethylene oxide. Sample E was obtained from sample B by calcination at 970 K for 4 h.

The surface acidity was determined by ammonia chemisorption at different temperatures and pressures (ref. 171) and by titration with different amines and indicators with different values of the Hammett constants (ref. 161), according to Benesi (ref. 241). Some of the later values are collected in Table 12.5. It was drawn that according to Kearby (ref. 237) these phosphates exhibit surface acidity exceeding that of silica-alumina cracking catalysts, although the number of acidic sites of great strength is not as high.

Recently, Marinas and co-workers (refs. 184,208,209,242) have described a

TABLE 12.5
Surface area (S), pore volume (V), main pore radius (r) and surface acidity titrated with butylamine (Benesi method)

Aluminium phosphate	S ($m^2 g^{-1}$)	V ($ml g^{-1}$)	r (nm)	Acidity (mol g^{-1} 10^3) $H_o = 6.8$	$H_o = 2.8$
A	312	0.64	2.5 - 5.0	1.31	0.64
B	325	0.96	2.5 - 5.0	1.20	0.54
C	10	0.02	2.5 - 5.0	0.50	—
D	36	0.06	2.5 - 7.0	0.20	0.10
E	316	0.61	2.5 - 7.0	1.26	0.94
F	156	0.52	2.5 - 7.0	0.98	0.42
P	224	0.49	1.5 - 5.0	1.43	1.16

method for the determination of the concentration of acidic, basic, oxidizing and reducing sites on the surface of $AlPO_4$-Al_2O_3 involving monitoring the adsorption of appropriate substances by a spectrophotometric method. A correlation between the acidity and the activity in the skeletal isomerization of alkenes has been demonstrated (ref. 187), and between the basicity and catalytic activity in the retroaldolization of diacetone alcohol (ref. 189). Moreover, the oxidizing-reducing properties of these surfaces are indeed connected to their acid-base nature and the basic sites coexist with the donor sites on the catalyst surface, the two being not entirely independent (ref. 242).

The results obtained in several previous studies (refs. 160,173,179,184,186, 189,208,209,242) are summarized in Tables 12.6 and 12.7, where the textural, acid-base and redox properties of eight solids are collected: three $AlPO_4$ (F, B and P, see Table 12.5); a chemically pure alumina AL; a commercial silica S; two $AlPO_4$-Al_2O_3 (75:25 mass%) systems BA and PA obtained by precipitation with ethylene and propylene oxide, respectively, on alumina AL, and $AlPO_4$-SiO_2 (20: 80 mass%) system BS obtained by precipitation with ethylene oxide on silica S. The addition of alkali-metal hydroxides to $AlPO_4$ leads to a change in the acidity-basicity balance with the formation of a new basic centre having no acceptor properties (ref. 186). For $AlPO_4$, the phenothiazine adsorption (ref. 242) has been correlated with the total acidity; the adsorption of 1,3-dinitrobenzene occurs on two different types of sites, one having only donor properties and the other both basic and donor properties. The results provide evidence that the surface of the $AlPO_4$ (and $AlPO_4$-Al_2O_3 and $AlPO_4$-SiO_2 systems) possesses oxidizing and reducing properties, and that these are indeed connected to the acid-base nature. It was noted that basic sites coexist with donor sites on the surface of

TABLE 12.6
Textural and acid-base properties of the catalysts[a]

Catalyst	S	V	d	acidity[b]			basicity[b]	
				CY	PY	AN	AA	PH
F	181	0.40	3.6	1042	260	76	294	81
B	256	0.60	4.5	789	320	86	298	55
P	210	0.42	2.5	732	380	124	473	71
AL	151	0.31	2.0	—c	60	—c	476	194
S	380	0.40	2.0	—c	206	10	95	13
BA	263	0.53	3.0	620	300	60	327	164
PA	237	0.70	2.5	684	385	35	518	129
BS	327	0.46	3.0	600	310	23	65	20

[a] Abbreviations and units used:
S = surface area ($m^2 g^{-1}$), V = pore volume ($ml\ g^{-1}$), d = main pore diameter (nm),
CY = cylohexylamine (pK_a = 10.6), PY = pyridine (pK_a = 5.3), AN = aniline
(pK_a = 4.6), AA = acrylic acid (pK_a = 4.25), PH = phenol (pK_a = 9.9).
[b] Uptake at equilibrium at 298 K ($\mu mol\ g^{-1}$).
[c] Not measured.

solids and that basic and reducing centres are not entirely independent.

Several authors (refs. 226-230) have recently discussed the catalytic properties of stoichiometric phosphates and, in particular, aluminium phosphates. Hagging (refs. 223,225) described a new family of microporous aluminophosphates developed by Union Carbide which have similar properties to those of zeolites, and will be probably have a variety of uses. Thus, initial investigations indicate that they have mildly acidic properties, and when used as supports for metals (platinum and palladium) they are active in hydrocracking, reforming and isomerization. Their synthesis seems to be according to the method of Kearby (ref. 237), although the aluminophosphate (obtained from equimolar portions of reactive hydrated alumina, phosphoric acid and water) is gelified in an organic amine or quaternary ammonium salt. As gelification agents, the tetrapropylammonium cation $(Pr_4N)^+$, tetramethylammonium hydroxide, di-n-propylamine and quinuclidine were employed. The organic gelification agent may be crucial because without it no sieves are formed. In brief, the idea is to employ an organic molecule which is physically and chemically compatible with the crystalline precursor, in order to obtain crystallization of the gel around the organic molecule (clathration), producing the pores and cages of a porous framework.

At the present time, only Marinas and co-workers (refs. 205-216) have studied

TABLE 12.7
Electron transfer properties of the catalysts[a]

Catalyst	Oxidizing centres[b]		Reducing centres[b]	
	ANT	PNTZ	DNB	TCNQ
F	0.019	2.1	3.3	0.4
B	0.021	1.8	3.5	0.8
P	0.018	1.0	2.5	0.6
AL	0.026	0.8	2.8	2.0
S	—c	0.4	1.1	—c
BA	0.022	1.8	4.3	0.4
PA	0.018	1.3	4.5	0.6
BS	0.010	0.5	3.1	—c

[a]Abbreviations:
ANT = anthracene (I.E. 7.55 eV), PNTZ = phenothiazine (I.E. 7.13 eV), DNB = 1,3-Dinitrobenzene (E.A. 2.10 eV), TCNQ = 7,7´,8,8´-Tetracyanoquinodimethane.
[b]Uptake at equilibrium at 298 K in μ mol g^{-1}.
[c]No adsorption.

the preparation and characterization of aluminium phosphate-supported nickel catalysts. They were prepared by impregnation to incipient wetness of the support with an aqueous solution of $Ni(NO_3)_2 \cdot 6H_2O$, to yield a nominal 5, 10 and 20 mass% metal. The impregnated dried samples were reduced in an ultrapure stream of hydrogen at 200 ml min^{-1} at 670 K for 4 h. Systems containing 20 mass% metal were also prepared by decomposition to the oxide form by heating in air at 573 K over 1 h, prior to reduction. Thus, for 20 mass% metal there are two series of catalysts: series I, not calcined, and series II, calcined prior to reduction. This is the result of optimizing the conditions in order to avoid incomplete nickel reduction and obtaining the highest metal surface area (ref. 206). Only bunsenite nickel oxide peaks were detected in the XRD diagrams of the calcined systems before reduction in the hydrogen stream. After reduction under the conditions indicated above, only the peaks of nickel metal were obtained. So, nickel oxide is reduced almost entirely to the metallic state.

The average crystallite diameter, D, was calculated by XRD measurements from the width of the (111) nickel peak at half the maximum peak height using Sherrer's relationship, according to the method of Moss (ref. 243). Furthermore, the surface basicity and acidity of supported nickel catalysts have been determined by a spectrophotometric method (refs. 208,209) which allows the titration of dark or coloured solids. Acrylic acid and pyridine were employed as titrants of the basic and acidic sites, respectively. In Table 12.8 the dispersion, acidity and basicity values are given, together with the rate, i.e., the activity per

TABLE 12.8
Values of surface acidity and basicity, average crystallite diameter and areal rate of supported nickel catalysts[a]

Catalyst	Acidity[b]	Basicity[c]	$\bar{D}$	r	r_A
$AlPO_4/Ni$					
5% I	306	288	5.4	1.43	0.39
10% I	295	274	7.6	1.51	0.58
20% I	192	164	7.7	3.17	1.20
20% II	238	159	18.4	1.26	1.18
Al_2O_3/Ni					
5% I	71	331	4.1	0.37	0.08
10% I	90	346	4.5	1.11	0.25
20% I	53	398	45.2	0.16	0.36
20% II	50	321	22.1	0.31	0.34
SiO_2/Ni					
5% I	81	300	4.2	5.33	1.12
10% I	96	311	7.3	3.40	1.26
20% I	81	186	12.6	3.10	1.98
20% II	94	171	15.8	2.37	1.89
$AlPO_4-Al_2O_3/Ni$					
5% I	287	237	7.5	0.66	0.25
10% I	170	220	7.3	0.66	0.25
20% I	127	252	7.9	0.62	0.25
20% II	143	190	9.3	0.47	0.25
$AlPO_4-SiO_2/Ni$					
5% I	149	166	3.7	1.42	0.27
10% I	237	96	4.4	2.93	0.66
20% I	267	263	10.0	3.98	2.01
20% II	189	154	11.3	3.77	2.16

[a] Abbreviations:
$\bar{D}$ = average crystallite diameter (nm), r = catalytic reaction rate with 0.5x10^{-3} g atom of nickel (mmol min^{-1}), r_A = areal rate (mmol min^{-1} m_{Ni}^{-2}).
[b] Uptake of pyridine (pK_a = 5.3) at equilibrium and 298 K (μmol g^{-1}).
[c] Uptake of acrylic acid (pK_a = 4.6) at equilibrium and 298 K (μmol g^{-1}).

unit surface area of nickel metal, in the liquid phase hydrogenation of 1-hexene. This reaction is taken as a probe and is carried out in such a way that the kinetic data are free from transport influences (ref. 210). According to the results, the deposition of nickel induce an important change in the acid-base characteristics, as compared to the supports. Owing to this, the catalytic behaviour of supported nickel systems should be correlated with their own acid-base characteristics instead of with those of the supports. On the other hand, the areal rate, r_A, was independent of the particle size for catalysts with the same support and nickel loading. However, it varied markedly with the support, decreasing as the nickel content decreased. This behaviour may be associated with a nickel-support interaction which decreased in the sequence

$$Ni/Al_2O_3 > Ni/AlPO_4-Al_2O_3 > Ni/AlPO_4 > Ni/AlPO_4-SiO_2 > Ni/SiO_2$$

for catalysts with the same nickel loading (ref. 210).

Marinas and co-workers (ref. 214) have also studied the effects of the carrier, nickel precursor and nickel loading on the particle size and catalytic behaviour of $AlPO_4$-supported nickel catalysts. Thus, the liquid-phase hydrogenation of 1-hexene, chosen as the probe reaction, was examined on $AlPO_4$-supported nickel catalysts prepared by several different means. The influence of several precursor materials and of different calcination treatments of the $AlPO_4$ support, were evaluated. The results are quoted in Tables 12.9, 12.10 and 12.11.

TABLE 12.9

Textural and acid-basic properties of $AlPO_4$ calcined at different temperatures
Abbreviations: T = calcination temperature in K (always for 3 h), S_{BET} = surface area by B.E.T. method (m^2 g^{-1}), S_t = surface area by Lecloux method, with n_3 according to monolayer value (m^2 g^{-1}), V_p = pore volume (ml g^{-1}), $\bar{r}$ = main pore radius (nm), PY = acidity obtained from pyridine adsorption (pK_a = 5.3) (µmol g^{-1}), AA = basicity obtained from acrylic acid adsorption (pK_a = 4.25) (µmol g-1).

T	S_{BET}	S_t	V_p	$\bar{r}$	PY	AA
770	190	188	0.89	2-4	161	61
920	228	236	0.94	2-4	180	80
1070	6	-	-	-	10	-

TABLE 12.10

Variation in the dispersion and catalytic activity with the support calcination temperature of 20 mass% $Ni/AlPO_4$ obtained by impregnation to incipient wetness in aqueous medium using $Ni(NO_3)_2 6H_2O$ as precursor salt

Abbreviations: T = calcination temperature of support in K, S_{Ni} = metal surface area determined from X-ray diffraction (m^2 g_{Ni}^{-1}), r_A = activity per unit surface of surface area of nickel metal (mol s^{-1} m_{Ni}^{-2}), r_g = activity per gram of catalyst (mol s^{-1} g^{-1} of catalyst), r_{Ni} = activity per gram of metal nickel supported (mol s^{-1} g_{Ni}^{-1}).

T	S_{Ni}	r_A 10^6	r_g 10^6	r_{Ni} 10^3
770	72	17.5	251	1.26
920	87	35.4	433	3.08
1070	18	46.7	171	0.84

TABLE 12.11

Relationship between the preparation procedure and nickel loading of Ni/AlPO$_4$-920 catalysts and their dispersion and catalytic activity in the liquid-phase hydrogenation of 1-hexene

S_{Ni}, r_A and r_{Ni} as in Table 12.10.

Ni (mass%)	Ni salt	Solvent	S_{Ni}	r_A 10^6	r_{Ni} 10^3
10	nitrate	ethanol	172	9.9	1.70
20	nitrate	ethanol	105	21.1	2.22
30	nitrate	ethanol	87	54.9	4.78
35	nitrate	ethanol	25	77.5	1.94
40	nitrate	ethanol	20	95.7	1.91
20	nitrate	water	87	35.4	3.08
20	chloride	ethanol	10	21.8	0.22
20	chloride	water	10	23.4	0.23
20	acetate	water	75	20.3	1.52

The calcination temperature has an important effect on the AlPO$_4$ surface which is manifested in its textural and acid-basic properties. Thus, in spite of the fact that the three AlPO$_4$ samples are identical from a chemical point of view, they show appreciable differences not only in their crystal form, surface area and acidity-basicity balance, but also in their catalytic activities.

On the other hand, there is a linear dependence between the metal surface area of nickel systems containing 20 mass% Ni and the specific surface area of the support S_{BET} (ref. 214). Accordingly, the surface area of supports seems to be an important factor in the dispersion control of supported metal catalysts. Its amorphous character provides the AlPO$_4$ with a porous texture, making it useful as a support for dispersed metals because of its large surface area.

As regards the effects of the nickel precursor and/or solvent, the results in Table 12.11 show that the particle size and catalytic activity are strongly dependent on the impregnation process. Thus, the catalysts are obtained by impregnation with nickel nitrate which exhibits higher metal surface area and catalytic activity per gram of nickel metal supported.

On the other hand, an increase in the nickel loading leads to an almost linear increase in the areal rate. This effect could be explained by a lessening of the metal-support interactions, as a result of the increase in the metal loading. However, the optimum activity per unit weight of supported nickel is obtained for catalysts with 30 mass% Ni. This fact is related to the saturation capacity of the AlPO$_4$ support, which is higher than the conventional supports like Al$_2$O$_3$ or SiO$_2$, containig 10 and 5 mass% Ni, according to the results in Table 12.8. Therefore, these supported nickel systems are useful as catalysts for the hydrogenation of alkenes. As the support possesses acidic and basic sites capable of taking part in the catalytic process, these metallic systems are bifunctional catalysts, with acidic and hydrogenation-dehydrogenation functions. Moreover, the previous impregnation of $AlPO_4$ with 5 mass% of alkali-metal ion (Li, Na or K) greatly decreases the metal surface area although the areal rate is modified only slightly (ref. 214).

- **$AlPO_4$-Al_2O_3 and $AlPO_4$-SiO_2 systems.** Marinas and co-workers have also studied the synthesis, characterization and application, as catalysts or metal supports, of $AlPO_4$-Al_2O_3 and $AlPO_4$-SiO_2 systems (refs. 170,173,175-185,191-216). These systems are complementary to $AlPO_4$ in heterogeneous catalysis. Marcelin and co-workers (refs. 60,221,231) subsequently described the preparation, characterization and use of $AlPO_4$-Al_2O_3 and MgO-Al_2O_3-$AlPO_4$ systems as nickel supports.

The preparation of $AlPO_4$-Al_2O_3 and $AlPO_4$-SiO_2 has been carried out in ammonium hydroxide as well as in ethylene or propylene oxides (refs. 173,175,176,179). In ethylene oxide the preparation of $AlPO_4$-Al_2O_3 involves the following reactions:

$$AlCl_3 \cdot 6H_2O + H_3PO_4 + H_2C\underset{O}{-}CH_2 \xrightarrow{H_2O} AlPO_4 + 3ClCH_2\text{-}CH_2OH \quad (12.3)$$

$$Al(NO_3)_3 \cdot nH_2O + 3NH_4OH \xrightarrow{H_2O} Al(OH)_3 + 3NH_4NO_3 \quad (12.4)$$

$$2Al(OH)_3 \xrightarrow{\Delta} \gamma\text{-}Al_2O_3 + 3H_2O \quad (12.5)$$

Table 12.12 shows the textural properties of several $AlPO_4$-Al_2O_3 and $AlPO_4$-SiO_2 systems; all them were dried at 390 K for 24 h and then calcined at 920 K for 3h. Their acid-base and redox characteristics have been determined by the Benesi method (refs. 173,175,176,179) as well as by spectrophotometry (refs. 184,185, 189,242), see Tables 12.6 and 12.7. Generally the $AlPO_4$-Al_2O_3 systems display a higher total acidity compared to $AlPO_4$. The acidity of $AlPO_4$-SiO_2 systems is less than that of $AlPO_4$.

Like pure $AlPO_4$, these systems have been used as supports for metals (refs. 192-216). The supported nickel catalysts are prepared under the same conditions as Ni/$AlPO_4$ catalysts shown in Table 12.8. The highest nickel dispersion is exhibited by the Ni/$AlPO_4$-SiO_2 systems, although all these catalysts show high activity in hydrogenation processes (linear and cyclic olefins, α,β-unsaturated carbonyl compounds, functionalized alkenes, etc.). Marcelin and co-workers (refs. 60,221,231) have reported the preparation of Ni/$AlPO_4$-Al_2O_3 and Ni/$AlPO_4$-OMg-Al_2O_3 systems containing 20 mass% Ni which are active in the hydrogenation of 2-ethylhexenal. Table 12.13 shows some characteristics of these supported nickel systems in comparison to other classic supported nickel catalysts (ref. 215). So, we can state that, together with Ni/$AlPO_4$, the Ni/$AlPO_4$-Al_2O_3 and Ni/$AlPO_4$-SiO_2 systems are effective catalysts of hydrogenation processes and an alternative to other nickel catalysts supported on conventional metal oxides such as SiO_2 or Al_2O_3.

TABLE 12.12

Surface area (S), pore volume (V) and main pore radius (r) of several $AlPO_4$-Al_2O_3 (refs. 173,179) and $AlPO_4$-SiO_2 (refs. 175,176) systems

System	Composition (mass%)	Gelification method	Synthesis procedure	S ($m^2 g^{-1}$)	V ($ml\, g^{-1}$)	r (Å)
$AlPO_4$-Al_2O_3	50:50	NH_4OH	Coprecipitation	109	0.20	21
$AlPO_4$-Al_2O_3	50:50	NH_4OH	Coprecipitation[a]	243	0.32	25
$AlPO_4$-Al_2O_3	25:75	NH_4OH	Coprecipitation[a]	241	0.34	25
$AlPO_4$-Al_2O_3	75:25	NH_4OH	Coprecipitation[a]	294	0.59	30
$AlPO_4$-Al_2O_3	75:25	$H_2C-O-CH_2$	Coprecipitation[a]	294	0.53	36
$AlPO_4$-Al_2O_3	75:25	CH_3-CH-O-CH_2	Coprecipitation[a]	250	0.29	25
$AlPO_4$-Al_2O_3	75:25	$NH_4OH, H_2C-O-CH_2$	Coprecipitation[a]	265	0.38	25
$AlPO_4$-Al_2O_3	75:25	NH_4OH, CH_3-CH-O-CH_2	Coprecipitation[a]	310	0.85	47
$AlPO_4$-SiO_2	20:80	NH_4OH	Precipitation[b]	280	0.46	35
$AlPO_4$-SiO_2	20:80	CH_3-CH-O-CH_2	Precipitation	410	0.52	35
$AlPO_4$-SiO_2	20:80	CH_3-CH-O-CH_2	Precipitation[c]	465	0.63	35
$AlPO_4$-SiO_2	20:80	$H_2C-O-CH_2$	Precipitation[c]	395	0.55	24

[a] Separate precipitation of system components.
[b] $AlPO_4$ precipitation onto commercial SiO_2 (Merck, Kieselgel 60, 70-230 mesh).
[c] Washed with isopropanol for 12 days.

TABLE 12.13

Hydrogen chemisorption and crystallite diameters of supported 20 mass% nickel catalysts (ref. 221)

Support	T^a (°C)	Hydrogen Chemisorption (μmol/g)	Percentage reduction	Apparent dispersion (%)	$\bar{D}^b$
SiO_2	300	160	100	10.0	
	500	170	100	10.5	45
TiO_2	300	30	45	3.9	
	500	10	100	0.6	85
Al_2O_3	300	110	80	8.2	
	500	130	100	7.5	90
$Al_2O_3 \cdot 2AlPO_4$	300	c	10	c	
	400	65	90	4.2	
	500	7	100	0.4	40
$4MgO \cdot 13Al_2O_3 \cdot 10AlPO_4$	300	78	40	12.5	
	400	45	90	2.9	
	500	19	100	1.1	40
$2MgO \cdot 3Al_2O_3$	300	-	0	-	
	400	241	85	17.0	
	500	384	100	21.0	30

For footnotes, see p. 434.

TABLE 12.13 continued

[a] Temperature of reduction.
[b] Average crystallite diameter (Å).
[c] Too small for accurate measurement.

12.3 METAL-SUPPORT INTERACTION

There is a growing interest in the role of the support in determining the adsorption and catalytic behaviour of supported nickel systems. The support effect comprises the appareance of new species, anomalous chemisorption as well as electron transfer.

In this section we discuss some significant data on the support effects in new supported nickel catalysts, although some references to them have been made in previous sections. The Ni/TiO$_2$ catalyst is perhaps the most studied system from this viewpoint. Support effects have been demonstrated in different research proyects on catalytic activity and selectivity in various hydrogenation processes, as well as in gas chemisorption or TEM and XRD studies. Thus, Baker et al. (refs. 123,124) employed high resolution electron microscopy to formulate a model to account for the strong metal-support interaction found in different metals on TiO$_2$, following reduction at $\geqslant 775$ K. One of the crucial steps in the model is the postulate that the dissociated hydrogen on the metal surface provides a source of hydrogen atoms, which then attack the support reducing it to a lower oxide Ti$_4$O$_7$. Mustard and Bartholomew (ref. 130) obtained TEM micrographs of Ni/TiO$_2$ which provide evidence for a "raft-like" metal structure attributed to SMSI. Moreover, it appears that this interaction is enhanced by treatment with hydrogen at high temperature, as evidenced by a new metallic phase on the surface of the support. The authors believe this could be intermetallic Ni-TiO$_x$ ($x < 2$) formed by the high temperature of reduction. Burch and Flambard (ref. 132) postulated that the adsorption of CO on partially reduced titania (Ti$_4$O$_7$ shear phase) is responsible for the increased activity in methanation. Fig. 12.1 shows the stages in the development of the Ni/TiO$_2$ systems as the temperature of activation increases. These authors (ref. 132) proposed a model, Fig. 12.3, which emphasizes the importance of the interface between the metal and the partially reduced support. New active sites created at the interface (concerned with interfacial metal-support interactions (IMSI)) are proposed as being responsible for the high specific activity of these catalysts for the CO/hydrogen reaction.

The SMSI effect is seen to be influential in the adsorption of CO and H$_2$ and the hydrogenation activity/selectivity of CO on Ni/TiO$_2$ catalysts (refs. 126,133,244). Thus, when these systems are reduced at moderate temperatures, 473 K, their chemisorption and catalytic properties are normal. However, their

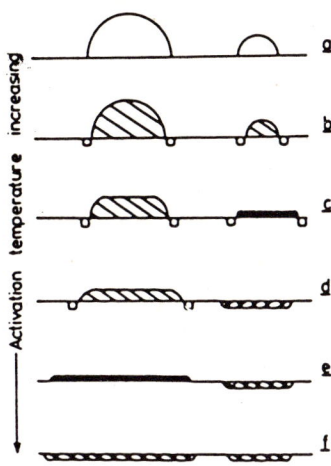

FIG. 12.1. Interactions in the Ni/titania system as the activation temperature is increased. ○ ,NiO; ⌬ ,bulk nickel metal; ▬ ,surface Ni(≤1 nm) in the SMSI state; ⌬ ,Ni in a partially ionized subsurface state; ☐ ,oxygen anion vacancy. Reproduced from ref. 132.

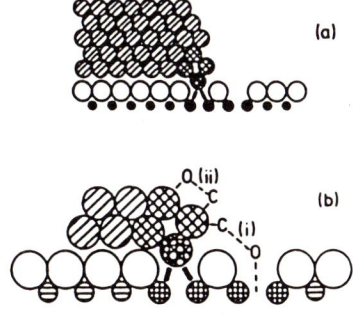

FIG. 12.2. Model of the active site in the CO/H_2 reaction over Ni/titania catalysts (corresponds to stage (b) in Fig. 12.1). ⊘ , normal nickel atom; ⊛ , $Ni^{\delta\delta-}$ atom; ● , $Ni^{\delta-}$ atom; ○ , oxide ion; ⊕ , Ti^{3+}; ⊖ , Ti^{4+}; ▌ , SMSI (ref. 132)

reduction at higher temperatures, 700-800 K, is accompanied by a decrease in both hydrocarbon conversion and hydrogen uptakes, which cannot be accounted for by metal sintering or encapsulation. Subsequent oxygen treatments followed by low temperature reductions restore most of the normal properties. Similar behaviour is found with other noble metals supported on TiO_2, such as Pt, Pd, Rh, Ru, Os or Ir (refs. 121-125,131,134,245-247).

Indeed, this effect has been investigated in several other reactions such as the hydrogenation of benzene (refs. 245,248), ethylene (refs. 245,249), styrene (ref. 245), the dehydrogenation of cyclohexane (ref. 245) and the hydrogenolysis of n-hexane (refs. 127,128,132,250), ethane (refs. 132,135,136) or butane (ref. 245). However, in these reactions the development of the SMSI is accompanied by a loss of catalytic activity, as expected since SMSI reduces the capacity of the supported metal to adsorb hydrogen. So, it is surprising, therefore, that titania-supported catalysts should exhibit exceptionally high activity in the reaction of CO with hydrogen, independently of the supported metal (refs. 121-132, 245-247). Such conflicting results have led several authors (refs. 127,128,251) to propose that the enhanced activity is specific to this reaction.

Table 12.4 shows the results obtained by Vannice (ref. 131) in a study of the

TABLE 12.14
Activity comparisons for supported metals (ref. 131)

Catalyst	Activity (μmol CO.s^{-1}.g metal^{-1})	N_{CO} at 275 °C (s^{-1} × 10^3)	
		a	b
2% Ru/TiO$_2$	120	320	12
5% Ru/Al$_2$O$_3$	190	325	--
1.53% Ni/TiO$_2$	270	75	16
1.5% Ni/TiO$_2$	720	130	41
10% Ni/TiO$_2$	1530	820	90
10% Ni/TiO$_2$[c]	830	330	50
5% Ni/Al$_2$O$_3$	84	38	--
2% Rh/TiO$_2$	110	430	11
1% Rh/Al$_2$O$_3$	80	17	--
2% Pd/TiO$_2$	5	100	0.5
1.86% Pd/TiO$_2$	10	27	1.1
2% Pd/Al$_2$O$_3$	14	13	--
2% Pt/TiO$_2$	17	130	3.4
1.75% Pt/Al$_2$O$_3$	8	3.4	--
2% Ir/TiO$_2$	30	45	5.8
2% Ir/Al$_2$O$_3$	6	2.6	--
1.53% Co/TiO$_2$	13	20	0.7
2% Co/Al$_2$O$_3$	20	28	--
1.46% Fe/TiO$_2$	0.13	0.02	0.0007
5% Fe/TiO$_2$	0.09	0.03	0.005
15% Fe/Al$_2$O$_3$	23	160	--

[a]Based on CO adsorption.
[b]Assuming 100% dispersion.
[c]Degussa P-25 titania.

catalytic activity of Group 8 metals supported on TiO_2 in the hydrogenation of CO. The highest activity is obtained with the Ni/TiO_2 system and, in general, with metals which tend to have more completely filled d bands (Ni, Pd, Pt, Rh, Ir). The increase in activity may be the result of higher surface concentrations of hydrogen (ref. 252) or a consequence of a strong metal-support interaction (ref. 253). On the other hand, the excellent correlation between the hydrogen chemisorption and ethane hydrogenolysis, shown by Ko et al. (ref. 136) with Ni/TiO_2 catalysts, provides direct evidence for a crystallite-size effect on the onset of metal-support interactions, since more severe reduction conditions are found necessary to induce SMSI in larger nickel crystallites, Table 12.15.

TABLE 12.15
Chemisorption and catalytic properties of titania-supported nickel catalysts in ethane hydrogenolysis (ref. 136)

mass%Ni	T^a (K)	V^b (ml g^{-1}Ni)	$r_a^c \times 10^6$ (mol m^{-2}Ni h^{-1})	Surface intermediate and power rate law
2	573	2	9	C_2H_4 (rate = $kp_E^{1.0} p_H^{-0.8}$)
10	573	16	50	C_2H_2 (rate = $kp_E^{0.9} p_H^{-1.7}$)
10	773	3	10	C_2H_4 (rate = $kp_E^{1.0} p_H^{-1.2}$)

[a] Temperature of reduction treatment in hydrogen by 1 h.
[b] Amount of hydrogen chemisorbed.
[c] Specific activity at 525 K.

Marinas and co-workers have inferred the existence of a metal-support interaction from the kinetic behaviour of $Ni/AlPO_4$, $Ni/AlPO_4$-SiO_2, $Ni/AlPO_4$-Al_2O_3, Ni/SiO_2 and Ni/Al_2O_3 in the hydrogenation of cyclohexene (ref. 205), 1-hexene (refs. 210,214), allyl alcohol (ref. 211), acrylic acid (ref. 211), α,β-unsaturated carbonyl compounds (ref. 212), styrene and α-methylstyrene (ref. 213,216) and other organic functional groups. The extent of interaction follows the sequence

Ni/Al_2O_3 > $Ni/AlPO_4$-Al_2O_3 > $Ni/AlPO_4$ > $Ni/AlPO_4$-SiO_2 > Ni/SiO_2

for catalysts with the same nickel loading.

The interaction between the support and the nickel crystallites is suggested to involve a transfer of electrons between the nickel atoms and the acidic (oxidizing) sites on the surface of the support, while its extent is related to the nature of the support and the metal support ratio (refs. 129,133,210). According to Vannice and Garten (ref. 254), oxidizing sites on the support surface may interact with the dispersed metal and reduce the concentration of electrons in the d band of the metal crystallites. This effect shifts the catalytic behaviour of supported nickel to a metallic behaviour more characteristic of cobalt. Consequently, this interaction can be expected to different degrees in all bifunctional metal catalysts, such as nickel-aluminium orthophosphate catalysts. The results of this study provide strong evidence of modifications in the catalytic behaviour of nickel due to the interaction with the support.

It is also reasonable to expect that the effect of the support on the catalytic activity of the nickel surface will be manifested in the apparent activation energy, Ea. The values of Ea obtained from Arrhenius plots in the temperature range 258-358 K, for the hydrogenation of 1-hexene, acrolein and methyl vinyl ketone (refs. 210,214) are collected in Table 12.16. As is seen, nickel supported on $AlPO_4$ is more active than nickel supported on Al_2O_3, but the most active is that supported on SiO_2.

TABLE 12.16
Apparent activation energies, Ea (in KJ mol^{-1}), and logarithm of the Arrhenius constant, A, for catalysts with 20 mass% Ni (refs. 210,214)

catalyst	1-hexene		acrolein		methyl vinyl ketone	
	Ea	log A	Ea	log A	Ea	log A
Ni/$AlPO_4$	44.0	6.54	55.5	5.54	37.8	2.83
Ni/Al_2O_3	71.0	9.00	--	--	--	--
Ni/SiO_2	16.1	0.10	--	--	--	--
Ni/$AlPO_4$-Al_2O_3	61.6	7.20	54.1	4.57	55.8	5.52
Ni/$AlPO_4$-SiO_2	36.3	3.65	50.4	5.14	31.0	1.50

In other words, according to Table 12.8, increasing the nickel content has two opposite effects on the catalytic activity for 1-hexene hydrogenation. One is a trend toward increasing r_A (areal rate), caused by lessening of the nickel-support interaction, and the other is a decrease in the catalytic activity caused by a reduction in the metal dispersion. For each nickel-support system there will thus be a nickel loading corresponding to the optimum catalytic activity per unit mass of nickel. For the hydrogenation of 1-hexene, the suitable nickel loading is 5 mass% in Ni/SiO_2, 10 mass% in Ni/Al_2O_3 and 20 mass% for the catalysts Ni/$AlPO_4$-SiO_2 and $AlPO_4$-Al_2O_3, respectively.

The same authors (ref. 215) by titrating the amount of poisoning using n-butanethiol as the poison, determined the number of active sites responsible for the liquid-phase catalytic hydrogenation of 1-hexene over a series of nickel catalysts supported on $AlPO_4$, SiO_2 and $AlPO_4$-SiO_2, with a wide range of dispersions and nickel loadings. These results, taken in conjunction with the metal surface area, obtained by XRD, yield the fraction of catalytically active surface nickel atoms. This fraction, which may be considered as a measure of the metal-support interaction, is independent of the particle size for catalysts with the same support and nickel loading, whereas it decreases as the nickel loading decreases. It decreases in the sequence:

$Ni/SiO_2 \simeq Ni/AlPO_4$-$SiO_2 > Ni/AlPO_4$

This behaviour is associated with a metal-support effect which may be explained by electronic effects.

Marcelin and Vogel (ref. 60) found that $Ni/AlPO_4$-Al_2O_3 catalysts (and other systems like Ni/TiO_2) exhibited behaviour typical of SMSI-type materials, i.e., suppresion of hydrogen chemisorption upon high temperature reduction. This behaviour is observed regardless of other components found in the support, such as MgO or Al_2O_3, but is absent whenever phosphate is deleted in the preparation, as evidenced by the normal hydrogen chemisorption results obtained for Ni supported on magnesia-alumina.

Thus, a metal-support interaction is evident in $Ni/AlPO_4$ and related systems. However this phenomenon is not restricted to the above mentioned systems. Thus, the reduction of Ni/ZrO_2 catalysts (ref. 52) above 700 K results in a decrease in hydrogen adsorption and in the catalytic activity towards the hydrogenolysis of ethane. Oxygen treatment, followed by reduction at low temperatures of samples reduced at high temperatures, restores the normal properties. This behaviour is similar to that observed for Ni/TiO_2 and Ni/SiO_2 and is typical of a strong metal-support interaction similar to that observed on Pt/TiO_2. Paradoxically, Ni does not interact strongly with ZrO_2 since it is easily reduced giving very large metallic particles.

Udrea et al. (ref. 58), using Ni/Al_2O_3-ZnO in methanation processes, have found that the different behaviours of the catalyst surface may be related to specific metal-support interactions. The role played by the support was evidenced by a relationship between the support acidity, as measured by IR spectroscopy of adsorbed pyridine, and the specific activity in methane synthesis. Such an effect is attributed to a change in electron density in the metallic phase due to metal-support interactions. Other authors (ref. 148), using Ni/MgO catalysts, found that an increase in the reduction temperature (770-1200 K) results in an increase in the catalytic activities in ethane hydrogenolysis and benzene hydrogenation, suggesting that no SMSI effect occurs. In the case of Ni/MgO, another kind of interaction could be based on the presence of unreduced nickel species which has nothing to do with the SMSI effects, although no bulk Ni-Mg alloy can be detected. This is related to an increase in the number of sites, due to the progressive reduction of Ni^{2+}, since complete reduction is obtained only at high temperatures.

In summary, the possibility that the support may modify the catalytic properties of supported nickel appear to be well established, although much work remains to be done in order to understand fully the SMSI effects.

12.4 CATALYZED REACTIONS

At the present time, most new nickel catalysts have been tested in three essential reduction processes: hydrogenolysis and hydrocracking, hydrogenation of organic functional groups, especially carbon-carbon double bonds, and methanation processes, Co and CO_2. In previous sections, different processes have been mentioned in which new supported nickel systems behave as catalysts. Sometimes a large number of catalysts are tested with respect to the same hydrogenation process. Thus, Hattori et al. (ref. 53) carried out the hydrogenation of N,N-dimethyl-2-propenylamine on nickel catalysts supported on various metal oxides, such as MgO, CoO, La_2O_3, ThO_2, ZrO_2, TiO_2, Al_2O_3 and SiO_2.

Table 12.17 summarizes the reactions which are known to be catalyzed on new supported nickel systems. As is seen, a large number of these catalysts have been tested in gas-phase processes and only a small number in liquid-phase hydrogenation. However, we think that most would be useful as catalysts in such liquid-phase processes. It is even conceivable that some could be the best catalysts for a particular reaction. The number of factors determining the activity and selectivity of a catalyst is very large. Thus, it is a fact that the catalytic behaviour of a catalyst is determined not only by the activity of the nickel surface but also by the acid-base properties, the average pore size and the pore-size distribution of the support.

In fact, all organic molecules, according to the support porosity, will be somewhat restricted from entering cavities of small size. Therefore, only the active sites found within sufficiently large pores will contribute significantly to the reaction. Taking this into account, the challenge now is to determine what supported nickel catalysts are specially suited to a particular reaction. Little research has been directed towards this end, although Marinas et al. have obtained promising results employing $Ni/AlPO_4$, $Ni/AlPO_4-SiO_2$ and $Ni/AlPO_4-Al_2O_3$ catalysts in the selective liquid-phase catalytic hydrogenation of E-cinnamaldehyde to hydrocinnamaldehyde (ref. 207) and acrolein to propanal (ref. 212). In both cases, the selectivity was very high ($\approx 100\%$) and furthermore these catalysts were able to reduce the aldehyde function to alcohol just after the olefinic double bond had been completely hydrogenated.

TABLE 12.17

Reactions catalyzed by new supported nickel systems

Catalyst	Tested reactions	Refs.
Ni/K-Graphite	alkene, carbonyl and nitrile hydrogenation	18,19
Ni/Montmorillonite	hydroisomerization and hydrocracking	36-38
Ni/Pumice	alkene hydrogenation	41
Ni/Vermiculite	ethylene hydrogenation	42
Ni/Ceramic	alkenes, acrylonitrile and benzene hydrogenation	43,44
Ni/Sepiolite	1-hexene hydrogenation	45
Ni/Silicone rubber	methyl acetoacetate hydrogenation	85,86
Ni/Polymer	alkenes and alkynes hydrogenation	87,88
Ni/NaF	2-propanol and 3-methyl-1-butanol dehydrogenation	46
Ni/CaF_2	acetone hydrogenation	47
Ni/SiO_2Mg	CO methanation	48
Ni/Cr_2O_3	citral and β-ionone hydrogenation	49,50
Ni/Silicon carbide	alkane hydrogenolysis	61-63
Ni/WO_3-Al_2O_3	alkene hydrogenation	55
Ni/ZnO	maleic and fumaric acids and ethylene hydrogenation	82-84,255
Ni/SiO_2ZnO	maleic acid hydrogenation	56
Ni/Al_2O_3-ZnO	CO methanation	57-59
Ni/ZrO_2	ethane hydrogenolysis and benzene and N,N-dimethyl-2-propenylamine hydrogenation	52,53
	CO methanation	64-73
Ni/ThO_2	N,N-dimethyl-2-propenylamine hydrogenation	53
	CO methanation	64-73
Ni/UO_2	CO methanation	64-73
Ni/La_2O_3	N,N-dimethyl-2-propenylamine hydrogenation	53
	CO methanation	74
Ni/CaO	1-butene and N,N-dimethyl-2-propenylamine hydrogenation	53
Ni/Zeolites	hydrocracking of paraffins	89,115,116
	Fisher-Tropsch synthesis	93
	benzene hydrogenation	105,117
	ethane and n-hexane hydrogenolysis	105
	CO hydrogenation	109
	butane hydrogenolysis	114
	1-hexene, cyclohexene, styrene and 4-phenyl-1-butene hydrogenation	117

(continued on p. 442)

TABLE 12.17 (continued)

Catalyst	Tested reactions	Refs.
Ni/TiO$_2$	D-fructose to D-manitol hydrogenation	256
	1-butene and N,N-dimethyl-2-propenylamine hydrogenation	53
	CO hydrogenation	126-133
	n-hexane hydrogenolysis	127,128,132,250
	ethane hydrogenolysis	132,135,136
	butane hydrogenolysis	245
	benzene hydrogenation	245,248
	ethylene hydrogenation	245,249
	styrene hydrogenation	245
	cyclohexane dehydrogenation	245
Ni/MgO	toluene and methylcyclopentane hydrogenolysis	6
	1-butene and N,N-dimethyl-2-propenylamine hydrogenation	53
	ethane hydrogenolysis	146
	benzene and methylbenzenes hydrogenation	150,152
	maleic acid hydrogenation	155,156
	steam reforming of butane	257
	cyclohexane dehydrogenation	258,259
	cyclopropane and propene hydrogenation	260
Ni/AlPO$_4$ and Ni/AlPO$_4$-SiO$_2$ and Ni/AlPO$_4$-Al$_2$O$_3$	cyclohexene hydrogenation	205
	1-hexene hydrogenation	210,214,215
	E-cinnamaldehyde hydrogenation	207
	acrolein and methyl vinyl ketone hydrogenation	212
	allyl alcohol and acrylic acid hydrogenation	211
	styrene and α-methylstyrene hydrogenation	213,216
	cycloalkenes and functionalized alkenes hydrogenation	45
Ni/AlPO$_4$-Al$_2$O$_3$ and Ni/AlPO$_4$	2-ethylhexanal hydrogenation	221

12.5 POISONING AND DEACTIVATION

One of the more important problems associated with the efficient use of dispersed metal catalysts is their deactivation, due to poisoning by chemical agents, coke or sintering. In this section, only the former points will be treated.

A large number of factors connected with both the chemical composition and preparation method affect not only the carbon deposition rate, but also the crystallite size for catalysts (ref. 257). Very little information concerning the relationship between the nickel crystallite size and the carbon deposition rate is available. Rostrup-Nielsen (ref. 261) found an effect of the nickel crystallite size on the equilibrium constant for methane decomposition.

On the other hand, Borowiecki (ref. 257) studied the influence of the mean nickel crystallite size on coking in steam reforming, using different supported nickel catalysts. The nickel crystallites differ widely in size, and the relationship between the coking rate and their average size is almost linear for a given ratio of reagents. The model proposed assumes two kinds of active sites on the catalyst surface with regard to the reagents, in that the hydrocarbon undergoes chemisorption on the nickel metal, while steam chemisorbs on the carrier. The amount of the steam chemisorbed is a function of the carrier-surface affinity towards it and its pressure in the gas phase. The model permits an understanding of the radical changes in resistence to coking which accompany the introduction of potassium compounds into the carrier (refs. 261, 262), see Fig. 12.3.

On the other hand, in the case of nickel metal supported on non-classical supports the poisoning with chemical agents has been little studied. Dalla-Betta et al. (ref. 51) studied the poisoning by H_2S of the hydrogenation of CO on Ni/ZrO_2, whereas Moldovan et al. (ref. 66) studied the same process in the presence of Ni/ZrO_2 and Ni/TiO_2, together with the coke deposition. They concluded that the Ni/ZrO_2 system is more sensitive to coke and sulphur poisoning than is Ni/TiO_2, although is more deactivated than is Ni/Al_2O_3 or Raney-nickel.

Also, Hattori et al. (ref. 53) showed that Ni/MgO catalyst is completely poisoned by CO_2 in the hydrogenation of N,N-dimethyl-2-propenylamine, whereas $Ni/SiO_2-Al_2O_3$ is not. However, it has been demonstrated (ref. 263) that the poisoning coefficient, α, calculated from the Maxted equation (ref. 264) linearly decreases with increasing electron density of nickel metal as determined by XPS, and the specific activity in the styrene hydrogenation, both parameters being related to the metal-support interaction. Therefore, we think that the resistance to poisoning and the regeneration of the new supported nickel catalysts is an interesting field of study. It is even possible that in some cases such resistance could be the best property of a catalyst.

In this context, some exploratory work has been done by Marinas and co-workers (ref. 215). Poisoning experiments were carried out using n-butanethiol in the liquid-phase catalytic hydrogenation of 1-hexene on a series of nickel catalysts supported on $AlPO_4$ and $AlPO_4-SiO_2$, with different loadings, dispersions, and preparation procedures. The results obtained were compared to those obtained with SiO_2 as the support and with bulk nickel, without a support. Furthermore, the catalytic behaviour of these catalysts was studied after they had been regenerated in flowing hydrogen (100 ml.min^{-1}) at 570 ± 5 K for 2 h.

The poisoning isotherms are shown in Fig. 12.4. There is a clear difference between bulk nickel and the supported nickel catalysts. With the former, a

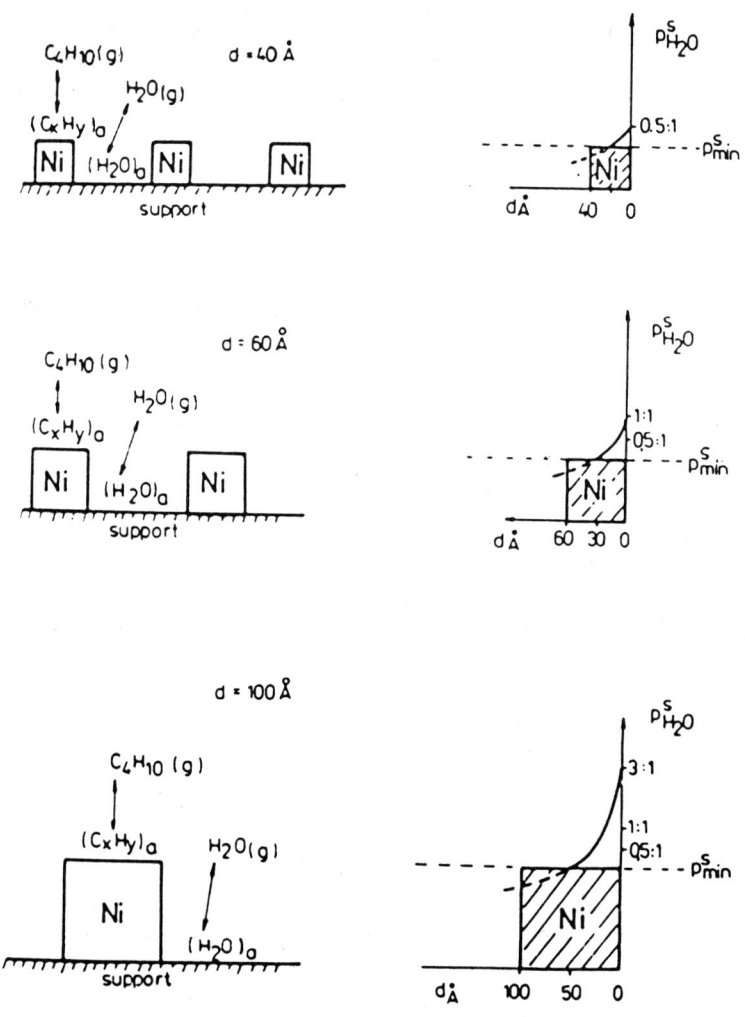

FIG. 12.3. A model of the influence of the mean size of nickel crystallites on the resistance to cocking, according to (ref. 257).

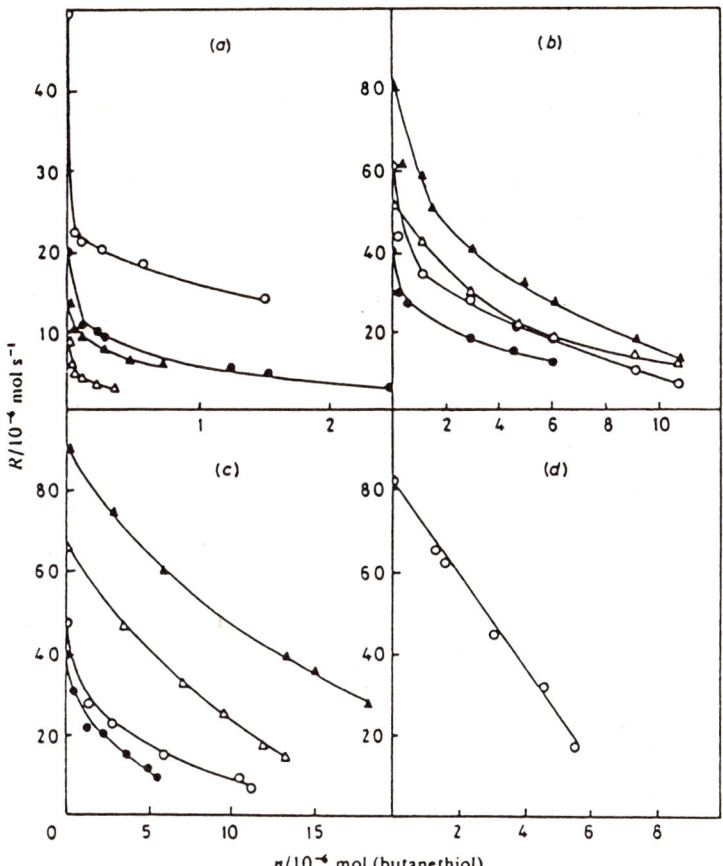

FIG. 12.4. Poisoning isotherms obtained with n-butanethiol on supported nickel catalysts (a) Ni/AlPO$_4$, (b) Ni/SiO$_2$ and (c) Ni/AlPO$_4$-SiO$_2$: △, 5 wt% I; ▲, 10 wt% I; ○, 20 wt% I; ●, 20 wt% II and (d) unsupported nickel (ref. 215)

linear reduction in hydrogenation activity is obtained with the amount of poison, according to the Maxted expression (ref. 264)

$$R = R_0(1 - \alpha n) \tag{12.6}$$

while with the latter the poisoning curves can be fitted to the expression

$$R = R_0 \cdot e^{-\alpha n} \tag{12.7}$$

were R_0 is the catalytic activity when the catalyst is clean, R is the activity when the catalyst is poisoned by an amount n, and α is the poisoning coefficient, which is a measure of the resistance to poisoning.

The α values of some catalysts indicated in Table 12.18 were obtained from the slopes of plots of ln R/R_0 or R/R_0 vs. n. The catalytic activities of clean and regenerated catalysts are also shown. The fraction of active sites decreased in the sequence:

TABLE 12.18
Poisoning coefficients, α, catalytic activity of clean (R) and regenerated (R_{reg}) catalysts and percentage of regeneration of catalysts, %Reg (ref. 215)

Catalyst	$\alpha \times 10^{6}$ [a]	$R \times 10^{6}$ [b]	$R_{reg} \times 10^{6}$ [b]	%Reg [c]
Ni(unsupported)	2.72	4.2	0.73	17.3
Ni/SiO$_2$				
5% I	0.15	51.6	51.44	99.7
10% I	0.14	80.9	45.00	55.6
20% I	0.16	51.7	20.37	39.4
20% II	0.14	39.0	19.30	49.5
Ni/AlPO$_4$				
5% I	1.27	9.6	0.00	0.0
10% I	0.66	13.6	1.02	7.5
20% I	0.26	27.8	6.20	22.3
20% II	0.50	19.1	3.44	18.0
Ni/AlPO$_4$-SiO$_2$				
5% I	0.10	67.2	9.07	13.5
10% I	0.05	90.5	39.82	44.0
20% I	0.12	48.2	20.20	41.9
20% II	0.20	40.5	32.85	81.1

[a] mol^{-1} of n-butanethiol per 0.5×10^{-3} g atom of Ni.
[b] Activity per 0.5×10^{-3} g atom of Ni (mol s^{-1}).
[c] (R_{reg}/R)100.

Ni/SiO$_2$ $\simeq$ Ni/AlPO$_4$-SiO$_2$ > Ni/AlPO$_4$

This behaviour is associated with a metal-support interaction which may be explained by electronic effects.

Taking into account that, according to the authors (ref. 215), the number of moles of n-butanethiol blocking the active sites present in 0.5×10^{-3} g atom of nickel is given by $n = \alpha^{-1}$, it is possible to establish that the supports SiO$_2$ and AlPO$_4$-SiO$_2$ enhance the catalytic activity of supported nickel metal by increasing the number of catalytically active surface nickel atoms. By contrast, when 20 mass% nickel is supported on AlPO$_4$, the fraction of surface nickel atoms available for catalytic reactions is only slightly higher than with unsupported nickel metal. However, at lower loadings, the surface density of the active sites is lower compared to bulk nickel.

The regeneration capacity of catalysts is related not only to the support but also to the nickel loading. Thus, in Ni/SiO$_2$ systems the regeneration capacity decreases from 100% to 35% when the nickel loading increases from 5 to 20%. However, in Ni/AlPO$_4$ the opposite behaviour occurs, the regeneration increasing from 0 to 25% on increasing the nickel loading from 5 to 20%. The systems Ni/AlPO$_4$-SiO$_2$ exhibit intermediate behaviours between Ni/AlPO$_4$ and Ni/SiO$_2$ catalysts, with the highest regeneration capacity being 10 mass%.

According to the results, the influence of the support on the resistance to poisoning of nickel catalysts is decisive. Therefore, if we consider the economic motivation to discover sulphur-tolerant catalysts, i.e., catalysts which retain sufficient methanation activity in the presence of sulphur contaminants, we have to conclude that it is very important to achieve not only a better understanding of the mechanisms of both sulphur poisoning and regeneration, but also to determine the tolerance to poisoning of new supported nickel catalysts. In this way, we think that it will be possible to obtain promising results.

12.6 PROMOTING EFFECTS

The catalytic properties of metals can be significantly affected by the addition of small amounts of foreign species. Some of these additives act as structural promoters, preventing sintering of the metallic phase; others, which are termed electronic or chemical promoters, change the catalytic activities or selectivities. Among chemical promoters, perhaps the alkali-metal ions have been the most studied.

Thus, the addition of alkali-metal ions to catalysts, especially to alumina--based catalysts, is widely used to control unwanted side reactions due to the carrier. This effect is mainly due to modifications in the acidity of the surface. However the action of an additive is not restricted to such alterations since other consequences of fundamental importance may take place. These pertain, as shown by Lo Jacomo et al (ref. 265), to the physico-chemical properties of the active phase deposited on the modified support, e.g., modification of the dispersion of the supported material and formation of a chemical association between the active phase and the support (ref. 265,266). This effect may proceed through two types of mechanisms:

(1) Via an alteration of the pH of the impregnating medium, during the deposition of the active phase, thus, promoting or inhibiting the adsorption, hydrolysis or precipitation of the cations.

(2) Via variations in the chemical properties of the carrier through solid--state reactions of the additive with the support, thus influencing the reactivity of the latter towards the deposited phase.

Many investigations have already been made of the influence of additives, but most have emphasized the first aspect, i.e., the variations of the catalytic activity with the acid-base properties of the modified carrier. Little attention has been made given to the influence of modifiers on the nature of the deposited phase and its dispersion over the carrier. moreover, most of the results are connected with the deposition of nickel oxide on modified alumina supports (refs.

267-270).

Only a few reports are available on the use of additives in nickel catalysts on unconventional supports. However, the role played by the alkali is equivocal. It has been suggested (ref. 271) that it might act as a strong base, transferring electrons to the metallic phase (refs. 272-274), thus increasing the bond strength of electron-acceptor molecules such as CO and N_2 via the metallic phase (ref. 275). Thus, sodium carbonate added during citral hydrogenation (ref. 50) increases the citronellol yield, through the adsorption of sodium on the surface of metallic nickel in Ni/Cr_2O_3 catalysts, thus modifying the adsorbability with respect to various unsaturated groups and the selectivity.

In this context, the modification of the surface of $AlPO_4$ with alkali-metal cations (Li, Na and K) leads to a change in the acidity-basicity balance of the catalytic surface, with the formation of a new basic site that does not possess reducing properties (ref. 186). When the nickel is impregnated until incipient wetness on $AlPO_4$ thus modified, the supported metal exhibits a lower metal surface area (ref. 214) although the areal rate remains approximately constant, Table 12.18. So, alkaly-metal impregnation, in principle, does not increase the metal-support interaction, but instead markedly decreases the activity per unit mass of nickel supported, especially in the case of Li, which is the alkali-metal ion with the smallest ionic radius.

TABLE 12.19

Effect of the addition of 5 mass% alkali-metal ion on the dispersion and catalytic activity of Ni/$AlPO_4$ (20 mass% Ni) in the liquid-phase hydrogenation of 1-hexene (ref.214)

Alkali ion	$^a S_{Ni}$	$^b r_a \cdot 10^6$	$^c r_{Ni} \cdot 10^3$
-^{d}	87	35.4	3.08
Li	17	32.5	0.55
Na	42	29.1	1.22
K	43	36.7	1.54

^{a}Metal surface area from XRD ($m^2 \, g_{Ni}^{-1}$).
^{b}Areal rate or activity per unit surface area of nickel metal (mol $s^{-1} m_{Ni}^{-2}$).
^{c}Activity per gram of metal supported (mol $s^{-1} g_{Ni}^{-1}$).
^{d}Without alkali-metal ion.

Other methods have been used to modify the activity and/or selectivity of supported nickel catalysts, although these only involve conventional supports with a few reports on other supports. These include the addition of oxides to the support, either insulator or conductor oxides, as well as alloy formation with the nickel, mainly by addition of copper metal. Thus, Nehering and co-workers (refs. 258-260) studied de addition of conductor oxides to Ni/MgO catalysts and found that n-conductor oxides improved the catalytic activity in the dehydrogenation of cyclohexane while the hydrogenations of cyclopropane and propene were inhibited. When p-conductor oxides are used the opposite behaviour is found. This is explained in terms of the conductivity of the support, due to an electronic nickel-support interaction. In another works, as has been stated in 12.2.3(iv), the addition of Al_2O_3 or SiO_2 to $AlPO_4$ supports leads to a supported nickel catalysts with a different metal-support interaction, which affects the catalytic activity in the hydrogenation of the carbon-carbon double bond of simple as well as of functionalized alkenes.

Supported alloy Ni-Cu catalysts have often been prepared, but only on conventional supports and from the data in the literature it appears that an increase in the surface area of the system can be obtained only at a very low amounts of introduced Cu (refs. 276-283). A very marked decrease in the surface area occurs when the content of Cu increases. The reasons for these phenomena are closely associated with the change in the energy characteristics of the nickel centres, manifested as a ligand or ensemble effect. In this way, previous results as yet unpublished have shown that small amounts of Cu (0.1-2.0 mass%) improve the areal rate of 20 mass% Ni/$AlPO_4$ catalysts in the hydrogenation of olefinic and acetylenic carbon-carbon bonds, although the active surface area of the supported catalysts is slightly modified. Some of these results are summarized in Table 12.19.

TABLE 12.20

Effect of copper addition to 20 mass% Ni/$AlPO_4$ on the dispersion and catalytic activity in the liquid-phase hydrogenation of 1-hexene

Cu (mass%)	S ($m^2 \, g_{Ni}^{-1}$)	$r_a \, 10^6$ (mol $s^{-1} m_{Ni}^{-2}$)
0	87	35.7
0.3	65	42.2
0.6	82	24.4
1.0	75	20.6
2.0	99	20.0

However, in order to draw conclusions in this area, more detailed experimental information is necessary.

12.7 CONCLUSIONS

This review of new supported nickel catalysts reveals the significant progress made in the characterization of supported catalysts and their use in catalytic processes. The increasing number of papers concerning new nickel catalysts reflects their importance in practice, although there still remains the difficulty of too little information available on the nickel-support interaction in systems other than Ni/TiO$_2$.

On the other hand, much attention is being given to the preparation and use of aluminium orthophosphate in catalysts and supports for metals. Initial investigations indicate that it has mildly acidic properties and appears useful as a support for nickel in the liquid-phase catalytic hydrogenation of several olefinic compounds. Furthermore, nickel supported on aluminium orthophosphate and related systems suppresses the hydrogen chemisorption which is typical of nickel-support interactions.

However, much more work needs to be done in the field of supported nickel catalysts, preferably on nickel-support interactions and their application to processes of industrial and laboratory interest. Thus, research in this direction will undoubtedly result in an even greater use of nickel catalysts in the near future.

REFERENCES

1. B. Delmon and M. Houalla, in B. Delmon, P. Grange, P. Jacobs and G. Poncelet (Editors), Preparation of Catalysts II, Elsevier, Amsterdam, 1979, p. 439.
2. B. L. Villemin, C. Hoang-Van and S. J. Teichner, Bull. Soc. Chim. Fr., 1 (1980) 16.
3. A. Merlim, B. Imelik and S. J. Teichner, C. R. Acad. Sci. Paris, Ser. C, 238 (1954) 254.
4. K. J. Klabunde, D. M. Ralston, R. W. Zoellner, H. Hattori and Y. Tanaka, J. Catal., 55 (1978) 213.
5. D. H. Ralston and K. J. Klabunde, Appl. Catal., 3 (1982) 13.
6. K. Matsuo and K. J. Klabunde, J. Catal., 73 (1982) 216.
7. Y. Yermakov, Catal. Rev. Sci. Eng., 13 (1976) 77.
8. M. Ichikawa, J. Chem. Soc., Chem. Commun., (1976) 26.
9. R. Kobo, J. Phys. Soc. Jpn., 17 (1962) 975.
10. M. Primet, J. A. Dalmon and G. A. Martin, J. Catal., 46 (1977) 25.
11. G. A. Martin and J. A. Dalmon, C. R. Acad. Sci. Paris, Ser. C, 286 (1978) 127.
12. G. A. Martin, J. Catal., 60 (1979) 345.
13. E. Zagli and J. L. Falconer, Appl. Catal., 4 (1982) 135.
14. H. Suzuki, S. Takasaki, F. Koga, A. Ueno, Y. Kotera, T. Sato and N. Todo., Chem. Lett.,(1982) 127.
15. A. Ueno, H. Suzuki and Y. Kotera, J. Chem. Soc., Faraday Trans. 1, 79 (1983) 127.
16. G. Gut, J. Kosiuka, A. Prabucki and A. Schuerch, Chem. Eng. Sci., 34 (1979) 1051.
17. J. Barcicki, D. Nazimek, W. Grzegorczyk, T. Borowiecki, R. Frank and M. Pielach, React. Kinet. Catal. Lett., 17 (1981) 169.
18. D. Savoia, E. Tagliavini, C. Trombini and A. Unami-Ronchi, J. Org. Chem. 46 (1981) 5340.

19 D. Savoia, E. Tagliavini, C. Trombini and A. Umani-Ronchi, J. Org. Chem., 46 (1981) 5344.
20 G. Dalmai-Imelik, C. Leclerq, J. Massardier, A. Maubert Franco and A. Zalhout, Jpn. J. Appl. Phys., Suppl. 2, Pt 2, (1974) 489.
21 G. A. Martin, B. Imelik and M. Prettre, C. R. Acad. Sci. Paris, Ser. C., 264 (1967) 1536.
22 G. Dalmai-Imelik, C. Leclerq and A. Maubert-Mugnet, J. Solid State Chem., 16 (1976) 129.
23 J. C. Vedrine, G. Hollinger and T. M. Duc, J. Phys. Chem., 82 (1978) 1515.
24 J. T. Richardson and R. J. Dubus, J. Catal., 54 (1978) 207.
25 J. A. Van Dillon, J. W. Geus, L. A. M. Hermans and V. der Mejden, Proc. 6th Intern. Congr. Catal., London, 1976, Chemical Society, London, 1977, p. 677.
26 D. L. Trimm. Pure Appl. Chem., 50 (1978) 1147.
27 D. L. Trim and E. R. Karal, in T. Seiyama and K. Tanabe (Editors), New Horizons in Catalysis, Proc. 7th Intern. Congr. Catal., Tokyo, 1980, Elsevier, Amsterdam, 1981, Part A, p. 517.
28 D. L. Trim. Design of Industrial Catalysts, Elsevier, Amsterdam, 1980.
29 J. W. Geus, in G. Poncelet, P. Grange and P. A. Jacobs (Editors), Preparation of Catalysts III, Elsevier, Amsterdam, 1983, p. 1.
30 J. M. Lalancette, U. S. Pat., 3,847,983 (1974).
31 K. Otto and M. Shelef, Carbon, 15 (1977) 117.
32 D. J. Smith, R, M. Fisher and L. A. Freeman, J. Catal., 72 (1981) 51.
33 D. L. Doering, J. T. Dickinson and H. Poppa, J. Catal., 73 (1982) 91.
34 W. T. Granquist, U. S. Pat., 3,252,757 (1966).
35 A. C. Wright, W. T. Granquist and J. Kennedy, J. Catal., 25 (1972) 65.
36 M. E. Swift and E. R. Black, Ind. Eng. Chem. Prod. Res. Dev., 13 (1974) 106.
37 J. P. Gianetti and D. C. Fisher, Amer. Chem. Soc. Div. Petr. Chem. Prepr., 20 (1975) 52.
38 H. E. Swift and R. F. Vogel, U. S. Pat., 4,065,380 (1977) and 4,138,326 (1979).
39 J. J. L. Heinerman, I. L. C. Freviks, J. Gaaf, G. T. Pott and J. G. F. Coolegen, J. Catal., 80 (1983) 145.
40 L. W. Covert, R. Connor and M. J. Adkins, J. Am. Chem. Soc., 54 (1932) 1651.
41 V. Mintsa-Eya, L. Hilaire, A. Choplin, R. Toroude and F. G. Gault, J. Catal., 82 (1983) 267.
42 M. Kermarec, M. Patel, P. Rabette, H. Pezerat and D. Delafosse, J. Chem. Soc., Faraday Trans. 1, 79 (1983) 599.
43 T. Tihanyi, K. Varga, I. Hannus, I. Kiricsi and P. Fajes, React. Kinet. Catal. Lett., 18 (1981) 449.
44 A. Parmaliana, A. Mezzapica, C. Crisafulli, S. Galvagno and N. Giordano, React. Kinet. Catal. Lett., 19 (1982) 155.
45 J. M. Campelo, A. Garcia, D. Luna and J. M. Marinas, unpublished results.
46 Yu. S. Mardashev and N. P. Karzhvina, Zhur. Fiz. Kim., 51 (1977) 2631; english transl., Russ. J. Phys. Chem. 51 (1977) 1531.
47 N. P. Karzhvina, F. L. Komolova, I. V. Razumovskaya and Yu. S. Mardashev, Zhur. Fiz. Khim., 57 (1983) 495; english transl., Russ. J. Phys. Chem., 57 (1983) 305.
48 P. J. Rencroft, H. Parekli, P. Ganesan, S. N. Russell and B. H. Davis, Appl. Catal., 3 (1982) 65.
49 D. V. Sokolskii, T. M. Omarkulov, U. Suyunbaev, M. A. Abishev, V. A. Shoshenkova, V. M. Mekeev and K. Kh. Nurgozhaev, Zh. Fiz. Khim., 56 (1982) 1773; Russ. J. Phys. Chem., 56 (1982) 1075.
50 A. M. Pak, S. R. Konuspaev, G. D. Zakumbaeva and D. V. Sokolskii, React. Kinet. Catal. Lett., 16 (1982) 339.
51 A. Dalla-Betta, A. G. Piken and M. Schelef, J. Catal., 40 (1975) 173.
52 P. Turlier and G. A. Martin, React. Kinet. Catal. Lett., 21 (1982) 387.
53 H. Hattori, H. Mai and K. Tanabe, Appl. Catal., 4 (1982) 87.
54 A. Hattori, H. Hattori and K. Tanabe, J. Catal., 65 (1980) 245.
55 J. Uchytil, E. Kocova and M. Kraus, Collect. Czech. Chem. Commun., 46 (1981) 2076.
56 R. Kudlacek and H. Solarova, Collect. Czech. Chem. Commun., 48 (1983) 2175.

57 I. Udrea, R. Morainu, M. Udrea and I. V. Nicolescu, Geterog. Katal., Tr. Mezhdunar Symp., 3rd 1975, 3 (1979) 133.
58 I. Udrea, M. Udrea and L. Frunza, React. Kinet. Catal. Lett., 21 (1982) 255.
59 I. Udrea, M. Udrea and L. Frunza, React. Kinet. Catal. Lett., 21 (1982) 261.
60 G. Marcelin, and R. F. Vogel., J. Catal., 82 (1983) 482.
61 C. J. Machielis and R. B. Anderson, J. Catal., 58 (1979) 253.
62 C. J. Machielis and R. B. Anderson, J. Catal., 58 (1979) 260.
63 C. J. Machielis and R. B. Anderson, J. Catal., 58 (1979) 268.
64 V. T. Coon, T. Takeshita, W. E. Wallace and R. S. Craig, J. Phys. Chem., 80 (1976) 1878.
65 A. Elattar, T. Takeshita, W. E. Gallace and R. S. Craig, Science, 196 (1977) 1093.
66 A. G. Moldovan, A. Elattar, W. E. Wallace and R. S. Craig, J. Solid State Chem., 25 (1978) 23.
67 V. T. Coon, W. E. Wallace and R. S. Craig, in G. J. McCarthy and J. J. Rhyne (Editors), The Rare Earths in Science and Technology, Plenum Press, New York, 1978.
68 A. Elattar, W. E. Wallace and R. S. Craig, Adv. Chem. Ser., 178 (1979) 2.
69 W. E. Wallace, in A. F. Andersen and A. J. Maeland (Editors), Hydrides for Energy Storage, Pergamon Press, Elmsford, N. Y., 1978, p. 33.
70 A. Elattar, W. E. Wallace and R. S. Craig, in A. F. Andersen and A. J. Maeland (Editors), Hydrides for Energy Storage, Pergamon Press, Elmsford, N. Y., 1978, p. 87.
71 C. A. Luengo, A. L. Cabrera, H. B. Mackey and M. B. Maple, J. Catal., 47 (1977) 1.
72 G. B. Atkinson and L. J. Nicks, J. Catal., 46 (1977) 417.
73 R. L. Chin, A. Elattar, W. E. Wallace and D. M. Hercules, J. Phys. Chem., 84 (1980) 2895.
74 H. Inamura and W. E. Wallace, J. Phys. Chem., 83 (1979) 2009.
75 H. Inamura and W. E. Wallace, J. Phys. Chem., 83 (1979) 3261.
76 H. Inamura and W. E. Wallace, J. Catal., 65 (1980) 127.
77 M. Houalla, C. L. Kibby, L. Petrakis and D. M. Hercules, J. Phys. Chem., 87 (1983) 3689.
78 H. C. Siegmann, L. Schlapbach and C. R. Brundle, Phys. Rev. Lett., 40 (1978) 972.
79 J. H. N. Van Vucht, F. A. Kuijpers and H. C. A. M. Bruning, Philips Res. Rep. 25 (1970) 133.
80 T. Takeshita, W. E. Wallace and R. S. Craig, Inorg. Chem., 13 (1974) 2282.
81 J. Barrault, D. Duprez and A. Guilleminot, Appl. Catal., 5 (1983) 99.
82 R. Kudlacek and J. Cabicar, Collect. Czech. Chem. Commun., 43 (1978) 1818.
83 R. Kudlacek and J. Cabicar, Collect. Czech. Chem. Commun., 43 (1978) 1838.
84 R. Kudlacek ans R. Jelinkova, Collect. Czech. Chem. Commun., 45 (1980) 1632.
85 T. Harada, A. Tai, M. Yamamoto, H. Ozaki and Y. Izumi, in T. Seiyama and K. Tanabe (Editors), New Horizons in Catalysis, Proc. 7th Intern. Congr. Catal., Tokyo, 1980, Elsevier, Amsterdam, 1981, p. 364.
86 A. Tai, Y. Imachi, T. Harada and Y. Izumi, Chem. Lett., (1982) 1651.
87 I. I. Schewager, R. J. Leak, E. L. Cole, Texaco, German Offen. 2,032,766 (1969); C. A.., 74 (1971) 76020e.
88 W. J. Jr. Leonard, H. V. A. Holler, Shell, German Offen. 2,213,485 (1972); C. A., 78 (1973) 44295.
89 K. G. Ione, P. N. Kuznetsov, D. M. Anufriev and V. N. Romannikov, Acta. Phys. Chem. (Szeged), 24 (1978) 153.
90 N. V. Klyneva, M. L. Valcheva, N. P. Davidova, K. G. Ione and D. P. Shopov, React. Kinet. Catal. Lett., 17 (1981) 315.
91 I. M. Kolesnikov and G. M. Panchelov, in Promoting Effect of the Reaction Mixture and Various Additives on Aluminosilicate Catalysts, (Ts N II Teneftekhin), Moskova, 1973.
92 N. Jaeger, U. Melville, R. Novak, H. Schribbers and G. Schulz-Ekloff, in B. Imelik, C. Naccache, Y. Ben Taarit, J. C. Vedrine, G. Coudurier and H. Praliaud (Editors), Catalysis by Zeolites, Elsevier, Amsterdam, 1980, p. 335.

93 H. H. Nijs, P. A. Jacobs, J. J. Verdonck and J. B. Uytterhoeven, in J. Bourdon (Editor), Growth and Properties of Metal Clusters, Elsevier, Amsterdam, 1980, p. 479.
94 N. P. Davidova, N. V. Peshev and D. M. Shopov, React. Kinet. Catal. Lett., 16 (1981) 201.
95 J. A. Rabo, C. L. Angell, P. H. Kasai and M. V. Shomaker, Discuss. Faraday Soc., 41 (1966) 328.
96 P. H. Kasai, R. J. Bishop Jr. and D. Mc Leod Jr., J. Phys. Chem., 82 (1978) 279.
97 E. Garbowski and M. V. Mathieu, Chem. Phys. Lett., 49 (1977) 247.
98 D. Olivier, M. Richard and M. Che., Chem. Phys. Lett., 60 (1978) 77.
99 D. Olivier, M. Richard, M. Che, F. Bozon-Verduraz and R. B. Clarkson, J. Phys. Chem., 84 (1980) 420.
100 M. F. Guilleux, D. Delafosse, G. A. Martin and J. A. Dalmon, J. Chem. Soc., Faraday Trans 1, 75 (1979) 165.
101 M. Brian-Faure, M. F. Guilleux, J. Jeanjean and D. Delafosse, Proc. Symp. Zeolites, Szeged, Hungary, 1978, p. 99.
102 D. Olivier, M. Richard, L. Bonneviot and M. Che, in J. Bourdon (Editor), Growth and Properties of Metal Clusters, Elsevier, Amsterdam, 1980, p. 193.
103 J. Jeanjean, D. Delafosse and P. Gallezot, J. Phys. Chem., 83 (1979) 2761.
104 M. Briand-Faure, J. Jeanjean, D. Delafosse and P. Gallezot, J. Phys. Chem., 84 (1980) 875.
105 J. T. Richardson, J. Catal., 21 (1971) 122.
106 H. Minchev, F. Steinbach and V. Penchev, Z. Phys. Chem. N. F. (Wiesbaden), 99 (1976) 223.
107 C. Minchev, V. Kanazirev, L. Kosova, V. Penchev, W. Gunsser and F. Schmidt, in L. V. Rees (Editor), Proc. Intern. Conference on Zeolites, Heyden, London, 1980, p. 255.
108 T. A. Egerton and J. C. Vickermann, J. Chem. Soc., Faraday Trans. 1, 69 (1973) 39.
109 D. Exner, N. Jaeguer, R. Nowak, H. Schrübbers and G. Schulz-Ekloff, J. Catal., 74 (1982) 188.
110 A. Andreev, D. Shopov, E. C. Hass, N. Jaeger, R. Nowak and R. Plath, Izv. Khim., 13 (1980) 108.
111 D. Olivier, M. Richard, L. Bonneviot and M. Che, Stud. Surf. Sci. Catal., 4 (1980) 193.
112 S. Briese-Gülban, H. Kompa, H. Schrübbers and G. Schulz-Ekloff, React. Kinet. Catal. Lett., 20 (1982) 7.
113 W. Romanowski, Polish J. Chem., 54 (1980) 1515.
114 G. N. Sauvion, M. Guilleux, J. F. Tempere and D. Delafosse, J. Chim. Phys., 79 (1982) 395.
115 K. G. Ione, P. N. Kuznetsov, N. V. Klyneva and G. V. Echevskii, Kinet. Katal., 18 (1977) 164.
116 P. N. Kuznetsov, D. M. Anufriev and K. G. Ione, Kinet. Katal., 19 (1978) 1236.
117 K. G. Ione, V. N. Romannikov, A. A. Davidov and L. B. Orlova, J. Catal., 57 (1979) 126.
118 P. N. Kuznetsov, D. M. Anufriev and K. G. Ione, Kinet. Katal., 19 (1978) 1250.
119 M. Briend-Faure, M. Guilleux, J. Jeanjean, D. Delafosse, G. D. Moriadassou and M. Bureau-Tardy, Acta Phys. Chem., 24 (1978) 99.
120 N. Davidova, N. Preshev and D. Shapov, J. Catal., 58 (1979) 198.
121 S. J. Tauster, S. C. Fung and R. L. Garten, J. Am. Chem. Soc., 100 (1978) 170.
122 S. J. Tauster and S. C. Fung, J. Catal., 55 (1978) 29.
123 R. T. K. Baker, E. B. Prestidge and R. L. Garten, J. Catal., 59 (1979) 293.
124 R. T. K. Baker, E. B. Prestidge and R. L. Garten, J. Catal., 56 (1979) 390.
125 S. J. Tauster, S. C. Fung, R. T. K. Baker and J. A. Horsley, Science, 211 (1981) 1121.
126 M. A. Vannice and R. L. Garten, J. Catal., 56 (1979) 236.
127 R. Burch and A. R. Flambard, J. Chem. Soc., Chem. Commun., (1981) 123.
128 R. Burch and A. R. Flambard, React. Kinet. Catal. Lett., 17 (1981) 23.
129 C. H. Bartholomew and R. B. Pannell, J. Catal., 65 (1980) 390.
130 D. C. Mustard and C. H. Bartholomew, J. Catal., 67 (1981) 186.

131 M. A. Vannice, J. Catal., 74 (1982) 199.
132 R. Burch and A. R. Flambard, J. Catal., 78 (1982) 385.
133 C. H. Bartholomew, R. B. Pannell and J. L. Butler, J. Catal., 65 (1980) 335.
134 C. H. Bartholomew, R. B. Pannell, J. L. Butler and D. G. Mustard, Ind. Eng. Chem. Prod. Res. Dev., 20 (1981) 296.
135 E. I. Ko and R. L. Garten, J. Catal., 68 (1981) 233.
136 E. I. Ko, S. Winston and Ch. Woo, J. Chem. Soc., Chem. Commun., (1982) 740.
137 M. A. Vannice and R. L. Garten, J. Catal., 66 (1980) 242.
138 R. Burch and A. R. Flambard, in G. Poncelet, P. Grange and P. Jacobs (Editors), Preparation of Catalysts III, Elsevier, Amsterdam, 1983, p. 311.
139 M. Che, C. Naccache and B. Imelik, J. Catal., 34 (1972) 328.
140 G. A. Zenkovets, E. A. Paukshtis, D. V. Tarasova and E. N. Yurchenko, Kinet. Katal., 26 (1981) 1266.
141 P. G. Menon and G. F. Froment, Appl. Catal., 1 (1981) 31.
142 X. Z. Jiang, T. F. Hayden and J. A. Dumesic, J. Catal., 83 (1983) 168.
143 J. S. Smith, P. A. Thrower and M. A. Vannice, J. Catal., 68 (1981) 270.
144 B. D. Flockhart, I. A. N. Scott and R. C. Pink, Trans. Faraday Soc., 62 (1966) 730.
145 H. Hattori, N. Yoshii and K. Tanabe, in J. W. Hightower (Editor), Proc. 5th Intern. Congr. Catal., Miami Beach, Florida, 1972, North-Holland/American Elsevier, 1973, Vol. 1, p. 1.
146 J. Kijenski and S. Malinowski, J. Chem. Soc., Faraday Trans. 1, 74 (1978) 250.
147 G. A. Martin, N. Ceaphalan, Ph. Mongolfier and B. Imelik, J. Chim. Phys., 70 (1973) 1422.
148 A. Maubert, G. A. Martin, H. Praliaud and P. Turlier, React. Kinet. Catal. Lett., 22 (1983) 203.
149 J. A. Lercher and H. Noller, J. Catal., 77 (1982) 152.
150 J. Volter, B. Lange and W. Khun, Z. Anorg. Allg. Chem., 340 (1965) 253.
151 G. A. Martin, B. Imelik and M. Prettre, J. Chim. Phys., 66 (1969) 1682.
152 V. Nikolajenko, V. Bosacek and V. L. Danos, J. Catal., 2, (1963) 130.
153 J. G. Highfield, A. Bossi and F. S. Stone, in G. Poncelet, P. Grange and P. A. Jacobs (Editors), Preparation of Catalysts III, Elsevier, Amsterdam, 1983, p. 181.
154 M. Pospisil and M. Tvrznik, Collect. Czech. Chem. Commun., 44 (1979) 1023.
155 R. Kudlacek and J. Rexova, Collect. Czech. Chem. Commun., 46 (1981) 2043.
156 R. Kudlacek and J. Lokoc, Collect. Czech. Chem. Commun., 47 (1982) 1780.
157 J. B. Moffat, Catal. Rev. Sci. Eng., 18 (1978) 199.
158 J. B. Moffat in M. Grayson and E. J. Griffith (Editors), Topics in Phosphorous Chemistry, Vol. 10, Wiley, New York, 1980, p. 286.
159 A. Alberola and J. M. Marinas, An. Quim., 65 (1969) 1001.
160 A. Alberola and J. M. Marinas, An. Quim., 65 (1969) 1007.
161 A. Alberola and J. M. Marinas, An. Quim., 66 (1970) 585.
162 A. Alberola and J. M. Marinas, An. Quim., 67 (1971) 631.
163 A. Alberola and J. M. Marinas, An. Quim., 67 (1971) 698.
164 A. Alberola, C. Fernandez and J. M. Marinas, An. Quim., 68 (1972) 1399.
165 A. Alberola, M. Blanco, J. M. Marinas and R. Perez-Osorio, Acta Cient. Venez., 24 (1973) 222.
166 A. Alberola, J. M. Marinas, E. Novella and R. Perez-Ossorio, An. Quim., 70 (1974) 648.
167 S. Esteban, J. M. Marinas, R. A. Perez-Ossorio and A. Alberola, An. Quim., 70 (1974) 944.
168 J. M. Marinas, C. Jimenez, R. Perez-Ossorio and J. V. Sinisterra, An. Quim., 70 (1974) 860.
169 M. Blanco, J. M. Marinas, R. Perez-Ossorio and A. Alberola, An. Quim., 71 (1975) 199.
170 C. Jimenez, J. M. Marinas, R. Perez-Ossorio and J. V. Sinisterra, An. Quim., 72 (1976) 254.
171 J. M. Marinas, R. Perez-Ossorio and J. V. Sinisterra, An. Quim., 72, (1976) 473.
172 J. M. Marinas, R. Perez-Ossorio and J. V. Sinisterra, An. Quim., 72, (1976) 185.
173 J. M. Campelo, J. M. Marinas and R. Perez-Ossorio, An. Quim., 72 (1976) 698.
174 A. Alberola, M. Blanco, S. Esteban, J. M. Marinas and R. Perez-Ossorio,

An. Quim., 72 (1976) 887.
175 C. Jimenez, J. M. Marinas, J. V. Sinisterra and R. Perez-Ossorio, An. Quim., 73 (1977) 736.
176 C. Jimenez, J. M. Marinas, J. V. Sinisterra and V. Borau, An. Quim., 73 (1977) 88.
177 A. Costa, S. Esteban, J. M. Marinas and R. Perez-Ossorio, An. Quim. 73 (1977) 1529.
178 J. M. Campelo, J. M. Marinas and R. Perez-Ossorio, An. Quim., 74 (1978) 86.
179 J. M. Campelo, J. M. Marinas and R. Perez-Ossorio, Rev. Port. Quim., 18 (1978) 84.
180 J. V. Sinisterra, C. Jimenez, V. Borau and J. M. Marinas, Rev. Port. Quim., 18 (1978) 359.
181 C. Jimenez, J. M. Marinas, R. Perez-Ossorio and J. V. Sinisterra, An. Quim., 70 (1979) 860.
182 J. V. Sinisterra, A. LLovera and J. M. Marinas, Afinidad, 36 (1979) 229.
183 J. V. Sinisterra, J. V. Morey and J. M. Marinas, Afinidad, 36 (1979) 305.
184 J. M. Marinas, C. Jimenez, J. M. Campelo, M. A. Aramendia, V. Borau and D. Luna, Proc. 7th Iberoamerican Symp. Catal., La Plata, Argentina, 1980, p. 79.
185 J. M. Campelo and J. M. Marinas, Afinidad, 38 (1981) 333.
186 J. M. Campelo, A. Garcia, J. M. Gutierrez, D. Luna and J. M. Marinas, J. Colloid Interface Sci., 95 (1983) 544.
187 J. M. Campelo, A. Garcia, J. M. Gutierrez, D. Luna and J. M. Marinas, Can. J. Chem., 61 (1983) 2567.
188 S. Esteban, J. M. Marinas, M. P. Martinez-Alcazar, M. Martinez and A. R. Agarrabeitia, Bull. Soc. Chim. Belg., 92 (1983) 715.
189 J. M. Campelo, A. Garcia, D. Luna and J. M. Marinas, Can. J. Chem., 62 (1984) 638.
190 M. A. Aramendia, J. M. Campelo, S. Esteban, C. Jimenez, J. M. Marinas and J. V. Sinisterra, Rev. Inst. Mex. Petrol., 12 (1983) 61.
191 M. A. Aramendia, J. M. Campelo, A. Costa, P. Deya, C. Jimenez, J. M. Marinas and J. V. Sinisterra, Rev. Inst. Mex. Petrol., in press.
192 M. A Aramendia, V. Borau, C. Jimenez and J. M. Marinas, Acta Cient. Venez., 32 (1981) 369.
193 M. A. Aramendia, V. Borau, C. Jimenez and J. M. Marinas, Acta Chim. Acad. Sci. Hung., 110 (1982) 97.
194 M. A. Aramendia, V. Borau, C. Jimenez and J. M. Marinas, Bull. Soc. Chim. Belg., 91 (1983) 743.
195 M. A. Aramendia, V. Borau, C. Jimenez and J. M. Marinas, React. Kinet. Catal. Lett., 18 (1981) 335.
196 M. A. Aramendia, V. Borau, C. Jimenez, J. M. Marinas and J. A. Pajares, J. Catal., 78 (1982) 188.
197 M. A. Aramendia, V. Borau, M. C. Gomez, C. Jimenez and J. M. Marinas, Appl. Catal., 8 (1983) 177.
198 A. Alba, M. A. Aramendia, V. Borau, A. Garcia-Raso, C. Jimenez and J. M. Marinas, Can. J. Chem., 62 (1984) 917.
199 J. M. Campelo, A. Garcia, D. Luna and J. M. Marinas, Gazz. Chim. Ital., 112 (1982) 221.
200 J. M. Campelo, A. Garcia, D. Luna and J. M. Marinas, React. Kinet. Catal. Lett., 21 (1982) 209.
201 J. M. Campelo, A. Garcia, D. Luna and J. M. Marinas, Colloids Surf., 5 (1982) 227.
202 J. M. Campelo, A. Garcia, D. Luna and J. M. Marinas, Afinidad, 40 (1983) 161.
203 J. M. Campelo, A. Garcia, D. Luna and J. M. Marinas, Appl. Catal., 10 (1984) 1.
204 J. M. Campelo, A. Garcia, D. Luna and J. M. Marinas, React. Kinet. Catal. Lett., 26 (1984) 73.
205 J. M. Campelo, D. Luna and J. M. Marinas, Rev. Roum. Chim., 26 (1981) 867.
206 J. M. Campelo, D. Luna and J. M. Marinas, Afinidad, 38 (1981) 299.
207 J. M. Campelo, A. Garcia, D. Luna and J. M. Marinas, React. Kinet. Catal. Lett., 18 (1981) 325.
208 J. M. Campelo, A. Garcia, D. Luna and J. M. Marinas, Afinidad, 39 (1982) 61.

209 J. M. Campelo, A. Garcia. D. Luna and J. M. Marinas, Afinidad, 39 (1982) 325.
210 J. M. Campelo, A. Garcia, D. Luna and J. M. Marinas, Appl. Catal., 3 (1982) 315.
211 J. M. Campelo, A. Garcia, D. Luna and J. M. Marinas, Proc. 8th Iberoamerican Symp. Catal., La Rabida, España, 1982, p. 125.
212 J. M. Campelo, A. Garcia, D. Luna and J. M. Marinas, Bull. Soc. Chim. Belg., 91 (1982) 131.
213 J. M. Campelo, A. Garcia, D. Luna and J. M. Marinas, Bull. Soc. Chim. Belg., 92 (1983) 851.
214 J. M. Campelo, A. Garcia, J. M. Gutierrez, D. Luna and J. M. Marinas, Appl. Catal., 7 (1983) 307.
215 J. M. Campelo, A. Garcia, D. Luna and J. M. Marinas, J. Chem. Soc., Faraday Trans. 1, 80 (1984) 659.
216 J. M. Campelo, A. Garcia, D. Luna and J. M. Marinas, Rev. Roum. Chim., 30 (1985) 427.
217 R. I. Stirton, U. S. Pat. 2,441,297 (1948).
218 L. A. Pine, U. S. Pat. 3,904,550 (1975).
219 W. L. Kehl, U. S. Pat. 4,080,311 (1978).
220 N. L. Cull., U. S. Pat. 4,233,184 (1980).
221 G. Marcelin, R. F. Vogel and H. E. Swift, J. Catal., 83 (1983) 42.
222 W. L. Kehl, U. S. Pat. 4,210,560 (1980).
223 J. Haggin, Chem. Eng. News, Dec. 13, 1982, p. 9.
224 J. Haber and U. Sybalska, Discuss. Faraday Soc., 72 (1981) 263.
225 J. Haggin, Chem. Eng. News, June 20, 1983, p. 36.
226 R. F. Vogel and G. Marcelin, J. Catal., 80 (1983) 492.
227 H. Itoh, A. Tada and H. Hattori, J. Catal., 76 (1982) 235.
228 H. Itoh, A. Tada and H. Hattori, J. Catal., 80 (1983) 494.
229 B. Gallace and J. B. Moffat, J. Catal., 76 (1982) 182.
230 J. B. Moffat and B. Gallace, J. Catal., 80 (1982) 496.
231 G. Marcelin, R. F. Vogel and W. L. Kehl, in G. Poncelet, P. Grange and P. A. Jacobs (Editors), Preparation of Catalysts III, Elsevier, Amsterdam, 1983, p. 169.
232 H. E. Swift, J. J. Stanolonis and E. H. Reynolds, U. S. Pat. 4,228,036 (1980).
233 M. Itoh, A. Tada and K. Tanabe, Chem. Lett., (1981) 1567.
234 P. Vitale, S. Cavallaro, R. Maggrore, G. Cimino, C. Caristi and S. Galvagno, Gazz. Chim. Ital., 112 (1982) 493.
235 J. B. Peri, Discuss. Faraday Soc., 52 (1971) 55.
236 P. L. Veltman, U. S. Pat. 2,301,013 (1942).
237 K. Kearby, Proc. 2th Intern. Congr. Catal., Paris, Technip, Paris, 1961, p. 2567.
238 K. Kearby, U. S. Pat. 3,342,750 (1967).
239 P. Deya, Thesis, Baleares University, Spain, 1980.
240 G. Wendt and C. F. Lindstrom, Z. Chem., 17 (1977) 118.
241 H. A. Benesi, J. Am. Chem. Soc., 78 (1956) 5490.
242 J. M. Campelo, A. Garcia, J. M. Gutierrez, D. Luna and J. M. Marinas, Colloids Surf., 8 (1984) 353.
243 R. L. Moss, in R. B. Anderson and P. T. Dawson (Editors), Experimental Methods in Catalytic Research, Vol. 2, Academic Press, N. Y., 1976, Ch. 2, p. 43.
244 X. Z. Jiang, T. F. Hayden and J. A. Dumesic, J. Catal., 83 (1983) 168.
245 P. Meriaudeau, O. H. Ellestad, M. Dufaux and C. Naccache, J. Catal., 75 (1982) 243.
246 R. T. K. Baker, E. B. Prestidge and L. L. Murrell, J. Catal., 79 (1983) 348.
247 R. T. K. Baker, E. B. Prestidge and L. L. Murrell, J. Chem. Soc., Chem. Commun., (1983) 993.
248 C. Hoang-Van, P. A. Compagnon, A. Ghorbel and S. J. Teichner, C. R. Acad. Sci. Paris, Ser. C, 285 (1977) 395.
249 D. Briggs, J. Dewing, A. G. Burden, R. B. Moyes and P. B. Wells, J. Catal., 65 (1980) 31.
250 R. Burch, J. Catal., 58 (1979) 220.
251 R. Burch and A. R. Flambart, J. Chem. Soc., Chem. Commun., (1981) 965.

252 S. Y. Wang, S. M. Moon and M. A. Vannice, J. Catal., 71 (1981) 167.
253 J. H. Sinfelt, Catal. Rev. Sci. Eng., 9 (1974) 147.
254 M. A. Vannice and R. L. Garten, J. Catal., 63 (1980) 255.
255 G. M. Schwab and G. Mutzbauer, Z. Phys. Chem., 32 (1962) 367.
256 J. R. Ruddlesden, A. Stewart, D. J. Thompson and R. L. Whelan, Discuss. Faraday Soc., 72 (1981) 397.
257 T. Borowiecki, Appl. Catal., 4 (1982) 223.
258 W. Langenbeck, D. Nehring and H. Dreyer, Z. Anorg. Chem., 307 (1960) 37.
259 D. Nehring and H. Dreyer, Z. Anorg. Chem., 315 (1962) 27.
260 W. Langenbeck, D. Nehring, H. Dreyer and H. Furmann, Z. Anorg. Chem., 314 (1962) 167.
261 J. R. Rostrup-Nielsen, Steam Reforming Catalysts, Teknisk Forlag A/S, Copenhagen, 1975.
262 S. P. S. Andrew, Ind. Eng. Chem. Prod. Res. Dev., 8 (1969) 331.
263 Y. Okamoto, Y. Nitta, I. Imanaka and S. Teranishi, J. Catal., 64 (1980) 397.
264 E. B. Maxted and V. Stone, J. Chem. Soc., (1934) 672.
265 M. Lo Jacomo, M. Schiavello and A. Cimino, J. Phys. Chem., 75 (1971) 1044.
266 H. Lafitau, E. Neel and J. C. Clement, in B. Delmon, P. A. Jacobs and G. Poncelet (Editors), Preparation of Catalysts I, Elsevier, Amsterdam, 1976, p.393.
267 M. Houalla and B. Delmon, C. R. Acad. Sci. Paris, Ser. C, 289 (1979) 77.
268 F. Delannay, M. Houalla, D. Pirotte and B. Delmon, Surf. Interface Anal., 1 (1979) 172.
269 M. Houalla and B. Delmon, C. R. Acad. Sci. Paris, Ser. C, 290 (1980) 301.
270 M. Houalla, J. Lemaitre and B. Delmon, J. Chem. Soc., Faraday Trans. 1, 78 (1982) 1389.
271 M. E. Dry and G. J. Oosthuizen, J. Catal., 11 (1968) 18.
272 A. Ozaki, K. Aika and Y. Morikawa in J. W. Hightower (Editor), Proc. 5th Intern. Congr. Catal., Miami Beach, 1972, North-Holland, Amsterdam, 1973, p. 1251.
273 K. Aika and A. Ozaki, J. Catal., 13 (1969) 232.
274 G. Ertl, in T. Seiyama and K. Tanabe, New Horizons in Catalysis, Proc. 7th Intern. Congr. Catal., Tokyo, 1980, Elsevier, Amsterdam, 1981, Part. A, p. 2.
275 R. J. Madon and H. Shaw, Catal. Rev., 15 (1977) 69.
276 F. L. Williams and M. Boudart, J. Catal., 30 (1973) 478.
277 J. J. Burton, E. Hyman and D. G. Fedak, J. Catal., 37 (1975) 106.
278 J. J. Burton and E. Hyman, J. Catal., 37 (1975) 114.
279 C. R. Helms and K. M. Yu, J. Vac. Sci. Technol., 12 (1975) 276.
280 W. M. H. Sachtler and R. A. Van Santen, Adv. Catal., 26 (1977) 69.
281 J. Barciczki, W. Grzegorczyk, T. Borowiecki, A. Machocki, A. Denis and D. Nazimek, React. Kinet. Catal. Lett., 8 (1978) 395.
282 A. Roman and B. Delmon, J. Catal., 30 (1973) 353.
283 H. Charcosset and B. Delmon, Ind. Chim. Belg., 38 (1983) 481.

Chapter 13

SUPPORTED METAL COMPLEXES AS HYDROGENATION CATALYSTS

YU.I. YERMAKOV and L.N. ARZAMASKOVA
Institute of Catalysis, Novosibirsk 630090 (USSR)

13.1 INTRODUCTION

The anchoring of metal complexes on support (or matrix) surfaces began a novel trend in catalyst preparation (refs. 1-16). After the first intensive attempts to develop this field in the 70s it was believed that the advantages of heterogeneous catalysts, e.g., the ease of separation of the reaction products and the thermal stability, and of homogeneous catalysts, e.g., mild conditions and the possibility of obtaining more definite information on the structure of the active centres and their mechanism of action, might be combined. The application of metal complexes on support surfaces enables a wide variation of the composition of the active centres because the choice of ligands is not restricted by the solubility of the metal compound.

The catalytic properties of materials prepared in this way have been reviewed (refs. 17-36). In this paper we analyze the application of the technique of metal complex anchoring to the preparation of hydrogenation catalysts. Hydrogenation seems to be the first catalytic reaction for which supported complexes were used as catalysts. As long ago as the 60s, Izumi (ref. 1,2) prepared hydrogenation catalysts by supporting transition metal complexes on polymers (natural silk). After the publications of Wilkinson et al. (refs. 37, 38) on new soluble hydrogenation catalysts, e.g., $RhCl(PPh_3)_3$ many research groups attempted to immobilize such complexes on organic polymer and mineral matrices (refs. 11,12,18,20,21,24,30,32,34).

13.1.1 General information on catalysts prepared by anchoring of metal complexes

(i) The use of non-functionalized oxide supports. The following reactions are possible during the anchoring of metal complexes, due to their interaction with the surface -OH, O^{2-} or

Lewis base and acid centres:

(a) The substitution of a ligand L' in the metal complex by a surface hydroxyl group

$$\underset{\equiv}{E}\text{-OH} + M_nL'_m \longrightarrow \underset{\equiv}{E}\text{-O-}\underset{H}{\overset{|}{M}}_nL'_{m-1} + L' \tag{13.1}$$

where L' = an organic ligand, anions or CO, n,m = stoichiometric coefficients (n ⩾ 1) and $\underset{\equiv}{E}$ is the support surface. There are no reliable data on such reactions.

(b) Oxidative addition of a surface hydroxyl group to the metal complex:

$$\underset{\equiv}{E}\text{-OH} + M_nL'_m \longrightarrow \underset{\equiv}{E}\text{-O}\underset{M}{\overset{M}{\diagup\!\!\diagdown}}H \quad M_{n-2}L'_{m-x} + xL' \tag{13.2}$$

This reaction may easily proceed for ions in zerovalent state and is characteristic of metal clusters, e.g., M = Os (refs. 29, 39,40), M = Ru, Fe (refs. 29,41).

(c) Protolysis of a M-L' bond by a surface hydroxyl group:

$$x(\underset{\equiv}{E}\text{-OH}) + M_nL'_m \longrightarrow (\underset{\equiv}{E}\text{-O})_xM_nL'_{m-x} + xHL' \tag{13.3}$$

This type of interaction was described for alkyl (ref. 17), allyl (refs. 17,19,22), chloride (refs. 27,42) and alkoxide (refs. 23, 27) complexes.

(d) Protolysis of some bonds of a ligand in the metal complex (ref. 43):

$$3(\underset{\equiv}{E}\text{-OH}) + (EtO)_3Si(CH_2)_2PPh_2Rh(acac)CO \longrightarrow (\underset{\equiv}{E}\text{-O})_3Si(CH_2)_2 \cdot$$

$$PPh_2Rh(acac)CO + 3EtOH \tag{13.4}$$

(e) The oxidation of a metal ion interacting with a hydroxyl group, with evolution of H_2:

$$2\underset{\equiv}{E}\text{-OH} + M_nL'_m \longrightarrow \underset{\underset{\equiv}{E}\text{-O}}{\overset{\underset{\equiv}{E}\text{-O}}{\diagdown\!\!\diagup}}M_nL'_{m-x} + H_2 + xL' \tag{13.5}$$

This type of interaction is typical of most carbonyl complexes when supported on Al_2O_3 (refs. 44,45).

(f) The formation of an anionic bond between the metal complex and the surface due to the nucleophilic attack of a surface hydroxyl upon coordinated CO, with evolution of CO_2:

$$\underset{\equiv}{E}-OH + CO-M_nL'_{m-1} \rightarrow \underset{\equiv}{E}^+ [HM_nL'_{m-1}]^- + CO_2 \qquad (13.6)$$

Examples are reactions of $M_3(CO)_{12}$ carbonyls (M = Fe, Ru) with MgO and Al_2O_3 (refs. 29,46,47).

(g) The substitution of coordinated CO by a surface O^{2-} group (refs. 48,49):

$$\underset{\equiv}{\equiv}O^{2-} + OC-M_nL'_m \rightarrow \underset{\equiv}{\equiv}-O^{2-}M_nL'_m + CO \qquad (13.7)$$

(h) The nucleophilic attack of a surface O^{2-} on coordinated CO, (h1) without removal of carbon monoxide

$$\underset{\equiv}{E}-O^{2-} + M_n(CO)_m \rightarrow \underset{\equiv}{E}\!\!<\!\!\overset{O}{\underset{O}{}}\!\!>\!\!C = M_n(CO)_{m-1} \qquad (13.8)$$

was described for iron carbonyls (refs. 29,47), and (h2) with removal of CO_2

$$\underset{\equiv}{E}-O^{2-} + M_n(CO)_m \rightarrow \underset{\equiv}{E}^{2+}[M_n(CO)_{m-1}]^{2-} + CO_2 \qquad (13.9)$$

e.g., M = Os (ref. 29).

(j) The interaction of a carbonyl ligand with a surface Lewis acid centre

$$\underset{\equiv}{E} + OC-M_nL'_m \rightarrow \underset{\equiv}{E} \leftarrow OC-M_nL'_m \qquad (13.10)$$

e.g., E = Al, La, Zr (refs. 50-52).

(k) Simultaneous interaction of the metal complex with surface Lewis acid and base centres:

$$\underset{\equiv}{\overset{O}{\underset{E}{}}} + \underset{CH_3}{OC-M_nL'_m} \rightarrow \underset{\equiv}{\overset{O-M_n\overset{L'_m}{}}{\underset{E-O}{}}}\!\!\!>\!C-CH_3 \qquad (13.11)$$

e.g., for the complex $Mn(CH_3)(CO)_5$, where E = Al (ref. 52).

(ii) <u>Functionalized organic polymer and oxide supports</u>. The functionalization of supports is usually the most time-consuming step in the preparation of catalysts containing "immobilized" metal complexes. For polymer supports, such functionalization occurs in several consecutive steps. A well known example is the introduction of phosphine groups into polystyrene (ref. 53):

$$\text{P-}\bigcirc \xrightarrow{Br_2} \text{P-}\bigcirc\text{-Br} \xrightarrow{LiPPh_2} \text{P-}\bigcirc\text{-PPh}_2 \qquad (13.12)$$

It is now possible to prepare various functionalities for the purpose of metal complex anchoring.

The following reactions are possible upon interaction of metal complexes with functionalized supports.

(a) The substitution of a ligand (L′) in the coordination sphere of the metal complex:

$$\text{P-L} + M_n L'_m \longrightarrow \text{P-}LM_n L'_{m-x} + xL' \qquad (13.13)$$

This is the most common way of metal complex anchoring. Examples are the heterogenization of Wilkinson's catalyst on polystyrene functionalized by $-PPh_2$, $-CH_2PPh_2$, $-C_5H_4$(Cp), 2,3-dihydroxy-1,4--bis(diphenylphosphino)butane (DIOP) (refs. 11,18,24,53)

$$\text{P-}\bigcirc\text{-PPh}_2 \begin{array}{c} \xrightarrow{RhCl(PPh_3)_3} \text{P-}\bigcirc\text{-PPh}_2RhCl(PPh_3)_2 \\ \\ \xrightarrow{RhH(CO)(PPh_3)_3} \text{P-}\bigcirc\text{-PPh}_2RhH(CO)(PPh_3)_2 \end{array} \qquad (13.14)$$

The soluble transition metal halogenides are anchored on polymers functionalized by phosphine (ref. 54), nitrile (ref. 55) or amine groups (ref. 15) due to the substitution of the solvent in the coordination spheres of the metal complexes.

$$2 \text{ P-L} + MCl_2 \longrightarrow (\text{P-L})_2MCl_2 \qquad (13.15)$$

where M = Co, Ni, L = a phosphine (ref. 54); M = Pd, L = a nitrile (ref. 55). In some cases the reduction of the metal ion occurs upon anchoring, e.g., the formation of Rh(II) upon anchoring of $RhCl_3 \cdot 3H_2O$ on polyacrylic acid in alkaline methanol (ref. 56).

   (b) Oxidative addition of the metal complex to the support (ref. 57).

   The preparation of functionalized polymer supports has been viewed (refs. 18,20,21,24,30,34,35).

   For functionalization of oxide supports (mainly silica) there are two general approaches (refs. 26,27,36,43,58-61):

   (a) Interaction of the surface hydroxyl groups with bifunctional compounds ("binding agents"):

$$x\overset{|}{\underset{|}{E}}-OH + Y_n E'R_m \longrightarrow (\overset{|}{\underset{|}{E}}-O)_x E'Y_{n-x} R_m + xHY \qquad (13.16)$$

Chloro- and alkoxysilanes $Y_n SiR_{4-n}$ (Y = Cl, OMe, OEt, R = hydrocarbon group) are usually used as "binding agents". This approach was first used in (refs. 59-61). The atom of silicon, attached to silica by siloxane bonds $(\overset{|}{\underset{|}{Si}}-O)_x Si$ is designated as SIL.

   (b) Consecutive synthesis of complex organic ligands on a silica surface using as starting groups the simple organic functionalities $SIL-CH_2CH=CH_2$, $SIL-(CH_2)_3Cl$, $SIL-(CH_2)_3NH_2$, $SIL-(CH_2)_2OH$, etc. (refs. 26,27,36). Considering that this technique involves many steps, the reactions used for the synthesis of surface ligands have to fulfil the following conditions:

   (1) High product yields.

   (2) The by-product which may remain on the support surface must not coordinate transition metal ions.

   (3) The carrier must not be destroyed during synthesis.
Some examples of surface functionalities prepared by the use of this approach are given in Fig. 13.1 of the next Section.

   The application of both approaches (a) and (b) provides a wide range of surface ligands available for anchoring complexes (see refs. 26,27,36,62-66).

   The anchoring of metal complexes on functionalized silica proceeds in a similar manner to that on functionalized polymers. For example, anchored complexes of Pd(II) and Pd(0) were prepared by treatment of SIL-L with solutions of $(PhCN)_2PdCl_2$ and Pd(dibenzylideneacetone)$_2$ (ref. 26).

   (iii) <u>Polynuclear metal complexes</u>. For the preparation of anchored complexes having the composition $\}-L_x M_n L'_{m-x}$, in general a solution of compounds $M_n L'_m$ having the same nuclearity is used.

This straightforward technique has been used for the attachment of many cluster complexes, containing such frameworks as Rh_4, Rh_6, Ir_4, Fe_2, Ru_3, Ru_4, Os_3, Os_6, Co_4, Co_2, Pt_{3-15}, Co_2Rh_2, Fe_3, Co_3Rh, $AuOs_3$, $FeCo_3$, Fe_2Pt, $RuPt_2$, etc. (for reviews see refs. 29,32,33,36,67).

Techniques have also developed for the preparation of anchored bi- and polynuclear complexes starting with compounds of lower nuclearity. Thus the interaction of $Ir(CO)_2Cl(p\text{-toluidine})$ and $Ir_2Cl_2(C_8H_{12})_2$ with a phosphine-functionalized poly(styrene-divinylbenzene) gave tetranuclear iridium clusters (refs. 68,69):

⌐-$PPh_2Ir_4(CO)_{11}$, ⌐-$(PPh_2)_2Ir_4(CO)_{10}$,

⌐$\genfrac{}{}{0pt}{}{(PPh_2)_2Ir_4(CO)_{10}}{PPh_2Ir_4(CO)_{11}}$, ⌐-$(PPh_2)_yIr_4(CO)_9(PPh_3)_{3-y}$ (y = 1,2)

Anchored binuclear complexes of Pd have been prepared by the reaction of surface palladium(0) compounds with palladium(II) complexes in solution (refs. 26,72,73):

$$\text{SIL-LPd(0)} + PdX_2 \longrightarrow \text{SIL-LPd}_2X_2 \qquad (13.17)$$

An IR and X-ray photoelectron spectroscopy (XPS) study of the structure of these complexes showed (refs. 70-72) that, depending on the nature of X, binuclear complexes containing a direct Pd-Pd bond or a bridged Pd-X-Pd bond may be prepared.

Polynuclear palladium complexes attached to phosphinated SiO_2 have been prepared (refs. 26,70,73-76) by the technique of "consecutive assembling":

$$\genfrac{}{}{0pt}{}{P}{P}\!\!>Pd(OAc)_2 \xrightarrow{HOCHO} \genfrac{}{}{0pt}{}{P}{P}\!\!>Pd\genfrac{}{}{0pt}{}{OH}{OCHO} \xrightarrow{Pd(OAc)_2} \qquad (13.18)$$

$$\genfrac{}{}{0pt}{}{P-Pd-OAc}{P-Pd-OAc} \longrightarrow \cdots \longrightarrow \genfrac{}{}{0pt}{}{P}{P}\!\!>Pd_n(OAc)_2$$

Anchored polynuclear complexes ⌐-$P_2Pd_n(OAc)_2$ (n ⩽ 5),

⌐-P_2 = SIL$\genfrac{}{}{0pt}{}{CH_2CH_2CH_2PPh_2}{CH_2CH(CH_3)CH_2PPh_2}$ have been studied by UV and IR spectroscopy, XPS and by X-ray diffraction and radial atomic distribution (RAD) techniques, and it was shown that the species have dimensions of less than 6 Å with Pd-Pd bond lengths of 2.8 Å (refs. 26,70).

13.1.2 The stability of anchored metal complexes

One of the major parameters determining the practical application of a catalyst is its stability under the reaction conditions. The stability of an anchored complex $\overset{|}{\underset{|}{E}}-LM_nL'_{m-x}$ depends on the nature of the support, ligand L, nature of the metal ion M and its oxidation number and the composition of the reaction mixture. The decomposition of the anchored complex may proceed at cleavage of $\overset{|}{\underset{|}{E}}-L$ and L-M bonds, so the strength of these bonds is of primary importance. The stability of complexes anchored to non-functionalized oxides is determined primarily by the strength of the M-O bonds. According to ref. 77, the M-O bond energies decrease in the sequence Os > Ru > Fe > Rh > Co > Ni > Pt > Ir > Pd. For example, the complexes $(\overset{|}{\underset{|}{E}}-O)_2Os(CO)_2$ (E = Al, Si, Ti) active in ethylene hydrogenation, contain osmium atoms bonded to two surface oxygen atoms and are stable to 400 °C (ref. 39). Supported catalysts of other Group 8 metals are less stable.

Metal complexes attached to surface organic ligands are quite sensitive to the temperature and to the composition of the reaction mixture. For example, $PhCH_2Mn(CO)_5$ and its analogues "heterogenized" on polymer ⌇-$PhCH_2Mn(CO)_5$ decomposed at 140 °C into $Mn_2(CO)_{10}$ and $PhCH_2CH_2Ph$ and ⌇-$PhCH_2CH_2Ph$-⌇ (ref. 78). In the hydroformylation of hexene-1 (45 atm, $CO:H_2$ = 1:1, 80 °C) catalyzed by rhodium complexes attached to silica by nitrogen-containing ligands a considerable part of the Rh was found in the reaction products. The detachment of the rhodium complex decreased with increasing donor properties of the nitrogen atoms (increasing Rh-N bond strength). For more stable rhodium complexes on phosphinated silica, reaction temperatures up to 140 °C could be employed. In this case also, rhodium was found in the reaction products. The breakage of Rh-P bonds is promoted by trace amounts of oxygen in the reaction mixture (ref. 43).

The surface organic ligands may also decompose under the reaction conditions. This complicates the stability analysis of the surface complexes. The thermolysis of surface nitriles catalyzed by metal ions results in the decomposition of anchored nitrile complexes of Pd upon heating to 200 °C (ref. 79). Hydrogenolysis of P-C bonds of the surface ligand L in the complexes $SIL-L_2Pd_2 \cdot (OAc)_2$ and $SIL-L_2Pd_5(OAc)_2$, which proceeds at 80 °C in the presence of hydrogen, results in a strengthening of M-L bonds due to the transformation of phosphine ligands to phosphides or phosphenes and the formation of more stable covalent Pd-P bonds (ref. 80).

The stability of anchored complexes depends on the degree of oxidation of the metal ion. During a catalytic cycle, a metal ion may undergo a change in oxidation number. Ions in one oxidation state may be weakly coordinated by the surface ligands and leave the surface, resulting in further cluster formation and the formation of dispersed metal particles (ref. 26):

$$
\begin{array}{c}
\text{\{-LPdX}_m \\
\text{\{-LPdX}_m \\
\text{\{-LPdX}_m
\end{array}
\xrightarrow[-HX]{H_2}
\begin{array}{c}
\text{\{-LPd(0)} \\
\text{\{-LPd(0)} \\
\text{\{-LPd(0)}
\end{array}
\longrightarrow
\begin{array}{c}
\text{\{-L} \\
\text{\{-L} \\
\text{\{-L(Pd(0))}_k
\end{array}
\qquad (13.19)
$$

It was concluded (ref. 26) that the stability of complexes $\text{\{-LPdX}_m$ in a reducing atmosphere increases with increasing coordination strength of L to Pd(II). The formation of strong Pd-L complexes seems to prevent the reduction of Pd(II) to Pd(0) due to the decrease in redox potential of the process $\text{\{-LPd(II)} \xrightarrow{2e} \text{\{-LPd(0)}$. The ability of the ligand L to coordinate to Pd(0) is important in retarding the decomposition of complexes in hydrogen (refs. 26,63,81).

The nature of the solvent (S) used for liquid-phase reaction may have a considerable effect on the stability of anchored complexes. It may act as a ligand, competing with the surface ligand for metal coordination:

$$\text{\{-LM}_n L'_m + S \rightleftharpoons \text{\{-L} + SM_n L'_m \qquad (13.20)$$

13.2 HYDROGENATION OF UNSATURATED HYDROCARBONS IN THE PRESENCE OF ANCHORED METAL COMPLEXES

13.2.1 The problem

Hydrogenation of unsaturated hydrocarbons is one of the most studied fields of catalysis by anchored complexes. One reason for this is the mild conditions required for this reaction which do not destabilize the surface complexes. Moreover, an increase in the temperature of olefin hydrogenation may be unfavourable due to the resulting decrease in the concentration of the intermediate olefin-metal complexes.(ref. 82).

The data accumulated in this field permit us to consider the effects of some parameters on the catalytic properties of anchored complexes. Examples of the application of anchored complexes in hydrogenation reactions are as follows:

(a) The same metal complex (e.g. $RhCl(PPh_3)_3$) has been used on various supports - phosphinated polystyrene, phosphinated cellulose, silica, alumina, activated carbon, diatomaceous earth (refs. 11, 18, 53, 83, 84).

(b) Some surface complexes differ only in the surface ligand L; for example, $RhCl(PPh_3)_3$ has been attached to polystyrene containing various functionalities L (refs. 11,18,24). Pd(II) and Pd(0) were attached to silica functionalized by various surface ligands L (refs. 26,62-65,85-90).

(c) Various mono- and polynuclear (including heteronuclear) complexes have been anchored on the same support. For example, $RhCl_3$, $RhCl_3+Y$ (Y= $PHPh_2$, PPh_3, C_2H_4), $RhCl(PPh_3)_3$, $RhCl(PHPh_2)_3$ (ref. 11), $RhCl(C_2H_4)(PPh_3)_2$ (ref. 16), $Ir(CO)(Cl)(PPh_3)_2$ (ref.91), $Rh_6(CO)_{16}$ (ref. 18), $Ir(CO)_2$(p-toluidine)Cl (ref. 69), $RuCl_2(CO)_2(PPh_3)_2$ (ref. 53), $H_4Ru_4(CO)_{12}$ (ref. 92), $Fe_2Pt(CO)_9 \cdot (PPh_3)$, $RuPt_2(CO)_5(PPh_3)_3$, $HAuOs_3(CO)_{10}(PPh_3)$, $ClAuOs_3(CO)_{10} \cdot (PPh_3)$, $H_2PtOs_3(CO)_{10}(PPh_3)_2$, $Co_2Pt_2(CO)_8(PPh_3)_2$ (refs. 93-95), etc., were anchored on phosphinated polystyrene.

The results allow the possibility to analyze the role of the substrate, solvent, carrier, the nature of the metal ion and its ligand environment in hydrogenation reactions. Such an analysis has already been performed (ref. 25) for data published before 1975, and in other reviews (refs. 20,26,31,32,35). Let us consider two main criteria for the catalyst activity - the accessibility of the active centre for the interaction with the substrate and the possibility of substrate activation in the coordination sphere of the anchored metal complex.

13.2.2 <u>Parameters influencing the accessibility of active centers</u>

(i) <u>Coordinative unsaturation as a necessary condition for catalyst activity</u>. This problem will be illustrated by examples of olefin hydrogenation in the presence of mono- and polynuclear anchored metal complexes.

Example 1. The use of such ligands as phosphine or C_2H_4 which are more strongly bound to the central metal ion than Cl^- results in a decrease in activity of rhodium(I) complexes anchored on phosphinated polystyrene in the hydrogenation of heptene-1. The yield of heptane noticeably (~ 16 times) decreases in the order (ref. 11):

$$\text{P-L-RhCl}_x > \text{P-L-RhCl}_p(\text{PHPh}_2)_r > \text{P-L-RhCl}_m(\text{PPh}_3)_n > \text{P-L-RhCl}_y \cdot$$
$$\cdot (C_2H_4)_z > \text{P-L-RhCl}(\text{PPh}_3)_2 > \text{P-L-RhCl}(\text{PHPh}_2)_2$$
$$(\text{P-L} = \text{P-}\langle\bigcirc\rangle\text{-CH}_2\text{PPh}_2).$$

The introduction of PPh_3 into the anchored complex $(SIL-(CH_2)_3NH \cdot (CH_2)_3NH_2)_2PdCl_x$ results in an inhibition of 1-heptene hydrogenation; as a result the catalyst may be used for selective hydrogenation of 1-heptyne to 1-heptene (ref. 96). These cases concern the effect of "outer" ligands L' on the activity and selectivity of anchored complexes.

Example 2. Triosmium clusters anchored on SiO_2 catalyze the hydrogenation of ethylene at 70-100 °C according to the mechanism (refs. 29,97)

$$(13.21)$$

CO ligands being omitted. According to ref. 29,

cluster (1) contains bridging hydride and 3-electron oxygen ligands; scission of the Os-O bond results in a vacant site (complexes 1', 2) for olefin coordination.

Example 3. Gates et al. (ref. 94), in a study of heteronuclear clusters anchored on phosphinated poly(styrene-divinylbenzene), showed that the clusters $ClAuOs_3(CO)_{10}(Ph_2P-\xi)$ and $Co_2Pt_2(CO)_8 \cdot (Ph_2P-\xi)_2$, with an open "butterfly" structure were active in ethylene hydrogenation (1 atm, ≤ 100 °C), whereas the clusters $HAuOs_3(CO)_{10}(Ph_2P-\xi)$ and $H_2PtOs_3(CO)_{10}(Ph_2P-\xi)_2$ with a closed tetrahedral structure to which olefin coordination is difficult were inactive.

In the first of the above examples the vacant coordination site is due to detachment of "outer" ligands L', in the second case to detachment of the surface ligand L and in the third case to the specific structure of the anchored metal complex.

(ii) <u>The role of the support and its functionalities</u>. In general the hydrogenation rate is decreased upon increasing the size of the substrate molecules due to steric hindrance between the substrate and the active centre. For anchored complexes, the presence of the support and the surface ligand L results in a more dramatic hindrance in comparison with homogeneous catalysts. The relative rates of hydrogenation of 1-hexene, cyclohexene, octadecene, cyclooctene, cyclododecene (cis- and trans-) and Δ^2-cholestene in the presence of soluble $RhCl(PPh_3)_3$ are 1.4:1:0.71:1:0.67:0.71, and in the case of phosphinated polystyrene anchored analogue the relative rates are 2.55:1:0.49:0.39:0.22:0.03 (ref. 12).

The nature of the support and the surface ligand may influence considerably the accessibility of the active centre toward olefin. For example, rhodium complexes on phosphinated polychlorovinyl show low activity for olefin hydrogenation (refs. 25,98) in comparison with complexes on phosphinated polystyrene (ref. 12). Rhodium(I) complexes on phosphinated SiO_2 are more active than on phosphinated polystyrene (ref. 20). It seems that on SiO_2 the active centre is less "shielded" by the carrier surface towards its interaction with a substrate. An increase in the accessibility of the active centre is probably also the reason for the increase in activity of carbonylphosphine cobalt complexes upon transfer from $SIL(CH_2)_2P(C_6H_{12})_2$ to $SIL(CH_2)_3P(C_6H_{12})_2$ as ligand (refs. 27, 99), i.e., when the distance between the support surface and the active centre is increased.

In the above example, when "shielding" of the active centre by the support surface is possible the activity of "heterogenized" complexes is less than that of soluble ones. However, in some cases the necessary coordinative unsaturation of the active centre can be achieved due to the anchoring on the support. For example, whereas $PdCl_2(PPh_3)_2$ is not acive in hydrogenation at 1 atm of H_2, Kaneda et al. (ref. 100) obtained a highly active catalyst by supporting $PdCl_2(PPh_3)_2$ on polymeric diphenylbenzyl phosphine. The autors (ref. 100) explained the high activity in terms of the formation of a coordinatively unsaturated palladium complex. Upon introduction of $SnCl_2$ to yield a ratio of Sn:Pd = 2:1, the hydrogenation was completely stopped. An increase in hydrogenation activity upon anchoring on a phosphinated styrene-divinylbenzene resin has been observed for iridium complexes (refs. 91, 101). At the ratio P:Ir < 5, the surface complex $(\text{\textphi-}PPh_2)_x Ir(CO)(Cl)(PPh_3)_{2-x}$ was about two orders of magnitude more active than its soluble analogue $Ir(CO)Cl(PPh_3)_2$.

Highly active catalysts for selective hydrogenation of conjugated dienes to olefins have been prepared by Novikova et al. (ref. 102) by supporting the complex $PdCl_2 \cdot N(C_9H_{19})_3$ on oxides. A moderate change in activity (factor of 2-3) was observed upon variation of the support. When surface complexes are directly bound to the oxygen of oxide carriers, a drastic change in activity is possible upon variation of the support. This was shown by Whyman et al. (refs. 103,104) for the osmium clusters $Os_3(CO)_{12}$, $Os_6(CO)_{18}$ and $H_4Os_4(CO)_{12}$ anchored on oxides of Si, Ti and Al. The molecular integrity of $H_4Os_4(CO)_{12}$ was intact upon anchoring, and the hydrogenation activity increased by about three orders of magnitude upon replacing Al_2O_3 by TiO_2. The effect of the support on the activity was explained by the change in electron transfer upon interaction of the cluster with the support.

(iii) <u>The role of the solvent</u>. As mentioned in Section 13.1.2, the solvent may influence the stability of the anchored metal complex by competing for coordination with surface ligands (see reaction 13.20). The solvent may also compete with a substrate for coordination at an active centre. Solvent polarity affects the swelling of polymer matrices (ref. 25) and the ability of a substrate to diffuse to the active centre (refs. 25,105, 106), thus influencing the catalytic activity.

For example, the activity of $PdCl_2 \cdot N(C_9H_{19})_3$ on Al_2O_3 increased by about a factor of 2 upon changing the aromatic solvents for aliphatic ones and upon changing cyclohexane for hexane (ref.102). The activity of $PdCl_2(PPh_3)_2$ on polymeric diphenylbenzyl phosphine in styrene hydrogenation at 25 °C and 1 atm of hydrogen increased about 90 times upon changing dimethylsulphoxide (DMSO) to dimethylformamide (DMF) as solvent. It was concluded that solvents with medium coordination ability to metal ions (for example, DMF, ethanol, tetrahydrofuran (THF), acetone) are favoured when solvents with low (cyclohexane) or high (DMSO) coordination ability produce an unfavourable effect on the activity (ref. 100). Similar effects have been found (ref. 99) for $Pd(OAc)_2$ and $(PhCN)_2PdCl_2$ anchored on SIL-L (L = N- and S-containing ligands) in the hydrogenation of cyclopentadiene, when heptane and dioxane were changed for ethanol.

The effect of the solvent on the selectivity for hydrogenation and isomerization of unsaturated compounds has been studied (ref. 107) for rhodium complexes anchored on polymers. The following mechanism of olefin transformation on this catalyst was considered:

$$\begin{array}{c} \text{PPh}_3 \\ | \\ \text{]-P-Rh-PPh}_3 \\ \text{Cl} \quad \text{I} \\ \downarrow +H_2 -PPh_3 \\ \text{H} \\ |/H \\ \text{]-P-Rh-PPh}_3 \\ \text{II} \end{array}$$

(13.22)

Isomerization of facilitated if the coordination site neighbouring to the alkyl group is vacant, but if this site is occupied by the solvent hydrogenation route is preferable.

13.2.3 Substrate activation in the coordination sphere of anchored palladium complexes

The effect of the ligand environment of the active centre on the coordination and activation of the substrate has been considered (refs. 26,62-65,70,85-90) for anchored palladium complexes.

(i) <u>Mononuclear anchored palladium complexes</u>. First we note that in hydrogenation reactions metallic palladium may be formed from complexes anchored on the support surface (see reaction 13.19). Metallic Pd is highly active in hydrogenation, but these are differences in its catalytic behaviour and that of its complexes, e.g., in the relative rates of hydrogenation of skeletal and positional olefin isomers (ref. 108), in the effect of substituents in the aromatic ring on azomethine hydrogenation (ref. 109) and in the relative rates of hydrogenation and isomerization (ref. 110). It has been shown (refs. 63,65) that hexene-1 and vinyl butyl ether are hydrogenated on anchored complexes with different rates, whereas on palladium metal the rates of hydrogenation of these substrates are the same.

The mechanism of hydrogenation on anchored palladium complexes involves the equilibrium formation of a π-olefin complex with subsequent insertion of olefin into a Pd-H bond (ref. 26). The formation of π-olefin complexes with $SIL-CH_2CH_2PPh_2 \cdot PdCl_2$ has been observed by IR spectroscopy (ref. 99). The oxidation state of Pd was not changed after hydrogenation (refs. 99,111, 112). Isotopic methods have shown (ref. 113) that hydrogenation on anchored amine complexes involves the formation of a Pd-H bond by heterolysis of molecular hydrogen. The solvent plays an important role in this step (ref. 114).

The nature of the "outer" ligands L' in anchored complexes $SIL-LPdL'_m$ also influences their catalytic activity. An increase in the activity of complexes $(SIL-CH_2CH_2PPh_2)_2Pd_2L'_4$ in the order L' = I, Br, Cl, OAc was explained by a facilitation of the molecular hydrogen heterolysis with increasing degree of interaction of the ligands with the proton in this order (ref. 26).

The rate of the catalytic reaction depends on the nature of the surface ligand L. For example, from a study of the reaction (13.23) it was concluded that ligands L forming a coordination type bond to Pd resulted in greater activity than those bound to Pd in a covalent manner (ref. 26). The properties of the sur-

$$\text{SIL-LPd} \begin{array}{c} Cl \\ \diagup \\ \diagdown \\ H \end{array} + BrCH_2Ph \xrightarrow[\text{pentane}]{0\ °C} \text{SIL-LPd} \begin{array}{c} Cl \\ \diagup \\ \diagdown \\ Br \end{array} + CH_3Ph \qquad (13.23)$$

face ligands influence the stability of the anchored complexes to poisoning. It has been shown (ref. 86) that an increase in the stability constants (K_s) for SIL-LPd(OAc)$_2$ complexes upon variation of L results in a decrease of the hydrogenation rate of sulpholene, but also in a decrease in the poisoning effect of thiophene (see Table 13.1). It seems that with increasing strength of the L-Pd bond the activation of hydrogen becomes more difficult and the hydrogenation rate decreases. However, such an increase in bond strength also makes difficult the coordination of the S atom of thiophene and results in an increase in stability of the palladium complex towards poisoning (ref. 26).

TABLE 13.1
Catalytic properties of palladium acetate anchored on SiO$_2$ in the hydrogenation of sulpholene-3 in isopropanol (25 °C, 1 atm H$_2$) (refs. 85,86,89,90)

Surface ligand	Activity (mol H$_2$/g-atom Pd·min)	Stability to poisoning (%)[a]	K_s
$\underset{\shortmid}{\overset{\shortmid}{\text{Si}}}$-OH	16	2	10^5
SIL-C$_2$H$_4$-⟨◯⟩-C$\overset{NH_2}{\underset{NOH}{\diagdown}}$	10	10	10^9
SIL-C$_3$H$_6$NHCS$_2$Na	0.8	32	10^{13}
SIL-C$_3$H$_6$SC$_9$H$_6$N	0.2	48	10^{14}
SIL-C$_3$H$_6$OCS$_2$Na	0.05	72	10^{15}
1% Pd/SiO$_2$[b]	22	1	–

[a] Stability to poisoning was characterized by the ratio A_s/A_o, where A_s = activity in the presence of 5 mol thiophene per g-atom of Pd, A_o = activity without thiophene.
[b] 50% dispersion of Pd.

The activity of the anchored complex depends on the oxidation state of palladium. The effect of the nature of the surface ligand L on hydrogen activation by anchored Pd(II) and Pd(0) has been studied (refs. 26,87) for methyl acrylate (MAC) activation. A two-step mechanism was considered:

$$\text{SIL-LPdCl}_2 + \text{CH}_2=\text{CHCOOCH}_3 \xrightleftharpoons{K_{rev}} \text{SIL-LPd}\underset{\text{Cl}}{\overset{\text{Cl}}{|}} \xleftarrow{} \overset{\text{CH}_2}{\underset{}{\|}} \text{CHCOOCH}_3$$

$$\text{SIL-LPdCl}_2(\text{CH}_2=\text{CHCOOCH}_3) + \text{H}_2 \xrightarrow{k} \text{SIL-LPdCl}_2 + \text{CH}_3\text{-CH}_2\text{COOCH}_3$$

(13.24)

As MAC formed strong π-complexes (K_{rev} was more than 100 for palladium(II) complexes and 10-80 for palladium(0) complexes), it was considered that the π-olefin complex concentration is practically equal to the total concentration of Pd. So the change in catalytic activity is due only to the change in rate constant, k, of the second step. The dependences of the catalytic activities of palladium(0) and palladium(II) complexes (see Fig. 13.1) on their stability constants was explained in terms of the different mechanisms of activation of the hydrogen molecule. With palladium(II) complexes the formation of palladium hydride upon rupture of the H-H bond may occur by the transfer of electron density from the bonding molecular orbital (MO) of hydrogen to the vacant atomic orbitals of Pd. With palladium(0) complexes the formation of hydrides occurs by the transfer of electron density from filled atomic orbitals of Pd to the vacant non-bonding MO of hydrogen (ref. 26). With increasing of K_s (strengthening of the Pd-L bond) the energy of the non-bonding E_g^* orbitals of Pd participating in hydrogen activation increases. When the energy of the E_g^* orbitals of Pd becomes high it facilitates the transfer of electron density from them to the non-bonding orbital of hydrogen in the case of Pd(0), and hinders the transfer of electron density from the bonding orbital of hydrogen to the E_g^* orbital of Pd(II).

It is worth noting that such dependences as shown in Figure 13.1 for a wide variation of ligands are not easy to obtain for homogeneous catalysts due to the impossibility of synthesizing soluble complexes with some of the ligands.

(ii) _Polynuclear anchored palladium complexes_. When the complexes SIL-LPd$_n$L$'_m$ (n > 1) (synthesized as mentioned in Section 13.1.1, see reactions 13.17, 13.18) were used for hydrogenation of cyclopentadiene (CPD) to cyclopentene (CPE) and cyclopentane (CP) (refs. 26,99,115,116) it was found that their catalytic properties depend on the number of palladium atoms and the structure of the -Pd$_n$L$'_m$ framework (see Table 13.2).

TABLE 13.2

Catalytic properties of anchored palladium complexes in hydrogenation of CPD and CPE (refs. 26,115,116)[a]

Nos	Anchored complex and its probable structure [b]	Catalytic activity[c] in hydrogenation (A)		X[d]
		CPD	CPE	
1	⌉-P$_2$Pd(OAc)$_2$	0	0	–
2	⌉-P$_2$Pd$_2$(OAc)$_2$			
	~P\ Pd–Pd(OAc)$_2$ ~P/	2.4	12.0	0.2
	AcO\ /OAc Pd–Pd ~P/ \P~	25.3	24.7	1.02
3	⌉-P$_2$Pd$_3$(OAc)$_2$	35.6	59.3	0.6
4	⌉-P$_2$Pd$_4$(OAc)$_2$	68.2	65.0	1.05
5	⌉-P$_2$Pd$_5$(OAc)$_2$	77.6	60.6	1.28
6	⌉-P$_1$Pd(0)			
	~P$_1$–Pd(dibenzylideneacetone)	109.6	52.2	2.1
7	⌉-P$_1$PdCl$_2$	32.7	17.3	1.9
8	⌉-P$_2$Pd$_2$Cl$_4$			
	Cl\ /Cl\ /Cl Pd Pd ~P/ \Cl/ \P~	13.8	2.9	4.8
9	⌉-P$_2$Pd$_2$Cl$_2$			
	/Cl\ ~P—Pd Pd—P~ \Cl/	92.6	17.9	5.3

[a] Reaction conditions: catalyst 0.02-0.25 g in 25 ml of ethanol, P_{H_2} = 1 atm, CPD - 0.5 ml, 30 °C, palladium content in the catalyst

[b] ⌉-P$_1$ = SIL-CH$_2$CH$_2$PPh$_2$; ⌉-P$_2$ = SIL$\big\langle^{CH_2CH_2CH_2PPh_2}_{CH_2CH(CH_3)CH_2PPh_2}$,

[c] Activity in mol of CPD or CPE per g-atom of Pd per min.

[d] The ratio of the activity for CPD hydrogenation to the A for CPE hydrogenation

The rate of hydrogenation of CPD to CPE increases with increasing number of palladium atoms in the anchored complex and the hydrogenation activity at n=5 was 75% of the activity of metallic Pd in a supported 1% Pd/SiO$_2$ catalyst. XPS revealed an in-

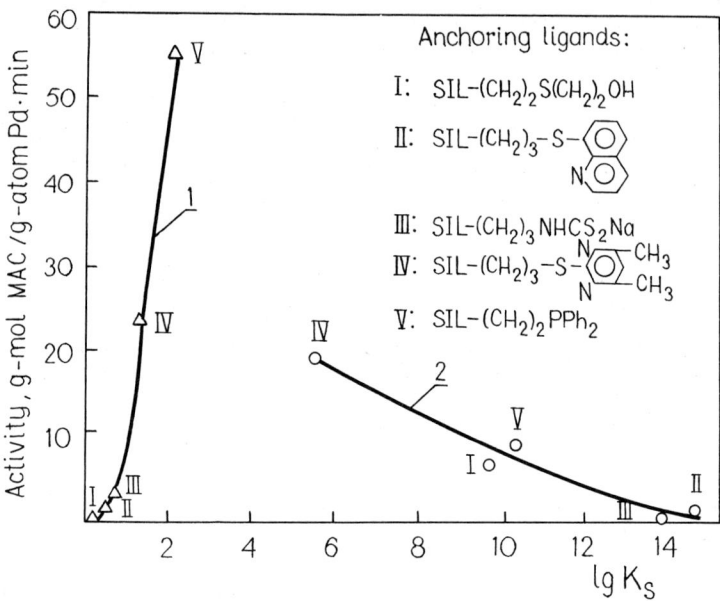

Fig. 13.1. The activity of anchored palladium complexes in methyl acrylate hydrogenation vs the stability constant K_s

$$SIL-L + PdL'_m \xrightleftharpoons{K_s} SIL-L-PdL'_m$$

1 - Pd(0), 2 - Pd(II). Reaction conditions: temperature, 60 °C, pressure of H_2 1 atm, concentration of methyl acrylate in dioxane ~10 vol.% (ref. 26).

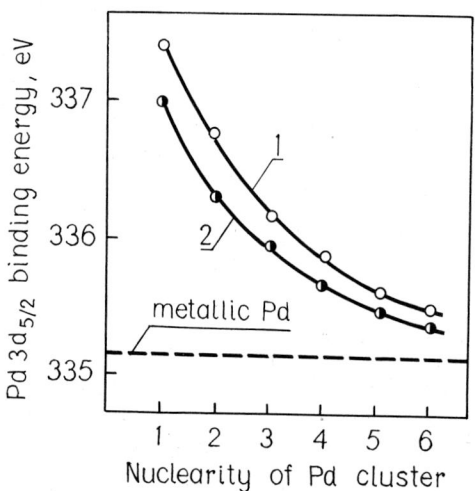

Fig. 13.2. Energy of the $Pd3d_{5/2}$ level vs. the number of palladium atoms in anchored clusters (ref. 26): 1 - I-$P_2Pd_n(OH)(CHO)$ in "initial" form; 2 - sample 1 after heating in vacuum at 110 °C, for 1 h.

crease in "metallic" character of the palladium atoms with an increase in their number in the anchored polynuclear complex. When this number was 10, the binding energy of the $Pd3d_{5/2}$ level in the anchored complex was practically the same as that in palladium foil (see Fig. 13.2). The increase in activity with increasing value of n (see complexes 1-5 in Table 13.2) was explained by a diminishing "poisoning" influence of the phosphine ligands on the metallic cluster (ref. 26). The structure of the active centres affects the ratio of the hydrogenation rates for CPD and CPE and have the selectivity for diene hydrogenation (see Table 13.2).

(iii) <u>Factors influencing the catalytic properties of anchoring palladium complexes</u>. A study of the anchored palladium complexes enabled some conclusions to be drawn on the factors influencing their catalytic properties (ref. 26). The reaction products did not depend on the nature of the surface ligand. So the nature of L in the complex $SIL-LPdL'_m$ in general did not affect the reaction mechanism, only the relative rates of the individual steps, resulting in quantitative (but not qualitative) change in the catalytic properties of the anchored complexes. The change in activity when a ligand L_1 was replaced by ligand L_2 may be attributed to the change in the free energy of activation for the reaction, $\lg (A_{L_1}/A_{L_2}) \sim \Delta G^{\#}_{L_1 \to L_2}$. The change in the stability constant, K, of the anchored complex is related to the corresponding change in the free energies of the complexes $\lg K_1/K_2 \sim \Delta G^{o}_{L_1 \to L_2}$. So the effect of the nature of the surface ligand on the catalytic activity may be explained on the basis of the Polyani-Semenov law

$$\Delta G^{\#}_{L_1 \to L_2} = \pm \beta_i \, \Delta G^{o}_{L_1 \to L_2}$$

where β_i is a correlation coefficient.

For MAC hydrogenation on anchored palladium(II) complexes, β_i has a minus, for palladium(0) complexes, a plus exponent. So, for $SIL-LPdL'_m$ complexes when L' does not vary, the catalytic activity is affected by thermodynamic factors such as the energy of the bonds Pd-L, Pd-olefin.

A change in the nuclearity of the anchored complexes $SIL-LPd_nL'_m$ in principle may result in a qualitative change in their catalytic properties, i.e. the reaction mechanism may alter from one typical for soluble and mononuclear anchored complexes to one typical for metallic palladium (ref. 26).

13.2.4 Some conclusions

It is possible to conclude that the anchoring of metal complexes enables the preparation of catalysts in which all the metal atoms participate in the catalytic reaction. These catalysts are easy to separate from the reaction mixture, and there are more possibilities of varying the composition of the active centre than in the case for soluble analogues.

After the first attempts to prepare and characterize heterogenized metal complexes, the difficulties which arose damped the initial enthusiasm of the pioneers in this field of catalysis. However, the subsequent accumulation of experience and knowledge has revealed ways of overcoming such difficulties in the preparation and application of these catalysts. For example, the restrictions in diffusion that often arise in the hydrogenation of complexes anchored on polymers may be avoided by appropriate choice of the polymer, e.g. see the results of Grubbs and co-workers (refs. 12,106) and Pittman et al. (ref. 53).

The homogeneity of catalysts containing anchored complexes in comparison with traditional heterogeneous catalysts facilitates the application of physical methods for catalyst characterization and the study of the mechanism of the catalytic reaction.

In general, the properties of the active centres - anchored - complexes - can be varied over a wide range due to the possibility of varying their compositions. By changing the nature of the carriers, oxides vs organic polymers, their structure (degree of cross-linking) and the length of the hydrocarbon chain linking L with the support surface it is possible to vary the accessibility of the active centres.

A promising application of anchored complexes is in the preparation of enantioselective catalysts. Effective catalysts for enantioselective hydrogenation of amino acids were prepared (ref. 117) on the basis of DIOP L-chiral phosphine ligands bound to polystyrene:

$$RCH=C\begin{smallmatrix}CO_2H\\NHCOCH_3\end{smallmatrix} \xrightarrow[\text{{\textstyle\frac{\lambda}{\lambda}}-LRhL'}]{H_2} RCH_2-CH^*\begin{smallmatrix}CO_2H\\NHCOCH_3\end{smallmatrix} \qquad (13.25)$$

The optical yield was 52-60% when R = H and 86% when R = Ph;

{{figure: structure of ${\textstyle\frac{\lambda}{\lambda}}$-LRhL' — copolymer of (CH$_2$–CH(C$_6$H$_4$–))$_{0.08}$ and (CH$_2$–C(CH$_3$)(CO$_2$C$_2$H$_4$OH))$_{0.92}$, with the phenyl group bearing a CH(O–CH(CH$_2$–PPh$_2$))$_2$ acetal coordinated to Rh(S)(Cl)}}

13.3 HYDROGENATION OF UNSATURATED HYDROCARBONS IN THE PRESENCE OF TRANSITION METAL IONS ON NON-FUNCTIONALIZED OXIDES

13.3.1 Catalysts prepared from anchored π-allyl complexes of Group 6 metals

Coordinatively unsaturated transition metal ions with low oxidation numbers bound to the surface of oxides may be prepared by treatment of anchored organometallic complexes with hydrogen:

$$({\textstyle\frac{\lambda}{\lambda}}E-O)_n MR_m + m/2\ H_2 \longrightarrow ({\textstyle\frac{\lambda}{\lambda}}E-O)_n M + mHR \qquad (13.26)$$

Samples prepared by this technique show catalytic activity for many reactions that usually proceed on partially reduced oxides, e.g. polymerization, metathesis, (de)hydrogenation, H-D exchange, etc. (refs. 27,118-126).

π-Allyl molybdenum complexes anchored on SiO_2 are reduced according to

$$({\textstyle\frac{\lambda}{\lambda}}Si-O)_2 Mo(C_3H_5)_2 \xrightarrow[>500\ °C]{H_2} ({\textstyle\frac{\lambda}{\lambda}}Si-O)_2 Mo + 2C_3H_8 \qquad (13.27)$$

The hydrogenation of ethylene on such catalysts proceeds with low (or close to zero) activation energy in the temperature range -78 to 200 °C. This reaction is first order with respect to C_2H_4 and H_2 (refs. 27,121,123). On metallic and oxide catalysts, this reaction is first order with respect to H_2 and zero (or negative) order with respect to ethylene.

In general, the preparation of catalysts from organometallic precursors results in more active hydrogenation catalysts than conventional techniques for catalyst preparation, e.g., impregnation by aqueous solutions. As regards the hydrogenation of ethylene at 25 °C, the activity of the catalyst prepared from $Mo(C_3H_5)_4$ anchored on SiO_2 is 6-10 times higher than for the catalyst prepared by impregnation of SiO_2 by $(NH_4)_2MoO_4$ solution. At higher temperatures (~ 200 °C) this difference becomes two orders of magnitude (refs. 27,123). The activity of samples prepared by supporting $Mo_2(\eta^3-C_3H_5)_4$ on Al_2O_3 and subsequent reduction by hydrogen at 585 °C is 3300 times higher in 1-butene and 1,3-butadiene hydrogenation than that of conventional molybdenum catalysts (ref. 124). Catalysts prepared by reduction of $(\overset{|}{Al}-O)_2Cr(C_3H_5)$ surface complexes at ~ 600 °C are considerably more active in C_3H_6 hydrogenation than impregnation catalysts (refs. 119,125).

It has been concluded (refs. 124-126) that the active and stable surface species in alkene hydrogenation at low temperatures are "paired structures" $\underset{\overline{///////////}}{\underset{O}{\diagdown}\underset{O}{\overset{M}{\diagup}}\underset{O}{\overset{M}{\diagup}}\underset{O}{\diagdown}}$ containing M^{2+} (M = Mo, Cr) on the surface of Al_2O_3 or SiO_2. A remarkable feature of the paired molybdenum structure on SiO_2 was its reversible coordination of oxygen at 0-155 °C (ref. 124).

13.3.2 Catalysts prepared from anchored complexes of Group 4 metals

Surface hydrides are formed when organometallic complexes of Group 4 metals anchored on oxide supports are treated with hydrogen. The $(\overset{|}{\underset{|}{Si}}-O)_2ZrH_2$ complexes anchored on SiO_2 prepared from tetraallyl- or tetraneopentylzirconium are active in the hydrogenation of 1-butene to butane (ref. 127). In the presence of surface hydrides of titanium prepared by hydrogen treatment of $Ti(CH_2C_6H_5)_4$ anchored on SiO_2, benzene and cyclohexene can be hydrogenated (ref. 128). From the correlation between the hydrogenation activity and content of surface hydrides it is possible to suggest that surface species with Ti-H bonds are the active centres for hydrogenation. The rate of benzene hydrogenation on anchored titanium complexes is four orders of magnitude lower than for cyclohexene, so only cyclohexane was obtained as the product of hydrogenation of benzene (ref. 128). However, zirconium hydrides prepared by the interaction of $ZrCl_4$ with an excess of LiC_4H_9 in organic solvents followed by treatment with hydrogen showed remarkable selectivity for benzene hydrogenation (85-93% of cyclohexene was obtained at 100 °C and P_{H_2} = 8.2 atm on some samples) (ref. 129). Note that moderate selectivities for cyclohexene have been observed for benzene hydrogenation on Pt/polyamide catalysts (10-67% C_6H_{10} (refs. 130-133)) or on modified ruthenium catalysts (11-56% C_6H_{10} (refs. 134,135)).

13.3.3 Catalysts prepared from anchored organometallic complexes of actinides

Very active hydrogenation catalysts have been prepared from $(CpMe_5)_2Th(CH_3)_2$, $(CpMe_5)_2U(CH_3)_2$ and $(CpMe_5)_2ThH_2$ ($CpMe_5$ = pentamethylcyclopentadienyl) anchored on Al_2O_3 (ref. 136). Alumina used as a support was first dehydroxylated at $\sim$ 950 °C and contained a low concentration of surface hydroxyl groups (0.12 OH/mm^2) (ref. 137). By addition of CO as a poison to the reaction mixture, it was found that only $\sim$ 2% of the supported U or Th are included in the active centres. However, these active centres showed remarkable activity for propylene hydrogenation: the turnover number at -63 °C was 30-50 s^{-1}. As was claimed (ref. 136), this activity is higher than that of supported rhodium catalysts under the same conditions.

13.4 ANCHORED ORGANOMETALLIC COMPLEXES AS PRECURSORS FOR THE PREPARATION OF HIGHLY DISPERSED METALLIC PARTICLES

In recent years supported metal carbonyls and other organometallic compounds have been extensively used for the preparation of supported metallic catalysts (refs. 27,138-158). The question may arise as to why these rather expensive and often quite unstable transition metal compounds are used for catalyst preparation. This may be justified as follows:

(a) The desire to obtain samples with unusual catalytic properties for practical applications.

(b) The possibility of obtaining "mixed" surface species or metal particles containing various elements with a composition that it is impossible to obtain by conventional techniques of catalyst preparation.

(c) The possibility of forming systems with more homogeneous surface compositions suitable for the study of the state of surface components by physical methods and for the study of theoretical problems of catalysis by supported metals.

13.4.1 Preparation of metallic catalysts by decomposition of surface organometallic complexes

Highly dispersed metallic catalysts may be prepared by the decomposition of anchored allyl complexes of Group 8 metals.

$$(\text{\textbardbl-O})_y M(C_3H_5)_x \xrightarrow{H_2} y(\text{\textbardbl-OH}) + \text{\textbardbl-M(0)} + xCH_3CH=CH_2 \longrightarrow$$

$$\xrightarrow[\text{of n atoms of M}]{\text{agglomeration}} \begin{array}{c}\text{\textbardbl-OH}\\ \boxed{M_n}\end{array} \qquad (13.28)$$

where M = Ni, Pd, Pt (refs. 27, 138-145, 148-150), Rh (refs. 151-153). The reduction of allyl complexes of Pd and Pt proceeds under mild conditions ($\sim$20 °C) (ref. 27). The data available on the reduction of allyl rhodium complexes are contradictory. In a study (ref. 154) of the activity of $Rh(C_3H_5)_3$ anchored on SiO_2 and TiO_2 in olefin hydrogenation at 23 °C in the presence of $P(CH_3)_3$, various anchored rhodium complexes were found. The hydrogenation activity of the catalysts varied in the order:

$$\text{Si-O-Rh}(C_3H_5)H > \text{Si-O-RhH}_2 > \text{Si-O-Rh}(C_3H_5)H(PMe_3)_2 > \text{Si-O-RhH}_2(PMe_3)_3$$

When six coordination sites in the complex $\text{Si-O-RhH}_2(PMe_3)_3$ were occupied by ligands no activity was observed. For all supported surface complexes, the hydrogenation of various olefins depended on the possible steric hindrance:

$$\sim\!\!\!\!\diagup \; > \; \diagup\!\!=\!\!\diagdown \; > \; \diagup\!\!=\!\!\diagup \; > \; \diagdown\!\!=\!\!\diagup \; > \; \diagdown\!\!=\!\!\diagup$$

Gates and co-workers (refs. 151,152) studied the reduction of $\text{Si-O-Rh}(C_3H_5)_2$ and found that, in an hydrogen atmosphere at 25 °C, metallic particles of Rh of 10-30 Å in size are easily formed on the surface of the support. It was shown that $\text{Si-O-Rh}(C_3H_5)_2$ does not react with hydrogen and reduction proceeds according to the scheme:

$$\xrightarrow[25\ °C]{H_2} Rh(0) + 3\ \underset{\overline{///Si///}}{|}^{OH}$$

$$CH_3CH=CH_2 \xrightarrow[25\ °C]{H_2,\ Rh(0)} CH_3CH_2CH_3 \qquad (13.29)$$

In the hydrogenation of toluene on this catalyst an induction period was observed during which active metal particles were formed from the initial inactive anchored complex (ref. 152).

The properties of supported catalysts prepared from anchored allyl complexes of Pt, Pd and Ni have been studied in detail (refs. 123,138,139,148-150,155-157). The formation of highly dispersed particles of these metals was observed. The particle size depended essentially on the number of hydroxyl groups of the support surface which is determined by the temperature of the preliminary dehydrogenation of the support, and on the concentration of the supported allyl complex (refs. 148-150). The catalytic properties of metallic Pt, Pd and Ni prepared from allyl complexes have been studied in the hydrogenation of phenylacetylene (ref. 149), 1-hexene (refs. 148,150), benzene, cyclopentadiene and sulpholene (refs. 155-157).

For the preparation of dispersed nickel hydrogenation catalysts supported on SiO_2 the application of $Ni_2Cp_2(CO)_2$ and $Ni_3Cp_3(CO)_2$ ($Cp = \eta^5-C_5H_5$) has been described (ref. 158).

13.4.2 Low-valent surface ions as anchoring sites for stabilization of dispersed metals and the problem of the strong metal-support interaction

Anchored organometallic compounds of Mo and W have been used (refs. 157,159-163) as precursors to obtain low-valent ions of these metals on the surface of SiO_2. The materials prepared were then treated with allyl complexes of Pt and Pd. After reduction by hydrogen at 200-800 °C, the catalysts obtained were characterized by the following features (refs. 22,157,159-163): (a) higher stability of metallic particles towards sintering; (b) decreased chemisorption of CO and H_2 on platinum and palladium particles; (c) modified catalytic properties in hydrocarbon conversion. After oxidation of samples at high (500 °C) temperature and subsequent reduction under mild conditions these specific features disappeared and the properties of $(Pt+M)/SiO_2$ and

(Pd+M)/SiO$_2$ (M = Mo or W) samples were similar to those of Pt/SiO$_2$ and Pd/SiO$_2$ catalysts.

In "bimetallic" platinum catalysts containing Mo and W the average oxidation numbers of Mo and W were close to 2. In catalysts containing Pd, reduction to oxidation numbers lower than 2 was possible. The dispersion of Pd and Pt in "bimetallic" catalysts prepared by the above technique and their activities in benzene hydrogenation were considerable higher than for monometallic Pt/SiO$_2$ and Pd/SiO$_2$ catalysts (ref. 157). XPS data showed a decrease in the electron density of Pt in the presence of Mo and W (ref. 157).

Some years ago the papers of Tauster et al. (ref. 164) and Baker et al. (ref. 165) drew attention to the unusual properties of metallic particles on reducible oxides (TiO$_2$, V$_2$O$_3$, Nb$_2$O$_5$, MnO). These unusual properties are similar to those described above for (Pt+Mo)/SiO$_2$ and (Pt+W)/SiO$_2$.

The reason for the change in the properties of metals on reducible oxides is believed to be a strong metal-support interaction (SMSI). Various hypotheses have been proposed to explain the effects of SMSI (ref. 166). One of them implies that this effect is determined by the interaction of metallic particles with low-valent ions formed on the oxide surface, e.g. TiO$_2$. To test this hypothesis, "bimetallic" catalysts containing titanium and zirconium ions on SiO$_2$ have recently been prepared (refs. 167,168). The preparation comprised anchoring of organometallic complexes of Ti and Zr and subsequent supporting of Ni, Pd or Pt. Such catalysts were characterized by SMSI and their activities were similar to those of Ni, Pd and Pt supported on TiO$_2$ and ZrO$_2$. So these preliminary data provide evidence in support of an interaction of the metallic particles with low valent Ti and Zr on TiO$_2$ and ZrO$_2$. For further elucidation of the nature of SMSI a detailed study of the structure of surface species in (M^8 + M^4)/SiO$_2$ (M^8 = metal of Group 8, M^4 = element of Group 4) catalysts is necessary.

13.4.3 The preparation of metallic catalysts by decomposition of anchored metal carbonyls

The decomposition of metal carbonyls on polymer and oxide carriers broadens the possible range of application of organometallic compounds for catalyst preparation because many transition metals form mono- or polynuclear carbonyls.

For the system "metal carbonyl - oxide support", as in the case of organometallic complexes anchored on oxides, the mechanism of interaction of the components and the types of surface species depend on the number of hydroxyl groups on the support surface and on the concentration of supported carbonyl. When the amount of carbonyl compounds is not higher than the content of surface hydroxyl groups, decarbonylation of samples results mainly in the formation of metal ions ($-O^-)_n M^{n+}$, but not metal particles. The oxidation of zerovalent metal is due to the removal of a surface proton in the form of hydrogen (see reaction 13.5). At high contents of supported metal carbonyl on the support, decarbonylation may result in formation of metal particles (refs. 44,45,169).

The conditions for decomposition of supported carbonyls are of primary importance. For example, highly dispersed superparamagnetic cobalt particles are formed upon rapid (50°/min) pyrolysis of $Co_2(CO)_8$ supported on various oxides (TiO_2, ZrO_2, SiO_2, MgO, Al_2O_3). However, when the increase in pyrolysis temperature is slow (1°/min), decomposition of the supported carbonyls results in the formation of large (>50 $\overset{o}{A}$) ferromagnetic cobalt particles as in the conventional impregnation technique of catalyst preparation (ref. 170). Superparamagnetic (<15 $\overset{o}{A}$) particles of metallic Fe have been prepared by thermal decomposition of $Fe(CO)_5$, $Fe_3(CO)_{12}$ and $[HFe_3(CO)_{11}]^-$ on dehydroxylated Al_2O_3 or MgO at <200 °C. At higher decomposition temperatures larger particles were formed. The thermal decomposition of iron carbonyls supported on a hydroxylated carrier results in a mixture of superparamagnetic particles and iron oxides (ref. 171). Iron particles of ~12 $\overset{o}{A}$ in size have been obtained (ref. 172) by decomposition of $Fe_3(CO)_{12}$ and $[HFe_3(CO)_{11}]^-$ on dehydroxylated MgO. The formation of metallic particles in the decomposition of carbonyls of Rh (refs. 173-179), Ru (refs. 51, 180-183), Os (refs. 184,185) and Ir (refs. 174,186,187) has been studied.

The rates of hydrogenation of 1-hexene and cyclohexene have been compared (ref. 188) for $Ru_3(CO)_{12}$ in solution, its analogue anchored on silica via a Ru-O-Si bond and for metallic Ru supported on SiO_2 prepared by pyrolysis of anchored carbonyl. The relative rates of hydrogenation of cyclohexene with these catalysts were 1:2.6:4.3.

It is worth mentioning the hydrogenation catalysts prepared

by decarbonylation of carbonyls of Group 6 metals $M(CO)_6$ (M = Cr, Mo, W) anchored on Al_2O_3 (refs. 44,45,136) (catalysts prepared by activation of $Mo(CO)_6/Al_2O_3$ were among the first examples of supported organometallic compounds (ref. 189)). On hydrated supports, the complete decarbonylation of $Cr(CO)_6$ at > 200 °C, $Mo(CO)_6$ and $W(CO)_6$ at > 300 °C results in oxidation of surface species by hydroxyl groups of the support and the formation of hard-to-reduce surface metal oxides having poor hydrogenation activity (refs. 44,45,169,190).

If Al_2O_3 is first dehydroxylated at $\sim$ 950 °C then used to anchor $Mo(CO)_6$ subsequent treatment with helium at 300 °C, with hydrogen at 650 °C results (ref. 136) in the formation of metallic Mo on Al_2O_3. This catalyst is extremely active in monoolefin hydrogenation: the turnover number for propylene hydrogenation at -46 °C is > 5 s^{-1}, whereas on Pt/SiO_2 it is only about 0.3 s^{-1} (refs. 136,191). $Mo(0)/Al_2O_3$ is three orders of magnitude more active than reduced MoO_3/Al_2O_3 (ref. 44).

Catalysts prepared from anchored carbonyls of Group 6 metals are 3-5 orders of magnitude more active in olefin hydrogenation than the hexacarbonyls of the same metals in solution (refs. 45, 192,193).

13.4.4 Some conclusions

The method of preparation of surface metallic catalysts via the decomposition of anchored organometallic complexes and carbonyls enables one to obtain the maximum metal dispersion for a given "metal-support" system and even to prepare surface clusters containing small numbers of metal atoms (refs. 157,158,171-174, 178,180,194-197). This is one of the advantages of the application of organometallic compounds for catalyst preparation in comparison with conventional techniques, e.g., impregnation.

The possibility to prepare superdispersed catalysts permits elucidation of the problem of the optimum size of metal particles for maximum catalytic activity or selectivity in a given reaction. With highly dispersed catalysts prepared from anchored $Rh_6(CO)_{16}$ clusters it was found (ref. 196) that the benzene hydrogenation rate on the size of the surface rhodium particles has a maximum at $\sim$ 17 Å, decreases for particles of $\sim$ 20 Å; particles of 8-12 Å were inactive.

The use of heteronuclear organometallic compounds for the preparation of surface species may be of special interest. It is possible that, by decomposition of mixed surface clusters, active species may be prepared that are difficult to obtain by the use of other techniques, e.g., impregnation. The most interesting examples of the application of supported bimetallic catalysts derived from mixed clusters are in the hydrogenation of CO (refs. 198-205) and in hydrocracking and reforming of hydrocarbons (ref. 206).

13.5 CONCLUSION

Several types of hydrogenation catalysts may be prepared via the anchoring of metal complexes on supports:

(a) Heterogenized metal complexes - analogues of the soluble metal compounds.

(b) Complexes attached to surface ligands, which have no direct analogues among soluble compounds. This is due to the possibility of using ligands which would normally form insoluble compounds with metals.

(c) Surface coordinatively unsaturated metal ions or hydrides formed by decomposition of organometallic compounds of metals in Groups 3-6.

(d) Superdispersed supported mono- and polynuclear catalysts prepared by decomposition of organometallic compounds anchored on organic or mineral carriers.

The systematic development and thorough study of anchored metal complexes and catalysts derived from them is at a relatively immature stage. In this review we have considered their use in only one reaction - the catalytic hydrogenation of unsaturated compounds. Interesting results have also been obtained in the hydrogenation of CO, isomerization, metathesis, hydroxylation, hydroformylation, carbonylation, oligomerization, polymerization and other reactions. However, we hope that we have succeeded in demonstrating the potential of catalysts prepared by the anchoring of metal complexes.

REFERENCES

1 Y. Izumi, Bull. Chem. Soc. Jpn, 32 (1959) 932-936.
2 Y. Izumi, Angew. Chem., 83 (1971) 956-966.
3 G.J.K. Acres, G.C. Bond, B.J. Cooper and J.A. Dawson, J. Catal., 6 (1966) 139-141.

4   P.R. Rony, Chem. Eng Sci., 23 (1968) 1021-1034.
5   P.R. Rony, J. Catal., 14 (1969) 142-147.
6   J. Manassen, Chim. Ind, 51 (1969) 1058-1062.
7   J. Manassen, Platinum Met. Rev., 15 (1971) 142-143.
8   W.O. Haag and D.D. Whitehurst, German Pat., 1, 800, 371 (1969), 1, 800, 379 (1969), 1, 800, 380 (1969); C. A. 71 (1969) 114951, 33823, 72 (1970) 31192, Belgian Pat., 721, 686 (1969).
9   V.L. Shmonina, N.N. Stefanovskaya, E.I. Tinyakova and B.A. Dolgoplosk, Vysomomolek. Soed., 12B (1970) 566.
10 Yu.I. Yermakov, in Yu.I. Yermakov (Editor), On the Possibility of Application of Organometallic Compounds of Transition Metals for the Synthesis of Supported Catalysts, Institute of Catalysis, Novosibirsk, 1971, pp. 5-13.
11 M. Capka, P. Svoboda, M. Cerny and J. Hetflejš, Tetrahedron Lett., 50 (1971) 4787-4790.
12 R.H. Grubbs and L.C. Kroll, J. Am. Chem. Soc., 93 (1971) 3062-3063.
13 K. Mosbach, Sci. Amer., 224 (1971) 26.
14 Yu.I. Yermakov, A.M. Lazutkin, E.A. Demin, Yu.P. Grabovskii and V.A. Zakharov, Kinet. Katal., 13 (1972) 1422-1427.
15 M. Capka, P. Svoboda, M. Kraus and J. Hetflejš, Chem. Ind. 16 (1972) 650-651.
16 J.P. Collman, L.S. Hegedus, M.P. Cooke, J.R. Norton, G. Dolcetti and D.N. Marquardt, J. Am. Chem. Soc., 94 (1972) 1789-1790.
17 D.G.H. Ballard, Adv. Catal., 23 (1973) 267-325.
18 C.U. Pittman and G.O. Evans, Chemtech., (1973) 560-566.
19 J.P. Candlin and H. Thomas, Adv. Chem. Ser., 132 (1974) 212-239.
20 Z.M. Michalska and D.E. Webster, Platinum Met. Rev., 18 (1974) 65-73.
21 J.C. Bailar, Cat. Rev. Sci. Eng., 10 (1974) 17-36,
22 Yu. I. Yermakov, Cat. Rev. Sci. Eng., 13 (1976) 77-120.
23 Yu. I. Yermakov, Relations entre Catalyse Homogène et Catalyse Hétérogène, Editions du CNRS, Paris, 1978, pp. 233-262.
24 D. Cornet, Bull. Union Phys., 71 (1977) 1183-1199.
25 F.R. Hartley and P.N. Vezey, Adv. Organomet. Chem., 15 (1977) 189-234.
26 Yu.I. Yermakov and V.A. Likholobov, Kinet. Katal., 21 (1980) 1208-1219.
27 Yu. I. Yermakov, B.N. Kuznetsov and V.A. Zakharov, Catalysis by Supported Complexes, Elsevier, Amsterdam, 1981, 522 pp.
28 A.K. Smith and J.M. Basset, J. Mol. Catal., 2 (1977) 229-241.
29 J.M. Basset and A. Choplin, J. Mol. Catal., 21 (1983) 95-108.
30 A.Ya. Yuffa and G.V. Lisichkin, Usp. Khim., 47 (1978) 1414-1443.
31 B.R. James, Adv. Organomet. Chem., 17 (1979) 319-405.
32 J. Evans, Chem. Soc. Rev., 10 (1981) 159-180.
33 B.S. Ramaswamy, J. Sci. Ind. Res., 40 (1981) 644-658.
34 M. Bartholin, C. Graillat and A. Guyot, J. Mol. Catal., 10 (1981) 361-375.
35 F. Ciardelli, G. Braca, C. Carlini, G. Sbrana and G. Valentini, J. Mol. Catal., 14 (1982) 1-17.
36 H.P. Boehm and H. Knözinger, in J.R. Anderson and M. Boudart (Editors), Catalysis, Science and Technology, vol. 4, Springer-Verlag, Berlin Heidelberg, New York, 1983 pp. 39-207.
37 F.H. Jardine, J.A. Osborn, G. Wilkinson and J.F. Young, Chem. Ind., (1965) 560.
38 J.F. Young, J.A. Osborn, F.H. Jardine and G. Wilkinson, Chem. Commun., (1965) 131.

39 M. Deeba and B.C. Gates, J. Catal., 67 (1981) 303-307.
40 R. Psaro, R. Ugo, G.M. Zanderighi, B. Besson, A.K. Smith and J.M. Basset, J. Organomet. Chem., 213 (1981) 215-247.
41 Y. Iwasawa, M. Yamada, S. Ogasawara, Y. Sato and H. Kuroda, Chem. Letters, (1983) 621-624.
42 C.G. Armistead, A.J. Tyler and J.A. Hockey, J. Phys. Chem., 73 (1969) 3947-3953.
43 R.D. Hancock, I.V. Howell, R.C. Pitkethly and P.J. Robinson, Proc. Int. Symp. on the Relations between Heterogeneous and Homogeneous Catalytic Phenomena, Brussels, 1974, D5, pp. 1-9.
44 A. Brenner, J. Mol. Catal., 5 (1979) 157-161.
45 T.J. Thomas and A. Brenner, J. Mol. Catal., 18 (1983) 197-202.
46 F. Hugues, A.K. Smith, Y.Ben Taarit, J.M. Basset, D. Commereuc and Y. Chauvin, J. Chem. Soc., Chem. Commun., (1980), 68-69.
47 F. Hugues, B. Besson, P. Bussière, J.A. Dalmon, J.M. Basset and D. Olivier, Nouv. J. Chim., 5 (1981) 207-210.
48 R.B. Bjorklund and R.L. Burwell, J. Colloid Interface Sci., 70 (1979) 383-391.
49 L. Bonneviot, D. Olivier and M. Che, J. Chem. Soc., Chem. Commun., (1982) 952-953.
50 J.L. Bilhou, A. Theolier, A.K. Smith and J.M. Basset, J. Mol. Catal., 3 (1977/78) 245-257.
51 V.L. Kuznetsov, A.T. Bell and Yu.I. Yermakov, J. Catal., 65 (1980) 374-389.
52 F. Correa, R. Nakamura, R.E. Stimson, R.L. Burwell and D.F. Shriver, J. Am. Chem. Soc., 102 (1980) 5112-5114.
53 C.U. Pittman, L.R. Smith and R.M. Hanes, J. Am. Chem. Soc., 97 (1975) 1742-1748.
54 K.G. Allum, R.D. Hancock, I.V. Howell, R.C. Pitkethly and P.J. Robinson, J. Organomet. Chem., 87 (1975) 189-201.
55 M. Kraus and D. Tomanova, J. Polym. Sci., Polym. Chem. Ed., 12 (1974) 1781-1785.
56 Y. Nakamura and H. Hirai, Chem. Lett., (1974) 645-650.
57 N. Kawata, T. Mizoroki, A. Ozaki and M. Ohkawara, Chem. Lett. (1973) 1165-1168.
58 K.G. Allum, R.D. Hancock, I.V. Howell, S. McKenzie, R.C. Pitkethly and P.J. Robinson, J. Organomet. Chem., 87 (1975) 203-216.
59 K.G. Allum, R.D. Hancock, S. McKenzie and R.C. Pitkethly, in J.W. Hightower (Editor), Proc. 5th Int. Congr. Catal., Florida, 1972, North-Holland/Elsevier, Amsterdam, New York, 1973, pp. 477-486.
60 H.-W. Kohlschütter, P. Best and G. Wirzing, Z. Anorg. Allg. Chem., 285 (1956) 236-245.
61 W. Stöber, Kolloid-Z., 149 (1956) 39-46.
62 Yu.I. Yermakov, in V.D. Sokolovskii (Editor), Actual Problems of Catalysis Science, Proc. Siberian Catal. Conf., 1977, Institute of Catalysis, Novosibirsk, 1978, pp. 73-103.
63 V.A. Semikolenov, V.A. Likholobov and Yu.I. Yermakov, Kinet. Katal., 18 (1977) 1294-1300.
64 V.A. Semikolenov, D.Kh. Mikhailova, Ya.V. Sobchak, V.A. Likholobov and Yu.I. Yermakov, React. Kinet. Catal. Lett., 10 (1979) 105-110.
65 V.A. Semikolenov, V.A. Likholobov and Yu.I. Yermakov, Kinet. Katal., 20 (1979) 269.
66 A.S. Lisitsyn, V.A. Likholobov, N.G. Maksimov, V.A. Zakharov and Yu.I. Yermakov, Izvest. Sib. Otd. Akad. Nauk SSSR, Ser. Khim. Nauk, 6 (1979) 96-99.

67 D.C. Bailey and S.H. Langer, Chem. Rev., 81 (1981) 109-148.
68 J.J. Rafalko, J. Lieto, B.C. Gates and G.L. Schrader, J. Chem. Soc., Chem. Commun., (1978) 540-541.
69 J. Lieto, J.J. Rafalko and B.C. Gates, J. Catal., 62 (1980) 149-156.
70 V.A. Semikolenov, Thesis, Institute of Catalysis, Novosibirsk, 1979.
71 V.A. Semikolenov, V.A. Likholobov, P.A. Zhdan, A.P. Shepelin and Yu.I. Yermakov, Kinet. Katal., 21 (1980) 429-435.
72 V.A. Semikolenov, V.A. Likholobov, P.A. Zhdan, A.P. Shepelin and Yu.I. Yermakov, in K.Kh. Razikov (Editor), Proc. 5th Japan-Soviet Seminar on Catalysis, FAN, Tashkent, 1979, pp. 209-212.
73 V.A. Semikolenov, V.A. Likholobov, P.A. Zhdan, A.I. Nizovskii and Yu.I. Yermakov, in B.N. Kuznetsov (Editor), Supported Metallic Catalysts for Conversion of Hydrocarbons, Proc. All-Union Conference, Institute of Catalysis, Novosibirsk, 1978, pp. 67-70.
74 V.A. Semikolenov, V.A. Likholobov and Yu.I. Yermakov, Kinet. Katal., 22 (1981) 1026-1030.
75 V.A. Semikolenov, V.A. Likholobov, P.A. Zhdan, A.I. Nizovskii, A.P. Shepelin, E.M. Moroz, S.V. Bogdanov and Yu.I. Yermakov, Kinet. Katal., 22 (1981) 1247-1252.
76 V.A. Semikolenov, V.A. Likholobov and Yu.I. Yermakov, in B.N. Kuznetsov (Editor), Metallic Catalysts for Conversion of Hydrocarbons, Proc. All-Union Conference, Institute of Catalysis, Novosibirsk, 1978, pp. 280-283.
77 V.N. Kondrat'ev (Editor), Energies of the Rupture of Chemical Bonds, Ionization Potentials and Affinity to Electron, Nauka, Moscow, 1974, 351 pp.
78 C.U. Pittman and R.F. Felis, J. Organomet. Chem., 72 (1974) 389-397.
79 V.A. Likholobov, B.N. Kuznetsov, V.A. Semikolenov, V.L. Kuznetsov, L.G. Karakchiev, V.G. Shinkarenko, Yu.I. Yermakov, in Ya.B. Gorokhovatskii (Editor), The Role of Coordination in Catalysis, Naukova Dumka, Kiev, 1976, pp. 18-23.
80 V.N. Zudin, V.A. Likholobov, V.D. Chinakov, V.M. Nekipelov and Yu. I. Yermakov, in Yu. I. Yermakov (Editor), Proc. 5th Int. Symp. on Relations between Homogeneous and Heterogeneous Catalysis, Novosibirsk, 15-19 July 1986, VNU Science Press BV, Utrecht, in press.
81 V.A. Semikolenov, V.A. Likholobov and Yu.I. Yermakov, in Yu. I. Yermakov (Editor), Catalysts Containing Supported Complexes, Proc. All-Union Meeting, Institute of Catalysis, Novosibirsk, 1977, pp. 43-46.
82 F.R. Hartley, Chem. Rev., 73 (1973) 163-190.
83 H. Pracejus and M. Bursian, East German Pat. 92031 (1972); C. A., 78 (1973) 72591.
84 P.R. Rony and J.F. Roth, J. Mol. Catal., 1 (1975/76), 13-25.
85 T.P. Voskresenskaya, V.A. Semikolenov, V.A. Likholobov, A.P. Shepelin, P.A. Zhdan, A.V. Mashkina, Kinet. Katal., 23 (1982) 382-387.
86 T.P. Voskresenskaya, V.A. Semikolenov, V.L. Kuznetsov, V.A. Likholobov, A.K. Trofimchuk, Yu.I. Yermakov and A.V. Mashkina, in A.V. Mashkina (Editor), Catalysts for the Processes of Preparation and Conversion of Sulphur Compounds, Institute of Catalysis, Novosibirsk, 1979, pp. 74-77.
87 I.I. Frolova, V.A. Semikolenov, V.A. Likholobov and Yu.I. Yermakov, in A.V. Mashkina (Editor), Catalysts for the Processes of Preparation and Conversion of Sulphur Compounds, Institute of Catalysis, Novosibirsk, 1979, pp. 81-84.

88   T.P. Voskresenskaya, V.A. Semikolenov, V.A. Likholobov, A.K. Trofimchuk, A.P. Shepelin, P.A. Zhdan and A.V. Mashkina, in Yu.I. Yermakov (Editor), Catalysts Containing Supported Complexes, Institute of Catalysis, Novosibirsk, 1980, part 2, pp. 22-25.

89   T.P. Voskresenskaya, V.N. Yakovleva, V.A. Semikolenov, V.A. Likholobov and A.V. Mashkina in M.V. Shimanskaya (Editor), Proc. 3rd All-Union Symposium, Zinatne, Riga, 1981, pp. 94-96.

90   A.P. Shepelin, P.A. Zhdan, T.P. Voskresenskaya, V.A. Semikolenov, V.A. Likholobov and A.V. Mashkina, in M.V. Shimanskaya (Editor), Proc. 3rd All-Union Symposium, Zinatne, Riga, 1981, pp. 99-101.

91   S. Jacobson, W. Clements, H. Hiramoto and C.U. Pittman, J. Mol. Catal., 1 (1975/76) 73-76.

92   Z. Otero-Schipper, J. Lieto and B.C. Gates, J. Catal., 63 (1980) 175-181.

93   R. Pierantozzi, K.J. McQuade, B.C. Gates, M. Wolf, H. Knözinger and W. Ruhmann, J. Am. Chem. Soc., 101 (1979) 5436-5438.

94   R. Pierantozzi, K.J. McQuade and B.C. Gates, in T. Seiyama and K. Tanabe (Editors), New Horizons in Catalysis, Proc. 7th Int. Congr. Catal., Kodansha LTD./Elsevier, Tokyo, Amsterdam, Oxford, New York, 1981, pp. 941-952.

95   B.C. Gates and J. Lieto, Chemtech., (1980) 248-251.

96   G.V. Kudryavtsev, A.Yu. Stakheev and G.V. Lisichkin, J. D.I. Mendeleev All-Union Chem. Soc., 27 (1982) 232-233.

97   B. Besson, A. Choplin, L.D' Ornelas and J.M. Basset, J. Chem. Soc., Chem. Commun., 15 (1982) 843-845.

98   British Petroleum Co. Ltd. British Pat., 1,295,675 (1972).

99   V.L. Kuznetsov, Thesis, Institute of Catalysis, Novosibirsk, 1977.

100   K. Kaneda, M. Terasawa, T. Imanaka and S. Teranishi, Chem. Lett., 10 (1975) 1005-1008.

101   C.U. Pittman, S.E. Jakobson and H. Hiramoto, J. Am. Chem. Soc., 97 (1975) 4774-4775.

102   A.V. Novikova, L.S. Kovaleva, O.P. Parenago and V.M. Frolov, Kinet. Katal., 25 (1984) 583-587.

103   S.D. Jackson, R.B. Moyes, P.B. Wells and R. Whyman, J. Catal. 86,(1984) 342-358.

104   D.J. Hunt, R.B. Moyes, P.B. Wells, S.D. Jackson and R. Whyman, Proc. 8th Int. Congr. Catal., Berlin (West), July 2-6, 1984, Verlag Chemie/Dechema, Weinheim, Deerfield Beach, Florida, Basel, pp. V27-V38.

105   S.D. Nayak, V. Mahadevan, M. Srinivasan, J. Catal., 92 (1985) 327-339.

106   R.H. Grubbs, L.C. Kroll and E.M. Sweet, J. Macromol. Sci., Chem., A7 (1973) 1047-1063.

107   T. Uematsu, T. Kawakami, F. Saitho, M. Miura and H. Hashimoto, J. Mol. Catal., 12 (1981) 11-26.

108   V.Z. Sharf, I.B. Slinyakova, L.H. Freidlin, V.N. Krutii, A.S. Gurovets and L.P. Finn, in Yu.I. Yermakov (Editor), Catalysts Containing Supported Complexes, Proc. All-Union Meeting, Institute of Catalysis, Novosibirsk, 1977, pp. 183-186.

109   G.V. Varnakova, E.I. Karpeiskaya, E.I. Klabunovskii and I.S. Shekoyan, in Yu.I. Yermakov (Editor), Catalysts Containing Supported Complexes, Proc. All-Union Meeting, Institute of Catalysis, Novosibirsk, 1977, pp. 159-162.

110　V.Z. Sharf, A.S. Gurovets, L.P. Finn, I.B. Slinyakova, V.N. Krutii and L.H. Freidlin, Izv. Akad. Nauk SSSR, Ser. Khim. (1979) 104-108.
111　V.L. Kuznetsov, M.R. MacLaury, B.N. Kuznetsov, J.P. Collman and Yu. I. Yermakov, React. Kinet. Catal. Lett., 3 (1975) 361-369.
112　B.N. Kuznetsov, V.L. Kuznetsov, M. MacLaury, M.I. Ioffe and Yu.I. Yermakov, Proc. 3rd Japan-Soviet Seminar on Catalysis, Kaz. NIINTI, Alma-Ata, 1975, preprint 21, pp. 1-15.
113　V.Z. Sharf, A.S. Gurovets, V.N. Krutii, I.B. Slinyakova, L.P. Finn and S.I. Scherbakova, Izv. Akad. Nauk SSSR, Ser. Khim., (1979) 2533-2535.
114　A.S. Berenblyum A.P. Aseeva and L.I. Lakhman, in Yu.I. Yermakov (Editor), Catalysts Containing Supported Complexes, Proc. All-Union Meeting, Institute of Catalysis, Novosibirsk, 1977, pp. 155-157.
115　V.A. Semikolenov, D.Kh. Mikhailova, Ya.V. Sobchak, V.A. Likholobov and Yu.I. Yermakov, Kinet. Katal., 21 (1980) 526-529.
116　V.A. Semikolenov, V.A. Likholobov and Yu.I. Yermakov, Kinet. Katal., 22 (1981) 1475-1479.
117　N. Takaichi, H. Imai, C.A. Bertelo and J.K. Stille, J. Am. Chem. Soc., 100 (1978) 264-268.
118　Y. Iwasawa, M. Yamagishi and S. Ogasawara, J. Chem. Soc., Chem. Commun., (1980), 871-873.
119　Y. Iwasawa, J. Mol. Catal., 17 (1982) 93-104.
120　Y. Iwasawa, Chem. Today, 159 (1984) 28-35.
121　Yu.I. Yermakov and B.N. Kuznetsov, Proc. 2nd Japan-Soviet Seminar on Catalysis, Science Council Hall, Tokyo, 1973, pp. 65-85.
122　Yu.I. Yermakov, B.N. Kuznetsov, L.G. Karakchiev and S.S. Derbeneva, React. Kinet. Catal. Lett., 1 (1974) 307-313.
123　B.N. Kuznetsov, Thesis, Institute of Catalysis, Novosibirsk, 1972.
124　Y. Iwasawa, S. Ogasawara and X. Kuroda, in A.A. Davydov (Editor), Proc. 7th Japan-Soviet Seminar on Catalysis, Institute of Catalysis, Novosibirsk, 1983, pp. 71-77.
125　Y. Iwasawa and S. Ogasawara, Chem. Lett., (1980) 127-130.
126　Y. Iwasawa, Y. Sasaki and S. Ogasawara, J. Chem. Soc., Chem. Commun., (1981) 140-142.
127　J. Schwartz and M.D. Ward, J. Mol. Catal., 8 (1980) 465-469.
128　Yu.I. Yermakov, O.S. Alekseev, V.A. Shmachkov, V.I. Sobolev, and Yu. A. Ryndin, Kinet. Katal., 26 (1985) 1270-1271.
129　G.P. Pez and R.K. Crissey, J. Mol. Catal., 21 (1983) 393-404.
130　P. Dini, D. Dones, S. Montelatici and N. Giordano, J. Catal., 30 (1973) 1-12.
131　S. Galvagno, P. Staiti, P. Antonucci, A. Giannetto and N. Giordano, React. Kinet. Catal. Lett., 21 (1982) 157-162.
132　P. Staiti, S. Galvagno, P. Antonucci, A. Rositani and P. Vitarelli, React. Kinet. Catal. Lett., 26 (1984) 111-116.
133　S. Galvagno, Z. Poltarzewsky, P. Staiti, R. Pietropaolo and N. Giordano, Proc. of the 1st Italian-Soviet Seminar on Catalysis Application to Energy Problems, Messina, 1984, pp. 214-218.
134　M.M. Johnson and G.P. Nowack, J. Catal., 38 (1975) 518-521.
135　J.A. Don and J.J.F. Scholten, Faraday Discuss. Chem. Soc., (1981) 145-156.
136　R.L. Burwell, J. Catal., 86 (1984) 301-314.

137 R.G. Bowman and R.L. Burwell, J. Catal., 63 (1980) 463-475.
138 Yu.I. Yermakov and B.N. Kuznetsov, Kinet. Katal., 13 (1972) 1355-1356.
139 Yu.I. Yermakov and B.N. Kuznetsov, Dokl. Akad. Nauk SSSR, 207 (1972) 644-646.
140 Yu.I. Yermakov, J. Mol. Catal., 21 (1983) 35-55.
141 Yu.I. Yermakov and B.N. Kuznetsov, Kinet. Katal., 18 (1977) 1167-1178.
142 M.S. Borisova, B.N. Kuznetsov, V.A. Dzisko, V.I. Kulikov and S.P. Noskova, Kinet. Katal., 16 (1975) 1028-1035.
143 Yu.I. Yermakov and B.N. Kuznetsov, React. Kinet. Catal. Lett., 1 (1974) 87-92.
144 Yu.I. Yermakov and B.N. Kuznetsov, J. Mol. Catal., 9 (1980) 13-40.
145 G.A. Domrachev, V.A. Varyukhin and B.A. Nesterov, React. Kinet. Catal. Lett., 22 (1983) 281-285.
146 G.A. Razuvayev, T.A. Sladkova, G.A. Domrachev, K.G. Shal'nova, B.G. Gribov and V.P. Maryin, Dokl. Akad. Nauk SSSR, 203 (1972) 848-851.
147 W. Hafner and E.O. Fisher, U.S. Patent, 2,953,586 (1960); C. A., 55 (1961) 4529.
148 G. Carturan, G. Cocco, L. Schiffini and G. Strukul, J. Catal., 65 (1980) 359-368.
149 G. Carturan, G. Facchin, G. Cocco, S. Enzo and G. Navazio, J. Catal., 76 (1982) 405-417.
150 G. Cocco, S. Enzo, L. Schiffini and G. Carturan, J. Mol. Catal., 11 (1981) 161-166.
151 S.J. DeCanio, H.C. Foley, C. Dybowski and B.C. Gates, J. Chem. Soc., Chem. Commun., 24 (1982) 1372-1373.
152 H.C. Foley, S.J. DeCanio, K.D. Tau, K.J. Chao, J.H. Onuferko, C. Dybowski and B.C. Gates, J. Am. Chem. Soc., 105 (1983) 3074-3082.
153 Y. Iwasawa and H. Sato, Chem. Lett., (1985) 507-510.
154 M.D. Ward and J. Schwartz, J. Mol. Catal., 11 (1981) 397-407.
155 Yu.I. Yermakov, B.N. Kuznetsov, Yu.A. Ryndin and A.M. Lazutkin, Kinet. Katal., 14 (1973) 1594-1595.
156 B.N. Kuznetsov, Yu.I. Yermakov, V.L. Kuznetsov, Yu.A. Ryndin, L.G. Karakchiev, V.G. Shinkarenko, E.K. Mamaeva and L.Ya. Startseva, Kinet. Katal., 16 (1975) 1356-1357.
157 Yu.A. Ryndin, Thesis, Institute of Catalysis, Novosibirsk, 1977.
158 M. Ichikawa, J. Chem. Soc., Chem. Commun., (1976) 26-27.
159 Yu.I. Yermakov, B.N. Kuznetsov, Yu.A. Ryndin and V.K. Duplyakin, Kinet. Katal., 15 (1974) 1093.
160 Yu.I. Yermakov, B.N. Kuznetsov and Yu.A. Ryndin, React. Kinet. Catal. Lett., 2 (1975) 151-161.
161 Yu.I. Yermakov, M.S. Ioffe, Yu.A. Ryndin and B.N. Kuznetsov, Kinet. Katal., 16 (1975) 807-808.
162 M.S. Ioffe, Yu.I. Shylga, Yu.A. Ryndin, B.N. Kuznetsov, A.N. Startsev, Yu.G. Borodko and Yu.I. Yermakov, React. Kinet. Catal. Lett., 4 (1976) 229-234.
163 M.S. Ioffe, B.N. Kuznetsov, Yu.A. Ryndin and Yu.I. Yermakov, in G.C. Bond, P.B. Wells and F.C. Tompkins (Editors), Proc. 6th Int. Congr. Catal., The Chemical Society, London, 1977, pp. 131-138.
164 S.J. Tauster, S.C. Fung and R.L. Garten, J. Am. Chem. Soc., 100 (1978) 170-175.
165 R.T.K. Baker, E.B. Prestridge and R.L. Garten, J. Catal., 59 (1979) 293-302.

166 G.C. Bond and R. Burch, in G.C. Bond and G. Webb (Editors), Catalysis, vol. 6, The Royal Society of Chemistry, Burlington House, London W1V OBN, 1983, pp. 27-60.
167 Yu.I. Yermakov, Yu.A. Ryndin, O.S. Alekseev and M.N. Vasilieva, J. Chem. Soc., Chem. Commun., (1984) 1480-1481.
168 Yu.I. Yermakov, Yu.A. Ryndin, O.S. Alekseev M.N. Vasilieva, G.K. Myakishev and V.V. Volkov, Kinet. Katal., 25 (1984) 1017-1018.
169 D.A. Hucul and A. Brenner, J. Phys. Chem., 85 (1981) 496-498.
170 A.S. Lisitsyn, A.V. Golovin, V.L. Kuznetsov and Yu.I. Yermakov, React. Kinet. Catal. Lett., 19 (1982) 187-191.
171 F. Hugues, P. Bussiere, J.M. Basset, D. Commereuc, Y. Chauvin, L. Bonneviot and D. Olivier, in T. Seiyama and K. Tanabe (Editors), New Horizons in Catalysis, Proc. 7th Int. Congr. Catal., Kodansha LTD/Elsevier, Tokyo, Amsterdam, Oxford, New York, 1981, pp. 418-431.
172 F. Hugues, J.A. Dalmon, P. Bussiere, A.K. Smith, J.M. Basset and D. Olivier, J. Phys. Chem., 86 (1982) 5136-5144.
173 K.L. Watters, R.F. Howe, T.P. Chojnacki, C.-M. Fu, R.L. Schneider and N.-B. Wong, J. Catal., 66 (1980) 424-440.
174 M. Ichikawa, Chemtech, (1982) 674-680.
175 J.L. Bilhou, V. Bilhou-Bougnol, W.F. Graydon, J.M. Basset and A.K. Smith, J. Mol. Catal., 8 (1980) 411-429.
176 J.L. Bilhou, V. Bilhou-Bougnol, W.F. Graydon, J.M. Basset, A.K. Smith, G.M. Zanderighi and R. Ugo, J. Organomet.,Chem. 153 (1978) 73-84.
177 A. Theolier, A.K. Smith, M. Leconte, J.M. Basset, G.M. Zanderighi, R. Psaro and R. Ugo, J. Organomet. Chem., 191 (1980) 415-424.
178 P. Bosch, D. Acosta, J. Zenith, B.M. Nicolson and B.C. Gates, J. Mol. Catal., 31 (1985) 73-80.
179 J. Evans and G.S. McNylty, J. Chem. Soc. Dalton Trans., (1984) 587.
180 A. Theolier, A. Choplin, L. D'Ornelas, J.M. Basset, G. Zanderighi, R. Ugo, R. Psaro and C. Sourisseau, Polyhedron, 2 (1983) 119-121.
181 A. Zecchina, E. Guglielminotti, A. Bossi and M. Camia, J. Catal., 74 (1982) 225-239.
182 E. Guglielminotti, A. Zecchina, A. Bossi and M. Camia, J. Catal., 74 (1982) 240-251.
183 E. Guglielminotti, A. Zecchina, A. Bossi and M. Camia, J. Catal., 74, (1982) 252-265.
184 H. Knozinger and Y. Zhao, J. Catal., 71 (1981) 337-347.
185 A. Choplin, M. Leconte, J.M. Basset, S.G. Shore and W.-L. Hsu, J. Mol. Catal., 21 (1983) 389-391.
186 K. Tanaka, K.L. Watters and R.F. Howe, J. Catal., 75 (1982) 23-28.
187 K. Tanaka, K.L. Watters, R.F. Howe and S.L.T. Anderson, J. Catal., 79 (1983) 251-258.
188 R.A. Sanchez-Delgado, I. Duran, J. Monfort and E. Rodriguez, J. Mol. Catal., 11 (1981) 193-203.
189 R.L. Banks and G.C. Bailey, Ind. Eng. Chem. Prod. Res. Dev. 3 (1964) 170-173.
190 T.J. Thomas, D.A. Hucul and A. Brenner, Am.Chem. Soc. Symp. Ser., 192 (1982) 267-279.
191 P.H. Otero-Schipper, W.A. Wachter, J.B. Butt, R.L. Burwell and J.B. Cohen, J. Catal., 50 (1977) 494-507.
192 M. Wrighton and M.A. Schroeder, J. Am. Chem. Soc., 95 (1973) 5764-5765.

193 M.S. Wrighton, D.S. Ginley, M.A. Schroeder and D.L. Morse, Pure Appl. Chem., 41 (1975) 671-697.
194 D.I. Kochubei, M.A. Kozlov, K.I. Zamaraev, A.N. Startsev and Yu.I. Yermakov, Khim. Fiz., 3 (1984) 1148-1155.
195 I.A. Ovsyannikova, V.L. Kraizman, A.N. Startsev and Yu.I. Yermakov, Kinet. Katal., 25 (1984) 446-451.
196 W.F. Graydon and M.D. Langan, J. Catal., 69 (1981) 180-192.
197 J.W.E. Coenen, R.Z.C. Van Meerten and H.Th. Rijnten, in J.W. Hightower (Editor), Proc. 5th Int. Congr. Catal., Florida, 1972, North-Holland/Elsevier, Amsterdam, New York, 1973, pp. 671-680.
198 P. Moggi, G. Albanesi, G. Predieri and E. Sappa, J. Organomet. Chem., 252 (1983) 89-92.
199 Z. Schay, L. Guczi, Acta Chim. Acad. Sci. Hung., 111 (1982) 607-615.
200 I. Boszörményi, S. Dobos, L. Guszi, L. Markó, K. Lázár, W.M. Reiff, Z. Schay, L. Takács and A. Vizi-Orosz, in Proc. 8th Int. Congr. Catal., Berlin (West), 2-6 July 1984, Verlag Chemie/Dechema,Weinheim, Deerfield Beach, Florida, Basel, pp.V183-194.
201 M. Kaminsky, Ki.J. Yoon, G.L. Geoffroy and M.A. Vannice, J. Catal., 91 (1985) 338-351.
202 J.R. Budge, B.F. Lucke, B.C. Gates and J. Toran, J. Catal., 91 (1985) 272-282.
203 L. Bruce, G. Hope and T.W. Turney, React. Kinet. Catal. Lett., 20 (1982) 175-180.
204 V.L. Kuznetsov, A.F. Danilyuk, I.E. Kolosova, Yu.I. Yermakov React. Kinet. Catal. Lett., 21 (1982) 249-254.
205 A.S. Lisitsyn, V.L. Kuznetsov and Yu.I. Yermakov, Kinet. Katal., 24 (1983) 764.
206 W. Keim, H.-J. Leuchs und B. Engler, Untersuchung zur Darstellung neuer Katalysatoren für Reforming und Hydrocracking, Westdeutscher Verlag, 1979, 36 pp.

Chapter 14

SUPPORTED ASYMMETRIC HYDROGENATION CATALYSTS

J. HETFLEJŠ
Institute of Chemical Process Fundamentals, Czechoslovak Academy of Sciences, Rozvojová 135, 165 02 Prague 6 - Suchdol (Czechoslovakia)

14.1 INTRODUCTION

Asymmetric catalysis is an efficient method of synthesis of optically active compounds from prochiral reactants. In the last 15 years, considerable research has been carried out on the development of this concept and impressive achievements have been reported especially in the area of asymmetric hydrogenations. Transition metal complexes containing optically active ligands have proved to be particularly useful asymmetric catalysts. For example, by using soluble rhodium complexes of chiral chelating diphosphines, optical yields higher than 90 % have frequently been attained in the hydrogenation of dehydro α-amino acids and their derivatives. The commercial synthesis of L-DOPA (3,4-dihydroxyphenylalanine) (refs. 1,2) by such a route represents an important practical application. In parallel to the design of highly stereoselective catalysts, remarkable progress has also been made in the understanding of the mechanism of homogeneously catalysed asymmetric hydrogenation (refs. 3-11). This area has recently been comprehensively reviewed, (refs. 12-15), and numerous specialized accounts have dealt with various aspects of the catalysis by soluble chiral transition metal complexes. Only those not included in above reviews will be cited here (refs. 16-25).

The use of soluble homogeneous catalysts has the disadvantage that their solubility in the reaction medium makes difficult their separation from the product, their recovery and repeated use. In general, this problem can be solved by performing the reaction in two-phase solvent systems, the catalyst being soluble in only one of the immiscible solvents, or by immobilizing the catalyst on an organic polymer or an inorganic support. So far, the first approach has not been widely applied in homogeneously catalysed hydrogenations (refs. 26-28) and no attempt has been made to adapt it to asymmetric hydrogenation. The first example of a water-soluble chiral diphosphine that would enable one to test this process has been reported only recently (ref. 29).

The immobilization of transition metal complexes has become an area of intense research, with the prospect of combining the relatively high activity and selectivity of homogeneous transition metal catalysts with the merits of heterogeneous ones. Polymer-supported catalysts, including illustrative examples of asymmetric hydrogenation catalysts (ref. 30), and transition metal catalysts attached to inorganic supports (ref. 31) have been comprehensively reviewed.

In addition to the advantages mentioned above in the case of asymmetric hydrogenation, the development of suitable supported analogues of homogeneous chiral catalysts is of interest also from other aspects. A perusal of the literature on homogeneous asymmetric hydrogenation shows that, with several exceptions, the highly stereoselective catalysts are not at the same time also highly active. To ensure a sufficient hydrogenation rate, a relatively low mol. ratio of the substrate to the metal is needed. Unless the catalyst properties are maintained over a long period, its application will be limited on economical grounds. Furthermore, the most successful homogeneous asymmetric hydrogenation catalysts for the processes of potential industrial interest, such as the hydrogenation of dehydroamino acids or dehydro dipeptides (refs. 32-34), are the complexes of the costly metal, rhodium, which contain expensive chiral ligands obtained by multistep syntheses. So far, there is no indication that their dominance could be endangered either by homogeneous catalysts based on less expensive metals or by the use of heterogeneous asymmetric hydrogenation catalysts. The latter catalysts are of limited value due to the fact that only a few chiral compounds act efficiently as modifiers of metallic surfaces. Thus, despite the recent progress in this area, as evidenced by the use of stereoselective nickel and platinum catalysts for asymmetric hydrogenation of β-ketoesters and β-diketones (refs. 35-39), the variability of transition metal complexes and their applicability to other types of substrates has hardly been exploited.

For the above reasons, the immobilization of homogeneous asymmetric catalysts could assist in the further development of asymmetric catalysts. In this chapter, the homogeneous asymmetric hydrogenation catalysts attached to organic polymers (Section 14.2) and inorganic supports (Section 14.3) will be surveyed.

14.2 POLYMER-SUPPORTED ASYMMETRIC HYDROGENATION CATALYSTS

Early attempts to immobilize chiral homogeneous hydrogenation catalysts have centered on the covalent attachment of the chiral ligand to the polymer with subsequent formation of the complex on the polymer surface by routes analogous to the synthesis of homogeneous analogues (Scheme 14.1), using polystyrenes as supports:

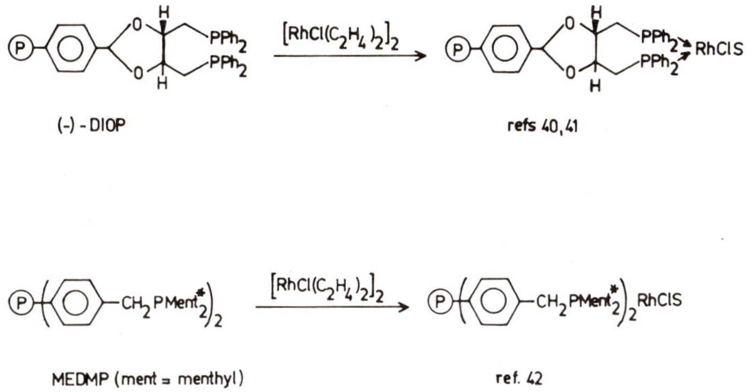

Scheme 14.1

The rhodium complex attached to lightly cross-linked styrene-divinylbenzene resin (2 % cross-linking) via covalently bonded 2,3-O-isopropylidene-2,3-dihydroxy-1,4-bis(diphenylphosphino)butane (DIOP) (Rh:DIOP mol ratio = 1:1) was found to be an efficient catalyst for the reactions of prochiral substrates soluble in non-polar solvents, such as the asymmetric hydrogenation (ref. 40) and hydroformylation (ref. 43) of alkenes and hydrosilylation of prochiral ketones (ref. 40). However, compared to the homogeneous Rh-DIOP catalyst, lower optical yields and lower hydrogenation rates were obtained with the supported catalyst. The attempted hydrogenation of α-acetamidoacrylic acid was unsuccessful because of shrinking of the polymer in the solvent system used (benzene/ethanol 1:1), blocking access of substrate molecules to the catalytic sites (ref. 40).

The effect of the polystyrene support on the catalytic behaviour of supported Rh-(-)-DIOP complexes has been investigated in more detail in the hydrogenation of several prochiral unsaturated acids, using a series of soluble polystyrenes (mean mol. weight 800-2800), insoluble non-cross-linked polystyrene (mean mol. weight 5000) and cross-linked styrene-2 % divinylbenzene copolymer (ref. 41). With all the catalysts, the hydrogenation of itaconic acid (1) in benzene-methanol(2:1) proceeded under relatively mild conditions (50°C, 0.1 MPa hydrogen pressure) to give R-(-)-α-methylsuccinic acid (2) in optical yields (16-38 %) depending on the type of polymer support (eqn. 14.1):

$$HOOCCH_2-\underset{CHCOOH}{\overset{\|}{C}}-COOH \xrightarrow{H_2} HOOCCH_2-\overset{*}{\underset{CH_2COOH}{C}H}-COOH \qquad (14.1)$$

$$(1) \qquad\qquad\qquad (2)$$

Soluble polystyrenes and the cross-linked styrene-divinylbenzene copolymer were found to be inferior supports compared to the insoluble non-cross-linked polystyrene, giving catalysts of low stereoselectivity and activity. Similar results were obtained also for α-methylcinnamic (3) and citraconic (4) acid. With all the conjugated acids examined, the best supported Rh-DIOP catalyst showed stereoselectivity comparable to that of the homogeneous Rh-DIOP complex, the activity of both catalysts being, however, very low. At least in the case of the homogeneous rhodium complex, the extraordinarily low efficiency is believed to result from the use of the chiral ligand in large excess with respect to the catalyst precursor (DIOP:Rh mol ratio = 4:1).

$$C_6H_5CH=\underset{CH_3}{\overset{|}{C}}-COOH \qquad CH_3-\underset{CHCOOH}{\overset{\|}{C}}-COOH$$

$$(3) \qquad\qquad\qquad (4)$$

In a kinetic study of the hydrogenation of Z-α-acetamidocinnamic acid (5) catalysed by homogeneous Rh-DIOP complexes prepared similarly to those used in ref. 41, it was found that the use of Rh:DIOP mol. ratios greater than 1:1 leads to a gradual decrease in the catalytic activity of the systems, essentially inefficient catalysts being obtained at mol ratios higher than 1:2 (ref. 44).

In contrast to microporous polystyrenes, the applicability of highly cross-linked macroporous styrene-divinylbenzene resins that possess high surface areas and are not so sensitive to solvent effects has not been thoroughly investigated. The Rh-DIOP (1:1) complex when attached to the commercial macroporous resin Amberlite has been reported to catalyse the

hydrogenation of acetophenone in benzene solution to give phenyl methyl
carbinol of the same configuration and in the same optical yield as that
obtained with the corresponding homogeneous systems (refs. 45,46). In the light
of the generally very low stereoselectivity and activity of neutral Rh-DIOP
complexes in the hydrogenation of prochiral ketones, the potential of this
catalyst is at present difficult to evaluate.

A series of cross-linked styrene-divinylbenzene resins (2.0-10 %
divinylbenzene) have been used to prepare supported chiral
dimenthylphosphinorhodium complexes (Rh-MEDMP, Scheme 14.1, ref. 42) and their
catalytic behaviour tested in the hydrogenation of Z-α-acetamidocinnamic acid
(5). Under rigorous reaction conditions (1 MPa hydrogen pressure, Rh:substrate
mol ratio = 1:100) the hydrogenation afforded (R)-N-acetylphenylalanine (6)
in 50-60 % optical yields (eqn. 14.2) comparable to those obtained with the
homogeneous catalyst (ref. 42) at the same Rh:P mol ratio (1:2.5).

$$C_6H_5CH=C-COOH \atop NHCOCH_3 \quad \xrightarrow{H_2} \quad C_6H_5CH_2-\overset{*}{C}H-COOH \atop NHCOCH_3 \qquad (14.2)$$

$$\underline{(5)} \qquad\qquad\qquad \underline{(6)}$$

An increase in the polymer cross-linking affected the stereoselectivity and
activity of the catalysts in a complex way, the best results being obtained
with the copolymer cross-linked with 5 % of divinylbenzene. Perhaps the most
interesting result in relation to the effect of the support is the finding
that the Rh-MEDMP complex anchored to the lightly cross-linked copolymer (2 %
divinylbenzene) catalysed the hydrogenation of acid (5) in both benzene and
ethanol apparently without any dependence on the polarity of the solvent.
However, similarly to non-chiral Wilkinson-type rhodium complexes containing
monodentate tertiary phosphines, the catalysts underwent dissociation during
hydrogenation, resulting in the gradual loss of their activity and
stereoselectivity on repeated use.

The disadvantages of microporous polystyrenes as catalyst supports for
reactions demanding polar solvents such as the process of most interest, namely
hydrogenation of dehydro-α-amino acids, could in principle be overcome either by
using a rigid support with an insensitive surface or by utilization of polymers
compatible with both polar and non-polar solvents. With the exception of
inorganic supports which will be discussed later (Section 14.3), the only
reported example of an asymmetric hydrogenation catalyst on an inert organic
support is the Rh-DIOP complex attached to graphitized carbon (ref. 47).

Discouraging results obtained in the hydrosilylation of ketones and hydrogenation of Z-α-acetamidocinnamic acid (5) makes this catalyst of little interest.

In contrast, the development of resins that swell in polar solvents has proved to be a promising route to efficient asymmetric hydrogenation catalysts. The supported analogues of (-)-DIOP (ref. 48) and
(2S,4S)-4-diphenylphosphino-2-diphenylphosphinomethylpyrrolidine (2S, 4S-BPPM, refs. 49-51) have been obtained by incorporation of monomers (7)-(9) into 2-hydroxyethyl methacrylates via copolymerization. The rhodium complexes derived from these polymers (Scheme 14.2) could readily be used in the hydrogenation of dehydroamino acids in ethanol or other polar solvents.

The hydrogenation in ethanol of the acid (5) and of α-acetamidoacrylic acid, using the catalyst (10), gave under mild conditions (0.1 MPa hydrogen, 25-30°C) the optically active amino acids in the same optical yields (60 and 80 % respectively) and with the same absolute configurations as were achieved with the homogeneous Rh-(-)DIOP complex (ref. 48). The catalyst could be reused without change in its properties providing that oxygen is excluded during recovery of the catalyst and subsequent handling. Similarly, the supported Rh-BPPM catalysts (11) and (12) retained the stereoselectivities of the soluble analogues. Nearly 90 % optical yields have been attained in the hydrogenation of N-α-acetamidocinnamic acid derivatives (refs. 49-52) and itaconic acid (1) (ref. 50), and the catalyst (11) (2S, 4S) has been used with success also in the asymmetric hydrogenation of ketopantolactone (13) to (R)-pantolactone (14) (ref. 53, optical yield 75.5 %), an important intermediate in the synthesis of pantheine and Coenzyme A (eqn. 14.3):

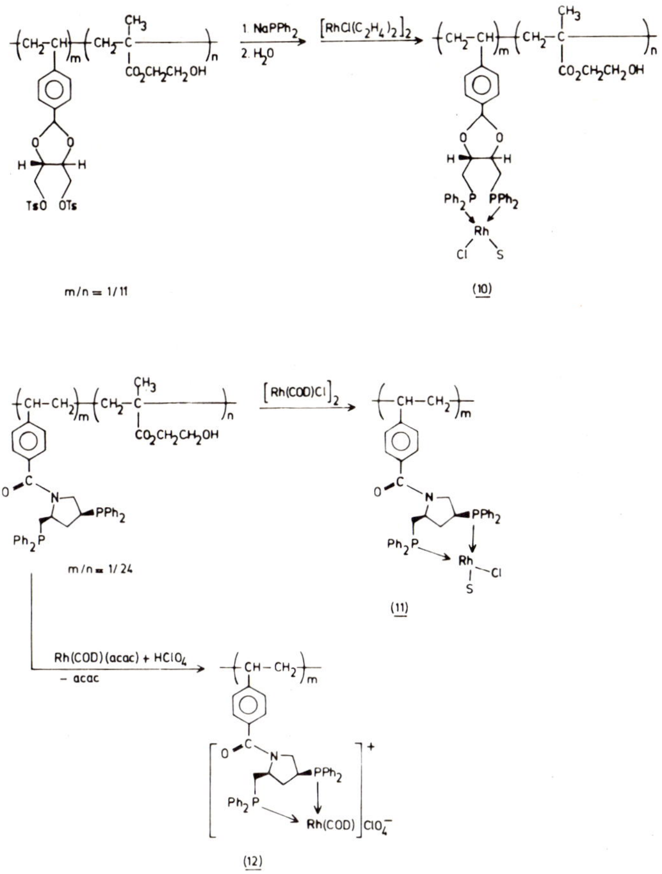

Scheme 14.2

The modification of the polymers by substitution of 2-dimethylacetamido for 2-hydroxyethyl groups affords catalysts of low stereoselectivity, presumably due to the competition of the amido group with the chiral phosphine for the catalytic site (ref. 51). The interaction between the catalytic site and ancillary group might be utilized to design supports in which the ancillary

$$\underset{(13)}{\text{[structure]}} \xrightarrow[\text{5 MPa } H_2, 50°C, 45h]{\text{cat.(11)}} \underset{(14)}{\text{[structure]}} \qquad (14.3)$$

group exerts a favourable effect, e.g., enhances the stereoselectivity of the catalyst. This concept has been tested by using cross-linked co-polymers of types (10) and (11) in which chiral 2-hydroxyethyl (15) (ref. 54) or 1-methyl-2-hydroxypropyl groups (16) were incorporated (ref. 55). When tetrahydrofuran was used as the solvent, the catalysts having opposite

(15), ref. 54 — (2R-) or (2S-)

(16), ref. 55 — (1R, 2R-) or (1S, 2S-)

chirality at the carbon of the alcohol group but otherwise identical showed different stereoselectivities in the hydrogenation of N-α-acetamidoacrylic acid. The optical yields obtained with the catalyst on the racemic support fell between those observed on the optically active supports (refs. 54,55). However in ethanol, the optical yields were independent of the configuration of the alcohol site, indicating predominant solvation of the catalytic site by the solvent (for a detailed discussion of this topic see ref. 56). This makes improbable the utilization of the effect of a second optical centre in hydrogenation of dehydroamino acids catalysed by supported rhodium complexes.

It has been pointed out that the synthesis of chiral polymers by copolymerization of suitable monomers has several advantages (ref. 56): it assures a high degree of ligand purity since few reactions are carried out on the support, and the comonomers can be chosen to give the optimum ligand density, cross-linking density and swelling characteristics for the supported catalyst. Notwithstanding, the application of this route is limited by the fact that only a few of the known chiral diphosphines could be converted into derivatives capable of polymerization (compare, e.g., the list of chiral phosphines in ref. 15).

Several attempts have been made to avoid multistep synthesis of chiral phosphine ligands by using rhodium complexes of readily available optically active amino acids. The copolymer containing (S)-phenylalanine (17) which swells in polar solvents has been obtained by copolymerization of styrene with the maleinimide of the acid (ref. 57). The rhodium complex derived from this polymer was found to catalyze hydrogenation of the acid (5) at a slow rate and with low optical yields (3-9 %). It is worth mentioning that low optical yields have been reported also for soluble rhodium-tertiary phosphine carboxylate complexes containing L-(+)-mandelate (ref. 58) and (S)-phenylalaninate (ref. 59) as chiral ligands.

$$\underset{(17)}{\left\{ \underset{\underset{\underset{C_6H_5CH_2\overset{*}{C}HCOOH}{|}}{N}}{\overset{O=C}{\underset{}{\diagdown}}\overset{}{\underset{}{\diagup}}\overset{C=O}{}}\text{CH-CH}\right\}_{50}\left\{\text{CH}_2\text{-}\underset{C_6H_5}{\overset{|}{\text{CH}}}\right\}_{50}} \quad \underset{(18)}{\left\{\text{CH-CH}_2\atop \underset{CONHCH_2CH=N-\overset{*}{C}H-COOH}{|}\atop \underset{CH_3}{|}\right\}_n}$$

In contrast, rhodium-phosphine carboxylate complexes derived from Schiff bases of optically active α-amino acids are highly enantioselective catalysts for hydrogenation of dehydroamino acid esters (ref. 60). Thus, the hydrogenation of methyl Z-α-acetamidocinnamate ctalyzed by the rhodium complex prepared in situ from $RhCl(PPh_3)_3$ and benzylidene-(S)-alanine in benzene-methanol (1:2) under mild conditions (0.1 MPa hydrogen, 50°C, Rh:substrate mol ratio 1:100) afforded S-phenylalanine in 88-93 % optical yields. Even higher optical yields (95-98 %) have been obtained in methanol when the rhodium complex is supported on the commercial polyaldehyde resin

Enzacryl (18). Although no further data about the properties of this catalyst are available, it is unlikely to be applied to the hydrogenation of dehydroamino acids because of the displacement of the chiral ligand from the coordination sphere of the metal by the substrate molecule (ref. 59).

Another way to avoid the synthesis of monomeric chiral ligands and their subsequent immobilization is to use an inherently chiral support. In this respect, polysaccharides are the supports of choice since their hydroxy groups can readily be phosphinated. The reaction of cellulose with PPh_2Cl yields 6-O-diphenylphosphinocellulose (19) to which a rhodium complex has been attached by an exchange reaction with $RhCl(PPh_3)_3$ (ref. 61). The rhodium

catalyst so obtained was found to catalyze the hydrogenation of methyl α-phthalimidoacrylate (21) (eqn. 14.4) but with very low rates and moderate optical yields (11-28 %, refs. 19,61).

(14.4)

An extremely low activity is displayed also by the rhodium complex derived from 2,3-O-bis(diphenylphosphino)-6-O-triphenylmethylcellulose (20). The interesting feature of this catalyst is its high enantioselectivity in the hydrogenation of some prochiral olefins (ref. 62). A high molar excess of the phosphinite groups with respect to rhodium (P/Rh = 11.5) is needed to ensure the optimum properties of the catalyst. However, even with such an excess, loss of the activity and enantioselectivity of the catalyst on repeated use could not be prevented (ref. 63).

Ion-exchange resins offer the possibility of anchoring the ligands by means of ionic bonding. Several examples of rhodium and ruthenium catalysts supported through such linkages with one of the functional groups of bifunctional phosphines on macroporous cation (eqns. 14.5 and 14.6) or anion-exchange resins (eqn. 14.7) have been reported (refs. 26,27,60-65). The leaching of the metal from the resin support is negligible, although the metal could readily be recovered from the ion exchanger by elution with acids (refs. 27,28).

$$\text{(P)}-SO_3H + Ph_2PCH_2CH_2NMe_3^+ X^- \xrightarrow[-HX]{\text{ref. 28}} \text{(P)}-SO_3^- Me_3N^+CH_2CH_2PPh_2 \qquad (14.5)$$

$$\text{(P)}-SO_3H + (p-Me_2NC_6H_4)_3P \xrightarrow{\text{ref. 64}} \text{(P)}-SO_3^- HP^+(C_6H_4 NMe_2-p)_3 \qquad (14.6)$$

$$\text{(P)}-NH_2 + HO_3S-\bigcirc-PPh_2 \xrightarrow{\text{ref. 65}} \text{(P)}-NH_3^+ \ ^-O_3S-\bigcirc-PPh_2 \qquad (14.7)$$

None of these routes has been utilized so far to immobilize asymmetric hydrogenation catalysts. Chiral bifunctional phosphines that display good asymmetric efficiency in the rhodium-catalysed enantioselective hydrogenation of prochiral unsaturated acids are now available (refs. 66-68) for testing the applicability of this method.

Recently, the high efficiency of homogeneous chiral cationic rhodium bisphosphine complexes of the type $Rh(diene)L^+ X^-$ (L = bidentate ligand; diene = 1,5-cyclooctadiene, norbornadiene, X = ClO_4^-, BF_4^-) (ref. 15) has led to a study of their immobilization through the exchange of their anion. The direct immobilization of the transition metal cation would be advantageous in that it avoids the modification and anchoring of chiral ligands and thus increases the variety of ligands that can be used to synthesize supported complexes. With the prospect of utilizing sulphonated resins as supports, a series of homogeneous arenesulphonate rhodium-bisphosphine complexes have been prepared (ref. 69, eqn. 14.8, chemical yields 87-93 %) and their catalytic properties investigated in the hydrogenation of alkenes and dehydro-α-acylamino acids (ref. 70). They were found to display the same catalytic activity and in

$$Rh(COD)(acac) + L_2 + p\text{-}CH_3\text{-}\langle O \rangle\text{-}SO_3H \xrightarrow[-acacH]{}$$

$$Rh(COD)L_2(O_3S\text{-}\langle O \rangle\text{-}CH_3\text{-}p) \qquad (14.8)$$

$L_2 = $ e.g. $Ph_2P(CH_2)_n PPh_2$ (n = 2-4), (-)-DIOP

the case of chiral rhodium complexes also the same enantioselectivity as similar cationic complexes of the above mentioned type. Thus, for example, the hydrogenation of Z-α-acetamidocinnamic acid (5) in benzene-ethanol (1:1) catalyzed by the arenesulphonate Rh-(-)-DIOP complex (eqn. 14.8) (0.1 MPa hydrogen, 20°C) afforded (R)-N-acetylphenylalanine in 83 % optical yield, while 82 % optical yield was obtained with $|Rh(COD)(-)\text{-}DIOP|^+ ClO_4^-$ under identical reaction conditions (ref. 71). The "innocent" role of the p-tosylate anion has been demonstrated on the basis of the kinetics of the reactions: in both cases, the hydrogenation was first order with respect to the catalyst and hydrogen and zero order with respect to the acid, and very similar activation parameters and dependences of optical yield vs. temperature were obtained with both catalysts (ref. 71).

The immobilization of the complexes on sulphonated microporous as well as macroporous styrene-divinylbenzene resins (ref. 72) and 2,3-epoxypropylmethacrylate-ethylene dimethacrylate copolymers (refs. 73,74) modified by sulphopropylation (ref. 75), in an analogous way to that illustrated in eqn. 14.8, affords efficient hydrogenation catalysts (refs. 76,77). The effect of the type of styrene-divinylbenzene resin (microporous vs. macroporous) and of its cross-linking on the properties of these catalysts is illustrated in Table 14.1 (ref. 78) for the hydrogenation of two substrates of different polarities. It is seen that nearly the same hydrogenation rate as with the homogeneous complex is achieved with the catalyst attached to the lightly cross-linked microporous resin (SDVB-2-Rh) in the hydrogenation of the alkene, while under the same conditions the dehydroamino acid (5) is hydrogenated only very slowly. This shows that sulphonation did not markedly affect the swelling properties of the polystyrene resin. Thus, in the solvent system used (benzene-ethanol 1:1), the non-polar resin will be swelled preferentially by benzene.

TABLE 14.1

Relative rates of hydrogenation of 1-octene and Z-α-acetamidocinnamic acid (5) catalysed by rhodium complexes supported on sulphonated styrene-divinylbenzene (SDVB) resins

Catalyst[a]	Surface area (m^2/g)	Ion exchange capacity (mmol/g)	1-Octene[b]	Z-α-acetamidocinnamic acid (opt. yield, %)[c]	
Soluble	-	-	118	1720	(72)
SDVB-2-Rh	0.1	4.5	103	4	
SDVB-8-Rh	0.1	4.8	13	2	
SDVB-15-Rh	0.1	4.1	10	1	
SDVB-15M-RH	35	3.8	17	375	(70)
SDVB-40M-Rh	110	3.0	1	328	(74)

[a] The number denotes cross-linking, M = macroreticular copolymer. Reaction conditions: benzene-ethanol 1:1, 40°C, 0.1 MPa hydrogen.
[b] Rh-1,2-bis(triphenylphosphino)ethane 1:1 complex.
[c] Rh-(-)-DIOP 1:1 complex.

As a result of the difference in the microenvironment of the catalytic site compared to that of the bulk solvent, a polarity gradient is set up, favouring a higher concentration of the alkene within the resin. An opposite trend can be expected for the polar acid (5) which is soluble in ethanol. In such a case, the best catalysts are those attached to macroreticular resins (compare SDVB-15 M-Rh and SDVB-40 M-Rh). Because of diffusion restrictions, their activity is not as high as that observed with the homogeneous catalyst. However, they retain the stereoselectivity of the soluble complex (Table 14.1) and no substantial loss of their activity and stereoselectivity is observed on repeated use (ref. 76). This method of immobilization thus seems to be of interest and deserves further attention.

14.3 ASYMMETRIC HYDROGENATION CATALYSTS ANCHORED TO INORGANIC SUPPORTS

Compared to the polymer-supported catalysts discussed in the preceding section, few studies have been devoted to the immobilization of chiral transition metal complexes on inorganic supports. This seems surprising both from the standpoint of the advantages of the dispersed inorganic materials (compatibility with solvents of different polarities due to the rigidity of the matrices, thermal stability and large scale availability of supports of the required texture) and with regard to the general progress made in heterogenization (cf. ref. 31).

Several attempts have been made to attach a chiral phosphine via covalent bonding to silica, utilizing the surface hydroxyl groups of the support as binding sites (eqn. 14.9).

$$\text{SiO}_2\text{-OH} + (\text{C}_2\text{H}_5)_3\text{Si}(\text{CH}_2)_n\text{Y} \xrightarrow{-\text{C}_2\text{H}_5\text{OH}} \text{SiO}_2\text{-O-Si}(\text{CH}_2)_n\text{Y} \begin{smallmatrix} \text{O-} \\ | \\ | \\ \text{O-} \end{smallmatrix} \quad (14.9)$$

Y = PPhMent*, n=3 (refs 78,79); PMent*$_2$, n= 1,3,5 (ref. 80)
-(-)-DIOP, n=4 (ref. 81)

The rhodium complex prepared by the reaction of (-)-Si-DIOP (Scheme 14.3) with /RhCl(cyclooctene)$_2$/$_2$ has been found (ref. 82) to catalyze the hydrogenation of Z-α-acetamidocinnamic acid (5) in benzene-ethanol (1:1) under mild conditions (0.11 MPa hydrogen, 25°C) to give (R)-acetylphenylalanine in 63 % optical yield. Its activity is comparable with that of the Rh-DIOP catalyst supported on poly(2-hydroxyethyl methacrylate) resin (ref. 48, cf. catalyst (10), Scheme 14.2). Although the catalyst could be reused without appreciable loss in its activity and stereoselectivity, in both these properties the supported catalyst was inferior to the soluble Rh-Si-DIOP complex (optical yield 73 %, approximately four times higher hydrogenation rate). It is noteworthy that the immobilization suppressed two unfavourable features of the homogeneous neutral Rh-DIOP system, i.e., its deactivation during hydrogenation due to formation of chloro-bridged dimeric species (ref. 83) and the dependence of the hydrogenation activity on the Rh:DIOP mol ratio (cf. Section 14.2, below Scheme 14.1), indicating that the catalytically active species formed even at relatively high Rh:DIOP mol ratios (up to 1:4) had one DIOP ligand attached to the metal.

Generally, the probability that a metal complex will be bonded to a surface by more than one anchored ligand depends on several factors such as the distribution of the ligands on the support surface, the mobility of the ligands, the metal loading and the metal to ligand mol ratio. The role of these factors has recently been investigated in more detail (ref. 80) in the asymmetric hydrogenation of Z-α-acetamidocinnamic acid (5) and itaconic acid (1) catalysed by chiral rhodium complexes supported on macroporous silica modified with silyl-substituted dimenthylphosphines (see eqn. 14.9). It was found that the formation of catalytically active species containing two phosphine ligands coordinated to the metal responsible for the stereoselectivity increases with increasing P:Rh mol ratio, decreasing metal loading and increasing length of the alkylene chain separating the phosphino group from the surface. As a result, the best anchored catalysts display

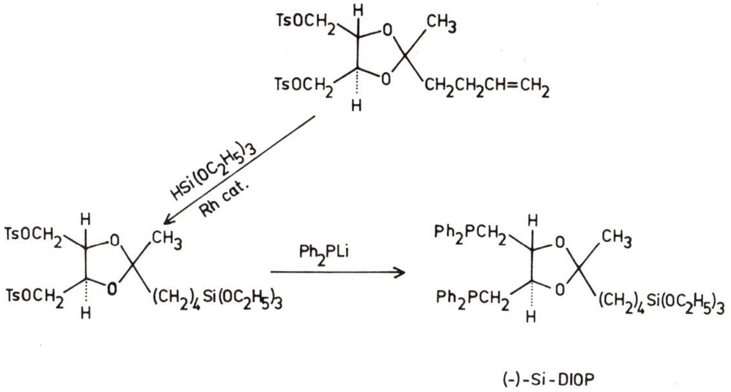

Scheme 14.3

stereoselectivities higher than those of the homogeneous analogues and the optical yields obtained in the hydrogenation of both acids (87 % for (5) and 83 % for (1)) are among the highest attained so far with immobilized catalysts. Unfortunately, as with the analogous complex attached to polystyrene resin (ref. 42, Scheme 14.1), metal leaching could not be avoided, thus limiting the recycling of these catalysts.

Immobilization through covalent bonding of a chiral ligand to an inorganic support requires the synthesis of a suitable silyl-substituted analogue of a given chiral ligand. This is usually a tedious task and together with the difficult purification represents the main restriction on the wider application of this method.

Efficient hydrogenation catalysts have been obtained by immobilization via ion exchange. Thus, for example, cationic rhodium-phosphine complexes intercalated by swelling of the layer-lattice silicate mineral hectorite have been reported to display activities twice that of the soluble catalyst in the hydrogenation of simple alkenes (ref. 84). A chiral analogue,

/Rh(COD)(PNNP)/$^+$ClO$_4^-$ (PNNP = N,N-bis(R(+)-α-methylphenyl)-N,N'-
-bis(diphenylphosphino)ethylenediamine), has also been prepared and tested in
the hydrogenation of some dehydro-α-acylamino acids (ref. 85). It was found
that the complex intercalated in hectorite shows the same stereoselectivity as
the homogeneous complex (optical yield 70 %) for hydrogenation of
α-acetamidoacrylic acid, while with Z-α-acetamidocinnamic acid,
N-acetylphenylalanine is obtained in much lower optical yield on the supported
catalyst, presumably due to the selective adsorption of the product on the
support. The use of other clay minerals such as bentonite, nontronite or
halloysite led to less efficient catalysts.

The first example of a chiral cationic rhodium complex directly immobilized
on silica has recently been reported (ref. 86). The supported rhodium complex
of N-benzyl-(3S,4S)-3,4-bis(diphenylphosphino)pyrrolidine was found to
catalyze the hydrogenation of N-acetamidocinnamic acid derivatives with the
same optical yields (95-99 %) as on the homogeneous catalyst.

This demonstrates that immobilization via ion exchange is a promising route
to supported asymmetric hydrogenation catalysts, the potential of which is
virtually unexplored.

14.4 CONCLUDING REMARKS

Compared with the achievements made in the asymmetric catalysis by soluble
chiral transition metal complexes, progress in the field of supported
asymmetric catalysts is much less impressive. Although several types of
immobilized catalysts have been reported that retain the stereoselectivity of
their homogeneous analogues, metal leaching and other deactivation processes
are an obstacle which has seldom been satisfactorily overcome. Although the
structure of the surface complexes depends on many factors, few thorough
studies of their effects have been made. Recent promising results lead the
author to believe that the present shortcomings of supported asymmetric
hydrogenation catalysts result from the lack of research effort devoted to
the solution of this problem rather than from any inherent limitations of the
method. Inevitably, much systematic study is needed to design supported
asymmetric hydrogenation catalysts of industrial interest.

REFERENCES
1   B.D. Vineyard, W.S. Knowles and M.J. Sabacky, J. Mol. Catal., 19 (1983)
    159-169.
2   R.E. Merill, Chem. Tech., (Washington, DC) (1981) 118-127.
3   J. Halpern, Inorg. Chim. Acta, 50 (1981) 11-19.
4   J. Halpern, Science, 217 (1982) 401-407.

5   J.M. Brown, P.A. Chaloner and D. Parker, in E.C. Alleya and D.W. Meek (Editors), Catalytic Aspects of Metal Phosphine Complexes, Adv. Chem. Ser., No. 196, American Chemical Society, Washington, DC, 1982, Ch. 22, p. 355.
6   B. Bosnich and N.K. Roberts, in E.C. Alleya and D.W. Meek (Editors), Catalytic Aspects of Metal Phosphine Complexes, Adv. Chem. Ser., No. 196, American Chemical Society, Washington, DC, 1982, Ch. 21, p. 337.
7   H. Brunner, B. Schoenhammer and C. Steinberger, Chem. Ber., 116 (1983) 3529-3538.
8   V.A. Pavlov and E.I. Klabunovskii, Dokl. Akad. Nauk SSSR, 269 (1983) 856-858.
9   J.M. Brown and D. Parker, J. Org. Chem., 47 (1982) 2722-2730.
10 C.R. Landis and J. Halpern, J. Organomet. Chem., 250 (1983) 485-490.
11 J.M. Brown and B.A. Murrer, J. Chem. Soc., Perkin Trans. 2 (1982) 489-497.
12 V. Čaplar, G. Commiso and V. Šunjič, Synthesis, (1981) 85-116.
13 E.I. Klabunovskii, Usp. Khim., 51 (1982) 1103-1128.
14 L. Marko and J. Bakos, in R. Ugo (Editor), Aspects of Homogeneous Catalysis, Vol. 4, D. Reidel Publishing Company, Dordrecht, 1981, p. 145.
15 H.B. Kagan, in G. Wilkinson (Editor), Comprehensive Organometallic Chemistry, Vol. 8, The Synthesis, Reactions and Structure of Organometallic Compounds, Pergamon Press, Oxford, 1982, Ch. 53, p. 463.
16 L. Achiwa, Fundam. Res. Homogeneous Catal., 3 (1979) 549-563.
17 H. Brunner, Kontakte, 3 (1981) 3-12.
18 U. Matteoli, P. Frediani, M. Bianchi, C. Botteghi and S. Gladiali, J. Mol. Catal., 12 (1981) 265-319.
19 H. Pracejus, Wiss. Z. Ernst Moritz Arndt Univ. Greifsw., Math.-Naturwiss. Reihe, 31 (1982) 5-13.
20 T. Hayashi and M. Kumada, Acc. Chem. Res., 15 (1982) 395-401.
21 W.S. Knowles, Acc. Chem. Res., 16 (1983) 106-112.
22 J.M. Brown and P.A. Chaloner, in L.H. Pignolet (Editor), Homogeneous Catalysis by Metal Phosphine Complexes, Plenum, New York, 1983, Ch. 13, p. 7.
23 G.P. Chiusoli, J. Mol. Catal., 19 (1983) 147-157.
24 I. Ojima, Pure Appl. Chem., 56 (1984) 99-110.
25 V.A. Pavlov and E.I. Klabunovskii, Izv. Akad. Nauk SSSR, Ser. Khim., (1983) 2015-2022.
26 F. Joo and Z. Toth, J. Mol. Catal., 8 (1980) 369-383.
27 R.T. Smith, R.K. Ungart and M.C. Baird, Transition Met. Chem. (Weinheim, Ger.), 7 (1982) 288-289.
28 R.T. Smith, R.K. Ungar, L.J. Sanderson and M.C. Baird, Organometallics, 2 (1983) 1138-1144.
29 Y. Amrani and D. Sinou, J. Mol. Catal., 24 (1984) 231-233.
30 C.U. Pittman, Jr., in G. Wilkinson (Editor), Comprehensive Organometallic Chemistry, Vol. 8, The Synthesis, Reactions and Structures of Organometallic Compounds, Pergamon Press, Oxford, 1982, Ch. 55, p. 553.
31 Yu.I. Yermakov, B.N. Kuznetsov and V.A. Zakharov (Editors), Catalysis by Supported Complexes, Elsevier, Amsterdam, 1981.
32 I. Ojima, N. Yoda and M. Yatabe, Tetrahedron Lett., 23 (1982) 3917-3920.
33 I. Ojima, T. Kogure, N. Yoda, T. Suzuki, M. Yatabe and T. Tamaka, J. Org. Chem., 47 (1982) 1329-1334.
34 I. Ojima, N. Yoda, M. Yatabe, T. Tanaka and T. Kogure, Tetrahedron, 40 (1984) 1255-1257.
35 A. Hoek, H.M. Woerde and W.M.H. Sachtler, in T. Seiyama and K. Tanabe (Editors), New Horizons in Catalysis, Elsevier, Amsterdam, 1981, Part A, p. 376.
36 Y. Orito, S. Imai and S. Niwa, Nippon Kagaku Kaishi, (1980) 670-677; CA 93 (1980) 113912.
37 A. Tai, K. Ito and T. Harada, Bull. Chem. Soc. Jpn., 54 (1981) 223-227.
38 T. Harada, M. Yamamoto, S. Onaka, M. Imaida, H. Ozaki, A. Tai and Y. Izumi, Bull. Chem. Soc. Jpn., 54 (1981) 2323-2329.

39 Y. Nitta, F. Sekine, I. Imanaka and S. Teranishi, J. Catal., 74 (1982) 382-392.
40 W. Dumont, J.C. Paulin, T.P. Dang and H.B. Kagan, J. Am. Chem. Soc., 95 (1973) 8295-8299.
41 K. Okhubo, K. Fujimori and K. Yohinaga, Inorg. Nucl. Chem. Lett., 15 (1979) 5-6.
42 H.W. Krause, React. Kinet. Catal. Lett., 10 (1979) 243-246.
43 S.J. Fritschel, J.J.H. Ackerman, T. Keyser and J.K. Stille, J. Org. Chem., 44 (1979) 3152-3157.
44 J. Vilím and J. Hetflejš, Chem. prum., 28/53 (1978) 135-140.
45 K. Okhubo, M. Haga, K. Yoshinaga and Y. Motozato, Inorg. Nucl. Chem. Lett., 17 (1981) 215-218.
46 K. Okhubo, M. Haga, K. Yoshinaga and Y. Motozato, Inorg. Nucl. Chem. Lett., 16 (1980) 155-158.
47 H.B. Kagan, T. Yamagishi, J.C. Motte and R. Setton, Isr. J. Chem., 17 (1979) 274-277.
48 N. Takaishi, H. Imai, C.A. Bertello and J.K. Stille, J. Am. Chem. Soc., 100 (1978) 264-268.
49 K. Achiwa, Y. Nakamoto and N. Ishizuka (Fuji Chemicals Industrial Co.), Jpn. Kokai Tokkyo Koho, 79, 158, 494; CA 93 (1980) 46192.
50 K. Achiwa, Chem. Lett., 8 (1978) 905-908.
51 G.L. Baker, S.J. Fritschel, J.R. Stille and J.K. Stille, J. Org. Chem., 46 (1981) 2954-2960.
52 P.D. Sybert, C. Bertelo, W.B. Bigelow, S. Varaprath and J.K. Stille, Macromolecules, 14 (1981) 502-511.
53 K. Achiwa, Heterocycles, 9 (1978) 1539-1545.
54 T. Masuda and J.K. Stille, J. Am. Chem. Soc., 100 (1978) 268-272.
55 G.L. Baker, S.J. Fritschel and J.K. Stille, J. Org. Chem., 46 (1981) 2960-2965.
56 G.L. Baker, S.J. Fritschel and J.K. Stille, in F.E. Bailey, Jr. (Editor), ACS Symposium Series, No. 212, Initiation of Polymerization, American Chemical Society, Washington, DC, 1983, p. 137.
57 V.K. Latov, V.M. Belikov, T.A. Belyaeva and A.I. Vinogradova, Izv. Akad. Nauk SSSR, Ser. Khim., (1978) 660-663.
58 Z. Nagy-Magos, S. Vastag, B. Heil and L. Markó, J. Organometal. Chem., 171 (1979) 97-102.
59 V. Vaisarová and J. Hetflejš, Collect. Czech. Chem. Commun., in press.
60 L.M. Koroleva, V.K. Latov and M.B. Saporovskaya, Izv. Akad. Nauk SSSR, Ser. Khim., (1979) 2390-2391.
61 H. Pracejus and M. Bursian, East Ger. Pat. 92 031; CA 78 (1973) 72591.
62 Y. Kawabata, M. Tanaka and I. Ogata, Chem. Lett., (1976) 1213-1214.
63 Y. Kawabata, M. Tanaka and I. Ogata, Tokyo Kogyo Shikensho Hokoku, 73 (1978) 25-31.
64 S.C. Tang, T.E. Paxson and L. Kim, J. Mol. Catal., 9 (1980) 313-321.
65 F. Joo and M.T. Beck, J. Mol. Catal., 24 (1984) 135-145.
66 G. Pracejus and H. Pracejus, J. Mol. Catal., 24 (1984) 227-230.
67 H. Brunner and F.M. Mokhlesur, Chem. Ber., 117 (1984) 710-724.
68 L. Horner and G. Simons, Z. Naturforsch., B: Anorg. Chem., Org. Chem., 39 B (1984) 512-516.
69 J. Reiss and J. Hetflejš, Czech. Pat., 224,134 (1985).
70 J. Reiss and J. Hetflejš, Collect. Czech. Chem. Commun., 51 (1986) 340-346.
71 J. Reiss and J. Hetflejš, React. Kinet. Catal. Lett., in press.
72 J. Reiss and J. Hetflejš, Czech. Pat., 231,113 (1985).
73 F. Švec, J. Hradil, J. Čoupek and J. Kálal, Angew. Makromol. Chem., 48 (1975) 135-143.
74 J. Lukáš, F. Švec and E. Votavová, J. Chromatogr., 153 (1978) 373-380.
75 J. Hradil and F. Svec, Polymer Bulletin, 6 (1982) 565-570.
76 J. Reiss and J. Hetflejš, unpublished results.
77 J. Reiss, J. Hetflejš, J. Hradil and P. Svec, Czech. Pat. 242 004 (1985).

78 M. Čapka, Synth. Inorg. Metal Org. Chem., 7 (1977) 347-354.
79 M. Čapka, Collect. Czech. Chem. Commun., 42 (1977) 3410-3416.
80 A. Kinting, M. Čapka and H. Krause, J. Mol. Catal., in press.
81 M. Černý, Collect. Czech. Chem. Commun., 42 (1977) 3069-3078.
82 I. Kolb, M. Černý and J. Hetflejš, React. Kinet. Catal. Lett., 7 (1977) 199-204.
83 J. Vilím and J. Hetflejš, Collect. Czech. Chem. Commun., 43 (1978) 122-133.
84 W.H. Quayle and I.J. Pinnavaia, Inorg. Chem., 17 (1979) 2840-2847.
85 M. Mazzei, W. Marconi and M. Riocci, J. Mol. Catal., 9 (1980) 381-387.
86 U. Nagel, Angew. Chem., 96 (1984) 425-426.

Chapter 15

LIQUID-PHASE HYDROGENATION: THE ROLE OF MASS AND HEAT TRANSFER IN SLURRY REACTORS

GUENTHER GUT, OEMER M. KUT, FUESUN YUECELEN and DANIEL WAGNER
Swiss Federal Institute of Technology (ETH), CH-8092 Zurich, Switzerland

15.1 INTRODUCTION

To perform liquid-phase hydrogenation, a slurry, containing a catalyst and a substrate, is contacted with gaseous hydrogen. The hydrogen has to be transferred from the gas to the liquid phase and then to the catalyst surface, on which the reaction takes place. The reaction product is desorbed and diffuses back into the bulk liquid.

To optimize such heterogeneous systems, as well as chemical parameters, also physical parameters of steps such as diffusion and sorption have to be considered. In recent years different authors have analysed the macrokinetics of such three-phase systems for cases with constant reaction orders (refs. 1-6). Here, hydrogenation will be modelled for Langmuir-Hinshelwood type kinetic behaviour. This implies that the reaction order will change during the reaction. Emphasis will be given to the effects of heat and mass transfer on the kinetics.

15.2 OVERALL MODEL

Fig. 15.1 shows schematically the concentration profiles of the reaction components for a simple hydrogenation reaction corresponding to eqn. 15.1:

$$A + \nu H_2 \rightarrow R \qquad (15.1)$$

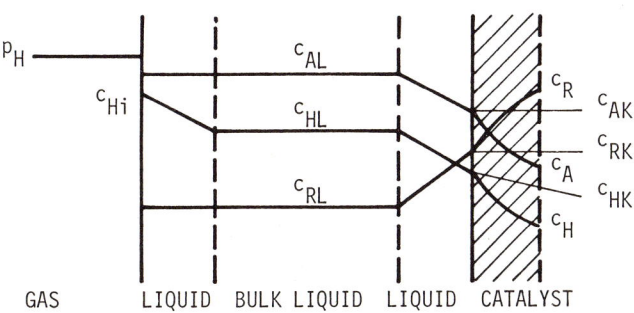

Fig. 15.1   Concentration profiles of reaction components.
A = Reactant, H = Hydrogen, R = Product.

Assuming that the organic species and hydrogen are adsorbed on different sites on the catalyst's surface under isothermal conditions, the following rate equations can be derived for the different reaction steps:

$$-r_H = k_{HL} a (c_{Hi} - c_{HL}) \quad \text{Hydrogen transfer from gas interface to bulk liquid} \quad (15.2)$$

$$= k'_{HL} a_K (c_{HL} - c_{HK}) \quad \text{Hydrogen transfer from bulk liquid to catalyst surface} \quad (15.3)$$

$$= \nu k'_{AL} a_K (c_{AL} - c_{AK}) \quad \text{Substrate transfer from bulk liquid to catalyst surface} \quad (15.4)$$

$$= m_K k_H \{c_{HK}(1 - \theta_H) - \frac{1}{K_H} \theta_H\} \quad \text{Adsorption of hydrogen on the catalyst} \quad (15.5)$$

$$= \nu m_K k_A \{c_{AK}(1 - \theta_A - \theta_R) - \frac{1}{K_H} \theta_A\} \quad \text{Adsorption of substrate A on the catalyst} \quad (15.6)$$

$$= \nu m_K k^* \eta \theta_A \theta_H \quad \text{Chemical surface reaction} \quad (15.7)$$

$$= \nu m_K k_R \{c_{RK}(1 - \theta_A - \theta_R) - \frac{1}{K_R} \theta_R\} \quad \text{Desorption of product R from the catalyst} \quad (15.8)$$

$$= \nu k'_{RL} a_K (c_{RK} - c_{RL}) \quad \text{Product transfer from catalyst to the bulk liquid} \quad (15.9)$$

At steady-state all these steps are occurring with the same rate and so it is possible to eliminate the variables c_{HL}, c_{HK}, c_{AK}, c_{RK}, θ_A, θ_R and θ_H, which cannot be measured directly from eqns. 15.2 - 15.9. The resulting general model is too complex to be useful. However, in practice it is often the case that only one or two of the above steps are rate-controlling, and accordingly this general reaction model can be simplified.

15.2.1 Surface reaction

If all the transport steps are fast compared to the chemical reaction steps and all the adsorption equilibria are established, then the chemical surface reaction is rate-controlling. For very fine or non-porous particles ($\eta = 1$), the rate equation is given by:

$$-r_{Ho} = -\nu r_{Ao} = \nu m_K k^* \{\frac{K_A c_{AL}}{K_A c_{AL} + K_R c_{RL}}\} \{\frac{K_H c_{Hi}}{1 + K_H c_{Hi}}\} \quad (15.10)$$

Eqn. 15.10 is based on the assumption that the catalyst surface is always covered with organic species, i.e. $K_A c_{AL} + K_R c_{RL} \gg 1$ (ref. 7). Consequently, only the ratio $Q = K_R/K_A$ and not the individual values of the adsorption constants K_A and K_R can be determined. Because of the low solubility of hydrogen in most organic liquids - typical values of $c_{Hi} = p_H/H$ are 0.03 to 0.3 mol/l -

normally the second term of eqn. 15.10 cannot be simplified.

At constant hydrogen pressure, eqn. 15.10 can be simplified further if the solubility of hydrogen is of the same order of magnitude in the reactant and product:

$$-r_{Ho} = -\nu\, r_{Ao} = \nu\, m_K\, k\, \left(\frac{K_A c_{AL}}{K_A c_{AL} + K_R c_{RL}}\right) = \nu\, m_K\, k\, \left(\frac{c_{AL}}{c_{AL} + Q c_{RL}}\right) \quad (15.11)$$

The dependency of the rate on hydrogen-pressure is now incorporated in the new kinetic constant, k:

$$k = k^* \theta_H = k^* \left(\frac{K_H c_{Hi}}{1 + K_H c_{Hi}}\right) \quad (15.12)$$

In general the more unsaturated substrate will be more strongly adsorbed than the hydrogenated product ($K_A c_A \gg K_R c_R$), hence the hydrogenation shows zero-order kinetic behaviour up to relatively high extents of conversion depending on the value of Q, as is demonstrated in Fig. 15.2.

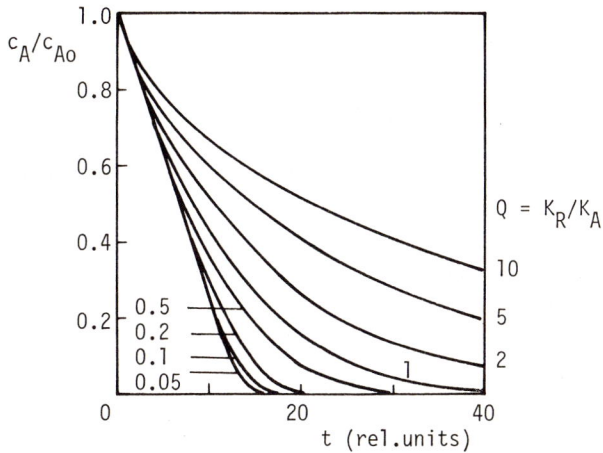

Fig. 15.2 Substrate concentration as a function of time during hydrogenation under constant pressure and isothermal conditions. Parameter: $Q = K_R/K_A$ (eqn. 15.11).

Therefore in many cases the initial hydrogenation rate can easily be determined:

$$-r_{Ho} = \nu\, m_K\, k = \nu\, m_K\, k^* \left(\frac{K_H c_{Hi}}{1 + K_H c_{Hi}}\right) \quad (15.13)$$

The relationship between the hydrogen pressure, p_H, and the saturation concentration, c_{Hi}, is given by Henry's law:

$$p_H = H \, c_{Hi} \tag{15.14}$$

The combination of eqns. 15.13 and 15.14 gives the relation between the hydrogen pressure and the initial hydrogenation rate:

$$-r_{Ho} = \nu \, m_K \, k^* \left\{ \frac{(K_H/H) \, p_H}{1 + (K_H/H) \, p_H} \right\} \tag{15.15}$$

Fig. 15.3 shows the initial hydrogenation rate of o-tert.-butylphenol on a palladium catalyst (5 % Pd/C; Engelhard) as a function of the hydrogen pressure at different temperatures.

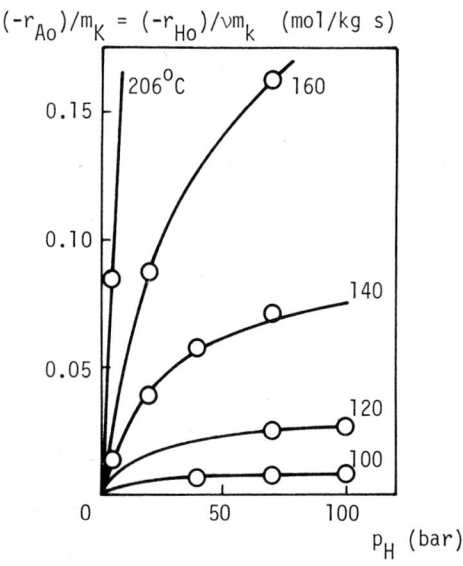

Fig. 15.3 Initial rate in the liquid-phase hydrogenation of o-tert.-butylphenol on a palladium catalyst as a function of hydrogen pressure at different temperatures (eqn. 15.15).

It is seen that the reaction order with respect to hydrogen changes from one to zero with increasing pressure. Such behaviour is characteristic for a great number of catalytic hydrogenations in the liquid phase, over a broad range of hydrogen pressures.

When the solubility of hydrogen in the feed and product differs widely, eqn. 15.16 may be used:

$$H = H_A \, (1 - x_A) + H_R x_A \tag{15.16}$$

Here H_A and H_R refer to the Henry constant of hydrogen in the reactant A and product R, and x_A is the fractional conversion of the reactant A. This simple equation may often be useful. For more precise calculations, Ramachandran and co-workers (ref. 8) proposed a method to correlate solubility data based on thermodynamic parameters for both a single solvent and mixtures. The effect of temperature as well as the heat of solution can also be predicted with this method.

15.2.2 Poisoning and inhibition

According to eqn. 15.10, the reaction rate is proportional to the fractional occupancy of the active sites of the catalyst by the reactants. Any further compound - also product or solvent - adsorbed at the surface of the catalyst diminishes the active area and thus leads to a decrease in the reaction rate. In the case of poisoning, the ratio of the reaction rate in the presence of the poison to that without poison, r_A/r_{Ao}, is an appropriate measure of the activity.

When a poison, which is strongly adsorbed, is in the feed, e.g., a sulphur-containing compound, the following model is found to fit the experimental activity very well. Its derivation and theoretical discussion have been published by Gut and co-workers (ref. 9) and an example can be found in ref. 10.

$$-r_A = m_K k^* \left(\frac{1}{1 + K_G c_G}\right) \theta_A \theta_H \tag{15.17}$$

The relation between the poison concentration in the reaction mixture, c_G, and the initial poison concentration, c_{Go}, is given by the mass balance of the poison:

$$c_{Go} = c_G + c_G^* c_K \tag{15.18}$$

In eqn. 15.18, c_G^* (mol/g) refers to the poison concentration of the catalyst, and c_K (g/l) to the catalyst concentration. Since the poison is strongly adsorbed on the catalyst surface and there is practically no interaction between the poison and the reacting species, the poison concentration, c_G, in the liquid remains approximately constant during the reaction. Therefore the value of the poisoning term, $\{1/(1 + K_G c_G)\}$, in eqn. 15.17 is nearly constant, smaller than one, which corresponds formally to a reduction in the amount of catalyst used. To account for this effect, instead of eqn. 15.17, eqn. 15.19 may be used:

$$-r_A = (m_K - m_{Ko}) k^* \theta_A \theta_H \tag{15.19}$$

The amount of catalyst poisoned, m_{Ko}, can be easily determined from experiments with different amounts of catalyst (see Fig. 15.4).

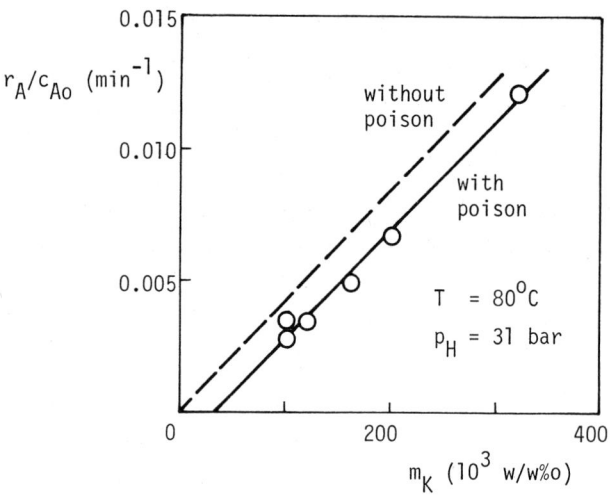

Fig. 15.4 Hydrogenation rate of 2,4-dimethylnitrobenzene on a palladium catalyst as a function of catalyst loading (ref. 11).

Since hydrogenations are highly exothermic reactions, experiments with different amounts of catalyst must be analyzed carefully for effects of insufficient heat transfer, resulting in a rise of temperature. As Cerveny and coworkers (ref. 12) pointed out, it is not permissible to conclude apriori on the basis of a linear relationship between the reaction rate and amount of catalyst - measured in an insufficiently cooled autoclave - that the experiments were done on the intrinsic kinetic regime. In such non-isothermal systems, a sigmoidal dependence of the reaction rate on the amount of catalyst will result. Hence, experimental results, based on measurements in a limited range of m_K, may be interpreted incorrectly as a proof of the presence of a poison in the feed.

If the product R is strongly adsorbed with increasing extent of conversion, an inhibition of the reaction rate will be observed. Thus in eqn. 15.11, $K_R > K_A$ and $Q = K_R/K_A$ has a value larger than one. This situation is also illustrated by Fig. 15.2, where conversion curves are plotted with Q as parameter.

To account for solvent effects, a term $K_L c_L$ may be introduced in the denominator of eqn. 15.11. However, it should be mentioned in relation to solution media or to other additives, that types of compounds, which exert an interaction with the unsaturated group, will thereby change the adsorption behaviour of the liquid substrate and hence the value of Q.

15.3 MASS TRANSFER AND KINETICS

In order to design a three-phase slurry system, chemical reaction steps have to be coupled to steps of mass transfer. Beginning with the transport of hydrogen, we will consider the steps of absorption and the inter- and intra-particle diffusion.

15.3.1 Transport of hydrogen

The transfer rate of hydrogen is proportional to the areas of the interphases and to the concentration gradients. For pure hydrogen, on the gas-phase side there exists no appreciable transfer resistance. This often also holds for systems with diluted hydrogen because the diffusion in the gas phase is about two orders of magnitude faster than in the liquid phase. The overall resistance is therefore only due to diffusion through the liquid films around the gas bubbles and around the catalyst particles. As long as the reaction rate is zero-order with respect to the substrate or using the initial rates, under steady-state conditions:

$$-r_H = k_{HL} a (c_{Hi} - c_{HL}) \tag{15.2}$$

$$= k'_{HL} a_K (c_{HL} - c_{HK}) \tag{15.3}$$

$$= \nu \, m_K \, k^* \, \eta \, \left(\frac{K_H c_{HK}}{1 + K_H c_{HK}}\right) \tag{15.7}$$

$$p_H = H \, c_{Hi} \tag{15.14}$$

Eqn. 15.2 corresponds to the rate of absorption, eqn. 15.3 to the transfer of hydrogen from the bulk liquid to the outer geometric surface of the catalyst particle and eqn. 15.7 to the surface reaction. The surface concentration of hydrogen may be calculated from eqns. 15.2 and 15.3:

$$c_{HK} = c_{Hi} \left\{ 1 - \frac{(-r_H)}{k_{HL} a \, c_{Hi}} - \frac{(-r_H)}{k'_{HL} a_K \, c_{Hi}} \right\} \tag{15.20}$$

In eqn. 15.20 the rate, $-r_H$, refers to the observable rate of hydrogen consumption, and $k_{HL} a \, c_{Hi}$ and $k'_{HL} a_K \, c_{Hi}$ are the maximum hydrogen-transfer rates to the liquid phase and the catalyst respectively. Introducing the Damköhler number, $\eta_H Da$, defined as (observed reaction rate)/(maximum hydrogen-transfer rate), eqn. 15.20 may be rewritten as:

$$c_{HK} = c_{Hi} (1 - \eta_H Da) \tag{15.20A}$$

and the rate eqn. 15.7 as:

$$-r_H = \nu\, m_K\, k^* \eta \left\{ \frac{K_H c_{Hi}(1-\eta_H Da)}{1+K_H c_{Hi}(1-\eta_H Da)} \right\} \tag{15.21}$$

Eqn. 15.21 also holds when the reaction is controlled only by the surface reaction, when $\eta_H Da \to 0$ and with $\eta = 1$ eqn. 15.21 becomes equal to eqn. 15.13.

15.3.2 Transport of substrate

If the hydrogenation is carried out at constant hydrogen pressure, and if the solubility of hydrogen is only a weak function of the extent of conversion, x_A, then eqn. 15.7 can be rewritten as:

$$-r_A = m_K\, k\, \eta \left(\frac{K_A c_{AK}}{K_A c_{AK} + K_R c_{RK}} \right) \tag{15.22}$$

where $k = k^* \theta_H$. The rate of transfer of component A to the catalyst is given by eqn. 15.4 and that of component R from the catalyst to the bulk liquid by eqn. 15.9:

$$-r_A = k'_{AL} a_K (c_{AL} - c_{AK}) \tag{15.4}$$

$$= k'_{RL} a_K (c_{RK} - c_{RL}) \tag{15.9}$$

Expressing the concentrations c_{AL} and c_{RL} in terms of the extents of conversion and initial concentrations, $c_{AL} = c_{ALo}(1-x_A)$ and $c_{RL} = c_{ALo} x_A$, and assuming that the approximation $k'_{AL} a_K = k'_{RL} a_K$ is valid (ref. 13), the following surface concentrations are obtained:

$$c_{AK} = c_{AL}\left\{1 - \frac{(-r_A)}{k'_{AL} a_K c_{ALo}(1-x_A)}\right\} = c_{AL}\left\{1 - \frac{\eta_A Da}{(1-x_A)}\right\} \tag{15.23}$$

$$c_{RK} = c_{RL}\left\{1 + \frac{(-r_A)}{k'_{AL} a_K c_{ALo} x_A}\right\} = c_{RL}\left\{1 + \frac{\eta_A Da}{x_A}\right\} \tag{15.24}$$

The rate, $-r_A$, in eqns. 15.23 and 15.24 corresponds to the observed rate of consumption of reactant A. The Damköhler number, $\eta_A Da$, is defined as (observed reaction rate)/(maximum substrate-transfer rate). On combining eqns. 15.22 - 15.24, the rate is given by:

$$-r_A = m_K\, k\, \eta \left\{ \frac{K_A c_{AL}\{1 - \eta_A Da/(1-x_A)\}}{K_A c_{AL}\{1 - \eta_A Da/(1-x_A)\} + K_R c_{RL}(1 + \eta_A Da/x_A)} \right\} \tag{15.25}$$

As long as $\eta_A Da \to 0$, the overall reaction is controlled by the surface reaction and eqn. 15.25 with $\eta = 1$ is equal to eqn. 15.11.

15.3.3 Estimation of the absorption coefficient of hydrogen

The value of $k_{HL}a$ is strongly dependent on the flow pattern of the reactor and in the case of a stirred tank reactor, strongly on the stirring rate or/and the hydrogen throughput. As is seen from Fig. 15.5, where the hydrogenation rate is plotted for different control regimes, the two extremes of chemical control and absorption control are easily distinguished.

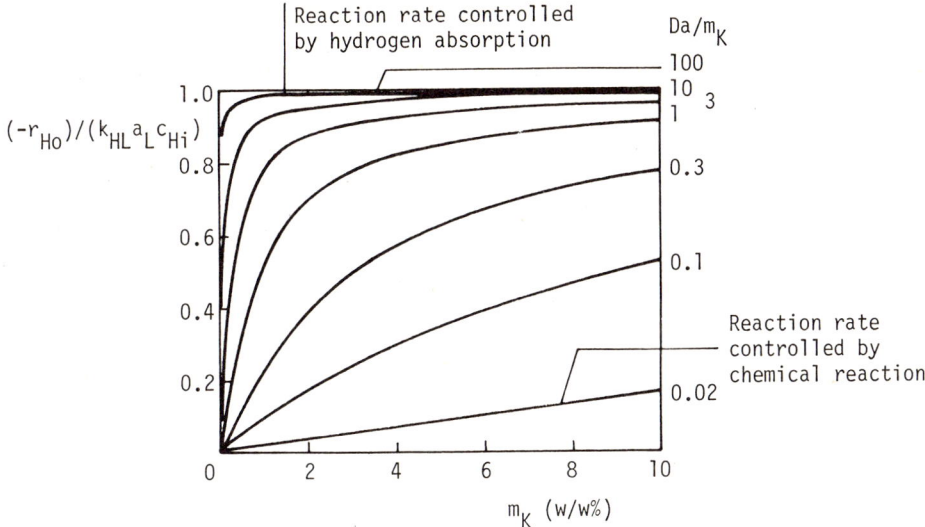

Fig. 15.5 Relationship between initial hydrogenation rate and catalyst loading for different control regimes.

If the chemical surface reaction is the only rate-controlling step ($\eta_H Da \to 0$), the hydrogenation rate is proportional to the amount of catalyst. At the other extreme, if the absorption of hydrogen is the slowest step, the rate is independent of the catalyst loading. To determine $k_{HL}a$, an appropriate fast hydrogenation reaction and a highly active catalyst may be used. For this limiting case, $c_{HL} \to 0$, and eqn. 15.2 simplifies to:

$$-r_H = k_{HL}a \, c_{Hi} \qquad (15.26)$$

Since hydrogenations are extremely exothermic reactions, in practice this method can be used only within limits, and often it is difficult to transfer the heat of reaction to the coolant fast enough (see also Section 15.2.2).

Another method is based on eqn. 15.21. Let us assume that a fine powdered and non-porous catalyst is used; hence practically no pore-diffusion resistance and no resistance to transfer from the bulk liquid to the catalyst is expected.

Let us furthermore assume the experiments are performed at relatively high temperature and low pressure, then the reaction may be first order in the hydrogen concentration (see Fig. 15.3). Based on these assumptions and for the chemical side limiting conditions, eqn. 15.21 can be simplified to:

$$-r_H = \nu\, m_K\, k^*\, K_H c_{Hi}\, \{1 - (-r_H)/k_{HL} a\, c_{Hi}\} \qquad (15.27)$$

or rewritten as:

$$\frac{c_{Hi}}{(-r_H)} = \frac{1}{k_{HL} a} + \frac{1}{\nu\, k^*\, K_H}\left(\frac{1}{m_K}\right) \qquad (15.28)$$

In order to estimate values of $k_{HL}a$, the observed hydrogenation rate, r_H, can be determined for different catalyst loadings, m_K, and calculated using eqn. 15.28. This method has found widespread application (refs. 1, 2, 14), but it gives good approximations only at low stirring rates and low pressures. A simulation (Fig. 15.6) using Langmuir-Hinshelwood kinetics shows that at higher pressures the assumed linear relationship is valid only for low catalyst loadings.

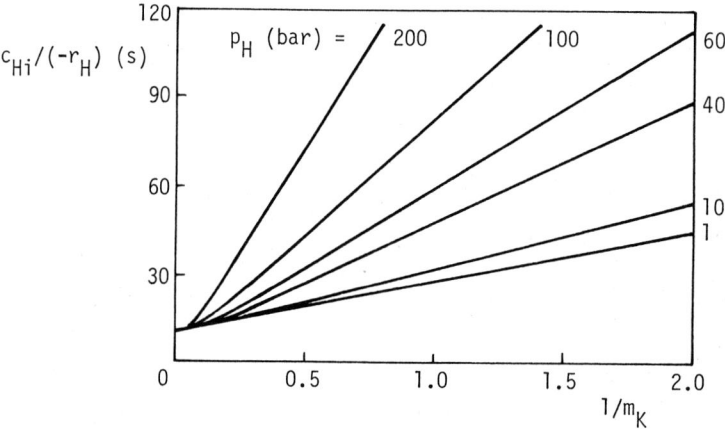

Fig. 15.6   Effect of catalyst loading on hydrogenation rate (eqn. 15.28).

Therefore, if only experimental data for low catalyst loadings are available, then the use of the value of the X-intercept of the extrapolated straight line, which passes through these points, will result in a large error in the estimation of $k_{HL}a$. In the literature, however, there are many empirical correlations of mass transfer parameter, $k_{HL}a$, with the stirring rate, N (refs. 15-18), or the power input (refs. 18-20). Thus the error mentioned above can be reduced substantially if the exact eqn. 15.21 is combined with an appropriate correlation equation for the mass-transfer parameter in order to calculate $k_{HL}a$. To account for $\eta_H Da$ in eqn. 15.21, eqn. 15.29 may be used:

$$\eta_H Da = \frac{(-r_H)}{c_{Hi}} \left\{ \frac{1}{k_{HL}a\,(N)} \right\} \tag{15.29}$$

Here $k_{HL}a\,(N)$ is the $k_{HL}a$ function corresponding to an appropriate correlation equation. As is seen from Fig. 15.7, in the case of the nickel-catalysed hydrogenation of o-cresol in a 500-ml unbaffled, mechanically agitated reactor with a six-bladed hollow-shaft turbine stirrer, there is good agreement between experimental points and calculated values (ref. 21).

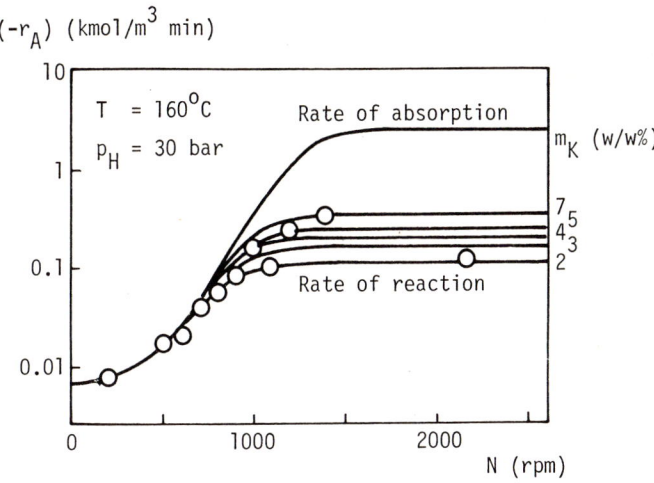

Fig. 15.7 Hydrogenation of o-cresol on nickel: Absorption and reaction rates as a function of stirrer speed. Curves calculated with model equations (ref. 21).

15.3.4 Estimation of the transfer coefficient to the catalyst

With decreasing particle size, the effectiveness factor, η, as well as the value of the liquid-solid transfer coefficient, $k'_L a_K$, will increase. Thus in practice, when working with particles smaller than 50 μm, preferably smaller than 10 μm, no limitations imposed by liquid-solid mass transfer or by pore-diffusion phenomena are expected.

To test, e.g., for the effects of substrate diffusion, a fast consecutive hydrogenation reaction may be convenient. If the rate is influenced by these phenomena, a decrease in selectivity will be observed. Thus, experiments with decreasing catalyst particle size will yield the required information. For example, Fig. 15.8 shows the effects of particle size on the selectivity of 2-amino-6-nitrotoluene formation in the hydrogenation of 2,6-dinitrotoluene on a catalyst, in which the palladium was homogeneously dispersed.

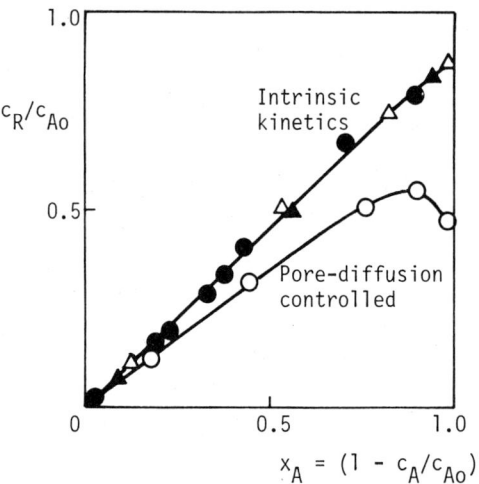

Fig. 15.8 Effects of particle size on the selectivity of a consecutive hydrogenation reaction. T = 60°C, p_H = 50 bar.
Particle sizes: ▲ 5 μm; △ 10-20 μm; ● 45-63 μm; ○ 345-500 μm.

It may be concluded that in this particular case the selectivity is constant when catalyst particles smaller than 63 μm are used. With particles of d_p > 345 μm, where the reaction is controlled by both pore diffusion and liquid-solid mass transfer, a decrease in selectivity occurs.

To estimate values of $k_L' a_K$ to a first approximation, eqn. 15.30 may be used:

$$k_L' a_K = \frac{0.24 \, D \, m_K \, \rho_L}{d_p^2 \, \rho_p} \tag{15.30}$$

It is based on the equation $k_L' = 4D/d_p$, proposed by Acres and Cooper (ref. 22), and is identical to eqn. 15.31 if the second term of the right-hand side is neglected. To account for a_K, sperical catalyst particles are considered. This equation gives quite satisfactory estimates for sufficiently small particles (d_p < 20 μm). For more accurate estimations, eqn. 15.31 and 15.32 may be used:

$$\left(\frac{k_L' \, d_p}{D}\right)^2 = 16.0 + 4.84 \left\{\frac{g \, d_p \, (\rho_p - \rho_L)}{18 \, D \, \mu}\right\}^{2/3} \tag{15.31}$$

In eqn. 15.31 the effects of the stirring rate have been taken into account (ref. 10). In eqn. 15.32, proposed by Brian and Hales (ref. 23), the specific power input, P, which can easily be measured, is considered:

$$\frac{k_L' d_p \rho_L^{1/3}}{D^{2/3} \mu^{1/3}} = f(P^{1/3} d_P^{4/3} \rho_L/\mu) \tag{15.32}$$

With estimates of k_L' and a_K, eqn. 15.33 and 15.36 or Figs. 15.9 and 15.10 may be used to account for the influence of the liquid-solid mass transfer of hydrogen and the substrate.

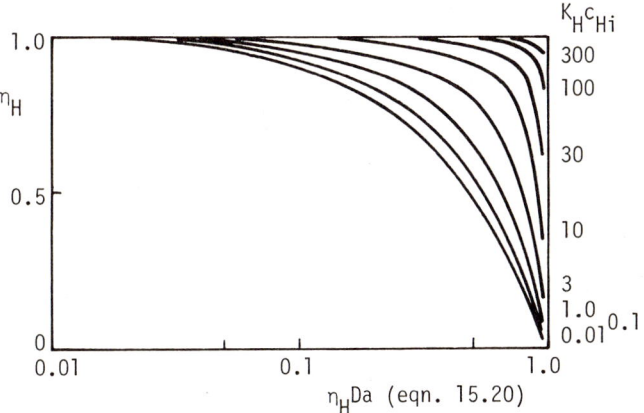

Fig. 15.9   Isothermal effectiveness factor for hydrogen as a function of observable reaction parameters (eqn. 15.33).

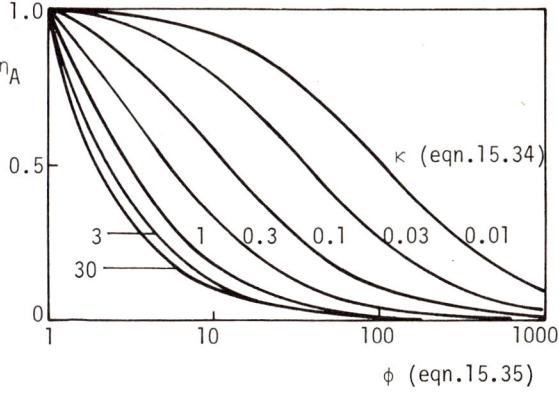

Fig. 15.10   Isothermal effectiveness factor for substrate A as a function of observable reaction parameters (eqn. 15.36).

15.3.5 Checking for effects of mass transfer on the kinetics

Let us assume that the absorption coefficient, $k_{HL}a$, and the liquid-solid mass-transfer coefficients, $k'_{HL}a_K$ and $k'_{AL}a_K$, are known or have been estimated by convenient methods (refs. 1, 2, 24). We can now calculate the effect of the overall mass transfer on the kinetics.

First the hydrogen is considered. The absorption rate is given by eqn. 15.2, the hydrogen transfer to the outer geometric surface of the catalyst by eqn. 15.3 and the chemical reaction rate by eqn. 15.7. From these equations, in Section 15.3.1, eqn. 15.21 was derived. Dividing eqn. 15.21 by 15.13 and assuming $\eta = 1$ leads to an expression for the effectiveness factor, η_H, under isothermal conditions:

$$\eta_H = \frac{(-r_H)}{(-r_{Ho})} = \frac{(1 + K_H c_{Hi})(1 - \eta_H Da)}{1 + K_H c_{Hi}(1 - \eta_H Da)} \tag{15.33}$$

In Fig. 15.9, η_H is given as a function of $\eta_H Da$ with $(K_H p_H/H) = K_H c_{Hi}$ as a parameter. When the kinetic parameters, K_H and c_{Hi}, and the mass-transfer parameters, $k_{HL}a$ and $k'_{HL}a_K$, are known, it is possible to estimate η_H and the undisguised chemical hydrogenation rate, r_{Ho} (ref. 13).

The transfer of substrate to the solid can be treated in a similar way. Dividing eqn. 15.25, derived from eqns. 15.9 and 15.22 (see Section 15.3.2), by eqn. 15.11 and introducing the parameters κ and ϕ as defined by eqns. 15.34 and 15.35, we obtain the expression for η_A under isothermal conditions (eqn. 15.36):

$$\frac{K_R}{K_A}\left(\frac{x_A}{1 - x_A}\right) = Q\left(\frac{x_A}{1 - x_A}\right) = \kappa \tag{15.34}$$

$$\left(1 + \frac{\eta_A Da}{x_A}\right)\left(1 - \frac{\eta_A Da}{1 - x_A}\right)^{-1} = \phi \tag{15.35}$$

$$\eta_A = \frac{(-r_A)}{(-r_{Ao})} = \frac{1 + \kappa}{1 + \kappa\phi} \tag{15.36}$$

Fig. 15.10 shows a plot of η_A versus ϕ with κ as a parameter. From this graph it is possible to evaluate η_A at any particular conversion level using the known kinetic and mass-transfer parameters.

15.3.6 Pore diffusion

Here, an estimation method will be given which is accurate enough to solve most of the problems connected with pore-diffusion phenomena. If the catalyst particles are small enough, the pores filled with liquid and the particles suspended in that liquid, it is likely that the temperature gradient between the liquid and solid is small and can be neglected. In such systems, η is a function only of two dimensionless parameters, namely the order of reaction, n, and the modulus, ϕ_S, which can be estimated from measurable data (refs. 1, 10):

$$\phi_S = \eta \phi^2 = \frac{d_p^2}{4D_e} \left\{ \frac{100 \, (-r_H) \, \rho_p}{m_K \, \rho_L} \right\} \frac{1}{c_{iK}} \tag{15.37}$$

Here D_e refers to the effective diffusivity of hydrogen or the substrate in the pores of the catalyst and c_{iK} is the concentration of the diffusing species at the pore mouth; D_e can be estimated using eqn. 15.38:

$$D_e = \varepsilon D / \tau \tag{15.38}$$

A typical value of the tortuosity factor, τ, is 4. Combining eqn. 15.38 with eqn. 15.37 and setting $\tau = 4$, we get:

$$\phi_S = \frac{100 \, d_p^2 \, \rho_p \, (-r_H)}{\varepsilon \, D \, m_K \, \rho_L \, c_{iK}} \tag{15.39}$$

All parameters on the right-hand side of eqn. 15.39 can be measured or calculated. If the gas-liquid and liquid-solid mass transfer is fast, then $c_{Hi} = c_{HL} = c_{HK}$; $c_{AL} = c_{AK}$ and $c_{RK} = c_{RL}$. If this is not the case, eqn. 15.20 or 15.23 and 15.24 may be used, providing the transfer coefficients are known or were previously estimated. After estimation of ϕ_S by eqn. 15.39, η may be calculated using diagrams given in the literature (refs. 1, 25).

15.4 EXAMPLES OF APPLICATION AND SIMULATION

Having established the main framework of the model, the effects of mass transfer on the reactor performance are now considered, together with the interaction between the mass transfer of hydrogen, rate of heat production and rate of heat transfer to the coolant.

15.4.1 Batch reactor: Effects of hydrogen and substrate transfer

This model is intended to include the mass-transfer steps required to bring the organic species and hydrogen to the outer geometric surface of the catalyst particle. In the absence of a solvent or in concentrated solutions, the concentrations of the organic species are normally much higher than that of the hydrogen dissolved in the liquid. Therefore, the first observable mass-transfer limitation is expected to be for the hydrogen. Hydrogen-transfer phenomena can also be expected at low pressures or low agitation intensity.

$$A + \nu H_2 \rightarrow R \qquad (15.1)$$

In diluted solutions and for hydrogenations at high pressures, the transfer of substrate to the particle may become rate-controlling. In this case, in the vicinity of the catalyst, there may be a relative decrease in the concentration of component A and an increase in the concentration of component R, as indicated in Fig. 15.1.

To account for these phenomena, we desire an expression in which the chemical rate and the mass-transfer rate are coupled to each other. In the model introduced here, it is assumed that all sorption equilibria are established and that a non-porous catalyst is used ($\eta = 1$). Considering a hydrogenation corresponding to eqn. 15.1, we obtain the expression:

$$-r_H = -\nu r_A = k_{HL} a (c_{Hi} - c_{HL}) \qquad (15.2)$$

$$= k'_{HL} a_K (c_{HL} - c_{HK}) \qquad (15.3)$$

$$-r_H = \nu k'_{AL} a_K (c_{AL} - c_{AK}) \qquad (15.4)$$

$$= \nu k^* m_K \theta_A \theta_H \qquad (15.7)$$

$$= \nu k'_{RL} a_K (c_{RK} - c_{RL}) \qquad (15.9)$$

Taking into account that $c_{RL} = c_{ALo} x_A$ and $c_{AL} = c_{ALo} (1-x_A)$, and using eqns. 15.4 and 15.9, the concentration of the product in the vicinity of the catalyst is:

$$c_{RK} = c_{ALo} - c_{AK} \qquad (15.40)$$

Hence, θ_A and θ_H are given by:

$$\theta_A = \frac{c_{AK}}{c_{AK} + Q(c_{ALo} - c_{AK})} \quad ; \quad \theta_H = \left(\frac{K_H c_{HK}}{1 + K_H c_{HK}}\right) \qquad (15.41)$$

From eqns. 15.2, 15.3 and 15.7, c_{HK} can be extracted:

$$c_{HK} = c_{Hi} \{1 - (\frac{\nu k^* m_K}{k_{HL} a \, c_{Hi}} + \frac{\nu k^* m_K}{k'_{HL} a_K c_{Hi}}) \theta_A \theta_H\} \qquad (15.42)$$

In a similar way, the substrate concentration, c_{AK}, can be found:

$$c_{AK} = c_{AL} \{1 - (\frac{k^* m_K}{k'_{AL} a_K c_{ALo} (1-x_A)}) \theta_A \theta_H\} \qquad (15.43)$$

Introducing modified Damköhler numbers, Da_H and Da_A (for definitions see eqns.

15.48 and 15.50), and taking into accoung eqn. 15.41, eqns. 15.42 and 15.43 may be rewritten as:

$$c_{HK} = c_{Hi} \{1 - Da_H \left(\frac{c_{AK}}{c_{AK} + Q(c_{ALo} - c_{AK})}\right)\left(\frac{K_H c_{HK}}{1 + K_H c_{HK}}\right)\} \quad (15.42A)$$

$$c_{AK} = c_{AL} \{1 - \frac{Da_A}{(1-x_A)} \left(\frac{c_{AK}}{c_{AK} + Q(c_{ALo} - c_{AK})}\right)\left(\frac{K_H c_{HK}}{1 + K_H c_{HK}}\right)\} \quad (15.43A)$$

From these equations the unknown concentrations, c_{AK} and c_{HK}, may be calculated. By rearrangement, two quadratic equations with two unknowns are obtained:

$$A z^2 + B z^2 y + C zy + D z + E y + F = 0 \quad (15.42B)$$
$$A'y^2 + B'y^2 z + C'zy + D'y + E'z + F' = 0 \quad (15.43B)$$

To simulate a stirred tank batch reactor, eqn. 15.7 in combination with eqn. 15.41 may be used, whereas for any fixed reaction time, c_{AK} and c_{HK} are calculated stepwise using eqns. 15.42B and 15.43B.

In this section no simulation examples are given because the effects of the mentioned transfer rates on the reactor performance can also be seen from a simulation of a continuous stirred tank slurry reactor (see Section 15.4.4).

15.4.2 Continuous stirred tank slurry reactor: Basic concept

In this section the performance of a continuous stirred tank slurry reactor (CSTR) is analyzed. The notation of the system is given in Fig. 15.11.

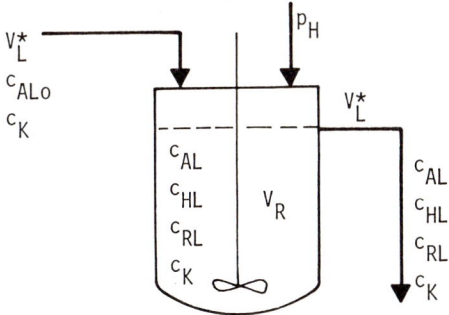

Fig. 15.11   Notation for a continuous stirred tank slurry reactor.

A hydrogenation reaction corresponding to eqn. 15.1 is considered. The mass balance of the CSTR is given by:

(input of A and H_2) = $\begin{pmatrix}\text{disappearance of A}\\\text{and } H_2 \text{ by reaction}\end{pmatrix}$ + (output of A and H_2)

$$V_L^* c_{ALo} = (-r_A) V_R + V_L^* c_{AL} \qquad (15.44)$$

$$V_R k_{HL} a (c_{Hi} - c_{HL}) = (-r_H) V_R + V_L^* c_{HL} \qquad (15.45)$$

Assuming pore diffusion to be negligible ($\eta = 1$), the rate of the surface reaction is given by eqn. 15.7, and the rate of transfer of species from the bulk liquid to the vicinity of the catalyst by eqns. 15.3 and 15.4. θ_A and θ_H are defined by eqn. 15.41 (see Section 15.4.1).

If we set $V_R/V_L^* = \tau$, eqn. 15.44 may be transformed to:

$$c_{AL}/c_{ALo} = (1 - x_A) = 1 - (-r_A)\tau/c_{ALo} \qquad (15.46)$$

Since the hydrogen concentration in the reaction mixture is normally relatively small, the output of hydrogen is smaller by a factor of 20-200 than its consumption by reaction. Thus we may state:

$$V_L^* c_{HL} \ll (-r_A) V_R \qquad (15.47)$$

As long as condition 15.47 holds, eqn. 15.45 simplifies to eqn. 15.2 (see Section 15.4.1).

15.4.3 CSTR: Combination of mass transfer with chemical reaction

Taking into account that $c_{RK} = c_{ALo} - c_{AK}$, and using the same procedure as described in Section 15.4.1, we obtain:

$$c_{AK} = c_{ALo} - c_{ALo} \left(\frac{\tau m_K k^*}{c_{ALo}} + \frac{m_K k^*}{k'_{AL} a_K c_{ALo}}\right) \theta_A \theta_H \qquad (15.48)$$

Introducing the Damköhler number, defined here as $Da_A = (m_K k^*)/(k'_{AL} a_K c_{ALo})$:

$$c_{AK} = c_{ALo} - c_{ALo} \left(\frac{\tau m_K k^*}{c_{ALo}} + Da_A\right) \theta_A \theta_H \qquad (15.48A)$$

$$c_{AK} = c_{ALo} - \beta \theta_A \theta_H \qquad (15.48B)$$

For τ = constant, the value of β is also a constant.

The hydrogen concentration, c_{HK}, may be expressed similarly:

$$c_{HK} = c_{Hi} - \alpha \theta_A \theta_H \qquad (15.49)$$

In eqn. 15.49, $\alpha = (p_H/H) Da_H = c_{Hi} Da_H$ and:

$$Da_H = \frac{\nu \, m_K \, k^*}{c_{Hi}} \left(\frac{1}{k_{HL} a} + \frac{1}{k'_{HL} a_K}\right) \quad (15.50)$$

Taking account of he surface coverage terms, eqns. 15.48B and 15.49 may be rewritten as:

$$c_{AK} = c_{ALo} - \beta \left(\frac{c_{AK}}{c_{AK} - Q c_{AK} + Q c_{ALo}}\right)\left(\frac{K_H c_{HK}}{1 + K_H c_{HK}}\right) \quad (15.51)$$

$$c_{HK} = c_{Hi} - \alpha \left(\frac{c_{AK}}{c_{AK} - Q c_{AK} + Q c_{ALo}}\right)\left(\frac{K_H c_{HK}}{1 + K_H c_{HK}}\right) \quad (15.52)$$

The desired concentrations, c_{AK} and c_{HK}, may be calculated from eqns. 15.51 and 15.52 using the same procedure as described in Section 15.4.1.

15.4.4 CSTR: Simulation of conversion profiles

In order to simulate conversion profiles in a CSTR, eqn. 15.7 is combined with eqn. 15.46, whereby for any time τ, eqns. 15.51 and 15.52 are used to calculate the appropriate values of c_{AK} and c_{HK}. Some examples are now given. The simulations are based on data estimated for the hydrogenation of a nitro compound to the corresponding aniline derivative (Fig. 15.12).

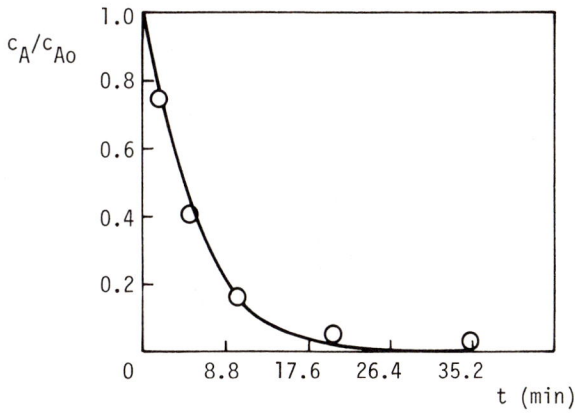

Fig. 15.12 Hydrogenation of an aromatic nitro compound in a batch stirred tank reactor. T = 75°C; catalyst = Pd/C, 0.125 w/w%; k = 1.41 mol/l min w/w%; Q = 0.65; K_H = 15.7 l/mol; p_H/H = 0.245 mol/l.

Fig. 15.13 shows the effect of the catalyst loading on the performance of a CSTR. It is seen that the conversion rate is fast at the beginning, and that all curves show a strong tailing. The Damköhler number is a measure of the combined effects of the resistances to mass transfer and chemical reaction on the overall reaction rate.

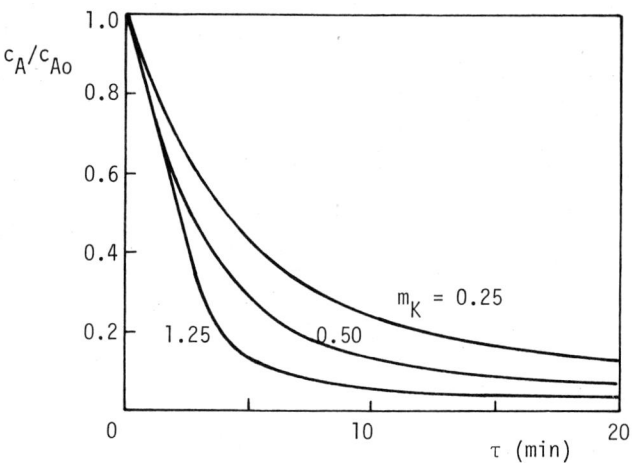

Fig. 15.13 CSTR: Effect of catalyst loading on the concentration profiles in the kinetically controlled regime. Simulation data as in Fig. 15.12. p_H = 40 bar.

In Fig. 15.14 the performance of a CSTR is given with the Da_H number as a parameter. The effects of the resistance to substrate transfer on the reactor performance are illustrated in Fig. 15.15.

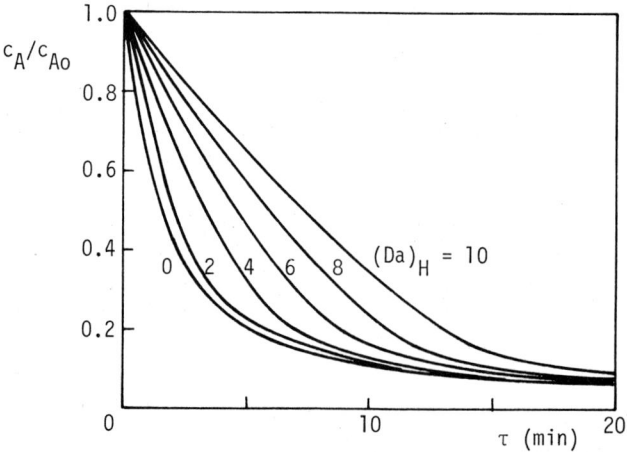

Fig. 15.14 CSTR: Effect of the rate of hydrogen transfer, Da_H number, on the concentration profiles. Kinetic data as in Fig. 15.12. m_K = 0.5 w/w%; p_H = 60 bar; Da_A = 0.

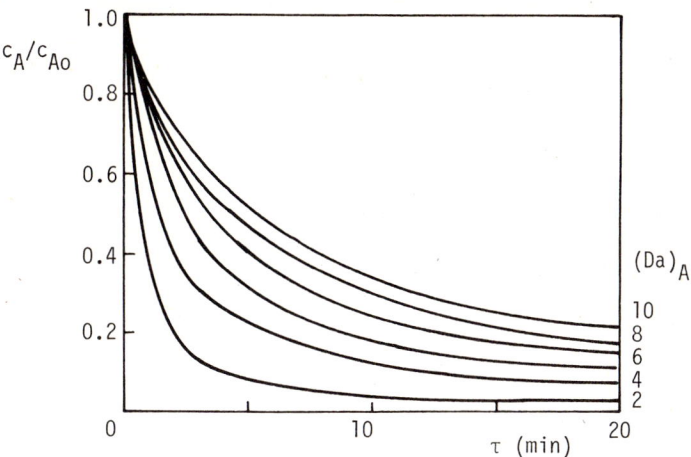

Fig. 15.15   CSTR: Effect of the rate of substrate transfer, Da_A number, on the concentration profiles. Kinetic data as in Fig. 15.12. $Da_H = 0$; $m_K = 1.5$ w/w%; $p_H = 60$ bar.

15.4.5   Batch reactor: Effects of the rates of heat production and heat transfer

In this model the effects of the rates of heat production and heat transfer on the kinetics are considered. It will be assumed that the chemical reaction rate is affected only by the rates of transfer of hydrogen and the substrate, not by pore-diffusion effects. The heat balance of the system is given by:

$$\begin{pmatrix}\text{heat released}\\ \text{by reaction}\end{pmatrix} - \begin{pmatrix}\text{heat removal}\\ \text{by coolant}\end{pmatrix} = \begin{pmatrix}\text{increase in internal}\\ \text{energy of the system}\end{pmatrix}$$

$$V(-r_A)(-\Delta H_R) - UA(T - T_a) = (m_i c_{Pi})\, dT/dt \tag{15.53}$$

By coupling the heat balance 15.53 to the mass-transfer kinetics, after introducing the usual simplifying assumptions (constant heat capacity, heat of reaction, heat-transfer coefficient and coolant temperature) (refs. 26, 27), the following system of differential and algebraic equations is obtained:

- The time-dependent temperature profile, where $\rho = m/V$

$$dT/dt = \frac{1}{\rho\, c_P}\{(-\Delta H_R)(-r_A) - \frac{UA}{V}(T - T_a)\} \tag{15.54}$$

- The concentration-time function (V = constant)

$$-r_A = m_K\, k^* \,\theta_A\, \theta_H \tag{15.7}$$

In eqn. 15.7, θ_A and θ_H are given by the Langmuir-Hinshelwood model (eqn. 15.41) and c_{HK} and c_{AK} (Section 15.4.1) by:

$$c_{HK} = c_{Hi} (1 - Da_H) \theta_A \theta_H \qquad (15.42)$$

$$c_{AK} = c_{AL} (1 - \frac{Da_H}{1 - x_A}) \theta_A \theta_H \qquad (15.43)$$

We note that the Damköhler number contains k^*, k_{HL}, k'_{HL} and k'_{AK} which show an Arrhenius-type temperature dependence with different activation energies, whereby $E_{chem} \gg E_{diff}$. The adsorption constant for hydrogen, K_H, will decrease with increasing temperature, while the solubility of hydrogen in the liquid increases so that the ratio K_H/H is not very temperature-dependent. This is also true for $Q = K_R/K_A$, which is the ratio of two adsorption constants with very similar temperature dependencies. If the cooling of the reactor is insufficient (UA/V = small) during the hydrogenation, the temperature will rise and the chemical reaction rate will increase. Simultaneously, Da_A and Da_H will increase. Because of this growing transport limitation, the reaction rate will increase less than for a homogeneous reaction with a similar activation energy. In Fig. 15.16 the concentration profiles for c_{AL}, c_{AK} and c_{HK} together with the corresponding temperature profiles are plotted for a given cooling capacity (UA/V = constant) with different initial levels of hydrogen-transfer limitations $\{Da_H(T_a) = variable\}$.

15.4.6 Batch reactor: Effects of film and pore diffusion

Let us assume the overall rate is affected by both the rate of mass-transfer to the particle and also by pore-diffusion phenomena. This problem of coupled external-internal diffusion with chemical reaction has been extensively analyzed for first-order reactions (refs. 24, 25, 28-30), for second-order reactions (ref. 31) and also for monomolecular reactions of general order (ref. 32). However, the case of the practically important Langmuir-Hinshelwood-type bimolecular reaction, such as hydrogenations, can be handled only by numerical integration of the appropriate differential equations.

It is not the objective of this paper to give an extensive mathematical formulation of this problem. This presentation will be restricted to isothermal systems (uniform temperatur throughout the particle and no temperature gradient between the particle and surrounding liquid) and to a short description of the method and criteria used to evaluate the concentration profiles. Further details can be found in refs. 33 and 34.

For our model reaction corresponding to eqn. 15.1 with Langmuir-Hinshelwood-type kinetics (eqn. 15.7), the individual concentration profiles in the liquid

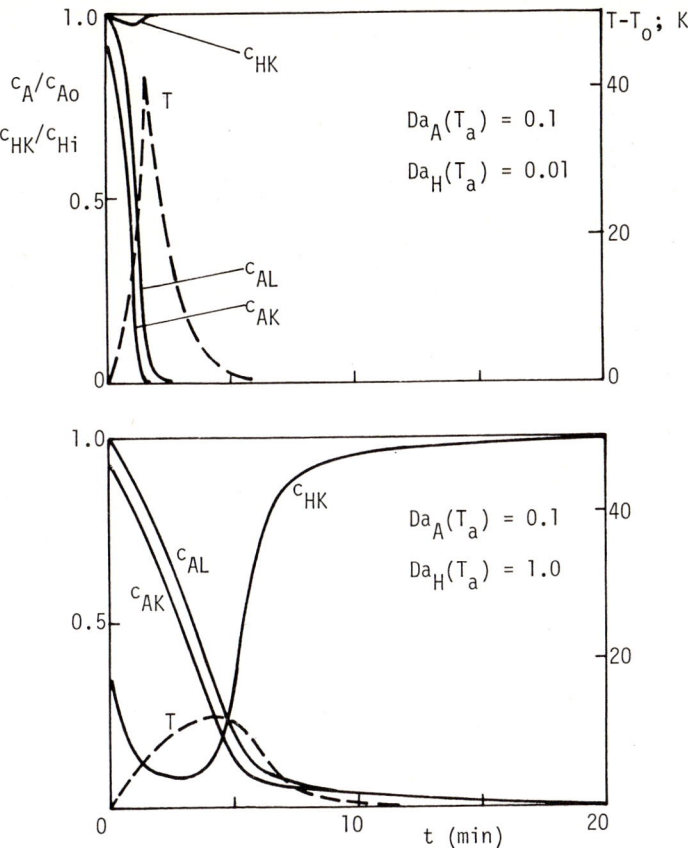

Fig. 15.16 Concentration and temperature profiles for a catalytic hydrogenation at constant pressure.

Simulation with: c_{Ao} = 1.5 mol/l; m_K k^* = 0.176 exp 4811 (1/T_a - 1/T) mol/l min; K_{HPH}/H = 4.8; Q = K_R/K_A = 0.65; E_{diff} = 10 kJ/mol; ρ c_p = 1.6 J/cm³ K; ΔH_R = 490 kJ/mol; T_a = 75°C.

film and in the pores at a given time can be calculated from the mass balance for a catalyst particle with radius R. It has become standard practice to use dimensionless variables. We use the following definitions:

Thiele moduli: $\phi_A = R \left(\dfrac{k_v}{c_{ALo} D_{Ae}}\right)^{1/2}$; $\phi_H = R \left(\dfrac{\nu_H k_v K_H}{D_{He}}\right)^{1/2}$ (15.55)

Biot number: $Bi = \dfrac{k'_{iL} R}{D_{ie}}$ (15.56)

$Q = K_R/K_A$; $C_A = c_A/c_{ALo}$; $C_H = K_H c_H$ (15.57)

Let us consider first the component A. In the steady-state the rate is equal to the mass flux at the particle surface (eqn. 15.58), and the concentration gradient in the film is coupled to the reaction rate by eqn. 15.4:

$$-r_A = a_K D_{Ae} \frac{dc_A}{dr}\bigg|_{r=R} \tag{15.58}$$

$$= k'_{AL} a_K (c_{AL} - c_{AK}) \tag{15.4}$$

Combining eqns. 15.4 and 15.58 gives:

$$c_{AL} - c_{AK} = \frac{D_{Ae}}{k'_{AL}} \frac{dc_A}{dr}\bigg|_{r=R} \tag{15.59}$$

A similar equation can be formulated for the hydrogen. Using dimensionless variables as indicated above, we obtain:

$$c_{AL} - c_{AK} = \frac{1}{(Bi)_A} \frac{dc_A}{dx}\bigg|_{x=1} \quad \text{(surface)} \tag{15.60}$$

$$c_{HL} - c_{HK} = \frac{1}{(Bi)_H} \frac{dc_H}{dx}\bigg|_{x=1} \quad \text{(surface)} \tag{15.61}$$

At the steady-state, the following material balances exist in the pore:

$$\frac{d^2 C_A}{dx^2} + \frac{2}{x} \frac{dC_A}{dx} = \phi_A^2 \{\frac{C_A}{C_A + Q(1-C_A)}\} \{\frac{C_H}{1+C_H}\} \tag{15.62}$$

$$\frac{d^2 C_H}{dx^2} + \frac{2}{x} \frac{dC_H}{dx} = \phi_H^2 \{\frac{C_A}{C_A + Q(1-C_A)}\} \{\frac{C_H}{1+C_H}\} \tag{15.63}$$

with the boundary conditions:

$$\begin{aligned} &x = 0; \quad dC_A/dx = dC_H/dx = 0 \\ &x = 1; \quad C_A = C_{AK}; \quad C_H = C_{HK} \end{aligned} \tag{15.64}$$

In order to evaluate the concentrations of substrate A and hydrogen H as a function of the pore length and hence be able to derive the expression for the rate in the pore, eqns. 15.60 - 15.63 must be solved. For this reaction model, the concentrations cannot be written explicitly and therefore an analytical solution of the equations is not possible. However, this set of differential equations can be solved simultaneously using numerical methods to find the concentration at any point in the film and the pore.

Concentration profiles for the conditions Bi = constant, $\phi_A/\phi_H = 5$, $Bi/\phi_A =$

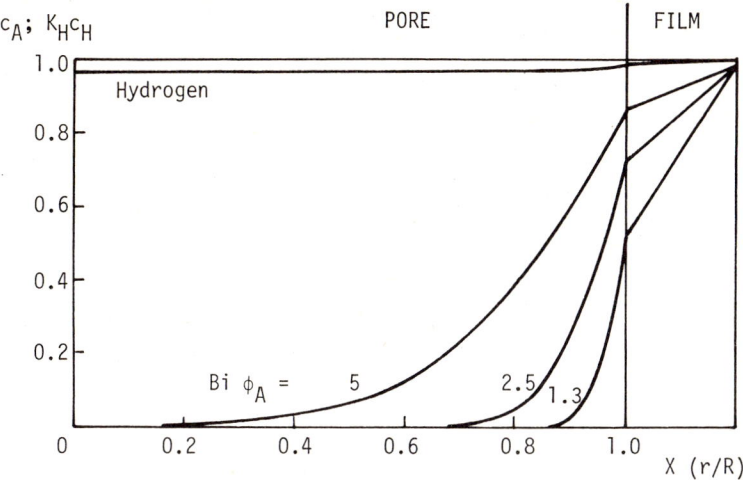

Fig. 15.17 Liquid-solid mass-transfer resistance and pore-diffusion limitation by the substrate (eqns. 15.58 - 15.62). $\phi_A/\phi_H = 5$; $Q = 0.10$; $Bi = 25$.

variable are presented in Fig. 15.17. Such calculations for different sets of mass-transfer parameters show that the relative importance of individual diffusion limitations is determined by the ratios of the Biot numbers and the Thiele modulus of the limiting component. As was shown for first-order kinetics, the effects of the film-transfer limitation can be neglected if $Bi \gg \phi$. A large concentration gradient in the film is expected only when $Bi \ll \phi$. In the case of hydrogenations on finely powered catalysts, the minimum liquid-solid mass-transfer rate for a single sphere resting in a large volume of stagnant liquid is given by (ref. 1):

$$Sh = k_L' R/D = 1 \qquad (15.65)$$

Using this boundary value for k_L', the lower limit of Bi for liquid-phase hydrogenation in slurries can be estimated:

$$(Bi)_{min} = k_L' R/D_e = D/D_e = \tau/\varepsilon \geq 1 \qquad (15.66)$$

Now the condition for severe film-transfer limitation can be rewritten and is given by $\phi \gg 1$. This means that, in liquid-phase hydrogenations using finely powdered catalysts, it is not expected that there will be important concentration gradients in the film around the particle without a simultaneous strong effect on the overall rate caused by pore-transport phenomena. The model presented here can be used to simulate conversion profiles, e.g., in a slurry batch reactor. It can be expanded to a competitive-consecutive reaction. In this case it may be used to describe the conversion and selectivity changes due to transfer limitations, e.g., as in Fig. 15.18.

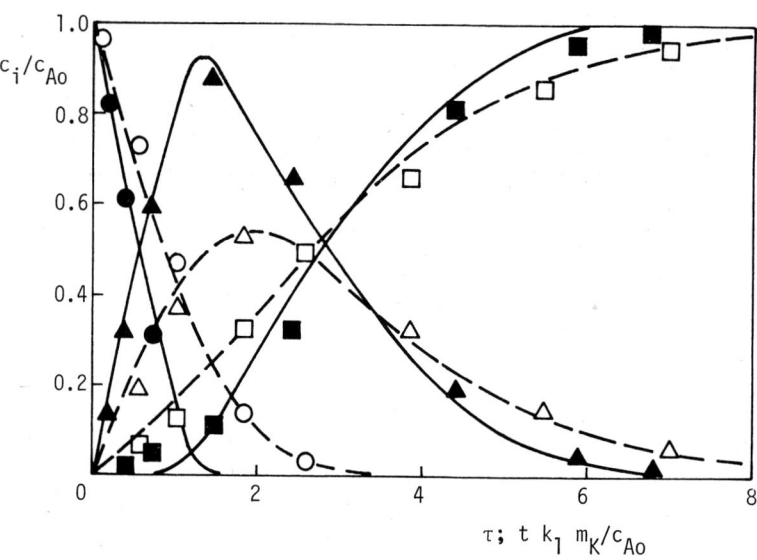

Fig. 15.18 Hydrogenation of 2,6-dinitrotoluene: Effects of pore-diffusion phenomena on selectivity. Curves calculated with model equations. T = 60°C; p_H = 50 bar. d_p: 2,6-dinitrotoluene (● 63 μm, ○ 177-245 μm), 6-aminotoluene (▲ 63 μm, △ 177-245 μm), 2,6-diaminotoluene (■ 63 μm, □ 177-245 μm).

15.5 NOTATION

A	Heat transfer area (m^2)
a	Specific surface area of gas-liquid interface (cm^2/cm^3)
a_K	Specific geometric surface area of catalyst particles (cm^2/cm^3)
Bi	Biot number for mass transfer (eqn. 15.56)
C_i	Dimensionless concentration of component i (eqn. 15.57)
c_i	Concentration of component i (mol/l)
c_G^*	Poison concentration on the catalyst surface (mol/g)
c_{Hi}	Saturation concentration of hydrogen in the liquid (mol/l)
c_p	Heat capacity (J/g K)
D	Molecular diffusion coefficient (cm^2/s)
D_e	Effective diffusion coefficient (cm^2/s)

Da	Damköhler number
d_p	Diameter of catalyst particle (cm)
E	Activation energy (kJ/mol)
g	Acceleration due to gravity (cm/s^2)
H	Henry's constant (1 bar/mol)
ΔH_R	Reaction enthalpy (kJ/mol)
K_i	Adsorption equilibrium constant (1/mol)
k	Pressure-dependent chemical rate constant (mol/l min, w/w%)
k*	Pressure-independent chemical rate constant (mol/l min, w/w%)
k_V	Pressure-dependent chemical rate constant based on catalyst volume (mol/cm$_K^3$ s)
k_i	Adsorption rate constant (one subscript) (min, w/w%)$^{-1}$
k_{iL}	Gas-liquid transfer coefficient (two subscripts) (cm/min)
k'_{iL}	Liquid-solid transfer coefficient (two subscripts) (cm/min)
m_i	Mass of component i (g)
m_K	Amount of catalyst (w/w%)
m_{Ko}	Amount of catalyst poisoned (w/w%)
N	Stirrer speed (rpm)
n	Order of reaction
P	Specific power input (erg/s g$_{slurry}$)
p_H	Hydrogen pressure (bar)
Q	Relative adsorptivity (eqn. 15.11)
R	Radius of catalyst particle (cm), gas constant (J/mol K)
$-r_i$	Reaction rate based on component i (mol/l min)
Sh	Sherwood number
T	Temperature (K, °C)
T_a	Coolant temperature (K, °C)
t	Time (s, min)
U	Overall heat-transfer coefficient (kJ/K min m^2)
V	Reactor volume (l, m^3)
V*	Volumetric feed rate (l/min)
x	Distance in direction of diffusion
x_A	Fractional conversion of component A
α, β	Constant used in eqns. 15.49 and 15.48B respectively
ε	Porosity of catalyst
η	Internal effectiveness factor
η_i	External effectiveness facor for component i
θ_i	Fractional occupancy of active sites by component i

μ	Dynamic viscosity (g/cm s)
ν	Stoichiometric coefficient
ρ	Density (g/cm^3)
τ	Tortuosity factor, mean residence time
ϕ_i	Thiele modulus (eqn. 15.55)
ϕ	Constant defined by eqn. 15.35
ϕ_S	Dimensionless modulus for pore diffusion based on particle diameter (eqn. 15.37)
κ	Constant defined by eqn. 15.34

Subscripts

A	Substrate
G	Poison
H	Hydrogen
K	Catalyst
L	Liquid phase, solvent
P	Particle
R	Product, reactor
o	Initial value

REFERENCES

1. C.N.Satterfield, Mass Transfer in Heterogeneous Catalysis, MIT Press, Cambridge, MA, 1970.
2. G.N.Roberts, in P.N.Rylander and H.Greenfield (Editors), Catalysis in Organic Synthesis, Academic Press, New York, 1976, p. 1-48.
3. S.Goto and J.M.Smith, AIChE J., 24 (1978) 286-293.
4. R.V.Chaudhari and P.A.Ramachandran, AIChE J., 26 (1980) 177-201.
5. R.V.Chaudhari and P.A.Ramachandran, Ind.Eng.Chem.Fundam., 19 (1980) 201-206.
6. P.A.Ramachandran and R.V.Chaudhari, Three-Phase Catalytic Reactors, Gordon and Breach, New York, 1983.
7. J.C.Jungers, Cinétique Chimique Appliquée, Soc.Ed.Technique, Paris, 1958.
8. K.Radhakrishnan, P.A.Ramachandran, P.H.Brahme and R.V.Chaudhari, J.Chem. Eng.Data, 28 (1983) 1-4.
9. G.Gut, R.U.Meier, J.J.Zwicky and O.M.Kut, Chimia, 29 (1975) 295-299.
10. J.J.Zwicky and G.Gut, Chem.Eng.Sci., 33 (1978) 1363-1369.
11. O.M.Kut, T.Bühlmann, F.Mayer and G.Gut, Ind.Eng.Chem.Process Des.Dev., 23 (1984) 335-337.
12. L.Cerveny, V.Heral and V.Ruzicka, Chem.Tech. (Leipzig), 35 (1983) 579-583.
13. G.Gut and T.Bühlmann, Chimia, 35 (1981) 64-68.
14. T.K.Sherwood and E.J.Farkas, Chem.Eng.Sci., 21 (1966) 573-582.
15. S.Sideman, O.Hortaçsu and J.W.Fulton, Ind.Eng.Chem., 58(7) (1966) 32-47.
16. T.Reith and W.J.Beek, Proc. 4th Europ.Symp.Chem.React.Eng., Bruxelles, 1968, Pergamon Press, Oxford, 1971, p.191-204.
17. L.L.van Dierendonck, J.M.H.Fortuin and D.Venderbos, Proc. 4th Europ.Symp. Chem.React.Eng., Bruxelles, 1968. Pergamon Press, Oxford, 1971, p.205-215.

18 T.Sridhar and O.E.Potter, Chem.Eng.Sci., 35 (1980) 683-695.
19 P.H.Calderbank, Trans.Inst.Chem.Eng., 36 (1958) 443-459.
20 M.Midoux and J.C.Charpentier, Entropie, 101 (1981) 3-31.
21 T.Bühlmann, G.Gut and O.M.Kut, Chimia, 36 (1982) 469-474.
22 G.J.K.Acres and B.J.Cooper, J.Appl.Chem.Biotechnol., 22 (1972) 769-785.
23 P.L.T.Brian and H.B.Hales, AIChE J., 15 (1969) 419-425.
24 J.J.Carberry, Chemical and Catalytic Reaction Engineering, McGraw-Hill, New York, 1976.
25 G.F.Froment and K.B.Bischoff, Chemical Reactor Analysis and Design, Wiley, New York, 1979.
26 G.Gut, SwissChem, 5/9a (1983) 13-23.
27 G.Gut, Proc. Eur.Conf.Evaluation of Thermic Hazards and Prevention of Runaway Chemical Reactions, Zurich, 1982, Oyez Scientific and Techn.Serv.Ltd., London, 1982, p.1-21.
28 R.Aris, Chem.Eng.Sci., 6 (1957) 262-268.
29 A.Ermakova and G.K.Ziganshin, Theor.Osn.Khim.Tekhnol., 4 (1970) 286-289. Eng.Transl.: Theor.Found.Chem.Eng., 4 (1970) 271-274.
30 B.D.Kulkarni and N.G.Karanth, Chem.Eng.Commun., 2 (1978) 265-269.
31 A.N.Namjoshi, B.D.Kulkarni and L.K.Doraiswamy, AIChE J., 30 (1984) 915-924.
32 B.N.Mehta and R.Aris, Chem.Eng.Sci., 26 (1971) 1699-1712.
33 F.Yücelen, Diss. No. 7544, Swiss Federal Institute of Technology, Zurich, 1984.
34 D.Wagner, Diss. No. 7761, Swiss Federal Institute of Technology, Zurich, 1985.

Chapter 16

APPLICATION OF FIXED-BED REACTORS TO LIQUID-PHASE HYDROGENATION

JIŘÍ HANIKA[a] and VLADIMÍR STANĚK[b]
[a]Department of Organic Technology, Prague Institute of Chemical Technology,
166 28 Prague 6 (Czechoslovakia)
[b]Institute of Chemical Process Fundamentals, Czechoslovak Academy of Sciences,
165 02 Prague 6 (Czechoslovakia)

16.1 INTRODUCTION

Multi-phase reactions may be performed either in slurry reactors usually in a batchwise manner, or continuously in fixed-bed reactors. Depending upon the method of addition of the liquid reaction mixture, we speak either of bubble columns (the liquid enters at the bottom concurrently with the gas) or trickle-bed reactors (TBRs), where the reaction mixture trickles down the elements of the catalytic bed, usually concurrently with the gaseous reactant. In the former system the liquid and gas flow-rates are limited by the necessity of avoiding loss of the catalyst. Moreover, a flooded bed reactor may be unstable, system runaway being initiated by hot spots formed near the gas distributor in the case of exothermic reactions.

Attention will be focused on the TBRs because of their great importance in hydrogenation and hydrotreatment processes in organic and petrochemical industries. A considerable advantage of such systems is the small liquid hold up in the reactor, which considerably reduces the possibility of homogeneous side reactions.

Three-phase catalytic hydrogenations will always occur when the volatilities of the two reactants, i.e., hydrogen and substrate, are very different, and thus two-fluid-phase reactors result from economic as well as technical considerations. Almost all natural feedstocks contain non-volatile heat--sensitive molecules and thus they must be processed in the liquid phase at low temperatures. Many hydrodesulphuration and new hydrocracking plants involve such processes. Other industrial applications of three-phase catalytic fixed-bed reactors have been reviewed (ref. 1).

A trickle-bed reactor is a concurrent downflow packed column. It provides a good means of carrying out a reaction in which gaseous and liquid reactants are to be contacted with catalyst particles or an inert packing. Three-phase TBRs have not been studied as intensively as the gas-phase reaction systems, despite their extensive use in petrochemical and organic technologies.

16.2 LITERATURE SURVEY

One of the first practical applications of a concurrent three-phase fixed-bed reactor was the synthesis of butynediol (ref. 2). Its importance in the chemical and petrol industries stimulated further research, and new papers and patents were published concerning petrol refining (refs. 1 - 11), various hydrogenations (refs. 12 - 17), selective hydrogenation (ref. 18), oxidation of organic compounds in waste waters (refs. 19 - 21), treatment of effluent gases (refs. 22 - 23), production of hydrogen peroxide (ref. 24), deuterium exchange reactions (refs. 25 - 26), reactions of immobilized enzymes (ref. 27), etc. The various applications have been reviewed (refs. 28 - 31).

The fundamental problem concerning the operation and design of TBRs is to obtain a reproducible relationship between the reactor productivity and the operating and system variables such as the temperature, pressure, reaction mixture composition and feed rate, reactor dimensions, catalyst particle diameter, etc. (ref. 32). An adequate model of the three-phase fixed-bed reactor should involve the interactions with the fluid flow dynamics as well as the resistance to mass and heat transfer between the individual phases.

Shah (ref. 29) has reviewed the advantages of fixed-bed reactors, and many other studies, e.g., (refs. 15, 33 - 36), have verified individual factors affecting reactor performance. In the last decade, several authors have critically reviewed (refs. 24, 32, 37, 38) the literature concerning the behaviour of multi-phase fixed-bed reactors. The advantages as well as the drawbacks of fixed-bed versus slurry reactors have also been reviewed (refs. 24, 27, 32, 33, 37 - 40). The need for the separation of the deactivated powdered catalyst and the higher installation costs of slurry reactors means that fixed-bed reactors are preferred. In addition, the flow in fixed bed reactors is close to the plug flow, which is very convenient if high degrees of conversion are to be achieved. On the other hand, the effectiveness factor of the catalyst pellets in packed beds is usually low due to their large size. The mechanical strength of the pellets must be also greater to avoid erosion by the liquid reaction mixture.

16.3 PARAMETERS AFFECTING THE PRODUCTIVITY OF MULTI-PHASE FIXED-BED REACTORS

The manner of operation of a fixed-bed reactor is affected by a number of factors which may be classified into several groups (ref. 38):

(a) Reactor construction parameters (diameter, length, inlet liquid distribution and location of redistributors for liquid, efficiency of heat exchange)

(b) Catalyst properties (diameter of pellet, activity, selectivity and long-term stability, radial distribution of the active component on the cross-section of the pellet, mechanical strength)

(c) Controlling factors affecting the reactor productivity (inlet temperature, operating pressure, feed composition and rate)

(d) Chemical reaction data (reaction rate, reaction orders with respect to individual components, reaction enthalpy, activation energy)

(e) Physico-chemical properties of the reaction mixture (heat capacities, diffusion coefficients, thermal conductivities, viscosities, etc.).
These quantities, however, are more or less fixed, but knowledge of them, in some cases, permits one to understand the specific behaviour of the reactor.

The instantaneous state of the system is affected by the character and/or value of all the parameters cited above. It is clear that the optimum operation and design of multi-phase fixed-bed reactors represents a complex problem.

16.4 MASS AND HEAT TRANSFER IN THREE-PHASE FIXED-BED REACTORS

The design and scale-up of multiphase reactors is generally very difficult (refs. 41, 42). The reactor performance depends not only on the reaction itself, but also on the fluid dynamics and transport processes of the reaction components between the individual phases of the reaction system.

In case of a hydrogenation reaction it is possible to distinguish the steps which may affect the operation of the reactor itself. The principal elementary steps of mass transfer between the three phases may be classified as follows (ref. 15):
- mass transfer of hydrogen from the bulk gas phase toward the external surface of the liquid film flowing down the external catalytic surface
- convective-diffusion transfer of hydrogen and the hydrogenated substrate present in the flowing liquid film and in liquid menisci, formed at the contact points of neighbouring catalyst pellets
- internal diffusion of the reactants and products within the pores of the catalyst pellet.

The mass transfer is also significantly affected by the liquid maldistribution in a randomly packed catalytic bed.

The reaction heat evolved, typically for hydrogenations carried out at high pressures in TBRs, markedly influences the thermal regime of the reactor.

Analyses of the basic relationships between the heat transfer and fluid flow in trickle-bed reactors have been published (refs. 43 - 47). In several papers (ref. 13, 16, 34, 48, 49) it has been reported that the reaction mixture may undergo changes in phase under certain conditions, with corresponding marked changes in the reaction rate. The increase in the reaction rate in the case of evaporation of the reaction mixture is the result of the lowered resistance to mass transfer in the gaseous phase (ref. 13). This may have an undesirable effect in an exothermic reaction, as it gives rise to a rather narrow reaction zone with steep temperature gradients (ref. 48, 49). Thus, the catalyst may be exposed to local overheating, which results in subsequent deactivation of the bed or the occurrence of undesirable side reactions. Furthermore, if the heat removed is insufficient, a hot spot may occur, resulting in the appearance of multiple steady states in the system. Also, in many hydroprocessing TBRs, such as the hydrocracker and the hydrodesulphuration reactors operated under extreme conditions, the extent of evaporation could be as high as 90 or 100 percent of liquid feed. The modelling of such reactors is therefore very complex.

16.5 STRUCTURAL PROPERTIES OF FIXED CATALYTIC BEDS

An important factor for effective operation of trickle-bed reactors is the creation of large catalyst/liquid and liquid/gas interfaces. The former mediates transfer of the reactants to the active catalytic surface, and in the opposite direction of products into the bulk liquid. The latter mediates transport of gaseous components participating in the reaction, i.e., hydrogen in the case of hydrogenation reactions. If this type of transport is weak, the overall rate of hydrogenation may be limited by the so-called external resistance to transfer.

Since the external surface of the catalyst is virtually never completely covered by the hydrogenated substrate, the presence of the catalyst/gas interface, (i.e., hydrogen plus vapours of the reaction mixture), allows the possibility of gas-phase hydrogenation. In this context, we note that the degree of wetting of the catalytic surface usually decreases with decreasing density of irrigation of the bed, and it is also affected by the physical properties of the liquid and the solid surface.

As far as hydrogenation reactions are concerned, the orientation of the three mentioned interfaces is usually immaterial. The catalyst particles are therefore simply freely dumped into the reactor and we may speak of a random layer of particles. In an ideal random layer of equal size particles, each particle has the same probability of occupying an arbitrary position in the bed.

Every volume element of a perfectly random layer thus has the same void fraction and the same catalyst/fluid interface or, with perfect wetting of the layer, the catalyst/hydrogenated substrate interface. These properties are important from the standpoint of the hydrogenation itself, as well as the physical phenomena affecting the hydrogenation. They are necessary, but not sufficient prerequisites for equal phase velocities, equal rates of energy dissipation and equal intensities of transfer in each volume element on an arbitrary horizontal level.

In real fixed-bed reactors, however, deviations from perfect randomness of the catalytic layer virtually always occur. Of these, the most significant for hydrogenation reactors appear to be deviations due to the presence of surfaces

confining the layer spatially (wall, supporting grid) and due to the different
methods of preparation of the layer (filling the reactor).

In catalytic reactors the layer of catalyst is confined in the horizontal
direction by the reactor walls and in the vertical direction by the supporting
grid or bottom. The radius of curvature of these surfaces is an order of
magnitude greater than that of the catalyst particles themselves. These
differences result in local structural inhomogenities of the bed and the
ultimate result is an oscillatory course of the void fraction in the proximity
of the confining surfaces and the preferential flow of both liquid and gas
through these regions. Details of the character of these oscillations may be
found, for instance, in papers by Staněk and Eckert (refs. 50, 51). Since also
the upper surface of the bed of catalytic particles is usually levelled,
analogous oscillations are observed here. Into the same category of structural
deviations fall the disturbance of randomness near the thermocouple wells,
cooling coils and other fixtures in the reactor.

The method of filling the reactor also affects the structure of the bed. It
has been observed that different filling methods lead to layers of different
compactness. For layers of spherical particles (not typical for catalyst
pellets), the void fraction ranges between 0.36 for "dense" and 0.40 for "thin"
layers.

For extrusion-moulded particles, as is generally the case for catalyst
pellets, the differences between freely dumped and dense layers may be as much
as 15 % in terms of catalyst volume (ref. 52). In this case the increased
volume of the catalyst per unit reactor volume not only increases the
productivity but also changes the prevailing orientation of individual catalyst
extrudates predominantly to horizontal. Beds with predominantly horizontally
located extrudate particles usually display superior properties from the
standpoint of liquid distribution. On the contrary, freely dumped beds of
extrudates tend to form isochronous (conical) surfaces slanting toward the
wall. Such beds are then prone to increased flow of the liquid substrate toward
the walls.

Another potential source of trouble is the formation of fines in the
originally monodisperse layers. Fines may appear as a consequence of mechanical
abrasion and/or thermal strain during operation. Deposits of fines increase the
local resistance to the flow of fluid phases, in turn increasing the probability
of additional deposition of fines. Deposits may also be caused by undesirable
polymerization reactions.

The relative magnitude of the preferential flow of liquid (wall flow) and
gas in the proximity of the confining wall is closely related to the relative
size of the catalyst particles, expressed by the ratio of the reactor diameter
to the nominal size of the catalyst pellets. Since, however, the gaseous

reactant, i.e., hydrogen, is generally in excess with respect to the liquid reactants, or in some cases is practically pure, a more significant flow inhomogeneity in trickle-bed hydrogenation reactors is that of the liquid.

As regards the size of the catalyst pellets, one should also keep in mind that a decrease in particle size would simultaneously lead to (i) an increase in the external catalyst/fluid surface, (ii) an increase in the resistance to the flow of gas, (iii) an increase in hold up of the liquid reaction mixture, (iv) an increase in the effective thermal conductivity of the bed and (v) to the already mentioned reduction in the effect of structural inhomogeneities, which influences primarily the distribution of the liquid substrate in the bed.

All these changes are not mutually independent, but, instead, are coupled through the interactions of the heat and mass flow and the course of the hydrogenation. A rational way of exploring these interactions is by the use of realistic mathematical models.

16.6 MATHEMATICAL MODELLING IN THE DESIGN AND OPERATION OF TRICKLE-BED REACTORS

From the standpoint of mathematical modelling, a trickle-bed catalytic reactor represents a very complex system, particularly because of the multitude of simultaneous transport mechanisms taking place and also in relation to the temperature dependence of the kinetics. A study of the individual transport mechanisms is often impossible.

Overly simplistic models are of limited significance as they fail to predict more complicated features of the course of hydrogenation which appear in real reactors. On the contrary, carefully formulated models, amenable to mathematical methods of solution, may become an effective tool in the design and operation of a hydrogenation reactor.

The following paragraphs summarize the results obtained with the basic model which regards the irrigated catalytic layer as a pseudohomogeneous medium with radially distributed properties (concentration, temperature, irrigation rate, etc.). The pseudohomogeneous approach, in contrast to reality, views the above properties as continuous functions of spatial coordinates.

Radial distribution is an important feature of the model, particularly because the transport properties of TBRs in the direction perpendicular to the principal direction of the flow are, as a rule, very poor. In the case of strongly exothermic hydrogenations, however, the problem of the removal of heat from the reactor may be critical and requires the design of relatively small diameter reactors (in terms of the size of the catalytic pellets). As already mentioned, in such reactors the structural effects on the homogeneity of the flow and, in turn on the course of the hydrogenation, are of particular significance.

The basic model of mutual interactions of the heat and mass transfer with the course of a heterogeneously catalyzed reaction (ref. 53) may be classified as one of pseudohomogeneous radially distributed plug flow for the trickling liquid with parallel wall flow. The principal idea of the model is that the sole mechanism responsible for radial dispersion of the reacting species (hydrogenated substrate) and heat in the reactor is the dispersion with the trickling liquid.

The model accounts for the fact that part of the liquid flows down the reactor wall in the form of the wall flow. Naturally, the wall does not exhibit catalytic properties. Mathematically, the model is represented by three partial differential equations for the concentration field of the hydrogenated substrate, temperature and the density of irrigation, plus two ordinary differential equations for the concentration of the hydrogenated substrate in the wall flow and its temperature. The analysis of these equations has shown that, under an adiabatic regime, a simple linear relationship may be written between the local temperature, T, and the local concentration, c, of the hydrogenated substrate

$$(T - T_0) \rho c_p + (c - c_0) (-\Delta H) = 0 \tag{16.1}$$

where T_0, c_0 designate the inlet temperature and concentration of the hydrogenated substrate, ρ and c_p are the density and specific heat capacity of liquid and $-\Delta H$ is the reaction heat per mole of the feed.

This equation is analogous to the expressions in the literature for the simple case of an adiabatic plug-flow reactor. Even for the radially distributed case with wall flow, it may be utilized either to reduce the numerical effort involved in the solution of the set of equations, or as a test of the numerical consistency of the computed results. In addition, it may be used to compute (after substituting c = 0) the maximum local temperature that may plausibly be reached in the reactor.

On the basis of an extensive set of numerical solutions of the model, one can divide the effects of flow inhomogeneities in the adiabatic trickle-bed reactor into two categories:

(a) Effects of the entrance region of the reactor, below the mouth of the distributor.

(b) Effects of the wall-flow region with more or less established dynamic equilibrium of the flow through the catalytic layer.

The effect of the inhomogeneities of the velocity field (a) in the entrance region becomes of particular importance for low order reactions and when the conditions in the feed are such that the reaction may commence immediately upon entering the catalytic section. Due to the flow inhomogeneities, markedly curved reaction zones may appear, which may eventually contribute to a slight increase in the overall reaction rate. A highly non-uniform flow, however, such as that arising, when feeding the liquid through a narrow centrally located nozzle, characterized by high local irrigation rates, tends to decrease the overall reaction rate. Yet, even with an uniform feed distribution, significant deviations from an uniform flow appear in the entrance region, as a consequence of the wall flow at the expense of the flow down the catalytic bed. This phenomenon, contributes to increased local reaction rates, and, may temporarily result in a slight increase in the overall reaction rate. The increase in the overall rate can be observed only at very low extents of conversion.

The effect of the wall flow (b), outside the entrance region of the reactor, where the wall flow is essentially in equilibrium, causes the part of the liquid flowing down the wall to react only gradually, due to the dynamic exchange of liquid between the wall flow and the trickle flow down the catalytic surface. In this region the wall flow always results in a decrease in the overall reaction rate.

The results of model calculations have been compared (ref. 54) with the course of hydrogenation of cyclohexene in a laboratory TBR, in the form of the dependence of the mean outlet concentration of cyclohexene on the depth of the catalytic bed. The experimental results and the predicted dependence were in relatively good agreement, considering that the predicted curves had not been optimized. The model is thus a useful tool for prediction of the extent of conversion, the necessary depth of the catalytic bed and for correct interpretation of peculiarities of the conversion curves.

An important assumption of the described model is the equivalence of the dispersion of heat and mass in the trickle-bed reactor. Even though there is evidence in support of this in the literature (agreement of the Peclet numbers for heat and mass transfer), other results reveal the contrary and show that, among possible mechanisms of radial dispersion of heat, the contribution of thermal conductivity need not be negligible. The importance of the conductivity contribution to the net heat transfer is relatively higher particularly at low densities of irrigation and low gas flow-rates. A significant role is also played by the material of the catalytic pellets, i.e., its thermal conductivity. Since in catalytic trickle-bed reactors the density of irrigation is, as a rule, very low, the conductivity contribution to the overall heat dispersion practically always represents a significant portion of the dispersion transport. At low densities of irrigation, the heat transported axially by convection together with the flowing gas may also be important.

Both of these additional transport mechanisms have therefore been incorporated into the model and used to modify the heat-balance equations (ref. 55). An analysis of the modified model shows that, with simultaneous conduction of heat through the catalyst particles themselves, eqn. (16.1) no longer holds. An analogous expression can be derived only for adiabatic operation of the reactor and zero gas velocity, and even then this is valid only between the integral mean temperature and integral mean concentration in the reactor. Nevertheless, it can be utilized to evaluate an upper limit for the local temperature in the reactor, while real values are expected to be lower.

The modified model was also applied to the hydrogenation of cyclohexene in a trickle-bed reactor. As the heat of this reaction (142 kJ/mol) considerably exceeds the heat of evaporation (30.5 kJ/mol), it is obvious that at normal pressures significant evaporation of the reaction mixture may occur due to the heat evolved. The two values quoted show that the evaporation can occur even in fairly diluted reaction mixtures.

Experimental studies have shown that under atmospheric pressure the rate of hydrogenation of cyclohexene increases with increasing temperature up to about $60^\circ C$. At higher temperatures the rate of liquid-phase hydrogenation should decrease, in accord with the significant decrease in partial pressure of hydrogen in the gas phase. In reality, the overall rate of hydrogenation remains approximately constant as the decrease in the liquid-phase reaction is offset by the gas-phase hydrogenation on the dry spots of the catalyst surface. Near the boiling point, the reaction mixture is evaporated with a correspondingly marked increase in the overall rate of hydrogenation.

The modified mathematical model has been solved (ref. 56), with the reaction term which expresses the just outlined properties of the kinetics of cyclohexene hydrogenation. A variety of conditions were employed and corresponding isotherms and contours of constant cyclohexene concentration have been obtained. In the region where the temperature reaches or even exceeds the boiling point of the reaction mixture, in a real situation are would expect the formation of a hot spot leading to considerably increased reaction temperature and eventual damage to the catalytic activity of the layer. The predicition of these spots by the model is only approximate, for the evaporation of the reaction mixture violates the validity of the model.

The analysis of a large number of model solutions of the course of the hydrogenation of cyclohexene under various conditions in an externally cooled reactor has led to the following conclusions.

If the reaction conditions (inlet temperature and concentration) permit evaporation of the reaction mixture under adiabatic conditions, the elimination of hot spots by means of external cooling is difficult. Even extremes of cooling are likely to achieve only restriction of the extent of the hot spot rather than its entire elimination.

Whether the hot spot forms in the reactor depends also on the method of

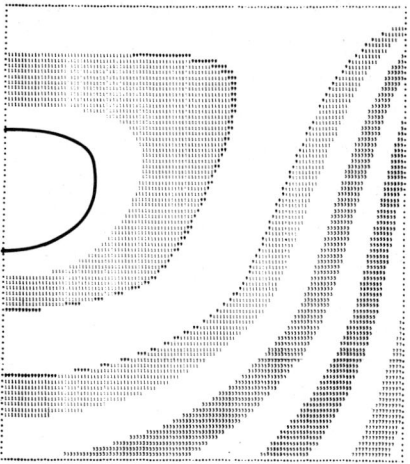

Fig. 16.1. Isotherms in a cooled reactor for hydrogenation of cyclohexene, with an uniform feed of the reaction mixture at 70°C. Temperature of the coolant: 20°C. Dimensionless effective thermal conductivity of the catalytic bed 0.1.

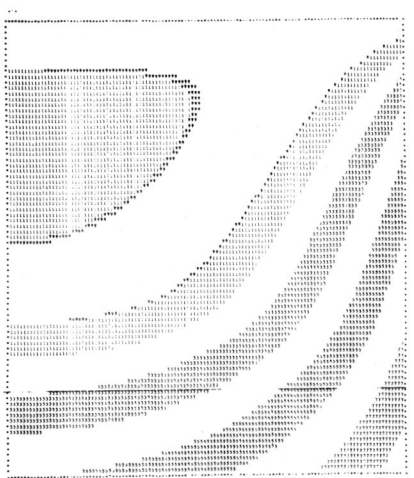

Fig. 16.2. Isotherms in a cooled reactor for hydrogenation of cyclohexene. Conditions as in Fig. 16.1., except that the dimensionless effective thermal conductivity of the catalytic bed was 0.29.

addition of the liquid to the reactor. If, for instance, the feed is brought into the reactor through a narrow central nozzle, evaporation may occur even though this would not be the case under an uniform feed of the reaction mixture under otherwise identical conditions. The inhomogeneity of the flow in the case of the central nozzle, though inducing, as mentioned earlier, lower overall reaction rates, hampers more significantly the radial transport of heat into the cooling jacket.

A relatively effective means of suppressing the appearance of hot spots in cooled trickle-bed reactors is increased effective thermal conductivity of the bed proper. The situation is illustrated graphically in Figs. 16.1. and 16.2. which depict isotherms computed from the model for a plane passing through the axis of a cylindrical reactor. In both cases the feed is supplied uniformly over the upper cross-section at a temperature of $70^{o}C$, while the temperature of the cooling liquid in the jacket is $20^{o}C$. They differ in the effective thermal conductivity of the catalytic layer, having values of 0.1 (see Fig. 16.1.) and 0.29 (see Fig. 16.2.). The left-hand vertical edge of each figure represents the axis of the reactor, the right-hand edge the wall of the reactor. In case of the lower effective thermal conductivity a limited hot spot appears in the reactor, shown in Fig. 16.1. by a solid contour. A more intensive radial transport of heat, typical for the case in Fig. 16.2., facilitated by the higher thermal conductivity of the catalyst particles, eliminated the hot spot. The differences between these cases are also apparent in the overall patterns of the isotherms, which are the less steep (lower temperature gradients) at higher intensities of radial transport. This is favourable for the operation of the reactor and the course of the hydrogenation.

An increase in the flow-rate of the gas phase also contributes to more intensive heat transfer. For example, increasing the flow-rate of hydrogen by a factor of four leads, according to the model calculations, to disappearance of the hot spot. Such increases also brought about significant changes in the overall pattern of isotherms.

The model calculations thus reveal relatively effective means of affecting the temperature regime of the reactor via modification of the feed, gas flow--rate and the non-catalytic properties of the catalyst particles.

The appearance of hot spots in a TBR has been systematically studied, again in the case of hydrogenation of cyclohexene in a laboratory reactor. Selected results are presented in the following paragraphs.

16.7 HYDROGENATION OF CYCLOHEXENE IN A TRICKLE-BED REACTOR

The degree of complexity of multi-phase fixed-bed reactors will be demonstrated in this chapter by presenting selected experimental results, obtained (ref. 55) for the hydrogenation of cyclohexene in a laboratory trickle-

-bed reactor. The progress of the hydrogenation was monitored from axial temperature profiles by a computing data logger, both in the steady state and the dynamic regimes.

The experiments were devised so as to determine the influence of changes in the depth of the catalytic bed on the productivity of the reactor with external cooling. Various feed rates and inlet temperatures of the feed were employed. Also presented will be results of measurements of the steady-state axial temperature profiles at various feed rates of the reaction mixture, concentrations of the hydrogenated substrate in the feed, types of distribution of the liquid feed over the upper surface of the bed, flow-rates of hydrogen, etc.

The experiments were also aimed at exploring the dynamic behaviour of the reactor: transient development of the productivity and transient temperature profiles following a change in the feed rate of the reaction mixture or a change in the concentration of the feed.

16.7.1 Experimental set-ups and reaction conditions

The behaviour of the reactor was studied using the hydrogenation of cyclohexene as a model reaction. The product, cyclohexane, was used as a solvent. The reaction was catalyzed by a commercially available carrier--supported palladium catalyst CHEROX 41-00 (3 % of Pd on activated carbon). Cylindrical extrudates 3.5 mm in diameter and 5 mm long were used.

The hydrogenation reaction is as follows:

$$\text{cyclohexene} + H_2 \longrightarrow \text{cyclohexane} \qquad -117.86^{198\ K}\ kJ/mol$$

At elevated reaction temperatures, particularly in hot reaction zones of the catalytic bed, the formation of benzene was observed. This by-product was produced by the following reaction:

$$3\ \text{cyclohexene} \longrightarrow \text{benzene} + 2\ \text{cyclohexane} \qquad -147.38^{298\ K}\ kJ/mol$$

This side reaction permitted us to ascertain the relationship between the formation of the hot spots in the catalytic layer and the selectivity of the hydrogenation.

According to earlier results (ref. 13), the liquid-phase hydrogenation of cyclohexene is zero-order with respect to cyclohexene. In the gas phase this reaction is first order with respect to the substrate. With respect to hydrogen, the reaction order is 0.63, regardless of the phase in which the hydrogenation takes place.

The experiments were carried out (ref. 55) under atmospheric pressure. The reactor proper was a glass tube, 30 mm in diameter and 350 mm long, provided with a cooling jacket. This tube was equipped with 16 side arms 20 mm apart, housing copper-constantan thermocouples. The measuring points were located along the axis of the reactor tube. The voltages of the thermocouples were monitored by an HP 9835 A desk-top calculator, an HP 3455 digital voltmeter and an HP 3495 A scanner.

Liquid reaction mixture was brought to the top of the catalyst layer either directly through a central nozzle, or indirectly. In the latter case, a 50-mm deep layer of catalytically inactive 2-mm glass beads was placed between the catalytic section and the central nozzle. The inactive packed section facilitated a more even distribution of liquid substrate over the catalytic section. This permitted us to observe the effects of different types of liquid distribution and hence the effect of non-uniformities in the liquid flow in the catalytic section on the productivity of the reactor.

16.7.2. Steady-state operation of the trickle-bed reactor

The principal parameters of the reactor are the amount of catalyst and the feed rate of the reaction mixture. The effect of the depth of the bed on the productivity of the reactor with heat exchange was assessed at three different feed rates of a 20 % solution of cyclohexene in cyclohexane: 82 ml/h, 125 ml/h and 238 ml/h. The temperature in the cooling jacket was kept equal to the temperature of the feed, i.e., $40^{\circ}C$. A central jet was used as a distributor. In all experiments the hydrogen flow-rate was maintained at a constant level of 0.25 l/min. The performance of the reactor was characterized by the degree of conversion of cyclohexene reached on a given depth of the catalytic layer. The experimentally observed values are plotted in Fig. 16.3.

It is apparent that the dependence of the outlet conversion on the depth of the catalytic section is, for a given feed of the substrate, linear over a certain range of depths. This confirms that the reaction order with respect to the substrate is close to zero (ref. 13). It is interesting that the straight line dependence does not start from the origin; instead, particularly for lower feed rates, it appears shifted toward higher performance. The cause of this is thought to be the presence of the dry surface of the pellet immediately below the mouth of the central nozzle, where the reaction proceeds in the gas phase.

The effect of temperature on the productivity of the reactor is presented in Fig. 16.4., which indicates clearly that, particularly at 60 and $70^{\circ}C$, close to

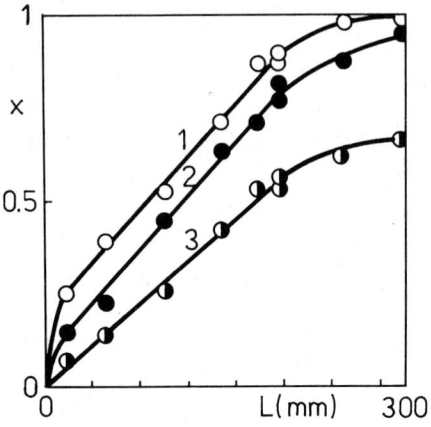

Fig. 16.3. The effect of bed depth on the extent of conversion at various liquid feed rates. T = 40°C, V_H = 0.25 l/min, C = 20 wt.%; distributor: central nozzle; (1) F = 82 ml/h, (2) F = 125 ml/h, (3) F = 238 ml/h.

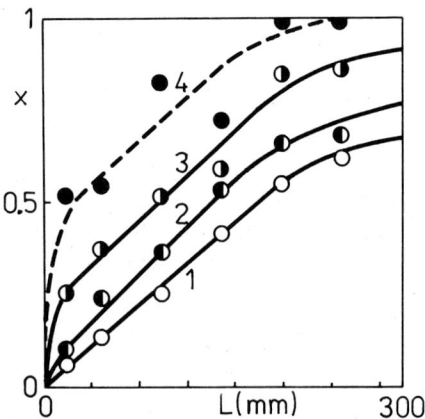

Fig. 16.4. The effect of bed depth on the extent of conversion at various feed temperatures. F = 238 ml/h, V_H = 0.25 l/min, C = 20 wt.%; distributor: central nozzle; (1) T = 40°C, (2) T = 50°C, (3) T = 60°C, (4) T = 70°C.

the boiling point of the reaction mixture (about 80°C), the hydrogenation of cyclohexene vapours becomes significant.

Nevertheless, the dependences plotted in Figs 16.3. and 16.4. reveal little about the nature of the processes of heat and mass transfer, which can be better elucidated from measurements of the temperature profiles in the catalyst bed.

The effect of the feed rate of the liquid substrate on the temperature profile is apparent from Fig. 16.5., which clearly reveals the evaporation of liquid at low feed rates in the hot spot formed near the inlet. At higher feed rates, the reactor operated in the liquid-phase regime along its entire depth. The unusual behaviour of the reactor is very important from the standpoint of its control. Evaporation causes a considerable decrease in the resistance to mass transfer both inside and outside the pellets and the reaction rate thus increases by as much as an order of magnitude (ref. 13).

Similarly, Fig. 16.6. shows the effect of the temperature of the cooling jacket, kept equal to the feed temperature. It is seen that the temperature profiles are flat as long as the reaction proceeds in the liquid-phase regime on perfectly wetted catalyst surface. The formation of a hot spot and evaporation of the liquid reaction mixture was observed near the inlet of the catalytic section when the temperature of the feed was 60°C. This again causes a marked overheating of the catalyst which may result ultimately in its deactivation or lead to undesirable side reactions.

An important parameter as regards the control of the operation of the TBR is the gas flow-rate. The effect of the hydrogen flow-rate on the outlet conversion was investigated with the central nozzle as the liquid distributor, a bed depth of 180 mm, a cooling-jacket temperature of 40°C and a liquid feed rate of 238 ml/h. The hydrogen flow-rate in the outlet gas was gradually adjusted to 0.05 l/min, 0.25 l/min and 0.50 l/min. These experiments were measured in the liquid-phase regime without hot-spot formation in the catalyst bed and the corresponding hydrogenation extent values 0.507, 0.518 and 0.547 were found.

It is apparent that an order of magnitude change in the hydrogen flow-rate has little effects on the extent of conversion obtained in the TBR. Its effect is seen in a greater turbulence of the laminar liquid film on the external surface of the catalyst pellets, which enhances the transfer of hydrogen toward the active surface of the catalyst and thereby increases the overall reaction rate. The limited effect of the hydrogen flow-rate on the operation of the reactor is also apparent from Fig. 16.7., which compares axial temperature profiles measured in the bed of catalyst under the above conditions. The reactor operated identically at various hydrogen flow-rates.

Another important parameter which may affect the operation and performance

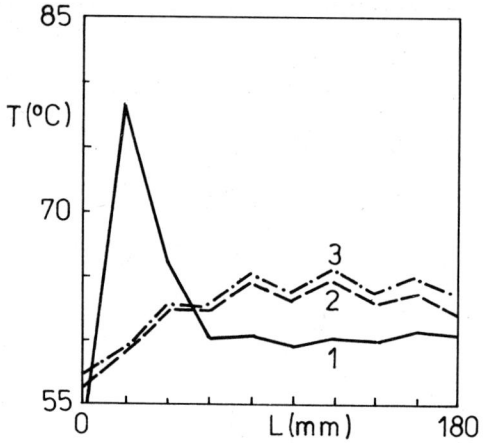

Fig. 16.5. Axial temperature profiles at different liquid feed rates. $T_o = 60°C$, $V_H = 0.25$ l/min, $C = 19$ wt.%; distributor: central nozzle; (———) (1) $F \cong 82$ ml/h, (– – –) (2) $F = 125$ ml/h, (–·–·–) (3) $F = 238$ ml/h.

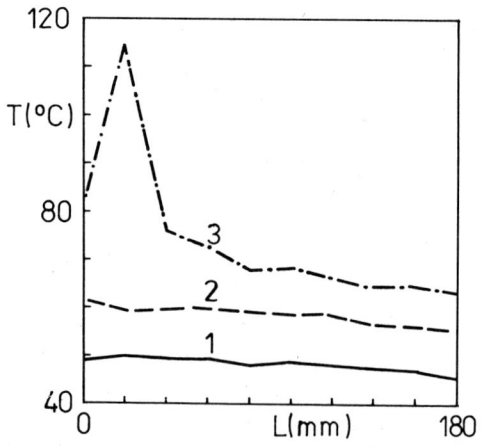

Fig. 16.6. Axial temperature profiles at different inlet and reactor-wall temperatures. $F = 238$ ml/h, $V_H = 0.25$ l/min, $C = 42$ wt.%; distributor: uniform; (———) (1) $T = 40°C$, (– – –) (2) $T = 50°C$, (–·–·–) (3) $T = 60°C$.

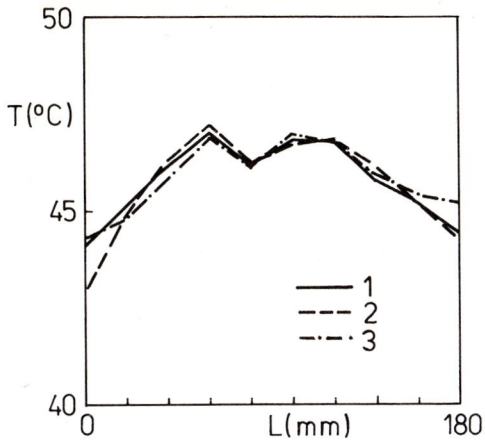

Fig. 16.7. The effect of the hydrogen flow-rate on the axial temperature profile. $T_o = 40°C$, F = 238 ml/h, C = 20 wt.%; distributor: uniform; (———) (1) V_H = 0.05 l/min, (– – – –) (2) V_H = 0.25 l/min, (–.–.–) (3) V_H = 0.5 l/min.

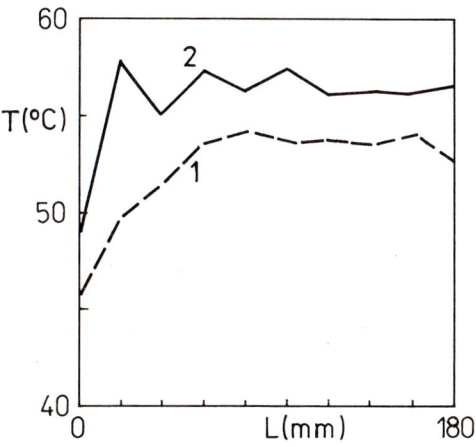

Fig. 16.8. The effect of the cyclohexene feed concentration on the axial temperature profile in the liquid-phase regime. $T_o = 50°C$, F = 238 ml/h, V_H = 0.25 l/min; distributor: central nozzle: (– – –) (1) C = 19 wt.%, (———)(2) C = 42 wt.%.

of the reactor is the concentration of the hydrogenated substrate. This is so in spite of the fact that the order of the reaction with respect to cyclohexene is close to zero. The effect of the concentration of cyclohexene on the axial temperature profile was investigated at two different concentrations: from Fig. 16.8. it is apparent that at the higher cyclohexene concentration the corresponding temperatures in the bed were higher. The reason for this is probably again the existence of the dry catalyst surface on which gas-phase hydrogenation takes place with a unit reaction order (ref. 13) with respect to the hydrogenated substrate. The dry surface results from the non-uniform flow distribution of the reaction mixture within the catalyst bed. The non-uniformities induced by the narrow central nozzle used are quite severe.

The results of the study of the effect of concentration on the performance of the reactor are summarized in Table 16.1, which shows the maximum temperatures detected in the catalytic bed, and, the rate and selectivity of the reaction.

TABLE 16.1

The effect of the cyclohexene concentration on the productivity of a trickle-bed reactor, on the maximum temperature of the bed and on the selectivity of the hydrogenation $T_0 = 70°C$, $F = 201$ ml/h, $V_H = 30$ l/h, $L = 180$ mm

Concentration (mass %)	T_{max}	Overall reaction rate (mol/sec.g_{cat}) 10^6	Selectivity
15.9	72.0	1.16	1.0
26.1	72.7	2.16	0.95
33.1	83.0	3.12	0.92
43.6	102.0	3.98	0.87
50.9	119.9	4.00	0.68
70.9	167.6	7.03	0.15
99.7	202.3	10.84	0.22

It is apparent that with increasing cyclohexene concentration in the feed mixture considerable overheating of the catalyst occurs, accompanied by an increased reaction rate and productivity. The values of the selectivity of the hydrogenation of cyclohexene, defined as the ratio of the reaction rate of the desired hydrogenation to the total rate of conversion of cyclohexene, shows that overheating of the catalyst is favourable for the disproportionation of cyclohexene to benzene and cyclohexane.

Detailed information about the reaction may be obtained from measurements of the temperatures in the catalytic bed. Temperature profiles measured along the reactor axis and near the wall under steady-state conditions and at various inlet concentrations are plotted in Fig. 16.9. It is seen that increasing the

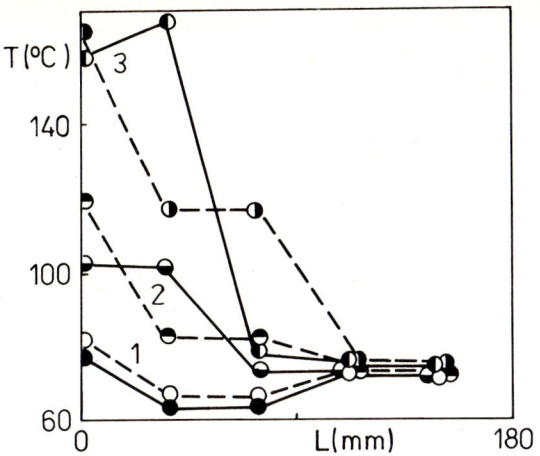

Fig. 16.9. Steady-state temperature profiles at various inlet cyclohexene concentrations. (———) on the axis, (– – –) near the wall; $T_g = 73°C$, $F = 201$ ml/h, $V_H = 0.50$ l/min; distributor: central nozzle; (1) C = 33.1 wt.%, (2) C = 50.9 wt.%, (3) C = 70.9 wt.%.

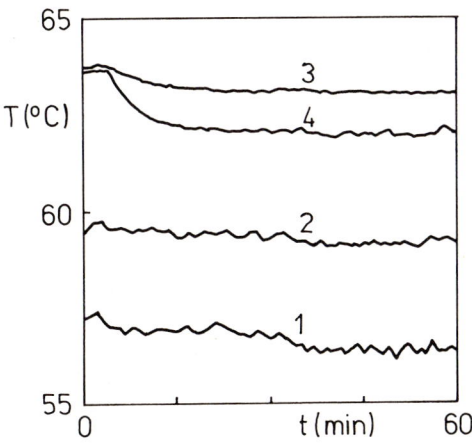

Fig. 16.10. Temperature monitoring at different axial points (bed depth is measured from the top of the bed). $T_g = 60°C$, $F = 125$ ml/h, $C = 19$ wt.%, $V_H = 0.25$ l/min; (1) L = 0 mm, (2) L = 20 mm, (3) L = 100 mm, (4) L = 180 mm; distributor: central nozzle.

concentration of cyclohexene in the feed mixture leads to an enlargement of the hot spot and the experimentally measured temperatures in this zone reach high values, considerably exceeding the boiling point of the reaction mixture. A comparison of the temperatures along the axis and along the wall of the reactor shows that at the maximum inlet concentration used the differences may be as large as $50^\circ C$.

16.7.3 Dynamic behaviour of the trickle-bed reactor

Continuous monitoring of the operation of the trickle-bed reactor by a data logger permitted examination of the temperature stability at a given point of the catalytic layer. The experimental results confirmed that, under certain conditions, fluctuations in temperature occur, in spite of the fact that the reactor is operating in the steady state. Fig. 16.10. shows the time dependence of the temperature at the inlet end (0 mm), in the inlet section (20 mm), in the middle (100 mm) and at the end (180 mm) of the catalytic section irrigated by the narrow nozzle. The maximum fluctuations (maximum $0.8^\circ C$) occur at the bed inlet and are probably due to the non-uniformities in the liquid intake. These fluctuations also influence the temperature in the upper part of the catalytic section, 20 mm from the inlet, though the amplitudes, are smaller. In the middle of the bed practically no temperature fluctuations are detected. At the end of the catalytic bed the extent of fluctuation is similar to that found 20 mm from the inlet.

A very interesting phenomenon is seen in Fig. 16.11., which shows a plot of the temperature 20 mm from the bed inlet. The temperature approaches the boiling point of the liquid phase. The steady state was not reached even after 170 min and temperature fluctuations amounted to almost $10^\circ C$. This part of the reactor operates alternatly in the liquid- and the gas-phase regimes. It is interesting that the temperature fluctuations in the upper part of the bed exercised no influence on the time course of the temperature in the middle of the bed (120 mm). This phenomenon apparently depends on the nature of the channelling of the liquid through the random catalyst bed, and it may be assumed that the spatial pattern of the individual rivulets in the bed changes with time.

Fig. 16.12. shows that the steady-state temperatures achieved under the gas-phase regime in the reactor exhibit no appreciable fluctuations. The reactor start-up can be conveniently monitored by measuring the axial temperature profiles in the bed. First one has to make sure that the whole surface of the freshly prepared bed of catalyst is wetted by liquid, usually an inert solvent. The process of wetting is accompanied by evolution of a considerable amount of heat (ref. 57). Interesting time-variable axial temperature profiles were observed experimentally during the start-up with a

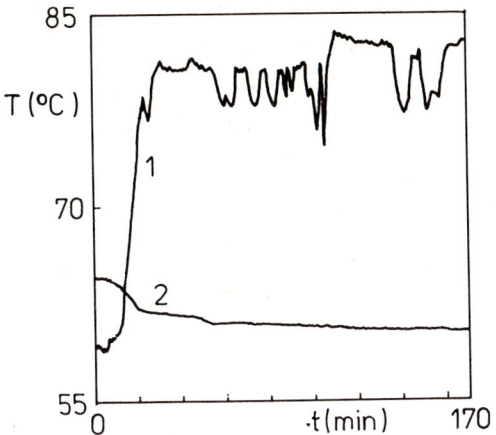

Fig. 16.11. Fluctuations with time of the temperature in a hot spot, at approximately the boiling point of the reaction mixture. $T_o = 60°C$, F = 82 ml/h, C = 19 wt.%, V_H = 0.25 l/min, distributor: central nozzle; (1) L = 20 mm, (2) L = 120 mm.

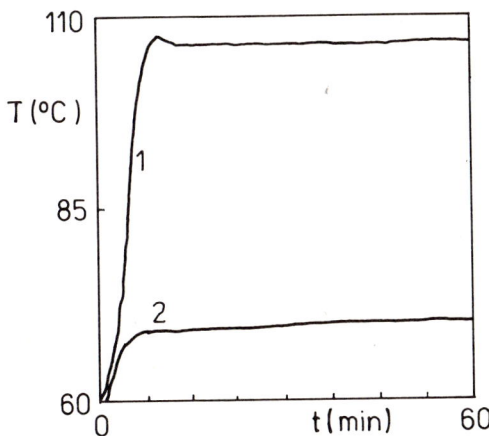

Fig. 16.12. Temperature - time dependence in the both liquid- and gas-phase regimes. $T_o = 60°C$, F = 238 ml/h, C = 41 wt.%, V_H = 0.25 l/min; distributor: central nozzle; (1) L = 20 mm, (2) L = 120 mm.

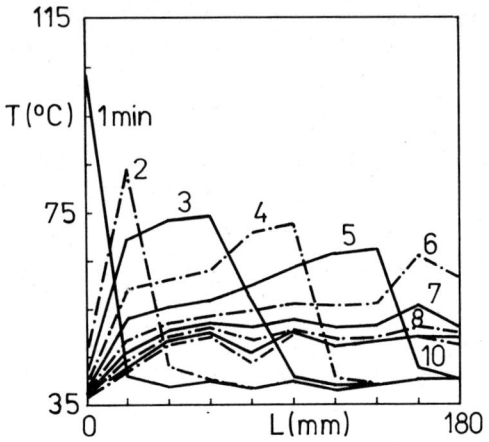

Fig. 16.13. Transient axial temperature profiles after the first contact of cyclohexane with the dry catalyst layer (numbers indicate the time in minutes from the start). $T_0 = 40°C$, $F = 125$ ml/h, $C = 0$ wt.%, $V_H = 0.25$ l/min; distributor: central nozzle.

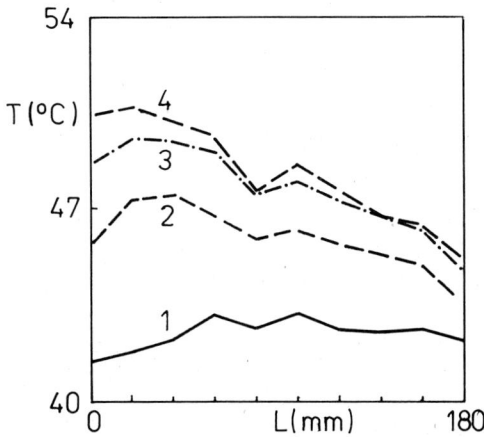

Fig. 16.14. Transient axial temperature profiles after the start of hydrogenation in the liquid-phase regime. $T_0 = 41°C$, $F = 125$ ml/h, $C = 42$ wt.%, $V_H = 0.25$ l/min; distributor: uniform; (———) (1) $t = 0$ min, (— — —) (2) $t = 5$ min, (—·—·—) (3) $t = 10$ min, (— —) (4) $t = 30$ min.

freshly prepared catalytic bed (see Fig. 16.13.). The temperatures reached in the bed testify to the evaporation of the liquid phase. After a certain period of time (about 40 min), maintaining the flow-rate of the feed liquid with simultaneous external cooling of the reactor, a steady state was again reached.

The changes in the axial temperature profiles during the start-up period after switching from the feed of inert solvent to the reaction mixture are shown in Figs. 16.14. and 16.15. The results indicate a gradual steadying of the temperature profiles during the first 30 min after the commencement of the reaction. The temperature profiles at the end of this period represent those of the steady-state regime. The curves in Fig. 16.14., depicting the situation when the reactor is operated in the liquid-phase regime, retain the same shape while shifting upwards with increasing time. The maximum rate of temperature increase occurs shortly after the beginning of the experiment. In contrast, between the 10th and the 30th minute the values increased only little. Fig. 16.15., illustrates the case of liquid evaporation due to the formation of a hot spot near the reactor inlet. The evaporation takes place within 5 min of the onset of the reaction.

16.7.4 Hysteresis properties of trickle-bed reactors

The importance of wetting of the catalytic layer prior to the addition of the substrate to be hydrogenated is demonstrated by the measurements illustrated in Figs. 16.16. and 16.17.

In the first case (Fig. 16.16.) the addition of the substrate to the dry bed was followed by the evolution of considerable heat of wetting and reaction. This caused an increase in the temperature of the inlet section and a greater part of the reactor operated in the "hot" steady state. The maximum temperature in the reactor was as high as $160^{o}C$ and the majority of the catalyst operated in the gas-phase regime. Further addition of the liquid phase did not result in a decrease in temperature. The high temperature is of course reflected in a high degree of conversion. In contrast, in the second case (Fig. 16.17.) the reaction mixture was added to a layer previously wetted by cyclohexane. Again evaporation occurred, but only as a consequence of the reaction heat. Consequently, the maximum temperature was only $108^{o}C$ and the greater part of the catalyst operated at a temperature above the boiling point of the reaction mixture. It is interesting that the steady state of the reactor in the two cases was the same since neither the temperature profiles nor the extents of conversion (0.92 and 0.81) differed. The reason for this observation lies in the hysteresis properties of the system during phase transitions in the reaction mixture. Such hysteresis had been observed previously (ref. 13). It plays, no doubt, an important role in the reactor start-up and operation. The steady state, established in the bed of catalyst, depends, for instance, on

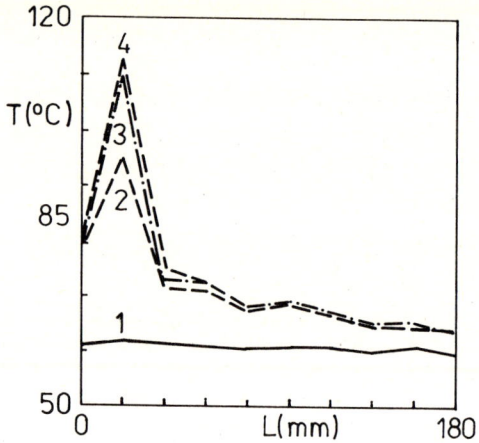

Fig. 16.15. Formation of a hot spot in the reactor after the start of hydrogenation. $T_o = 60°C$, $F = 238$ ml/h, $C = 42$ wt.%, $V_H = 0.25$ l/min; distributor: uniform; (———) (1) $t = 0$ min, (– – –) (2) $t = 5$ min, (–.–.–) (3) $t = 10$ min, (— —) (4) $t = 30$ min.

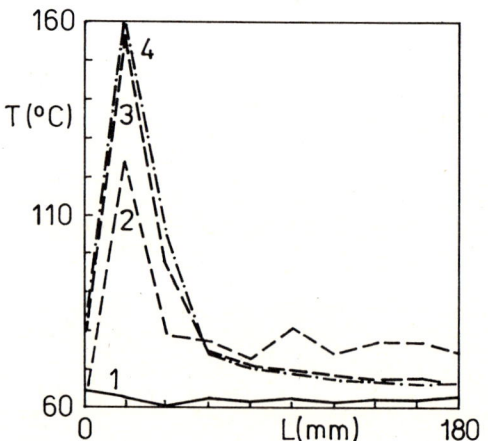

Fig. 16.16. Transient axial temperature profiles after the start of hydrogenation with a dry catalyst bed. $T_o = 60°C$, $F = 238$ ml/h, $C = 41$ wt.%, $V_H = 0.25$ l/min; distributor: central nozzle; (———) (1) $t = 1$ min, (– – –) (2) $t = 5$ min, (— —) (3) $t = 10$ min, (–.–.–) (4) $t = 30$ min.

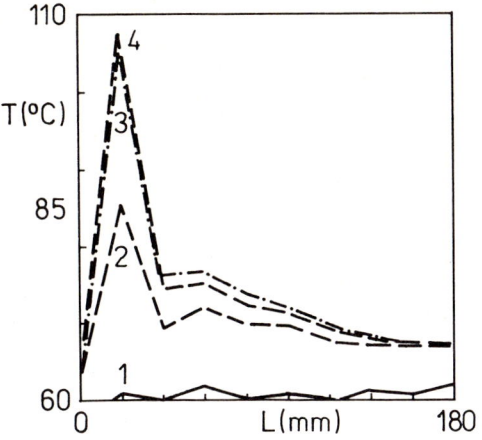

Fig. 16.17. Transient axial temperature profiles after the start of hydrogenation with a catalyst wetted by solvent. $T_o = 60°C$, $F = 238$ ml/h, $C = 41$ wt.%, $V_H = 0.25$ l/min; distributor: central nozzle; (———) (1) $t = 1$ min, (– – –) (2) $t = 5$ min, (— —) (3) $t = 10$ min, (–·–·–) (4) $t = 30$ min.

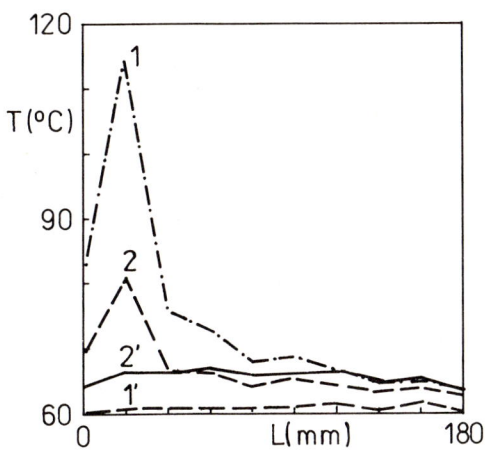

Fig. 16.18. Two different pairs of both starting and steady-state temperature profiles under the same reaction conditions. $T_o = 60°C$, $F = 238$ ml/h, $C = 19$ wt.%, $V_H = 0.25$ l/min; distributor: central nozzle; (–·–·–) (1) T'_{start}, (– – –) (2) T'_{steady}, (— —) (1') T'_{start}, (———) (2') T'_{steady}.

its initial temperature.

The hysteresis behaviour of the studied reaction system, i.e., the effect of the initial temperature in the reactor on the steady state attained is illustrated in Fig. 16.18. The corresponding data are summarized in Table 16.2.

TABLE 16.2
Extent of conversion in the reaction shown in Fig. 16.18.

Curve	Extent of conversion	Comments
1	0.803	Initial temperature profile prior to a change in feed concentration from 42 to 19 % mass.
2	0.816	Steady state following a change in feed composition from 42 to 19 % mass.
3	-	Temperature profile after the bed was wetted by inert solvent.
4	0.750	Steady state following a change in feed concentration from 0 to 19 % mass.

Curves 2 and 4 in Fig. 16.18. represent two steady-state temperature profiles in a bed of catalyst under otherwise identical conditions. The profile (———) was obtained after starting the reactor with fresh catalyst wetted by pure solvent, while the initial temperature profile corresponds to the curve (— —). In contrast, the profile (---) was obtained after a changing the feed concentration of cyclohexene from 42 to 19 mass %, during the addition of the more concentrated substrate, hot spots (curve (—.—.—) in this figure) were observed. It is thus apparent that the steady-state temperature profile depends on the reactor's history (the initial state), provided that phase transitions in the reaction system are possible. The hysteresis behaviour of the TBR is associated with the existence of a temperature gradient between the particle and the bulk stream of the reaction mixture in the case when the reaction takes place in the gas phase (ref. 13).

A very interesting result, demonstrating the hysteresis of the system, is illustrated in Fig. 16.19. The reaction rate, expressed in terms of the disappearance of cyclohexene, is plotted as a function of the initial composition of the feed. The experiment was carried out at a constant hydrogen flow-rate, i.e., 30 l/h, and the liquid feed rate was 201 ml/h. In the concentration range between 50 and 80 mass % of cyclohexene in the feed, two steady states of the reactor were observed: One, with a lower reaction rate, corresponded to a temperature lower than that of the boiling point of the mixture, i.e., predominantly a liquid-phase regime. The other corresponded to

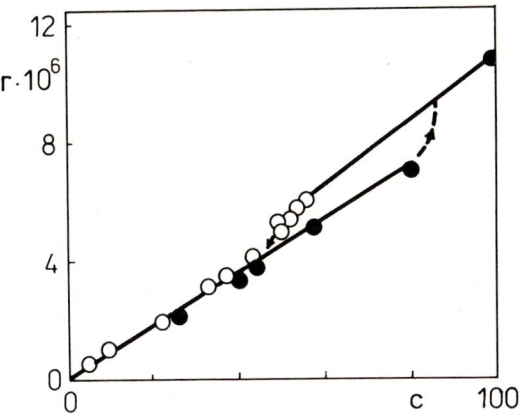

Fig. 16.19. The effect of a change in the feed concentration of cyclohexene on reaction rate; direction of concentration increase ●, decrease ○. $T_0 = 73°C$, $F = 201$ ml/h, $V_H = 0.50$ l/min, $L = 180$ mm; distributor: central nozzle.

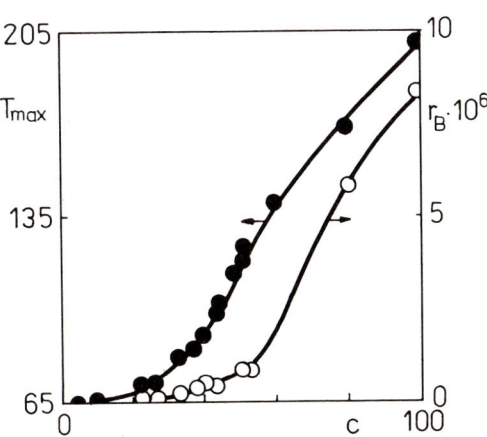

Fig. 16.20. Dependence of the temperature in the hot spot of the reactor and the rate of cyclohexene disproportionation on the cyclohexene concentration in the feed. $T_0 = 73°C$, $F = 201$ ml/h, $V_H = 0.50$ l/min, $L = 180$ mm; distributor: central nozzle.

reaction solely in the gas-phase regime.

The state in which the reactor steadied down also depended on its immediate history during the experiment. For this reason the above dependence exhibits a significant hysteresis loop. Unlike the experiments at lower temperatures, with a correspondingly different reaction order (0 or 1) in the gas- or liquid-phase regime, the apparent kinetics for the lower branch of the loop corresponded approximately to a first-order reaction. This is associated with the fact that in the lower branch of the loop the reaction takes place largely on the dry surface of the catalyst. This reaction was facilitated by the relatively high temperature of the experiment (73^oC), close to that of the boiling point of the reaction mixture (82^oC).

At a high concentration of cyclohexene, a transition occurs a gas-phase regime, accompanied by considerable overheating of the catalyst. Fig. 16.20. shows the temperature of the hot spot in the bed and the rate of formation of benzene in the side disproportionation reaction as a function of the cyclohexene feed concentration. A comparison of these dependences leads to the conclusion that the disproportionation of cyclohexene takes place primarily in the hot spot of the bed.

At high temperatures (experiments with high inlet cyclohexene concentrations) the disproportionation reaction is, in fact, preferred to hydrogenation of cyclohexene, as is apparent from the values of selectivity summarized in Table 16.1.

Axial temperature profiles measured near the reactor wall after reaching the steady state are shown in Fig. 16.21. at different concentrations. The figure also shows the temperature profiles in the reactor which stabilized after a previous increase/decrease of the feed concentration of cyclohexene (solid/broken line). The experimental results indicate that the steady-state temperature profile, after gradual increase of the feed concentration of the hydrogenated substrate, is somewhat lower than the temperatures measured along the reactor axis after a decrease in the cyclohexene content in the feed mixture. This histeresis behaviour of the reaction system is significant, particularly at medium concentrations.

16.8 CONCLUSION

The large number of papers dealing with trickle-bed reactors that have recently been published is no doubt a sign of the industrial importance of these reactors.

The role of the hydrodynamics and heat- and mass-transfer phenomena are now much better understood, but there is a lack of pertinent physicochemical and transport data, necessary for adequate modelling, design and scale-up of these systems. The simple pseudohomogeneous reactor model can now be recommended for

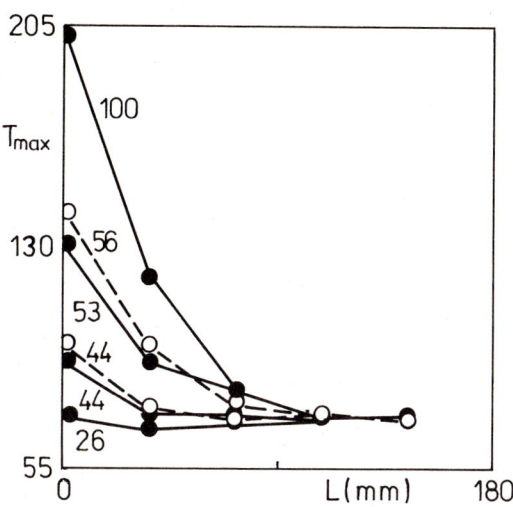

Fig. 16.21. Axial temperature profiles near the reactor wall for various feed concentrations of cyclohexene; direction of concentration increase ●, decrease ○. $T_0 = 73°C$, $F = 201$ ml/h, $V_H = 0.50$ l/h, $L = 180$ mm; distributor: central nozzle (numbers on profiles represent cyclohexene concentrations in feed in % wt.).

the description of most hydrotreatments of petroleum fractions or organic hydrogenations. Experimental work has improved our understanding of the complex phenomena such as phase transitions and the associated "pathological" effects occurring in these systems.

So far, inadequate attention has been paid to the role of the liquid maldistribution and the wall flow in reactor operation, i.e., the effects on productivity and selectivity of reactions and the thermal regime of the reactor, including the appearance of hot spots.

SYMBOLS

c	Concentration (mol/l)
C	Concentration of cyclohexene in the feed (mass %)
C_p	Specific heat capacity (J/g K)
F	Volume flow-rate of liquid (ml/h)
ΔH	Reaction heat (kJ/mol)
L	Depth of catalytic bed (mm)
r	Reaction rate (mol/s kg_{cat})

t   Time (min)
T   Temperature ($^\circ$C)
V_H Flow-rate of hydrogen (l/min)
W   Mass of catalyst in the reactor (g)
x   Extent of conversion
ρ Liquid density (g/ml)

REFERENCES

1. A. Germain in E. Alper (Editor), Proc. NATO Adv. Study Inst. on Mass Transfer with Chemical Reaction in Multiphase Systems, Martinus Nijhoff, The Hague, 1983, pp. 19-72.
2. H.P. Hofmann, Chem.-Ing.-Tech., 47 (1975) 823-868.
3. A. Bondi, CHEMTECH, (1971) 185-188.
4. H. Hoog, H.C. Klinkert and A. Schaafsma, Petrol. Ref., 32 (5) (1953) 137-141.
5. D.E. Mears, Chem. Eng. Sci., 26 (1971) 1361-1366.
6. A.A. Montagna and Y.T. Shah, Ind. Eng. Chem. Proc. Des. Dev., 14 (1975) 479-483.
7. E.V. Murphree, A. Voorhies Jr. and F.X. Mayer, Ind. Eng. Chem. Proc. Des. Dev., 3 (1964) 381-386.
8. J.A. Paraskos, J.A. Frayer and Y.T. Shah, Ind. Eng. Chem. Proc. Des. Dev., 14 (1975) 315-322.
9. L.D. Ross, Chem. Eng. Prog., 61 (No. 10) (1965) 77-82.
10. Y.T. Shah and J.A. Paraskos, Ind. Eng. Chem. Proc. Des. Dev., 14 (1975) 368-372.
11. N.D. Sylvester and P. Pitayagulsarn, AIChE J., 19 (1973) 640-644.
12. B.D. Babcock, G.T. Mejdell and O.A. Hougen, AIChE J., 3 (1957) 366-372.
13. J. Hanika, K. Sporka, V. Ružička and J. Krausová, Chem. Eng. Commun., 2 (1975) 19-25.
14. S. Morita and J.M. Smith, Ind. Eng. Chem. Fundam., 17 (1978) 113-120.
15. C.N. Satterfield, A.A. Pelossof and T.K. Sherwood, AIChE J., 15 (1969) 226-234.
16. W. Sedriks and C.N. Kenney, Chem. Eng. Sci., 28 (1973) 559-569.
17. F. Turek, R. Lange, A. Busch and R. Löve, Chem. Tech. (Leipzig), 28 (1976) 149-152.
18. W. Krönig, Erdöl Kohle, Erd-gas Petrochem., 16 (1963) 520-523.
19. J. Klassen and R.S. Kirk, AIChE J., 1 (1955) 488-495.
20. S. Goto and J.M. Smith, AIChE J., 21 (1975) 706-720.
21. Levec J. and J.M. Smith, AIChE J., 22 (1976) 159-168.
22. S. Goto and J.M. Smith, AIChE J., 24 (1978) 286-293.
23. M. Hartman and R.W. Coughlin, Chem. Eng. Sci., 27 (1972) 867-881.
24. C.N. Satterfield, AIChE J., 21 (1975) 209-228.
25. J.T. Earight and T.T. Chuang, Can. J. Chem. Eng., 56 (1978) 236-250.
26. M. Shimizu, A. Kitamoto and Y. Takashima, J. Nucl. Sci. Technol., 20 (1983) 36-47.
27. S. Goto, J. Levec and J.M. Smith, Catal. Rev. Sci. Eng., 15 (1977) 187-247.
28. H.I. Joschek, in Ullmanns Encyklopädie der technischen Chemie, Vol. 3, Verlag Chemie, Weinheim, 4th ed., 1973, pp. 494-518.
29. Y.T. Shah, Gas-Liquid-Solid Reactor Design, McGraw Hill, New York, 1979, Chs. 4 and 6, pp. 105-148 and 180-229.
30. P.A. Ramachandran and R.V. Chaudhari, Three-Phase Catalytic Reactors, Gordon and Breach, London, 1982, Ch. 12.
31. M.O. Tarhan, Catalytic Reactor Design, McGraw Hill, New York, 1983, Chs. 6 and 7, pp. 155-240.
32. H.P. Hofmann, Catal. Rev. Sci. Eng., 17 (1978) 71-117.
33. A. Gianetto, G. Baldi, V. Specchia and S. Sicardi, AIChE J., 24 (1978) 1087-1104.

34 J. Hanika, V. Vosecký and V. Ružička, Chem. Eng. J., 21 (1981) 108-114.
35 J.C. Charpentier, Adv. Chem. Eng., 11 (1981) 1-133.
36 C.S. Tan and J.M. Smith, Lat. Am. J. Chem. Eng. Appl. Chem., 11 (1981) 59-69.
37 M. Herskowitz and J.M. Smith, AIChE J., 29 (1983) 1-18.
38 J. Hanika and V. Staněk, in N. Cheremisinoff (Editor), Handbook of Heat and Mass Transfer Operations, Vol. II, Gulf Publ., Co., Houston, 1986, Ch. 24, pp. 1029-1080.
39 P.A. Ramachandran and R.V. Chaudhari, Chem. Eng., 87 (No. 24) (1980) 74-85.
40 H. Van Landeghem, Chem. Eng. Sci., 35 (1980) 1912-1949.
41 P.L. Mills and M.P. Dudukovic, Symp. Catalytic Reactor Design, Seattle Meeting, Seattle, March 20-25, 1983. ACS Symp. Ser., 237 Chem. Catal. React. Model 237 (1984) 37-59.
42 Y.T. Shah and D. Smith, 8th Internat. Symp. Chemical Reaction Engineering, Edinburgh, September 10-13, 1984.
43 M. Crine, Chem. Eng. Commun., 19 (1982) 99-114.
44 V. Specchia and G. Baldi, Chem. Eng. Commun., 3 (1979) 483-499.
45 A. Matsuura, Y. Hitaka, T. Akehata and T. Shirai, Heat Transfer Jpn. Res., 8 (1979) 44-52.
46 A. Matsuura, Y. Hitaka, T. Akehata and T. Shirai, Heat Transfer Jpn. Res., 8 (1979) 53-60.
47 K. Hashimoto, K. Muroyama, K. Fujiyashi and S. Nagata, Internat. Chem. Eng., 16 (1976) 720-727. (See also Kagaku Kogaku Ronbunshu, 2 (1976) 53-59).
48 J. Hanika, K. Sporka, V. Ružička and J. Hrstka, Chem. Eng. J., 12 (1976) 193-197.
49 J. Hanika, K. Sporka, V. Ružička and R. Pištěk, Chem. Eng. Sci., 32 (1977) 525-528.
50 V. Staněk and V. Eckert, Coll. Czech. Chem. Commun., 44 (1979) 829-840.
51 V. Staněk and V. Eckert, Chem. Eng. Sci., 34 (1979) 933-940.
52 A.I. Snow and M.P. Grosboll, Oil Gas J., 75 (No. 21) (1977) 61-65.
53 V. Staněk, J. Hanika, V. Hlaváček and O. Trnka, Chem. Eng. Sci., 36 (1981) 1045-1067.
54 V. Staněk and J. Hanika, Chem. Eng. Sci., 37 (1982) 1283-1288.
55 J. Hanika, 8th Congress CHISA'84, Prague, September 2-7, 1984.
56 V. Staněk and J. Hanika, 8th Congress CHISA'84, Prague, September 2-7, 1984.
57 J. Hanika and V. Ružička, Sb. Vys. Sk. Chem. Technol. Praha, Org. Chem. Technol., C26 (1980) 79-86.

Chapter 17

CONTROL OF HYDROGENATION AUTOCLAVES

JOSEF HORÁK
Prague Institute of Chemical Technology, 166 28 Prague (Czechoslovakia)

17.1 INTRODUCTION

Autoclaves are high pressure chemical reactors, thus their control is a special case of chemical reactor control. Despite the fact that chemical reactors are basic equipment of all chemical plants, the literature dealing with practical aspects of reactor control is not as extensive as expected. Chapters on reactor control are included in several textbooks and monographs on control engineering (refs. 1 - 4). However, attention is focused on reactor analysis and design and discussions on reactor control are limited to the simplest control mode - that achieved by stabilization of the reactor inputs (the reactant feed and the coolant inlet temperature and flow-rate). The dynamic aspects considered are limited to systems with stabilized inputs (autonomous systems). More sophisticated control methods, for example feedback and feedforward temperature controls, are not included. Similar comments can be made regarding the theoretical treatments (refs. 5 - 7). In the last three decades the meaning of the terms "Reactor dynamics" and "Reactor control" has been reduced to the analysis of autonomous systems with multiple steady states, autonomous oscillating systems and related phenomena. Many reviews have been published on this subject (refs. 8 - 15).

There is thus a lack of literature dealing with the practical aspects of advanced reactor control. Perhaps, most progressive in the application of modern non-traditional control methods has been made in production of polyvinyl chloride (refs. 16 - 18). A feedback temperature control both for stable and unstable pseudostationary states has become a standard tool in polyvinylchloride polymerization reactors. As these reactors are operated at high pressures, they can be considered as autoclaves.

Consequently, the experience ganed in the control of polymerization reactors can be utilized in designing control systems for large hydrogenation autoclaves. However, no monographs or specialized chapters in textbooks are available on this subject.

(i) <u>Problems in the application of non-traditional control methods</u> (refs. 19,20). Recent developments in electronics and the present state of mathematical modelling have opened the way to the aplication of advanced control methods. These methods are not yet widely utilized for two reasons:

(ii) <u>Technical difficulties in chemical industries. Delay in receiving information</u> about the process caused by the necessity to evaluate the composition of the reaction mixture. Thus, knowledge of the instantaneous composition is obtained too late to be used for feedback control. <u>Stochastic changes</u> in the feedstock composition and in the catalyst activity. These changes are often difficult to evaluate before commencement of the process. <u>Unmeasurable quantities</u>, for example the feedstock quality in hydrogenations of natural products, the product quality. <u>Lack of satisfactory process models</u>.

(iii) <u>Conservatism of chemists and chemical engineers</u>. Most chemists consider that isothermal operation achieved by simple stabilization of the coolant inlet temperature is the only method of controlling the reactor operation. It may be very effective in the control of small autoclaves, but may fail with largescale reactors. In the control of industrial reactors, the advanced control methods feedback temperature or a non-isothermal adaptive control, may be profitable.

In this paper guidane is provided in the design of control systems. The scope is limited to smaller plants with a changing product spectrum. In such plants the control system is usually designed by the technological engineer developing the process. The design of special control systems for hydrogenations in petrochemical industries is not included.

17.2 FORMULATION OF THE CONTROL TASK

17.2.1 Control objectives

Autoclave control in research laboratories is aimed at achieving defined reaction conditions, i. e., the reaction temperature and hydrogen partial pressure. Simultaneously the safe operation of the autoclave must be ensured. In industry, technological aspects must also be taken into account. These aspects are listed below in order of their decreasing importance:

(i) <u>Process safety</u>. Since hydrogenations are exothermic processes the control system must avoid an uncontrollable temperature and pressure increase in the autoclave.

(ii) <u>Process selectivity</u>. If the yield of the main product is reduced by undesirable side reactions, the autoclave regime should be optimized so as to suppress such reactions.

(iii) <u>Conversion of reactants</u>. In most cases the extent of conversion should be as high as possible in order to reduce the amount of unreacted starting material which needs to be recycled. For some reaction systems, e.g., which needs to consecutive reactions, the extent of conversion must be controlled exactly in order to maximize the process selectivity.

(iv) <u>Reactor performance</u>. The reactor performance may or may not be important. Autoclaves being expensive equipments the process intensification may be desirable.

17.2.2 Characteristics of the reaction regime

The technological aspects listed above must be transformed into data defining the reaction regime:
- The set value of the reaction temperature or the set function if the reaction temperature is to be changed according to a prescribed programme.
- The desired reaction pressure.
- The desired value of the extent of conversion.

The choice of these values represents one of the most important steps in developing a control system. Let us define the following categories of variables:

(1) <u>Strictly controlled variable</u> - a process variable the value of which is to be kept at a prescribed level

(2) <u>Constrained controlled variable</u> - a process variable the value of vhich is permitted to change within a certain interval and must not exceed the regim corresponding to safe and economic operation.

(3) <u>Disturbing variable</u> - a process variable which changes, irrespective of the control system.

(4) <u>Manipulated variable</u> - a quantity which can be varied so as to change the values of controlled variables.
The choice of the controlled and manipulated variables must reflect the control objective, the properties of the reaction and local limitations and constraints.

(i) <u>When must be the extent of conversion strictly controlled?</u>
If the process selectivity is not important or if it does not depend on the extent of conversion, the latter should be chosen as the constrained variable, the value of which must exceed a certain minimum determined by economic considerations. The cost of raw material, and of the product separation decrease with increasing extent of conversion, whereas the autoclave performance decreases.

If the selectivity or the product quality depend on the extent of conversion the latter must be strictly controlled in order to optimize the process selectivity. This type of behaviour is typical of systems of consecutive reactions where if an intermediate is the desired product. In the control of such reactions, any change in the reaction conditions, e.g., the temperature, catalyst activity or pressure, must be compensated by a corresponding change in the reaction time in order to optimize the extent of conversion.

(ii) <u>When must the reaction temperature be strictly controlled?</u>
Chemists are used to specifying the reaction temperature as a fixed, strictly controlled variable, thus constraining the

process optimization. Fixation of the temperature is necessary and useful in the following cases:

(1) When a high degree of process reproducibility is to be achieved. Of course, in hydrogenations, stabilization of the temperature may not be effective due to changes in the catalyst activity.

(2) When the catalyst selectivity is extremely sensitive to temperature changes.

In all other cases the reaction temperature should be the constrained controlled variable which must not exceed a certain maximum. Typically, the maximum is determined from an analysis of the process safety. Another phenomenon limiting the maximum allowable temperature may be the catalyst deactivation.

17.2.3 Optimization of the reaction temperature

If the reaction temperature is not strictly controlled, it can be changed in such a way as to optimize the reactor operation. The chemical engineering literature dealing with the control of batch processes, including optimization of the reaction temperature in batch reactors, has been reviewed by Rippin (ref. 21)

The following situations may occur:

(i) <u>Reaction kinetics versus equilibrium</u>. For exothermic reversible reactions the rate constants increase with increasing temperature, whereas the equilibrium constants decrease. As a result, for each degree of conversion an optimum reaction temperature can be found ensuring the maximum production rate.

(ii) <u>Production rate versus selectivity</u>. If the process selectivity decreases with increasing temperature, an economic optimum must be evaluated because the production rate increases and the product yield decreases with increasing temperature.

(iii) <u>Instantaneous reaction rate versus rate of catalyst deactivation</u>. The rate of the catalyst deactivation depends on the reaction temperature. If catalyst deactivation is rapid, the long-term production rate depends on both the instantaneous reaction rate and the rate of catalyst deactivation. In most cases the deactivation rate increases with increasing temperature. Consequently, a compromise is necessary.

(iv) <u>Production rate versus process safety</u>. The hazard of a temperature runaway increases with increasing reaction temperature, simultaneously, the production rate increases. The maximum attainable production rate is affected by the reliability of the control system (refs. 4, 20 - 22). The closer the reaction temperature to the safety limit, the higher is the reactor performance. A similar problem must be solved when the process selectivity increases with increasing temperature. Then the process safety decreases while the production rate and selectivity increase with increasing reaction temperature.

17.2.4 <u>Control constraints</u>

Control constraints may play a dominant role in the design of the control system. They may be derived from the properties of the reaction and catalyst, equipment and instrumentation or from local specific plant conditions.

(i) <u>Measurements and instrumentation</u>. For safe and reliable control, the process monitoring should be as complex as possible. The following state variables should be measured:
(a) The reaction temperature, by a rapidly responding sensor, directly in the reaction mixture.
(b) The temperatures of the coolant inlet and outlet streams.
(c) The reaction pressure inside the autoclave.
(d) The hydrogen feed rate and outflow rate, with the possibility to integrate the instantaneous values to obtain the total hydrogen consumption.
(e) The instantaneous composition of the reaction mixture, if possible without delay.

(f) The content of inert gases in the hydrogen inside the autoclave.

Unfortunately, the instrumentation for many autoclaves is poor. The reaction temperature is measured by sensors in massive tubes having unfavourable dynamic properties, in some cases in the heating jacket which may cause a significant delay in the measurement. Many autoclaves are not equipped with a flowmeter to measure the rate of hydrogen consumption. Sometimes, sampling of the liquid reaction mixture is not possible during hydrogenation.

(ii) <u>Actuators</u>. Sophisticated control methods cannot be applied if the autoclave is not equipped with the corresponding actuators:
(a) Valves to manipulate the hydrogen inlet stream and the purge stream in order to remove inert gases from the system.
(b) A valve to manipulate or stabilize the pressure inside the reactor.
(c) A valve to manipulate the coolant flow.
(d) A heat exchanger to manipulate or stabilize the coolant inlet temperature.
(e) A pump to control the reactant feed in continuous systems.

As the instrumentation of many autoclaves is poor, the application of modern control methods may be limited.

(iii) <u>Technological constraints</u>. When formulating the control task, local interests and limitations may be important. In continuous processes the substrate feed rate may be affected by the operation of the supply equipments. Consequently, the feed rate cannot be employed as a manipulated variable. The minimum attainable coolant temperature and the maximum temperature of the heated medium may be determined by a central cooling and heating system.

17.3 CONTROL OF SMALL LABORATORY AUTOCLAVES

Technological aspects are not important in the control of laboratory autoclaves. The aim is to achieve defined reaction conditions.

17.3.1 Temperature control

The control of the temperature in small autoclaves is simple. The reaction heat is easy to remove from the reaction mixture due to the favourable ratio of the heat-transfer area to the reactor volume. As the heat capacity of the autoclave wall is comparable with that of the reaction mixture, the autoclave walls are able to absorb the reaction heat. Thus, the thermal effects of the reaction are suppressed and any simple control system can be effective in stabilizing the reaction temperature.

(i) <u>Compensation of heat loss by electrical heating</u>. If the heat loss represents the prevailing item in the heat balance of the autoclave, the reaction temperature can be stabilized by electrical heating compensating for the heat loss. The massive autoclave walls act as a thermostat, damping temperature fluctuations. Any laboratory controller can be employed to control the heating. The temperature measured in the heating jacket is used as the controller input. If the temperature measured inside the autoclave is used, the dynamic properties of the control loop are poor due to the inertia of the reactor walls. Difficulties in controller tuning can avise.

(ii) <u>Stabilization of the coolant inlet temperature</u>. In small jacketed autoclaves the reaction temperature can be controlled by stabilizing the coolant inlet temperature. The smallest autoclaves can be immersed in a thermostatted bath. Stabilization of the coolant temperature results in stabilization of the reaction temperature at a value close to that of the coolant temperature.

17.3.2 Pressure control

The decisive state variable affecting the hydrogenation rate is the partial pressure of hydrogen. If the hydrogen feed contains inert components or if inert components are formed as by products, they will tend to accumulate in the gas space thus reducing the hydrogen partial pressure. Then a small purge stream must be removed from the gas space to withdraw inerts. A pseudostationary state will be reacted, at which the concentration of their components is a function of the purge flow, hydrogen feed rate and their content in the hydrogen feed.

Since very pure hydrogen is used in a laboratory the control of partial pressure can be replaced with control of the total pressure. The system of pressure control must be modified according to the instrumentation available.

(i) <u>Batch system</u>. When using this system, the autoclave is not connected with a hydrogen-supply vessel during hydrogenation. The autoclave is filled with hydrogen at the beginning of the experiment and then it is closed. The gas space of the autoclave serves as the hydrogen source. The hydrogen consumption can be evaluated from the pressure decrease. This simple system can be effective if the ratio of the gas space to the liquid volume is large and the initial pressure is high.

(ii) <u>Batch system with a supply tank</u>. The autoclave is connected with a supply tank to form a batch system with a large hydrogen inventory. Consequently, the changes in the hydrogen pressure due to hydrogenation are small and nearly isobaric data are obtained experimentally. Nevertheless, the hydrogen consumption can be evaluated from the pressure decrease in the system.

(iii) <u>Semibatch system with stabilized pressure</u>. The autoclave is connected with a hydrogen-supply tank by means of an automatic valve stabilizing the pressure in the autoclave. The isobaric data thus obtained are simple to evaluate. The instantaneous rate of hydrogen consumption can be determined from the hydrogen flow-rate, the hydrogen consumption from the pressure changes in the supply tank.

(iv) <u>Autoclaves operated between two pressure levels</u>. For this control mode two set values of pressure are defined, the upper, P_1, and the lower, P_2. The autoclave is filled with hydrogen to the value P_1, the autoclave is closed until the pressure falls to value P_2 due to hydrogen consumption. Then, the procedure is repeated. This simple control mode can be used for manual control.

17.4 INDUSTRIAL AND PILOT-PLANT BATCH AUTOCLAVES

Control of large autoclaves may be difficult due to problems with heat exchange. The surface-to-volume ratio is unfavourable with such vessels, consequently the rate of cooling may limit the production rate and the process safety. The temperature regime may become non self-regulating and require application of a feedback controller.

17.4.1 Process safety

Safety control is of increasing interest. Regenass has published a series of papers devoted to the safe design and control of exothermic processes (refs. 23 - 25). Theofanous analyzed the causes of the accident in Seveso (refs. 26, 27). A runaway in an industrial hydrogenation trickle-bed reactor was described and analyzed by Eigenberger (refs. 28, 29). Symposia have been devoted to process safety inculding the control of exothermic reactions (refs. 24, 30 - 32), and a collection of papers dealing with systems and facilities for automation of potentially dangerous processes has been published in the USSR (ref. 33).

There are two sources of hazard in hydrogenations:
- High pressure. Control of high-pressure equipment is not a specific problem of hydrogenations. Experience with the design and safe operation of high-pressure facilities has been summarized by Livingstone (ref. 34). The operating of a small autoclave was described by Riemenschneider (ref. 35), and the safe design of emergency venting systems was solved by Davies, (ref. 36), Woods and Thornton (ref. 37), Friedel and Purps (ref. 38).

(i) Classification of hydrogenations from the point of view of their potential hazard. Hydrogenations are exothermic reactions. For safe operation the reaction heat must be removed from the reaction mixture. From the point of view of the potential hazard of a thermal explosion, hydrogenations can be classified into three groups:

(a) Simple hydrogenations. The main reaction is the only source of reaction heat.
(b) Selective hydrogenations. Side reactions (total hydrogenation, hydrogenolysis) represent an additive source of heat.
(c) Dangerous substances are involved which are able to decompose exothermically in the absence of hydrogen.

For group (a), safety is easy to achieve. The potential danger can be diminished by avoiding the accumulation of reactants in the autoclave, for example by operating the autoclave continuously at high extents of conversion. In batch autoclaves the potential hazard decreases with increasing extant of conversion. A temperature runaway can be dealt with shutting off the hydrogen feed. Catalytic hydrogenations cannot reach explosive rates due to the resistances to mass transfer, furthermore, most hydrogenations are reversible and possess autoregulative properties. Since a temperature runaway is connected with the hydrogen consumption, the increase of pressure in the autoclave is damped.

For group (b) a temperature runaway can caused by both the main and the side reactions. As both the reactants and the products may take part in the side reactions, this type of runaway cannot be eliminated by avoiding accumulation of reactants. Shutting off the hydrogen feed may be effective in damping the runaway.

Group (c) is the most dangerous one. It includes hydrogenations of nitro-compounds (ref. 40). In order to control these processes a thorough analysis is necessary to determine the temperature limits over which the exothermic decomposition takes place. Relevant data can easily and safely be obtained by means of preparative calorimetry and microthermoanalytic instrumentation (refs. 40 - 46). If an exothermic decomposition is found every care must be taken to keep the decomposable material within the allowable temperature range (refs. 23, 24). In this case, performance optimization is not recommended. The lowest reasonable reaction temperature is the best choice in designing the autoclave regime.

(ii) __Thermodynamic characteristics of potential hazards__. A basic parameter used in chemical reaction engineering to characterize the thermal effects of chemical reactions is the adiabatic temperature rise. This is defined as the ratio of the potential chemical heat to the heat capacity of the system. It can be interpreted as the temperature increase due to the reaction heat if no heat if removed.

For a hydrogenation described by the stoichiometric equation

$$a A + b H_2 = \text{Products} + \Delta H \tag{17.1}$$

the maximum adiabatic temperature rise can be limited by either the reactant or the hydrogen inventory inside the autoclave. The heat capacity of the system is comprised of the heat capacity of the equipment which can absorb heat (the stirrer, the autoclave walls, the catalyst). The effective heat capacity of the autoclave will be denoted by the symbol C.

In the analysis of the potential hazards, several modified values of the adiabatic temperature rise can be utilized:

(a) __The initial adiabatic temperature rise limited by the reactant content__.

$$\Delta T_{ad,Ai} = \frac{c_{Ai}(-\Delta H) V}{a C} \tag{17.2}$$

This adiabatic temperature increase can be attained if hydrogen is present in excess (the hydrogen feed is not closed). A small value of this characteristic (10 - 20 K) indicates that the system is thermally indifferent and is not dangerous. High values (100 - 500 K) are characteristic of dangerous systems the control of which is impossible without effective cooling.

(b) The instantaneous adiabatic temperature rise limited by the reactant content.

$$\Delta T_{ad,A} = \frac{c_A (-\Delta H) V}{a\, C} \qquad (17.3)$$

where c_A is the instantaneous concentration of the reactant. This adiabatic temperature rise indicates whether a runaway starting from the instantaneous state may be dangerous. It corresponds to the maximum temperature runaway. In trickle-bed reactors its value can be exceeded due to local and temporary flow fluctuations occurring inside the reactor (refs. 28, 29).

(c) The instantaneous adiabatic temperature rise limited by the hydrogen content.

$$\Delta T_{ad,H} = \frac{c_H (-\Delta H) V_g}{b\, C} = \frac{(-\Delta H) P V_g}{b\, C\, R\, T} \qquad (17.4)$$

This adiabatic temperature rise indicates what would happen if the hydrogen feed was shut off, the hydrogen accumulated inside the reactor being the only hydrogen available.

Generally, all hydrogenations carried out in solvents are thermally indifferent due to the low concentration of the reactant. Hydrogenations in small autoclaves are safe due to the large heat capacity of the autoclave walls. Industrial hydrogenations of concentrated substrates may be dangerous; in most cases hydrogen is the limiting substance. This means that a temperature runaway can be effectively damped by shutting off the hydrogen feed. Exceptionally, in high pressure autoclaves whith large gas spaces, the hydrogen inventory inside the autoclave can be large enough for a dangerous temperature runaway to occur even after the hydrogen feed has been shut off.

17.4.2 Temperature control

When designing a control system it is useful to realize that difficulties encountered in the control of temperature are strongly dependent on the production rate. One should begin by answering the question: "Is it necessary and useful to maximize the autoclave performance?" If not, the temperature control can be simplified by using smaller portions of the catalyst or a less active catalyst. A simple self-regulating control mode can be effective at smaller production rates. At high production rates even a very sophisticated control method may fail.

(i) Stability of temperature regimes. Hydrogenations are exothermic reactions, therefore, removal of the reaction heat presents one of the most important control objectives. To achieve a pseudostationary state, the reaction heat must be removed from the reaction mixture at exactly the same rate as it is formed. An important parameter in the design of the control system is the stability of the pseudostationary state. A detailed explanation and reviews on stability can be found in the literature (refs. 8 - 15). In this paper only a brief description of stability will be presented.

Let us analyse the effect of the reaction temperature on both the rate of heat generation and the rate of cooling in an autoclave with a stabilized coolant inlet temperature and flow-rate. The reaction rate is an exponential function of the reaction temperature. The total rate of heat generation in the autoclave is related to the rate of hydrogen consumption by the equation:

$$R_h = R_H(-\Delta H)/b = r_o W(-\Delta H/b) \exp \frac{E(T-T_o)}{R T T_o} \qquad (17.5)$$

The rate of cooling is a linear function of the temperature difference between the reaction mixture and the cooling system. Supposing that both the reaction mixture and the cooling system can be described by a model of an ideally mixed system and that the cooling system is in a pseudostationary state, the following equations are obtained:

$$R_c = A k_h (T-T_{ce}) \qquad (17.6)$$

$$A k_h (T - T_{ce}) = F_c \rho_c c_{pc} (T_{ce} - T_{ci}) \qquad (17.7)$$

Combining eqns. 17.6 and 17.7, the cooling rate can be expressed as a linear function of the raction temperature and the coolant inlet temperature. This difference will be called <u>the cooling driving force</u>:

$$R_c = \frac{A k_h F_c \rho_c c_{pc}}{A k_h + F_c \rho_c c_{pc}} (T - T_{ci}) = k_c (T - T_{ci}) \qquad (17.8)$$

The pseudostationary states of the temperature regime correspond to intersections of the heat production curve and the cooling line. An example is given in Fig. 17.1.

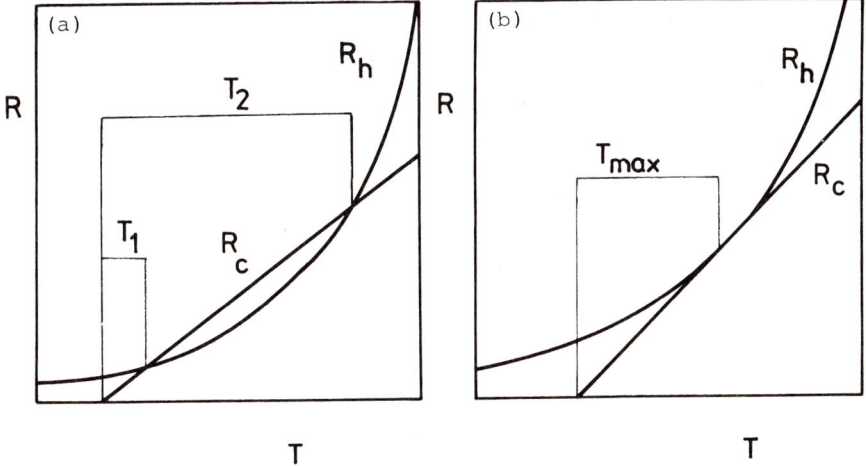

Fig. 17.1. Graphical representation of the heat balance of an autoclave.
(a) Two pseudostationary states, the state with the smaller cooling driving force is stable, the one with the larger cooling driving force is unstable.
(b) The limit case indicating the largest cooling driving force for stable pseudostationary states.

(ii) <u>Self-regulating ability of pseudostationary states.</u> Self regulation is the property of being able to reach a

pseudostationary state (relative to a stationary state for
continuous systems) without application of a feedback controller.
If a pseudostationary state of an autoclave is self regulating
(stable), the temperature control can be accomplished by stabilizin
stabilizing all inlets, i. e., the inlet coolant temperature and
flow and the hydrogen pressure. If a pseudostationary state is
non self-regulating (unstable), a feedback controller must be
applied to manipulate one of the inlet variables in such a way to
keep the reaction temperature with in the unstable regime.
Fig. 17.1 indicates that for chemical reactors only the
pseudostationary states with small cooling driving forces are
self-regulating. In these states the cooling rate is more
sensitive to temperature changes than is the rate of heat
generation. Consequently, if the reaction temperature is higher
than that corresponding to the pseudostationary state, more heat
is removed than is produced and the reaction temperature will
decrease. In unstable states, exceeding the pseudostationary
temperature results in an uncontrollable temperature increase
because more heat is produced than is removed.

The stability of the temperature regime represents one of the
most important constraints imposed on the reaction temperature
If a simple temperature-control method is used, safety requires
that the cooling driving force be maintained below some limiting
value so as to avoid the unstable region. The maximum allowable
cooling driving force depends on the activation energy. Examples
ar given in Table 17.1. When designing control systems for
industrial autoclaves, a control margin must be left below the

TABLE 17.1

Maximum cooling driving forces of stable pseudostationary states
of temperature regimes as a function of activation
energy.

E/RT	0.01	0.02	0.03	0.04	0.05	0.06	0.07	0.08
$(T_s - T_{ci})$ max	80	40	30	22	20	15	13	12

limit value so as to safeguard the process against a temperature runaway.

If a reliable feedback controller can be employed for temperature control, both stable and unstable states can be utilized. Consequently, the production rates may be higher.

(iii) <u>Kinetic characteristics of potential hazards</u>. Let us assume that the aim of the autoclave control is to reach a desired production rate as measured by the hydrogen consumption, $R_{H,des}$. As the reaction heat must be removed from the reaction mixture, the value of the corresponding cooling driving force can be calculated from eqn. 17.8. This value will be called <u>the necessary cooling driving force</u>. It provides important information indicating whether the temperature control is simple or difficult.

$$(T - T_{ci})_{nec} = R_{H,des}(-\Delta H/b)/k_c \qquad (17.9)$$

According to the value of the necessary cooling driving force, the following principles of temperature control can be employed:

(a) <u>Temperature control by stabilization of the coolant inlet temperature</u>. This simplest control method can be used if the necessary cooling driving force is small. The difference 1 - 2 K can be chosen as a limit:

$$(T - T_{ci})_{nec} < 2 \text{ K} \qquad (17.10)$$

For application of this method the autoclave must be equipped with a thermostat stabilizing the coolant inlet temperature. As the thermal effects of the reaction are week, heat loss may represent an important part of the heat balance. Consequently, both cooling and heating must be available for attaining the reaction temperature and for compensating the heat loss.

(b) <u>Temperature control by stepwise changes of the coolant inlet temperature</u>. If the necessary cooling driving force is smaller than the stability limit (10 - 15 K) and is of the same magnitude

$$2 \text{ K} < (T - T_{ci})_{nec} < 10 - 15 \text{ K} \qquad (17.11)$$

the reaction heat can be removed by employing the stable pseudostationary states. In principle, stabilization of the coolant inlet temperature could suffice. Unfortunately, variation of the catalyst activity may represent a serious problem. In batch processes the activity of the catalyst may change from batch to batch due to varying quality and quantity. To keep the procesess safe, it must be protected against a temperature runaway due to high catalyst activity. One a control method, is based on stepwise changes of the coolant inlet temperature (refs. 47, 48). The aim is to keep the cooling driving force within the safe, stable region, close to the stability limit.

Principle of the method: As the activity of a new catalyst portion is not known before the batch cycle is started, the reaction is started at a low inlet coolant temperature to keep the process safe. The coolant temperature is stabilized until a temperature maximum is reached, corresponding to a pseudostationary state. Then the experimentally measured temperature difference between the reaction mixture and the coolant is used for evaluation of the adaptive parameter R_H/k_c.

$$\frac{R_H}{k_c} = \frac{(T - T_{ci})_{max}}{(-\Delta H/b)} \qquad (17.12)$$

The overheating, $(T - T_{ci})_{max}$, is the controlled variable. A mathematical model of the system has been used for prediction of a new coolant inlet temperature which corresponds to the desired value of the overheating (refs. 47, 48).

(c) <u>Temperature feedback control by manipulating the coolant flow</u>. If the necessary cooling driving force exceeds the stability limits (10 - 20 K), temperature control based on stabilization of the coolant temperature is impossible:

$$(T - T_{ci})_{nec} > 10 - 20 \text{ K} \qquad (17.13)$$

The control system must be able to operate in both stable and unstable operating states, which is possible only with a feedback controller.

Jacketed autoclaves can be controlled by manipulating the coolant flow-rate (refs. 1 - 4). To achieve effective temperature control a cascade system is used. The coolant exit temperature is stabilized by manipulating the coolant flow, while the desired value of this temperature is derived from the measured and the set value of the reaction temperature.

For maximum stability and intensive heat transfer, coolant must be recirculated with fresh coolant added into the circuit. Hot water or steam is necessary for heating during the reactor start-up. The safety depends on the dynamic properties of the control system.

(d) <u>Temperature control by manipulating the hydrogen feed</u>. If the necessary cooling driving force exceeds the largest available difference between the reaction temperature and the coolant due to the fact that both the minimum inlet coolant temperature and the maximum permissible reaction temperature are limited, the desired production rate cannot be reached. The reactor must be operated at a lower production rate.

There are two ways to solve the problem:
(1) The production rate is diminished by using less catalyst per batch. The necessary cooling driving force will decrease and the corresponding control method given in the previous chapters can be applied.
(2) The autoclave is operated at the maximum attainable cooling driving force and the production rate is kept at a safe level by manipulating the hydrogen pressure.

The safety and reliability achievable by manipulating the hydrogen pressure is dependent on the dynamic properties of the system.

17.4.3 <u>Dynamic properties of autoclaves</u>

The utilization of unstable steady states of chemical reactors may be very effective in improving the reactor performance (refs. 47, 49, 50). The control reliability depends on the dynamic properties of the system (ref. 51); if a rapidly responding manipulated variable is used the control is simple and safe, whereas if a slowly responding variable is used the control is dangerous and difficult (ref. 52). Poor dynamic properties of the control system are difficult to compensate by applycation of a sophisticated algorithm (refs. 53, 54). The dynamic properties of autoclaves can change according to their size and construction. Few papers have described dynamic experiments with hydrogenation autoclaves (refs. 55, 56).

From the dynamic point of view, an autoclave can be analyzed in terms of five capacities:
(1) The volume of the liquid reaction mixture incorporating the reactants.
(2) The gas space inside the autoclave incorporating hydrogen.
(3) The heat capacity of the reaction mixture and of autoclave bottom and catalyst.
(4) The heat capacity of the cooling system (the cooling jacket, its walls, the wall separating the coolant from the reaction mixture).
(5) The heat capacity of the sensor delaying the sensor response.

Generally, the heat capacities augmenting the heat capacity of the reaction mixture improve temperature control by damping temperature changes. The capacities augmenting the heat capacity of the cooling system make the temperature control by manipulating the coolant flow more difficult and less reliable, but they may improve the control if the hydrogen feed is used as the manipulated variable. Augmentation of the sensor heat capacity increases the sensor inertia and may be dangerous, especially, in unstable operating states.

(i) Testing of control algorithms. Direct experiments with the chemical reaction in an industrial autoclave may be hazardous. Simulations may be helpful in dynamic analysis and algorithm testing or controller tuning. Two types of simulations can be utilized:

(a) Experimental modelling in the autoclave to be controlled. A safe experimental test is based on using an electrically heated element producing heat inside the autoclave. For sophisticated experiments the heating element input can be controlled by a computer to simulate the behaviour of the chemical reaction (refs. 57, 58).

(b) *Mathematical simulation*. A model for simulating the dynamic behaviour of the autoclave can be developed from experimental measurements under safe conditions, and the properties of the chemical reaction (the reaction heat, activation energy and the rate equation).

In this paper only the simplest dynamic model is presented. The autoclave is described as a system of two heat capacities, the reaction mixture with an effective heat capacity C and the cooling system with an effective heat capacity C_c. The relevant equations for a zero-order chemical reaction at constant hydrogen pressure transformed into the semi-dimensionless form are given in Scheme 17.1.

An example of a simulation and an algorithm test is presented in Fig. 17.2. The control algorithm tested is the bang-bang controller (the simple relay) described by:

$$F_c/F_{cs} = 0 \quad \text{if} \quad T < T_s \tag{17.14}$$

$$F_c/F_{cs} = 2 \quad \text{if} \quad T > T_s \tag{17.15}$$

The reaction temperature is controlled by closing and opening the coolant flow. The feedback control of the autoclave is simple and safe if this simple algorithm results in small amplitudes of temperature oscillations.

The reliability of control is dependent on two parameters:
(a) *The cooling system inertia, D*. Small values are dangerous ($D < 0.03$), delaying the adion of cooling.

(b) *Reserve in the cooling ability of the cooling system, B*. Small values are unfavourable, indicating that the autoclave regime approaches the cooling limit. Near this limit the cooling rate is insensitive the coolant flow-rate and the control by manipulating it is not efficient.

Another simulation is given in Scheme 17.2. and in Fig. 17.3. The aquations represent the simplest model simulating the temperature control by manipulating the hydrogen feed.

Scheme 17.1. Simple dynamic model for simulation of the temperature control by manipulating the coolant flow.

$$\frac{d(T - T_s)}{d\tau} = (R_H/R_{H,s}) - R_c$$

$$\frac{d R_c}{d\tau} = -\left\{ D \frac{R_c}{1-B} + \frac{F_c}{F_{cs}} (1/B) \left[R_c - 1/(1-B) \right] \right\}$$

$\tau = T\, R_{H,s}(-\Delta H)/C$ $\qquad R_c = (T_s - T_{ce})/(T_s - T_{ces})$

$B = (T_{ces} - T_{ci})/(T_s - T_{ci})$ $\qquad D = C/C_c (T_s - T_{ci})$

$R_H/R_{H,s} = \exp E(T-T_s)/RTT_s$

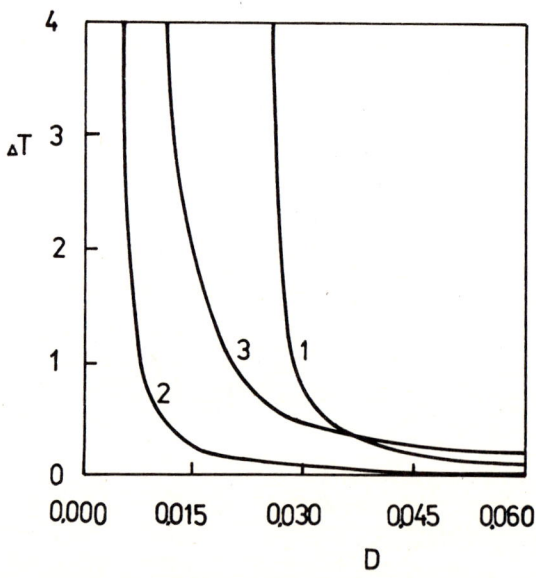

Fig. 17.2. Examples of the relationships betwen the temperature overregulation and dynamic properties of the cooling system in a bang-bang manipulation of the coolant flow (simple relay).
(1) E=50 kJ mol^{-1}, B=0.5; (2) E=50 kJ mol^{-1}, B=0.9;
(3) E=80 kJ mol^{-1}, B=0.9.

Scheme 17.2. Simple dynamic model for simulation of the temperature control by manipulating the hydrogen feed.

$$\frac{d(T - T_s)}{d\tau} = (R_H/R_{H,s}) - (T - T_s/(T_s - T_{ci}) - 1$$

$$\frac{d(P/P_s)}{d\tau} = D_1 (F - R_H/R_{H,s})$$

$$\tau = t\, R_{H,s}(-\Delta H)/C \qquad F = F_H/R_{H,s}$$

$$D_1 = RT\, C/V_g P_s (-\Delta H) \qquad (R_H/R_{H,s}) = (P/P_s)\,\exp E(T-T_s)/RTT_s$$

The algorithm tested:
$F = 0$ if $T > T_s$
$F = 2$ if $T < T_s$ maximum $(P/P_s) = 2$

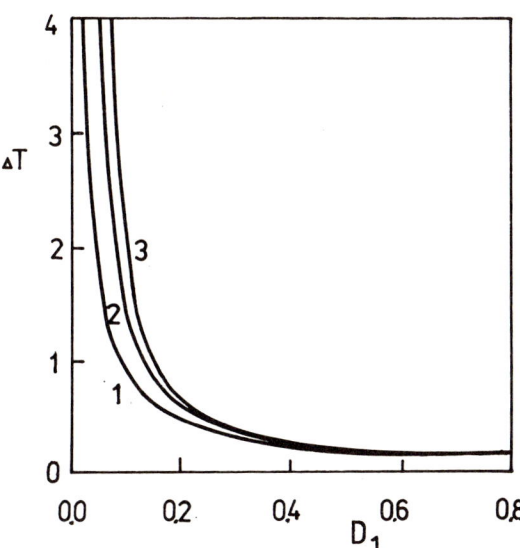

Fig. 17.3. Examples of the relationships between the temperature overregulation and dynamic properties of the autoclave in the temperature manipulation of the hydrogen feed. (1) $E=50$ kJ mol^{-1}, $(T_s-T_{ci})=20$; (2) $E=50$ kJ mol^{-1}, $(T_s-T_{ci})=50$; (3) $E=80$ kJ mol^{-1}, $(T_s-T_{ci})=50$.

The reliability of control is dependent on two parameters:
(a) The gas space inertia, D_1. Small values are dangerous ($D_1 < 0.4$), delaying the pressure changes in the gas space.
(b) The cooling driving force, $(T_s - T_{ci})$. Large values indicate that the operation is less stable.

17.4.4 Conversion control

In batch autoclaves the extent of conversion is controlled by stopping the process after an appropriate reaction time. The control may be complicated by the necessity to take samples and analyse them in the laboratory. Therefore, the degree of conversion may be known only ofter significant delay.

The following procedures can be applied to solve the problem:
(a) A substitute variable is used to indicate the reaction time (viscosity, density).
(b) The delayed information on the sample analyses is used for model identification. Then the course of the conversion is predicted by integrating the model equations.
(c) The reactant conversion is evaluated from the hydrogen consumption.
(d) The conversion is evaluated from the amount of heat removed from the autoclave (ref. 40)

17.5 INDUSTRIAL AND PILOT-PLANT CONTINUOUS AUTOCLAVES

Control of continuous autoclaves may be more complicated than control of batch systems because two variables must be controlled simultaneously, the reaction temperature and the extent of conversion. In most cases the substrate feed rate is determined by the desired reactor performance, consequently, the reaction time cannot be employed as a manipulated variable for the conversion control. The behaviour of continuous reactors with stabilized inputs may be very complicated, multiple steady states, instabilities and autonomous oscillations may occur, especially if a large cooling driving force must be utilized to remove the reaction heat (refs. 8 - 15).

The most reliable way to solve the control task consists

in independent stabilization of the reaction temperature. After the reaction temperature is stabilized, another manipulated variable is selected for control of the extent of conversion.

17.5.1 Temperature control

The problems with selecting the system for temperature control of continuous processes are similar to those in the temperature control of batch autoclaves. In continuous autoclaves a part of the reaction heat is removed by the feed if it is not preheated. The following characteristics may be helpful:

(a) <u>The cooling driving force necessary to remove the excessive reaction heat</u>.

$$(T - T_{ci})_1 = \frac{R_H(-\Delta H/b)}{k_c} \left[1 - \frac{F c_p (T-T_i)}{R_H(-\Delta H/b)} \right] \qquad (17.16)$$

The second term in the brackets represents the fraction of the reaction heat removed by the cold feed. The driving force, $(T - T_{ci})_1$, indicates whether the reaction heat can be removed from the autoclave or not. A small value of this characteristic does not mean that simple stabilization of inputs may be effective in controlling the reaction temperature, because the utilization of the cold feed may cause instabilities.

(b) <u>The difference between the reaction temperature and the temperature of zero cooling rate</u>. The temperature corresponding to the zero cooling rate is expressed as:

$$T_{zc} = (k_c T_{ci} + F \varrho c_p T_i)/(k_c + F \varrho c_p) \qquad (17.17)$$

For the stability considerations, the cooling rate can be expressed as a function of the reaction temperature and the temperature of zero cooling rate:

$$(T_s - T_{zc})_{nec} = R_H (-\Delta H/b)/(k_c + F \varrho c_p) \qquad (17.18)$$

The difference $(T_s - T_{zc})_{nec}$ can aid in selecting the method of temperature control, analogously to the cooling driving force for batch autoclaves.

(c) <u>The cooling driving force necessary to remove all reaction heat</u>

$$(T - T_{ci}) = R_H(-\Delta H/b)/k_c \qquad (17.19)$$

A small value of this difference indicates that the reserve in the cooling ability is large and that seef-regulative regimes can be achieved.

(i) <u>Autoclaves operated at high extents of conversion</u>. The extent of conversion is the constrained controlled variable, the value of which has to exceed a certain minimum. The temperature control is safe due to the low concentration of reactants in the autoclave. A reaction temperature runaway is not hazardous during the normal regime. However, any accumulation of unreacted reactants inside the autoclave must be avoided. Any decrease in reaction rate resulting in a decrease in the extent of conversion must be observed and compensated by stopping the reactant feed and slowing down the cooling to avoid subcooling of the autoclave. The accumulation of reactants caused by subcooling, catalyst deactivation or a failure in the catalyst may result in a serious temperature runaway.

(ii) <u>Autoclaves operated at medium degrees of conversion</u>. The temperature control may be difficult, especially if a large driving cooling force is necessary to remove the reaction heat. The control system must be able to avoid both an uncontrollable temperature increase and a decrease (extinction). The coolant flow must be employed as the manipulated variable for the temperature stabilization. The hydrogen pressure and the catalyst concentration cannot be used for this purpose due to their effect on conversion.

After the reaction temperature has been stabilized another manipulated variable is used to keep the extent of conversion at the set value. The following control modes can be applied:

(a) <u>The floating reaction pressure.</u> The reaction pressure is stabilized at a changing level. This level is slowly changed to compensate the effects of the changing catalyst activity and the reactor load.

(b) <u>The floating catalyst concentration</u>. If the catalyst is fed into the autoclave and removed from the autoclave with the reaction mixture, the slowly changing catalyst concentration can be utilized as the manipulated variable.

(c) <u>The floating reaction temperature</u>. In this case the set reaction temperature is slowly changed.

The control loop manipulating the floating variable must operate slowly in order that the loop stabilizing the reaction temperature be able to eliminate the disturbance resulting from the changes in the floating variable.

(iii) <u>Hydrogen feed as the manipulated variable</u>. If all inputs to the autoclave are stabilized and hydrogen is added in the stoichiometric ratio corresponding to the set extent of conversion, the reaction temperature and the reaction pressure will tend to reach a stationary state. As the hydrogen is consumed at the same rate as it is added, the extent of conversion is determined by the stoichiometric ratio of the feeds. This method can be applied both in stable and unstable stationary states. A dynamic analysis of the system is necessary before application because oscillatory behaviour may appear if the hydrogen inventory inside the autoclave exceeds a certain critical value (ref. 59).

17.6 EARLY RECOGNITION OF HAZARDOUS STATES

An early recognition of the start-of a temperature runaway is an important component of the control system. If recognized early, the hazard can be dealt with shotting of the hydrogen feed or by switching to full cooling.

The following alerting systems can be applied (ref. 24):

(a) <u>A fixed alarm temperature</u>. This simplest criterion may be effective for continuous and simple batch processes. The critical temperature must be related to that of the coolant, overheating with respect to the coolant inlet temperature indicating a hazardous state.

(b) <u>A black box mathematical model</u>. A method which is independent of specific settings, because it detects progressive heat evolution, has been proposed by Hub (ref. 60). A runaway situation is likely when the following conditions are met:

$$\frac{d^2 T}{dt^2} > 0 \quad \text{and} \quad \frac{d(T - T_{ci})}{dt} > 0 \qquad (17.20)$$

An on-line warning device (OLIWA) based on this reasoning is commerically available (ref. 61). The method is simple and no additional information on the process in necessary. When applied in digital form, the smoothing of the measured data represents the crucial problem (ref. 62).

(c) <u>An on-line deterministic model</u>. This most sophisticated method was suggested by Gilles and Schuler (refs. 63,64). It is based on a comparison of the measured and simulated temperature courses and an evaluation of the actual process parameters, for example the catalyst activity. The model is then used to predict the reactor behaviour.

There are three obstacles to a general introduction of this method:

The high parametric sensitivity of processes which tend to runaway, and consequently the high accuracy of the data required.

The complex nature of the theory, which is not easily understood by the practitioners.

The necessity to introduce information on the system into the model. The non-adaptive part of the model may result in identification failure in the case of an unexpected change in the system properties.

<u>List of symbols</u>

a   Stoichiometric coefficient of the reactant, dimensionless.
A   Reactant A.
b   Stoichiometric coefficient of hydrogen, dimensionless.
B   Dimensionless parameter defined in scheme 1.
c   Concentration, mol m^{-3}.

c_A Instantaneous concentration of reactant A, mol m^{-3}.
c_{Ai} Initial concentration of reactant A, mol m^{-3}.
c_H Concentration of hydrogen, mol m^{-3}.
C Effective heat capacity of the reaction mixture, J K^{-1}.
C_c Effective heat capacity of the cooling system, J K^{-1}.
c_p Specific heat capacity of the raction mixture, J kg^{-1}K^{-1}.
c_{pc} Specific heat capacity of the coolant, J kg^{-1}K^{-1}.
D Parameter characterizing the dynamic properties of the cooling system, defined in scheme 1, K^{-1}.
D_1 Parameter characterizing the dynamic properties of the gas space inside the autoclave, defined in scheme 2, K^{-1}.
E Activation energy, J mol^{-1}.
F Relative feed rate related to the pseudostationary value defined in scheme 2, dimensionless.
F_H Feed rate of hydrogen, mol s^{-1}.
F_c Flow-rate of the coolant, m^3 s^{-1}.
F_{cs} Flow-rate of the coolant in the pseudostationary state, m^3 s^{-1}.
ΔH Reaction heat, J mol^{-1}.
k_c cooling coeficient defined by equation 8, W K^{-1}.
k_h Heat-transfer coefficient, W m^{-2}K^{-1}.
P Total pressure, Pa.
P_s Total pressure corresponding to the pseudostationary state, Pa.
r_o Reaction rate at a standard temperature T_o, mol s^{-1}kg$_{kat}^{-1}$.
R Gas constant, J mol^{-1} K^{-1}.
R_c Relative rate of cooling defined in scheme 1, dimensionless.
R_h Total rate of heat production, W.
R_H Production rate measured in terms of the hydrogen consumption, mol$_H$s^{-1}.
$R_{H,des}$ Desired value of the production rate, mol$_H$s^{-1}.
$R_{H,s}$ Hydrogen consumption corresponding to the pseudostationary state, mol$_H$s^{-1}.
t Time, s.
T Temperature of the reaction mixture, K.
ΔT_{ad} Adiabatic temperature rise, K.
$\Delta T_{ad,A}$ Adiabatic temperature rise limited by the instantaneous concentration of reactant A, K.
$\Delta T_{ad,Ai}$ Adiabatic temperature rise limited by the initial concentration of reactant A, K.

$\Delta T_{ad,H}$ Adiabatic temperature increase limited by the hydrogen content inside the autoclave, K.
T_c Temperature of the coolant, K.
T_{ce} Coolant exit temperature, K.
T_{ces} Coolant exit temperature in the pseudostationary state, K.
T_{ci} Coolant inlet temperature, K.
T_o A standard temperature, K.
T_s Set value of the reaction temperature, K.
T_{zc} Temperature of the reaction mixture at which the rate of cooling equals zero, K.
V Volume of the reaction mixture, m^3.
V_g Volume of the gas space inside the autoclave, m^3.
W Catalyst mass inside the autoclave, kg_{kat}.
ϱ Specific mass of the reaction mixture, $kg\ m^{-3}$.
ϱ_c Specific mass of the coolant, $kg\ m^{-3}$.

REFERENCES

1 F.F. Shinskey, Process Control Systems, McGraw-Hill, New York, 1967.
2 M. Rezsö and F. Béla, Chemical Process Dynamics, Elsevier, Amsterdam, Akademiai Kiado, Budapest, 1982.
3 J. Hengstenberg, B. Sturm and O. Winkler (Editors), Messen, Steuern and Regeln in der chemischen Technik, Band III, Messwertverarbeitung zur Prozessführung I, Analoge und binäre Verfahren, Springer, Berlin, 1981.
4 F.G. Shinskey, Controlling Multivariable Processes, Instrument Society of America. Research Triangle Park, N C, 1981.
5 W.H. Ray, Comput. Chem. Eng., 7 (1983) 367-394.
6 J. Horák, Paper Int.Summer School "Modelling of Heat and Mass Transfer Processes and Chemical Reactors", Bulgarian Acad.Sci., Burgas, 1985.
7 J. Horák, Model-aided Control of Chemical Reactors, State of the Art Lecture, International CHISA Congress, Section Reaction Engineering, Prague, 1984.

8   L. Lapidus and N.R. Amundson (Editors), Chemical Reactor Theory, A Review, Prentice Hall, Engelwood Cliffs, N Y, 1977.
9   G. Eigenberger, Chem.-Ing.-Tech., 50 (1978) 924-933.
10 I. Endo and T. Furusawa, Catal.Rev.-Sci.Eng., 18 (1978) 297-335.
11 V.V. Kafarov and V.A. Chetkin, Itogi Nauki Tekh., Protsessy Appar.Khim.Tekhnol., 8 (1980) 77-151.
12 V. Hlaváček and P. Van Rompay, Chem.Eng.Sci., 36 (1981) 1587-1597.
13 V.B. Volter and I.E. Salnikov, Stability of the Working Regime of Chemical Reactors, Chimia, Moscow, 1981.
14 W.H. Ray, Springer Ser.Chem.Phys., 18 (1981) 337-354.
15 M. Kubíček and M. Marek, Computational Methods in Bifurcation Theory and Dissipative Structures, Springer, New York, Berlin, Heidelberg, Tokyo, 1983.
16 H. Gran, H. Grande and S. Lange, 4th International Symp.Chem. React.Eng., Heidelberg, Dechema, Frankfurt, 1976, Preprints VIII, pp. 363-371.
17 G. Piras and C.A. Bestetti, Consideration on Computer Automation of PVC Plants, Int.Congr.Proc.-Chem.Tech. 80, Symp. 6., Paper 3., Edited by Shah, Jasn, Bombay, 1980.
18 A.E. Hamielec and J.F. Mac Gregor, Latex Reactor Principles. Design, Operation, and Control, in Piirma, Irja (Editor), Emulsion Polym. 1982, Academic Press, New York, 1982, pp. 319-355.
19 B.G. Freedman, AICHE Symp. Ser., 72 (1979) 206-218.
20 R.K. Pearson, Adv.Instrum., 39 (1984) 885-893.
21 D.W.T. Rippin, Comput.Chem.Eng., 7 (1983) 137-156.
22 J. Horák and F. Jiráček, Collection Czech.Chem.Commun., 48 (1983) 711-721.
23 W. Regenass, U. Osterwalder and F. Brogli, Inst.Chem. Eng. Symp.Ser., 87 (1984) 369-376.
24 W. Regenass, Symp.Pap."Protection of Exothermic Reactors" Chester, 1984, 15 pp.
25 W. Regenass, Swiss.Chem., 5,9a (1983) 37.
26 V.G. Theofanous, Chem.Eng.Sci., 38 (1983) 1615-1629.
27 V.G. Theofanous, Chem.Eng.Sci., 38 (1983) 1631-1636.
28 G. Eigenberger, Springer Ser.Chem.Phys., 18, (1981) 284-304.
29 G. Eigenberger and U. Wegerle, ACS Symp.Series. (1982) (Chem. React.Eng.-Boston), 133-143.

30 H. Fierz, P. Finck, G. Giger and R. Gygyx, 4th Int.Symp.on Loss Prevention and Safety Promotion in the Process Industries, Vol. 3, Chemical Process Hazards, The Institution Of Chemical Engineers, Rugby, 1983, pp. A12-A21.
31 O. Klais and T. Grewer, ibid, pp. C24-C34.
32 V. Pilz, H. Schacke and N. Schulze, ibid, pp. B1-B9.
33 P.A. Obnovlenskii (Editor), Systems and Facilities for Automating Potentially Dangerous Chemical Engineering Processes. Leningr.Tekhnol.Inst., Leningrad, 1982.
34 E.H. Livingstrone, Chem.Eng.Prog., 80, (1984) 70-75.
35 W. Riemenschneider, Chem.-Ing.-Tech., 54 (1982) 1065-1066.
36 R. Davies, Inst.Chem.Eng.Symp.Ser., 87 (1984) 361-368.
37 W.A. Woods and E.R. Thornton, I.Mech.Comp.Publ.,(1983) 1-10.
38 L. Friedel and S. Purps, 4th Int.Symp.on Loss Prevention and Safety Promotion in the Process Industries, Vol. 3, Chemical Process Hazards, The Institutions of Chemical Engineers, Rugby, 1983, pp. B18-B33.
39 T. Grewer and E. Duch, ibid. pp. A1-A11.
40 W. Regenass, Thermochim.Acta, 20 (1977) 65-75.
41 J. Hakl, Thermochim.Acta, 38 (1980) 253-258.
42 J. Schildknecht, Thermochim.Acta, 49 (1981) 87-100.
43 C.F. Coates and W. Ridell, Chem.Ind. (London), (1981) 84-88.
44 J. Hakl, Chem.Rundschau, 34 (1981) 1-2.
45 L.F. Whiting and J.C. Tou, J.Thermal.Anal., 24 (1982) 111-120.
46 J. Hakl, Thermochim.Acta, 80 (1984) 209-219.
47 J. Horák, P. Beránek and D. Maršálková, Collect.Czech.Chem. Commun., 49 (1984) 2566-2578.
48 J. Horák, P. Beránek ane D. Maršálková, Collect.Czech.Chem. Commun., in press.
49 J. Horák, F. Jiráček and L. Ježová, Collect.Czech.Chem.Commun., 47 (1982) 251-261.
50 J. Horák and F. Jiráček, Collect.Czech.Chem.Commun., 47 (1982) 454-464.
51 W.L.Luyben and M.Melcic, Hydrocarbon Process., 57 (1978) 115-117.
52 J. Horák and F. Jiráček, Chem.Eng.Sci., 35 (1980) 483-491.
53 J. Horák, F. Jiráček and L. Ježová, Collection Czech.Chem. Commun., 48 (1983) 2627-2635.
54 J. Horák, F. Jiráček and L. Ježová, Collection Czech.Chem. Commun., 49 (1984) 1642-1652.

55 J. Hanika, K.Sporka, L. Kráčmar and V. Růžička, Chem.Prum., 30 (1980) 31-34.
56 J. Hanika, I. Kult and V. Růžička, Chem.Prum., 30 (1980) 390-393.
57 D. Mukesh and A.R. Cooper, Ind.Eng.Chem.Fundam., 22 (1983) 145-149.
58 J. Horák and P. Beránek, Collect.Czech.Chem.Commun., in press.
59 R. Geike, J. Horák and F. Turek, Chem.Tech. (Leipzig), 36 (1983) 283-286.
60 L. Hub, Chem.-Ing.-Tech., 54 (1982) 181.
61 OLIWA System Manufacturer, Systag CH-8803 Rüschlikon.
62 J. Horák and L. Králiková, Chem.Prum., 35 (1985) 169.
63 H. Schuler, VDI-Z, 125 (1983) 167-168.
64 E.D. Gilles and H. Schuler, Chem.-Ing.-Tech., 53 (1981) 673-682.

Chapter 18

SELECTIVE HYDROGENATION APPLIED TO THE REFINING OF PETROCHEMICAL RAW MATERIALS PRODUCED BY STEAM CRACKING

MICHEL L. DERRIEN, Institut Français du Pétrole, Rueil Malmaison, 92506 Cedex (France)

18.1 INTRODUCTION

The field of petrochemicals has undergone considerable expansion since the early 1950s. The replacement of natural products such as cotton, wood and rubber by plastics, synthetic fibres and synthetic rubbers, has required the production of larger and larger quantities of raw materials. A substantial proportion of these raw materials, whose essential characteristic is their chemical reactivity and hence an unsaturated nature, is produced by steam cracking.

Steam cracking is a process that takes place at high temperature (about 800°C) and in the presence of steam. It is mainly used for producing ethylene as well as a series of important coproducts (propylene, 1-butene, butadiene, isoprene, benzene) from a wide variety of hydrocarbon fractions ranging from ethane and propane via naphtha to vacuum distillates (b.p. 300-550°C).

The world capacity for ethylene production is currently 40 million metric tons per year, corresponding to a steam-cracking capacity of about 100 million tons per year. The production of the basic raw materials for petrochemistry is thus a very large-scale industry.

The cracking of hydrocarbon fractions in the presence of steam gives rise to an entire range of olefins in which ethylene predominates, but also to other families of hydrocarbons such as paraffins, diolefins, acetylenics and aromatics. The number of carbon atoms in the hydrocarbons obtained ranges from one to greater than 15 and for each carbon-atom number there is actually a dominant hydrocarbon, i.e., olefin, diolefin or aromatic, together with small amounts of hydrocarbons from other families. For example, the C_2 hydrocarbons break down as follows (in mol.%) : acetylene, 1.2 ; ethylene, 83.5 ; ethane, 15.3. The same is true of the C_3 fraction in which propylene dominates, but where propane is also found together with smaller amounts of more unsaturated hydrocarbons, i.e., methylacetylene and propadiene. For the C_4 fraction, in addition to the dominant 1,3-butadiene, small amounts of all possible acetylenics are found along with vinylacetylene. Likewise, for the mixture of C_5 and heavier hydrocarbons, because of the possibility of greater diversity, large numbers of more and less unsaturated hydrocarbons are found together with dominant cyclopentadiene, isoprene, benzene and styrene.

The raw material must be concentrated to maximize the productivity of the downstream processes. Likewise, it must be pure, i.e. devoid of other reactive hydrocarbons so as to prevent during downstream processing the formation of secondary products that might be detrimental to the quality of the end-product.

Very briefly, refining of steam cracking products includes the following series of operations :

A first fractionation by low-temperature distillation, resulting in different fractions or cuts, i.e., the ethylene-rich cut, the propylene-rich cut, the C_5, C_6C_8, ... cuts. For light fractions this step will be accompanied by desulphuration using amine wash and caustic wash processes.

A second stage called refining, which generally consists of catalytic hydrorefining, during which selective hydrogenation is used to remove from each fraction compounds more strongly unsaturated than the main hydrocarbon. A typical example is the hydrogenation of acetylene to purify the ethylene-rich C_2 cut. Other processes, e.g., selective adsorption may be used, but only when hydrogenation is not effective enough.

A third superfractionation stage (simple distillation or with a solvent) during which the required concentration is achieved. An example is the separation of ethylene and ethane.

Refining by selective hydrogenation has only gradually increased in importance. The first steam crackers were designed for cracking ethane. The only important product was ethylene, and it is entirely logical that acetylene hydrogenation was used as a deacetylenization method, especially since the purification level required was relatively modest. This enabled acceptable ethylene recovery rates to be obtained (98% and higher) with mundane catalysts (Ni-Co-Mo) working in the presence of H_2S.

The development of naphtha steam cracking, producing all the coproducts already mentioned in addition to ethylene, led to a whole range of selective hydrogenation processes :

selective hydrogenation of methylacetylene and propadiene in the propylene-rich cut,

selective hydrogenation of acetylenics in the butadiene-rich C_4 cut,

selective hydrogenation of diolefins in the benzene-rich gasoline cut.

This development is continuing as illustrated by the appearance of two new hydrorefining processes in the last five years. On one hand the establishment of a polymerization market for 1-butene has resulted in a specific process for the hydrogenation of the residual butadiene contained in 1-butene-rich cuts : in addition to the need to prevent the hydrogenation of olefins, an additional property is required, i.e., prevention of 1-butene losses by migration of the double bond (isomerization). On the other hand, since most applications of n-butenes, exclusive of polymerization favour 2-butenes, processes for the isomerization of 1-butene into 2-butenes have also been developed.

All in all, in most cases selective hydrogenation has replaced other refining methods. The reasons for this are numerous :

Hydrorefining is relatively simple to implement. The fixed-bed catalyst technology is commonly used. Therefore, the refining stage requires low investment and operating costs.

Hydrorefining is efficient. There is actually no thermodynamic limitation to purification under the low-temperature and medium-pressure conditions chosen. Furthermore, the constant improvement of catalysts has simultaneously enabled excellent yields to be maintained while requirements of purity increase gradually over the years. Since hydrorefining often makes use of consecutive reactions (acetylene ⟶ ethylene ⟶ ethane), there is even some gain of the main hydrocarbon in many cases.

Hydrorefining is easy to operate. It is easy to control the hydrogenation rate. Very active catalysts enable operation at low temperature. The real dangers of polymerizing polyunsaturated hydrocarbons are almost entirely eliminated, and the cycle length of these catalysts are six months or longer. Likewise, catalyst regeneration is easy, and the actual catalyst life may be several years.

Currently, most naphtha steam-cracking units involve several selective hydrogenation refining units. A total of more than 400 industrial units are in operation. The development of selective hydrogenation for the refining of steam-cracking products has resulted in original technologies (especially for limiting temperature rises in the reactor) as well as new generations of catalysts.

In this chapter, the thermodynamic, kinetic, catalytic and technological aspects of the hydrorefining of different fractions will be discussed according to the following logical order of increasing complexity :

hydrorefining of the ethylene-rich cut (C_2 cut),
hydrorefining of the propylene-rich cut (C_3 cut),
hydrorefining of the C_4 cut,
hydrorefining of the gasoline cut (C_5-C_{10} cut).

18.2 C_2 CUT HYDROREFINING

18.2.1 The different processes for the hydrorefining of the ethylene-rich cut (C2 cut)

The C_2 cut hydrorefining, i.e. the selective hydrogenation of acetylene contained in the ethylene-rich fraction, is performed in a unit which is closely integrated in the low-temperature fractionation section of the steam-cracking plant, i.e., the section where the narrow hydrocarbon fractions (the acetylene-ethylene-ethane cut, in other words the C_2 cut ; the C_3 cut ; the C_4 cut) are produced.

They are many alternative process-schemes for this section because the composition of the cracked product, more specially the light/heavy hydrocarbon ratio, changes with the boiling-range of the steam-cracking feedstock and because each engineering company has its own optimization of the fractionation (ref. 1, 2). In addition, the development of more selective catalysts, usually also more sensitive to contaminants, results in modifications of fractionation processes.

The main variants of the ethylene-rich cut hydrorefining are :

the hydrorefining of the cracked gases as a whole, i.e. the hydrorefining of the C_5 minus fraction, the fraction containing C_5 and lighter hydrocarbons (H_2, CO, CH_4, C_2's, C_3's, C_4's, C_5's),

the hydrorefining of the C_2- cut (H_2, CO, CH_4, C_2 hydrocarbons) or of the C_3- cut,

the hydrorefining of the C_2 cut, i.e. after removal of the lighter (H_2, CO, CH_4) and heavier hydrocarbons (C_3, etc ...).

(i) <u>Hydrorefining of the cracked gases</u>. This is historically the first case of hydrorefining of steam-cracking. 2 alternative implementations are feasible :

It is implemented in the compression section which means before the caustic scrubbing. Therefore it operates at medium pressure, about 15 bar and employs H_2S-resistant catalysts, i.e. catalysts such as NiCoMo or NiCrMo active in the sulfide form. The resulting relatively high operating temperature (around 200°C) leads to a poor selectivity which may be acceptable if the tolerance of residual acetylene is fairly high : around 50 ppm. This process is almost obsolete nowadays.

It is implemented after the compression section which means also after the caustic-scrubbing. Hence it operates at a higher pressure (ca. 25 bar) and may employ non H_2S-resistant, more active, palladium catalysts. In this case it is in fact applied to all the C_2- fraction mixed with the uncondensed part of the C_3 and C_4 hydrocarbons. Due to the lack of selectivity for the broad cut, only the acetylene specification is targeted and the C_3+ fraction is reprocessed separately or burned-off which is sometimes still carried out for ethane or propane steam-crackers.

(ii) <u>C2- cut or C3- cut hydrorefining</u>. This implies the presence of the entire production of H_2 and CO, i.e., a large excess of H_2 and the concomitant overhydrogenation risks, and a high content of CO, a reaction inhibitor, entailing long contact-time and therefore large amount of catalyst. The standard catalyst is again based on palladium on an alumina carrier. For this and all subsequent cases the operating pressure is around 25 bar. A single reactor is sometimes sufficient.

<u>C2- cut hydrogenation</u>. The reaction product is then sent to the demethanizer and the isolated C2 cut is sent to the ethylene/ethane splitter.

<u>C3- cut hydrogenation</u>. In this case, the reaction product is sent to the demethanizer, then to the C_2-C_3 splitter, the bottom stream (C_3 cut) is then subjected to a final hydrogenation to obtain a C_3 cut meeting methylacetylene/propadiene specifications.

(iii) <u>C2 (or C2/C3) cut hydrorefining</u>. This hydrogenation also takes place in the gas phase, on the stream taken from the head of the de-ethanizer or depropanizer. The pressure is also about 25 bar. The essential feature of this hydrogenation is that it is carried out by controlled make-up of high-purity hydrogen from the purification-section and after methanation. The carbon monoxide content of the make-up hydrogen is adjusted to a value that is low but not nil (50 < CO < 500 ppm), because the presence of carbon monoxide traces improves the selectivity. This is obtained by partly bypassing the methanator, i.e. the catalytic converter where carbon monoxide contained in the hydrogen-rich gas, is converted into methane.

This hydrorefining is normally carried out on the C_2 cut only, but may also be performed on the C_2/C_3 cut if the depropanizer is intentionally placed before the de-ethanizer. This specific hydrorefining will be discussed in further detail below.

The foregoing processes all take place in the gas phase ; only one liquid-phase process exists, without present industrial application, namely the Bayer process. This is in contrast to the very clear preponderance of liquid-phase operation in the hydrorefining of C_3 and heavier cuts. The reason is essentially the low temperatures (around -20°C) at which C_2 liquid-phase hydrogenation would have to be carried out in order to preserve the integration of the hydrorefining unit into the fractionation section, which means operating at around 25 bars. Heat removal at -20°C is obviously a costly operation because it demands the use of low-temperature refrigeration. This is why Institut Français du Pétrole is developing a "solvent phase" process based on the use of a heavy solvent, circulating in a closed loop in the reaction section in order to profit from the advantages of the liquid phase (continuous removal of the oligomers by washing of the catalytic bed and improved heat transfer, thus resulting in a much longer cycle) (ref. 3).

18.2.2 Industrial catalysts

At the present time, practically all such catalysts are based on palladium supported on an alumina carrier. They have a low palladium content, about 0.04% on the average, because, by reducing the reaction rate, mass-transfer limitations are reduced, thus improving the selectivity and reducing the risk of a runaway.

Many manufacturers offer similar catalysts. Front-end hydrogenation : United Catalysts Inc. (UCI) (USA) C 35-2-01, Girdler-Südchemie (F.R.G.) G 83, Imperial Chemical Industries (ICI) (UK) 38-1, Leuna-Werke (G.D.R.) 7741. Tail-end hydrogenation : United Catalyst Inc. (UCI) (USA) C 31-1A-01, Girdler Südchemie G58, Procatalyse (F) LT 261, Imperial Chemical Industries ICI 38-2, Engelhard Industries (USA) HPN II, Leuna-Werke 7746, Procatalyse LT 279 promoted catalyst (ref. 4).

Palladium-based catalysts promoted by a metal are now becoming available. The promotor improves selectivity or stability (Procatalyse LT 279, for example (ref. 3)).

18.2.3 Selective hydrogenation of acetylene in the C2 cut, detailed study

The most common case will be described in detail. If the C_2 cut is taken from the top of the tail-end de-ethanizer, its composition is shown in Table 18.1.

TABLE 18.1
Composition of the C_2 cut (mol.%)

	Typical composition	Range of composition	Commercial specification of ethylene
C_1-	< 0.01	< 0.01	
Acetylene	1	0.7 - 2	2 ppm max.
Ethylene	79	70 - 90	99.5 +
Ethane	20	qsp 100	
C_3+	< 0.01	< 0.01	

The reactions involved in C_2 hydrorefining, apart from intensive acetylene hydrogenation, are the hydrogenation of ethylene to ethane as well as spurious oligomerization, with the formation of C_6-C_{12} olefins commonly called green-oils, which cause operating problems (catalyst fouling and need for regeneration, plugging of piping). The reaction scheme can therefore be written as follows.

$$\begin{array}{c} C_2H_2 \xrightarrow{+H_2} C_2H_4 \xrightarrow{+H_2} C_2H_6 \\ \searrow {\scriptstyle +H_2} \\ \quad C_6\text{-}C_{12} \text{ olefins} \end{array} \qquad (18.1)$$

All these reactions are exothermic :

$C_2H_2 + H_2 \longrightarrow C_2H_4 + 42.3$ Kcal/mol.

$C_2H_4 + H_2 \longrightarrow C_2H_6 + 32.6$ Kcal/mol.

and result in a reduction of the number of molecules : at low temperature and under pressure, they are almost complete.

Given the large excess of ethylene in the feed (ethylene/acetylene ratio $\geqslant 60$), the main difficulty is to hydrogenate the acetylene completely (99.99% conversion) with minimum, if any, ethylene losses.

(i) <u>Thermodynamic and kinetic features</u>. The hydrogenation reactions implemented can be considered as thermodynamically absolutely complete at temperatures $< 250°C$, even with low hydrogen pressures. This enabled the technique to keep pace with the increasing severity of specifications, which lowered the acetylene content of the ethylene product from 50 ppm to 5 ppm, and then to 2 ppm over the last 20 years. The reaction is extremely fast and operating conditions are chosen so as not to increase the reaction rate and thus decrease the catalyst volumes employed, but to improve the selectivity (maximum ethylene yield) and to integrate hydrogenation fully into the fractionation scheme from a pressure standpoint (no recompression, hence operation at about 25 bar).

In refining C_2 as well as heavier cuts, selectivity is the key objective. Hence operating conditions, catalyst and type of process (reaction section configuration) are selected so as to achieve maximum selectivity. Resulting greater capital investment or higher operating cost if any are rapidly recovered if the ethylene yield is improved even slightly. The selectivity in hydrorefining is routinely expressed by the ethylene yield, or ethylene recovery for a given acetylene conversion.

Ignoring the formation of oligomers, because it is actually negligible in comparison with other reactions, reaction 18.2 is a standard two-step consecutive scheme, with the presence, in the feed, of the intermediate product in high concentration and a common co-reactant (H_2) :

$$C_2H_2 \xrightarrow[r_1]{+H_2} C_2H_4 \xrightarrow[r_2]{+H_2} C_2H_6 \qquad (18.2)$$

r_1 and r_2 are the rates of hydrogenation, respectively of acetylene and ethylene.

A simple mathematical treatment can be used to obtain an expression giving the ethylene yield of hydrorefining for a feedstock with a given composition, and for different values of the residual acetylene content.

Let b and v be the ethylene contents respectively of the feed and the product, a and u the acetylene contents of the feed and the product, and S be the ratio r_1/r_2, i.e. the kinetic selectivity a characteristic term for each process, or at least for each set of parameters (temperature, pressure, hydrogen, etc) and each catalyst.

The mathematical treatment (ref. 5) culminates in the expression :

$$v = \frac{S}{1-S} u + \mu u^{1/S} \quad \text{with} \quad \mu = \frac{b - \left(\frac{S}{1-S}\right)a}{a^{1/S}} \qquad (18.3)$$

Knowing the "performance level" of the process as measured by the selectivity factor S, one can then calculate the ethylene yield for any type of feed and product. A graphic of representation in the form of a normograph, as in Fig. 18.1 enables one to determine S by comparison with theoretical curves. In Fig. 18.1 a logarithmic abscissa scale has been employed in order that the changes in the ethylene content in the interesting zone, marked by very low residual acetylene contents, may be discerned more readily.

In C_2 hydrorefining, on recent palladium catalysts, S may be as high as 1500. As shown in Fig. 18.1, this helps to eliminate acetylene with practically no loss of ethylene.

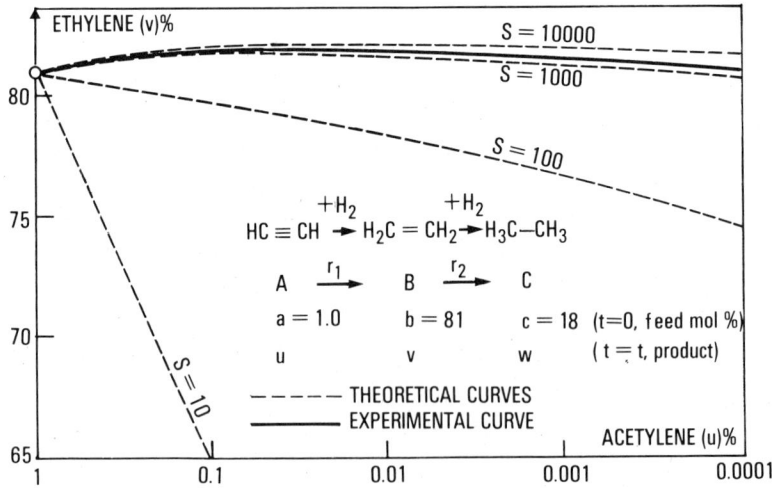

Fig. 18.1 - Selectivity for acetylene hydrogenation over a typical palladium-based catalyst (Procatalyse LT 261).

Increase in selectivity by reduction of the reaction rate. The selectivity of 1500 indicated above, which gives a 100% yield for a residual acetylene content of 5 ppm, is obtained if the rate-determining step of the hydrogenation is the reaction on the catalyst surface itself. If the reaction is too fast, diffusion in the catalyst particles becomes limiting, resulting in a reduction in selectivity (selectivity tends toward $S' = \sqrt{S}$, i.e. $S' = 40$). If the reaction rate is such that the intergranular limitations intervene, S will tend toward 1. This is why it is better to have a relatively low reaction rate on the catalyst surface. This is achieved by using slightly active catalysts and inhibitors (carbon monoxide) and operating at low temperature.

a) Use of slightly active catalysts. Palladium catalysts for C_2 hydrogenation have a low palladium content, 0.03 to 0.05%, as compared with 0.2 to 0.5% for hydrogenations of less reactive hydrocarbons (C_4 and heavier).

b) Use of the inhibiting effect of carbon monoxide. Carbon monoxide is a highly effective activity inhibitor or moderator. By adjusting its content in the hydrogen (partial bypass of the methanator) to between 50 and 500 ppm, the activity can be reduced by a factor of 2 to 4, and selectivity is improved.

c) Low-temperature operation. Given the cost of cooling, operations are carried out at the lowest possible temperature compatible with water cooling, i.e., a reactor inlet temperature of 40°C at the start and 60 to 80°C at the end. This rise in temperature, which is uniform during a cycle, is intended to compensate for deactivation of the catalyst. It is also essential to take account of the exothermicity of the reactions and the resulting temperature rises in the catalyst beds, causing decreases in selectivity.

(ii) Specific features of hydrogenation processes. If performed in an adiabatic reactor, the hydrogenation of a C_2 cut containing 1% acetylene, assuming an ethylene yield of 99.5% would give a ΔT of 70°C and hence reactor outlet temperatures of 110°C at the start and 150°C at the end of the run. This would result in a substantial loss of selectivity with two consequences : deterioration of ethylene yield and loss of acetylene specification due to the lack of available hydrogen.

The first alternative to avoid such high exothermicities in the catalytic bed is the splitting of the reaction between two or more reactors in series, separated by an intercooler. With two reactors in series and the circumstances discussed above, this would give the following mid-cycle profile (in °C) :

50 (R1 inlet)   85 (R1 outlet)   50 (R2 inlet)   85 (R2 outlet)

The sharing of the conversion and hence of the exothermicity between the two reactors is controlled by adjusting the make-up hydrogen at the inlet of each reactor.

The second alternative consists of using cooled reactors. The multitube reactor (catalyst in the tubes) conveys a closed-circuit stream of low-boiling-point hydrocarbon (isopentane, methanol), which is itself cooled in a heat-exchanger in the loop. The flow diagram in Fig. 18.3 illustrates this principle (ref. 6). The temperature increase across the catalytic tubes is reduced to 10 to 20°C and selectivity is accordingly improved.

Both technologies can be combined by placing a quasi-isothermal multitube reactor at the head, followed by a chamber-type adiabatic reactor.

(iii) <u>Industrial implementation of C2 hydrorefining (tail-end hydrogenation on C2 cut only) (Fig. 18.2)</u>. The feed to the hydrogenation unit is the gas phase of the reflux drum of the deethanizer (C_2-C_3+ splitter). The C_2 hydrocarbons are raised to the reactor inlet temperature (35 - 80°C) by feed/effluent heat exchange followed by a preheat using low-pressure steam. After the addition of hydrogen, the CO content of which is adjusted by partial bypass of the methanator, the H_2/HC ratio is used to control the progress of the reaction and hence the exothermicity and selectivity. For a feed containing 1.2% acetylene, the H_2/HC ratio is adjusted to get about 0.3% acetylene at the outlet. The effluent from the first reactor is cooled in a water cooler to allow to keep the outlet temperature of the second reactor below the 130°C limit. After heat exchange, the effluent from the second reactor is cooled and sent to the green-oil removal column.

This column is a counter-current liquid-gas absorption column. The absorption liquid is a circulating C_2 reflux obtained by condensation at the top of the reactor of all the effluent from the reactor in a propylene cooler. The green-oils are obtained at the bottom as a solution in an ethylene-ethane mixture, and this solution is sent to the de-ethanizer, where the green-oils are drawn off at the bottom with the C_3+. After the cascade of fractionations of C_3+, the green-oils (C_6, C_8, C_{10}, C_{12} and olefins) are found at the bottom of the debutanizer and, as a mixture with C_5+, are subjected to gasoline hydrogenation. This tower has a reflux rate of only 0.2.

The top stream from the green-oil tower is then sent to a stripper or secondary demethanizer (removal of light compounds : excess of H_2, CO and CH_4 introduced with the hydrogen), and then to the ethane/ethylene splitter. The ethane is recycled to the steam cracker.

At least three interchangeable reactors are employed, two in service and one on stand by. A furnace is provided to heat the steam-and-air mixture used for in-situ regeneration of the catalyst.

<u>Adjustment of operating conditions</u>. At the start of the cycle (new or regenerated catalyst), the unit is operated at the lowest possible temperature (35°C) by bypassing the preheater. Hydrogen is introduced stoichiometrically into each reactor, on the basis of a total conversion of acetylene (1.2 to 0.3% and 0.3% to 2 ppm) and an ethylene yield of 99 to 100%.

Carbon monoxide is added to slow down the reaction to a rate compatible with the product specification. The reaction "spreads" in the catalyst bed and the selectivity improves. With time, the catalyst deactivates, particularly by the green-oils laydown and it fouls. The CO content is then progressively reduced to maintain the extent of conversion at low temperature, then, in a next step, the reactor temperature is increased, first by heat exchange and then by heat input from the preheater at the end of the cycle.

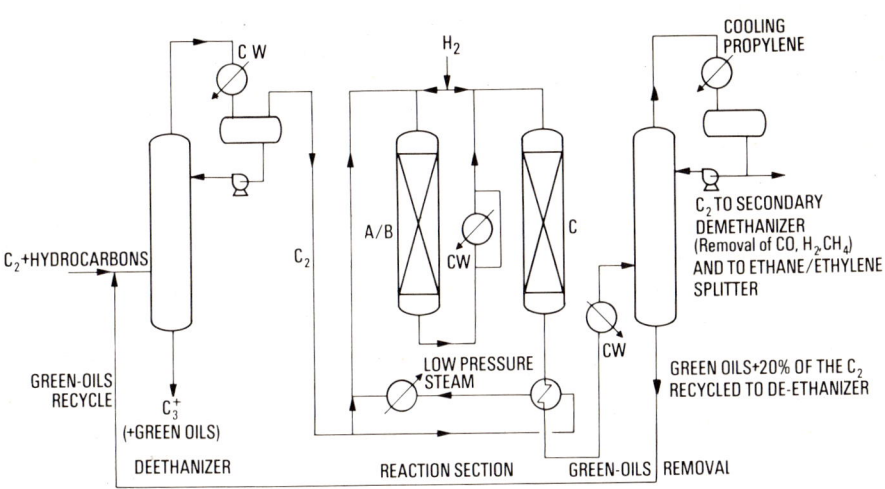

Fig. 18.2 - C_2 cut selective hydrogenation. Simplified flow diagram of the process.

If the acetylene content or the feed rate varie. The H_2/HC ratio is adjusted (the slight excess of hydrogen can be decreased or increased as the contact time varies when changing the feed rate or acetylene content) and the CO content is adjusted so as to operate always at the lowest possible temperature, and to avoid an excessively fast reaction.

Catalyst cycle, regeneration and lifetime. Reactor-inlet temperature is gradually increased during the cycle. Reactor-outlet increases accordingly. The cycle length, i.e. the time to reach the temperature limit of 120 to 130°C at the reactor-outlet is about six months. The deactivated reactor is taken offline and replaced by the stand by reactor. The catalyst is regenerated by in situ combustion at 400°C of the deposits (heavy oligomers, coke) using an air/steam mixture containing 3 wt% air. The catalyst is then ready for reuse, reduction taking place during the first few hours of operation. Oxidative regeneration is effective and the catalyst is mechanically strong enough to withstand many regenerations. Catalyst service life is about five years.

TABLE 18.2

C_2 cut hydrorefining - Typical operating conditions - Temperature profile in catalytic beds - Typical performances.

Space velocity (in m^3 STP[a] per m^3 and per hour) = 1000 to 1500[b] (2000 to 3000 per reactor).

Service pressure 25 bar ; H_2/C_2H_2 mol. ratio : first reactor = 1.1, second reactor = 5.

CO/C_2 (vpm) : start of cycle = 5 to 10, end of cycle = 0

Typical temperatures (°C)	Start of cycle Inlet	Outlet	End of cycle Inlet	Outlet
Reactor 1	35	80	80	125
Reactor 2	35	75	80	120

Typical performance wt%	Feed	Effluent
Acetylene	1	2 ppm
Ethylene	80	79.60
Ethane	19	20.37
Oligomers	--	0.03

[a] STP, standard temperature and pressure, i.e. 20°C and atmospheric pressure.
[b] About 60 m^3 of catalyst installed (three reactors of 20 m^3) for 400,000 t/year ethylene steam cracker.

(iv) Other C2 hydrorefining methods. We will only discuss briefly the processing of the other ethylene-rich fractions.

C2/C3 hydrogenation (tail-end hydrogenation). This is carried out according to the same flow-diagram. The gaseous feed is taken from the depropanizer. Two reactors are still sufficient despite the higher exothermicity (augmented by the hydrogenation of methylacetylene and part of the propadiene, causing a possible temperature increase of 85°C instead of 70°C), provided lower reactor inlet temperatures in end-of-run are planned, and hence shorter cycles are accepted.

C2- hydrogenation (front-end hydrogenation, de-ethanizer in front). The high hydrogen and methane content provides a substantial means of heat dilution with two consequences.

a) There is never any need for more than two reactors and for acetylene contents of ca 1%, a single reactor is even adequate. The isothermal reactor system can also be employed (Fig. 18.3) (ref. 6).

b) With two reactors, the temperature rise per reactor is small, around 25°C, and, insofar as the contact time selected is sufficient, operations can be conducted at a lower average reactor temperature, and the cycle times can be lengthened accordingly.

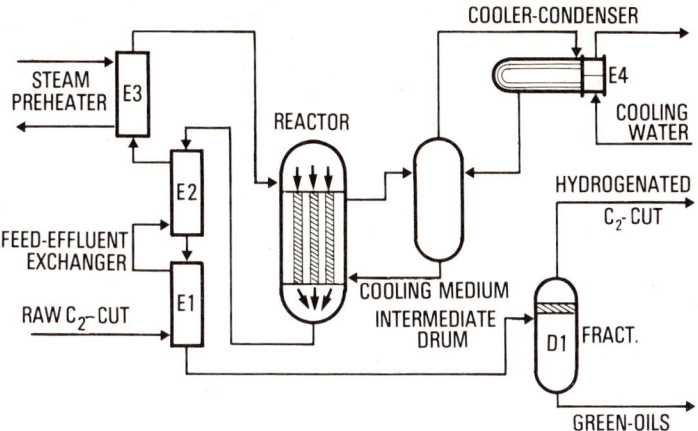

Fig. 18.3 - Front-end isothermal acetylene converter (feed = C_2- cut). Simplified flow diagram.

The reaction scheme is simplified. The de-ethanizer overhead-stream is sent to the fifth compression stage and then into a standard reaction section. The reactor effluent is cooled and sent to the demethanizer to remove light ends (H_2, CH_4, CO) and then to the ethylene-ethane splitter. The ethane collected at the bottom is flashed in an evaporator to isolate the green-oils, and then recycled to the steam cracker.

18.3 C_3 CUT HYDROREFINING

18.3.1 Purpose

The C_3 cut produced by steam cracking typically contains 90% propylene, but also about 4% methylacetylene + propadiene (MAPD) which obviously put the propylene out of polymerization specifications.

Table 18.3 summarizes the composition of typical feeds available from steam cracking or fluidized bed catalytic cracking (FCC), which represent the second current source of propylene, as well as the standard propylene grades.

TABLE 18.3

Composition of propylene cuts and commercial propylenes

	C3 cuts		Propylene	
	SC[a]	FCC	Chemical grade	Polymerization grade
Propylene (%)	92	65	92 - 94	95 - 99.9
Methylacetylene (MA) + propadiene (PD) (%)	4	0.01	20 - 30 ppm	10 ppm
C_4^+ (wt. ppm)	2000	2000	200	50
C_6^+ (wt. ppm) (polymers)			1000 ppm	10 ppm

[a]From naphtha steam cracking.

The MAPD content of the C_3 cut naturally varies according to the severity of the steam cracking operation and the design of the furnaces, but it depends above all on the type of feed cracked, as shown by Table 18.4.

TABLE 18.4
Composition of propylene cut for different steam-cracking feeds

Type of feed cracked	Ethane	Propane	Butane	Naphtha	Gas-oil
Propylene yield (wt %)[a]	2	16	15	13.2	13.1
Composition of C_3 cut					
C_3H_4 (MAPD)	1.6	3.2	4.2	3.6	
C_3H_6 (propylene)	64	94	92	93	
C_3H_8 (propane)	34.4	2.8	3.8	3.4	

[a] On the basis of 100 of feed to the cracking furnaces.

MAPD was initially eliminated by distillation ; another purification method is solvent extraction, but it is scarcely used. The application of catalytic hydrogenation has gradually increased and is now widely used because of the excellent propylene yields, resulting usually in propylene gains.

Comparative propylene recoveries of the various deacetylenization methods on the following bases : C_3 cut containing 4% MAPD, end-propylene purity 99.9%, recycling of the bottom-stream of the propylene/propane splitter back to the cracking furnaces.

Propylene recovery with fractionation = 95% ; with extraction of MAPD followed by fractionation = 100% ; with selective hydrogenation followed by fractionation = 103%. It is worth in addition to notice that in the fractionation case, the recycle is a cut containing 20% MAPD that fouls the furnaces and shortens their cycle length.

18.3.2 Thermodynamic and kinetic aspects of hydrogenation

Hydrorefining of the C_3 cut involves three hydrogenation reactions plus one spurious oligomerization reaction with the formation of hexenes and nonenes :

$$CH_3 - C \equiv CH + H_2 \longrightarrow CH_3 - CH = CH_2$$
$$CH_2 = C = CH_2 + H_2 \longrightarrow CH_3 - CH = CH_2$$
$$CH_3 - CH = CH_2 + H_2 \longrightarrow CH_3 - CH_2 - CH_3$$
$$2C_3H_4 + H_2 \longrightarrow C_6H_{10}$$

These reactions are exothermic and result in a reduction in the number of molecules. They are therefore favoured thermodynamically by low temperatures, under pressure. In fact, they are complete under standard operating conditions (20°C < T < 120°C, P 25 bar, H_2/HC 6% mol.).

The isomerization reaction

$$CH_3 - C \equiv CH \longrightarrow CH_2 = C = CH_2$$

which is thermodynamically feasible under the operating conditions, does not take place with usual catalysts.

Considering that oligomerization is a minor reaction in comparison with hydrogenation (it corresponds to about 5% of the MAPD), the reaction mechanism reduces to a competitive/consecutive scheme of the type :

$$C_3H_4 \text{ (MA)} \xrightarrow{r_{12}} C_3H_6 \xrightarrow{r_2} C_3H_8 \qquad (18.4)$$
$$C_3H_4 \text{ (PD)} \xrightarrow{r_{11}}$$

Although this is more complex than the simple consecutive scheme representing the hydrorefining of the ethylene cut, it is also amenable to similar mathematical treatment (ref. 8).

Let b and v be the propylene content of the feed and of the hydrogenated product, a_1 and u_1 the propadiene content of the feed and of the product and a_2 and u_2 the methylacetylene content of the feed and of the product. Since experience shows that propadiene is the least reactive C_3H_4 isomer, it is taken as a hydrogenation reference.

The change in the MA content may be evaluated by considering the hydrogenation of the two isomers in a system of competitive reactions. $S_{MA/PD}$ the relative hydrogenation reactivity is a constant given by :

$$\frac{a_2}{u_2} = \frac{a_1}{u_1}^{S_{MA/PD}} \quad \text{with } S_{MA/PD} = \frac{r_{12}}{r_{11}} \qquad (18.5)$$

In the consecutive mechanism already presented (reaction 18.4) 2 other relative rates of reaction, the 2 actual selectivities, are to be defined :

$$S_{PD} = S_{PD/C_3H_6} = \frac{r_{11}}{r_2} \quad \text{and} \quad S_{MA} = S_{MA/C_3H_6} = \frac{r_{12}}{r_2}$$

with the relationship $S_{MA} = S_{PD} \cdot S_{MA/PD}$

Then, the calculation of the relationship between the change in propylene content and the progress of hydrogenation leads to the following formula :

$$v = \frac{S_{PD}}{1-S_{PD}} u_1 + \frac{S_{MA}}{1-S_{MA}} u_2 + \lambda u_1^{1/S_{PD}} \qquad (18.6)$$

λ is determined by applying equation 18.6 to the feed, i.e. to the C_3 cut to be processed, giving :

$$b = \frac{S_{PD}}{1-S_{PD}} a_1 + \frac{S_{MA}}{1-S_{MA}} a_2 + \lambda a_1^{1/S_{PD}}$$

Thus it appears that, provided two constants are known, namely S_{PD}, the propadiene/propylene reactivity ratio, and $S_{MA/PD}$, the methylacetylene/propadiene reactivity ratio which are characteristic for each catalyst under specific processing conditions, one can, for any C_3 cut, predict the propylene yield for different product specifications.

The following values can be taken as a first approximation:

$S_{PD} = 400$, and $S_{MA/PD} = 1.8$

Fig. 18.4 shows how the propylene yield, as defined by equation 18.6 varies with the two main variables, the MAPD content of the feed and the MAPD content of the product, for the above values of the constants.

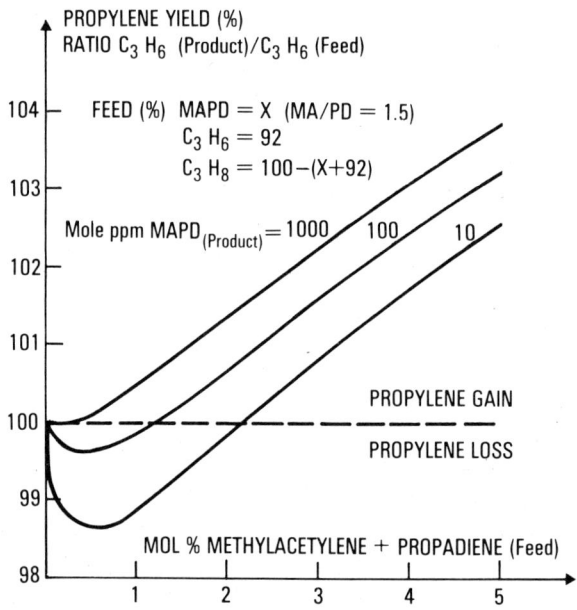

Fig. 18.4 - Hydrorefining of the propylene-rich cut. The recovery of propylene (propylene yield) as function of the MAPD content of the feed and the hydrogenated product. Propylene recovery data according to IFP technology.

Two conclusions may be drawn : hydrorefining is an excellent method for refining the propylene cut. Considering the standard specification of 20 ppm MAPD in the hydrogenated product refining takes place without propylene losses for feeds containing over 1.5% MAPD, which corresponds to all C_3 cuts from steam cracking (Table 18.4). Even substantial production of propylene occurs if the feed contains more than 3% MAPD, and one can consider that, with naphtha steam cracking (4% MAPD), the propylene gain corresponds to 70% of the MAPD contained. In conclusion, hydrorefining produces propylene.

The propylene yield, for a given feed, increases with decreasing severity of the MAPD specification (within the range 0 to 5000 ppm residual MAPD). Hence it is important to avoid hydrogenation levels beyond the quality requirements of commercial propylene. Moreover, since it is well known that the propylene/propane fractionation of a C_3 raw cut is accompanied by a reduction in the MAPD content of the propylene, it is advisable to combine the effects of hydrogenation and fractionation, i.e. to perform only partial hydrogenation. This combination results in increasing propylene gains.

18.3.3 Technological aspects of C3 hydrorefining ; Liquid-phase or gas-phase operation

The hydrogenation of the C_3 cut as a mixture with the C_2 cut has been considered in the section dealing with C_2 hydrogenation. Only the case of separate treatment of the C_3 cut will therefore be considered here.

As already observed, C_2 hydrorefining takes place logically in the gas phase, because a liquid-phase operation would require costly refrigeration at the conventional operating pressures of about 25 bar. Hydrorefining of the C_4 and heavier cuts is performed in the liquid phase unless vaporization results in a significant gain in performance.

The C_3 cut occupies an intermediate position and the two techniques of gas-phase hydrogenation and mixed-phase hydrogenation (referred to subsequently as liquid-phase hydrogenation), have been developed and are offered as valid alternatives. Gas-phase hydrogenation is the older technique and obviously stems directly from C_2 hydrogenation (similar scheme to Fig. 18.2). Liquid-phase hydrogenation has been developed with naphtha steam cracking and the front-end demethanizer scheme. It is currently predominant, being more suitable for high MAPD contents in the C_3 cut (see Fig. 18.5).

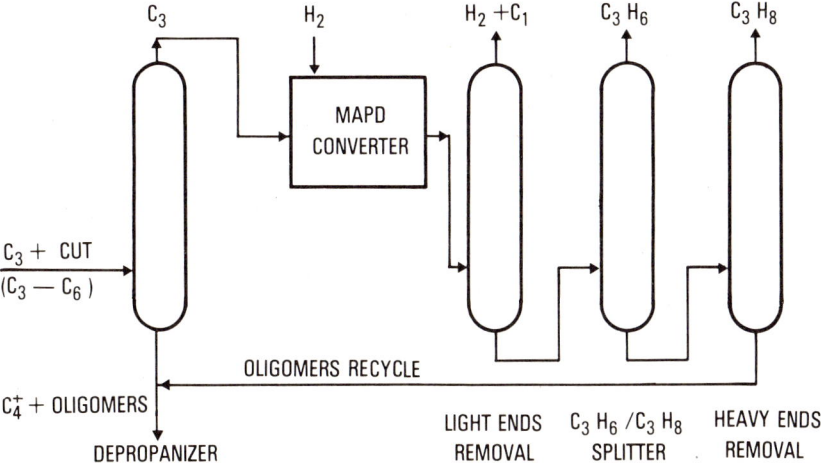

Fig. 18.5 - Hydrorefining of the propylene rich cut in liquid phase. Simplified flow diagram, from the depropanizer to the propane-green oils splitter.

Gas-phase hydrogenation. This is analogous to C_2 hydrogenation (see Fig. 18.2) and will not be described in detail. The C_3 cut is taken as the gas-phase from the depropanizer reflux drum. The hydrogen/hydrocarbon mixture is raised to the reaction temperature (about 60°C) and then selectively hydrogenated in two or three reactors in series, with intercoolers. The effluent from the last reactor is cooled and sent to a washing column, where the oligomers or green-oils are eliminated from the C_3 cut by absorption in a liquid reflux of the C_3 cut itself. The bottom product is sent back to the depropanizer for C_3/oligomer separation. The C_3 compounds are recycled. The condensed top product from this washing column is sent to a secondary demethanizer (elimination of light fractions : excess of H_2, CO and methane) and then to the propylene/propane splitter.

If the MAPD content exceeds 3%, even with good selectivity (slight formation of propane), the exothermicity is higher than in C_2 hydrogenation, because of the higher chemical consumption of hydrogen. Since the temperature limit of 120 to 130°C also apply, it is necessary to increase the number of reactors and intercoolers or to use multitube reactors with internal cooling or, as it is generally the case, to reduce the MAPD content at the reactor inlet and hence the rise in temperature, by dilution. This dilution is achieved by increasing the recycling of the bottoms from the washing column, or by adding a recycle of part of the overhead-stream of the same column : this recycle is re-evaporated. At least 25% of the raw C_3 cut feed rate is recycled. This need for "long" recycling of large quantities of the hydrogenated C_3 cut represents, with the associated loss in yield and the additional operating cost, one of the major handicaps of the gas-phase hydrogenation of MAPD-rich C_3 cuts.

Typical operating conditions are : P = 18 bar, temperature = 40°C (min) / 120°C (max) ; Space velocity (STP m^3/m^3 h) = 500 to 1000 (total for two reactors) ; H_2/HC (mol) = 1.5 to 2.

18.3.4 Industrial catalysts

These are all based on palladium and are of two types, depending on whether gas-phase or liquid-phase hydrogenation is applied. Stability and selectivity promoters are also employed.

(i) Catalysts for gas-phase hydrogenation. These are closely comparable to catalysts used for tail-end hydrogenation of C_2 (or C_2/C_3) cuts : United Catalyst Incorporated (UCI), C 31-1-01 ; Girdler Südchemie, G 55 ; Imperial Chemical Industries, ICI 38-3 ; Procatalyse, LT 261/LT 279 ; Engelhard, HPN III.

(ii) Catalysts for liquid-phase hydrogenation. Procatalyse LD 265 and LD 273, Bayer K 8330, Leuna-Werke 7751, BASF (F.R.G.) H012.

18.3.5   Industrial liquid-phase processes

Several processes are currently available, the "Kalt hydrierung" process of the Bayer Company (F.R.G.), the process developed by Institut Français du Pétrole (I.F.P.), as well as the process developed by Leuna-Werke (D.R.G.).

(i)   Bayer process (Fig. 18.6). The Bayer process is distinguished by its operation at low temperature, around 15°C, in a so-called trickling phase (the liquid trickles on the catalyst in an hydrogen atmosphere) and the use of a multitube reactor with propylene cooling (Fig. 18.6) (ref. 9, 10).

The C_3 cut to be processed is taken as a liquid, at a temperature of about 35°C, from the top of the depropanizer. It is cooled by feed/effluent heat exchange, and then in a propylene evaporation cooler, where it reaches a temperature of about 15°C. If the feed is not dry, a coalescer is installed to eliminate the water released by cooling. The hydrocarbon feed and methanated hydrogen are introduced separately at the top of a reactor similar to a tube-type exchanger. The catalyst is contained in the tubes, and the reaction mixture is cooled by controlled evaporation of propylene in the shell.

The reactor effluent is sent to a separator drum. To avoid losses, the hydrocarbon-rich gas phase is sent to the cracked-gas compressor. The liquid-phase is sent to downstream fractionation. For high MAPD contents, the same pump recycles part of the liquid product to the inlet of the reactor, to dilute the fresh feed and to limit the exothermicity. This recycling of hydrorefined product has the disadvantage to reduce the propylene yield.

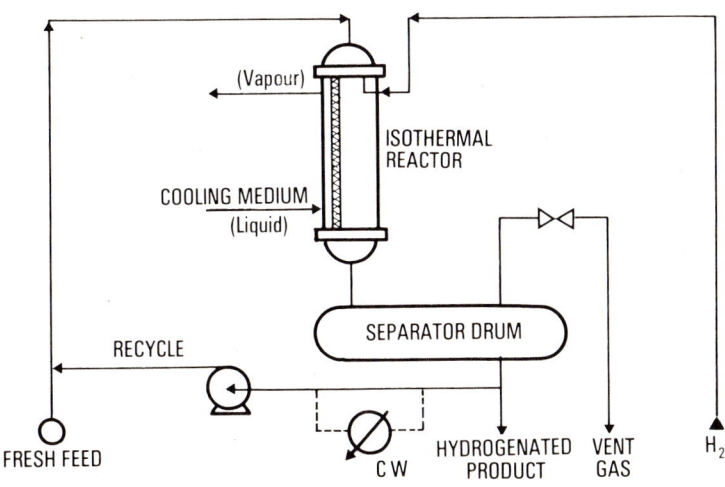

Fig. 18.6 - Hydrorefining of the propylene-rich cut. Simplified flow diagram according to the Bayer liquid-phase process. Isothermal multitube reactor with catalyst in the tubes and cooling medium evaporating in the shell. The Leuna-Werke process is similar.

The trickling phase is considered essential, and, to ensure the trickling of the C_3 cut on the catalyst, operations are carried out in a continuous hydrogen atmosphere. A special device at the head of the reactor distributes the liquid through nozzles above each catalyst tube, while the hydrogen is introduced separately. The pressure, reactor temperature and C_3 cut feed rate are kept within narrow ranges to avoid excessive hydrocarbon evaporation and channeling in the catalytic tubes.

The high activity of the palladium-rich catalyst allows high liquid hourly space velocities (LHSV, m^3 of liquid C_3/m^3 of catalyst x hour, 20-50) and consequently a small-size reactor. The high linear velocity of the liquid (v, cm/sec, above 1) in the tubes, obtained by the use of 50 mm-diameter-tubes, enhances the mixing of the gas and liquid phases thus favouring high conversion of MAPD. The unit operates with a slight excess of hydrogen, 10-20% of chemical consumption.

The catalyst's activity tends to decrease with time due to deposition of polymers, adsorption of catalyst poisons such as arsines, etc. MAPD conversion is maintained by progressively raising the partial pressure of hydrogen, by increasing the operating pressure.

The reactor is usually not mechanically designed for temperatures above 100°C. If catalyst regeneration proves necessary, this is conducted ex situ so that a stand by reactor is provided to guarantee continuous propylene production. Cold hydrogen flushings can be carried out in situ to eliminate the accidental accumulation of water on the catalyst.

Table 18.5 summarizes the operating conditions, and provides examples of performance.

TABLE 18.5
Selective hydrogenation of the C_3 cut by the Bayer process
Range of operating conditions :
Pressure : 10-20 bar ; LHSV : 20-50 liquid m^3/m^3.h ;
H_2/C_3H_4 : 1-2 mol/mol ; Temperature : 5-25°C.

| | Performances (mol. %) | | | |
| | Case 1 | | Case 2 | |
	Feed	Product	Feed	Product
Propane	3.34	9.30	8.12	9.89
Propylene	85.09	90.69	88.31	90.11
Propadiene	4.58	<20 ppm	3.05	<20 ppm
Methylacetylene	6.98	<10 ppm	0.52	<10 ppm
MAPD converted into propylene (%)		48		50

Case 1 corresponds to the selective hydrogenation of a C_3 cut produced in a very high severity NGL (natural-gas liquids) cracker. Case 2 corresponds to the finishing hydrogenation of a C_3 cut, after a first hydrogenation stage in a mixture with the C_2 cut.

(ii) <u>IFP process (Fig. 18.7)</u>. The IFP concept of selective hydrogenation of the C_3 cut takes place in liquid phase, or rather mixed phase, to allow evaporation, which is an excellent method to eliminate the heat released by the reaction and hence to avoid large increases in temperature. This is achieved under the most economical conditions by operating with conventional chamber-type reactors, without internal cooling but rather the reactor effluent is condensed and cooled with water and not with propylene. The use of conventional chamber-type reactors allows in situ catalyst regeneration and hence greater hydrorefining reliability.

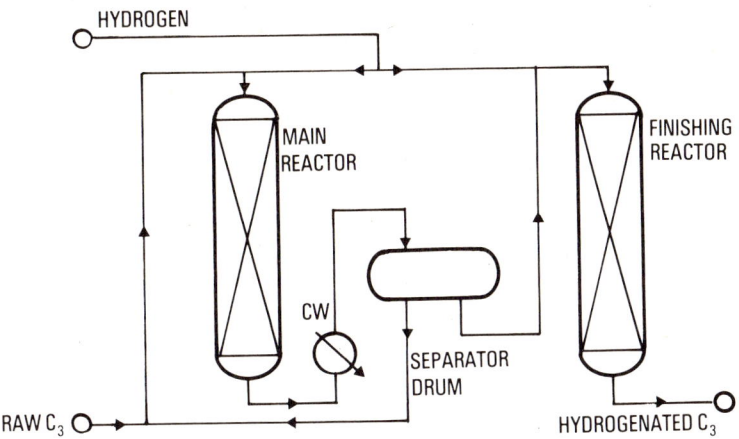

Fig. 18.7 - Hydrorefining of the propylene-rich cut in liquid phase. IFP technology employing chamber-type reactors. Simplified process flow diagram.

High MAPD contents require dilution to limit the exothermicity in the catalyst bed and to maintain a large liquid phase, in order to guarantee continuous washing of the catalyst which is indispensable for long cycle length. Thus part of the product must be recycled. To avoid the inevitable propylene losses, a system of two reactors in series is recommended with recycling of the intermediate product at the inlet of the first reactor. The recycling of a product containing 3000 ppm MAPD has practically no detrimental effect on propylene yield. The finishing reactor operates in a single pass.

The liquid C_3 cut is taken from the top of the depropanizer. It is diluted by intermediate recycling, mixed with a first hydrogen make-up and sent to the top of the first reactor. The heat liberated by the reaction causes a slight temperature rise and partial vaporization. The progress of the reaction is controlled by the hydrogen make-up, to keep MAPD in the effluent above 3000 $\pm$ 1000 ppm. The reactor effluent is sent to the separator drum, where the pressure is maintained by a second H_2 make-up, avoiding hydrocarbons losses or the need for gas recycling to the cracked gas compressor. The pressure of the first reactor is adjusted by a booster pump, to optimize the operation of the water cooler.

The liquid effluent from the separator drum is partly recycled. The net production is recompressed in order to operate the second reactor at the maximum pressure compatible with hydrogen availability, and a third hydrogen make-up is provided before the inlet to this second reactor, which operates entirely in the liquid phase to guarantee the most severe specification (possibly MAPD $<$ 1 ppm) without any excess of hydrogen. This may eliminate the need for the tail-end stripper to eliminate the light compounds (CO, CH_4, H_2).

The H_2/HC ratio controls the progress of the reaction for any feed composition. In addition, this ratio is raised slightly during the cycle, increasing the hydrogen pressure in the first reactor, in order to offset catalyst deactivation. The catalyst was specifically designed to improve selectivity. The monomodal pore-size distribution, large pore diameter and heterogeneous location of the palladium towards the exterior of the bead are favourable to the production of propylene. The low acidity of alumina reduces the formation of oligomers. In situ regeneration of the catalyst is infrequent (average cycle longer than fifteen months), but a two-function stand by reactor guarantees continuous production of high-purity propylene, even in the case of poisoning, particularly by arsenic.

This method combines the simplicity of the equipment of the gas-phase process with the inherent advantages of liquid-phase operation. Typical operating conditions and an example of performance are presented in Table 18.6.

18.3.6 <u>Conclusions on C3 hydrogenation</u>. It is clear that many methods are available for this selective hydrogenation, because they depend on at least three parameters : the layout of the cracked gas fractionation train, itself depending on the type of feed, the type of technology adopted and finally the target grade of the propylene-product. The layout of the cracked gas fractionation train determines whether hydrogenation takes place separately or in a mixture with the C_2 fraction, in an operation with a large excess of hydrogen and CO, or with controlled hydrogen addition (and of CO in the gas-phase process). The choice of the gas or liquid-phase technology is most frequently a factor in separate C_3 hydrogenation. The target propylene grade determines the complexity and cost of hydrogenation as the effects of fractionation and hydrogenation are combined to eliminate MAPD.

TABLE 18.6

Operating conditions and performances of the IFP Process

		Operating conditions	
		Reactor 1	Reactor 2
Pressure (bar)		20	28
LHSV (m^3 (liquid) C_3/m^3 cata.h)		15-30	30
Temperature (°C) (inlet outlet)		40 ➚ 60	40 ➚ 60
H_2/C_3H_4 (mol ratio)		1.2	6
		Performances	
wt %	Feed	Intermediate product	End-product
Methylacetylene	2.2	0.05	1 ppm
Propadiene	1.8	0.25	10 ppm
Propylene	92.0	95.30	94.40
Propane	4.0	4.10	5.50
C_6^+	0	0.30	0.30

18.4 C_4 CUT HYDROREFINING

Selective hydrogenation is also a factor in the upgrading of the C_4 cut. Given the complexity of the mixture (butadiene, isobutene, 1-butene and 2-butenes), and the large number of potential uses (raw material for the production of polymers and elastomers or of first-generation intermediates such as octenes and ketones), no longer a single hydrorefining but several hydrorefinings must be considered, each with its special features.

Hydrorefining of the 1,3-butadiene/butene cut (C4 raw cut). Selective hydrogenation of vinylacetylene to improve the butadiene recovery rate.

Hydrorefining of the butene cut (raffinate 1). The butene cut is available after butadiene extraction. This involves selective hydrogenation of the residual butadiene contained in the olefin cut, with an effort to achieve maximum butene yield (in this case, the distribution of n-butenes between 1-butene and 2-butenes - cis and trans - does not matter).

Hydrorefining of the 1-butene-rich cut (raffinate 2). The 1-butene-rich cut is available after butadiene and isobutene extraction. This hydrorefining involves selective hydrogenation of the residual butadiene in the n-butene-rich cut with the specific objective to avoid 1-butene losses. Therefore, in this case, apart from prevention of the formation of n-butane, additional losses of 1-butene by isomerization to 2-butenes (shift of the double bond), must be avoided, and this reaction develops easily.

Hydroisomerization of 1-butene. Apart from their use as monomers, 2-butenes are preferable to 1-butene for most synthesis of petrochemical intermediates. In other respects, n-butene/isobutene separation is easier if the n-butenes are present in the form of 2-butenes. The hydroisomerization of 1-butene to 2-butenes is therefore a specific variant of hydrorefining and can be considered as the reverse of the previous case.

The hydrorefining of the butene cut will not be examined in detail because it can be considered as a variant of hydroisomerization.

18.4.1 Hydrorefining of the C4 butadiene-rich cut (Selective hydrogenation of vinylacetylene)

(i) Purpose. The C_4 raw cut produced in a steam cracker is a complex-mixture of hydrocarbons with typically the following composition (wt %) : vinylacetylene (VAC) 1,2 ; ethylacetylene (ETAC) 0.2 ; 1,2-butadiene 0.2 ; methylacetylene + propadiene (MAPD) 0.2 ; 1,3-butadiene 45 ; butenes 45 ; butanes 8.2. The usual processing is butadiene extraction. The typical objective of hydrorefining is to improve the economy of the butadiene extraction by raising the butadiene yield or by making the extraction process simpler. The first generation of butadiene solvent-extraction processes, such as the Esso process using a cupro-ammoniacal complex, demanded the prior elimination (case 1 of the scheme Fig. 18.8) of acetylenics at a level of 50 ppm to avoid prohibitive solvent consumption. Hydrorefining meets this requirement but at the cost of high butadiene losses (4-8%). The second generation of extraction processes and solvents (BASF with the N-methylpyrrolidone ; Nippon Zeon with the dimethylformamide and Shell with the acetonitrile, in particular) are designed on the following basis : they separate the C_4 cut into 2 main fractions (high-purity 1,3-butadiene and a butenes-butanes mixture either called the raffinate n° 1) and 3 purges (the C_4 acetylenics extract, the methylacetylene concentrate, the 1,2-butadiene concentrate) that are 1,3-butadiene-rich fractions, thus causing 1,3-butadiene losses. These new processes no longer require prior hydrorefining, but hydrorefining can still serve to :

raise the butadiene yield by sharply reducing the purges that are mainly an acetylenic/butadiene mixture,

upgrade the acetylenics that have a fuel value at the most into olefins that are at least more valuable LPG,

eliminate a risky cut because for concentrations above 40%, vinylacetylene is explosive.

With this type of extraction, currently the most widespread, hydrogenation is no longer necessarily located at the front end, at feed preparation (ref. 11). In fact, the hydrorefining/extraction combination can be implemented by three different processes, as shown in Fig. 18.8.

Process 1, hydrorefining is applied to the raw C_4 cut only.
Process 2, hydrorefining is applied downstream from extraction, on the C_4 acetylenics extract and the hydrorefining product is recycled to the extraction unit.
Process 3, hydrorefining is applied to the mixture of the raw cut and the recycled acetylenics extract.

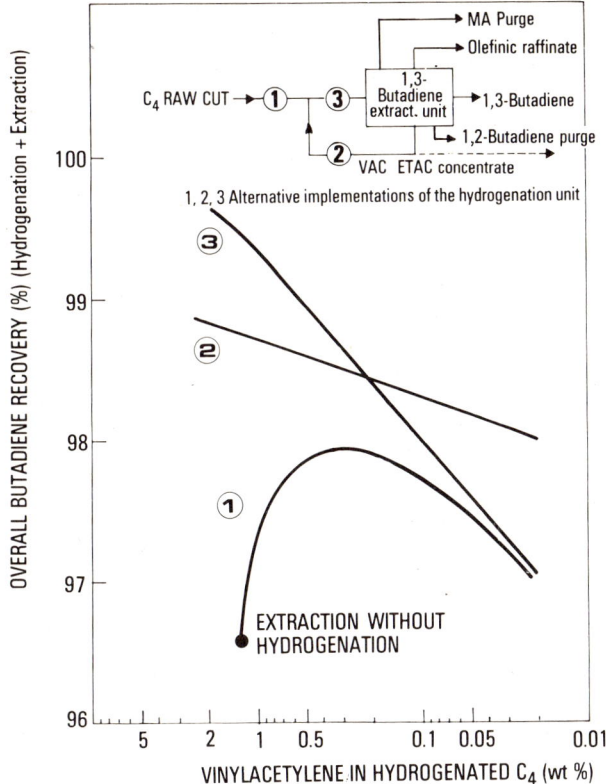

Fig. 18.8 - The selective hydrogenation of C_4 acetylenics increases the overall (hydrogenation + extraction) 1,3-butadiene recovery ratio. For a given feedstock (VAC 1.2%, ETAC 0.2%, 1,3-butadiene 45%) a recovery of 96.5% in an extraction plant is feasible. In case 1, hydrogenation of the raw C_4 cut, low conversion of the VAC results in an improvement of this recovery, but at higher conversion the recovery reaches a maximum then decreases because butadiene losses during hydrogenation becomes higher than butadiene savings obtained in the extraction unit by reduction of the acetylenics purge. Cases 2 and 3, involving recycle of unconverted C_4 acetylenics, allow to operate at moderate conversion thus avoiding hydrogenation losses, and lead to a much higher recovery ratio of 1,3-butadiene. The comparison of the 3 alternates is made on the following basis : 1,3-butadiene in the olefinic raffinate = 0.3%, acetylenics in 1,3-butadiene 15 ppm ; composition of C_4 acetylenics concentrate : VAC + ETAC + 1,2-butadiene 50% ; 1,3-butadiene 50%.

Fig. 18.8 compares the different possible locations of hydrorefining from the standpoint of butadiene yield, on the basis of the same hydrogenation selectivity of the reaction system. A number of conclusions can be drawn :

Process 3 is the most interesting, slightly more than Process 2. Processes 3 and 2 are much more interesting than Process 1. Total hydrogenation of the acetylenics in a single pass (first generation extraction) is not economically valid, with processes currently available.

The butadiene yield for Processes 2 and 3 rises with lower acetylenic conversion in the reactor. It is essential to operate with partial conversion, even if this slightly increases the recycled acetylenic extract rate, still anyway in the range of 5-10% of the C_4 raw cut.

Butadiene recovery rates close to 100% can be attained.

(ii) <u>Kinetic features of the reaction</u>. The predominant acetylenic is vinylacetylene. Since it is also the one that displays the highest relative reactivity to butadiene and is at least partly converted into 1,3-butadiene, vinylacetylene is taken as the basic compound for hydrogenation.

The reaction mechanism is rather complex, because this acetylenic can yield several isomers upon partial hydrogenation : 1,3-butadiene, 1,2-butadiene and ethylacetylene :

$$C_4H_4 \xrightarrow{+ H_2} C_4H_6 \xrightarrow{+ H_2} C_4H_8$$

vinylacetylene → 1,2-butadiene, 1,3-butadiene, ethylacetylene → 1-butene + 2-butenes → n-butane

Hopefully, this reaction scheme can be simplified because the isomerization between these three isomers, which is thermodynamically possible and favours 1,3-butadiene is not catalysed by standard catalysts and because the formation of n-butane is negligible.

The reaction scheme can thus be simplified to :

vinylacetylene $\xrightarrow{r_1}$ 1,3-butadiene $\xrightarrow{r_2}$ n-butenes; vinylacetylene $\xrightarrow{r'_1}$ n-butenes

Using once again the same symbols, let b and v be the butadiene content of the feed and the hydrogenated product, and a and u the vinylacetylene content of the feed and of the product. Here come several ratios of hydrogenation rates in other words several selectivity factors S :

$$S_{12} = \frac{r_1}{r_2} \qquad S_{11} = \frac{r'_1}{r_1} \qquad \text{and constants} \qquad \beta = \frac{1}{1 + S_{11}} \qquad \text{and } \alpha = \frac{\beta}{S_{12}}$$

The relationship between 1,3-butadiene and VAC contents is given by the following expression, taken from a mathematical treatment (ref. 8) as above :

$$v = \lambda u^\alpha - \frac{\beta}{(1-\alpha)} u \qquad (18.7)$$

is determined from equation 18.7 applied to initial conditions, i.e. to the feed :

$$b = \lambda a^\alpha - \frac{\beta}{(1-\alpha)} a$$

The approximate values of S_{12} and S_{11}, for a typical palladium-based catalyst, are 80 and 0.4 respectively. Fig. 18.8 is drawn using equation 18.7 and the above values for S_{12} and S_{11}.

The hydrogenation of the C_4 acetylenics and vinylacetylene in particular on a palladium catalyst is an extremely fast reaction. Hence it can be carried out "cold" and benefit from the advantages of the liquid phase (energy savings, longer cycle length) and medium pressure (6 to 12 kg/cm^2).

(iii) <u>Special features of the catalyst, instability of palladium catalysts</u>. The standard catalyst is again based on palladium. This hydrogenation displays an important feature not possessed by other hydrogenations. The catalyst life is no longer virtually infinite, and may be only a few months. Palladium is eluted from the catalyst! (ref. 3, 12).

As shown above, hydrogenation selectivity is achieved by preferential adsorption of the acetylenic. Vinylacetylene shows such an affinity for palladium that it combines with it, at temperatures below 100°C. The vinylacetylene/palladium complex formed is soluble in the hydrocarbon cut, and the catalyst thus progressively loses its palladium. This also means that the catalyst performance, especially its activity and selectivity, deteriorates with time. All conventional palladium-based catalysts suffer from this effect. Three solutions have been proposed and implemented :
The reactor is operated at a higher temperature, above 100°C, to avoid formation of the palladium/acetylene complex.
The palladium is stabilized by an additive. This has led to bimetallic palladium catalysts (ref. 3).
Other active elements not giving rise to this combination, such as copper, may be employed. This raises the problem of poisoning by sulphur, which is solved by the use of a guard bed or a caustic wash to get a sulphur-free C_4 cut or by more frequent replacement of the catalyst.

(iv) <u>Industrial processes : BASF Process (Fig. 18.9) (Process 2) (ref. 13)</u>. This process is applied according to case 2, hence on the C_4 acetylenic extract, i.e. on a cut containing 20 to 40% vinylacetylene. Given the very high heat of reaction and operation in the liquid phase, the process operates as follows. The acetylenic extract, by-product of the extraction unit, is condensed, rid of free water and traces of solvent by passage in a coalescer. The feed to the unit is diluted by high recycling of the product and/or extraction raffinate, and is injected with hydrogen make-up into a multitube water-cooled reactor. The reactor outlet gas phase is sent to a propylene-cooled drum to recover C_4 hydrocarbons. This cold liquid phase is mixed with the hot reactor-outlet liquid phase. Part of the mixture is cooled and used as a diluent for the reactor feed, and the remainder is recycled directly to the feed of the extraction unit. Patents indicate that the catalyst is palladium/zinc on barium carbonate/alumina.

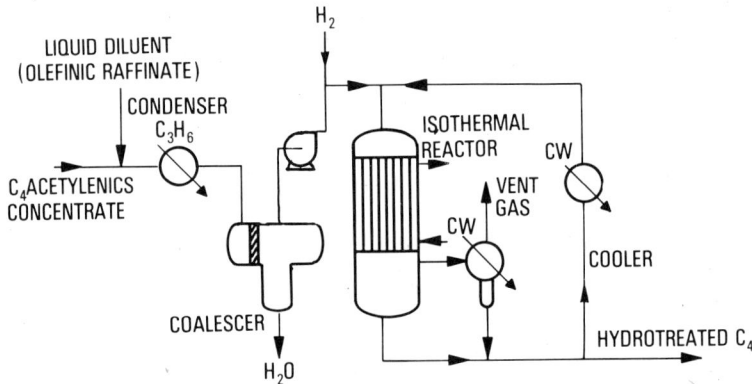

Fig. 18.9 - Selective hydrogenation of the C_4-acetylenics concentrate to improve the butadiene recovery. Simplified process flow-diagram of the BASF technology.

Arco-Engelhard process (applied to Process 1) (HPN IV-A process) (ref. 14). This process is the fruit of Engelhard's experience with palladium catalysts in general, and in particular with those operating in liquid phase on C_4 hydrocarbons, as well as Arco's (Atlantic Richfield) experience with petrochemicals. It involves hydrorefining of the raw C_4 cut (case 1).

Its essential feature is a reactor in up-flow operation (Fig. 18.10). The C_4 cut, i.e., the debutanizer overhead-stream, is picked up by a pump, cooled and sent to the reactor bottom. Hydrogen is injected separately at the reactor bottom, by a special distributor to perfect the gas-liquid, i.e., hydrogen-hydrocarbons, mixing. The reactor contains a single catalytic bed and the effluent is cooled and sent as such to the battery limit of the unit, almost all the hydrogen being consumed. To account for the high exothermicity and limitations of the up-flow reactor, which must operate virtually in the liquid phase, the reaction is divided between two reactors in series. Applied to Process 3, the Engelhard-Arco process employs more than two reactors.

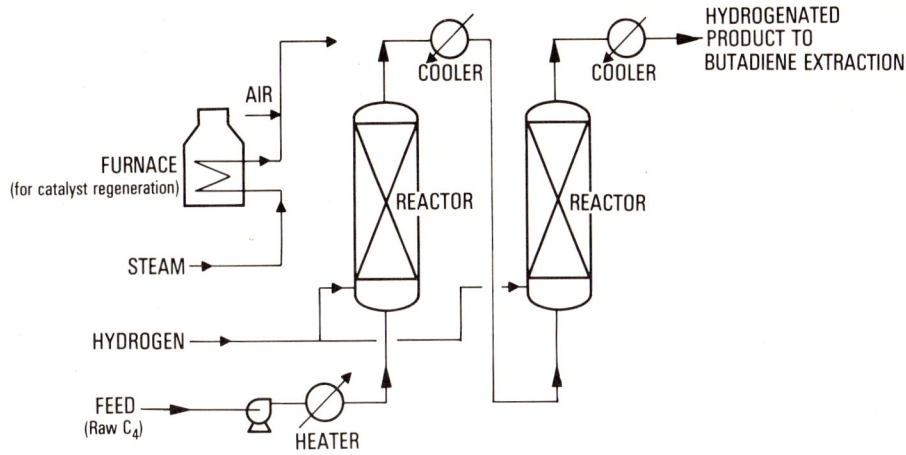

Fig. 18.10 - Hydrorefining of the C_4 butadiene-rich cut (selective hydrogenation of acetylenics). Process flow diagram according to the Arco-Engelhard technology. 2 in-series up-flow reactors operating at ambient temperature. The hydrogenated product is fed to the butadiene-extraction unit.

The reactor is specially designed to improve selectivity and hence the butadiene yield. The use of a catalyst based on active noble metal (palladium), Engelhard reference HPN IV-A, helps to reduce reactor size. The catalyst is regenerated by burning in an air/stream mixture. The up-flow reactor is considered to allow a wide variation in feed rate, the reduction and even elimination of risks of hot spots and optimization of gas-liquid distribution and hence selectivity.

IFP process (Fig. 18.11). The IFP process can be applied to the three types of feed and data in table 18.7 refer to the hydrogenation of the C_4 raw cut (process scheme 1), but, as shown by the comparative study (Fig. 18.8), it is specially recommended for feed 3 (fresh feed + acetylenic recycling), operating with partial conversion (50-80%). This results in a fairly high (2-3%) mean vinylacetylene concentration in the reactor, and this operation is only possible through the use of a stable catalyst (negligible elution of palladium with time, thanks to a promoter), developed by IFP and Procatalyse.

Application to Case 3. The C_4 acetylenic concentrate obtained by extraction is condensed, the water and solvent entrained allowed to settle and the hydrocarbon cut is mixed with the debutanizer-head C_4 cut. The feed thus prepared is mixed with make-up hydrogen, preheated by exchange, with a stream of warm solvent from the extraction unit. This preheating helps to lower the hydrogen pressure, and hence the reaction rate, which improves selectivity. The H_2/HC mixture is then introduced at the top of the reactor. The heat liberated by the hydrogenation reactions causes partial vaporization of the hydrocarbons.

The reactor effluent as such is subjected to extraction (if extraction is gas-liquid process, the reactor effluent is fed to the extraction-feed vaporization-tank). Nearly 100% overall 1,3-butadiene recovery can be achieved.

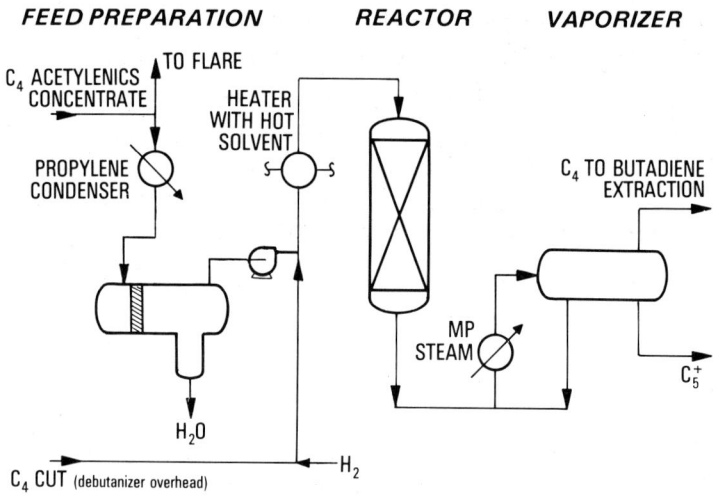

Fig. 18.11 - The hydrorefining of the C_4 butadiene-rich cut (selective hydrogenation of C_4 acetylenics and 1-2 butadiene). Simplified process flow diagram according to IFP liquid phase technology. In this case, all the acetylenics or 1,2-butadiene concentrates obtained as purges in the butadiene extraction plant are recycled, condensed and mixed with the C_4 raw cut. C_4 acetylenics concentrate is mixed downstream of the debutanizer while methylacetylene and 1,2-butadiene concentrates are mixed upstream of this splitter.

TABLE 18.7

Selective hydrogenation of raw C_4 cut - Performance of IFP Process

Composition (wt%)	C_4 feed	C_4 "optimum" product
Propane + propylene	0.01	0.11
Methylacetylene	0.11	0.038
Propadiene	0.04	0.016
Isobutane + n-butane	1.15	1.15
Isobutene	22.98	22.98
1-butene	15.40	16.36
cis-2-butene trans-2-butene	8.18	8.98
1,2-butadiene	0.13	0.11
1,3-butadiene	50.54	50.04
Ethylacetylene	0.184	0.0900
Vinylacetylene	1.27	0.127

Dow-K gas-phase process (applied to Process 1) (ref. 15). The basic concept consists of operating at a temperature such that the vinylacetylene/palladium complex is not formed (T > 100°C), and, on the other hand, using a very selective catalyst, at least for hydrogenation (palladium/copper catalyst) as well as a process adaptable to the short cycles that inevitably result from operation around 150 to 200°C (requiring a "swing" system of reactors).

The C_4 cut is completely vaporized for operation at a relatively high temperature, while retaining reasonable operating pressures, mixed with hydrogen and introduced at the top of the reactor. Three reactors are available, one in operation, one on stand by and one in regeneration, because the operating cycles are limited by oligomerization and fouling, and only last a few days! This process is designed for intensive acetylenic conversion (total remaining acetylenics = 50 ppm maximum), and has frequently been employed to prepare the feed for extraction units in the Exxon process with the cupro-ammoniacal complex as solvent.

The butadiene yield is satisfactory (94-97%). Deacetylenization is partly achieved by polymerization. This process offers usefull yields but at a prohibitive operating cost. It has recently undergone considerable changes, being now conducted in the liquid phase and at temperature close to the ambient and on a stabilized catalyst claimed to be a copper-alumina catalyst (ref. 16, 17). The resulting performances, namely the selectivity, are claimed to be very good. Commercial data are not yet published.

Amongst other processes, it is worth to mention the Showa-Denko (J) process.

18.4.2 Hydrorefining of 1-butene-rich cuts (selective hydrogenation of butadiene)

(i) Purpose. A demand for 1-butene suddenly grew a few years ago with the success of the new linear-low-density polyethylene (LLDPE), whose manufacture requires the incorporation of 5-10% 1-butene. The size of the world market is approximately 400,000 t/year.

When hydrorefining is specifically aimed at polymerization-grade 1-butene, an attempt is made to promote the hydrogenation of the residual butadiene with the minimum formation of butane, as usual but also with the minimum formation of 2-butenes so to avoid shifting of the double bond, also called hydroisomerization, which takes place very easily :

1-butene ⟶ 2-butenes

Thus arose the need to develop specific hydrorefining processes in which this secondary reaction, reducing the recovery rate of 1-butene, is avoided. These processes are distinguished by the use of catalysts that are specifically non-isomerizing or whose selectivity is improved by additives.

(ii) _Thermodynamic and kinetic aspects_. The essential feature is a double-bond shift, which is strongly favoured thermodynamically at medium temperatures, as shown by Fig. 18.12.

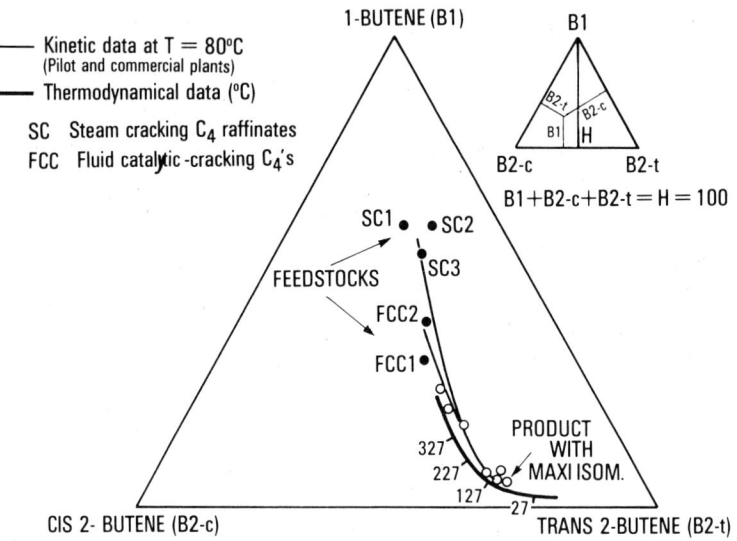

Fig. 18.12 - Thermodynamical and kinetic (at T = 80°C) distribution of n-butenes. At low temperature, isomerization of 1-butene contained in n-butenes-rich cut, into 2-trans-butene is favoured.

1-butene is the most stable of the butenes at high temperatures (around T > 700°C). At temperatures below 200°C, a classic hydrogenation range, the stable form is the 2-butenes (cis-trans) with a 2-butenes/1-butene ratio of 15.

$$\text{1-butene} \xrightarrow{\text{2 cis-butene}} \text{2 trans-butene}$$

The 1-butene-rich cuts normally available, e.g., de-isobutanized C_4 cuts, representing raffinate 2 from steam-cracking and the de-isobutanized cut from FCC displaying 2-butenes/1-butene ratios of 0.75 and 1.3 respectively, are therefore likely to be strongly isomerized. This reaction takes place slowly, in the absence of hydrogen, on alumina and silica/alumina catalysts around 250°C, but its rate rises sharply by hydrogen activation on conventional hydrogenation catalysts. All in all, the hydrorefining of 1-butene-rich cuts is described by the following reaction mechanism :

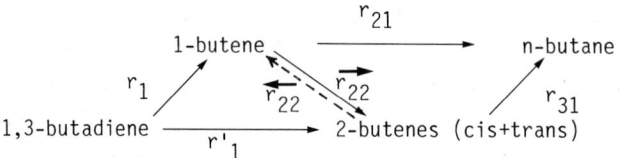

Given the relatively low hydrogenation and isomerization rates of 2-butenes, r_{31} and r_{22} are negligible ; the mechanism can thus be simplified as follows :

$$
\begin{array}{c}
\text{1,3-butadiene} \xrightarrow{r_1} \text{1-butene} \xrightarrow{r_{21}} \text{n-butane} \\
 \searrow_{r'_1} \text{2-butenes} \nwarrow^{r_{22}}
\end{array}
$$

It becomes close to the one representing the hydrogenation of vinylacetylene in 1,3-butadiene-rich cuts, equation 18.7 applies again and 1-butene as a function of 1,3-butadiene in the product, is given by :

$$v = \lambda u^\alpha - \frac{\beta}{(1-\alpha)} u \qquad (18.7)$$

where u and v are respectively the 1,3-butadiene and 1-butene content of the product, λ is given by the equation 18.7 applied to the feed composition, and β and α are given by :

$$\beta = \frac{1}{1+S_{11}} \quad \text{and} \quad \alpha = \frac{\beta}{S_{12}}$$

the selectivity factors being determined as follows :

$$S_{12} = \frac{r_1}{r_{21}+r_{22}} \; ; \; S_{11} = \frac{r'_1}{r_1} \; ; \; S_2 = \frac{r_{21}}{r_{22}}$$

For specific catalysts the performances can be predicted on the basis of :

$$S_{12} = 165 \; ; \; S_{11} = 0.4 \; ; \; S_2 = 0.5$$

whereas for conventional palladium catalysts, S_{12} is only 50-80, thus resulting in a significantly lower 1-butene recovery, as illustrated by Fig. 18.13.

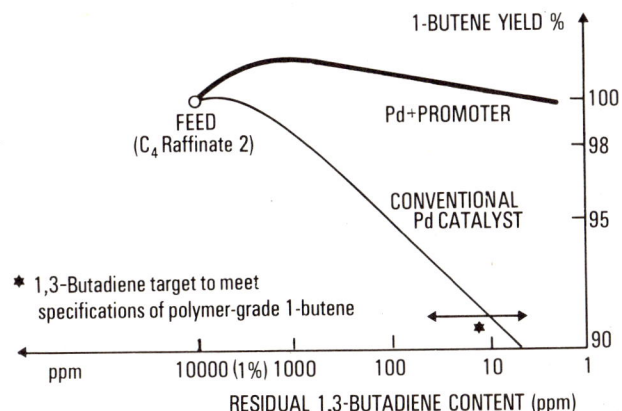

Fig. 18.13 - 1-butene yield (1-butene in product/1-butene in feed ratio) as a function of 1,3-butadiene hydrogenation for a given feedstock (1,3-butadiene = 1%, 1-butene = 27%). Comparison between conventional Pd-based catalysts and specific Procatalyse LD 271 (palladium + promoter). A log-scale is used to plot the 1,3-butadiene content of the product.

(iii) <u>Specific aspects of the catalyst and technology</u>. The catalyst is of fundamental importance for the performance, as evidenced by the possibility that the selectivity, S_{12}, can vary by a factor of 3. A first generation of catalysts developed in the 1960s (Ni-Cu-Cr catalysts) had outstanding selectivity, isomerization apparently having been eliminated, but they were not very reliable because of poisoning by traces of sulphur, even below 1 ppm, a routine content in C_4 cuts. These were replaced by conventional palladium catalysts thanks to their excellent stability towards sulphur poisoning, but their still unsatisfactory selectivity sparked major progress in two directions :

Use of selectivity-enhancing effect of carbon monoxide, which reduces the rate of isomerization but also the rate of 1,3-butadiene hydrogenation, thus resulting in the need for a much larger catalyst volume with concomitant higher capital investment (ref. 18).

Development of catalysts based on palladium + promoter, endowing the catalyst with very high intrinsic selectivity (Fig. 18.13) and avoiding the drawbacks of carbon monoxide (pollution of "polymerization grade" product, delicate adjustment of CO content, increased investment) (ref. 19).

<u>Process</u>. Isomerization is reduced by operating at low temperature ($20 < T < 80°C$) and low hydrogen pressure, thus decreasing the reaction rate on the catalyst surface and accordingly minimizing the intervention of diffusional mechanisms. The reactor dimensions and design are selected to avoid channeling and back-flow effects that are detrimental to selectivity (ref. 19).

(iv) <u>Industrial processes ; Chemische Werke Hüls process (SHP process)</u>. An initial process was developed in 1965 and implemented industrially. Characteristics : gas phase (3 bar pressure, 100°C, Ni-Cu-Cr catalyst), excellent selectivity with inhibition of the formation of 2-butenes (ref. 20). The process is handicapped by its extreme sensitivity to sulphur compounds and to water.

A new process emerged in the early 1980s, with hydrogenation carried out at low temperature, in the liquid phase, in a down-flow reactor. The H_2/butadiene ratios are almost stoichiometric, and isomerization is limited by the controlled addition of CO (ref. 18). A palladium-based catalyst is used. This process, combined with the synthesis of methyl tertio-butyl ether (MTBE) and superfractionation, helps to produce 1-butene to the most severe specifications. The 1-butene yields are stated to be close to 100% (98-100%).

<u>Arco-Engelhard process</u>. Up-flow process like almost all the liquid phase processes developed by this license holder. Low temperature process (30-60°C) on palladium catalyst.

<u>IFP process (Fig. 18.14)</u>. This is a low-temperature process (30-60°C) in the liquid phase on a special catalyst (palladium + promoter), the promoter making it possible virtually to eliminate isomerization of 1-butene to 2-butenes.

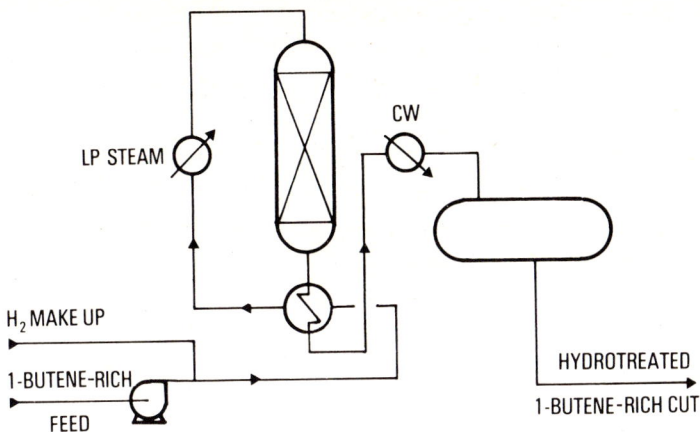

Fig. 18.14 - Selective hydrogenation of 1,3-butadiene in 1-butene-rich cut (formation of n-butane and 2-butenes is minimized). IFP technology. Simplified process flow diagram.

Hence the feed is an olefin C_4 cut after de-isobutenization by conventional processes (MTBE, sulphuric acid, selective oligomerization). The feed is mixed with carbon monoxide-free hydrogen, preheated by exchange and in a preheater, to guarantee better control of the operating conditions without depending on the battery-limit conditions (constant temperature, controlled hydrogen pressure). The H_2-HC mixture is introduced at the top of the reactor. The reactor effluent is subjected to different treatments depending on the location of the hydrogenation unit (before or in between the 2 fractionation columns used to separate 1-butene from other remaining C_4 hydrocarbons : isobutane, 2-butenes and n-butane). Typical performance figures are given in Table 18.8 and show that it is possible to nearly avoid any 1-butene losses.

TABLE 18.8
Typical performances on a raffinate-2

wt %	Feed	Hydrotreated product
Isobutane	2.75	2.75
Propane + Propylene	0.17	0.17
n-Butane	8.10	8.70
Butene-1	49.95	49.20
Isobutene	0.10	0.10
cis-2-Butene	16.23	16.66
trans-2-Butene	21.67	22.42
1,3-Butadiene	1.03	10 ppm

18.4.3 Hydroisomerization of 1-butene cuts

(i) **Purpose.** Apart from the polymerization objective, 2-butenes are usually more valuable than 1-butene. In petrochemistry, for example, methyl ethyl ketone and 1,3-butadiene are more easily synthesized from 2-butenes. Furthermore n-butenes/isobutene separation is considerably facilitated if the n-butene is 2-butenes, particularly cis-2-butenes. In refining, in the production of octane dopes by aliphatic alkylation of n-butenes, the octane of the 2-butenes alkylate is higher than the octane of the 1-butene alkylate.

Given the fact that, together with isomerization which upgrades the n-butenes hydrogenation of residual butadiene takes place simultaneously and thus eliminates operating problems in many downstream processes (gums, fouling, excessive consumption of acid), it appears that hydroisomerization is a valid method to supplement the range of C_4 hydrorefining processes (ref. 21).

(ii) **Thermodynamic aspects.** The thermodynamic aspects were discussed in the previous section dealing with selective hydrogenation of 1-butene cuts. Referring to the triangular diagram for n-butenes (Fig. 18.12), isomerization is favoured by low temperatures, especially the formation of trans-2-butenes. The equilibrium composition at 100°C is : 1-butene, 4.8% ; Cis 2-butene, 28.6% ; trans 2-butene, 66.6%. This means that, if complete conversion of 1-butene contained in industrial cuts is not feasible, the conversion level nevertheless may exceed 80% as shown by the Table 18.9.

TABLE 18.9
Isomerization : industrial performances - IFP process

(mol.%)	Steam cracked C_4 cut		FCC C_4 cut	
	Feed	Product	Feed	Product
Isobutane	2	2	37	37
Isobutene	47	47	13	13
1-Butene	25	2.5	12	2.2
Butadiene	0.3	<10 ppm	0.2	<10 ppm
Butane	6	6.5	12	12.3
trans-2-butene	11.7	12.8	15	25.0
cis-2-Butene	8	29.2	11	10.7
1-Butene conversion (%) actual		90		82
(thermodynamic limitation)		(91.5)		(85)

(iii) **Kinetic aspects.** Hydroisomerization is obviously accompanied by hydrogenation of the butadiene present, as well as slight hydrogenation of n-butenes.

Hydrogenation of isobutene is negligible. Hence the reaction scheme is as follows :

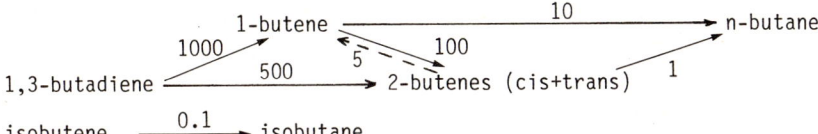

isobutene $\xrightarrow{0.1}$ isobutane

The numbers 0.1, 1, 10, ..., are reaction-rate ratios (S) to the rate of hydrogenation of 2-butenes, assumed to be equal to 1. The reaction mechanism thus appears to be highly complex and the precise mathematical relationship between the yield of 2-butenes and the 1-butene conversion difficult to use.

However, an accurate idea of what occurs in the reaction can be derived by means of a few simplifications that are realistic on the basis of the values given for the reaction-rate ratios (see mechanism above) valid for a palladium catalyst and for typical feedstocks and operating conditions : isobutene hydrogenation-level and butadiene-content are negligible. The reaction scheme thus reduces to :

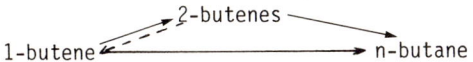

This again represents a consecutive scheme with leakage, complicated in this case by an equilibrium.

The relationship 2-butenes = f(1-butene) can thus also be expressed graphically by a curve with a maximum as demonstrated by experimental data (Fig. 18.15).

It appears that olefin losses are negligible provided that thermodynamic equilibrium is not too closely approached, making it industrially feasible to proceed with selective isomerization. The very fact that 1-butene is converted into 2-butenes, less active in hydrogenation, helps to avoid excessive hydrogenation.

(iv) <u>Catalytic and technological aspects</u>. All metallic catalysts (Ni, Pd, Pt) and sulphide catalysts (Ni-W) activate the shift of the double bond and the order of activity is : Pd > Ni > Ni-W $\gg$ Pt.

The palladium catalyst is therefore once again the most suitable catalytic metal. Anyway, attempts are made to increase its intrinsic selectivity by finally increasing the ratio :

$$\frac{\text{1-butene isomerization rate}}{\text{1-butene hydrogenation rate}}$$

Only one method is known to be applied industrially : presulphuration of the catalyst which, by sharply reducing the hydrogenation rate, improves the yield in 2-butenes (ref. 22) (Table 18.9).

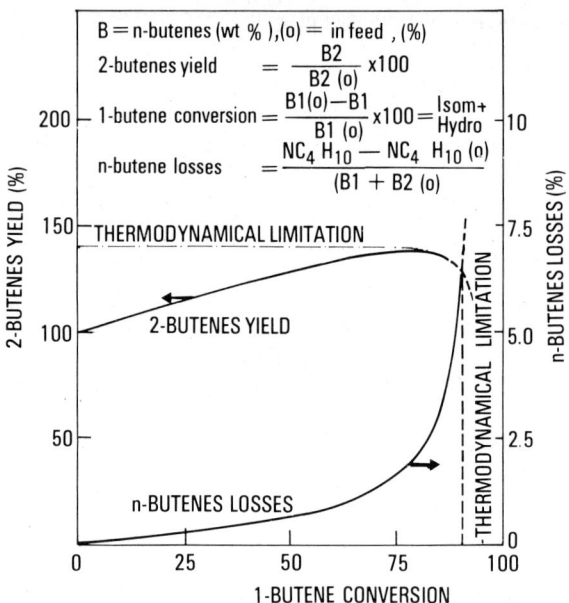

Fig. 18.15 - Hydroisomerization of an FCC C_4 cut. Typical performances obtained in conversion of 1-butene into 2-butenes, regarding the 2-butenes yield and the n-butenes losses. Operating conditions : T = 90°C ; P = 20 bar, 2-butenes yield increases by isomerization and in the final stage, slightly decreases by hydrogenation.

Since hydroisomerization is a relatively slow reaction, compared with butadiene hydrogenation, it is carried out at above-ambient temperatures. A temperature around 100-120°C, reached by processing the C_4 cut through a steam-heater, is ususally used. This temperature level conciliates the requirements of thermodynamics (operation at low temperature) with those of kinetics (operation at high temperature).

(v) <u>Industrial processes</u>. The first processes were developed for refining, i.e., the preparation of aliphatic alkylation feeds. These processes were operated at around 150-200°C, hence in the gas phase, using sulphide catalysts (Co-Mo, Ni-W) which resist the poisons contained in C_4 cuts from catalytic cracking (NH_3, HCN, RSSR). Improvements in FCC-cut refining techniques and the availability of thoroughly decontaminated steam-cracked C_4 cuts have allowed the use of considerably more active catalysts, making the liquid-phase operation more possible at a pressure of about 15-25 bar, thus leading to more economical hydroisomerization processes.

<u>Arco-Engelhard process (HPN IV-B process) (ref. 21)</u>. This process employs the usual Engelhard concept of the up-flow reactor. The feed is preheated by exchange with low pressure steam and introduced, separately to hydrogen at the bottom of the reactor. For high capacities parallel reactor trains are employed.

An Engelhard palladium-based catalyst is used.

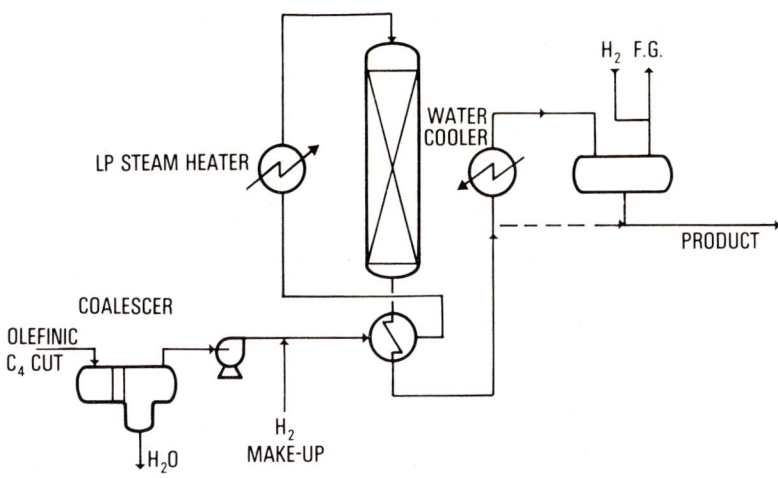

Fig. 18.16 - Hydroisomerization of 1-butene into 2-butenes in order to upgrade an alkylation feedstock. Simplified process-flow diagram according to IFP technology.

<u>IFP process (Fig. 18.16)</u>. It uses the concept of the down-flow reactor. The feed is mixed with hydrogen and preheated by feed/effluent exchange, and then by low-pressure steam. The H_2/HC mixture circulates in a co-current mixed phase downwards on a fixed bed of palladium catalyst (specially activated to improve selectivity). The reactor effluent is cooled and sent to the battery limit. Given the long cycles (1 year), no stand by reactor is provided. The catalyst can be regenerated by a standard combustion method (air + steam).

With a single reactor, a maximum of 85% conversion is feasible thermodynamically. For more intensive conversions, causing greater exothermicity, two reactors are required : one main reactor as described and one finishing reactor operating at a lower temperature to minimize hydrogenation losses and to be in a more favourable range from the thermodynamic standpoint. The progress of the reaction in each reactor, and hence isomerization, is controlled by adjusting the hydrogen make-up.

<u>Phillips process (Hydroisom) (ref. 23)</u>. This gas-phase process operates at around 120°C. One or two reactors are used with an intercooler. The catalyst displays a long cycle time and life. Regeneration is carried out by a standard combustion method (air + steam). See Table 18.10 for performance details.

<u>Bayer Process</u>. This liquid-phase trickle process designed according to Bayer principles, comprises an isothermal multitube reactor with ex situ regeneration of the palladium catalyst.

UOP process and U.C.I. processes. They are gas-phase processes employing nickel-tungsten or nickel catalysts.

TABLE 18.10
Performance of Phillips process (liquid vol.%)

	Steam-cracked C_4		FCC C_4	
	Feed	Product	Feed	Product
Isobutane	7.5	7.55	27.9	28.05
Isobutene	36.4	36.35	14.0	13.90
1-Butene	20.4	4.50	20.2	5.70
n-Butane	19.9	20.60	12.0	12.40
Butadiene	2.0	10 ppm	0.1	10 ppm
2-Butenes	13.8	31.00	25.8	39.95
Total	100.0	100.00	100.0	100.00

18.5. GASOLINE CUT HYDROREFINING

18.5.1 Characteristics of steam-cracked gasolines and specifications of hydrogenated products

The steam-cracked gasoline cut differs from straight-run and catalytically-reformed gasolines in its high content of diolefins such as isoprene and cyclopentadiene, and of alkenyl aromatics such as styrene and indene. For any use, these polymerizable hydrocarbons, also called dienes, require specific refining by selective hydrogenation.

The high content of unsaturated compounds (olefins and aromatics) in the steam-cracked gasoline gives it a high octane rating, as well as providing an attractive source of aromatics. However, due to its instability (induction period, gums, colour), it fails to meet commercial specifications. Specific refining is necessary, and hydrogenation appears to be ideal and effective as shown by Fig. 18.7.

The elimination by hydrogenation of hydrocarbons responsible for the instability (dienes) is followed by the reduction of the diene value of the gasoline. The selectivity of the operation (hydrogenation of dienes without hydrogenation of the other unsaturated compounds : olefins and aromatics) results in preservation of the octane rating. Stabilization is confirmed by the drastic improvement in the induction period and the elimination of potential gums.

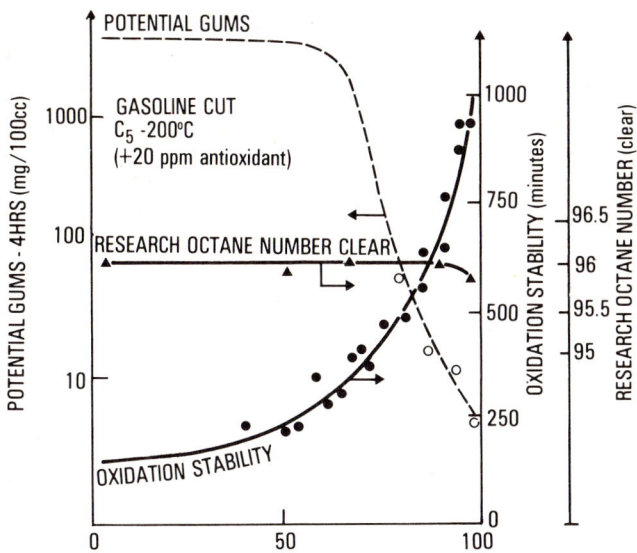

Fig. 18.17 - Hydrorefining of steam-cracking gasolines. Selective hydrogenation of gum-forming dienes, as measured by reduction of the diene value, increases drastically the stability without octane losses.

Two alternative processing schemes are possible, depending on the type of final product desired. If the purpose is to obtain an automotive fuel (gasoline), or, more precisely, a component of the gasoline pool, meeting stability and non-corrosiveness specifications, selective hydrogenation of dienes, called first-stage hydrogenation or hydro 1, is employed. If the purpose is to obtain an aromatic cut in order to recover one or more high-purity aromatics, the first-stage hydrogenation is followed by fractionation to isolate the aromatic cut and second-stage hydrogenation (hydrogenation of olefins and hydrodesulphurization), called hydro 2.

This discussion will deal only with the selective hydrogenation of diolefins and alkenyl aromatics (hydro 1).

18.5.2 Refining by selective hydrogenation

(i) Principle. Fig. 18.18 shows the main chemical reactions involved :
Partial hydrogenation of diolefins to olefins, mainly internal olefins, possibly with isomerization of the external olefins present to internal olefins, which helps to limit the complete hydrogenation to paraffins (internal olefins hydrogenate slowly).
Partial hydrogenation of cyclodiolefins to cycloolefins. Here also cycloparaffin saturation is very low, so that the selectivity is high except in the

specific case of cyclopentene which hydrogenates into cyclopentane. The reaction cyclopentene + H_2 ⟶ cyclopentane represents the only non-aimed hydrogenation that cannot be avoided, even with a high-performance process. Hydrogenation, perfectly selective in this case, of the alkenyl aromatics (styrene, methylstyrene, indene) to aromatics.

Non-aimed polymerization reactions generally determining the catalyst cycle time, with the formation of oligomers that may require redistillation of the hydrogenated gasoline.

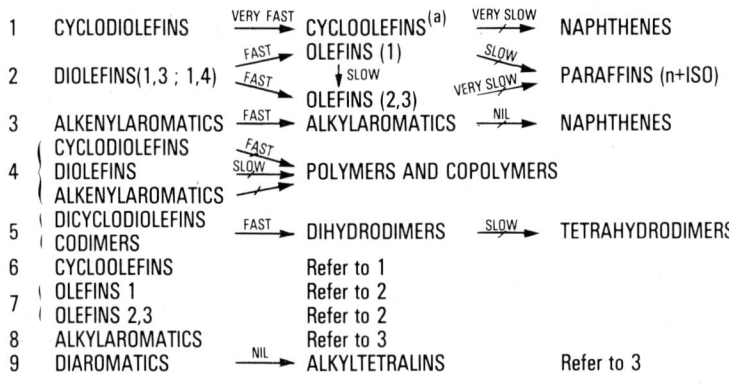

(a) Cyclopentene hydrogenation is fast, other cycloolefins hydrogenate very slowly

Fig. 18.18 - Synopsis of the various hydrogenation reactions involved in the hydrorefining of steam-cracking gasolines. Indicative (fast, slow, very slow, nihil) relative rate of reaction. ⟶ means that the corresponding reaction is undesirable.

Two kinds of selectivities that are essential for catalyst design, the choice of operating conditions and the type of process can be identified : hydrogenation selectivity i.e., formation of saturates and non-polymerization selectivity are minimized and, if possible, virtually avoided. Table 18.11 illustrates the effectiveness of existing technologies in reaching these goals : detailed gas chromatographic analysis of both a commercial feed (C_5 to 215°C end-point fraction) and the corresponding hydrogenated product were performed. The hydrocarbons whose disappearance or formation would demonstrate the selectivity if any, were monitored : the dienes (diolefins, dimers, alkenyl aromatics) whose hydrogenation must be almost 100% ; the saturates whose content should not significantly increase and the alkyl aromatics whose content should increase as much as the content of the corresponding alkenyl aromatics is decreased. Olefins are too numerous to be considered in details.

TABLE 18.11

Selectivity of the hydrogenation (commercial data, IFP process, palladium catalyst, G.C. analysis)

wt %	Feedstock	Hydrogenated Product
Saturates		
iso-Pentane	0.35	0.45
n-Pentane	4.28	4.50
Cyclopentane	0.65	2.30
Cyclohexane	0.44	0.51
Methylcyclopentane	1.00	1.02
Tetrahydrodicyclopentadiene	<0.10	0.80
Total paraffins (n + iso)	8.40	10.06
Total naphthenes (cyclopentane excluded)	2.15	2.48
Total saturates (cyclopentane excluded)	10.65	13.24
Diolefins, alkenyl aromatics, dimers		
Isoprene	2.50	0.30
Pentadiene	3.77	0.30
Cyclopentadiene	6.69	0.30
Methylcyclopentadiene	0.55	0.30
Cyclohexadiene	1.60	0.30
Styrene	4.19	0.50
2-Methyl- + 3-Methylstyrene	0.75	0.50
α-Methylstyrene	0.20	0.50
Indene	1.68	0.60
Dicyclopentadiene	2.50	0.30
Alkyl aromatics		
Ethylbenzene	0.61	4.60
o- + m-Ethyltoluene	0.23	1.14
Cumene	0.21	0.45
Naphthalene	1.55	1.60

The hydrogenation selectivity is very good. For ≥ 95% conversion of dienes, the formation of saturates is highly limited : starting from 24% dienes, 10% olefins and 55% alkyl aromatics (benzene included) the saturates content increases only by 4.24, which means a saturation of 5% only. Table 18.11 shows that alkenyl aromatics are selectively hydrogenated to alkyl aromatics (ethylbenzene increase corresponds to styrene hydrogenation) which means that cycloparaffins are formed by the hydrogenation of cyclodiolefins. A new calculation excluding alkenylaromatics results in a saturation, even including cyclopentane limited to 15%. Hence the only significant side-reaction is the hydrogenation of cyclopentene (and dihydrodicyclopentadiene) which confirms research data that the rate of hydrogenation of cyclopentene is much higher that for all the other C_5+ olefins.

18.5.3 Processes

(i) Characteristics and schemes. Selectivity ultimately emerges as the most important aspect of the hydrogenation of pyrolysis gasolines : hydrogenation selectivity to preserve the value of the gasoline (octane rating or aromatics content), and selectivity to avoid the undesired formation of oligomers resulting in fouling of catalytic reactors. The optimization of these two types of selectivity determines the choice of operating conditions, type of process scheme and catalyst, aimed to achieve two objectives known to favour these selectivities, i.e., keep the temperatures in the reactor as low as feasible and keep a liquid phase to wash the catalyst.

Since hydrogenation selectivity increases with decreasing operating temperature, operations are conducted between 40 and 100°C, a technically and economically accessible temperature range (air or water cooling). The need for highly active catalysts therefore eliminates sulphides (Co-Mo and Ni-Mo) in favour of metals, except for specific cases in which Ni-W sulphide is retained. Since platinum is not selective enough, the catalysts will generally be based on nickel or palladium. Operations will also be carried out under pressure in order to increase the activity : around 25-30 bar, the available pressure of steam-cracker hydrogen, or above.

Since polymerization remains the main reason for frequent regenerations, a mixed phase is generally used (hydrogen partially dissolved in the liquid hydrocarbon feed) to extract the polymers from the catalyst by a washing effect. Extremely inert catalyst carriers are also used (low area alumina, silica), which are quite different from those employed in other hydroprocessing methods, such as hydrodesulphurization.

Another specific feature is the need to control an highly exothermic and hence self-accelerating reaction. Various technological alternatives are available and these characterize the processes : use of cooled multitube reactors, catalytic beds with intermediate quenching or high dilution of reactive gasoline by recycling the hydrogenated product.

Fig. 18.19 shows a typical scheme for the selective hydrogenation of steam-cracked gasoline with two sections, the reaction section and the fractionation section. For most processes, the latter section features the following standard components :

a column for stabilization, to eliminate light compounds and to obtain the initial ASTM boiling point as specified for automotive-fuel.

a second redistillation column, to remove heavy products (present in the feed or formed in the reactor) and obtain the final ASTM boiling point.

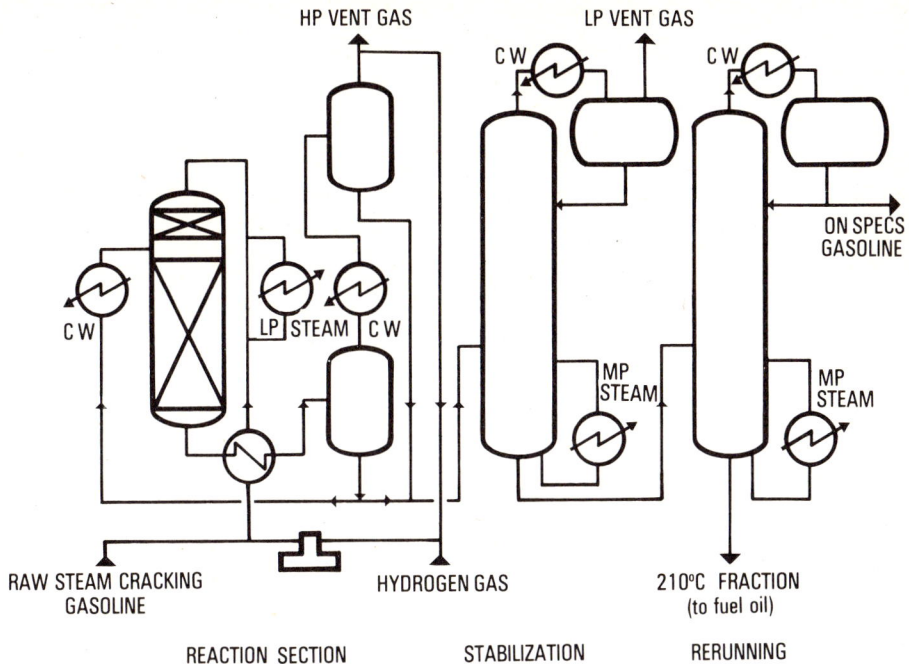

Fig. 18.19 - Hydrorefining of steam-cracking gasoline. Selective hydrogenation of dienes in order to get a stable, olefin and aromatic-rich fraction with a high octane-value. Simplified process flow diagram according to IFP technology.

The reaction section reflects IFP technology : the feed and fresh hydrogen are mixed then heated by heat-exchange with the reactor effluent (or by steam for start-up) and sent to a fixed-bed reactor. The condensed effluent is separated into two phases. An inter-bed liquid quench, taken from the separator drum, serves to limit the temperature rise in the reactor. If necessary, gas recycling is used to reach the hydrogen partial pressure required in the reactor with the make-up hydrogen barely more than that consumed. A stand by reactor is often present (refs. 24, 25). A furnace is normally provided for intermittent regeneration (burning of carbonaceous deposits by air/steam mixture) and for catalyst reduction

Selective hydrogenation is only carried out alone for steam-cracked gasolines, for the production of automotive fuels. For the production of aromatics, as with the refining of other pyrolysis gasolines, selective hydrogenation is employed as a pretreatment before hydrodesulphuration/hydrogenation of the olefins, referred to as second-stage or hot hydrogenation as opposed to first-stage or cold hydrogenation.

The process has many variants according to whether fractionation is intermediate (aromatic scheme for pyrolysis gasolines) or not, and also depending on the degree of integration required in the two reaction stages. Fig. 18.20

provides an example of UOP technology for a highly integrated two stages scheme, which is interesting from the capital investment and utilities standpoint (only one gas-recycling compressor, one feed-dilution pump, one furnace for both stages) (ref. 2). However, it is unfavourable for the catalyst used in the first stage, because the reaction mixture contains not only contaminants from the fresh feed (hydrocarbons + hydrogen), but also their second-stage decomposition products, e.g., H_2S formed in the second-stage out of the gasoline sulphur compounds. Thus longer contact times and hence larger quantities of catalyst must be employed, requiring larger reactors.

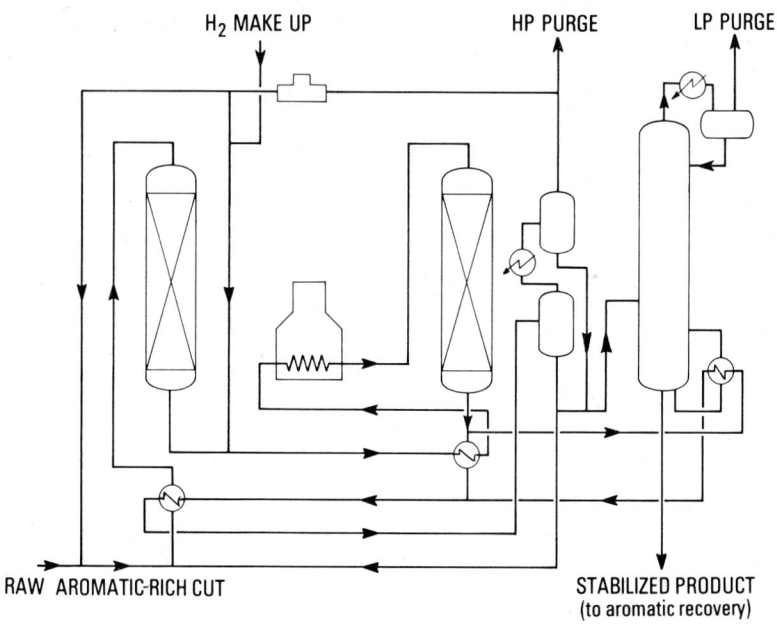

Fig. 18.20 - Hydrofining of steam-cracking gasoline. 2-stages hydrogenation in order to produce an olefin and sulfur-free aromatic-rich fraction. UOP technology integrating the 2 stages of reaction in one single unit. Simplified process flow diagram.

(ii) <u>Industrial processes</u>. Many specific processes have been developed in the past twenty years, for the hydrogenation of pyrolysis gasolines and particularly the hydrogenation of steam-cracked gasolines. They may be considered either as extension of existing refining processes (UOP process on Co-Mo catalysts, based on hydrodesulphuration catalysts) or as brand-new inventions for a new need (British Petroleum process and IFP process).

These processes are too numerous to describe all of them and, besides IFP and UOP processes, we will only mention the essential features of the other most important.

B.P. (British Petroleum) process (refs. 27, 28). The earliest process of this type is distinguished by the search for minimum capital investment : operation at low pressure to avoid the hydrogen make-up compressor, use of a nickel catalyst on sepiolite, a cheap natural clay ; the catalyst is made selective by the sulphur compounds, thiophene mainly, contained in the gasoline. Temperature rise in the reactor is limited by diluting the fresh feed with a recycle of the product.

Bayer process (ref. 29). The reaction section is similar to the reaction-section already described for C_3 or C_4 cuts selective hydrogenation : the main feature is the use of isothermal, heat-exchange-type multitube reactor. The process looks for long operating cycles but capital investment is high : the feed is predistilled to eliminate gums, thus almost avoiding catalyst fouling. An hydrogen make-up compressor leads to an operating pressure of 50-60 bar. Active Pd-catalysts are used.

Lummus process (ref. 30). Lummus-Crest, an engineering company active in the ethylene market has developed its own process, based on the use of Pd-catalyst. The concept favours a moderate capital cost at the expenses of the selectivity. Operation at high pressure (50-60 bar) and high dilution ratio of the feed with a recycle of product favour long cycles but increase the hydrogenation of olefins.

It is also worth to mention other processes developed and licensed by Engelhard (U.S.A.), B.A.S.F. (F.R.G.), U.O.P., Houdry, Shell, Kellog (U.S.A.).

18.5.4 Industrial performance

Many hydrogenation catalysts, which are quite different from standard hydro-desulphuration catalysts, have been developed and are commercially available. Since processes are not usually associated with the use of a specific catalyst beyond the life-time of the first catalyst load, a "free market" for hydrogenation catalysts exists. Palladium alumina catalysts dominate the market, followed by nickel-based catalysts, and mixed sulphide (Ni-W) based catalysts, which are less specific for hydrogenation than the formers.

Many manufacturers offer similar catalysts : Bayer K 8317 B, Engelhard PGC3, BASF H022, Procatalyse LD 265, UCI G-68 and Harshaw Pd-0501 T, Calsicat (Mallinckrodt Inc., USA) E-143, UOP LT Unibon are Pd catalysts ; Laporte, Shell 204 and Procatalyse LD 241 are Ni catalysts and Procatalyse LD 155 is a Ni-W catalyst (ref. 31).

Here catalyst performance will be examined in terms of product stability, gasoline yield, catalyst resistance to impurities in the feed, regenerability and mechanical properties.

(i) _Catalyst activity and selectivity_. Table 18.12 compares the characteristics of a steam-cracked gasoline before and after selective hydrogenation, and reveals that stabilization, the basic objective of the hydrotreatment performed, is attained.

TABLE 18.12

Characteristics of raw (unhydrogenated) and hydrogenated steam-cracking gasoline

Characteristics			Raw Gasoline	Hydrotreated Gasoline
Density sp.gr.	D 20/4		0.821	0.819
ASTM distillation (°C)	IBP		30	40
	5		42	42
	10		49	48
	50		98	100
	90		164	166
	95		174	179
	EBP		195	204 [a]
Diene value (UOP 326-65)	$gI_2/100$ g		27	0.8
Bromine number	$gBr_2/100$ g		75	47
Total sulphur	ppm		400	380
Acidic sulphur	mg/l		18	< 5 [b]
Doctor test			> 0	< 0 [b]
Copper test				1 [a]
Actual gums before/after heptane wash	mg/100 cc		30/25	25/< 2 [a]
Copper-beaker gums	mg/100 ml		>500	10
Induction time (mn)	0		100	200
(with anti-oxidant ppm)	20		180	480
	70		200	960
Potential gums (4-H test)	mg/100 ml		3500	10
Colour Saybold			Yellow	+ 26 (16) [a]
Research octane number	Clear		97	97
	Lead 0.15 g/l		97.5	98
Motor octane number	Clear		85	85
	Lead 0.15 g/l		85.5	86

[a] Before redistillation
[b] With nickel-based catalysts

The diolefins and alkenyl aromatics, whose content is determined by measuring the diene value, have been hydrogenated, so that the stability of the gasoline as measured by the existing and potential gum content, the induction period and anti-oxidant susceptibility, has been achieved. Other properties of the gasoline can be improved simultaneously, such as its passivity, as measured by its corrosive mercaptan content and related tests (doctor test, copper test) and its susceptibility to antiknock agents (lead additives).

These properties are related to the catalyst's activity. Selectivity (non-hydrogenation of the other unsaturated hydrocarbons) is measured by the slight reduction in the bromine number, once the effect of hydrogenation of alkenyl aromatics has been taken into account, as well as the preservation of the gasoline octane rating. Whereas, by definition, all the processes and catalysts serve to produce equally hydrogenated products, consequently of equivalent stability, this cannot be said of the corrosiveness and the octane rating. This points to one of the essential differences in performance between palladium catalysts and nickel catalysts.

Palladium catalysts on the one hand, actually appear to be more selective (lower olefin hydrogenation rate) for intrinsic reasons, and also because their greater stability makes it possible to maintain low operating temperatures, favourable to high and stable selectivity during the life of the catalyst. For instance, the average cyclopentene hydrogenation ratio is 50% on nickel or on sulphides, while it is only 25% on palladium. This higher hydrogenation selectivity reduces one of the causes of octane loss and contributes to the overall slight octane gains on palladium as opposed to slight losses on nickel.

On the other hand, nickel catalysts offer a definite advantage in terms of gasoline sweetening : Nickel and Ni-W based catalysts can destroy mercaptans in the feed, unlike palladium catalysts which tend to produce mercaptans. The reactions involved are as follows :

$$RSSR' + H_2 \xrightarrow{Pd, Ni \ (?)} RSH + R'SH$$

$$H_2S + R-CH = CH_2 \xrightarrow[Pd, Ni]{H_2} R-SH$$

$$RSH + R'-CH = CH_2 \xrightarrow[Ni]{H_2} R-S-R' \text{ (accompanied by other reactions)}$$

These features of nickel and palladium catalysts are summarized in Table 18.13.

With respect to the "sweetening" of steam-cracked gasolines, the following reference points can be employed.

"Copper test pass" or copper specification = 1 (Europe). Accessible on all types of catalyst, for steam-cracked gasolines containing less than 120 ppm RSH, and, in practice, gasolines from all steam crackers except those fed with high-sulfur fractions (straight-run vacuum distillates).

"Doctor test pass" or mercaptan specification = 10 ppm max (Japan and U.S.A.). Accessible on palladium if the steam-cracked gasoline contains less than 5 ppm RSH. Accessible on nickel if the load contains less than 120 ppm RSH. The operating conditions of the process are adjusted (temperature is increased and LHSV decreased) to achieve this performance.

(ii) <u>Resistance to feed impurities, stability and cycle time</u>. By definition, pyrolysis gasolines contain many gum-generating compounds, and are therefore unsuitable for long cycle times. Moreover, the feed undergoes no preparation and, due in particular to the low operating temperature, the catalysts employed are very sensitive to many natural impurities (sulphur compounds, peroxides and possibly arsenic) and also to artificial pollutants (lead in antiknock additives, water, silicon in antifoamers). The hydrogen used may or may not contain carbon monoxide and sometimes hydrogen sulphide.

All these contaminants cause more or less rapid deactivation of the catalyst, which may be offset by raising the operating temperature, but at the expense of speeding up the polymers formation. Hence contamination results in a reduction in the cycle time and possibly in the catalyst service life. The sensitivity to contaminants, of the various catalytic species is therefore of great importance as regards the performance of the reactor.

Table 18.13 presents a sort of classification of the harmfulness of the most widely observed impurities to the three categories of catalyst species employed : Ni, Pd and sulphided Ni-W. Nickel-based catalysts appear to display greater sensitivity, reflected by their shorter life. Nickel combines irreversibly (at least in relation to industrial regeneration procedures) with a large number of metalloids (sulphur for instance) and metals. Palladium, on the other hand, whose oxide form is only stable at high temperatures, is generally only inhibited, allowing for the possibility of compensation by decreasing the LHSV or increasing the operating temperature, and is rarely poisoned. Mixed sulphides (specially Ni-W sulphides) display good resistance to poisons, but much lower activity. Hence larger reactors or higher operating temperatures, causing excessively short cycle times, must be employed ; anyway they are easily regenerated.

The selection criteria are thus extremely numerous and the most appropriate catalyst is rarely obvious. For example, the rate of poisoning of nickel may be sufficiently low to be acceptable compared to the very rapid inhibition of palladium, i.e., a medium catalyst life may be preferable to very short cycle times.

(iii) <u>Mechanical properties and regeneration strength</u>. For the various reasons discussed above, cycle times are generally moderate (a few months) and consequently regenerations are frequent. The quantities of polymers deposited may be high, and consequently the heat stresses during high-temperature regenerations are also high.

The mechanical properties are measured either on the catalyst particle (grain crushing) or on a catalyst bed (attrition in a rotating tube, bed crushing) and are determined at the end of the catalyst service life. The latter is related to the number of regenerations possible, and also to the frequency of catalyst unloadings for screening to eliminate fines that generate pressure drops or to replenish the bed with fresh catalyst. The mechanical properties are vitally important, and often point to the superiority of a given commercial catalyst. The type of carrier and preparation method are decisive, and generally depend on the manufacturer's know-how (ref. 32).

TABLE 18.13
Summary of the specific features of palladium and nickel catalysts

Activity : Palladium-based catalysts are 2-3 times more active than nickel-based catalysts (LHSV is 3 to 5 instead of 1 to 2).

Sweetening : Nickel-based catalysts provide a sweetening effect. Doctor test of the gasoline becomes negative. Palladium-based catalysts show no sweetening effect ; instead they sour the gasoline.

Selectivity : Palladium-based catalysts give a higher selectivity than LD 241 catalyst.

Sensitivity to contaminants : Lower for palladium-based catalysts.

Stability : Life of palladium-based catalysts is 2-4 times longer than that of nickel-based catalysts.

Optimum temperature (°C) for	Pd	Ni
Activation (by H_2)	100	400
Reactivation (by H_2 stripping)	430	410
Regeneration (by air burning)	390	400

Effect of typical contaminants of steam-cracked gasoline and hydrogen-rich gas on catalyst performance

Impurity	Catalyst type		
	Palladium	Nickel	Nickel-tungsten
Carbon monoxide	Inhibitor	Inhibitor	Inhibitor
Sulphides + thiophenes	Inhibitor	Inhibitor[a]	Inhibitor
Acidic sulphur (existing and potential) (RSH + H_2S + CS_2 + RSSR)	Inhibitor	Temporary Poison[a,b]	Inhibitor
Organic chloride compounds	Inhibitor	Permanent Poison	Inhibitor
As and derivatives	Temporary Poison	Temporary Poison	Temporary Poison
Pb (antiknock additive) Si (antifoaming additive)	Permanent Poison	Permanent Poison	Permanent Poison

[a] Permanent poison above 180°C
[b] With an effect on life-time.

(iv) <u>Restoration of the activity of selective hydrogenation catalysts : reactivation and regeneration</u>. The causes of deactivation are many and varied, and regeneration is frequently necessary. Apart from conventional procedures involving burning, similar to those conducted on HDS catalysts, specific techniques have been developed. The end of a cycle is rarely due to a carbon deposit that can only be eliminated effectively by combustion, but more frequently to deposits of polymers and gums, or to the build-up of poisons (sulphur, arsenic, chlorine compounds, etc ...) which are more easy to eliminate from the catalyst. This has led to three procedures, conventional regeneration (burning in air/steam mixtures), reactivation by stripping in hot hydrogen and extraction of soluble deposits by washing with water.

<u>Hydrogen reactivation</u>. Treatment in an hydrogen-rich gas at about 350-450°C is generally effective in eliminating build-up of polymers, generally located at the top of the bed and causing high pressure drops. This is achieved by stripping, possibly accompanied by hydrocracking. A number of impurity/catalyst combinations are simultaneously destroyed with the formation of volatile hydrides, so that most of the initial activity of the catalyst is restored.

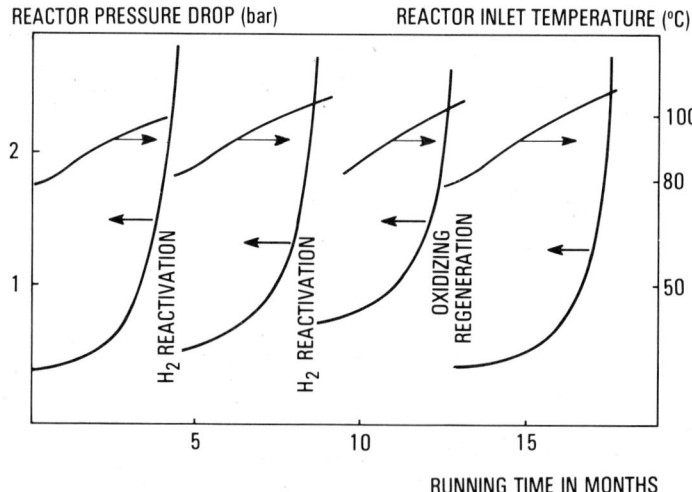

Fig. 18.21 - Hydrogenation of steam-cracking gasoline. Restoring the activity of a nickel-based catalysts (Procatalyse LD 241) by a hot-hydrogen stripping (reactivation) or by a burning with steam-air mixture (regeneration). Activity of catalyst (as measured by reactor inlet temperature), and feasibility of operation (as measured by pressure drop in reactor) are recovered after each treatment.

Fig. 18.21 shows the effectiveness of this treatment for a nickel catalyst. It is possible, by rapid (24-48 h), easy (simple shutting of the feed and heating in hydrogen) and safe (no risks of runaway by uncontrolled combustion) treatment to obtain cycle times, i.e. operating times between two regenerations, up to one year. The same high-temperature hydrogen treatment can be applied to other catalysts, but less effectively, because the hydrocracking activity of nickel (formation of CH_4) around 400°C helps to eliminate polymers. Hence hydrogen reactivations are also carried out on nickel/tungsten catalysts, more rarely on palladium catalysts (requiring a higher temperature).

18.6 CONCLUSION

Research on the selective hydrogenation of hydrocarbons began in 1900, and it is worth noting the works by P. Sabatier and J.B. Senderens as well as by V.N. Ipatieff. Nevertheless, it has developed considerably only in the early 60's, due to the demand by the petrochemical industry, particularly with the boom of olefin production by steam cracking.

The need to purify cuts rich in olefins or diolefins, to achieve the purity level required for producing constantly improving qualities of polymers and elastomers, has led to the development of a great many hydrogenation processes and catalysts as well as to great improvements in scientific knowledge and know-how in the field of catalytic hydrogenation. However, the possibilities of cracking and polymerization technologies have as yet been only partially exploited, as shown, for example, by the marketing in 1980 of a new polyethylene : the linear low-density polyethylene (LLDPE). Therefore, further progress in hydrorefining will be necessary, and the selective hydrogenation of hydrocarbons will continue to be improved and diversified.

REFERENCES

1 S.D. Zdonik, E.J. Green and L.P. Hallee, Manufacturing ethylene, reprinted from Oil and Gas Journal, 66, 67, 68, 69, 70. The Petroleum Publishing Cy, Tulsa, Oklahoma, USA.
2 L.K.Ng, C.N. Eng. and R.S. Zack, Hydrocarbon Processing, 66, 12 (1983), pp. 99-103.
3 J.P. Boitiaux, J. Cosyns, M. Derrien and G. Léger, Hydrocarbon Processing, 63, 4 (1985), pp. 51-59.
4 Commercial brochures of catalyst manufacturers. UCI (United Catalyst Incorporated, USA) ; ICI (Imperial Chemical Industries, U.K.) ; Girdler Südchemie (Fed. Rep. of Germany) ; Leuna-Werke (Democratic Rep. of Germany) ; Procatalyse (France).
5 J.C. Jungers, Cinétique Chimique Appliquée, Technip, Paris 1958, pp. 172-181.
6 Commercial brochure on Hydrogenation Processes, 1981, V.E.B. Leuna-Werke Walter Ulbricht, DDR-422, Leuna 3.
7 M.L. Derrien, Revue de l'Association Française des Techniciens du Pétrole, 314 (1985), pp. 9-15.
8 J. Thomas, M.L. Derrien, personal communication.
9 U.S. Pat. 3,075,917, W. Krönig, A. Meckelburg, B. Schleppinghoff, and G. Shärfe, 1963.
10 W. Krönig, Hydrocarbon Processing, 49, 3 (1970), pp. 121-126.
11 M.L. Derrien, C. Bronner, J. Cosyns and G. Léger, Hydrocarbon Processing, 58, 5 (1979), pp. 175-179.
12 H. Lauer, Erdoel und Kohle, Erdgas Petrochemie, 36, 6 (1983), pp. 249-251.

13 U.S. Pat. 4,049,742, H.M. Weitz, V. Wagner, K. Volkhammer, E. Schubert and F. Bandermann, 1977.
14 Arco-Engelhard, U.S.A., Commercial brochure, Acetylene Hydrogenation Process, 1981, Engelhard Ind. Div., Newark, N.J. 07105, U.S.A.
15 U.S. Pat. 2,426,604.
16 U.S. Pat. 4,440,956, M.C. Couvillion, 1983.
17 U.S. Pat. 4,493,906, M.C. Couvillion, 1983.
18 Eur. Pat. 0,081,041, F. Obenaus, F. Nierlich, O. Reitemeyer, B. Scholtz, 1982.
19 J.P. Boitiaux, J. Cosyns, M. Derrien and G. Léger, Proc. American Institute of Chemical Engineers Meeting, Houston March 1985, paper 79b.
20 U.S. Pat. 3,481,999, M. Reich, 1969.
21 R.M. Heck, G.R. Patel, W.S. Breyer and D.S. Merril, Oil and Gas Journal, 81, 3 (1983), pp. 103-113.
22 U.S. Pat. 4,132,745, P. Amigues, J. Gaillard, J.F. Le Page, R. Stern, 1979.
23 C.L. Rogers, Oil and Gas Journal, 69, 45 (1971), pp. 60-61.
24 M.L. Derrien, J.W. Andrews, P. Bonnifay and J. Léonard, Chem. Eng. Prog. 70, 1 (1974), pp. 74-80.
25 M.L. Derrien, J. Léonard and R. Stern, Proc. American Institute of Chemical Engineers Meeting, Houston 1975.
26 C.H. Watkins, Proc. 7th World Pet. Congress, Mexico 1967, Panel 20, pp. 85-92.
27 P.T. White, F.W.B. Porter and A.A. Yeo, Proc. 5th World Pet. Congress, New York 1960, pp. 335-345.
28 K.H. Bourne, P.D. Holmes and R.C. Pitketly, Proc. Int. Congress in Catalysis 2, (1965), pp. 1400-1415.
29 U.S. Pat. 4,097,370, W. Krönig, A. Meckelburg and B. Schleppinghoff, 1961.
30 Anonymous, Hydrocarbon Processing, 55, 9 (1976), p. 144.
31 Commercial brochures of Catalysts Manufacturers and Process Licensors.
32 J. Cosyns, M.T. Chenebaux, J.F. Le Page and R. Montarnal, in B. Delmon, P. Jacobs and G. Poncelet, Preparation of Catalysts I, Proc. 1st Int. Symp. on the Scientific Preparation of Heterogeneous Catalysts, Louvain 1976, Elsevier, Amsterdam 1976, pp. 459-472.

Subject Index

Absorption coefficient of hydrogen 525
Acceptor 53
Activation energy 438
Adiabatic temperature rise 590
Adsorbed species
 - in carbonyl compounds hydrogenation 94
 - of hydrocarbons on metal surfaces 155
Adsorption
 - of carbonyl compounds on metal catalysts 79
 - of metals 376, 379
Alloy catalyst 60, 159
Amines formation 106
Ammonia synthesis 304
Anchored complexes 459, 465, 474, 480
Asymmetric hydrogenation 281, 497, 509
Auger electron spectroscopy (AES) 321, 417
Autoclave
 -, control of 579, 585, 588
 - dynamic properties of 597
 -, industrial continuous 602
Axial temperature profiles 562, 568, 575

Balandin's multiplet theory 4, 35
Batch reactor 531, 537
Bimetallic catalyst 145, 313, 337, 373, 376

Biot's number 539

Calcination in a platinum catalyst preparation 340
Carbonyls of metal 484
Carrier effect 277
Catalyst
 -, alloy 60, 159
 -, bimetallic 145, 313, 337, 373
 -, cobalt based 111, 114, 206, 266, 270, 293
 -, copper based 112, 119
 - for asymmetric hydrogenation 497
 - for carbon oxide hydrogenation 285, 289, 290, 293
 - for hydrodesulphurization 58, 206, 257
 - for nitriles hydrogenation 105
 -, highly dispersed 155
 -, iron based 293, 322
 -, low dispersed 154
 -, molybdenum based 206, 266, 270, 295
 -, nickel based 111, 114, 206, 266, 270, 285, 315, 348, 390, 411
 -, palladium based 56, 290, 326, 348, 381, 394
 -, platinum based 56, 66, 111, 121, 145, 289, 318, 330, 338, 376, 385, 400
 -, promoted 60, 112

-, rhenium based   398
-, rhodium based   292, 348
-, ruthenium based 291, 319, 331
-, Wilkinson's   462, 501
-, wolfram based   270
Chemisorption of hydrogen   346
Chloroplatinic acid   338
Cobalt catalyst
- for carbon oxide hydrogenation   293
- for nitrile hydrogenation 111, 114
-, molybden and sulphur based 206, 266, 270
Coking   69, 62, 353, 367, 444
Competitive hydrogenation 35, 92
Complexes
-, π   5, 375, 472, 474
-, π-allyl   479
-, anchored   375
-, bimetallic   376
-, metallic   459, 481
- of palladium   472, 474
Continuous stirred tank reactor (CSTR)   517, 533
Control
- of hydrogenation autoclave 579, 585, 588
- of pressure   586
- of selectivity   67
- of temperature   586, 591, 603
Controlled surface reaction 373

Conversion
- of n-hexane   393
- profiles simulation   535
Coordinatively unsaturated sites 314, 467
Copper catalyst   112, 119
Cracking   71, 156

Damköhler's number   524, 532, 538
Deactivation
- of nickel catalyst   442
- of supported catalysts   305
Dealkylation   417
Deamination   113
Decomposition
- of anchored metal carbonyls 484
- of surface organometallic complexes   482
Dehydrocyclization   351
Dehydrogenation
- of butane   351
- of cyclohexane   318, 320, 349, 360, 365, 436, 441
- of cyclohexanone   351
- of cyclohexylamine   351
- of dodecane   351
- of formic acid   302
- of 3-methyl-1-butanol   441
- of 2-propanol   441
- on bimetallic catalysts   349, 351
Dehydroisomerization   352
Denitrogenation   201
Deuteration of carbonyl compounds 83, 85, 87
Diffusion   44, 528, 530, 538

Dimerization   323
Disproportionation   421
Dynamic behaviour of autoclave   597

Effect
- -, electronic   146, 284, 318, 332, 439, 446
- -, geometric   146, 284, 330
- of ammonia on nitriles hydrogenation   132
- of carbon oxide on acetylene hydrogenation   327, 329
- of carrier on hydrogenation properties of metals   277
- of catalyst loading   536
- of entrance region   553
- of film and pore diffusion   538
- of heat transfer   537
- of hydrogen pressure on hydrodesulphurization   262
- of hydrogen pressure on nitriles hydrogenation   130
- of hydrogen transfer   531, 536
- of mass transfer on kinetics   523, 530
- of matrix   322
- of nitrile structure on hydrogenation   125
- of promoting   447
- of reaction conditions on nitriles hydrogenation   128
- of ring size on cyclo-alkanones hydrogenation   93
- of particle size   315, 322
- of structure on carbonyl compounds hydrogenation   92
- of structure on hydrodesulphurization   262
- of substrate transfer   531
- of support on hydrogenation properties of metals   277
- of synergy on hydrodesulphurization   71
- of temperature on hydrodesulphurization   262
- of temperature on nitriles hydrogenation   128
- of wall-flow region in a trickle bed reactor   553
- of water on nitriles hydrogenation   135

Effective thermal conductivity   556
Effectiveness factor   523, 529
Electrochemical metal adsorption   379
Electron
- nuclear double resonance (ENDOR)   419
- paramagnetic resonance (EPR)   419
- spectroscopy for chemical analysis (ESCA)   417
- spinning resonance (ESR)   419

Electronic effect   146, 284, 318, 332, 439, 446
Electrophilic attack   219
Enamine   107
Enantioselective hydrogenation   34, 302
Ensemble size effect   315, 322

Equilibrium of hydrogenation reactions 80
Extended X-ray absorption fine structure (EXAFS) 319, 341
External diffusion 44

Farkas and Farkas' mechanism 4
Ferromagnetic resonance 420
Film diffusion in a batch reactor 538
Fischer-Tropsch synthesis 60, 441
Fixed-bed reactor 547, 550
Functionalized support 462

Geometric effect 146, 284, 330

Hammett's equation 92
Hazardous states 605
Heat
 - production in a batch reactor 537
 - transfer 517, 537, 549
Henry's law 520
Highly dispersed catalysts 155
Horiuti-Polanyi mechanism 4, 82
Hot spots in a trickle bed reactor 557
Hydrides of palladium 326
Hydrocarbon
 - adsorbed on metals 155
 -, conversion of 358
Hydrocracking of paraffins 415, 420, 441

Hydrodenitrogenation
 -, mechanism of 201, 246
 - of acridine 237
 - of aniline 211, 222
 - of carbazole 243
 - of dipentylamine 221
 - of indole 241
 - of isoquinoline 233
 - of nitrogen in aliphatic amines 218
 - of nitrogen in aromatic amines 222
 - of nitrogen in aromatic heterocycles 225, 239
 - of nitrogen in saturated monocyclic rings 244
 - of piperidine 245
 - of pyridine 225
 - of pyrrole 239
 - of pyrrolidine 244
 - of quinoline 229
 - of o-toluidine 205
Hydrodeoxygenation 205, 209
Hydrodesulphurization
 -, catalyst for 58, 257
 -, effect of reaction conditions on 71, 262
 -, kinetics of 259
 - of benzothiophene 260, 266, 271
 - of dibenzothiophene 261, 268, 271
 - of 2-methylthiophene 205, 207
 - of thiophene 259, 265, 269
Hydroformylation 499
Hydrogen
 -, chemisorption of 346
 - deuterium exchange 6, 21

-, effect of pressure 130, 262
-, para-ortho conversion of 21
-, reactivity of 376
-, solubility of 518
- spill-over 53, 62, 68, 69, 73
- transfer 531, 536

Hydrogenation
-, asymmetric 281, 497, 509
-, competitive 35, 92
-, enantioselective 34, 302
- of α-acetamidoacrylic acid 499
- of α-acetamidocinnamic acid 500, 510
- of acetone 23, 33, 84
- of acetonitrile 123, 126
- of acetophenone 89, 501
- of acetylene 15, 62, 326, 330, 441, 618
- of acroleine 438
- of acrylic acid 437
- of acrylonitrile 405, 415, 441
- of adiponitrile 114
- of alicyclic ketones 95
- of alkylphenols 26
- of allylalcohol 437
- of aromatic compounds 7, 18, 205, 297, 299
- of aromatic nitro compounds 535
- of benzene 20, 32, 57, 282, 296, 320, 348, 361, 392, 415, 420, 435, 441, 480, 483, 484
- of benzoic acid 33
- of benzonitrile 123, 126
- of 1-butene 282, 417, 441, 480
- of butyronitrile 114, 122, 126, 131
- of carbon oxides 61, 284, 289, 320, 324, 417, 421, 487
- of carbonyl compounds 79, 94, 416, 432, 437, 441
- of p-chloronitrobenzene 279
- of cinnamaldehyde 280, 440
- of citraconic acid 500
- of citral 441
- of o-cresol 527
- of crotonaldehyde 25, 37, 280
- of cycloalkanones 53
- of cyclohexanone 26, 37
- of cyclohexene 420, 441, 437, 469, 480, 485, 554, 557
- of cyclooctene 469
- of cyclopentadiene 41, 300, 474, 483
- of cyclopentanone 85
- of cyclopropane 442
- of dehydro-α-acylamino acids 507
- of dehydro-α-amino acids 501
- of dienes 14, 470
- of 2,4-dimethylnitrobenzene 522
- of N,N-dimethyl-2-propenyl-amine 417, 440
- of 2,6-dinitrotoluene 528
- of diolefins 299
- of ethyl acetoacetate 34
- of ethylene 5, 10, 62, 68, 299, 327, 348, 361, 435, 441, 465, 468, 479
- of 2-ethylhexanal 442

- of fatty oils 301
- of formaldehyde 301
- of D-fructose 442
- of fumaric acid 418, 441
- of furfural 29
- of 1-heptene 37, 468
- of 1-heptyne 37, 468
- of heterocyclic compounds 28
- of 1,5-hexadiene 300
- of 1-hexene 348, 361, 415, 420, 429, 437, 441, 443, 448, 469, 472
- of isoprene 41, 61
- of ketones 86, 95, 501
- of lauronitrile 108, 115, 117, 126, 138
- of maleic acid 418, 441
- of methylacetoacetate 302, 441
- of methylacrylate 474
- of α-methylcinnamic acid 500
- of methyl α-phthalimido-acrylate 506
- of α-methylstyrene 437
- of methylvinylketone 438
- of mixtures of organic compounds 35, 92
- of nitriles 105, 441
- of nitrobenzene 27
- of nitrogen oxide 304
- of octadecene 469
- of 1-octene 420, 509
- of olefins 10, 37, 299, 415, 441, 506, 507
- of palmitonitrile 114
- of 1,3-pentadiene 61
- of phenol 25, 37
- of phenylacetylene 483
- of 4-phenyl-1-butene 420, 441
- of propionaldehyde 413
- of propionitrile 126
- of propylene 442, 481, 486
- of stearonitrile 112, 120, 128, 134, 138
- of styrene 420, 432, 435, 437, 441, 443
- of sulpholene 33, 473, 483
- of tert. butyl phenol 302, 520
- of thymol 302
- of toluene 483
- of unsaturated acids 499
- of unsaturated hydrocarbons 466, 479
- of valeronitrile 123
- of vinylbutyl ether 472
- of xylenes 281
-, thermodynamics of 80

Hydrogenolysis
- of alkanes 417, 441
- of 1-aminotrialkylamines 109
- of n-butane 315, 320, 323, 420, 436, 441
- of C-C bond 16, 145, 157, 441
- of cyclohexane 315, 320
- of ethane 320, 323, 332, 417, 436, 441
- of n-hexane 363, 442, 436
- of hydrocarbons 16
- of methylcyclopentane 441
- of toluene 442
- on bimetallic catalysts 145, 353
-, thermodynamics of 150

Hydroisomerization 415, 441

Hydrorefining
- of C_2 cut   616
- of C_3 cut   625
- of C_4 cut   635
- of gasoline cuts   652
- of petrochemical raw materials   613

Impregnation of alumina by chloroplatinic acid   338
Infrared spectroscopy (IR) 14, 94, 345, 413, 439, 464, 472
Inhibition   521
Interfacial metal - support interaction (IMSI) 284, 434
Internal diffusion   44
Iron catalyst
-, bimetallic   322
- in carbon oxide hydrogenation   293
Isokinetic reactions   40
Isomerization   325, 415, 441
Isotopic effect   16, 21

Kinetics
- of carbonyl compounds hydrogenation   79, 91
- of hydrodesulphurization   257
- of hydrogenation   1, 10, 31
- of hydrogenation of nitriles   105
- of hydrogenolysis   152

Langmuir-Hinshelwood kinetics 3, 13, 29, 517, 526, 537
Low dispersion catalysts   154

Mass transfer
- and kinetics   523, 530
- in combination with chemical reaction   534
- in fixed bed reactors   549
- in slurry reactors   517
Matrix effect   322
Maxted equation   443
Mechanism
-, Farkas and Farkas'   4
-, Rideal and Eley's   3, 5
-, Twigg and Rideal's   5, 10
- of carbonyl compounds hydrogenation   79
- of cracking   156
- of hydrodenitrogenation   246
- of hydrodesulphurization   257
- of hydrogenation   3, 7
- of hydrogenolysis   152
- of nitriles hydrogenation 105
Metal
-, adsorption of   376, 378, 379
- carbonyls   484
- complexes   459, 474, 480
-, dispersed   481, 483
- support interaction   62, 283, 305, 330, 342, 421, 434, 479
Methanation   60, 282, 306, 414, 439, 441
Molecular sieve zeolites   418
Molybdenum catalysts
- for carbon oxide hydrogenation   295
- for hydrodesulphurization 290
Mössbauer spectroscopy   322, 344, 346
Multiplet theory   4, 35

Nernst potential 376
Nickel catalysts 411, 441
-, bimetallic 315, 348, 390
- for nitrile hydrogenation 111, 114
- for carbon oxide hydrogenation 285
- for hydrodesulphurization 206, 266, 270
-, poisoning of 442
Nitrogen compounds in liquid fuels 201, 203
NMR spectroscopy 16
Non-functionalized oxide support 459
Non-stationary regime 43
Nuclear magnetic resonance (NMR) 16
Number
-, Biot's 539
-, Damköhler's 524, 532, 538
-, Peclet's 554
-, turnover 287, 305, 321

Optimization of reaction temperature 583

Para-ortho hydrogen conversion 21
Palladium
- complexes 472, 474
- hydrides 326
Palladium catalysts
-, bimetallic 326, 328, 348, 381, 394
- for carbon oxide hydrogenation 290

-, supported 56
Peclet's number 554
Pilot-plant batch autoclave 588
Platinum catalysts
-, alloyed 159
-, bimetallic 145, 318, 330, 343, 376, 385, 400
- for carbon oxides hydrogenation 289
- for nitriles hydrogenation 111, 121
- with aluminium oxide support 66, 338
- with inorganic oxides support 56
Poisoning
- by thiophene 473
- by n-butanethiol 443
- coefficient 443, 446, 521
- isotherms 445
- of nickel catalysts 442
Polymer supported asymmetric hydrogenation catalysts 498
Polynuclear anchored palladium complexes 474
Pore diffusion 530, 538
Pressure control in laboratory autoclaves 586
Promoted catalysts 60, 112, 447
π-allyl complexes 479
π-complexes 5, 375, 472, 474
π-oxaallylic species 87

Radial atomic distribution (RAD) 465
Radial distribution 552
Raney catalysts 111

Reactivity
- factor 157
- of aromatic nitrogen compounds in hydrogenation 247
- of nitrogen compounds in C-N bond splitting 249

Reactors
-, batch 531, 537, 588
-, fixed bed 547, 549
-, slurry 517, 533
-, trickle bed 547, 552, 557

Refining
- of liquid fuels 201
- of oils 216
- of petrochemicals 613

Reflectance spectroscopy 419

Reforming 351, 442

Regeneration
- of platinum catalysts 342
- of surface sites 70

Remote control 65

Resonance
-, electron nuclear double (ENDOR) 419
-, electron spinning (ESR) 419
-, ferromagnetic 420

Rhenium catalysts 398

Rhodium catalysts 292, 348

Rigid band model 146

Ruthenium catalysts 291, 319, 331

Safety control of autoclave 588

Segregation 330

Selectivity of hydrogenation 38, 42, 67, 326

Self-poisoning reaction 330, 332

Sintering 342

Sites
-, acidic 219
-, coordinatively unsaturated 314, 467

Slurry reactor 517, 533

Small-angle X-ray scattering (SAXS) 413

Small particles 149

Spectrophotometric methods 426, 428

Spectroscopy
-, Auger electron 321, 417
-, electron for chemical analysis (ESCA) 417
-, infrared (IR) 14, 94, 345, 413, 439, 464, 472
-, Mössbauer's 322, 344, 346
- nuclear magnetic resonance (NMR) 16
-, photoelectron X-ray (XPS) 320, 329, 342, 345, 416, 443, 464, 484
-, reflectance 419
-, ultraviolet (UV) 99, 342, 419, 464

Spill-over hydrogen 53, 62, 68, 69

Stability
- of dispersed metals 483
- of temperature regimes 592

Strong metal-support
   interaction (SMSI)   62, 283,
   342, 421, 434, 479, 483
Substrate transfer in batch
   reactor   531
Support
 -, acidic properties of   290
 - effect on metal properties
   277
 -, oxidic   459, 462
 -, polymeric   462
Supported asymmetric hydrogenation catalysts   497
Supported bimetallic
   catalysts   380, 385, 391
Supported metal complexes   459
Surface
 - area   448
 - controlled reaction   373
 - intermediates in acetylene
   hydrogenation   327
 - mobile species   53, 67, 74
 - reaction   518
 - segregation   149, 315
 - site   71
Synergistic effect   53, 262,
   272
Synergy
 - in bifunctional catalysts
   59
 - in bimetallic catalysts
   148

Taft's equation   92
Temperature
 -, adiabatic   590
 - axial profiles   562, 568,
   575

 - control in autoclaves   586,
   591, 603
 - effect on hydrodesulphurization   262
 - effect on nitriles
   hydrogenation   128
 -, optimization of   583
 - programmed desorption (TPD)
   98, 318, 383, 421
 - programmed reduction (TPR)
   320, 345, 387
 - regimes stability   592
Thermodynamics
 - of hydrogenation   80
 - of hydrogenolysis   150
Thiele moduli   539
Tortuosity factor   531
Transfer
 - of heat   517, 537, 549
 - of mass   517, 523, 549
Transmission electron
   microscopy (TEM)   347, 356,
   413, 419, 420
Trickle-bed reactor
 -, dynamic behaviour of   566
 -, hydrogenation of
   cyclohexene in   557
 -, hysteresis properties of
   569
 -, mathematical modelling of
   552
 -, steady-state operation in
   559
Turnover number   287, 305, 321
Twigg and Rideal's mechanism
   5, 10

Ultrahigh vacuum 99
Ultraviolet spectroscopy (UV)
  99, 342, 419, 464

Wilkinson's catalyst 462, 501
Wolfram catalysts 270

X-ray
 - absorption spectra 319, 342
 - diffraction 345, 419, 423, 465
 - photoelectron spectroscopy (XPS) 320, 329, 342, 345, 416, 443, 464, 484

Zeolites 418

RAYMOND H. FOGLER LIBRARY
DATE DUE